ELEMENTS OF BIOLOGICAL SCIENCE

Elements

W·W·NORTON & COMPANY·INC NEW YORK

# of Biological Science

SECOND EDITION

WILLIAM T. KEETON

CORNELL UNIVERSITY

ILLUSTRATED BY PAULA DI SANTO BENSADOUN

SECOND EDITION

Library of Congress Cataloging in Publication Data
Keeton, William T.
    Elements of biological science.
    Shortened version of the author's Biological science,
2d ed., published in 1972.
    Includes bibliographies.
    1. Biology. I. Title.
QH308.2.K442 1973   574   73–2569
ISBN 0–393–09346–8

  2 3 4 5 6 7 8 9 0

Contents

Energy transformations 71

PART II   THE BIOLOGY OF ORGANISMS

Nutrient procurement and processing 95

Gas exchange 116

Internal transport 125

Preface

Biological science can and should be one of the most stimulating subjects a college student encounters. Nothing else, after all, has such immediate personal relevance as the phenomenon of life; and biological science, as the study of life, sheds light on what every individual experiences in himself and observes around him. Given the inherent excitement of the subject, there is no excuse for an introductory biology course to be dull.

A textbook for today's students, who are much better prepared than those of a decade ago, must do justice to the newer areas of biological science without neglecting the older that provide the foundation upon which the new rests. I have tried for an accurate and honest picture of the current state of our knowledge (and ignorance) in each of the major areas, without prejudice to any one of them; thus such topics as cellular ultrastructure, photosynthesis and respiration, molecular genetics, and developmental biology have been carefully examined, but not at the expense of "whole organism" and population biology.

Elements of Biological Science is designed especially for courses two quarters or one semester long. However, it is also appropriate for use in full-year courses when the instructor wishes to combine a text less extensive than my longer book, *Biological Science,* with his own selection of supplemental readings and lecture material.

The book does not follow the once-traditional phylum-by-phylum organization; nor does it put plant and animal biology in separate sections, as is commonly done. Instead, it discusses the basic problems faced by all forms of life, whether plant, animal, or microbe (I have tried hard to avoid slighting the plants, a tendency to which integrated texts are prone), and compares, in the light of modern evolutionary theory, the various alternative "solutions" to these problems seen in a variety of organisms. In this way, students are constantly reminded of the unity in diversity that characterizes life.

There is much disagreement among teachers about the sequence in which the various topics in an introductory biological science course should be taught. The disagreement is a legitimate one, since a good case can be made for each of several sequences. The plain fact is that each topic can best be understood if all the other topics have preceded it. This being impossible, every teacher must choose the sequence that best suits his conceptual approach to the subject. Thus, although strong arguments can be given for placing the discussions of cell division, genetics, and development with the other cellular topics at the beginning of the sequence, I have deferred them until later (Part III) for two reasons: First, I feel that study of genetics benefits more from prior knowledge of anatomical and physiological characteristics than study of anatomy and physiology benefits from prior knowledge of genetics. Second, since evolutionary theory rests upon genetics, there is considerable advantage in having the chapters on these two subjects close together. Those who find these justifications unconvincing can easily move Part III forward and follow the sequence I, III, II, IV, V. Since each chapter can be studied as an independent entity, diverse sequences of chapters can be followed.

It is a particularly difficult decision whether to cover the major groups of organisms near the beginning of a textbook, so that students will be familiar with them when they are mentioned in other chapters, or whether to leave them until the end of the course, when students may approach their examination with more insight. I have followed the latter alternative here (although the major groups are briefly introduced in Chapter 3) for the sake of a more meaningful discussion of evolutionary patterns in the plant and animal kingdoms.

Given the rate at which our knowledge is advancing, no biological science textbook can be fully up to date at the time a student reads it. It is important for him to understand this as one aspect of the dynamic, searching, inquiring nature of science. For this reason I have not hesitated to warn repeatedly that more research is needed before some of the ideas discussed can be reliably evaluated. Often evidence is cited both for and against two or three opposing hypotheses, and it is made clear that a final decision is not yet possible. This approach, which encourages the student to evaluate the evidence for himself, can bring it home to him that much remains to be done and that the effort will be exciting.

This second edition of *Elements of Biological Science* incorporates many of the major changes in content and improvements in presentation of the second (1972) edition of *Biological Science*. Thus many topics have been updated (e.g. the mechanism of photophosphorylation, the ATP yield from the complete respiration of glucose, the regulation of gene action), and material on recent discoveries (e.g. the role of cyclic AMP in hormonal action, that of calcium in muscle contraction) and on new hypotheses (e.g. concerning the origin of eucaryotic cells) has been added. Like its predecessor, this edition omits some of the relatively specialized topics covered in *Biological Science* and gives a somewhat simplified treatment of others. More than its predecessor, it emphasizes the biology of man, treating human reproduction at greater length and dealing with such subjects as immunology, biological clocks, the mode of action of neurological drugs, overpopulation, and the degradation of the environment. It also tries out different approaches to a few subjects that students often find difficult (e.g. the incorporation of energy into ATP in photosynthesis and respiration). Finally, it includes innumerable small changes that I hope will make for greater clarity and readability.

The references and readings at the end of each chapter have of course been updated. The student should be encouraged to consult this material, which is accessible to the beginner, for no one book can hope to give adequate coverage to all the diverse aspects of biological science. Many of the suggested readings are articles from *Scientific American;* the majority of these are available as offprints.

Once again the index may err on the side of inclusiveness, having been compiled on the cautious principle that too many entries are better than too few. As before, italicized page numbers refer the student to basic explanations and definitions in context. The glossary, designed to refresh his memory about important recurrent terms and to give him some sense of the etymology of the biological vocabulary, has been expanded.

Paula DiSanto Bensadoun, whose illustrations were an essential part of the first edition, has reworked a major portion of her drawings and diagrams to help the student toward a more vivid and realistic conception of organisms and their parts and a readier grasp of processes and interrelationships. Use of an accent color has permitted greater clarity and a more precise emphasis on important points. The new presentation has greatly benefited from the suggestions of Hugh O'Neill, Art Director of W. W. Norton, who also designed the book in a new format that allows the use of larger type and permits illustrations to be placed closer to text discussion.

Thirty-two color plates, containing 79 illustrations, explore themes that lend themselves especially well to color presentation. If the selection of subjects seems weighted in favor of birds on the one hand and microscopic organisms and structures on the other, the reason lies in a happy chance to choose among the superlative photographs in these fields by David G. Allen and Roman Vishniac.

The book owes a great debt to others also, who contributed valuable photographs both in color and in black and white: my colleagues at Cornell University Thomas Eisner, J. G. Franclemont, Herbert Israel, J. M. Kingsbury, D. J. Paolillo, Verne N. Rockcastle; E. R. Lewis of the University of California at Berkeley; and the Carolina Biological Supply Company. The legends identify these and other contributors—individuals, museums, business firms—who graciously allowed reproduction of their photographs.

Many persons read parts of this book in manuscript and offered valuable advice and criticism. Others graciously provided me with information from their special fields that I might not have obtained in any other way. I wish to acknowledge especially the aid of my colleagues Karen Arms, David W. Bierhorst, W. C. Dilger, D. J. Paolillo, Efraim Racker, Richard B. Root, and Stanley Zahler. Numerous teachers who used the first edition of *Biological Science* or *Elements* in their courses sent me suggestions on specific topics; these were of great help to me during preparation of the revisions.

Melvin Kreithen aided me throughout the preparation of the revised editions by searching references, running down obscure facts, commenting on the manuscript, preparing a first draft of the new material in Chapter 18, and helping in other ways too numerous to mention.

Bertha Blaker, my secretary, assisted me in many ways during preparation of my books. It is a pleasure to acknowledge her help.

To Dr. Horst Mittelstaedt I am indebted for pleasant quarters and stimulating companionship during my sabbatical visit to his department in the Max-Planck-Institut für Verhaltensphysiologie at Seewiesen, Bavaria, where I completed this revision.

Once more I wish to express my gratitude to my wife for her endless patience, interest, and help.

Finally, I thank the many students throughout the world who have taken the time to write to me. Their kind comments have made the long hours of labor worthwhile.

W. T. K.

Seewiesen, Bavaria
January, 1973

ELEMENTS OF BIOLOGICAL SCIENCE

CHAPTER 1

Introduction

The subject of this book is biological science—the study of life. But what is life? We all have some idea. After all, we ourselves are manifestations of life, and endlessly varied forms of it abound around us. But when asked to say explicitly what life is, we find that a satisfying answer eludes us. One dictionary defines life as "the condition which distinguishes animals and plants from inorganic objects and dead organisms," and defines dead as "deprived of life"; in other words, life is what animals and plants have when they are not dead, and dead is what those same organisms are when they lack life. Clearly, such circular definitions get us nowhere. Here is a word for something at the very center of our concern, and yet we are at a loss to offer a meaningful definition of it.

In lieu of a definition, many biology texts give a list of attributes that, taken together, are said to characterize life. The list usually includes metabolism, reproduction, growth, responsiveness, and movement. And there can be no doubt that investigation of these attributes does help toward an understanding of what life is. But merely setting down one-paragraph descriptions of each of these properties with the assertion that they are peculiar to life seems a singularly inadequate way to describe something so rich and complex, so varied and changeful. This entire book is about life; yet when you have finished it, chances are that you still won't be able to define life convincingly or describe it satisfactorily, although you will perhaps have a deeper appreciation and understanding of what it is to be alive.

BIOLOGY AS A SCIENCE

We live in a science-conscious age. Radio and television advertise every product as the latest result of "scientific" research. Commercials are full of "scientists" in starched white lab coats. The news media deal every day with some fresh development in space science or medical science or agricultural science. The President and Congress have scientific advisers, and millions of dollars are spent every year by both the national and the state governments on scientific research and development. Yet there is a surprising lack of understanding of what science is. It is viewed as something akin to magic or as a cold and relentless mathematical game. There is a belief—less prevalent now than in the past—that this game is played exclusively by strange men so narrow and specialized that they have only the most simple-minded notions about anything falling outside their specialty.

The advent of space vehicles and moon probes has changed the public image of at least some scientists. But the new conception of scientists as glamorous adventurers is as erroneous as the older one. Most scientists are just people who have a deep curiosity about the world they live in. They want to know more about the earth and stars, rocks and rivers, atoms and molecules, plants and animals. For many scientists, the need to know is in itself sufficient motivation. For others, the possibility that what they discover may benefit mankind is additional motivation. But whatever motivates them, they are usually neither naïve recluses nor glamorous adventurers. And they are not a strange breed of modern magician waving a wand called "scientific method."

The scientific method

So much has been said about the powers of the scientific method that many suspect it involves some formula too complicated for ordinary people to understand. It doesn't. It is used to some extent by almost everyone every day. Its power in the hands of a good scientist stems from the rigor of its application. Let us briefly examine the elements of this method.

Science is concerned with the material universe, seeking to discover facts about it and to fit those facts into conceptual schemes, called theories or laws, that will clarify the relations among them. Science must therefore begin with observations of objects or events in the physical universe. The objects or events may occur naturally, or they may be the products of planned experiments, but the important point is that they must be observed, either directly or indirectly. Science cannot deal with anything that cannot be observed.

Once a scientist has made careful observations, he must do something with his data. It is not enough simply to amass data; the data must be fitted into some sort of generalization. The observation that a particular fly has three pairs of legs is interesting, but it will remain an essentially useless isolated fact until it is incorporated into a generalization such as, "All flies have three pairs of legs," or the even broader generalization, "All insects have three pairs of legs." The step from isolated observations to generalization can be made with confidence only if enough observations have been made to give a firm basis for generalization and only if the individual observations have been reliably made.

After the scientist has reasoned inductively from the specific to the general (i.e. from specific observations to a general statement), he must reverse his field and reason deductively, from the general to the specific. Suppose, for example, he has observed thirty kinds of flies, ten kinds of beetles, four kinds of wasps, and six kinds of grasshoppers and has found that all have three pairs of legs. And suppose he has arrived at the following generalization from these observations: "All insects have three pairs of legs." He may now reason deductively that if all insects have three pairs of legs, and if cockroaches, crickets, moths, and bees are insects, then they must have three pairs of legs. In other words, he uses his generalization to formulate a hypothesis about things he has not yet observed. Then he proceeds to test his hypothesis, in this case by examining as many kinds of cockroaches, crickets, moths, and bees as he can to check if they really do have three pairs of legs. When he finds that all these insects do indeed have three pairs of legs and that his hypothesis was therefore a good one, he feels more confidence in his generalization. But he doesn't stop there; he makes new hypotheses designed to test his generalization still further. Perhaps he will predict that since immature flies and moths are insects they too must have three pairs of legs. But observation shows him that many grubs (immature flies) and caterpillars (immature moths) have no legs at all. His generalization must therefore take the more restricted form: "All adult insects have three pairs of legs."

Let us summarize the steps followed up to this point by our hypothetical scientist. He began, as all good scientists must, with observations. He used these observations to formulate a generalization, which he then tested by basing simple predictions on it and checking if the predictions were accurate. When he found a discrepancy between his generalization and observable fact, he changed his generalization. Any good scientist must be ready to alter or even abandon his most cherished generalizations when new facts contradict them. He must always remember that his generalizations, his theories, even his physical laws, are dependent upon observable facts and not vice versa.

We have here dealt with a case involving very simple data and a very elementary generalization, but the same conditions would apply to data obtained by the most elaborate experiment and to the most sophisticated and complex generalization based on those data. And they would apply also to higher levels of generalization, such as purported explanations of biological events. At this higher level of generalization, the steps one must follow may be less clear. For example, hypotheses may not flow so obviously from previous generalizations; they may often depend more on educated guesses or hunches than on strict deduction. But in the final analysis the basic rules are the same. The hypothesis must be testable, or it is of no value, and as testing proceeds it must be altered when necessary to conform with the evidence. The scientist's query must always be: "What is the evidence?"

Limitations of the scientific method. The insistence on testability in science imposes limitations on what it can do. For example, the hypothesis that there is a God working through the natural laws of the universe is not testable and hence cannot be evaluated by science. Science can say neither that there is such a God, nor that there is not. Yet it can legitimately say something about certain attributes ascribed to God. Throughout history men have made statements about the physical universe in the name of their gods, and have insisted that denial of their statements is a denial of their gods. If this is so, then science may well have to deny those gods. Any aspect of the physical universe can be studied by science, and anyone making the existence of his God stand or fall on some supposed fact about the universe risks having science destroy his God.

Science cannot make value judgments; it cannot say that a painting or a sunset is beautiful. And science cannot make moral judgments; it cannot say that war is immoral. Science can, however, analyze the technique of a painting, and it can analyze the biological, social, and cultural implications of war. It can, in short, provide people with information that they can use in making value or moral judgments. But the act of making the judgment itself is not science.

THE RISE OF MODERN BIOLOGICAL SCIENCE

The scientific approach discussed above is of relatively recent origin. It may seem obvious to you that the way to learn about nature is to look at nature, to seek evidence by observation, but for many centuries such an approach was not in the least obvious. True, major scientific discoveries were made in several of the ancient civilizations, especially those of Egypt, Babylonia, and Greece, but this early scientific activity began a steady decline more than a century before the birth of Christ. Between about 200 and 1200 A.D., there were almost no important scientific advances, and much of what the ancients had known was forgotten. Superstition reigned during this period; greater reliance was placed on religious dogma and on the writings of a few venerated ancient scholars than on observation of the universe itself. Haggling over the exact meaning of a sentence in one of Aristotle's books on plants or animals took the place of studying the plants or animals themselves; the possibility that Aristotle might sometimes have erred seems not to have been entertained. No real distinction was made between science and theology, and the one was questioned as little as the other.

Then came a period of intellectual reawakening in Europe. Two very influential theologians and philosophers, Albertus Magnus (1193–1280) and Thomas Aquinas (1225–1274), both of whom taught at the University of Paris, recognized a distinction between natural truth and revealed truth. This rationalistic approach, separating large segments of human knowledge and speculation from theology, prepared the way for a relatively independent development of science. At about the same time, Roger Bacon (ca. 1210–1293) at Oxford University was calling for an end to unthinking acceptance of traditionally authoritative writings such as Aristotle's: "Cease to be ruled by dogmas and authorities; look at the world!" Three centuries later, Francis Bacon (1561–1626) vigorously championed the experimental approach to knowledge, urging men to trust no statements without verification, to examine all things with the utmost rigor.

The new era in the physical sciences. The intellectual climate was changing; it was no longer beneath the dignity of an educated man to look at the material objects of nature. Thus Nicolaus Copernicus (1473–1543), a Polish astronomer, analyzed the movements of the heavenly bodies and announced that the earth moves around the sun rather than the sun around the earth. Now, the intellectual climate may have been changing and becoming more friendly to science, but it was not yet ready for such a proposition as this. After all, if the earth moved around the sun, then the earth was not the center of the universe and man did not stand at the center of all creation. An outrageous suggestion! When the great Galileo Galilei (1564–1642) embraced Copernicus' theory, he was forced to recant publicly under threat of excommunication.

But the ideas of Copernicus, Galileo, and the other great scientists who followed them could not be suppressed. Educated men gradually became convinced that the physical universe could be understood in terms of orderly relationships, of universal laws, that physical events have impersonal causes which men can hope to understand, that capricious whims of gods and magicians and evil spirits need no longer be invoked to explain a physical event. No other name stands out so prominently during this period as that of Isaac Newton (1642–1727), who was born in the year of Galileo's death. His discovery of the Law of Gravitation and his explanation in 1685 of the movements of the planets caused a revolution in human thought and carried physical science into a new era. In a very real sense, the work of Newton marks the birth of modern physics.

The science of the new era was largely restricted to physics, astronomy, and chemistry, i.e. to the physical sciences. If men were now ready to give up their place at the center of the universe and admit that the earth circles the sun, they were nevertheless not yet ready to admit that they themselves and other living creatures could be understood in terms of impersonal forces. Life seemed too full of purpose, of design, to be studied in the same way as chemicals and moving particles. If Copernicus was a threat to human dignity and pride, how much greater a threat lay in an explanation of life processes in mechanistic terms! It was not until 1859, nearly two hundred years after Newton, that biological science experienced its own revolution and entered its modern era.

The new era in the biological sciences. The year 1859 is taken as the beginning of the modern era of biology because it was in that year that Charles Darwin (1809–1882), the great British naturalist, published *The Origin of Species,* in which he proposed his theory of evolution by natural selection.[1] We do not wish to imply that no important biological work was done before the time of Darwin—that would be nonsense. But important as the work of Darwin's predecessors was, it did not spark the sort of explosive growth of biological science that Newton's work stimulated in the physical sciences.

An explosion is precisely what Darwin's book did cause. Men had

[1] Actually, the theory was first announced in a short paper read before the Linnaean Society of London in 1858. A paper by Alfred Russel Wallace (1823–1913) containing essentially the same conclusions was also presented on this occasion. Darwin had conceived the theory first and had labored many years to prepare a massive tome that would provide convincing proof. For this reason and because what he published the following year was far more complete than anything Wallace ever published on the theory of evolution, the theory is usually attributed to him. It is, however, a good indication that the times were ripe for such a theory that two men advanced it independently at almost the same time. In fact, earlier in the nineteenth century several other workers had published comments that foreshadowed the ideas of Darwin and Wallace.

long since reconciled themselves to living on a small planet far from the center of the universe, but they were not prepared to accept the ultimate indignity—that they had descended from some lowly form of life in the distant past and shared common ancestors with monkeys and apes and even worms. Again, the old cry of heresy was raised. Again, some men felt that the very basis of all they held dear had been challenged, that if Darwin's views should prevail religion would be destroyed. The outcry was loud and anguished and has not fully subsided yet, although most major Western religions and denominations now accept the theory of evolution and no longer consider it a threat to their existence.

Copernicus' theory was not generally accepted until long after his death, as we have seen. But delayed recognition was not to be the fate of Darwin's theory. And despite violent denunciations from some quarters, Darwin never had to recant like Galileo. The times were ripe for Darwin. A major part of the scientific community welcomed and promptly championed his views. In fact, he himself seldom debated the merits of his theory in public. He didn't need to. Some of his greatest contemporaries eagerly acted as his defenders. His spark had ignited an excitement in the scientific community that was not to be extinguished. Biology was never again to be the same. Almost immediately, interest in biological research began to grow rapidly. Whether to prove or to disprove Darwin, men began to investigate the phenomena they had so long considered beyond the scope of science. The dynamic growth of biological research, begun more than a century ago, has never slackened; in fact, the rate of growth is greater now than ever before. And to this day, the theory of evolution by natural selection remains one of the most important unifying principles in all biology. We shall have cause to refer to it in every chapter of this book.

DARWIN'S THEORY

The theory of evolution, as modified in the years since Darwin, will be treated at some length in Chapter 17. But since we shall have to refer to this all-important unifying principle of biology in interpreting much of the material covered in earlier chapters, let us briefly examine the central concepts of Darwin's theory here.

The theory consists of two major parts: the concept of evolutionary change and the concept of natural selection. First, Darwin rejected the notion that living creatures are the immutable products of a sudden creation, that they exist now in precisely the form in which they have always existed. He insisted that, on the contrary, change is the rule, that the organisms living today have descended by gradual changes from ancient ancestors quite unlike themselves. Second, Darwin said that it is *natural selection* that determines the course of the change, and that this guiding factor can be understood in completely mechanistic terms, without reference to conscious purpose or design. Let us examine the two parts of Darwin's theory separately.

The concept of evolutionary change

In the mid-twentieth century, the idea that lineages of organisms change with time seems far from revolutionary. We are used to change. We should probably be surprised to find anything that remained the same for any very long period of time. But in Darwin's day things moved more slowly. The idea of a world in constant flux had only few adherents. The vast majority accepted without question the proposition that the universe was created a few thousand years before the birth of Christ, and that all the species of plants and animals were put on the earth at that time and had perpetuated themselves without change ever since. What sorts of evidence could Darwin bring forward to combat this view of a static universe?

First, he could point to the fossils. During the latter part of the eighteenth and the first half of the nineteenth century, geologists had unearthed many fossils and realized that most of them represented species no longer living on the earth and, conversely, that few living species were represented in the fossil record. In other words, forms of life different from those known today inhabited the earth in past ages. This is a point now familiar to grade-school children, who are aware that dinosaurs once roamed the earth in vast numbers, but are not encountered nowadays. They can go to museums and see dioramas of ancient seas filled with strange fish and shelled creatures unlike anything living today, or they can see reconstructions of weird coal-age forests with plants that became extinct millions of years ago. They study about the early cavemen, who, though clearly human, were also clearly different in many ways from modern men. To us the existence of fossils seems convincing evidence that the history of life on earth has been marked by change. But when it was first suggested that the extinct creatures whose remains are preserved in the rocks as fossils represented the ancestors from which the organisms living today are descended, it was urged instead that these extinct species indicated the occurrence of catastrophic extinctions at various times in the history of the earth, followed by new episodes of divine creation. According to this hypothesis, each species would have remained unchanged from the time of its creation until the time of its extinction; there would have been no evolution.

Soon, however, the fossil record itself made this hypothesis untenable. As more and more fossils were discovered and studied, it became evident that gradual shifts in characters (physical traits) could be traced through time. If an investigator studied the fossils in one rock layer and then studied the fossils in a slightly more recent layer, he would often find that those in the more recent layer, though very similar to the older ones, showed slight differences. If he then studied a third layer slightly more recent than the second, he would again find that slight changes in the characters of the fossil species could be detected. In this way, by studying a series of successive rock layers, he could reconstruct the sequence of changes through which a given lineage had passed. He could even predict what the fossils in some intermediate layer not yet studied would be like and then test his hypothesis by locating and studying such a layer. The notion of catastrophic extinctions and repeated creations hardly seemed adequate to explain such fossil sequences. It was far more likely that the changes seen in the fossils were the result of accumulation of many small alterations as the generations passed.

Second, Darwin could point to resemblances between living species. If one looks at the forelimbs of a variety of different mammals, for

example, one will find essentially the same bones arranged in the same order. The basic bone structure of a man's arm, a dog's front leg, or a seal's flipper is the same; the same bones are even present in a bird's wing. True, the size and shape of the individual bones vary from species to species, and some bones may be missing entirely in one or another species, but the basic construction is unmistakably the same. To Darwin, the resemblance suggested that all these species had descended from a common ancestor from which each had inherited, with distinctive modifications, its forelimb. The fact that some species possess in reduced and nonfunctional form structures that in other species have important functions further convinced Darwin of the validity of his theory. Why would the Creator have given pigs, which walk on only two toes per foot, two other toes that dangle uselessly well above the ground? Why would he have given human embryos gill pouches and well-developed tails only to make them disappear before the time of birth? It seemed much simpler to assume that such structures were inherited vestiges of structures that functioned in ancestral forms and that still function in other species descended from the same ancestor.

Third, and particularly convincing, Darwin could point to changes produced in domesticated plants and animals. How could anyone confronted with the historical evidence of the changes in domesticated forms doubt that great changes can occur in organisms with time? Where were French poodles and Mexican Chihuahuas two thousand years ago? Where were Guernsey cattle and Leghorn chickens? Where were the modern strains of tomatoes and corn and roses? None of them existed. Their ancestors existed, but those ancestors bore little resemblance to poodles or Chihuahuas or Guernseys or Leghorns or garden tomatoes, corn, and roses. Obviously, radical changes have occurred in a few thousand or even a few hundred years. The ancestors of the poodles and Chihuahuas were wolves. The ancestors of modern corn were small wild plants with ears less than an inch long. Let anyone who would still insist that species cannot change explain these facts.

The concept of natural selection

It was easy for Darwin to see that evolutionary change had occurred. But it took him many years to figure out what caused the changes. His first clue came from the breeding of domesticated plants and animals. When pigeon breeders, for example, are developing a new strain, they exploit the variation always seen among individuals by selecting the ones best endowed with the characteristics they want to propagate and using them as the parents for the next generation. The same procedure is followed in each successive generation; those individuals that most nearly approximate the desired type are selected as breeders, and individuals that deviate markedly from the desired type are eliminated. After many generations of such selection, the pigeons will be very different from the ones with which the breeders began (Fig. 1.1). Essentially the same procedure is used in developing a new breed of dog or horse or wheat or chrysanthemum. Since individual variation occurs in all populations of wild organisms, just as it does in populations of domesticated ones, Darwin reasoned that evolutionary change in these populations must be caused by some sort of natural selection for individuals with certain characteristics and elimination of individuals with other characteristics. But what sort of selective force might be at work in nature? The answer eluded Darwin for several years.

Then in 1838 he read a book entitled *An Essay on the Principle of Population*, written by Thomas R. Malthus (1766–1834) in 1798. This book suggested to him how he could account for the selection he felt sure must be operating in nature. Consider for a moment a population of gray squirrels. If this population is to be perpetuated at a stable level, each pair of squirrels must leave enough offspring to replace itself —two, if we assume that all the offspring survive to reproduce. If the average number of progeny per pair were more than two, then the population density would rise; if the average number were less than two, then the population density would fall. Now, even a casual study of actual populations will reveal that the average number of offspring per pair is always more than two, usually far more. A single female frog may lay many thousands of eggs each year; a single pair of robins usually has two clutches per year of four to five eggs each; and a pair of gray squirrels usually has two litters per year containing two to four young each. A single oak tree may produce millions of seeds during its lifetime. Very large reproductive potentials are, in short, the rule in all types of organisms. Yet natural populations usually remain relatively stable over long stretches of time; they may fluctuate noticeably, but they never even approach the level that would be expected if all their progeny survived to reproduce. It is obvious, therefore, that a very high percentage of the young of any species fail to survive and reproduce.

Once Darwin recognized that in nature the majority of the offspring of any species die before they reproduce, he had the clue he needed to explain natural selection. If survival of the young organisms were totally random, if each individual in a large population had exactly the same chance of surviving and reproducing as every other individual, then there would probably be no significant evolutionary change in the population. But survival and reproduction are never totally random. Some individuals are born with such gross defects that they stand almost no chance of surviving to reproduce. And even among individuals not so severely afflicted, differences in the ability to escape predators or obtain nutrients or withstand the rigors of the climate or find a mate, etc., ensure that survival will not be totally random. The individuals with characteristics that weaken their capacity to escape predators or obtain nutrients or withstand the rigors of the climate, etc., will have a poorer chance of surviving and reproducing than individuals with characteristics enhancing these capacities. In each generation, therefore, a slightly higher percentage of the well-adapted individuals will leave progeny. If the characteristics are inherited, those favorable to survival will slowly become more common as the generations pass and those unfavorable to it will become less common. Given enough time, these slow shifts can produce major evolutionary changes.

Now let us compare the propagation of favorable characteristics in nature, as outlined above, with their propagation in domesticated organisms. In each case far more offspring are born than will survive and reproduce; i.e. in each case there is differential reproduction, or selection. In the breeding of domesticated plants and animals, the selection (differential reproduction) results from the deliberate choice of the breeder. In nature the selection (differential reproduction) results

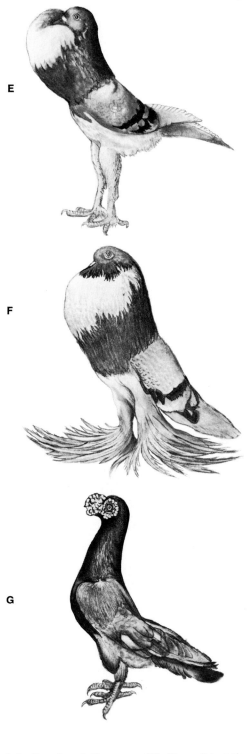

1.1. Breeds of Pigeons. (A) The wild rock pigeon of Europe is thought to be the ancestor of the domesticated breeds shown here. (B) Fantail. (C) Frillback. (D) Satinette oriental frill. (E) English pouter. (F) Pomeranian pouter. (G) Carrier. [Based on photographs in W. W. Levi, *The Pigeon*, Levi Publishing Co., 1957.]

simply from the fact that individuals with different inherited characteristics have unequal chances of surviving and reproducing. Both sorts of selection, artificial and natural, cause some inherited characteristics to become more prominent in the population and others to become less so as the generations pass. Notice that individuals, once born, are not changed by selection. An individual cannot evolve. The change is in the makeup of the population.

One difference between evolutionary change in nature and the change produced by breeders should be noted. That is the rate of the change. Breeders can practice very rigorous selection, eliminating all undesirable individuals in every generation and allowing only a few of the most desirable to reproduce. They can thus bring about very rapid change, as we all know. Natural selection is usually much less rigorous. Some poorly adapted individuals manage to survive and reproduce, and some well-adapted individuals are eliminated. Hence evolutionary change is usually rather slow; major changes may take many thousands or even millions of years. Fortunately for Darwin and his theory, the geologists of his day, particularly Charles Lyell (1797–1875), one of the greatest geologists of all time and a close friend of his, had provided evidence that the earth could not possibly have been created in 4004 B.C., as many churchmen insisted, but must be incomparably older—according to current estimates, several billion years older. Without the geologists' gift of immense spans of time, Darwin's theory of natural selection could not adequately have explained evolutionary change.

In summary, we see that Darwin's explanation of evolutionary change in terms of natural selection depends upon five basic assumptions:

1. Many more individuals are born in each generation than will survive and reproduce.
2. There is variation among individuals; they are not identical in all their characteristics.
3. Individuals with certain characteristics have a better chance of surviving and reproducing than individuals with other characteristics.
4. At least some of the characteristics resulting in differential reproduction are heritable.
5. Enormous spans of time are available for slow, gradual change.

All the known evidence supports the validity of these five assumptions.

PART I

The chemical and cellular basis of life

CHAPTER 2

Some simple chemistry

One of the most common complaints leveled against introductory courses in biology is that they contain too much chemistry. Why, in a course that is supposed to teach biology and not chemistry, should the student be asked, right at the start, to plunge into a chapter devoted almost exclusively to chemistry?

Admittedly, some aspects of biology can be understood reasonably well without reference to chemistry or physics. Someone with an exclusively descriptive approach to living things and little or no concern with underlying mechanisms might conceivably dispense with knowledge of the physical sciences and yet make meaningful contributions to biological knowledge. It is certainly true that not many years ago much of the fundamental biological research was done without reference to the physical sciences, and many unanswered questions and undiscovered facts could still be approached in this manner. But it is also true that in recent years biologists have gained a new appreciation and awareness of the contributions chemistry and physics can make to biology. More and more, living organisms are being viewed as integral parts of the physical universe, to which the fundamental physical laws have as important an application as to atoms and molecules, rocks and minerals, planets and stars. It is becoming increasingly plain that a person without background in the physical sciences is severely limited in his choice of biological pursuits. Even if he chooses an area of biology in which work can be done without this background, his outlook is likely to be so narrow and restricted that he will never achieve the insight and productivity his abilities might otherwise gain him.

But perhaps you are studying biology only because you believe some knowledge of it is part of a liberal education (or because the course is required and you couldn't get out of it), and you may object that for your purposes a simplified version of biology, omitting all chemistry, would be sufficient. Such an objection misses the basic point of taking an introductory science course, which is to get some insight into what the science is today, what questions it asks, how it attempts to answer them, and what its probable future lines of development will be. If all chemistry were omitted from this book, you would get an untrue picture of modern biology, for biology without the physical sciences is not the biology of today or of tomorrow.

THE ELEMENTS

The matter of the universe is composed of a limited number of basic substances called elements. Elements are substances that cannot be decomposed into simpler substances by chemical reactions. There are 92 naturally occurring elements.

Each element is designated by a chemical symbol of either one or two letters that stands for its English or Latin name. Thus H is the symbol for hydrogen, O for oxygen, C for carbon, Cl for chlorine, Mg for magnesium, K for potassium (the Latin name is *kalium*), Na for sodium (the Latin name is *natrium*), etc.

Matter is not continuous; i.e. it cannot be subdivided without limit. Progressive subdivision leads ultimately to units indivisible by ordinary chemical means. These units are called *atoms*. A single atom is customarily represented by the chemical symbol of the element concerned; e.g. N stands for a single atom of nitrogen.

Six elements play such a salient role in the phenomenon of life that they should be given special mention. They are hydrogen, carbon, oxygen, nitrogen, phosphorus, and sulfur. They occur in all living creatures and are indispensable to life as we know it on this planet. We shall have countless occasions to refer to them in the course of our examination of life processes. Do not get the impression, however, that these six are the only elements essential for life. Others such as calcium, sodium, potassium, magnesium, and iron are of great importance in most organisms. Still others, commonly called trace elements, are present in minute amounts in many cells and, despite the extremely small quantities in which they occur, may be indispensable for the maintenance of life.

Atomic structure

Although atoms can be considered the basic units of matter, anyone living today, in the so-called atomic age, is well aware that they are composed of still smaller particles and that, if appropriate methods are used, these particles can be separated from each other. Atoms, then, are complex entities, and if we are to understand their chemical behavior we must consider their detailed structure.

The atomic nucleus. The positive charge and almost all the mass of an atom are concentrated in its center, or *nucleus,* which contains two kinds of primary particles, *protons* and *neutrons.* Each proton carries an electronic charge of $+1$. The neutrons, as their name implies, have no charge. Protons and neutrons have roughly the same mass,[1] which is close to one (see Table 2.1).

[1] Mass is measured in atomic mass units. They are relative units based on the arbitrary assignment of 12 atomic mass units as the atomic weight of C^{12}.

Table 2.1. Fundamental particles

Particle	Mass (atomic mass units)	Charge (electronic charge units)
Electron	0.00055	−1
Proton	1.00728	+1
Neutron	1.00867	0

The number of protons in the nucleus is unique for each element. This number, called the **atomic number,** is usually symbolized by the letter Z and is sometimes written as a subscript immediately before the chemical symbol. Thus $_1$H indicates that the Z number of hydrogen is one; i.e. its nucleus contains only one proton. Similarly, $_8$O indicates that oxygen nuclei contain eight protons. It is often desirable to indicate the total number of protons and neutrons in a nucleus; this number is called the **mass number,** because it approximates the total mass of the nucleus. The mass number is commonly written as a superscript immediately following the chemical symbol. For example, most atoms of oxygen contain eight protons and eight neutrons; the mass number is therefore 16, and the nucleus can be symbolized as O^{16} or, if we wish to show both Z number and mass number, as $_8O^{16}$.

We have indicated that the number of protons (Z) is the same for all atoms of the same element. But the number of neutrons is not always the same, and neither, consequently, is the mass number. For example, most oxygen atoms, as we have said, contain eight protons and eight neutrons and have a mass number of 16; some, however, contain nine neutrons and thus have a mass number of 17 (symbolized as O^{17}), and still others have ten neutrons and a mass number of 18 (symbolized as O^{18}). Atoms of the same element that differ in mass, because they contain different numbers of neutrons, are called **isotopes.** Thus O^{16}, O^{17}, and O^{18} are three isotopes of oxygen. Some elements have as many as 20 isotopes; others have as few as two. Modern nuclear reactors have made it possible to create many new artificial isotopes that do not occur in nature. Apparently the number of neutrons in the nucleus does not affect the chemical properties of an atom, for all isotopes of the same element have essentially the same chemical characteristics.

The number of neutrons does, however, affect the physical properties of a nucleus. Some isotopes are unstable and tend to break down to more stable forms, emitting high-energy radiation in the process. This radiation from the so-called radioactive isotopes is increasingly important in the modern world. As we shall have occasion to note again and again, the radioactive isotopes are invaluable research tools.

The electrons. The portion of the atom outside the nucleus contains only negatively charged particles, the electrons. Electrons have very little mass (see Table 2.1). As a result, almost the total mass of the atom is contributed by the protons and neutrons in the nucleus, even though the extranuclear region constitutes most of the volume of the atom. Each electron carries a charge of −1; i.e. its charge is exactly the opposite of that of a proton.

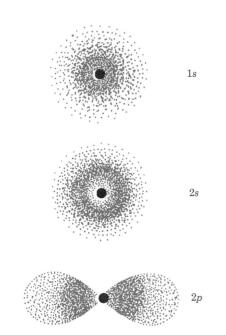

2.1. Representations of electron orbitals. Although the orbitals may look two-dimensional here, they are actually three-dimensional. The density of the dots is proportional to the probability of finding the electron in the region around the nucleus (black). The orbitals of *s* electrons are approximately spherical, those of *p* electrons roughly dumbbell-shaped. The numerals before *s* and *p* indicate the energy level or shell. Thus the 1*s* electron is in the first electron shell, that nearest the nucleus; the 2*s* and 2*p* electrons are in the second electron shell, at a higher energy level and hence at a greater average distance from the nucleus than the 1*s* electron. Note that, despite the very different shapes of their orbitals, the 2*s* and 2*p* electrons are in the same shell; i.e. their most probable distances from the nucleus are nearly the same.

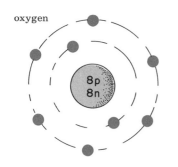

2.2. Atomic structure of oxygen. There are two electrons in the first shell, six in the second.

In a normal neutral atom, the number of electrons orbiting around the nucleus is exactly the same as the number of protons in the nucleus of that atom; the positive charges of the protons and the negative charges of the electrons cancel each other. Consequently, in a neutral atom, the Z number represents both the number of protons inside the nucleus and the number of electrons outside the nucleus. Thus the symbol $_{17}Cl^{35}$ tells us that a neutral atom of this isotope of chlorine has 17 protons, 18 neutrons, and 17 electrons. Similarly, the symbol $_{19}K^{39}$ means that this isotope of potassium contains 19 protons, 20 neutrons, and 19 electrons.

The electrons are not in fixed positions outside the nucleus. Each is in constant motion, and it is impossible to know exactly where a given electron is at any particular moment. All that can be specified is the region around the nucleus in which it is most likely to be. This region, known as the ***orbital*** of the electron, can be described in terms of its distance from the nucleus and its shape (Fig. 2.1).

The distance of an electron from the nucleus is a function of its energy; the higher the energy, the farther from the nucleus will be the probable location of the electron. The average energy levels of electrons in an atom correspond to a series of so-called electron shells, which can conveniently be represented by concentric circles around the nucleus. In an atom of oxygen, for example, there are two electrons in the first shell (the one closest to the nucleus) and six in the second shell (Fig. 2.2).

As shown in Fig. 2.1, the orbitals of electrons may have different shapes. In the first electron shell, this shape is always approximately spherical (it is symbolized by *s*). In the second electron shell, both the spherical shape (*s*) and a dumbbell shape (symbolized by *p*) occur. Additional shapes occur in succeeding electron shells.

It has been shown that there is a maximum number of electrons each shell can contain. The maximum for the first shell is 2, for the second 8, for the third 18, for the fourth 32, etc. Although the third and successive shells can hold more than eight electrons, they are in a particularly stable configuration when they contain only eight. For our purposes, then, the first shell can be considered complete when it holds two electrons and every other shell can be considered complete when it holds eight electrons.

Electron distribution and the chemical properties of elements. When the elements are arranged in sequence according to their Z numbers—beginning with hydrogen, which has the number one, and proceeding to uranium, the last of the natural elements, which has the number 92—it can be seen that elements with very similar properties occur at regular intervals in the list. For example, fluorine, number 9, is more like chlorine, number 17, and bromine, number 35, than like oxygen, number 8, or neon, number 10, the two elements immediately adjacent to it in the list.

The explanation for this periodicity is that the chemical properties of elements are largely determined by the number of electrons in their outermost shell. If that shell is complete, as in helium (Z = 2) or neon (Z = 10), the element has very little tendency to react chemically with other atoms. If the outermost shell has one electron less than the full complement, the element has certain characteristic chemical properties; if it lacks two electrons the element has somewhat different properties; if it lacks seven electrons the element has very different properties.

We have said that fluorine is chemically similar to chlorine and bromine. The critical characteristic shared by these three elements is that each has seven electrons in its outer shell (Fig. 2.3). Oxygen (Fig. 2.2), the element immediately preceding fluorine in the chemists' list, has six electrons in its outer shell and hence is chemically quite different from fluorine. Neon, the element just after fluorine, has a full eight electrons in its outer shell and therefore lacks the reactivity of fluorine.

A convenient way to represent the electronic configuration of the outer shell is to symbolize each electron by a dot placed near the chemical symbol for the element under consideration. Thus fluorine and chlorine, which, as we have said, have seven electrons in their outer energy level, would have the electronic symbols

$$: \overset{..}{\underset{..}{F}} \cdot \qquad : \overset{..}{\underset{..}{Cl}} \cdot$$

Similarly, hydrogen with one electron in its shell, carbon with four in its outer shell, nitrogen with five, and oxygen with six would be shown as follows:

$$H \cdot \qquad \cdot \overset{.}{C} \cdot \qquad \cdot \overset{..}{N} \cdot \qquad : \overset{..}{O} \cdot$$

It goes without saying that the placement of the dots in no way indicates the actual positions of the electrons concerned.

CHEMICAL BONDS

The atoms of most elements possess the property of binding to other atoms to form new and more complex aggregates. When two or more atoms are bound together in this fashion, the attraction that holds them together is called a chemical bond. Each bond represents a certain amount of potential chemical energy. The atoms of a particular element characteristically can form only a certain precise and limited number of such bonds; atoms of some other element may be capable of forming a different but equally precise and limited number of bonds. Our previous examination of atoms should now help you understand how atoms bond together and why each element has its own characteristic bonding capacity.

Ionic bonds

We have said that atoms are in a particularly stable configuration when their outer electron shell is complete, i.e., in most cases, when it contains eight electrons. There is consequently a general tendency for atoms to form complete outer shells by reacting with other atoms. These reactions are the stuff of chemistry, and it is clear that the tendency of atoms to gain complete outer shells forms the basis upon which all chemistry is built.

Consider, for example, an atom of sodium ($Z = 11$). This atom has two electrons in its first shell, eight in the second, and only one in the third. One way sodium might gain a complete outer shell would be to acquire seven more electrons from some other atom or atoms. But the

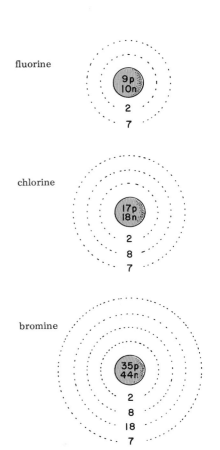

2.3. Atomic structure of fluorine, chlorine, and bromine. The three elements have similar chemical properties because each has seven electrons in its outermost shell.

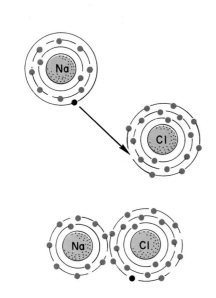

2.4. Ionic bonding of sodium and chlorine.
Sodium has only one electron in its outer shell, while chlorine has seven. Sodium acts as an electron donor, giving up the lone electron in its outer shell; the complete second shell then functions as its new outer shell. Chlorine acts as an electron acceptor, picking up an additional electron to complete its outer shell. But after sodium has donated an electron to chlorine (top), the sodium, left with one more proton than it has electrons, has a positive charge. Conversely, the chlorine, with one more electron than it has protons, has a negative charge. The two charged atoms, called ions, are attracted to each other by their unlike charges (bottom). The result is sodium chloride (NaCl).

sodium would then have an enormous excess of negative charge, and, since like charges repel each other, the electrons would tend to push each other away from the sodium. In point of fact, sodium cannot obtain a full outer shell by appropriating seven additional electrons. An alternative way, and the way actually followed in nature, is for the sodium atom to give up the lone electron in its third shell to some electron acceptor, leaving the complete second shell as the new outer shell (Fig. 2.4).

Next, consider an atom of chlorine ($Z = 17$). This atom has two electrons in its first shell, eight in its second shell, and seven in its third shell. In other words, its outer shell is almost complete, lacking only a single electron. Chlorine can acquire a complete outer shell by gaining an extra electron from some electron donor.

If a strong electron donor like sodium (i.e. an atom with a strong tendency to get rid of an electron) and a strong electron acceptor like chlorine (i.e. an atom with a strong tendency to acquire an extra electron) come into contact, an electron may be completely transferred from the donor to the acceptor. The result, in the present example, is a sodium atom with one less electron than normal and a chlorine atom with one more electron than normal. Once the sodium has lost an electron, it is left with one more proton than it has electrons, and it therefore has a net charge of $+1$. Similarly, the chlorine atom that gained an electron has one more electron than it has protons and has a net charge of -1. Such charged atoms (as well as charged aggregates of atoms) are called *ions,* and are symbolized by the appropriate chemical symbol followed by a superscript indicating the charge. Sodium and chlorine ions are written Na^+ and Cl^-.

A sodium ion with its positive charge and a chlorine ion with its negative charge tend to attract each other, since opposite charges attract. Consequently the two ions are held together by an electrical attraction and form the compound we know as table salt, or sodium chloride, NaCl (Fig. 2.4). Such a bond, involving the complete transfer of an electron from one atom to another and the binding together of the two ions thus formed, is termed an *ionic bond.*

Ionic bonding may entail the transfer of more than one electron, as in calcium chloride, another common salt. Calcium ($Z = 20$) has two electrons in its outermost shell, and it loses both to form the calcium ion Ca^{++} (Fig. 2.5). Chlorine, however, need gain only one electron to complete an octet in its outer shell, as we have already seen. Hence it takes two chlorine atoms to act as acceptors for the two electrons from a single calcium atom, and the calcium chloride formed involves the bonding together of a total of three ions, symbolized as $CaCl_2$, where the subscript 2 indicates that there are two chlorine atoms for each calcium atom in this compound. We can say, then, that calcium has a bonding capacity, or valence, of $+2$, while sodium has a valence of $+1$ and chlorine a valence of -1.

Ionic bonds occur between strong electron donors (typically configurations with one or two electrons in their outer energy level) and strong electron acceptors (typically configurations with six or seven electrons in their outer energy level). Ionic bonding is not common between configurations that have intermediate numbers of electrons in the outer shells, or between units both of which are strong electron donors or electron acceptors.

In many instances *ionization* (i.e. the transfer of one or more electrons from one atom to another to form ions) occurs without true molecular formation. A *molecule* is generally defined as an electrically neutral aggregate of atoms bonded together strongly enough to be regarded as a single entity. Substances like sodium chloride (NaCl) or calcium chloride ($CaCl_2$), in which the bonds are almost exclusively ionic, have a pronounced tendency to dissociate into separate ions when in solution. When they are dissociated or ionized in this manner, they do not exist as molecules. In solution, NaCl forms two separate entities, a Na^+ ion and a Cl^- ion (Fig. 2.6). Similarly, $CaCl_2$ in solution forms three separate entities, a Ca^{++} ion and two Cl^- ions.

Ions, being charged particles, behave differently from neutral atoms or molecules in living systems, and substances wholly or partly ionized in water play many important roles in the functioning of biological systems. In later chapters we shall be concerned with the effects of charge on the movements of materials through the membranes of living cells, and with the partitioning of positive and negative ions that gives rise to the electric potential differences essential for nerve and muscle activity.

Two classes of ionic compounds are of such immense importance that they deserve special mention. These are the acids and the bases. There are a number of different definitions of acids and bases, but for our purposes an *acid* can be defined simply as a substance that increases the concentration of hydrogen ions (H^+) in water, a *base* as a substance that increases the concentration of hydroxyl ions (OH^-) in water. The degree of acidity or basicity (usually called alkalinity) of a solution is usually measured in terms of a scale known as the *pH scale.* A solution is neutral, neither acidic nor basic (i.e. it contains equal concentrations of H^+ ions and OH^- ions), if its pH is exactly 7. Substances with a pH of less than 7 are acidic (i.e. contain a higher concentration of H^+ ions than of OH^- ions); the lower the pH, the more acidic the substance. Conversely, substances with a pH higher than 7 are basic (i.e. contain a higher concentration of OH^- ions than of H^+ ions); the higher the pH, the more basic the substance.

Covalent bonds

Ionic bonds, as we have just seen, involve complete transfer of electrons from one atom to another. But in many, indeed most, cases bonding occurs without complete transfer, by a sharing of electrons between the atoms involved. Bonds of this sort, based on shared electrons, are called *covalent bonds.*

Consider the first element, hydrogen (Z = 1). An atom of hydrogen has only one electron. A complete first shell would contain two electrons. If the hydrogen gained an electron, it would have a full shell, but there would be twice as much negative charge as positive charge in the atom (one proton and two electrons). Hydrogen does not, in fact, ionize in this manner. It tends to do the reverse, losing its single electron and forming H^+ ions, which are simply isolated protons since the hydrogen nucleus contains no neutrons. But suppose there is no strong

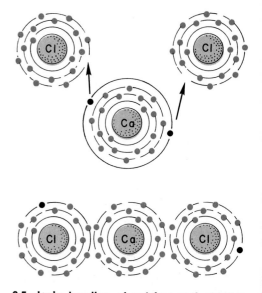

2.5. Ionic bonding of calcium and chlorine. Calcium has two electrons in its outer shell. It donates one to each of two chlorine atoms, and the two negatively charged chlorine ions thus formed are attracted to the positively charged calcium ion (Ca^{++}) to form calcium chloride ($CaCl_2$).

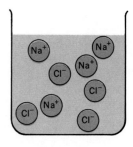

2.6. Ionization of sodium chloride. When in solution, the NaCl dissociates into separate Na^+ and Cl^- ions.

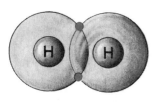

2.7. Covalent bonding of two hydrogen atoms.
The electrons are shared, so that, in a sense, each atom has two electrons in its shell.

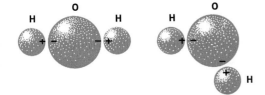

2.8. Molecular structure of water. Two hydrogen atoms are bonded covalently to one oxygen atom, but the shared electrons are pulled closer to the oxygen than to the hydrogen. The bonds are thus polar, with the oxygen side more negative. If the three atoms were arranged linearly, as at left, the charge distribution within the whole molecule would be symmetrical and the molecule would be nonpolar. But since in fact the atoms are arranged as shown at right, the charge distribution is asymmetrical and the molecule is polar.

electron acceptor available and the hydrogen cannot ionize. One possible reaction is for two atoms of hydrogen to bond with each other to form what we call molecular hydrogen, H_2

$$H \cdot + H \cdot \rightarrow H : H$$

In this molecule, each atom shares its electron with the other atom, so that each hydrogen has, in a sense, two electrons in its shell (Fig. 2.7). Clearly, when two atoms of the same element bond together, as in this case, one will not have a greater attraction for the electrons than the other. Instead, the electrons will be shared equally by the two atoms. In other words, there is no more probability that the negative charges will be nearer one atom at any given time than that they will be nearer the other atom. Such a bond is said to be a *nonpolar* covalent bond.

Suppose now that two hydrogen atoms, instead of being bonded to each other, are covalently bonded to an oxygen atom, forming water, H_2O

$$H \cdot + H \cdot + : \overset{..}{\underset{.}{O}} \cdot \rightarrow H : \overset{..}{\underset{..}{O}} : H$$

Oxygen ($Z = 8$) has only six electrons in its outer shell (Fig. 2.2) and needs two more. By sharing electrons with two hydrogen atoms, the oxygen atom can obtain a full outer octet, while at the same time each hydrogen obtains a complete first shell of two electrons. A covalent bond between a hydrogen atom and an oxygen atom is somewhat different from one between two hydrogen atoms or between two oxygen atoms, however. No two elements have exactly the same affinity for electrons. Consequently, when a covalent bond forms between two different elements, the shared electrons tend to be pulled closer to the one element than to the other. Such a bond is called a *polar* covalent bond; i.e. the charge is distributed asymmetrically within the bond. In water, the electrons are closer to the oxygen than to the hydrogen, and the atoms are arranged in such a way that the whole molecule is polar (Fig. 2.8). Molecular polarity, generally a result of the polarity of the bonds within the molecule, often helps explain the structural organization of living systems. In cell membranes, for example, the polarity of the molecular components is an important factor in determining how these components are arranged.

Once we realize that many covalent bonds are polar, and that there are all degrees of polarity—from bonds where the electrons are much closer to one atom than to the other to bonds where they are only slightly closer to one than to the other—we begin to see that there is actually no sharp distinction between ionic bonds and covalent bonds. Ionic bonds are simply one extreme, where the electrons are pulled completely from one atom to the other, and nonpolar covalent bonds are the other extreme, where the electrons are pulled exactly equally by two atoms. Polar covalent bonds represent the middle ground between these two extremes; the electrons are pulled closer to one atom than to the other, but not all the way. Since electrons do not remain in one position but are constantly moving, a bond may be essentially ionic one instant and covalent the next instant. Thus even a compound that is primarily covalent may be very slightly ionized.

In our discussion of covalent bonds so far, we have mentioned only the sharing of one electron pair between two atoms, i.e. the formation of a single bond. Sometimes two atoms share two or three electron pairs and form double or triple bonds. When two atoms of oxygen bond together, they form a double bond (remember that an oxygen atom needs two electrons to complete its outer shell), and when two atoms of nitrogen ($Z = 7$) bond together, they form a triple bond because each nitrogen atom needs three additional electrons to fill its outer shell:

$$: \overset{..}{O} : : \overset{..}{O} : \qquad : N : : : N :$$

So far, we have been diagraming covalent bonds by a pair of dots representing a pair of electrons. Often, however, each pair of shared electrons, which constitutes a covalent bond, is indicated simply by a line between the two atoms, and the other electrons in the outer shells are ignored. Shown in this manner, H_2, H_2O, O_2, and N_2, the molecules we have discussed, appear as follows:

$$H-H \qquad H-O-H \qquad O=O \qquad N \equiv N$$

Hydrogen bonds

One other type of bond, which is extremely important in biological systems, should be mentioned here. This is the so-called **hydrogen bond.** It forms only between a few small, very electronegative atoms like oxygen, fluorine, and nitrogen. It is a low-energy bond in which a hydrogen atom acts as though it were bonded simultaneously to two other atoms; i.e. the hydrogen atom is shared between two other atoms and forms a bridge between them. For example, water molecules are commonly linked together by hydrogen bonds between the oxygen atoms of the water, so that it is difficult to say where one H_2O molecule ends and another begins. Similarly, hydrogen bonds often serve to bind water molecules loosely to the molecules of many other compounds. We shall refer to hydrogen bonds again when we discuss such critically important substances as proteins and nucleic acids.

SOME IMPORTANT INORGANIC MOLECULES

Chemists have traditionally referred to complex molecules containing the element carbon as **organic** compounds. All other compounds are called **inorganic,** a designation that should not mislead you into assuming that these compounds play no role in life processes. Many inorganic substances are, in fact, basic to the chemistry of life. We shall examine a few of the most important below, for without some knowledge of them, it is hardly possible to understand the more complex organic compounds to which we will soon turn our attention.

Water

Let us review some of the facts about water that we have accumulated so far: A molecule of water is composed of two atoms of hydrogen bonded to one atom of oxygen and is therefore symbolized as H_2O. The bonds between the hydrogen and oxygen are primarily polar covalent (although there is always a small degree of ionization: roughly one molecule out of 554 million is ionized). The water molecule itself is polar, and in the liquid and solid states many water molecules are bound together by hydrogen bonds. What more can we say about water here?

Hydrocarbon chains may be

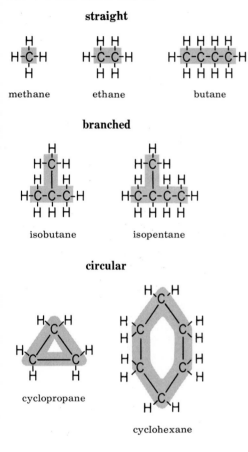

straight

methane   ethane   butane

branched

isobutane   isopentane

circular

cyclopropane

cyclohexane

Carbon-to-carbon bonds may be single,

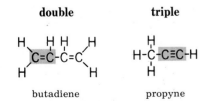

double   **triple**

butadiene   propyne

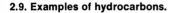

2.9. Examples of hydrocarbons.

Living things are composed largely of water; about 80 to 90 percent of living material is water. Not only is water the major internal component of organisms, but it is also one of the principal external environmental factors affecting them. The chemistry of life, and indeed much of the chemistry of the earth, is water chemistry (i.e. the chemical reactions take place within a water medium). One of the main reasons is that water is among the best solvents known; more different substances will dissolve in water, and in greater quantity, than in almost any other liquid. It is sometimes even called the "universal solvent"—something of an exaggeration, as there are actually many materials, such as gasoline and oils, for which water is a poor solvent. Nonetheless, because of its excellent properties as a solvent in general, and its tendency to produce ionization of substances in solution, it plays a most important role in facilitating chemical reactions. But this ubiquitous substance does more than simply facilitate reactions; it is itself often a reactant or a product of chemical reactions. Water is the principal source of the hydrogen incorporated into the many organic compounds in the bodies of living things.

Aside from its essential chemical roles, water also plays many important physical roles in the world of living things. It is a medium of transport for substances within the bodies of plants and animals. It is almost certainly the medium within which life arose, and vast numbers of organisms still live all or part of their lives in water. Furthermore, the great amount of heat necessary to increase the temperature of water or to change liquid water into water vapor and, conversely, the great amount of heat liberated into the environment when liquid water changes into ice all play crucial roles in stabilizing temperatures on the earth; since the quantity of water on the earth's surface is enormous, the warming or cooling of this water is the requisite for any major widespread temperature changes.

Carbon dioxide

Carbon ($Z = 6$) has only four electrons in its outer electron shell and, as a result, has a covalent bonding capacity of four. Carbon dioxide (CO_2) is the compound formed when two atoms of oxygen bond to one atom of carbon. Though this substance contains carbon, it is generally thought of as inorganic, because it is simpler than all but a few of the compounds classified as organic.

Only a very small fraction of the atmosphere, roughly 0.033 percent, is carbon dioxide;[2] yet atmospheric carbon dioxide is the principal inorganic source of carbon, and carbon is the principal structural element of living matter. Carbon dioxide and water are the raw materials from which green plants manufacture many complex organic compounds essential to life, as we shall see in detail in a later chapter. And when these complex compounds have run their course in the life system, they are broken down again to carbon dioxide and water, and the carbon dioxide is eventually released into the atmosphere. The simple compound carbon dioxide, then, is the beginning and the end of the immensely complex carbon cycle in nature.

[2] The value of 0.033 percent is based on data for 1965. Prior to 1900, the figure was 0.029 percent. The increase is due primarily to man's burning of fossil fuels.

Oxygen

Molecular oxygen (O_2) constitutes approximately 21 percent of the atmosphere. It is a necessary material for maintenance of life in most organisms, although a few can live without it. It can be utilized directly, without change, by both plants and animals. Although the majority of organisms require O_2, most of the oxygen incorporated into the structures of living things comes, not from O_2, but from H_2O. The principal function of O_2 in the life system is seen to be—once the intricacies of its reactions are stripped away—essentially janitorial; it combines with waste hydrogen to form water, a substance of which organisms can easily rid themselves.

SOME SIMPLE ORGANIC CHEMISTRY

Organic compounds are based on the element carbon ($Z = 6$), which has a covalent bonding capacity of four. This element is present in an enormous number of known chemical compounds; of all the elements, only hydrogen has more known compounds. Although carbon can and does bond to a variety of different elements, it is most commonly bonded to hydrogen, oxygen, nitrogen, or more carbon.

Of central importance in organic chemistry are the compounds containing carbon and hydrogen; the number of different compounds of this kind, called hydrocarbons, is immense. A basic reason for the great variety of hydrocarbons is the readiness with which carbon-to-carbon bonds can form, producing chains of varying lengths and shapes (Fig. 2.9). These chains may be simple, as in methane, ethane, butane, and other still longer substances; or they may be branched, as in isobutane and isopentane; or they may form circles of varying numbers of carbons, as in cyclopropane and cyclohexane. Obviously, the more atoms there are in a molecule, the most different arrangements of those atoms are possible. Compounds of the same atomic content and molecular formula but of differing structures and hence differing properties are called *isomers* (Fig. 2.10). Very large organic molecules may have hundreds of different isomers.

Another factor adding variety to hydrocarbon compounds is the capacity of adjacent carbon atoms to form single, double, or triple bonds (Fig. 2.9). And, of course, substitution of other elements or groups of elements for hydrogen atoms makes possible an almost infinite number of derivative hydrocarbons.

Obviously, we cannot even attempt in these few pages to cover the enormous field of organic chemistry, but as the word "organic" suggests, organisms are composed of organic substances, and it is impossible to understand many of the most important attributes of life unless one is familiar with a few simple facts of organic chemistry. We shall mention only four major classes of organic compounds: carbohydrates, lipids, proteins, and nucleic acids.

Carbohydrates

Carbohydrates are compounds composed of carbon, hydrogen, and oxygen.[3] Characteristically, the hydrogen and oxygen are present in the

[3] Derivative carbohydrates may contain other elements in addition to these three.

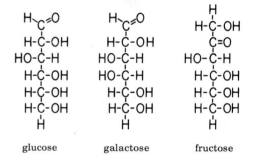

| glucose | galactose | fructose |

2.10. Three isomeric hexoses. Each of these six-carbon sugars has the same molecular formula, $C_6H_{12}O_6$; hence each is an isomer of the others.

same proportions as in water; i.e. there are usually two hydrogen atoms and one oxygen atom for each carbon atom. Consequently the grouping CH_2O, diagrammed H—C—OH, recurs frequently in carbohydrate molecules.

Some carbohydrates, like starch and cellulose, are very large and complex molecules. Fortunately for our grasp of them, they, like most very large organic molecules, are composed of many simpler "building-block" compounds bonded together. If we understand the constituent or building-block compounds, then we shall be able to understand much about the more complex substances.

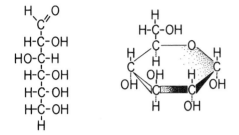

2.11. Two forms of glucose. Glucose may exist in the straight-chain form shown at left or as a ring structure, as shown at right. The ring structure (of which there is an alternative form, not shown here) is the more common.

Simple sugars. The basic carbohydrate molecules are simple sugars, or monosaccharides. All sugars, when in straight-chain form, contain a C=O group (Fig. 2.10). The carbon chain that forms the backbone of the sugar can be of different lengths. Some sugars contain as few as three carbons (trioses); others contain five carbons (pentoses), six carbons (hexoses), or more. Though both trioses and pentoses play significant biological roles and will be mentioned in later chapters, it is the hexoses, six-carbon sugars, that constitute the major building-block compounds for more complex carbohydrates.

There are many six-carbon sugars. Having the proportions of oxygen and hydrogen typical of carbohydrates, all have the same molecular formula, $C_6H_{12}O_6$, even though they differ structurally. Glucose and fructose are two of the most important six-carbon sugars. Glucose exists sometimes as a straight-chain compound, more often in ring form, generally a ring composed of five carbons and one oxygen (Fig. 2.11). Glucose plays a unique role in the chemistry of life. It is in a very real sense the crossroads of the chemical pathways in the bodies of plants and animals. Other six-carbon monosaccharides, among them fructose and galactose, are constantly being converted into glucose or synthesized from glucose. The more complex carbohydrates such as disaccharides and polysaccharides are composed of monosaccharides bonded together in sequence. And even such classes of compounds as fats and proteins can be converted into glucose or synthesized from glucose in the living body.

Disaccharides. Disaccharides, or double sugars, are compound sugars composed of two simple sugars bonded together through a reaction that involves the removal of a molecule of water. This kind of reaction is called a *condensation* reaction or a dehydration reaction.

Let us first examine the double sugar maltose, or malt sugar. This compound is synthesized by a condensation reaction between two molecules of glucose, described by the following equation:

$$2C_6H_{12}O_6 \rightarrow C_{12}H_{22}O_{11} + H_2O$$

This equation tells us very little, however. Any simple six-carbon sugar has the formula $C_6H_{12}O_6$, and any double sugar synthesized from such building blocks will have the formula $C_{12}H_{22}O_{11}$. Consequently the above equation can describe many different reactions involving a variety of reactants and products. What actually happens in the condensation synthesis of maltose is best understood by reference to the structure of the molecules (Fig. 2.12). The process—to give a simplified version of

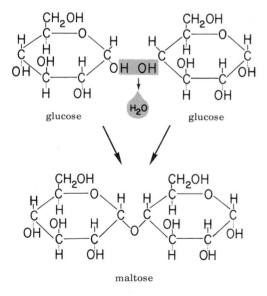

2.12. Synthesis of a double sugar. Removal of a molecule of water between the two molecules of glucose results in formation of a bond between the two. The double sugar produced is maltose.

it—is one where the hydrogen atom from a hydroxyl group (OH) of one molecule of glucose combines with a complete hydroxyl group from the other molecule of glucose to form water. The oxygen valence vacated by removal of hydrogen and the carbon valence vacated by removal of OH are filled by the bonding together of the oxygen of one glucose molecule with the carbon of the other glucose molecule. As a result, the two glucose units are connected by way of an oxygen atom shared between them. The product is the double sugar maltose. *Sucrose,* our common table sugar, is also a disaccharide. It is synthesized by a condensation reaction between a molecule of glucose and a molecule of fructose.

Having seen how double sugars are synthesized, we should also consider the reverse, the breaking of the disaccharide into its constituent simple sugars. This reaction, called *hydrolysis,* involves addition of a water molecule:

$$C_{12}H_{22}O_{11} + H_2O \rightarrow 2C_6H_{12}O_6$$

We will focus particular attention on hydrolysis reactions when we discuss digestion in a later chapter.

We are now in a position to define a simple sugar, or monosaccharide, more precisely than we have so far done. Unlike compound sugars, a monosaccharide is a sugar that cannot be hydrolyzed into smaller carbon-containing substances.

Polysaccharides. The prefix *poly-* means many, and polysaccharides are complex carbohydrates composed of many simple-sugar building blocks bonded together in long chains (Fig. 2.13). They are synthesized by exactly the same kind of condensation reactions as the disaccharides. And like the disaccharides, they can be broken down to their constituent sugars by hydrolysis.

A number of complex polysaccharides are of great importance in biology. *Starches,* for example, are the principal carbohydrate storage products of higher plants; they are composed of many glucose units bonded together. *Glycogen* is the principal carbohydrate storage product in animals and is sometimes called "animal starch"; its molecules are more branched than those of starch. *Cellulose* is a highly insoluble polysaccharide occurring widely in plants, where it is a major supporting material; the bonds between its sugars are somewhat different from those of starch and glycogen, and are more resistant to hydrolysis.

2.13. Two polysaccharides. Small segments of molecules of starch and cellulose are shown. Each of these polymers consists of a very large number of glucose units bonded together, but the units are arranged differently in the two.

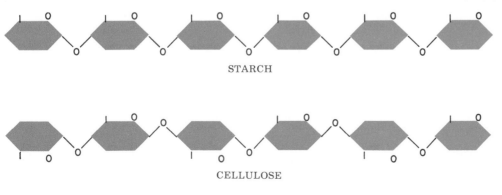

STARCH

CELLULOSE

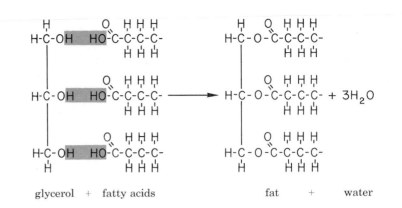

glycerol + fatty acids   fat   +   water

2.14. Synthesis of a fat. Removal of three molecules of water results in the bonding of three molecules of fatty acid to a single molecule of glycerol. The carbon chains of the fatty acids are usually longer than shown here.

Reactions like those that form polysaccharides—i.e. reactions in which small molecules bond together to form long chains—are called polymerization reactions. The products formed are called *polymers.* Polymers of several types play many critical roles in life processes, as we shall see.

Lipids

Lipids constitute a second major group of important biological compounds. Like carbohydrates, they are composed principally of carbon, hydrogen, and oxygen, but may sometimes also contain other elements, particularly phosphorus and nitrogen. They differ from carbohydrates in that they contain a much smaller proportion of oxygen. They are insoluble in water.

Fats. Among the best-known lipids are the fats. Each molecule of fat is composed of two different types of building-block compounds, an alcohol called glycerol and fatty acids.

Glycerol (also sometimes called glycerin) has a backbone of three carbon atoms; attached to each carbon is a hydroxyl (OH) group (Fig. 2.14).

Fatty acids, like all organic acids, contain a COOH group (called a carboxyl group):

There are many different fatty acids, varying in carbon-chain length, in the number of single or double carbon-to-carbon bonds, and in other characteristics. Most of the fatty acids that enter into the formation of edible fats have relatively long carbon backbones, usually from 4 to 24 carbons, or more.

Organic acids and alcohols have a tendency to combine through condensation reactions. Since glycerol has three alcoholic (OH) groups, it can combine in this fashion with three molecules of fatty acid (Fig. 2.14); the result is a fat. A molecule of fat, then, is composed of one molecule of glycerol and three molecules of fatty acid.

Various fats differ in the specific fatty acids, or types of fatty acids, of which they are composed. You have doubtless read of the controversy over the relative health value of saturated and unsaturated fats.

UNSATURATED FAT

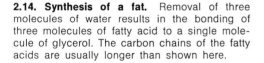

SATURATED FAT

2.15. Unsaturated and saturated fats compared. The fatty acids of an unsaturated fat contain less than the maximal amount of hydrogen.

Saturated fats are simply those incorporating fatty acids that have the maximum possible number of hydrogen atoms attached to each carbon and that therefore have no carbon-to-carbon double bonds (Fig. 2.15). The fatty acids in unsaturated fats (or perhaps we should say oils, since they are usually liquid at room temperature) have at least one carbon-to-carbon double bond; i.e. they are not completely saturated with hydrogen.

Since fats are synthesized by condensation reactions (removal of water), they, like complex carbohydrates, can be broken down to their building-block compounds by hydrolysis, as happens in digestion.

Other lipids. Other important lipids (fatlike substances) include waxes (in which the alcohol is larger than glycerol) and *phospholipids,* which are composed of glycerol, fatty acids, phosphoric acid, and usually a nitrogenous compound (Fig. 2.16). The part of the phospholipid molecule where the phosphate group is located is electrically charged and thus water-soluble, whereas the two long fatty acid chains are not charged and are not soluble in water. Thus the molecule is polar and its two ends behave quite differently in the living cell. Phospholipids are important components of many cellular membranes.

Another group of compounds, the *steroids,* are commonly classified as lipids because their solubility characteristics are similar to those of fats, oils, waxes, and phospholipids; all are insoluble in water but soluble in ether. As you can see from Fig. 2.17, the structure of steroids is very different from the structures of the other lipids we have discussed. Steroids are not based upon a bonding together of fatty acids and an alcohol; they are complex molecules composed of four interlocking rings of carbon atoms, with various side groups attached to the rings. Steroids are very important biologically; some of the vitamins and hormones are steroids, and steroids are frequently important structural elements in living cells, particularly in cellular membranes.

Proteins

Far more complex than either carbohydrates or lipids, proteins are basic to both the structure and function of living material. The number whose structure is known is still very small. Indeed, one of the most active fields of research in biochemistry today is the study of protein structure.

In spite of the gaps in our knowledge about proteins, some basic facts about them can be explained even in such an elementary book as this, because proteins, like carbohydrates and lipids, are composed of relatively few simple building-block compounds.

The structure of proteins. All proteins contain four essential elements: carbon, hydrogen, oxygen, and nitrogen. Atoms of these four elements are bonded together to form compounds called *amino acids,* which, being organic acids, contain the COOH group; in addition, they each have an amino group, NH_2

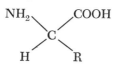

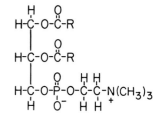

2.16. A phospholipid. One of the fatty acids has been replaced by a group containing phosphorus and nitrogen. Because the group is charged, this portion of the molecule is water-soluble. (R stands for the long hydrocarbon chains of the fatty acids.)

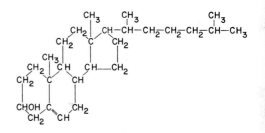

2.17. A steroid. All steroids have the same basic unit of four interlocking rings. They differ in the side groups attached to those rings. This particular steroid is cholesterol.

28

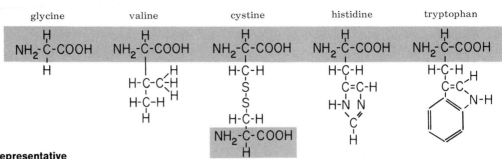

glycine valine cystine histidine tryptophan

2.18. The structure of some representative amino acids. All the amino acids that enter into the formation of protein have the same NH_2—CH—COOH unit at one end (color). They differ in their subsidiary groups (usually denoted as R). Glycine is the simplest, R being a single hydrogen atom. Some amino acids have straight-chain R groups and others have rings (by convention a hexagon means a six-carbon ring). Notice particularly cystine, a symmetrical molecule in which the two identical ends are held together by a disulfide linkage.

The various amino acids differ in their side chains, shown as R in the structural formula. R may be very simple, as in glycine, where it is only a hydrogen atom, or it may be very complex, as in tryptophan, where it includes two ring structures. Twenty-odd different amino acids are commonly found in proteins; the structural formulas of some of them are shown in Fig. 2.18.

Proteins are long and complex polymers of the twenty-odd amino acids. The amino acid building blocks form bonds by condensation reactions between the COOH groups and the NH_2 groups (Fig. 2.19). Such bonds are called *peptide bonds,* and the chains they produce are called *polypeptide chains.* The number of amino acids in a single protein molecule may vary from about 50 to 50,000 or more. Since proteins may differ in the total number of amino acid units they contain, in the relative quantities of the different amino acids, and in the specific sequence in which the amino acids are bonded, enormous variation is obviously possible. The number of different proteins is almost endless.

But protein molecules are even more complex than our discussion so far may indicate. We have mentioned amino acid content (number, type, and sequence). Now we must also mention the spatial arrangements of the polypeptide chains in which the amino acids are combined. Let us begin by examining the structure of one of the more unusual of the amino acids, cystine, shown in Fig. 2.18. Notice that this curious molecule is a sort of double molecule, a molecular twin as it were. At each end is a carbon to which are attached a carboxyl (COOH) group and an amino (NH_2) group. The two ends of the molecule are joined together by two atoms of sulfur, forming what is called a disulfide bond (–S–S–). Each half of a cystine molecule can, as a result of its structure, enter into reactions as though it were an

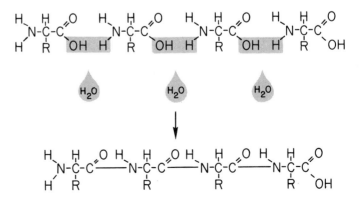

2.19. Synthesis of a polypeptide chain. Condensation reactions between the —COOH and —NH₂ groups of adjacent amino acids result in peptide bonds between the acids.

independent amino acid.[4] If one half of a cystine molecule were to enter into the formation of one polypeptide chain and the other half into the formation of a different polypeptide chain, the two polypeptide chains would be linked together by the disulfide bond in the cystine molecule and would constitute a single protein (Fig. 2.21). In a similar manner, one half of a cystine molecule might be incorporated into a polypeptide chain at one point and the other half into the same chain at a different point, causing the single polypeptide chain to fold back on itself (Fig. 2.21). Disulfide bonds, then, play a very important role in linking together the constituent polypeptide chains of complex proteins and in determining the folding pattern of proteins.

Proteins are not laid out simply as two-dimensional chains of amino acids, as our discussion so far might seem to imply. Instead, they are coiled and folded into very complex spatial patterns. These three-dimensional patterns are apparently maintained primarily by hydrogen bonds between oxygen and nitrogen atoms.

When proteins are exposed to excessive heat, radiation, electricity, excessive acidity, various chemical reagents, etc., their structure may become disorganized. One type of disorganization is called denaturation; apparently the hydrogen bonds are ruptured, causing the normally regular and rigid arrangement to unfold into a more diffuse formation (Fig. 2.20). Denaturation is partly or wholly reversible in some cases. Sometimes the process of breakdown and disorganization of the protein goes so far that there is loss of solubility. The protein is then said to be coagulated; coagulation cannot be reversed. Denatured or coagulated proteins ordinarily lose their normal properties and activity—a clear indication that the three-dimensional architecture is an essential aspect of protein structure; ultimately, any attempt to explain the biological properties of proteins must take this aspect into account.

Determining the structure. As we have mentioned, the structures of only a few proteins are known so far. It will be instructive to trace some of the steps by which Frederick Sanger and his colleagues at Cambridge University determined the structural formula for insulin, a hormone produced by the pancreas and the first protein for which a structural formula could be written (Fig. 2.21). It took some ten years of exhaustive experimentation before Sanger, in 1954, could feel confident that he had finally established the full sequence of amino acids in insulin. His achievement, a milestone in the history of biochemical knowledge, won him the Nobel Prize in 1958.

Insulin was well suited to be the object of pioneering work on protein structure; it can easily be obtained in pure form, and it is one of the smallest proteins known. Sanger's method of determining the sequence of amino acids in insulin involved breaking the molecules into fragments and then trying to establish how the pieces fitted together. If a protein is heated in an acid solution for about 24 hours, all its peptide bonds are hydrolyzed; i.e. a water molecule is added at the site of each peptide bond, the bond thus breaking and the amino acids being uncoupled. The results can then be analyzed, and it can be determined which amino acids are present and in what amounts.

[4] Actually the cystine molecule usually forms only after the two ends, as free molecules of cysteine, are incorporated into polypeptide chains.

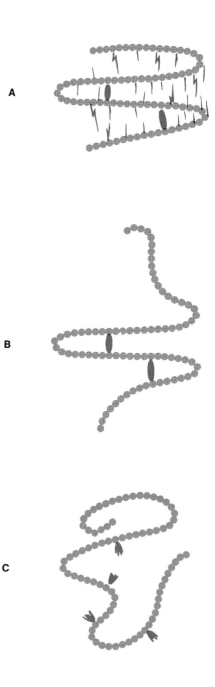

2.20. Denaturation of a protein. First the weak attractions (A) between adjacent parts of the folded molecule are destroyed (B), and then the disulfide linkages are ruptured (C). The three-dimensional configuration of the molecule is completely changed.

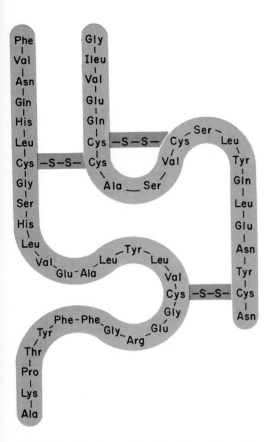

2.21. The structure of beef insulin. The molecule consists of two polypeptide chains joined by two disulfide linkages. There is also one disulfide linkage within the shorter chain (right). The amino acids are abbreviated as follows: Ala, alanine; Arg, arginine; Asn, asparagine; Cys, cysteine; Glu, glutamic acid; Gln, glutamine; Gly, glycine; His, histidine; Ileu, isoleucine; Leu, leucine; Lys, lysine; Phe, phenylalanine; Pro, proline; Ser, serine; Thr, threonine; Tyr, tyrosine; Val, valine.

By less drastic treatment of insulin with hydrolyzing agents, Sanger could preserve some of the peptide bonds and thus obtain many protein fragments consisting of two, three, four, five, or more amino acids. He analyzed these fragments for their amino acid content. After analyzing vast numbers of such pieces, he attempted to fit them together in proper sequence by looking for fragments with regions of apparent overlap. For example, he found the following two fragments that seemed to have overlapping sequences at their ends:

Leu–Val–Cys–Gly–Glu–Arg–Gly–Phe–Phe
Gly–Phe–Phe–Tyr–Thr–Pro–Lys

The reasonable inference was that one part of the insulin molecule contained the sequence

Leu–Val–Cys–Gly–Glu–Arg–Gly–Phe–Phe–Tyr–Thr–Pro–Lys

He then hunted for other fragments that overlapped this sequence, so that he could extend it. After long and laborious investigations of this sort, he finally determined the entire amino acid sequence.

Since Sanger's analysis of insulin, several other proteins have been similarly elucidated. The more recent work has been aided, as you would expect, by many technological advances. Of particular importance has been the development of a method for splitting off amino acids one at a time from the end of the polypeptide chains.

The three-dimensional configuration. Our knowledge of the three-dimensional structure of proteins has also moved forward. The use of X rays as analytical tools has been fundamental to this work. Proteins can be crystallized, an indication that all the molecules have the same detailed shape and that they are arranged in a regular three-dimensional array. If X rays are sent through a crystal while it is being turned in various directions, the X rays do not travel in a straight line, but are scattered by the electrons of the atoms in the crystal. If these scattered X rays fall on a photographic plate, they produce a characteristic pattern determined by the angle at which the X rays strike the planes of the crystal. Since crystals of different substances have different atomic and electron arrangements, they produce different X-ray patterns, from which the structure of the crystal can be deduced.

This technique of X-ray crystallography has been applied to several proteins, among them myoglobin, a protein similar to but smaller than hemoglobin, the red pigment in blood. John C. Kendrew and his associates at Cambridge University recorded thousands of reflections of X rays through myoglobin crystals at various angles and fed them into a computer for analysis. They then used the computer results as a guide in constructing a three-dimensional model of myoglobin (Fig. 2.22). As the figure shows, this protein contains a group different from the characteristic polypeptide chains, a so-called heme group, which is a flat grouping of atoms with an iron atom at its center. Hemoglobin contains four such heme groups. A nonproteinaceous grouping of this kind, known as a *prosthetic group*, is attached to many proteins and is usually essential to their characteristic chemical reactivity. For example, the heme group in myoglobin and hemoglobin is necessary to the interactions of those substances with oxygen.

Nucleic acids

Nucleic acids, a fourth major class of organic compounds, are long polymers composed of building-block units called nucleotides (see Fig. 15.3, p. 322). The nucleotides contain carbon, hydrogen, oxygen, phosphorus, and nitrogen. It is the sequence in which the different nucleotides are arranged that gives each nucleic acid its own distinctive properties (see Figs. 15.6 and 15.7, p. 324).

Nucleic acids play a fundamental role in the transmission of hereditary traits, in the control of the functioning of living cells, and in the synthesis of proteins. We shall study them in detail in a later chapter and so shall not examine them further here.

GENERAL CONDITIONS FOR CHEMICAL REACTIONS

In this chapter we have mentioned several types of chemical reactions that take place within the organism, but have said nothing about the conditions under which these reactions will take place. It is time to examine those conditions briefly.

At this point, instead of talking about condensation reactions or hydrolysis reactions or any other specific type of reaction, let us consider some hypothetical generalized reactions. Suppose, for example, that two substances, A and B, can react with each other to form two new substances, C and D:

$$A + B \rightarrow C + D$$

What factors determine the rate at which such a reaction occurs? First, we must recognize that if A and B are to react with each other, their molecules must come into contact. It follows that anything that increases the probability of contact increases the rate of reaction. One obvious way to increase the probability of molecular contact is to increase the concentration of the reactants; the more molecules per unit volume, the more will by chance bump into each other per unit time.

Increasing the concentration of the reactants is not the only way to speed up a reaction. Raising the pressure on the reactants is another way. Heating the reactants is still another. Heat is a form of energy, and all forms of energy involve motion of some sort. For heat, the motion is that of the atoms and molecules of which the substance in question is composed. At all temperatures above the theoretical absolute zero ($-273°C$), all particles possess motion, ordinarily a vibratory randomly directed type of motion. This motion increases as the temperature increases. Clearly, the more of this sort of random heat motion (or thermal agitation, as it is often called) the molecules of the reactants possess, the greater will be the chance that they will bump into each other. Thus, applying heat to the reactants will increase the rate of the reaction. Application of heat is a standard and frequently used technique of chemists in their laboratories.

So far, we have not mentioned the energy changes that take place in reactions. All substances contain energy. Conversion of one substance into another, or of two substances into new ones, ordinarily involves changes in energy content. In some reactions, the products contain less

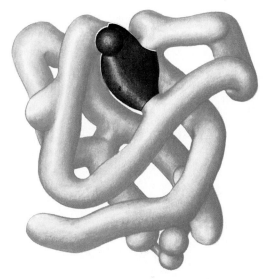

2.22. Spatial configuration of a molecule of myoglobin. The molecule consists of a single complexly folded polypeptide chain of 151 amino acid units. Attached to the polypeptide chain is a nonproteinaceous prosthetic group, called heme, which incorporates an atom of iron, shown here as a large ball. [Modified from J. C. Kendrew, *Science*, vol. 139, 1963.]

energy than the reactants, the extra energy being released by the reaction:

$$A + B \rightarrow C + D + energy$$

Such reactions are said to be *exergonic.* In other reactions, the products contain more energy than the reactants; an external source of energy is necessary if these reactions, called *endergonic* reactions, are to take place:

$$W + X + energy \rightarrow Y + Z$$

It should be clear that just increasing the concentrations of the reactant substances in an endergonic reaction is not sufficient effectively to increase the rate of the reaction; the energy supply must also be increased. In living things, it is very common for exergonic and endergonic reactions to be coupled, the energy released by the one being used to drive the other.

In summary, concentration, pressure, temperature, and energy changes all exert an influence on reactions. But these are not all the factors we should consider. Many exergonic reactions do not begin spontaneously even when the conditions of concentration, pressure, and temperature are adequate. They need, so to speak, a push to get them going. This push must come from some outside source and is called *activation energy.* For example, a sheet of paper can burn (i.e. combine with oxygen) easily once ignited, but that same paper may exist in direct contact with a plentiful supply of oxygen for centuries without ever beginning to burn. Activation energy, usually in the form of extra heat from a match or some other hot object, must be applied to start the reaction of paper with oxygen. The magnitude of activation energy needed is different for each reaction. It is a measure of the stability of the reactants under the given set of conditions.

Catalysis

In the laboratory, activation energy is generally provided by applying heat. But how can activation energy be supplied for reactions that take place within the bodies of living things? For example, how do hydrolysis reactions involving carbohydrates, fats, and proteins occur? We all know from everyday experience that the simple mixing of starch and water is not followed by hydrolysis and the rapid production of sugar. Hydrolysis reactions require activation energy of fairly high magnitude. Clearly, when such reactions occur in our bodies, the activation energy cannot be applied in the form of heat, or our bodies would be destroyed. The same is true for all other living things. Yet hundreds or thousands of reactions that in a test tube would require high activation energy take place every instant within an organism at relatively low temperatures. What is the explanation?

The chemist knows that many reactions can be made to take place much more readily if a small quantity of catalyst is added to the mixture. For example, a simple mixture of hydrogen gas and oxygen gas shows no appreciable reaction. If, however, a small quantity of platinum is added, an explosive reaction takes place, and water is formed. At the end of the reaction, the platinum is still present, unchanged. This lack of alteration is a general property of *catalysts,* which are substances that speed up reactions but are unchanged themselves when the reactions

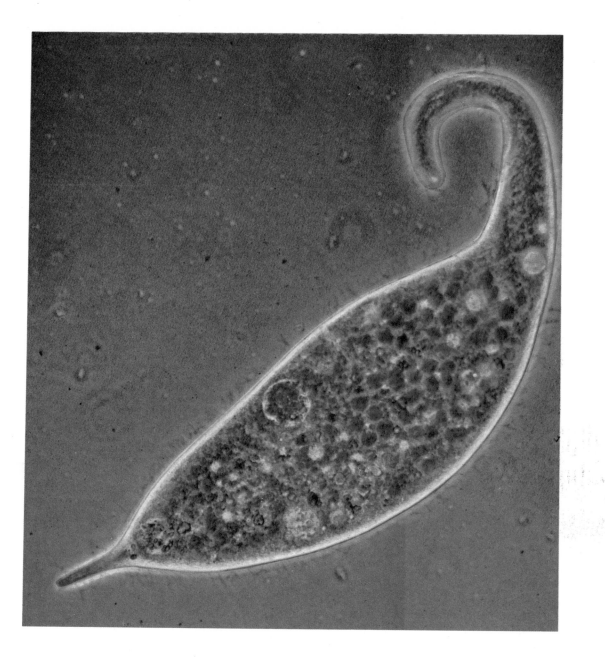

I-1 A SINGLE-CELLED ORGANISM

All living things are made up of cells. At the lowest level of biological organization are the single-celled organisms—protozoans (literally, "first animals"), bacteria, some algae. As single-celled organisms go, *Dileptus*, the protozoan shown here, is fairly complex, possessing many special structures. Cilia, faintly discernible around the edge of the cell, cover the whole surface; they serve in locomotion and in sweeping food particles into the mouth, which is at the base of the long flexible proboscis. Food vacuoles (large brownish spots) digest the food and distribute it throughout the cell. There may be as many as 80 nuclei in the undivided, granular cytoplasm of *Dileptus*.

Photo, Roman Vishniac.

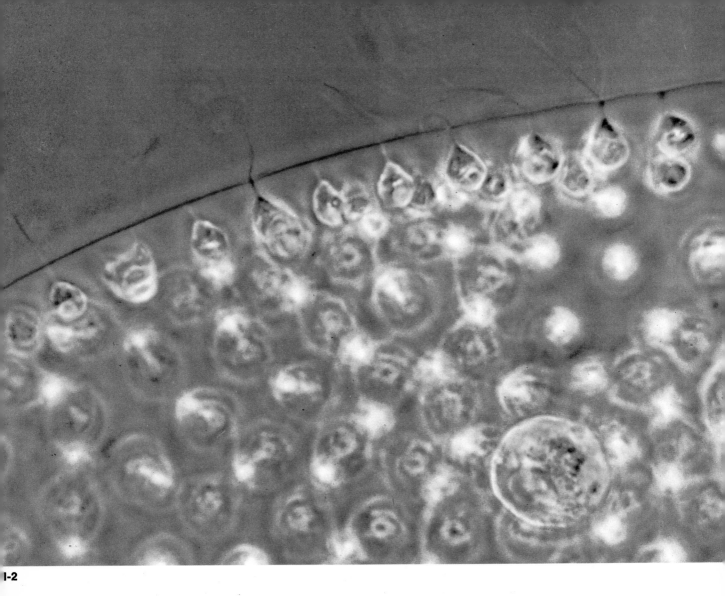

I-2, I-3 COLONY OF UNICELLS...OR MULTICELLULAR ORGANISM?

A *Volvox* colony is a sphere (3A) with a single layer of cells embedded in gelatinous material; the interior of the sphere is filled with a watery mucilage. The cells are plantlike in having chlorophyll and cell walls, animal-like in that each has two flagella (see above). There is some coordination of movement among the flagella, and the colony can thus swim as a unit. Although the cells are not actually in contact, they are interconnected by protoplasmic strands (3C). Certain cells are specialized for sexual reproduction, others for asexual reproduction. The latter give rise to daughter colonies, which are released upon disintegration of the parent colony; small daughter colonies can be seen in 3A, and larger ones, ready to lead an independent existence, in 3B. Intercellular connections, coordination of movement, and specialization among cells are all features of multicellularity, and *Volvox* might be considered a very early stage of multicellular life. Indeed some biologists have speculated that multicellular animals arose from a colonial flagellate like *Volvox* but differing from it in lacking chlorophyll and cell walls.

3B, J. M. Kingsbury. All others, Roman Vishniac.

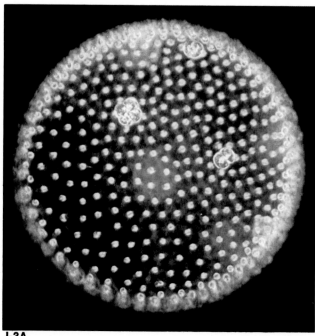

I-3A
I-3C

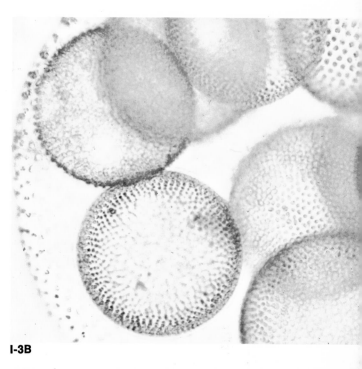

I-3B

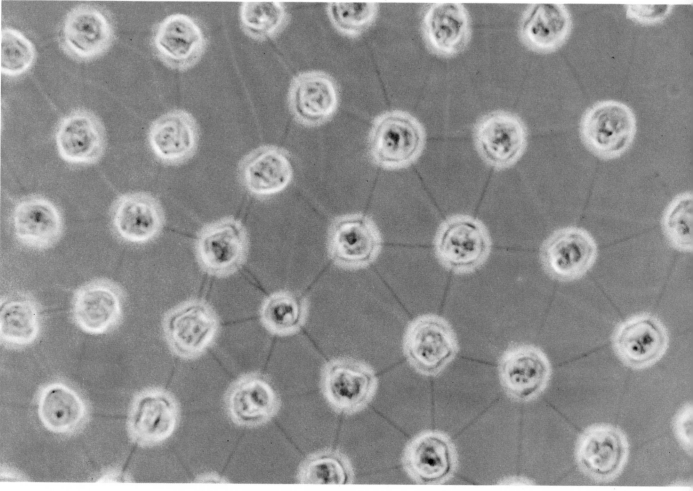

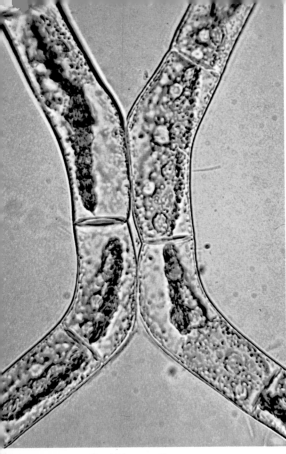

I-4A A green alga, *Mougeotia*

I-4B A horsetail

I-4C A red alga, *Grinnellia*

I-4, I-5 PLANTS, AUTOTROPHIC AND HETEROTROPHIC

Autotrophic plants, plants that can manufacture their own food, possess the green pigment chlorophyll, which enables them to use light energy to synthesize high-energy compounds from carbon dioxide and water. The examples here are (4A) a filamentous green alga, showing chloroplasts in each cell; (4C) a red alga, whose red pigment masks the green chlorophyll; (5A) a brown alga, whose brown pigments mask the green; (4D) moss, exhibiting two life stages: the green gametophyte and, growing from it, the red-stalked, white-capsuled sporophyte; and (4B) a horsetail, a primitive vascular plant. Some plants, however, lack chlorophyll and are heterotrophic, resembling animals in their dependence on ready-made high-energy compounds. The examples here are (5B) a mold of the species that produces penicillin; (5C) an edible mushroom and (5D) a poisonous one; (5E) orange fungi and lichens, which are composed of a heterotrophic fungus and an autotrophic alga living in symbiosis.

4A, B, courtesy Carolina Biological Supply Co.; 4C, 5A, J. M. Kingsbury; 4D, 5D, D. J. Paolillo; 5B, Pfizer Inc.; 5C, E, David G. Allen.

I-4D *Polytrichum* moss

I-5A A brown alga, *Fucus*

I-5B A mold colony, *Penicillium chrysogenum*

I-5C Shaggymane mushrooms

I-5D A poisonous mushroom, *Amanita*

I-5E Velvet-stemmed collybias and lichen

I-6A
I-6B

I-6, I-7 SOME INVERTEBRATE ANIMALS

(6A) The file shell, *Lima,* is a scalloplike bivalve mollusc with numerous tentacles along the margin of its mantle. In the background at left is a coelenterate, *Bartholomea.* (6B) *Dirona,* a nudibranch mollusc, is a shell-less marine animal, related to the snails. (7A) Centipeds are many-segmented terrestrial arthropods with a pair of walking legs on nearly every segment. (7B) Sea urchins are echinoderms with numerous spines on the outer surface of their hard boxlike skeleton. (7C) *Goni-onemus* is a marine coelenterate that spends most of its life as a medusa—a weakly swimming bell-like creature with numerous tentacles.

6A, 7B, Roman Vishniac. 6B, 7A, 7C, courtesy Carolina Biological Supply Co.

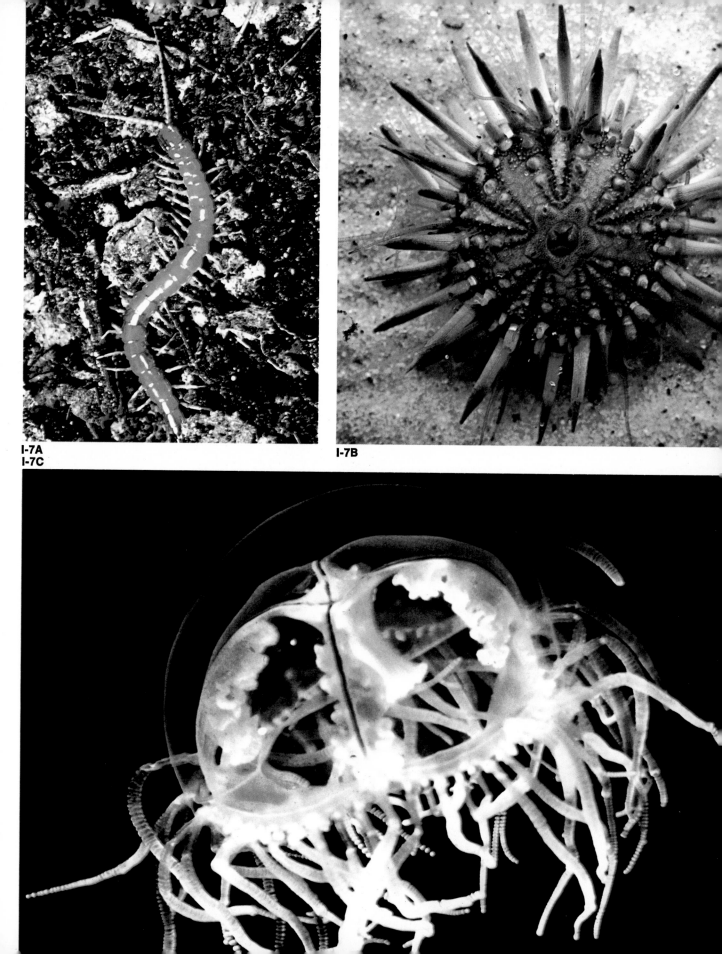

I-7A
I-7C

I-7B

I-8   SEDENTARY LIVES

Animals get about under their own power, plants stay put—that is the most obvious distinction between them, and by and large a valid one. But some animals, especially aquatic ones, are hardly less sedentary than plants, remaining in one spot most or all of their lives. The corals (A, lower left) have no more tendency to alter their position than the red algae attached to them, for the limy skeletons that coral polyps secrete around themselves are rooted to the substratum. Peacock worms (B) are segmented worms that never leave the tubes they build for themselves; their finely branched radioles serve in respiration and feeding. Of the worms seen here among underwater rocks, some are expanding their crowns, others have half retired, and still others have retired entirely.

A, Roman Vishniac. B, Douglas P. Wilson.

A
B

are over (they may have been temporarily changed during the reactions). In the terms of our discussion, catalysts decrease the activation energy needed for reactions to take place. Just how they do this is not fully clear. Apparently, part of the answer is that they tend to adsorb molecules of the reactants to their surfaces, thereby increasing molecular contact or putting unusual strain on chemical bonds and facilitating the reactions. Note that catalysts can speed up only those reactions that are possible to begin with.

The principle of reduction of activation energy by catalysts is fundamental to life. Here is the explanation for the "cold" chemistry in living things. The vast majority of chemical reactions taking place within organisms are catalyzed by special catalysts called **enzymes.** All enzymes are proteins, and some scientists think that all proteins may play some enzymatic role. At any rate, our earlier discussion of the nature and structure of proteins can now be considered a description of enzymes.

Current theory holds that the key to enzyme function—as to the operation of simpler inorganic catalysts—is surface activity. Proteins, as we have seen, are enormously complex molecules with intricate three-dimensional contours. Each different kind of protein has its own distinctive surface geometry. It seems probable that an enzyme combines briefly with reactants whose molecular surfaces "fit" the enzyme surfaces (Fig. 2.23). In this manner, two or more reactants may be brought into close proximity, a state facilitating their interaction; or a reactant molecule, by being attached to an enzyme, may be altered or deformed in its configuration and thus become more reactive.

We can now begin to understand some of the distinctive properties of enzymes. Enzymes are inactivated at temperatures well below that of boiling water (100°C). The explanation, in accordance with the theory of surface activity of enzymes, is that the surface is simply not the same at high temperatures and reactivity is lost; we have already seen that high temperatures break the hydrogen bonds in protein molecules and disrupt their three-dimensional shape. Similarly, we can understand the high sensitivity of enzymes to acidity changes (some work best when the solution is acid, others when the solution is basic, but very few are active over a wide pH range); such changes affect the three-dimensional configurations of proteins.

A characteristic of enzymes critical to the functioning of living systems is their specificity. Most enzymes are highly specific with regard to the reactants they affect (often called substrates) and with regard to the type of reactions they catalyze. Thus, for example, the enzymes that catalyze the hydrolysis of starch are different from those that catalyze the hydrolysis of glycogen, despite the fact that starch and glycogen are two very similar polysaccharides. Similarly, some protein-hydrolyzing enzymes are so specific that they will act only on peptide bonds involving one or two particular amino acids and not on those involving any other amino acids; i.e. they will hydrolyze proteins at only a few specific points along the polypeptide chains. The specificity exhibited by enzymes confers upon living systems enormous potential for very precise control of chemical reactions; this precise control is a basic feature of life.

Enzyme specificity can be seen as a reflection of the distinctiveness of the proteins themselves. We have already seen that each different

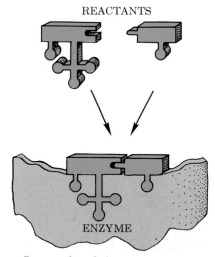

REACTANTS

ENZYME

Reactants brought in contact on enzyme

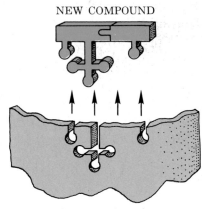

NEW COMPOUND

Enzyme unchanged

2.23. Activity of enzymes. It is thought that enzymes have characteristic surface configurations that "fit" their specific substrates. The enzyme probably combines briefly with the reactant (substrate) molecules in such a way as to bring them close together and facilitate their interaction. When the reaction is complete and the enzyme–substrate complex dissociates, the enzyme is left unchanged.

protein has a surface geometry that is uniquely its own. Since it can catalyze only those reactions that involve substrates capable of "fitting" its surfaces, its catalytic potential is severely limited and its specificity is the result.

Some enzymes function only in conjunction with other organic substances, called *coenzymes.* It seems probable that in some cases the enzyme and its coenzyme combine during the course of a reaction; the coenzyme, which is smaller and less complex than a protein, may cause configurational changes in the enzyme that render it more reactive and thus enhance the catalytic function. In other cases, the coenzyme may function as a substrate for one part of a reaction, being regenerated by the time the entire reaction is complete. Since coenzymes, like enzymes, are not used up or permanently altered by the reactions in which they participate, only very tiny amounts are needed. However, if the supply falls below normal, the health, or even the life, of the organism may be endangered.

REFERENCES

DICKERSON, R. E., and I. GEIS, 1969. *The Structure and Action of Proteins.* Harper & Row, New York.

FIESER, L. F., and M. FIESER, 1957. *Organic Chemistry,* 3rd ed. Heath, Boston.

SIENKO, M. J., and R. A. PLANE, 1966. *Chemistry,* 3rd ed. McGraw-Hill, New York.

WHITE, A., P. HANDLER, and E. L. SMITH, 1968. *Principles of Biochemistry,* 4th ed. McGraw-Hill, New York.

WHITE, E. H., 1970. *Chemical Background for the Biological Sciences,* 2nd ed. Prentice-Hall, Englewood Cliffs, N.J.

SUGGESTED READING

BAKER, J. J. W., and G. E. ALLEN, 1970. *Matter, Energy, and Life,* 2nd ed. Addison-Wesley, Reading, Mass.

GREEN, D. E., 1960. "The Synthesis of Fat," *Scientific American,* February. (Offprint 67.)

KENDREW, J. C., 1961. "The Three-Dimensional Structure of a Protein Molecule," *Scientific American,* December. (Offprint 121.)

ROBERTS, J. D., 1957. "Organic Chemical Reactions," *Scientific American,* November. (Offprint 85.)

SPEAKMAN, J. C., 1966. *Molecules.* McGraw-Hill, New York.

STEIN, W. H., and S. MOORE, 1961. "The Structure of Proteins," *Scientific American,* February. (Offprint 80.)

THOMPSON, E. O. P., 1955. "The Insulin Molecule," *Scientific American,* May. (Offprint 42.)

CHAPTER 3

Cells: Units of structure and function

We saw in the last chapter that a great variety of chemicals, some simple and some complex, enter into the composition of organisms. But the molecules do not of themselves possess the properties we recognize as life. This fact strongly suggests that some kind of order is imposed on them, that they are not simply dispersed in random fashion in an aqueous medium. As we shall see, the components of living matter are indeed elaborately organized. The fundamental unit of organization, to which we shall refer throughout this book, is the cell. It is the purpose of this chapter to examine the cell in some detail.

THE CELL THEORY

The discovery of cells, and of their structure, is bound up with the development of magnifying lenses, particularly the microscope. Some of the optical properties of curved surfaces were known as long ago as 300 B.C., but it was not until the seventeenth century that previous observations were brought to fruition in the development of the microscope. Antoni van Leeuwenhoek (1632–1723) and his contemporaries refined the production of lenses and made microscopes satisfactory for simple scientific observations. Thus in 1665 Robert Hooke was able to present before the Royal Society of London the results of his investigations on the texture of cork. Hooke's work can be taken as the beginning of the study of cells.

It was not until the early nineteenth century, however, that intensive work on cells began. The first clear enunciation of the so-called cell theory is commonly credited to two German investigators, the botanist Matthias Jakob Schleiden and the zoologist Theodor Schwann, who published their ideas in 1838 and 1839 respectively. The *cell theory,* one of the most important generalizations of modern biology, says that all living things are composed of cells. Schleiden and Schwann were not the first to enunciate this principle, but they stated it with particular clarity, and it was their support of the idea that helped it gain general credence.

A very important extension of the cell theory was contributed in 1858 by Rudolf Virchow of Germany. Virchow said that all living cells arise from pre-existing living cells, that there is no spontaneous creation of cells from nonliving matter. Virchow's theory of *biogenesis,* life from life, soon received convincing support from a classic series of experiments by Louis Pasteur in France, which finally laid to rest the old and widespread belief in the spontaneous generation of life. As we shall see in a later chapter, the principle of biogenesis has been somewhat modified in recent years. Current theory holds that spontaneous generation of life from nonliving matter does not occur under present conditions, but that it probably did occur under the conditions existing on the primitive earth when life first arose.

The two components of the cell theory—that all living things are composed of cells and that all cells arise from other cells—give us the basis for our first definition of living things: Living things are chemical organizations composed of cells and capable of reproducing themselves. Notice we have said that this is our first definition of living things. It is a working definition that will allow us to proceed further, but it will not be the last definition we shall consider in this book. We shall see later that any attempt to draw a sharp line between the nonliving and the living becomes essentially arbitrary. It is often convenient, for example, to treat viruses as living things, even though they are not composed of cells in the usual sense.

CELL STRUCTURE

The years since Schleiden, Schwann, and Virchow laid the foundations of the cell theory have witnessed many major advances in the knowledge of cells. Most of the later work has been made possible by the development of better and more powerful microscopes. The electron microscope (Fig. 3.1), in particular, has opened up whole new vistas in the study of cells. This microscope, as its name implies, uses a beam of electrons instead of light as its source of illumination. The electrons pass through the specimen and fall upon a photographic plate, where they produce an image of the specimen. Electron microscopes are capable of resolving[1] objects about 10,000 times better than the unaided human eye. Many of the details of cellular structure discussed in this chapter would not be known but for the electron microscope.

Cell size

Most cells are very small and can be distinguished only with a microscope. Some, however, such as the egg cells of birds, can be seen with the naked eye. Others, like nerve cells, may be very small in some of their dimensions, but extremely long; a single human nerve cell may be as much as 3 or 4 feet long, and an elephant's nerve cell

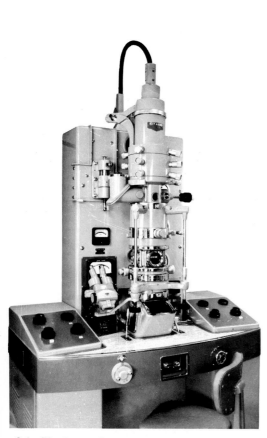

3.1. Electron microscope. [Courtesy Perkin-Elmer Corp.]

[1] Resolution is the capacity to render distinct (show as separate) two objects that are very close together.

may be even longer. To say that cells are generally small is not saying much, however, because even among microscopic cells there is a wide range in size. The diameter of a human red blood cell is about 35 times greater than that of some very tiny microorganisms, while that of a human egg cell is about 14 times greater than that of a red blood cell; the diameter of an ostrich egg cell, in turn, is about 1,500 times greater than that of a human egg cell. Most cells, however, have a diameter of 0.5–40 microns.[2]

Colloids and protoplasm

Before we examine the internal characteristics of cells, we must consider the physical nature of the cellular material. When substances are mixed, there are several possible distribution patterns they may assume.

At one extreme, a *solution* may form. A solution is a homogeneous mixture of two or more components, in which the particles of the different substances are so small that they cannot be distinguished in the mixture. Consider, for example, the solution of a solid in a liquid. If the solid is progressively subdivided, its particles are eventually broken down to individual atoms or molecules or very tiny clusters of these. At this point, a true solution is obtained. The two phases, solid and liquid, can no longer be distinguished. The solution is uniform throughout. The individual molecules cannot be detected even with the strongest microscope. No matter how long the solution is left standing, the dispersed particles do not settle out, nor can they be removed from the solution by filtration (Fig. 3.2).

True solutions represent one extreme—the subdivision of the dissolved substance into minute particles of atomic or molecular size. At the other extreme is a *suspension.* Here the dispersed substance consists of particles so large that it is only by constant agitation that they remain suspended within the liquid. A mixture of sand grains and water is an example. The sand can be kept dispersed in the water as long as the mixture is agitated. When agitation ceases, the sand grains soon settle to the bottom. A suspension, then, is a heterogeneous mixture. The two phases remain distinct.

Between the extremes of a true solution and an obvious suspension are many possible gradations. The transition from homogeneity to heterogeneity is not an abrupt one. A substance may break down to particles so small that they cannot be seen and do not form an obviously separate phase; yet these particles may still be too large to form a true solution. Often the substance has not broken down all the way to its constituent molecules; its particles remain molecular aggregates. Particles in this intermediate size range do not settle out at an appreciable rate. Such a system is called a *colloid.* Cigarette smoke is a colloid of solid dispersed in air. Ordinary milk is a colloid involving a liquid dispersed in another liquid, in this case fat globules dispersed through an aqueous solution.

Let us consider in more detail colloids in which a solid or a liquid is dispersed in water. The obvious initial question is: What prevents the small particles from settling to the bottom over a period of time?

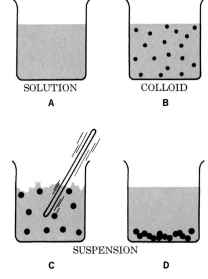

3.2. Solutions, colloids, and suspensions. The tiny particles in a solution cannot be detected and they do not settle out (A). The large particles of a suspension remain distinct from the liquid and will settle out rapidly (D) unless agitated (C). The particles of a colloid are of intermediate size and do not settle out at an appreciable rate (B).

[2] A micron (μ) equals 0.001 millimeter (mm.).

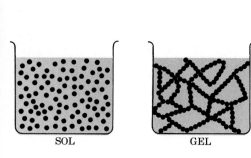

3.3. Sol and gel states of a colloid. The particles in a sol are dispersed in a random and discontinuous fashion. The particles in a gel form a more orderly network.

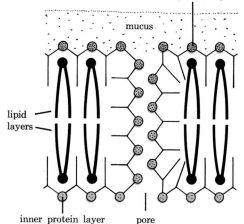

3.4. Model of the cell membrane. Danielli suggested that there are two layers of lipids sandwiched between two layers of protein (as indicated here, the outer and inner protein layers are probably not identical). The molecules of lipid are oriented with their polar heads (circles) near the two surfaces and their nonpolar tails projecting at right angles from the surfaces. The polar portions of the protein molecules are oriented parallel to the surfaces while their nonpolar side groups project at right angles from the surfaces. Tiny pores probably penetrate the membrane at some points. Many cells have a coating of carbohydrate-containing mucoprotein on their outer surface.

One factor is the constant motion of the particles and the water molecules, a motion that is due to their heat energy; the moving particles are constantly jostling one another, which helps keep them in suspension. Another factor is the tendency of particles with like charges to repel each other. Still another is a mutual attraction between particles of the dispersed phase and the aqueous continuous phase.

Colloidal particles often exhibit a capacity for interconversion between sol and gel states. In the *sol* state, the colloidal particles are dispersed throughout the medium in random fashion; they constitute the dispersed phase, while the liquid in which they are dispersed is the continuous phase. When in the sol state, the colloid is essentially fluid. In the *gel* state, the colloidal particles interact and form a more orderly three-dimensional spongy network extending throughout the medium, so that both phases, in effect, become continuous (Fig. 3.3). When in the gel state, the colloid forms a semisolid. Changes between the sol and gel states are brought about by changes in such factors as temperature, concentration, pH, salt concentration, pressure, and agitation. Think of gelatin, for example. At high temperatures, it is a sol, a fluid; when the temperature is lowered, it is converted into a gel, a semisolid.

An understanding of the properties of colloidal systems is important in understanding the properties of living material. You will remember that protein molecules, which form the principal structural basis of living systems, are extremely large. In aqueous media the individual protein molecules often separate to form a true solution, but being larger than the molecular aggregates of many other substances, they behave physically like colloidal particles. Protein solutions, then, have some of the properties of a true solution and some of the properties of a colloidal system; they have the greater reaction potential of a solution, but the sol-gel potential of a colloid.

Cells are generally said to be composed of *protoplasm,* "living substance" according to the conventional definition. The statement could hardly be less precise, living material being composed of a wide variety of substances, intricately organized. Protoplasm may be regarded as both a complex solution and a heterogeneous colloid. The colloidal material is mostly protein molecules and fat globules. Various other materials such as inorganic salts, simple sugars, and amino acids are in solution. The sol-gel states to which protoplasm is subject because of its colloidal nature are a source of many of its unusual properties. Thus the different regions or structures revealed on close examination of cells are in part explained by the fact that given areas of the cell are in different sol or gel conditions. Through such organization, a system composed largely of water can maintain a high degree of structural integrity. Keeping in mind the properties of colloids, let us now examine the subcellular components.

Structure of the cell membrane

Long before any direct evidence was available, biologists took it for granted that cells are bounded by a membrane—the *plasma membrane,* as it is often called. They deduced its characteristics from various characteristics of the cells themselves, for the membrane is usually not visible even under the most powerful light microscopes.

It had been known for a long time that lipids and many substances soluble in lipids move with relative ease between the cell and the surrounding medium. From this fact it was deduced that the outer boundary of the cell, the cell membrane, must contain lipids, and that fat-soluble substances could move across the membrane by being dissolved in it. The lipid model of the cell membrane was not sufficient to explain all the results of permeability studies, however. Because it was observed that many water-soluble substances also move quite freely between the inner portion of the cell and its external environment, it was postulated, in addition, that the cell membrane is a kind of sieve, containing pores or nonlipid patches. But still another observation had to be accounted for. Many small water-soluble ions do not exhibit the same facility for crossing the cell boundary; some move rather freely, others only very slightly. These different ions often possess different electric charges. It was therefore assumed that the cell membrane possesses charge and tends to attract some ions and repel others. Finally, the physical properties of the cell boundary, especially its wettability and elasticity, made it necessary to hypothesize the presence of protein in the membrane.

These various ideas about the cell membrane—its lipid and protein components and pores—all stemming primarily from permeability studies, were put together in 1940 by J. F. Danielli of King's College, London. He concluded that the membrane consists of inner and outer layers of protein with two layers of lipid between them (Fig. 3.4), the whole membrane being about 80 angstroms thick (an angstrom is a hundred millionth of a centimeter).

In recent years, both electron microscopy and X-ray diffraction studies have provided more direct information about the membrane. For example, electron micrographs made by J. David Robertson of the Harvard Medical School indicate that the cell membrane is composed of two electron-dense layers separated by a somewhat wider lighter area (Fig. 3.5). His measurements of the total thickness of the membrane are very close to the 80 angstroms hypothesized by Danielli in 1940. X-ray diffraction studies indicate that the molecules of the material in the lighter middle portion of the cell membrane are oriented perpendicular to the surface of the cell. Again, this evidence supports Danielli's hypothesis of the arrangement of the lipid molecules (see Fig. 3.4). X-ray studies also seem to indicate that the outermost and innermost portions of the membrane are composed of molecules oriented parallel to the surface of the cell—a confirmation of Danielli's idea of the arrangement of the proteins. Danielli's model of the cell membrane has been supported amazingly well by later experimental approaches, and furnishes an excellent example of the value of model construction in science. Conceptual models often help us assess the current state of our knowledge; they may point up problems yet to be tackled and thus provide an impetus to further research.

Robertson, impressed with the close agreement between his early electron-microscope studies and Danielli's lipoprotein-sandwich model, proposed this model as the basic structure of all cellular membranes, including both the plasma membrane and the various intracellular membranes, to be discussed later; he called this structure—a bimolecular

3.5. Electron micrograph showing cell membrane of human red blood cell. The cytoplasm of the cell is in the upper right half of the picture. The membrane consists of two dark lines (probably protein) separated by a lighter area, which is probably lipid. × 280,000. [Courtesy J. David Robertson, Harvard University.]

lipid layer between two layers of protein—the *unit membrane.* However, more recent research has cast doubt on the idea of a uniform structure for all cellular membranes. Though most membranes appear to be composed of lipids and proteins, they vary in thickness, from about 50 to 100 angstroms; in lipid content, from about 20 to 45 percent; in the ratios of the various types of lipids they contain; etc. It seems likely that these differences reflect corresponding differences in structural detail. Thus a variety of alternatives to and modifications of the unit-membrane model have been suggested: e.g. that the proteins constitute a matrix within the lipid bilayer rather than coatings on the surfaces; that the lipid is in the form of a mosaic of lipid globules rather than a bilayer; or, according to one widely accepted modification, that there are places along the membrane where the lipids are in direct contact with the surface (i.e. where they are not coated by proteins) and other places where parts of protein molecules extend deep into the interior of the membrane (possibly even penetrating through it). It is not yet clear which if any of the alternative models accurately describe actual membranes in the living cell. For the moment, the unit-membrane model—with various minor modifications, depending on the membrane being examined—remains the one best supported by experimental evidence. But there is much still to be learned.

Movement through the cell membrane

The cell membrane is not simply an envelope giving mechanical strength and shape and some protection to the cell, though these functions are important. It is an active component of the living cell, playing a complex and dynamic role in life processes. It regulates the traffic in materials between the precisely ordered interior of the cell and the essentially unfavorable and potentially disruptive outer environment. All substances moving between the cell's environment and the cellular interior in either direction must pass through a membrane barrier.

Diffusion.   Before examining in more detail the role of the membrane, we must discuss some of the factors affecting the movement of materials from one place to another in general. Several times already, we have mentioned the motion that characterizes small particles as a consequence of heat energy. This thermal agitation, which, as we saw, affects the speeds of chemical reactions and the behavior of colloidal particles, is also a major factor in the tendency of particles to move from one place to another.

Let us consider a small rectangular box containing 20 marbles, all placed in a tight cluster near one end of the box (Fig. 3.6A). The marbles will remain there as long as they are without motion. But suppose we now cause the marbles to move randomly. In a very short time, they will be distributed with relative uniformity throughout the box; the tight cluster at one end will have disappeared (Fig. 3.6B). Why? We said that the motion of the marbles was random; they were as likely to begin moving in one direction as in any other. But look at the marbles in Fig. 3.6A; among all the possible directions in which a given marble might move, there are more leading away from the center of the cluster than toward it. It is therefore more probable that random movement will result in disruption of the cluster than in its maintenance. Or, to word

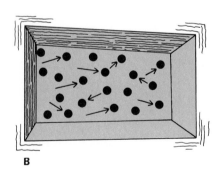

3.6. Mechanical model for diffusion. (A) All 20 marbles are placed in a cluster at one end of a rectangular box.   (B) When the box is shaken to make the marbles move randomly, they become distributed throughout the box in nearly uniform density. For fuller explanation, see text.

it another way, in the absence of any counteracting external influence, a dynamic system will tend to move toward the more probable disorganized state rather than toward the less probable organized state.

Besides the more numerous directions of movement away from the cluster than toward its center, there is another factor that favors the dispersion of the marbles. Movement toward the cluster has a high probability of resulting in a collision of two or more marbles, with the effect of stopping the forward motion of the colliding marbles and causing them to head off in other directions. Movement away from the cluster carries much less probability of collision; a marble has a good chance of continuing on an uninterrupted path to the outer portions of the box. On the average, then, more marbles move away from the center of concentration than toward it.

Notice that the above arguments are both statistical. It is possible that, as a result of random motion, 20 scattered marbles will all come to form a tight cluster at one end of the box. This result has a finite probability, but an extremely small one, so small that we generally feel justified in disregarding it. This kind of reasoning is typical of most scientific reasoning. Science cannot make absolute statements. Its facts and laws are statistical statements. They describe nature in terms of the probabilities of various events as scientists see them.

We can now make a generalization based on our example of the marbles in the box and on others like it: *The net movement of the particles of a particular substance is from regions of higher concentration of that substance to regions of lower concentration of that substance, provided that all regions are at the same temperature and pressure.* Note that we said the *net* movement. There will always be some particles moving, by chance, in the opposite direction, but when the movements of all the particles are considered jointly, the net movement is away from the centers of concentration. When a uniform density is reached, the system is in equilibrium; the particles continue to move, but there is little net change in the system.

Movement of particles from one place to another in the manner we have been discussing is called *diffusion.* Diffusion is fastest by far in gases, where there is much space between the molecular particles and hence relatively little chance of collision. Diffusion in a liquid is much slower; in the absence of convection currents, it takes a very long time, years in fact, for a substance to move in appreciable quantity only a few feet through water.

Thus far we have discussed diffusion in terms of concentration gradient, but diffusion is not strictly a function of concentration. More accurately, it is a function of a gradient in the free energy of the particles. *Free energy* is roughly similar to potential energy in a mechanical system; it is a potential for producing change. In a sense, free energy is a measure of the orderliness or improbability of a system; an orderly, improbable arrangement is, of course, more likely to undergo net change than a disorderly, more probable arrangement. Look again at Fig. 3.6. The marbles in Fig. 3.6A are arranged in a tight cluster—an arrangement most unlikely to arise purely by chance. Because the tight cluster is an orderly, improbable arrangement, the particles in it possess more free energy than particles in the more disorganized arrangement shown in Fig. 3.6B. Thus we can rephrase our earlier generalization as follows:

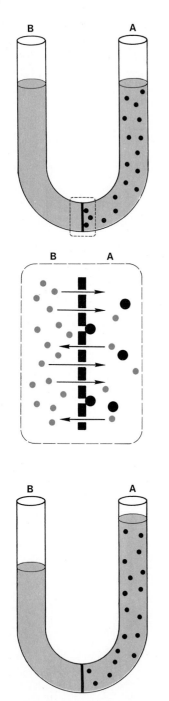

3.7. U tube divided by a semipermeable membrane at its base. The membrane is permeable to water but not to sugar molecules (black balls). Side A contains a sugar solution; side B contains only water. Initially the quantity in the two sides is the same (top). More water molecules (colored balls, center) bump into the membrane per unit time on side B than on side A. Hence there is a net movement of water from B to A, and the level of fluid on side B falls while that on side A rises (bottom).

The net movement of the particles of a particular substance is from regions of greater free energy of that substance to regions of less free energy of that substance.

You may ask why we have bothered to discuss diffusion in terms of free energy; after all, in the example given, the concentration gradient sufficed to explain the observed phenomenon, and concentration is an easier concept to grasp than free energy. Our reason for invoking free energy is that it is not only the more correct basis for understanding diffusion but also the more broadly applicable. Consider a situation where there is a slight concentration gradient from point Y to point Z and a pronounced temperature gradient in the reverse direction. If we regarded only concentration, we would predict net diffusion from Y to Z, but the net diffusion may actually be from Z to Y if the temperature gradient overweighs the concentration gradient. Remember that diffusion is a consequence of the heat motion of particles. Since that motion becomes faster with rising temperature, the rate of diffusion increases as the temperature increases. Now, obviously, rapidly moving particles possess more energy than slowly moving particles. We conclude, then, that free energy is a function of several variables, of which concentration and temperature are two (pressure is a third). And we see that diffusion from Z to Y in our hypothetical example is in full accord with our generalization that the net movement will be down the free-energy gradient. Here an explanation in terms of free energy is effective where one based only on concentration fails.

Osmosis. Now let us consider another situation, diffusion through a membrane. Imagine a chamber divided into two halves by a membrane partition through which particles of some substances can pass while particles of other substances cannot. Such a membrane is said to be differentially permeable, or *semipermeable.* How will the semipermeable membrane affect the diffusion of materials between the two halves of the chamber? Suppose our chamber is a U tube like that shown in Fig. 3.7, divided in half by a semipermeable membrane. Suppose side A contains a solution of sugar in water and side B an equal quantity of pure water, both sides being subject to the same initial temperature and pressure. If the membrane is permeable to water but not to sugar, water molecules will be able to pass in both directions, from A to B and from B to A. Since water is already present on both sides of the membrane, we might at first suppose that the net effect of the diffusion of water molecules across the membrane would be zero, but such a supposition would be wrong.

Consider the differences between the pure water and the sugar solution more carefully. On side B, all the molecules that bump into the membrane during a given interval of time are water molecules and, because the membrane is permeable to water, many of these molecules will pass through the membrane from B to A. By contrast, on side A, some of the molecules bumping into the membrane during the same interval will be water molecules, which may pass through, and some will be sugar molecules, which cannot pass through because the membrane is impermeable to them. At any given instant, then, part of the membrane surface on side A is in contact with sugar molecules and part is in contact with water, whereas on side B all the membrane surface is in contact with water. Consequently, more water molecules will move across the membrane from side B to side A per unit time than in the opposite direction.

Think of the matter another way. If somehow we could dip periodically into the pure water and remove at random a sample consisting of one molecule, we would know beforehand that every dip would yield a water molecule. If, on the other hand, we followed this procedure with the sugar solution, we could not be certain whether a given dip would yield a water molecule or a sugar molecule, because the two kinds of molecules are distributed randomly within the mixture. In other words, the arrangement of water molecules in pure water is orderly and we know that every molecular location will be occupied by a water molecule, whereas the arrangement in the sugar solution is disorderly and we cannot know whether a given molecular location is occupied by a water molecule or by a sugar molecule. Now, we have said that free energy is a measure of orderliness. It follows that the orderly water molecules in the pure water (side B) have more free energy than the disorderly water molecules in the sugar solution (side A). There is a free-energy gradient for water from side B to side A, and, according to our generalization concerning diffusion, we would expect a net diffusion of water down this gradient, from B to A.

We are now in a position to make some additional generalizations: (1) *The free energy of water molecules is always decreased if other substances are dissolved (or colloidally suspended) in the water.* (2) *The decrease in the free energy of water molecules is proportionate to the number of dissolved particles per unit volume.* From these generalizations, and from our earlier generalization that net diffusion is always from the region with greater free energy to the region with less free energy, it follows logically that if two different solutions are separated by a membrane permeable only to water, the net movement of water will be from the solution with the fewer dissolved particles per unit volume to the solution with the more dissolved particles per unit volume.

Another question arises in connection with the U-tube example of Fig. 3.7. If, as we have seen, the net movement of water is from side B to side A, the volume of fluid will increase on side A and decrease on side B. How long can this process continue? Will an equilibrium point be reached?

Clearly, conditions on the two sides of the membrane will never be equal, no matter how many water molecules move from B to A, because the fluid in A will remain a sugar solution, though an increasingly weak one, and the fluid in B will remain pure water if the membrane is completely impermeable to sugar molecules. We might therefore expect that the net movement of water from B to A would continue indefinitely. However, this is not in fact what happens. Under normal conditions, the fluid level in A will rise to a certain point and then cease to rise further. Why? The column of fluid is, of course, being pulled downward by gravity; i.e. it has weight. As the column rises, therefore, it exerts increasing hydrostatic pressure upon the membrane at its base. Eventually the column of sugar solution becomes so high that the pressure it exerts against the membrane is great enough for water molecules to be forced across the membrane from A to B as fast as they move from B to A. When this point is reached—when water is passing through the membrane in opposite directions at the same rate—the system is in equilibrium. Under given conditions of temperature and pressure, the

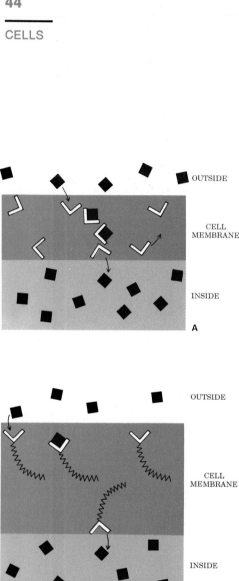

3.8. Model of active transport across cell membrane. A given substance, S, here represented by black squares, is in lower concentration outside the cell than inside; yet it continues to enter the cell. (A) At the outer surface of the membrane, molecules of S combine with lipid-soluble carrier molecules (wedges). The newly formed complex molecules diffuse across the membrane from the region of their higher concentration near the outer surface to the region of their lower concentration near the inner surface, where the S molecules are released into the interior of the cell. The free carrier molecules then diffuse back across the membrane to the outer surface. (B) In an alternative version of the model, the carrier is anchored in the membrane but can undergo conformational changes that move S across the membrane. In both (A) and (B), the chemical reactions involved require expenditure of energy.

equilibrium point is determined by the difference in solute-particle concentration on the two sides of the membrane; the greater the difference, the higher the column will rise before equilibrium is reached. It is important to note that by concentration we here mean not concentration by weight, but rather molecular concentration, i.e. the number of particles per unit volume.

The movement of water through a semipermeable membrane is called *osmosis,*[3] and the hydrostatic pressure exerted by the column of sugar solution at equilibrium in our example is known as **osmotic pressure.** Osmotic pressure is *the pressure that must be exerted on a solution or colloid to keep it in equilibrium with pure water when the two are separated by a semipermeable membrane.* Obviously, that pressure does not have to be exerted by a column of liquid; pressure upon the solution by any other means would also be effective.

It is possible to speak of the potential osmotic pressure of a solution even when the solution is not actually separated from pure water by a semipermeable membrane. What is meant, of course, is the pressure that would have to be put upon that solution to keep it from gaining water if it were separated from a supply of pure water by a semipermeable membrane. Clearly, then, *the osmotic pressure of a solution or colloid is a measure of the tendency of water to move by osmosis into it.* The more dissolved or colloidal particles in a solution, the more water will tend to move into it, and the higher the value of the solution's osmotic pressure will be.

We have discussed diffusion and osmosis and the role played by semipermeable membranes at such length because the cell membrane is semipermeable and the processes of diffusion and osmosis are fundamental to cell life. The membranes of different types of cells vary widely in their permeability characteristics, but a few rough generalizations are possible: Cell membranes are relatively permeable to water and to certain simple sugars, amino acids, and lipid-soluble substances; they are relatively impermeable to polysaccharides, proteins, and other very large molecules; their permeability to small inorganic ions differs greatly depending on the particular ion, but in general negatively charged ions can cross cellular membranes more rapidly than positively charged ions, though neither can do so as readily as uncharged particles.

Regulating the exchange of materials. The movement of substances through the cell membrane is to some extent a passive process, the result of concentration gradients between the cell contents and the

[3] Actually, osmosis is the movement of the principal solvent of a system—whether water or another substance—through a semipermeable membrane, but since the principal solvent in living systems is water, this book will always use the term with reference to water.

It should also be noted that osmosis is more than a process of simple diffusion through a semipermeable membrane. Even though both the direction of movement and the equilibrium conditions can be understood in terms of simple diffusion, the movement of water in a membrane is far too rapid to be explained by the random molecular motions upon which diffusion depends. Apparently there is bulk flow of water in the membrane, specifically through the pores.

It seems that an extremely steep concentration (free-energy) gradient at one end of each pore results in such rapid diffusion at that point that the density of the water within the pore decreases; the consequent hydrostatic-pressure gradient along the length of the pore causes bulk flow of water through it.

external medium, or of electric-potential gradients between the inside and the outside of the cell, or (less often) of bulk flow through the pores. But this is far from the complete story. The membrane is a part of the living cell, and as such it is an active, dynamic entity. We can, therefore, speak of *active transport,* movement of substances across cell membranes through the doing of work by the cell; doing work, of course, demands expenditure of energy by the cell. While active transport often involves moving into or out of the cells materials that passive diffusion would cause to move in the same direction, though not as rapidly, it may also involve moving substances against their concentration gradient.

The exact mechanism of active transport is not known. Most hypotheses proposed to explain this phenomenon, so important to the maintenance of life, assume that some sort of carrier molecule is involved. The carrier is presumed to react chemically with the molecule to be transported, forming a compound that is soluble in the lipid portion of the membrane. According to one version of this model, the compound then moves through the membrane along a concentration gradient to the other side, where it is broken apart by enzymes (Fig. 3.8A). The transported molecule is released, and the carrier diffuses back through the membrane, again along a concentration gradient, to pick up another load. This model of active transport depends on passive diffusion of a compound molecule in a direction in which neither the transported molecule alone nor the carrier molecule alone would diffuse. The energy expenditure is in the chemical reactions involved, which make possible the pickup of a substance by a carrier and its release from the membrane on the side where it is already in high concentration.

An alternative version of the above model suggests that after the carrier reacts with the molecule to be transported, it undergoes conformational changes that carry that molecule across the membrane (Fig. 3.8B). In this model, the carrier molecule, presumably a protein, is assumed to be relatively fixed in the membrane (i.e. the carrier as a whole does not move about in the membrane as it is assumed to do in the diffusion model). Energy expenditure would be needed either to produce the conformational change that effects transport across the membrane, or to produce the reverse conformational change that restores the carrier molecule to its original condition, or both.

Which if either of these models corresponds to the actual transport mechanisms in living cells remains to be determined by future research.

Carrier molecules may be involved in some passive transport as well. Molecules of sugar, for example, may be in higher concentration outside a liver cell than in it and hence tend to diffuse into the cell. But in order to cross the membrane, they may have to combine with a carrier and form a lipid-soluble compound. Since the movement is with the concentration gradient, however, no energy might be required.

Another way in which materials may enter cells involves the active engulfing of the material by the cell. When the material engulfed is in the form of large particles or chunks of matter, the process is called *phagocytosis* (Fig. 3.9). Usually portions of the cell flow around the material, enclosing it within a chamber. The membrane of this chamber, or vesicle, then becomes detached from the membrane of the cell surface, and the vesicle migrates into the interior of the cell. When the engulfed material is liquid or consists of very small particles, the process is often termed *pinocytosis* (Fig. 3.10). Pinocytosis is much like phago-

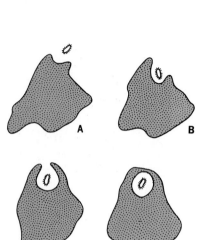

3.9. Phagocytosis. Pseudopodia, armlike processes of the cell, flow around the food particle until it is entirely enclosed within a vacuole.

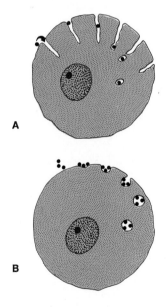

3.10. Pinocytosis. (A) Membrane with an adsorbed particle flows into a deep narrow channel, at the end of which the particle is enclosed in a vesicle that detaches and moves into the interior of the cell. (B) Alternatively, adsorbed particles may be enclosed in vesicles detached directly from the cell surface.

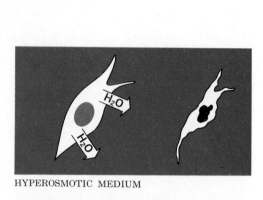

HYPEROSMOTIC MEDIUM

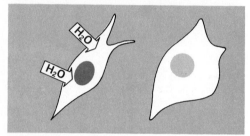

HYPOOSMOTIC MEDIUM

ISOSMOTIC MEDIUM

3.11. Osmotic relationships of an animal cell.
Cell in a hyperosmotic medium loses water (left) and shrivels (right). Cell in hypoosmotic medium gains water and swells. Cell in isosmotic medium has no net gain or loss of water.

cytosis except that the material is not surrounded to the same extent by arms of the cell. Instead, the material apparently becomes adsorbed on the cell surface. Then the loaded membrane either flows inward into a deep narrow channel, at the end of which vesicles are formed, or the small vesicles are simply detached directly from the membrane of the cell surface.

When material is enclosed within vesicles formed through phagocytosis or pinocytosis, it has not yet entered the cell in the fullest sense. It is still separated by a membrane from the cellular substance, and it must eventually cross that membrane (or the membrane must disintegrate) if it is to become incorporated into the cell. Usually the material is first acted upon by enzymes in the vesicles and broken down to smaller, simpler substances that can move more easily across the vesicular membranes. In this manner, substances that cannot penetrate the plasma membrane can enter the cell, usually in the form of their breakdown products.

If we have given the impression that the cell membrane can completely regulate the exchange of materials between the cell and the surrounding medium and always maintain optimum conditions within the cell, we must point out that such is not the case. Some poisons can apparently move freely across the cell membrane, much to the detriment of the cell. And some beneficial substances are lost to the cell because the membrane cannot prevent them from diffusing out. Furthermore, the membrane's great permeability to water can sometimes result in harmful or even fatal effects upon the cell. When a cell is in a medium that is *hyperosmotic* relative to it (i.e. in a medium that has a higher concentration of osmotically active dissolved or colloidal particles and thus a higher osmotic pressure), the cell tends to lose water by osmosis to the medium (Fig. 3.11). As it loses water it shrinks, and if the process goes too far, the cell may die. Conversely, when a cell is in a medium *hypoosmotic* relative to it (i.e. in a medium that has a lower concentration of osmotically active particles and thus a lower osmotic pressure), the cell tends to gain water by osmosis from the medium. As it gains water it swells, and unless it has special mechanisms for expelling the excess water, or special structures that prevent excessive swelling (as in most plant cells), it may burst. A cell in an *isosmotic* medium (i.e. one that has the same concentration of osmotically active particles and thus the same osmotic pressure) neither loses nor gains appreciable quantities of water by osmosis. Obviously, the osmotic relationship between the cell and the medium surrounding it is a critical factor in the life of the cell.

Plant cell walls

One of the most striking differences between the cells of plants and those of animals is that the former possess a very conspicuous cell wall, while the latter do not. The cell wall is located outside the cell membrane. It is generally not considered part of the cellular protoplasm, although it is a product of the protoplasm.

The cell wall is composed primarily of the complex polysaccharide *cellulose* and other similar compounds. The cellulose is generally present in the form of long threadlike structures called fibrils. The spaces between the fibrils, though partly filled with other compounds, usually allow water, air, and dissolved materials to pass freely through the cell wall. The wall does not usually exert a discriminating effect as to which materials can enter the cell and which cannot. That function is reserved to the cell membrane, located below the cell wall.

The first portion of the cell wall laid down by a young growing cell is the *primary wall.* The fibrils within the primary wall are arranged in a loose random network (Fig. 3.12A). Where the walls of two cells abut, an intercellular layer known as the *middle lamella* is located between them (Fig. 3.13). The middle lamella is common to both cells. *Pectin,* a complex polysaccharide, is one of its chief constituents. It is this middle lamella between the primary walls of the two cells that binds the cells together.

Cells of the soft tissues of the plant have only primary walls and intercellular middle lamellae. The cells that eventually form the harder, more woody portions of the plant add further layers to the cell wall, forming what is known as the *secondary wall.* Since this wall, like the primary wall, is deposited by the protoplasm of the cell, it is located internal to the earlier-formed primary wall, lying between it and the membrane (Fig. 3.13). The secondary wall is often much thicker than the primary wall and is composed of a succession of compact layers, or lamellae. The cellulose fibrils in these lamellae are not arranged in a loose network like those of the primary wall, but lie parallel to each other and are generally oriented at angles of about 60 degrees to the fibrils of the next lamella (Fig. 3.12B). This arrangement gives added strength to the cell wall. In addition to cellulose, secondary walls usually contain other materials such as *lignin,* which make them stiffer.

Deposition of the secondary wall usually begins when the cell is nearing the end of its growth. When deposition of the secondary wall has been completed, many cells die, leaving the hard tube formed by their walls to function in mechanical support and internal transport for the body of the plant.

Cell walls generally do not form completely uninterrupted boundaries around the cells. There are often tiny holes in the walls through which delicate protoplasmic connections between adjacent cells may run. These connections are called *plasmodesmata* (see Fig. 3.19). Thus the protoplasm of an individual cell in a multicellular plant body is not isolated, but is in contact and communication with the protoplasm of other cells by way of the plasmodesmata. A large portion of the intercellular exchange of such materials as sugars and amino acids probably takes place through protoplasmic connections.

The presence of cell walls means that plant cells can withstand very dilute external media without bursting. In such media, they are of course in a condition of turgor (distention). Water tends to move by osmosis into the cell as a result of the high osmotic concentration of the protoplasm. The cell swells, building up *turgor pressure* against the cell walls, which exert an equal opposing pressure against the swollen cell. The wall of a mature cell can usually be stretched only a minute amount. Equilibrium is reached when the resistance of the wall is so great that no further increase in the size of the cell is possible and, consequently, no more water can enter the cell. Because of the wall pressure, plant cells can withstand much wider fluctuations in the osmotic makeup of the surrounding medium than animal cells.

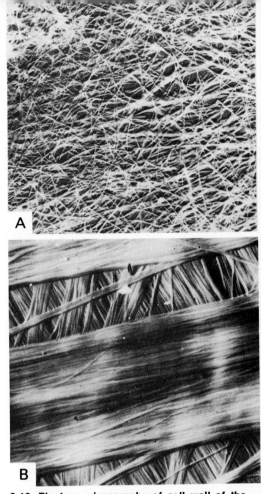

A

B

3.12. Electron micrographs of cell wall of the alga *Valonia ventricosa.* (A) Young primary wall. The cellulose fibrils are randomly arranged and form a loose network. × 10,000. (B) Secondary wall. The cellulose fibrils lie parallel to each other in compact layers. Three layers, each oriented at 60 degrees to the one below it, can be seen here. × 14,200. [Courtesy F. C. Steward and K. Mühlethaler, *Ann. Botany (London),* vol. 17, 1953.]

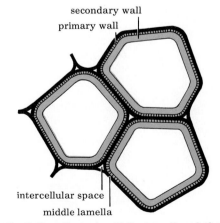

secondary wall
primary wall

intercellular space
middle lamella

3.13. Cell walls and middle lamella of three adjacent plant cells.

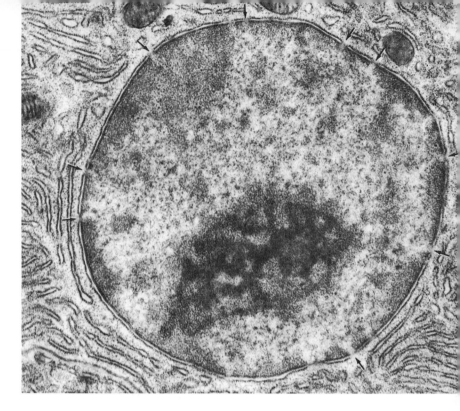

3.14. Electron micrograph of the nucleus of a pancreatic cell. The dark area inside the nucleus is a nucleolus. Numerous "pores" (arrows) can be seen in the double nuclear membrane. × 15,300. [Courtesy D. W. Fawcett, *A Textbook of Histology,* Saunders, 1962.]

3.15. Nuclei of dividing cells in root tip of onion, showing chromosomes. Separate chromosomes cannot be distinguished in the dark-stained nuclei of the nondividing cells at top and bottom of photograph. [Courtesy General Biological Supply House, Inc., Chicago.]

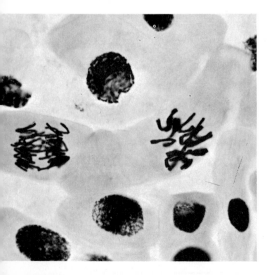

The nucleus

Within the cells of most organisms (though not of bacteria and blue-green algae), the largest and one of the most conspicuous structural areas is the nucleus (Fig. 3.14), the control center of the cell. The nucleus plays the central role in cellular reproduction, the process whereby a single cell undergoes cell division and forms two new cells. It also plays a central part, in conjunction with the environment, in determining what sort of differentiation a cell will undergo and what form it will exhibit at maturity. And the nucleus directs the metabolic activities of the living cell. In short, it is from the nucleus that the "instructions" emanate that guide the life processes of the cell as long as it lives. Biologists attach so much importance to the nucleus that they even have a special name, *cytoplasm,* for all protoplasm other than the nucleus; the nucleus and the cytoplasm can thus be thought of as the two major subdivisions of protoplasm.

Within the nucleus are several relatively distinct types of structures, notably the chromosomes and the nucleoli. The entire nucleus is bounded by a nuclear membrane.

The *chromosomes* (Fig. 3.15) are elongate, threadlike bodies clearly visible only when the cell is undergoing division. They are composed of nucleic acid and protein. The chromosomes bear, apparently in linear arrangement, the basic units of heredity, called *genes.* It is the genes that determine the characteristics of cells, as they are passed from generation to generation, and that act as the units of control in the day-to-day activities of living cells. They are the code units, if you will, for the transmission of bits of information from parent to offspring, and for the determination of the enzymes that so precisely regulate the myriad interdependent chemical reactions of life. Indeed, the genes are at the very hub of life processes, and they and the chromosomes on which they lie will demand much discussion in later chapters.

Nucleoli are dark-staining, generally oval bodies usually clearly visible within the nuclei of nondividing cells (Fig. 3.14). There may be one or more per nucleus, depending on the species. The current view is that they play an important role in sending instructions for protein synthesis to the cytoplasm.

The presence of a *nuclear membrane* surrounding the nucleus makes possible different environments within the nucleus and in the surrounding cytoplasm. The nuclear membrane differs from the plasma membrane in that it is double, enclosing a distinct space (Figs. 3.14, 3.16). Electron-microscope studies indicate that it is interrupted at intervals by fairly large pores where the outer and the inner membrane are continuous. But the pores do not make an indiscriminate sieve of the nuclear membrane; on the contrary, permeability studies show that this membrane is highly selective. Some substances that can cross the plasma membrane—including substances with molecules smaller than the pores—are restricted to the cytoplasm because they apparently cannot easily cross the nuclear membrane. Yet some macromolecules readily pass through the pores; these seem to be primarily substances that act as chemical messengers between the nucleus and the cytoplasm.

The electron microscope has revealed another particularly interesting fact about the nuclear membrane—the continuity, at some points, of this double membrane with an extensive cytoplasmic membrane system called the endoplasmic reticulum (Fig. 3.16). The connection may

3.16. Electron micrograph showing nuclear membrane of a cell from corn root. The nucleus is the large structure filling the upper left quarter of the picture. The unlabeled arrow indicates a point where the endoplasmic reticulum and the double nuclear membrane interconnect. ER, endoplasmic reticulum; G, Golgi apparatus; M, mitochondrion; N, nucleus; NM, nuclear membrane; P, pore in nuclear membrane; W, cell wall. × 15,600. [Courtesy W. G. Whaley, H. H. Mollenhauer, and J. H. Leech, *Am. J. Botany*, vol. 47, 1960.]

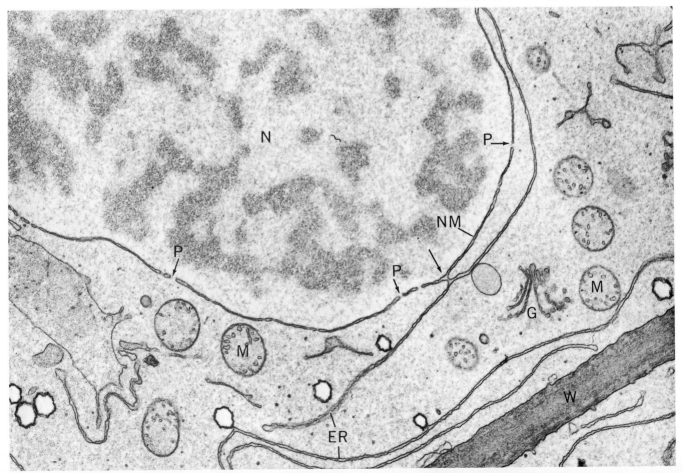

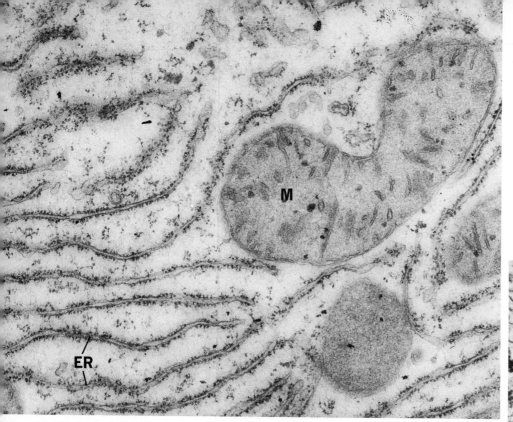

3.17. Electron micrograph of rough endoplasmic reticulum of a rat liver cell. The dark particles studding the surface of the vesicles of the endoplasmic reticulum are ribosomes. Several mitochondria can be seen. × 41,600. [Courtesy K. R. Porter, in *The Cell,* Academic Press, 1961.]

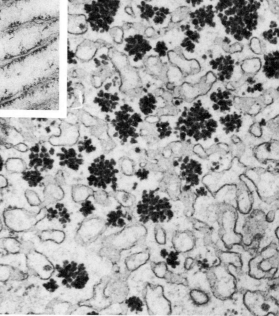

3.18. Electron micrograph of smooth endoplasmic reticulum from liver of a three-day-old rat. There are no ribosomes on the surfaces of the numerous membranous vesicles. Dense glycogen particles are clustered in spaces between the vesicles. × 48,000. [Courtesy G. Dallner, P. Siekevitz, and G. E. Palade, *J. Cell Biol.,* vol. 30, 1966.]

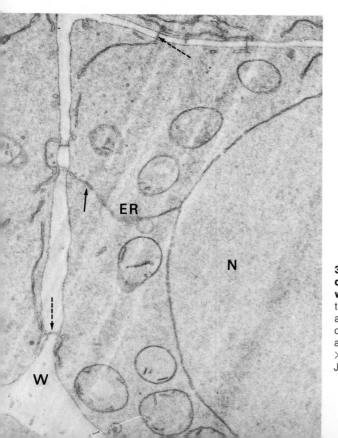

3.19. Electron micrograph of cell from corn root showing a canal of the endoplasmic reticulum (solid arrow) continuous with the nuclear membrane. In fact, this canal appears to run through the cell wall and join the ER of the adjacent cell. Two additional places where ER canals in the two cells seem to be continuous through plasmodesmata are indicated by dashed arrows. N, nucleus; ER, endoplasmic reticulum; W, wall. × 12,920. [Courtesy W. G. Whaley, H. H. Mollenhauer, and J. E. Kephart, *J. Biophys. Biochem. Cytol.,* vol. 5, 1959.]

be of great significance for the functioning of the nucleus. We turn next, therefore, to an examination of the intracellular membrane system.

The endoplasmic reticulum and ribosomes

Biologists long wondered if the cytoplasm had some sort of invisible structural organization, and faint traces of such a cytoplasmic "skeleton" were reported at various times. Finally, in 1945, Keith R. Porter of the Rockefeller Institute, using a phase microscope, which has much greater resolution than the usual light microscope, described a complex system of membranes forming a network in the cytoplasm. This system, called the *endoplasmic reticulum* by Porter, has since been extensively studied with the electron microscope. It has been shown to be present in all nucleated cells.

Although the endoplasmic reticulum (ER) varies greatly in appearance in different cells, it always forms a system of membrane-enclosed spaces. In many cells, though not all, these spaces are interconnected, forming a true reticulum. Sometimes the membranes of the ER are lined on their outer surfaces by small particles called *ribosomes,* in which case the ER is spoken of as "rough" (Fig. 3.17); when no ribosomes line the membranes, the ER is described as "smooth" (Fig. 3.18). Just as the ER may exist without associated ribosomes, ribosomes, which are present in almost all cells, may occur independently of the ER. There is now abundant evidence that ribosomes are the sites of protein synthesis.

The ER is almost always closely associated with the nuclear membrane. In fact, electron micrographs have repeatedly shown connections between the outer portion of the double nuclear membrane and adjacent elements of the ER (Figs. 3.16, 3.19). Thus the spaces between the two components of the nuclear membrane are continuous with the membrane-enclosed spaces and channels of the ER. It seems, therefore, that the nuclear membrane is only a specialized part of the general cell-membrane system.

One possible function of the ER is immediately apparent. Its channels could serve as routes for transport of materials between various regions of the cytoplasm or between the various parts of the cytoplasm and the nucleus, forming a communications network, as it were, between the nuclear control center and the rest of the cell.

Another function of the ER is related to its large protein content; proteins, you will remember, may act both as structural elements in cells and as enzymes catalyzing chemical reactions. There is now abundant evidence that at least some of the many protein molecules of which the ER membranes are composed act as enzymes and that the ER functions as a cytoplasmic framework providing manufacturing surfaces for the cell; its complex folding provides an enormous surface area.

The Golgi apparatus

The subcellular elements identified as Golgi apparatus characteristically consist of a system of membrane-delimited vesicles arranged approximately parallel to each other (Figs. 3.16, 3.20). There is evidence that the smooth membranes of the Golgi apparatus often have connections (probably transient) with the membranes of the endoplasmic reticulum and therefore constitute another portion of the complex cellular membrane system.

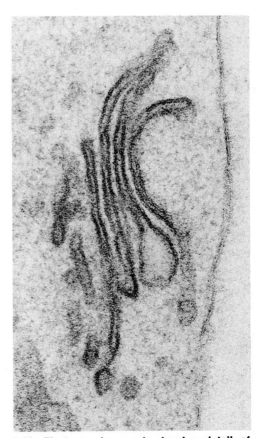

3.20. Electron micrograph showing detail of Golgi apparatus. The Golgi apparatus is composed of vesicles, each bounded by a unit membrane. × 140,700. [Courtesy Herbert W. Israel, Cornell University.]

Many years ago, it was noticed that the Golgi apparatus was particularly prominent in cells thought to be involved in the secretion of various chemical products; as the level of secretory activity of these cells changed, corresponding changes were observed in the morphology of the Golgi apparatus. The inference was that the Golgi apparatus played some part in the secretory process. And, in fact, more recent research has established that the Golgi is involved in storage, modification (e.g. removal of water or emulsification of lipids), and packaging of secretory products. These products are enclosed in vesicles formed by the outer portion of the Golgi. The vesicles move from the Golgi to the cell surface, where they become attached to the plasma membrane and then rupture, releasing their contents to the exterior. The membranes of the ruptured vesicles may remain as permanent additions to the plasma membrane.

Mitochondria

Among the formed inclusions of the cytoplasm that can be detected with the light microscope are small bodies, known as mitochondria, which occur in almost all types of cells from most kinds of organisms (Figs. 3.16, 3.17). These organelles are, in a sense, the powerhouse of the cell. They are the site of many crucial chemical reactions, indispensable to life, that extract energy from foodstuffs and make it available to the cell for its innumerable energy-demanding activities. These chemical reactions will be one of the main subjects of the next chapter.

Electron microscopy reveals that each mitochondrion is a double-walled vessel. The outer wall is a smooth membrane, and the inner wall is a membrane with many inwardly directed folds (Fig. 4.15, p. 91). These folds, called *cristae*, extend into a semifluid amorphous matrix.

Lysosomes and peroxisomes

First described in the 1950's, lysosomes (Fig. 3.24) are membrane-enclosed bodies slightly smaller than mitochondria, with which they were often confused in the past. They contain many powerful digestive (hydrolytic) enzymes and are thought to function, in a sense, as the digestive system of the cell, enabling it to process some of the bulk materials taken in by phagocytosis or pinocytosis.

The lysosome membrane, which is single (unlike the double-membrane envelope of mitochondria), must be both impermeable to the outward movement of the hydrolytic enzymes and capable of resisting their digestive action. Packaged in the lysosomes, the enzymes are prevented from digesting the material of the cell itself. If the lysosome membrane is ruptured, they are released into the surrounding cytoplasm and immediately begin to break it down.

Lysosomes are probably produced by the Golgi apparatus. First discovered in rat liver cells, they have since been found in many kinds of animal cells, and may occur in all of them. They have also been found in some fungi, but it is not yet clear how widely they occur in other plant cells.

Improved cell-fractionation methods have recently revealed in some cells, particularly those of kidney and liver, a type of membrane-bounded organelle previously confused with lysosomes. These organelles, known as peroxisomes, have been found to contain an assortment of powerful enzymes—not digestive enzymes as in the lysosomes, but oxidative enzymes, which catalyze the oxidative removal of amino groups from amino acids, for example.

Plastids

Plastids are large cytoplasmic organelles found in the cells of most plants (fungi are an exception), but not in animal cells. They can easily be observed with an ordinary light microscope. There are two principal categories of plastids: *chromoplasts* (colored plastids) and *leucoplasts* (white or colorless plastids).[4]

Chloroplasts, which are chromoplasts containing the green pigment *chlorophyll,* are extremely important to all life. Energy from sunlight is trapped in them by chlorophyll and utilized in the manufacture of complex organic molecules (particularly sugar) from simple inorganic raw materials. Chloroplasts contain, in addition to chlorophyll, various yellow or orange pigments called *carotenoids.*

The electron microscope reveals that, like the mitochondrion, the typical chloroplast is bounded by a double membrane. Though the inner membrane of a chloroplast is not folded into cristae, there is a complex internal membranous organization (Fig. 3.21); numerous double-membrane lamellae are embedded in a rather homogeneous matrix called the stroma. In most higher plants, these lamellae are differentiated into two varieties, separate lamellae running through the stroma and stacks of platelike lamellae forming regions known as *grana.* The chlorophyll is located in the lamellae, apparently bound to the proteins and lipids of the membranes.

Chromoplasts lacking chlorophyll are usually yellow or orange (occasionally red) because of the carotenoids they contain. It is these

[4] Some classifications of plastids recognize three principal categories: chloroplasts, chromoplasts, and leucoplasts. However, chloroplasts, or green plastids, can appropriately be considered a subcategory of chromoplasts.

3.21. Electron micrograph of chloroplast of tobacco. The stacks of disclike lamellae are the grana. The less tightly packed lamellae running between the grana are the stroma lamellae. × 33,250. [Courtesy Herbert W. Israel, Cornell University.]

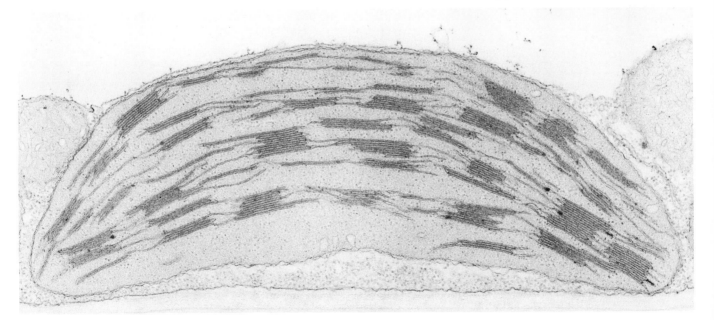

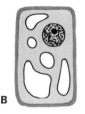

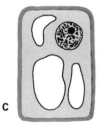

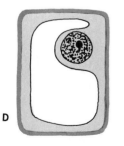

3.22. Development of plant-cell vacuole. The immature cell (A) has many small vacuoles. As the cell grows, these fuse to form the large vacuole occupying most of the space in the mature cell (D). The cytoplasm is pushed to the outer circumference of the cell.

kinds of chromoplasts that give the characteristic yellow or orange color to many flowers, ripe fruits, and autumn leaves.

The colorless plastids, or leucoplasts, are primarily organelles in which materials such as starch, oils, and protein granules are stored. Plastids filled with starch are particularly common in storage roots and stems (e.g. carrots, potatoes) and in seeds.

Vacuoles

Membrane-enclosed, fluid-filled spaces called vacuoles are found in both animal and plant cells, though they have their greatest development in plant cells. There are various kinds of vacuoles with a corresponding variety of functions. In some Protozoa, specialized vacuoles, called contractile vacuoles, play an important role in expelling excess water and some wastes from the cell; we shall discuss them in greater detail in a later chapter. Many Protozoa also possess food vacuoles, chambers that contain food particles. Similar vacuoles, or vesicles, are formed by many kinds of cells when they take in material by phagocytosis or pinocytosis, processes discussed earlier in this chapter.

In most mature plant cells, a large vacuole occupies much of the volume of the cell. The immature cell usually contains many small vacuoles (Fig. 3.22). As the cell matures, the vacuoles take in more water and become larger, eventually fusing to form the very large definitive vacuole of the mature cell. This process pushes the cytoplasm to the periphery of the cell, where it forms a relatively thin layer.

Centrioles

Centrioles, small dark bodies located just outside the nucleus of most animal cells in a region of specialized cytoplasm, have been seen for many years and have been known to play a role in cell division. They do not occur in the cells of most higher plants, although they are found in some algae and fungi and in a few reproductive cells of higher plants. We shall discuss them more fully when we describe the process of cell division in Chapter 13.

Cilia and flagella

Some cells of both plants and animals have one or more movable hairlike structures projecting from their free surfaces. If there are only a few of these appendages and they are relatively long in proportion to the size of the cell, they are called *flagella* (Fig. 8.4, p. 157). If there are many and they are short, they are called *cilia.* Actually the basic structure of flagella and cilia is the same, and the terms are often used interchangeably. Both usually function either in moving the cell or in moving liquids or small particles across the surface of the cell. They occur commonly on unicellular and small multicellular organisms and on the male reproductive cells of most animals and many plants, in both of which they may be the principal means of locomotion. They are also common on the cells lining many internal passageways and ducts in animals, where their beating aids in moving materials.

"Typical" cells

By now you realize that a living cell, small as it is, is an extraordinarily complex unit, containing numerous components that are themselves relatively complex. The extent of this complexity was not truly appreciated until the advent of the electron microscope and modern biochemical techniques, which combined to change our whole picture of the cell. Not too long ago, biology texts regularly included a diagram of a so-called typical cell that showed only five or six simple internal components, a diagram easily memorized by students (Fig. 3.23). No simple diagram of this sort can be given today. In the first place, there is no such thing as a typical cell. Not only do plant and animal cells differ in many important ways, but the various cells of the body of any one plant or animal are often strikingly different from one another in shape, size, and function. This much, of course, has been known for a long time. But now that the number of known cellular components has grown so large and that their great variability has been demonstrated, it becomes even more obvious that no single diagram, or even series of diagrams, can really portray a "typical" cell. Nevertheless, in an effort to help you visualize the arrangement of the organelles discussed in the preceding pages, two composite diagrams (Figs. 3.24 and 3.25) are given here. As you examine them, keep in mind that not all the components shown always occur together in any one real cell.

Procaryotic cells

You will have noticed that throughout the discussion of cellular components, we have had to caution you that the descriptions do not apply to bacteria or blue-green algae. The cells of these unicellular organisms differ in numerous basic ways from the cells of all other organisms. In recognition of their distinct characteristics, these cells are customarily termed *procaryotic cells,* while the cells of all other organisms, both plant and animal, are termed *eucaryotic cells.* The most

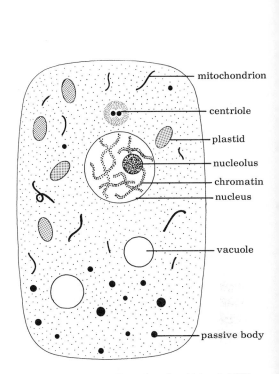

3.23. A "typical" cell as visualized about 1925. [Modified from E. B. Wilson, *The Cell in Development and Heredity*, Macmillan, 1925.]

Table 3.1.   Procaryotic and eucaryotic cells compared

Characteristic	Procaryotic cells	Eucaryotic cells
Nuclear membrane	Absent	Present
Chromosomes	Composed only of nucleic acid	Composed of nucleic acid and protein
Photosynthetic apparatus	May contain chlorophyll, but not in chloroplasts	Chlorophyll, when present, contained in chloroplasts
Mitochondria, Golgi apparatus, endoplasmic reticulum, etc.	Absent	Present
Flagella	Lack 9–2 fibrillar structure	Have 9–2 fibrillar structure
Cytoplasmic streaming or amoeboid movement	Does not occur	May occur
Cell wall	Contains muramic acid	When present, does not contain muramic acid

striking difference between the two is that procaryotic cells lack the nuclear membrane characteristic of eucaryotic cells. For a long time it was thought that procaryotic cells have no nucleus at all, but in recent years it has been shown that they have a nuclear region containing genetic material; however, this region is not enclosed by a membrane. The genetic material is not organized into chromosomes of the sort seen in eucaryotic cells, though it does apparently have an analogous structure. Some bacteria and all blue-green algae contain chlorophyll, which is associated with membranous vesicles, or lamellae (see Fig. 19.1, p. 444); but the lamellae do not lie within organelles, and hence there are no distinct plastids. Procaryotic cells have no mitochondria, no endoplasmic reticulum, no Golgi apparatus, no lysosomes, and no peroxisomes. In short, they lack the internal membranous structures characteristic of eucaryotic cells. There is evidence, however, that the membrane that bounds the cell is folded inward at various points, and that this membrane carries out many of the enzymatic functions associated with the membranes of the mitochondria, endoplasmic reticulum, and other internal membranes of eucaryotic cells. Some bacteria have flagella, but these structures do not show the characteristic internal structure found in the flagella of other cells. Table 3.1 gives a summary of some of the most important differences between procaryotic and eucaryotic cells.

3.24. Current conception of "typical" animal cell. The cell should be visualized as three-dimensional.

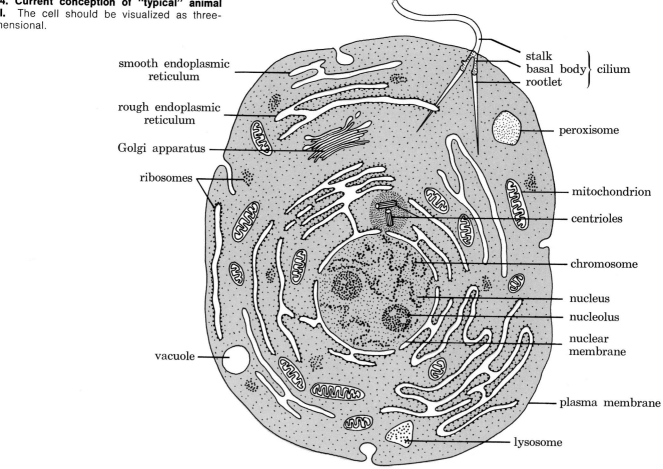

smooth endoplasmic reticulum

rough endoplasmic reticulum

Golgi apparatus

ribosomes

vacuole

stalk
basal body } cilium
rootlet

peroxisome

mitochondrion

centrioles

chromosome

nucleus

nucleolus

nuclear membrane

plasma membrane

lysosome

MULTICELLULAR ORGANIZATION

Some animals and some plants are unicellular (i.e. composed of a single cell), but most organisms are multicellular (i.e. composed of many cells). Ordinarily the bodies of multicellular organisms, particularly animals, are organized on the basis of tissues, organs, and systems. A *tissue* is composed of many cells, usually similar in both structure and function, that are bound together by intercellular material. An *organ*, in turn, is composed of various tissues (not necessarily similar) grouped together into a structural and functional unit. Similarly, a *system* is a group of interacting organs that "cooperate" as a functional complex in the life of the organism.

The following sections will introduce you briefly to some of the basic plant and animal tissues, organs, systems, and organisms. We shall examine them in more detail in later chapters.

PLANT TISSUES

Plant tissues have been classified in a variety of ways by different botanists. The system used here is not necessarily better than other possible systems; it is simply one of several acceptable ones. The lack of

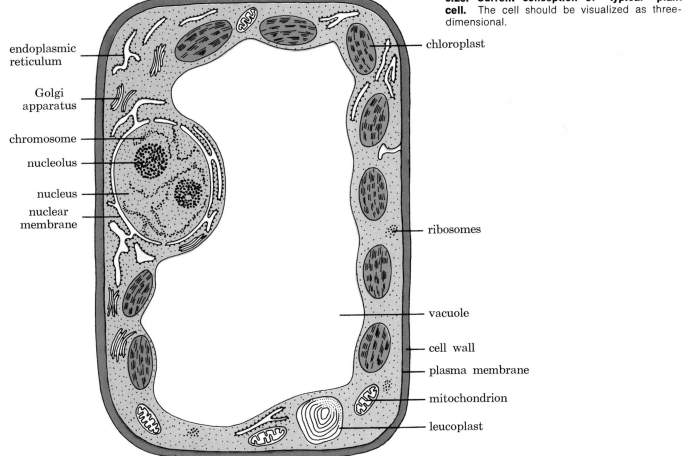

3.25. Current conception of "typical" plant cell. The cell should be visualized as three-dimensional.

endoplasmic reticulum

Golgi apparatus

chromosome

nucleolus

nucleus

nuclear membrane

chloroplast

ribosomes

vacuole

cell wall

plasma membrane

mitochondrion

leucoplast

full agreement on any one classification springs from characteristics of the plant cells themselves. The different cell types intergrade, and a given cell may even change from one type to another during the course of its life. Consequently the tissues formed from such cells also intergrade and may share structural and functional characteristics. Furthermore, plant tissues may contain cells of only one type, or they may be complex, containing a variety of cell types. In short, plant tissues cannot be fully characterized or distinguished on the basis of any single criterion such as structure, function, location, or mode of origin.

All plant tissues can be divided into two major categories: meristematic tissue and permanent tissue. *Meristematic tissues* are composed of immature cells and are regions of active cell division; *permanent tissues* are composed of more mature, differentiated cells. This distinction is not an absolute one, however, for some permanent tissues may revert to meristematic activity under certain conditions.

The permanent tissues fall into three subcategories: surface tissues, fundamental tissues, and vascular tissues. Each of these, in turn, contains several different tissue types. The classification used here may be summarized as follows:

I. Meristematic tissue
II. Permanent tissue
   A. Surface tissue
      1. Epidermis
      2. Periderm
   B. Fundamental tissue
      1. Parenchyma
      2. Collenchyma
      3. Sclerenchyma
      4. Endodermis
   C. Vascular tissue
      1. Xylem
      2. Phloem

It must be emphasized that this classification is based on the higher land plants, the vascular plants. It has little relevance for other plants, where multiple tissue types seldom occur.

Meristematic tissue

Meristematic tissues are composed of embryonic, undifferentiated cells capable of cell division. Cell division occurs throughout the body of the very early embryo, but as the young plant develops, many regions of the body become specialized for other functions and cease playing a primary role in the production of new cells. Consequently cell division becomes restricted largely to certain undifferentiated tissues in localized regions; these tissues are the meristems.

Regions of meristematic tissue are found at the growing tips of roots and stems. These apical meristems are responsible for increase in length of the plant body. In many plants, there are also meristematic areas toward the periphery of the roots and stems, and these lateral meristems are responsible for increase in girth.

Surface tissue

As the term implies, surface tissues form the protective outer covering of the plant body. In young plants and herbaceous adult plants

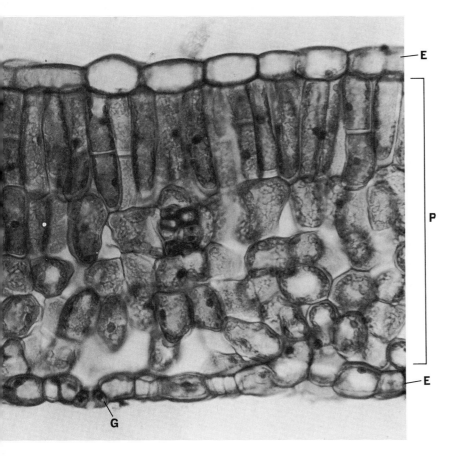

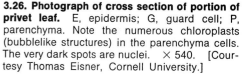

3.26. Photograph of cross section of portion of privet leaf. E, epidermis; G, guard cell; P, parenchyma. Note the numerous chloroplasts (bubblelike structures) in the parenchyma cells. The very dark spots are nuclei. × 540. [Courtesy Thomas Eisner, Cornell University.]

lacking active lateral meristems, the principal surface tissue of roots and stems is the *epidermis;* epidermis is also the surface tissue of all leaves (Fig. 3.26). Often the epidermis is only one cell thick, though it may be thicker, as in some plants living in very dry habitats, where protection against water loss is critical. Epidermal cells on the aerial parts of the plant often secrete a waxy, water-resistant *cuticle* on their outer surface, which aids in protection against loss of water, mechanical injury, and invasion by parasitic fungi.

As the stems and roots of plants with active lateral meristems increase in diameter, the epidermis is slowly replaced by another surface tissue, the *periderm,* which constitutes the corky outer bark so characteristic of old trees (Fig. 7.5, p. 130). The functional cork cells forming the periderm are dead; it is their cell walls, waterproofed with suberin, that function as the protective outer covering of the plant.

Fundamental tissue

Most of the fundamental tissues are simple tissues; i.e. each is composed of a single type of cell. Often these same types of cells also occur as components of the complex vascular tissues. The various fundamental tissues are not necessarily structurally similar, although they form from the same embryonic regions. They are often defined simply as those tissues that are neither surface tissues nor vascular tissues.

Parenchyma. Parenchyma tissue occurs in roots, stems, and leaves. The parenchyma cells, like the cells that make up almost the

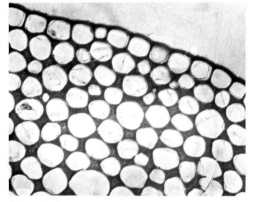

3.27. Collenchyma cells from petiole of beet leaf. Notice the particularly thick walls at the corners of the cells. [Used by permission from J. D. Dodd, *Form and Function in Plants*, © 1962 by The Iowa State University Press.]

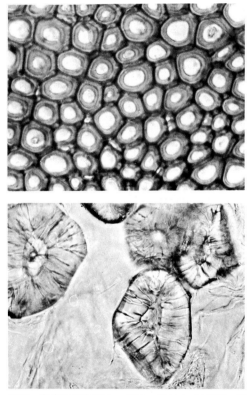

3.28. Sclerenchyma. Top: Cross section of fibers from corn stem. Bottom: Stone cells of pear fruit. [Used by permission from J. D. Dodd, *Form and Function in Plants*, © 1962 by The Iowa State University Press.]

whole body of lower plants, are relatively unspecialized. They have not lost the capacity for cell division, and in some circumstances take on meristematic activity; they also sometimes undergo further specialization, forming other cell types. Parenchyma cells usually have thin primary walls and no secondary walls. They generally have a large vacuole surrounded by a peripheral layer of cytoplasm. Since they are ordinarily loosely packed, intercellular spaces are abundant in parenchyma tissue (Fig. 3.26). Most of the chloroplasts of leaves are in the cells of parenchyma tissue, and it is largely here that photosynthesis occurs. Parenchyma of stems and roots functions in storage of nutrients and water.

Collenchyma. Like parenchyma, collenchyma is a simple tissue whose cells remain alive during most of their functional existence. Though collenchyma cells are characteristically more elongate, they are structurally similar to parenchyma cells, except that their walls are irregularly thickened. The thickened areas are usually most prominent at the edges (the "corners" when viewed in cross section) (Fig. 3.27). Collenchyma functions as an important supporting tissue in young plants, in the stems of nonwoody older plants, and in leaves.

Sclerenchyma. Sclerenchyma is a type of simple fundamental tissue that, like collenchyma, functions in support. However, sclerenchyma cells are far more specialized than collenchyma cells; at functional maturity, most are dead, and their uniformly very thick, heavily lignified secondary walls give strength to the plant body. Often these walls are so thick that the lumen (internal space) of the cell has been nearly obliterated.

Sclerenchyma cells are customarily divided into two categories: fibers and sclereids. *Fibers* (Fig. 3.28, top) are very elongate cells with tapered ends. They are tough and strong, but flexible; commercial flax and hemp are derived from strands of sclerenchyma fibers. *Sclereids* are of variable, often irregular, shape. The simpler, unbranched sclereids, frequently called stone cells, are common in the shells of nuts and in the hard parts of seeds, and are scattered in the flesh of hard fruits; the gritty texture of pears is due to small clusters of stone cells (Fig. 3.28, bottom).

Endodermis. Endodermis is a type of tissue difficult to place in any classification. It occurs as a layer surrounding the vascular-tissue core of roots (Fig. 5.4, p. 99) and, less frequently, of stems. Young endodermal cells are much like elongate parenchyma cells, except that a band of chemically distinctive thickening runs around the cell on its radial (side) and end walls. This lignified and suberized ("water-proofed") band is called the *Casparian strip.* The cells of endodermal tissue occur in a single layer and are compactly arranged without intercellular spaces. Possible functions of the endodermis will be discussed in a later chapter.

Vascular tissue

Vascular, or conductive, tissue is a distinctive feature of the higher plants, one that has made possible their extensive exploitation of the terrestrial environment. It incorporates cells that function as tubes or ducts through which water and numerous substances in solution move from one part of the plant body to another. There are two principal types of vascular tissue: xylem and phloem. Both of these are complex tissues; i.e. they consist of more than one kind of cell.

Xylem. Xylem is a vascular tissue that functions in the transport of water and dissolved substances upward in the plant body. It forms a continuous pathway running through the roots, the stem, and appendages of the stem such as leaves. In its most advanced (evolved) form, in the flowering plants, it commonly includes two types of cells unique to xylem: *tracheids* and *vessel elements.* It also includes numerous parenchyma and sclerenchyma cells. The parenchyma cells are the only living cells in mature functioning xylem, inasmuch as both the cytoplasm and the nucleus of tracheids, vessel elements, and sclerenchyma cells disintegrate at maturity, leaving the thick cell walls as the functional structures.

Transport is not the only function of xylem. Another important one is support, particularly of the aerial parts of the plant. The numerous sclerenchyma fibers in the xylem function almost exclusively in this way. One need only remember that the common name for xylem is wood to understand the enormous strength characteristic of this tissue.

Phloem. The second vascular tissue, phloem, is unlike xylem in that materials can move both up and down in it. Phloem functions particularly in the transport of organic materials such as carbohydrates and amino acids. For example, newly synthesized organic molecules are moved in the phloem from the leaves to the roots and stem for storage or to the growing points of the plant for immediate use. Like xylem, phloem is a complex tissue and contains both parenchyma and sclerenchyma cells in addition to the cells unique to it: *sieve cells* and *companion cells.* The sieve cells are the vertical transport units.

PLANT ORGANS

The body of the higher land plants is customarily divided into two major parts: the *root* and the *shoot* (Fig. 3.29). These two fundamental organ systems can be distinguished on the basis of numerous morphological characteristics, principally the arrangement and mode of origin of the vascular tissue, differences in the way branch roots and branch stems are formed, and the presence of leaves on the shoot but not on the root. Yet note that many of the tissues are essentially continuous throughout the entire axis, both root and shoot. For example, the vascular tissue of root and shoot, despite a somewhat different arrangement in each, forms an uninterrupted transport system.

The roots of a plant function mainly in procurement of inorganic nutrients such as minerals and water, in transport, in nutrient storage, and in anchoring the plant to the substrate. The shoot is somewhat more complex structurally. It consists of the stem and the appendages of the stem, particularly the foliage leaves and the reproductive organs. The stem, of course, functions in internal transport and in support, while

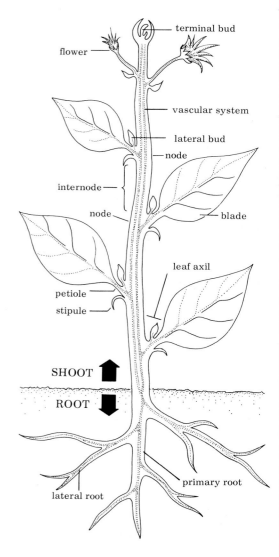

3.29. Diagram of flowering plant body.

the foliage leaves are organs in which the critical process of photosynthesis takes place.

You will notice that the number of distinct organs mentioned here —root, stem, foliage leaf, and reproductive organs—is much smaller than it would be for higher animals. In general, the plant body is simply not as clearly divided into readily distinguishable functional components, or organs, as the animal body. The parts of the plant grade more imperceptibly into each other and, in some ways, form a more continuous whole. This fundamental difference between plants and animals is doubtless correlated with the different modes of life that characterize them, particularly their different feeding methods and their different adaptations for a sedentary existence on the one hand (plants) and active locomotion on the other (animals).

ANIMAL TISSUES

It is traditional to divide all animal tissues into four categories: epithelium, connective tissue, muscle, and nerve. Each of these, particularly the second, is a diverse assemblage containing numerous subtypes. It should be emphasized that the subtype classification is based primarily on the vertebrate animals, especially man, and that its application to other animals, particularly the lower invertebrates, is difficult and sometimes even meaningless.

The classification of animal tissues used here can be summarized as follows:

   I. Epithelium
      A. Simple epithelium
         1. Squamous
         2. Cuboidal
         3. Columnar
      B. Stratified epithelium
         1. Stratified squamous
         2. Stratified cuboidal
         3. Stratified columnar
  II. Connective tissue
      A. Vascular tissue
         1. Blood
         2. Lymph
      B. Connective tissue proper
         1. Loose connective tissue
         2. Dense connective tissue
      C. Cartilage
      D. Bone
 III. Muscle
      A. Skeletal muscle
      B. Smooth muscle
      C. Cardiac muscle
 IV. Nerve

Epithelium

Epithelial tissue forms the covering or lining of all free body surfaces, both external and internal. The outer portion of the skin, for example, is epithelium, as are the linings of the digestive tract, the lungs, the blood vessels, the various ducts, the body cavity, etc. Epithelial cells are packed tightly together, with only a small amount of cementing material between them and almost no intercellular spaces. Thus they provide a continuous protective barrier between the underlying cells and the external medium. Since anything entering or leaving the body must cross at least one layer of epithelium, the permeability characteristics of the cells of the various epithelia play an exceedingly important role in regulating the exchange of materials between different parts of the body and between the body and the external environment.

It is customary to group epithelial cells into three categories: squamous, cuboidal, and columnar. *Squamous cells* are much broader than they are thick and have the appearance of thin flat plates (Fig. 3.30). *Cuboidal cells* are roughly as thick as they are wide and, as their name implies, have a rather square or cuboidal shape when viewed in a section perpendicular to the tissue surface. *Columnar cells* are much thicker than they are wide and, in vertical section, appear as rectangles set on end.

Epithelial tissue may be only one cell thick, in which case it is called *simple epithelium,* or it may be two or more cells thick and is then known as *stratified epithelium.* There is, in addition, a third cate-

3.30. Epithelial tissues.

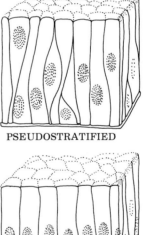

SIMPLE SQUAMOUS

SIMPLE CUBOIDAL basement membrane SIMPLE COLUMNAR PSEUDOSTRATIFIED

STRATIFIED SQUAMOUS STRATIFIED CUBOIDAL STRATIFIED COLUMNAR

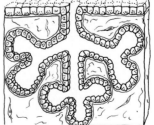

UNICELLULAR EPITHELIAL GLANDS MULTICELLULAR SIMPLE GLAND MULTICELLULAR COMPOUND GLAND

gory, called ***pseudostratified epithelium,*** in which the tissue looks stratified but actually is not; whereas in true stratified epithelium only the cells in the lowest layer are in contact with the underlying membrane, in pseudostratified epithelium all the cells are in contact with it. The various types of epithelia are named on the basis of cell type and number of cell layers; it is the cells in the outermost layer of stratified epithelia that determine the name. All types of epithelium are usually separated from the underlying tissue by an extracellular fibrous ***basement membrane*** (Fig. 3.30).

Epithelial cells often become specialized as gland cells, secreting substances at the epithelial surface (Fig. 3.30). Sometimes a portion of the epithelial tissue becomes invaginated, and a multicellular gland is formed.

Connective tissue

In connective tissue, the cells are always embedded in an extensive intercellular matrix. Much of the total volume of connective tissue is

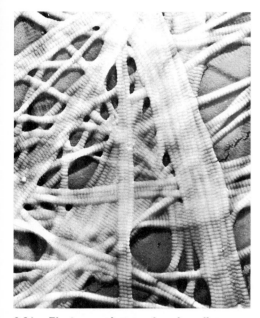

3.31. Electron micrograph of collagenous fibers from human skin. One of the three types of fibers commonly found in connective tissue proper, collagenous fibers are flexible but resist stretching. × 12,300. [Courtesy Jerome Gross, Harvard University.]

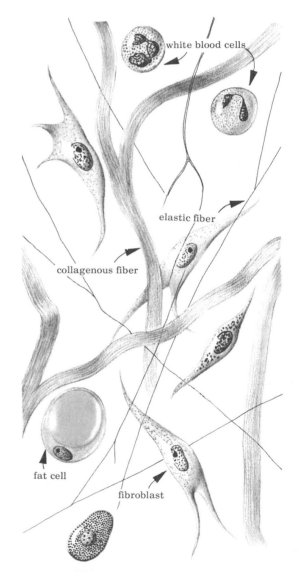

white blood cells

elastic fiber

collagenous fiber

3.32. Loose connective tissue. The several varieties of cells are embedded in an extensive extracellular matrix of fibers and ground substance. The fibroblasts are cells that are thought to secrete the proteins from which the fibers form.

fat cell

fibroblast

matrix, the cells themselves often being widely separated. The matrix may be liquid, semisolid, or solid. Connective tissue is often divided into four main types: (1) blood and lymph, (2) connective tissue proper, (3) cartilage, and (4) bone. The last three are sometimes collectively described as supporting tissues.

Blood. Blood and lymph are rather atypical connective tissues with liquid matrixes. They will be discussed in some detail in Chapter 7.

Connective tissue proper. Connective tissue proper is very variable, but its intercellular matrix always contains numerous fibers (Fig. 3.31). Both cells and fibers are embedded in a rather amorphous ground substance, which is a mixture of water, proteins, carbohydrates, and lipids. Associated with the ground substance is the *tissue fluid,* a liquid derived from the blood.

Connective tissue proper is customarily subdivided into two basic types—loose connective tissue and dense connective tissue—though there is no rigid separation between them, and intermediate types sometimes occur.

Loose connective tissue is characterized by the loose, irregular arrangement of its fibers, the extensive amount of ground substance, and the presence of numerous cells of a variety of types (Fig. 3.32). It is very widely distributed in the animal body. Because of its flexibility, loose connective tissue allows movement between the units it binds or connects.

Dense connective tissue is characterized by the compact arrangement of its many fibers, the limited amount of ground substance, and the relatively small number of cells (Fig. 3.33A). The fibers may be irregularly arranged into an interlacing network, as in the dermis of the skin, or they may be arranged in a definite pattern, usually in parallel bundles oriented to withstand tension from one direction, as in tendons connecting muscle to bone, or ligaments connecting bone to bone.

Cartilage. Cartilage (gristle) is a specialized form of dense fibrous connective tissue in which the intercellular matrix has a rubbery consistency (Fig. 3.33B). The relatively few cells are located in cavities in the matrix. Cartilage can support great weight; yet it is often flexible and somewhat elastic.

Cartilage is found in the human body in such places as the nose and ears (where it forms pliable supports), the larynx and trachea (you can feel the rings of cartilage in the front of your throat), intervertebral discs, surfaces of skeletal joints, and ends of ribs. Most of the skeleton of the early vertebrate embryo is composed of cartilage; the developing bones follow this model and slowly replace it. Some vertebrate groups, the sharks for example, retain a cartilaginous skeleton even in the adult.

Bone. Bone has a hard, relatively rigid matrix impregnated with inorganic salts such as calcium carbonate and calcium phosphate. This inorganic material may constitute as much as 65 percent of the dry weight of an adult bone. The few bone cells are widely separated and are located in spaces in the matrix (Fig. 11.3, p. 237). We shall discuss the histology of bone in more detail in Chapter 11.

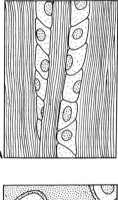

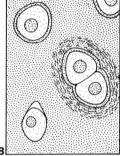

3.33. Dense connective tissue and cartilage. (A) Rows of cells and collagenous fibers alternate in tendon. (B) The nearly spherical cells of cartilage are widely spaced in an extracellular matrix that looks almost homogeneous under an ordinary microscope. Actually, the matrix contains a dense network of thin fibrils.

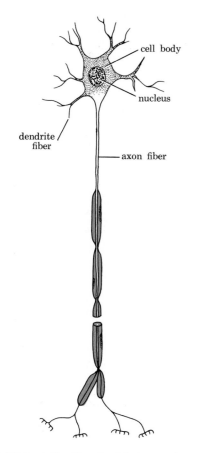

cell body

nucleus

dendrite
fiber

axon fiber

3.34. Nerve cell. The dendrites carry impulses toward the cell body. The axon carries impulses away from the cell body.

Muscle

The cells of muscle have greater capacity for contraction than most other cells, although all protoplasm probably possesses this capacity to some extent. Muscles are responsible for most movement in higher animals. The individual muscle cells are usually elongate and are bound together into sheets or bundles by connective tissue. Three principal types of muscle tissue (Fig. 11.5, p. 239) are recognized in vertebrates: (1) skeletal or striated muscle, which is responsible for most voluntary movement; (2) smooth muscle, which is involved in most involuntary movements of internal organs; and (3) cardiac muscle, the tissue of which the heart is composed. This classification does not hold for many invertebrate animals, which may lack one or more of these types altogether and nearly always show a different distribution of them.

Nerve

To some extent, all protoplasm possesses the property of irritability —the ability to respond to stimuli—but nerve tissue is highly specialized for such response. Nerve cells are easily stimulated and can transmit impulses very rapidly. Each cell is composed of a cell body, containing the nucleus, and one or more long thin extensions called fibers (Fig. 3.34). Nerve cells are thus admirably suited to serve as conductors of messages over long distances. An individual nerve cell may be 3 or 4 feet long, or even longer; no other kinds of cells even approach such length.

The functional combination of nerve and muscle tissue is fundamental to all multicellular animals except sponges. It is these tissues that confer upon animals their characteristic ability to move rapidly in response to stimuli.

THE VARIETY OF ORGANISMS

In our discussion in later chapters of the many different problems faced by all living things and the numerous alternative solutions to them evolved in different kinds of organisms, we shall need to refer repeatedly to representative organisms drawn from a variety of plant and animal groups. The following brief synopsis of a selected few such groups is given here so that you will be familiar with them when they are mentioned. A much more complete treatment of these and other groups will be found in Part V.

Bacteria

Bacteria do not fit well into either the plant or the animal kingdom, and it is meaningless to insist that they must belong to one or the other. Bacteria are very small single-celled procaryotic organisms. They are not usually green (though a few have a distinctive type of chlorophyll and carry out photosynthesis). They can be seen only under magnification. As we shall see, bacterial cells differ in many significant ways from the cells of most other organisms. Many bacteria are notable as disease-producing agents. Others are important scavengers, destroying the dead bodies of other organisms.

Plant groups

Green algae.   Green algae are relatively simple plants that live only in the water or in very moist environments on land. Some are unicellular, others multicellular. In multicellular species, most of the cells are similar to one another and are not highly specialized; consequently there are no separate layers or regions of different tissue types. The essentially parenchymatous cells form what might be considered a single continuous tissue, which is the plant body (Pl. I-4A). These algae possess chlorophyll, as the name "green algae" implies. They are of special interest because it is believed that many (if not all) of the higher plant groups evolved from ancestral flagellated green algae (see Pls. I-2, 3).

Brown algae and red algae.   Both the brown algae and the red algae are primarily marine and are commonly known as seaweeds. Their bodies are fairly complex and are always multicellular. Some are very large. In most cases, there is relatively little differentiation into distinct tissue types. A few species of brown algae do show such differentiation and have evolved many other characteristics resembling those of the higher land-plant groups, but this evolution is generally considered entirely independent and not an indication of close phylogenetic relationships. The common names of these two groups refer to the presence of brown and red pigments, which frequently mask the green chlorophylls that the plants also contain (see Pls. I-4C, I-5A).

Fungi.   Fungi are a group of relatively primitive plants whose members lack chlorophyll and are therefore unable to manufacture their own food. Like animals, they must obtain the complex high-energy nutrients they need in an already synthesized form. Some fungi are unicellular (e.g. yeast), but most are multicellular (e.g. bread mold, fruit molds, mushrooms, toadstools, bracket fungi; see Pls. I-5B–E). Their bodies, like those of algae, show little differentiation into distinct tissue types.

Mosses.   Green and photosynthetic, mosses (Pl. I-4D) live on land, but are not entirely independent of the ancestral aquatic environment; thus they occur only in very moist habitats. They are restricted to such habitats because they have not evolved highly specialized vascular tissues such as xylem and phloem, and therefore lack an efficient internal transport system.

Vascular plants.   These plants, as their name implies, all possess vascular tissue and thus have an effective internal transport system. As a result, they are less dependent on water in the surrounding environment than the previously discussed plants, and they constitute the dominant plant group on land today. Of all members of the plant kingdom, this group shows the greatest internal specialization into tissues and organs, and it is to the higher members of the group, particularly the flowering plants, that our earlier description of plant tissue types and body form applies.

The vascular-plant group is usually subdivided into several sections, three of which are doubtless familiar to you: the ferns, the conifers and their allies (gymnosperms), and the flowering plants (angiosperms). Of these three groups, the ferns are the most primitive; they appeared on the ancient earth before the other two groups, dominated the land for a long time, but eventually gave way to the other groups and are now largely overshadowed by them.

The gymnosperms and flowering plants are known collectively as the *seed plants;* they are more highly specialized for a terrestrial existence than the ferns. Some commonly encountered gymnosperms are pine, cedar, spruce, fir, and hemlock; all of these bear cones and have needlelike leaves, though not all gymnosperms do.

The angiosperms, or flowering plants, are the most advanced of the three groups. The majority of the land plants familiar to you belong to this group, which includes plants of every shape and size from grasses to cactuses, and from tiny herbs and wildflowers to large oaks and maple trees. This huge and very diverse group is customarily divided into two subgroups, the dicots and the monocots. The *dicot* subgroup includes such plants as beans, buttercups, privets, dandelions, oak trees, maple trees, roses, potatoes, and a great variety of others. The *monocot* subgroup includes the grasses and other grasslike plants such as corn, lilies, irises, and palm trees.

Animal groups

Protozoa. The Protozoa are single-celled organisms, but in many cases the single cell is so highly specialized and complex that it resembles a multicellular organism (Pl. I-1). They are generally much larger than bacteria and very mobile, swimming rapidly through the water in which they live or crawling along the bottom or on submerged objects. Some of them, like *Euglena* (Fig. 8.4, p. 157), possess long flagella and propel themselves by their whiplike motion. Others, like *Paramecium* (Fig. 5.6, p. 105), bear many cilia, which may function in both locomotion and feeding. Still others, like *Amoeba* (Fig. 22.2, p. 492), have neither flagella nor cilia, but move by a complex flowing of the cytoplasm as the cell, constantly changing shape, sends out extensions into which the rest of the cytoplasm then flows. We shall refer to representative Protozoa many times as we examine their fascinating evolutionary adaptations.

Coelenterates. The coelenterates constitute a large group of aquatic animals. Their saclike bodies are composed of two distinct tissue layers with a much less distinct third layer between them. They have a digestive cavity, but it has only one opening, which must serve as both mouth and anus. Tentacles are often present around the mouth and are used in capturing prey. The nerves and muscles of these simple animals are of an exceedingly primitive type, and no circulatory system is present. The body plan, or symmetry, is radial rather than bilateral (Pl. I-7C), a reflection of the relatively sedentary existence of many coelenterates during at least part of their life cycle. The group includes jellyfish, sea anemones (Pl. II-3C), and corals (Pl. I-8A). A fresh-water form to which we shall often refer as representative is the hydra (Fig. 5.8, p. 107).

Flatworms.   The flatworms (Platyhelminthes) are more complex than the coelenterates in some ways, but they, too, have a digestive tract with only one opening. The body is composed of three primary tissue layers, and the symmetry is bilateral. Many flatworms, such as flukes and tapeworms, are parasites and show numerous interesting specializations for this mode of existence. Others, like planaria (Fig. 5.9, p. 107), a small animal to which we shall refer repeatedly, are free-living aquatic organisms.

Molluscs.   Fairly complex animals, most of which possess shells, the molluscs include snails, clams (Pl. I-6A), oysters, and scallops, as well as octopuses (Pl. II-3B) and squids, which do not have obvious shells. These animals are particularly abundant in the oceans, as anyone who has collected their shells along the seashore will know. They are also common in fresh water. Some snails have evolved lungs and become fully terrestrial.

Annelids.   The annelids are often called the segmented worms. As this term implies, the bodies of these highly evolved worms are divided into a series of units or segments, which are often clearly visible externally. Though most of these animals are aquatic, some, like the earthworm, occur on land, though always in moist places. We shall generally take the earthworm as representative of this group; occasionally, however, we shall mention *Nereis,* a marine worm with large lobes growing from each side of the body segments (Fig. 6.5, p. 120), or the various tube-dwelling worms common in the oceans (Pl. I-8B).

Arthropods.   The arthropods constitute an immense group of very advanced animals that includes more different species than all other animal groups combined. All arthropods have jointed legs and a hard outer skeleton. Spiders, scorpions, crabs, lobsters, crayfish, centipeds, millipeds, and insects all belong to this major group; of these, the insects are by far the largest subgroup and are among the most successful of all land animals, rivaled only by the mammals and man himself (see Pls. II-4, 5).

Echinoderms.   All echinoderms are strictly marine; they have apparently never been able to invade either the fresh-water or the terrestrial habitats. Though their symmetry is radial, they are fairly advanced in many ways. The group includes sea stars (starfish), sand dollars, sea urchins (Pl. I-7B), sea cucumbers, and a variety of other forms. Though their appearance hardly suggests it, the echinoderms are regarded by most biologists as the group of animals most closely related to the next group, the chordates, to which man himself belongs.

Chordates.   This very important group includes a major subgroup called the ***vertebrates,*** which comprises all animals possessing an internal bony skeleton, particularly a backbone. Fish, amphibians (e.g. frogs, salamanders), reptiles (e.g. snakes, lizards, turtles, alligators), birds, and mammals (including man) belong to this group. We shall pay particular attention to the chordates throughout this book.

REFERENCES

AREY, L. B., 1968. *Human Histology,* 3rd ed. Saunders, Philadelphia.

BLOOM, W., and D. W. FAWCETT, 1968. *A Textbook of Histology,* 9th ed. Saunders, Philadelphia.

BRACHET, J., and A. E. MIRSKY, eds., 1961. *The Cell,* vol. 2. Academic Press, New York.

BROWN, R., 1960. "The Plant Cell and Its Inclusions," in vol. 1A of *Plant Physiology,* ed. by F. C. Steward. Academic Press, New York.

DEROBERTIS, E. D. P., W. W. NOWINSKI, and F. A. SAEZ, 1970. *Cell Biology,* 5th ed. Saunders, Philadelphia.

DUPRAW, E. J., 1968. *Cell and Molecular Biology.* Academic Press, New York.

ESAU, K., 1965. *Plant Anatomy,* 2nd ed. Wiley, New York.

SUGGESTED READING

BRACHET, J., 1961. "The Living Cell," *Scientific American,* September. (Offprint 90.)

DEDUVE, C., 1963. "The Lysosome," *Scientific American,* May. (Offprint 156.)

FOX, C. F., 1972. "The Structure of Cell Membranes," *Scientific American,* February. (Offprint 1241.)

GROSS, J., 1961. "Collagen," *Scientific American,* May. (Offprint 88.)

HOKIN, L. E., and M. R. HOKIN, 1965. "The Chemistry of Cell Membranes," *Scientific American,* October. (Offprint 1022.)

HOLTER, H., 1961. "How Things Get into Cells," *Scientific American,* September. (Offprint 96.)

JENSEN, W. A., 1970. *The Plant Cell,* 2nd ed. Wadsworth, Belmont, Calif.

LOEWENSTEIN, W. R., 1970. "Intercellular Communication," *Scientific American,* May. (Offprint 1178.)

LOEWY, A. G., and P. SIEKEVITZ, 1969. *Cell Structure and Function,* 2nd ed. Holt, Rinehart & Winston, New York.

NEUTRA, M., and C. P. LEBLOND, 1969. "The Golgi Apparatus," *Scientific American,* February. (Offprint 1134.)

PORTER, K. R., and M. A. BONNEVILLE, 1968. *Fine Structure of Cells and Tissues,* 3rd ed. Lea & Febiger, Philadelphia.

ROBERTSON, J. D., 1962. "The Membrane of the Living Cell," *Scientific American,* April. (Offprint 151.)

SOLOMON, A. K., 1960. "Pores in the Cell Membrane," *Scientific American,* December. (Offprint 76.)

CHAPTER 4

Energy transformations

It is a basic principle of physics that all systems have a natural tendency toward increasing disorder. The more orderly any arrangement of matter is, the less probable it is, and the less likely it is to be maintained if energy is not expended to counteract the tendency toward greater disorder. For example, the organization of wood, bricks, and other materials in the form of a house is one of an infinitely large number of possible arrangements of those materials, but it is so orderly an arrangement that there is almost no likelihood it would occur spontaneously. Similarly, if all the type used to print this page were dropped from a box onto the floor, one possible arrangement in which it could come to rest is the one you see before you, but this orderly arrangement of letters into words and sentences is so very unlikely to occur by chance that you doubtless feel certain it could be achieved only if energy were expended by a typesetter. The living cell, whose complex structure we examined in the last chapter, is also an inherently unstable and improbable organization. Only by constant use of energy can it maintain itself and keep from falling into the more stable and more probable random and disorganized state. The acquisition of energy in usable form is thus a necessity for every living cell.

CHARACTERISTICS OF ENERGY

Energy is generally defined as the capacity to do work. It may be stored as potential energy. But before work can actually be done, the energy must become active, which is to say that there must be motion of some sort. Thus a car parked on a steep hill has potential energy by virtue of its position; when its brake is released, the car rolls down the hill, converting its potential energy into the energy of motion. Similarly, a lump of coal contains potential energy, which is released as heat energy when the coal burns.

Energy can occur in many different forms. Light is energy, and so is electricity; there is heat energy, mechanical energy, chemical energy, etc. All forms of energy are interconvertible, at least partly. According to the *First Law of Thermodynamics* (also called the Law of Conservation of Energy), when energy is converted from one form into another no energy is either gained or destroyed (for this law to be strictly valid, matter must be considered a form of energy).

Living cells draw primarily on chemical energy; they do work by utilizing the potential energy in chemical bonds. Every bond in every molecule represents an amount of chemical-bond energy equal to the amount of energy that was necessary to link the atoms together originally. Living cells, then, are transducers that turn other forms of energy into chemical-bond energy, or the reverse.

We have already seen that the molecules of many of the compounds of living cells are highly complex, often being composed of thousands of atoms. Many of the bonds holding these atoms together are relatively rich in chemical-bond energy. Thus a living organism is a storehouse of potential chemical energy, which can be used, when necessary, to do work. But as this stored energy is converted into other forms, less and less remains in reserve. A source of usable energy outside the organism must be available to replenish its supply of chemical-bond energy. For many organisms, that outside energy source is other organisms; i.e. one living thing obtains new supplies of energy-rich molecules by eating the bodies of other living things. Since according to the First Law of Thermodynamics all these energy conversions are accomplished without reduction of the total amount of energy, it might seem at first glance that the same energy could be passed continuously from organism to organism and that no source of energy outside the system composed of all living things would be required. A little further thought shows that this is not true, since energy is constantly passed from organisms to nonliving matter, as when you throw a rock or move a pencil or when heat from your body warms the air; such energy is lost to the life system. Furthermore, the molecules of substances that leave the body retain some energy, and this energy, too, may be lost. But there is another basic reason why life on earth would run down if there were no non-living source of energy to be tapped. Living systems, like nonliving systems, obey the *Second Law of Thermodynamics,* which says that every energy transformation results in a reduction in the usable or free energy of the system; or, to put it another way, there is a steady increase in the amount of energy that cannot be used to do work.

PHOTOSYNTHESIS

As you doubtless know, the ultimate energy source for most living things is sunlight, and the organisms that transform the light energy into chemical-bond energy are primarily the green plants. The process by which this transformation is carried out is called photosynthesis. There is still much about the process that is not known, but in the last decades enormous strides have been made in analyzing the chemical pathways involved. It is not our purpose here to discuss in detail all the chemistry of photosynthesis, or to mention all reactions and compounds now thought to be involved in it; but photosynthesis is so fundamental to life that you ought to understand at least the broad outlines of the process and some of the principles of energy transformations in living systems exemplified by it.

Historical development

As early as 1772, an English clergyman and chemist, Joseph Priestley, demonstrated that green plants affect air in such a way as to reverse the effects of breathing (Fig. 4.1). His experiments were the first to show that plants produce oxygen, though he himself did not realize that this was what was happening; nor did he realize that light was essential for the process he observed. But his findings stimulated interest in photosynthesis, as we now call it, and led to further work. Only seven years later, the Dutch physician Jan Ingen-Housz demonstrated the necessity of sunlight for oxygen production (though, like Priestley, he knew nothing about oxygen at that time and explained his results in other terms), and he showed also that only the green parts of the plant could photosynthesize. In 1782 a Swiss pastor and part-time scientist, Jean Senebier, showed that the process was dependent upon a particular kind of gas, which he called "fixed air" (and we call carbon dioxide). Finally, in 1804, another Swiss worker, Nicolas Théodore de Saussure, found that water is necessary for the photosynthetic production of organic materials.

Thus, early in the nineteenth century, all the important materials of the photosynthetic process were at least vaguely known, and could be summarized by the following equation:

$$\text{carbon dioxide} + \text{water} \xrightarrow[\text{light}]{\text{green plants}} \text{organic material} + \text{oxygen}$$

Later, scientists came to believe that light energy splits carbon dioxide, CO_2, and that the carbon is then combined with water, H_2O, to form the unit (CH_2O),[1] which is the atomic grouping upon which carbohydrates are based. According to this view, the oxygen released by the plant during photosynthesis comes from CO_2. This idea received a severe blow about 1930, when C. B. van Niel of Stanford University showed that some photosynthetic bacteria that use hydrogen sulfide (H_2S) instead of water as a raw material for photosynthesis give off sulfur instead of molecular oxygen as a by-product. Now, H_2S and H_2O have obvious chemical similarities, and if the sulfur produced by the bacteria during photosynthesis came from H_2S, the oxygen produced by higher plants during photosynthesis might well come from H_2O and not CO_2. This was at last conclusively shown by use of a heavy isotope of oxygen (O^{18} instead of the usual O^{16}). If photosynthesizing plants are given normal carbon dioxide plus water incorporating heavy oxygen, the heavy isotope appears as molecular oxygen:

$$CO_2 + 2H_2O^{18} \rightarrow O_2^{18} + (CH_2O) + H_2O$$

[1] The parentheses indicate that this combination of atoms does not represent an actual molecule, but only a grouping within some larger compound; thus it takes six (CH_2O) groupings to make a simple six-carbon sugar, $C_6H_{12}O_6$.

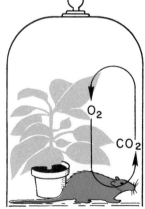

4.1. Priestley's demonstration that plants and animals "restore" the air for each other. A plant alone in a closed jar died, and a mouse alone in another closed jar died, but when plant and mouse were together in the same jar, both lived.

Chlorophyll, the green pigment of plants, traps the energy required to split water. The currently accepted equation for green-plant photosynthesis is now given in greater detail; the dashed lines indicate the fates of all the atoms involved:

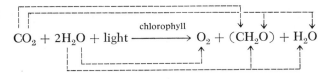

$$CO_2 + 2H_2O + light \xrightarrow{\text{chlorophyll}} O_2 + (CH_2O) + H_2O$$

Multiplying this summary equation by 6 has the advantage of showing that glucose, a six-carbon simple sugar, is often the end product:

$$6CO_2 + 12H_2O + light \xrightarrow{\text{chlorophyll}} 6O_2 + C_6H_{12}O_6 + 6H_2O$$

It may seem curious that water should appear on both sides of the equation. The reason is that the water produced by the photosynthetic process is new water; it is not the same water as that used as a raw material.

Although the above equation is a convenient summary of photosynthetic carbohydrate synthesis, it tells us nothing about how the synthesis is actually achieved. The process is certainly not one gross chemical reaction, as the summary equation might imply. It was recognized years ago that many reactions are involved and that some of these are "dark" reactions—ones that can take place without light. It is only recently, however, that scientists have begun to learn the actual reactions of photosynthesis and have been able to clarify the role of light.

The chemical nature of photosynthesis

Carbon dioxide is an exceedingly energy-poor compound; sugar is energy-rich. Photosynthesis, then, is a process that converts light energy into chemical-bond energy and stores it by synthesizing energy-rich sugar from energy-poor carbon dioxide. In chemical terms, the energy is stored by the reduction of carbon dioxide.

Basically, *reduction* means the addition of an electron (e^-), and its converse, *oxidation,* means the removal of an electron. Since an electron added to one molecule must have been removed from some other molecule, it follows that whenever one substance is reduced another is oxidized:

$$A^{e-} + B \rightarrow A + B^{e-}$$

electron donor   electron acceptor   has been oxidized (lost energy)   has been reduced (gained energy)

Among the many ways a compound may be reduced, two of the most common involve removal of oxygen or addition of hydrogen; similarly, oxidation often involves addition of oxygen or removal of hydrogen:

$$A + BO \rightarrow AO + B$$

electron donor   electron acceptor   has been oxidized   has been reduced

or:

$$AH + B \rightarrow A + BH$$

electron donor   electron acceptor   has been oxidized   has been reduced

In biological systems, removal or addition of hydrogen constitutes the most frequent mechanism of oxidation-reduction reactions. In general, then, when we speak of reduction we shall mean addition of hydrogen, and when we speak of oxidation we shall mean removal of hydrogen. Reduction reactions store energy in the reduced compound, while oxidation reactions liberate energy.

It now becomes clear why the synthesis of sugar from carbon dioxide constitutes reduction of the carbon dioxide. Hydrogen obtained by splitting water molecules is added to the CO_2 to form compounds based on (CH_2O) units, and energy from light is stored in the process. Our attention must therefore be focused on two key points: the mechanism of trapping and handling energy and the mechanism of transferring hydrogen from water to carbon dioxide.

Cyclic photophosphorylation

What happens when light of a proper wavelength strikes a chlorophyll molecule? The exact answer is not known, but some well-informed guesses can be made. Light energy comes in discrete units or packets called photons. When a photon strikes a chlorophyll molecule and is absorbed, its energy is apparently transferred in some manner to an electron of the chlorophyll. This electron is raised from its normal stable energy level to a higher energy level (Fig. 4.2). This higher level is relatively unstable, and the "excited" electron will show a marked tendency to fall back to its normal level, giving up the absorbed energy. Isolated chlorophyll in a test tube promptly loses the energy it traps by re-emitting it as visible light; thus the chlorophyll alone is incapable of converting light energy into chemical-bond energy. Once light energy has raised a chlorophyll electron to a high-energy state, other cellular components must come into play if the energy is to be captured and utilized. Let us look at this process in more detail.

In the chloroplasts of a living photosynthesizing cell, the pigment molecules are organized into functional groups called **photosynthetic units.** Two types of photosynthetic units occur in most plants; collectively these make up two different **photosystems.** Photosystem I will be considered here, photosystem II in the next section.

Each unit of photosystem I contains molecules of chlorophyll *a*, the most widespread type of chlorophyll, molecules of carotenoid pigment, and one molecule of a highly specialized form of chlorophyll *a* designated P700. When a photon of light strikes one of the pigment molecules in a photosynthetic unit of photosystem I, the energized electron may be passed from pigment molecule to pigment molecule within the unit until eventually it reaches the molecule of P700, which traps the electron. In the functioning chloroplast, the excited electron of P700 does not simply fall back to its normal lower energy level. Instead, a molecule of a substance designated Z (not yet characterized chemically), which has a very great affinity for electrons, leads the high-energy electron away from the P700. Substance Z then passes the electron to a second electron-acceptor molecule, which passes it to still another acceptor molecule, and so forth; some of the acceptor molecules are iron-containing pigments called **cytochromes.** After being passed from molecule to molecule, the electron finally arrives back in the chlorophyll whence it started. But when the electron leaves the chlorophyll it is energy-rich, and when it finally returns it is energy-poor (Fig. 4.3).

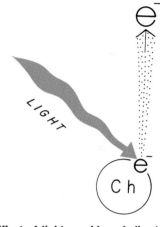

4.2. Effect of light on chlorophyll. Light striking a chlorophyll molecule (Ch) causes an electron (e^-) of the chlorophyll to be raised to a higher energy level.

To say, in the customary phrase, that the electron is "passed" from acceptor molecule to acceptor molecule may do an injustice to the complexity of the process and obscure its chemical nature. The "passing" is actually accomplished by a series of oxidation-reduction reactions, each catalyzed by a different enzyme. Imagine a series of electron acceptors, A, B, C, D, each of which—owing to its particular molecular configuration—has a somewhat higher energy content than the next one in the series. If A now passes an electron to B, A will be oxidized and B reduced; if B then passes the electron to C, B will be oxidized and C reduced; and so on. Because of the relative energy levels of the molecules, the reactions are energy-yielding:

$$A^{e-} + B \rightarrow A + B^{e-} + energy$$
$$B^{e-} + C \rightarrow B + C^{e-} + energy$$
$$C^{e-} + D \rightarrow C + D^{e-} + energy$$

It is clear that the alternating reduction and oxidation of each molecule allows it to function as an electron acceptor over and over again.

By taking part in a series of energy-releasing reactions of this kind, the light-energized electron from chlorophyll is eased back step by step to its normal state. By the time it returns to the chlorophyll, it has lost all its extra energy, but it has not lost it all at once, as would happen in the case of isolated chlorophyll in a test tube. Instead, the energy has been released in a series of small portions of manageable size.

Step-by-step release of energy is observable in many everyday situations. The potential energy in a waterfall can be harnessed by interposing a mill wheel. The potential energy in a tank of gasoline can be harnessed to propel a car, not by putting a match to the whole tankful—which, while a sure way of releasing the energy, wouldn't do the car much good!—but by burning small quantities at a time. In each case —waterfall, gasoline, or high-energy electron—the total amount of energy released is unaffected by whether the change in energy levels is abrupt or gradual, but the results of the release are considerably affected. Step-by-step reactions are a general property of biological chemical systems. In living cells, almost all major chemical conversions are ac-

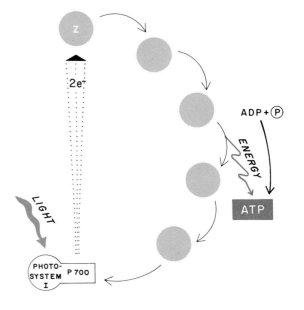

4.3. Cyclic photophosphorylation. Light striking molecules of chlorophyll in photosystem I causes electrons to be raised to a high energy level. The electrons are picked up by an electron-acceptor molecule, Z, which passes them to a series of further acceptor molecules (gray circles), each at a slightly lower energy level. Eventually the electrons return to the chlorophyll from which they started. Some of the energy released as the electrons are eased step by step down the energy gradient is used in the synthesis of ATP from ADP and inorganic phosphate.

complished by a series of smaller conversions, each catalyzed by its own enzyme.

The energy released as electrons are lowered down an energy gradient is not used by the organism directly. Instead, that portion which is not lost or transformed into heat is incorporated into a kind of energy-storage compound on which the organism can draw at need to power its multiple activities. This compound, found in every living thing and one of the most important substances of life, is called adenosine triphosphate, or **ATP** for short. Before describing ATP in some detail, we should point out that not all reactions by which an electron is transferred from one acceptor molecule to another yield enough energy for the synthesis of this high-energy compound; in the electron-transport system illustrated in Fig. 4.3, for example, only one of the reactions yields enough energy for the synthesis of an ATP molecule.

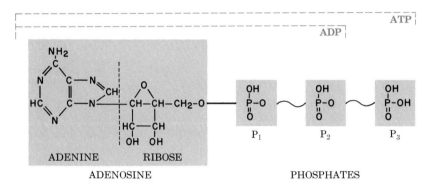

4.4. The ATP molecule. As its full name, adenosine triphosphate, implies, this molecule is composed of an adenosine unit (a complex of adenine and ribose sugar) combined with three phosphate groups arranged in sequence, with the last two attached by high-energy bonds (wavy lines). When the last bond is broken, the result is a release of energy and production of adenosine diphosphate, ADP.

The ATP molecule (Fig. 4.4) is composed of a nitrogen-containing compound (adenosine) plus three phosphate[2] groups bonded in sequence:

$$\text{adenosine} - ⓟ \sim ⓟ \sim ⓟ$$

The bonds by which the second and third phosphate groups are attached are not the usual kind of chemical bond, but are energy-rich bonds (customarily represented by the symbol $\sim$). If these bonds are broken by hydrolysis, far more calories of energy are released than if the other bonds in the ATP molecule are broken. Actually, it is often only the terminal high-energy phosphate bond of ATP that is involved in energy conversions. The exergonic reaction by which this bond is broken and the terminal phosphate group removed leaves a compound called adenosine diphosphate or **ADP**, consisting of adenosine plus only two phosphate groups:

$$\text{ATP} \xrightarrow{\text{enzyme}} \text{ADP} + ⓟ + \text{energy}$$

New ATP can be synthesized from ADP and phosphate if adequate energy is available to force a third phosphate group onto the ADP. Addition of phosphate is termed **phosphorylation:**

$$\text{ADP} + ⓟ + \text{energy} \xrightarrow{\text{enzyme}} \text{ATP}$$

---

[2] The symbol ⓟ is customarily used to designate the entire phosphate (actually phosphoryl) group:

$$-\overset{\displaystyle OH}{\underset{\displaystyle OH}{P}}=O$$

ATP is often called the universal energy currency of living things, and this characterization is fully justified. It is the phosphate-bond energy of ATP that is used to do all manner of work: synthesis of more complex compounds, muscular contraction, nerve conduction, active transport across cell membranes, light production, etc. Whenever any organism is doing work of any kind, you can be certain that ATP is involved; the energy price of the work is paid in ATP's (i.e. the energy is obtained through the breaking down of ATP to ADP).

ATP molecules can be synthesized from ADP in a variety of ways in living organisms, as will be seen later. But although neither synthesis of ATP nor step-by-step electron transport is unique to photosynthesis, the light-induced production of the high-energy electron that starts the process is. Any organism can manufacture energy currency by using up other energy-rich compounds, but only photosynthetic organisms (and a few chemosynthetic bacteria) can manufacture such currency from energy-poor inorganic raw materials. In short, it is the unique capacity of chlorophyll to absorb light energy and act as a donor of high-energy electrons that is critical for photosynthesis.

In the photosynthetic phosphorylation process we have so far described, chlorophyll acts both as electron donor and as the ultimate electron acceptor, donating excited electrons and eventually accepting the electrons in a low-energy state. Because the same electrons can be carried round and round the system and no outside source of electrons is involved, this method of synthesizing ATP, which was discovered in the mid-1950's, is called *cyclic photophosphorylation.*

Noncyclic photophosphorylation

Under certain conditions (as when carbon dioxide is not available), it seems likely that green plants use light for cyclic photophosphorylation alone. But most photosynthetic organisms generally draw on light energy for another critical reaction also—one involving transfer of hydrogen from water to an intermediate compound, which can then act as a hydrogen donor in the reduction of carbon dioxide.

Like the process of cyclic photophosphorylation already described, this process begins when photons of light strike molecules of chlorophyll in photosynthetic units of photosystem I and raise electrons to an excited state (Fig. 4.5). Again, the excited electrons are led away from the P700 molecule by a strong electron acceptor, probably the one we have called Z. But here the similarity to cyclic photophosphorylation ends. Instead of passing the electrons to the cytochrome carrier chain and hence back to the chlorophyll, Z donates the electrons, via several intermediate compounds, to an extremely important substance called nicotinamide adenine dinucleotide phosphate, or **NADP.** The NADP retains the energized electrons, and the now energy-rich compound, displaying a great affinity for hydrogen, apparently pulls two hydrogen protons (H^+) away from water[3] and forms $NADPH_2$. This utilization of light energy in the production of $NADPH_2$ by the splitting of water is crucial

[3] The assumption here of a direct movement of protons from water to NADP may not be correct. Some workers think that the water is first ionized ($H_2O \rightarrow H^+ + OH^-$) and that the energized NADP then picks up free hydrogen ions. According to this model, the electrons donated by water to photosystem II (discussed on pp. 79–80) come from hydroxyl ions.

to photosynthesis by green plants, for $NADPH_2$ can act as a hydrogen donor in the reduction of carbon dioxide to carbohydrate.

Since the energized electrons from the chlorophyll of photosystem I are retained by $NADPH_2$ and eventually incorporated into carbohydrate, photosystem I is left short of electrons. These "electron holes" are filled indirectly, by electrons derived from water as a result of a second light event. This second light event involves photosynthetic units of photosystem II, each composed of molecules of chlorophyll *a*, molecules of chlorophyll *b*, *c*, or *d*, depending on the species of plant,[4] and presumably one molecule of a specialized electron-trapping molecule (not yet isolated or identified) analogous to P700.

When light of the proper wavelength strikes a pigment molecule of photosystem II, the energy, in the form of excited electrons, eventually reaches the molecule of specialized electron-trapping pigment. This molecule, in turn, donates the high-energy electrons to a strong electron acceptor designated Q (Fig. 4.5). Substance Q then passes the electrons

[4] Chlorophyll *b* is found in green algae, bryophytes, and vascular plants. Chlorophyll *c* occurs in brown algae, and chlorophyll *d* in red algae.

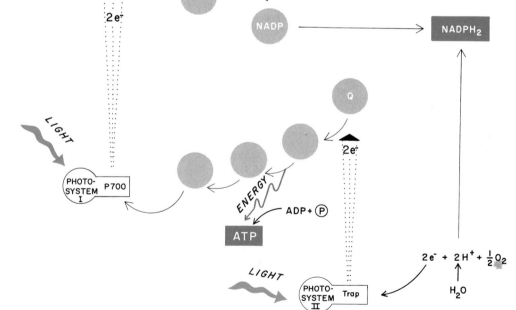

4.5. Noncyclic photophosphorylation. Light strikes pigment molecules of photosystem I (left). The electrons thus energized eventually reach the molecule of P700, a specialized form of chlorophyll *a*. The energized electrons are pulled away from P700 by Z, a very strong electron acceptor. Z then passes the electrons, via an intermediate compound, to NADP. NADP thereby gains enough energy to pull two hydrogen protons (H$^+$) away from water and form $NADPH_2$. A second light event raises electrons from photosystem II (bottom) to a higher energy level. These electrons are captured by an acceptor molecule, Q, which then passes them to a chain of acceptor compounds (gray circles), which lower the electrons step by step down an energy gradient until they reach the pigment molecules of photosystem I, where they replace the electrons lost as a result of the first light event; energy released from the electrons as they move down the energy gradient is used in synthesis of ATP from ADP and inorganic phosphate. Electrons to replace those lost from photosystem II come from water. Oxygen atoms from water combine to form molecular oxygen, O_2 (two molecules of water must be split for one complete O_2 molecule to be formed). The two important products of noncyclic photophosphorylation, ATP and $NADPH_2$, make possible the synthesis of carbohydrate from CO_2.

to a chain of acceptor molecules (some of them cytochromes), which lead the electrons to the electron holes in photosystem I. As the electrons move along the chain of carrier molecules, they are eased step by step down an energy gradient. Some of the energy thus released is utilized in the synthesis of ATP from ADP and inorganic phosphate.

Thus the electron holes created in photosystem I by the first light event are refilled by electrons moved from photosystem II by the second light event. But this process alone would leave electron holes in photosystem II. It is at this point that the electrons from water, mentioned earlier, play their role. As Fig. 4.5 indicates, it is thought that water is split into protons, electrons, and molecular oxygen:

$$H_2O \rightarrow 2H^+ + 2e^- + \tfrac{1}{2}O_2$$

(Note that two molecules of water must be split to yield a complete molecule of O_2.) The protons ($2H^+$), as we saw previously, are picked up by NADP. Photosystem II receives the electrons as replacements for those it lost in the second light event. The electrons involved in the second light event thus move from water to photosystem II to Q to carrier chain to photosystem I.

If we combine these steps with the electron movement associated with the first light event, as traced above, we obtain the following abbreviated sequence showing the overall electron movement:

$$H_2O \rightarrow \text{photosystem II} \rightarrow Q \rightarrow \text{carrier chain} \rightarrow \text{photosystem I} \rightarrow$$
$$Z \rightarrow NADPH_2 \rightarrow \text{carbohydrate}$$

This sequence reinforces our earlier statement that the hydrogen and electrons necessary to reduce CO_2 to carbohydrate come from water, but it shows that the movement of electrons from water to carbohydrate is an indirect and complex process.

Since electrons are not passed in a circular chain in this process, some leaving the system via $NADPH_2$ and others entering the system from water as replacements, this series of reactions is termed *noncyclic photophosphorylation*. The whole process results in the formation of both ATP and $NADPH_2$ and in the release of molecular oxygen.

We should point out that the details given above apply only to photosynthesis by green plants. As we have already indicated, some bacteria possess a form of chlorophyll and can utilize light energy in synthesis of ATP and $NADPH_2$, but they do not use water as the source of hydrogen ions and electrons, and they do not release oxygen as a by-product.

Carbohydrate synthesis

Early in this chapter we said that in studying photosynthesis attention should be focused on two key points: the mechanism of trapping and handling energy and the mechanism of transferring hydrogen from water to carbon dioxide. Both of these key mechanisms have now been explained in terms of ATP and $NADPH_2$. Still to be shown is how these two energy-rich compounds make possible the synthesis of carbohydrates from CO_2.

Actually we have already discussed the reactions unique to photosynthetic organisms and at the heart of photosynthesis: the synthesis of ATP and $NADPH_2$ by means of light energy. The utilization of these two key compounds in the synthesis of carbohydrate can be carried out in the dark; the dark reactions of carbohydrate synthesis are fully separable from the light-driven synthesis of ATP and $NADPH_2$ on which they depend. One way to separate the light and dark phases of photosynthesis experimentally in both time and space is to expose a suspension of chloroplasts to light under conditions of plentiful supply of ADP, inorganic phosphate, and NADP, but absence of CO_2. Under such conditions, the chloroplasts will synthesize large quantities of ATP and $NADPH_2$. If the chloroplasts are then fractionated and the solid green portion in which photophosphorylation takes place is discarded, and if CO_2 is added to the remaining mixture in the dark, the CO_2 will be assimilated into carbohydrate at the expense of the earlier-synthesized ATP and $NADPH_2$. In other words, the production of ATP and $NADPH_2$, on the one hand, and the assimilation of CO_2, on the other, may occur at different times, under different conditions, and even in different parts of the chloroplasts.

Now, carbohydrate contains much chemical-bond energy, while CO_2 contains very little. As we might predict from what we have already learned about the chemistry of living cells, the reduction of CO_2 to form glucose proceeds by many steps, each catalyzed by an enzyme. In effect, CO_2 is pushed up an energy gradient through a series of intermediate compounds, some of them unstable, until the stable carbohydrate end product is formed. An analogy would be a man moving a large and very heavy chest up a flight of stairs from the first to the second floor of his home. The man might be able to lift the chest just high enough to get it up one step at a time, balancing it on each step just long enough to marshal his strength before the next heave. If he let go (i.e. stopped applying energy) at any point between the stable level of the first floor and the stable but higher energy level of the second floor, the chest would come crashing to the bottom. The steps, then, make it possible to move the chest up an energy gradient, but they themselves are unstable intermediate levels. In this case, the energy necessary to move the chest through the series of unstable intermediate levels to the stable high energy level at the top is supplied by the man. In the case of synthesis of carbohydrate from CO_2, the energy comes from light via ATP and $NADPH_2$ (Fig. 4.6).

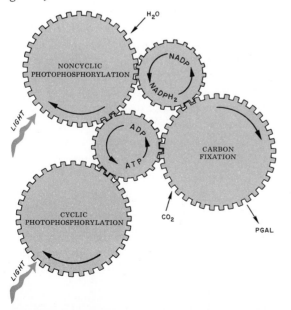

4.6. The relationship between photophosphorylation and carbon fixation. The entire photosynthetic process can be visualized as a series of interlocking gears. Energy from light turns the two photophosphorylation gears. The turning of the cyclic-photophosphorylation gear causes the gear of ATP synthesis to turn, and the turning of the noncyclic-photophosphorylation gear causes both the ATP-synthesis and the $NADPH_2$-synthesis gears to turn. These two gears cause the carbon-fixation gear to turn, with resultant production of carbohydrate (PGAL) from CO_2.

If there are many sequential steps in the reduction of CO_2 to carbohydrate, and if many of the intermediate compounds occur also in other processes leading to different end products, you may well wonder how the exact sequence of steps could be discovered. The tool that made such discoveries possible was a radioactive isotope of carbon, designated C^{14}. Carbon 14 was utilized by Melvin Calvin and his associates at the University of California at Berkeley in an intensive long-term investigation of carbon dioxide fixation in photosynthesis, begun in 1946. They exposed algal cells to light in an atmosphere of $C^{14}O_2$ for a few seconds and then killed the cells by immersing them in alcohol. The alcohol not only killed the cells but also inactivated the enzymes that catalyze the reactions of photosynthesis. With the enzymes inactivated, whatever amount of each intermediate compound existed in the cell at the moment of inactivation was, in effect, locked in. Calvin and his co-workers could then determine which of these locked-in intermediate compounds contained C^{14}. The length of time during which the algal cells were exposed to the $C^{14}O_2$ before being killed determined the number of compounds in which C^{14} was detected; if the time was very short, the C^{14} would reach only the first few compounds in the synthetic sequence, while, if the time was longer, the isotope would have moved through more steps in the sequence and would appear in a great variety of compounds. After years of painstaking research, Calvin worked out a sequence of reactions outlined in very abbreviated form in Fig. 4.7.

4.7. Synthesis of carbohydrate. The CO_2 combines with ribulose, a five-carbon sugar, to form an unstable six-carbon compound, which promptly splits into two molecules of a three-carbon compound. After phosphorylation by ATP, these molecules are reduced by hydrogen from $NADPH_2$ to form PGAL, a three-carbon sugar. Much of the PGAL is used in synthesis of more ribulose by a complicated series of reactions (not shown separately here) driven by ATP. But some of the PGAL molecules combine to form glucose, which the organism can use in a variety of ways.

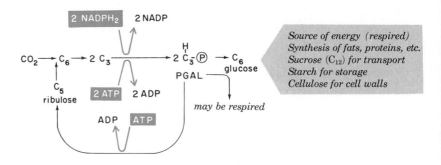

Under experimental conditions, the dark reactions of photosynthesis begin with the combination of CO_2 with a five-carbon sugar called **ribulose.** The stable end product of the dark reactions is, in a sense, the energy-rich three-carbon sugar phosphoglyceraldehyde, **PGAL** for short. Most of the molecules of PGAL are used in the formation of new ribulose with which more CO_2 can be processed. Some of the PGAL is used as a source of metabolic energy by the cell in which it is synthesized, and some is combined and rearranged in a series of steps to form the six-carbon sugar glucose, traditionally considered the end product of photosynthesis.

The leaf as an organ of photosynthesis

Photosynthesis is not restricted to the leaves; it occurs in all green parts of the plant. In most higher vascular plants, however, it is the leaves that expose the greatest area of green tissue to the light and are therefore the principal organs of photosynthesis. As you consider the structural features of a representative leaf, ask yourself how these contribute toward making it an efficient organ for carrying on photosynthesis. Keep in mind such requisites of the process as exposing large numbers of chloroplasts to sunlight and much moist cell-membrane surface to air, preventing excessive water loss, and moving the products of photosynthesis from their site of production to other parts of the plant.

Do not conclude, because attention here will be focused on the leaves of the more advanced land plants, that only these are of importance as photosynthetic organisms. Indeed, it has been estimated that only about 10 percent of all photosynthesis is conducted by the land plants most familiar to us. The other 90 percent is carried out by algae, principally in the ocean. Many of these algae are microscopic, and all lack true leaves.

Most dicot leaves consist of a stalk, or *petiole*, and a flattened *blade* (Fig. 4.8). The blade is usually broad and thin and contains a complex system of veins. Because of the flatness of the blade, the leaf exposes to the light an area that is very large in relation to its volume.

If a transverse section of a leaf is examined microscopically (Fig. 4.9), the outer surfaces are seen to be formed by layers of epidermis, usually only one cell thick, but sometimes two, three, or more cells thick. A waxy layer, the *cuticle*, usually covers the outer surfaces of both the upper and lower epidermis; it is generally thicker on the former. The chief function of the epidermis is protection of the internal tissues of the leaf from excessive water loss, from invasion by fungi, and from mechanical injury. Most epidermal cells do not contain chloroplasts.

The entire region between the upper and lower epidermis contains parenchyma cells, which constitute the *mesophyll* portion of the leaf. The mesophyll is commonly divided into two fairly distinct parts: an upper *palisade* mesophyll consisting of cylindrical cells arranged perpendicular to the epidermis and a lower *spongy* mesophyll composed of irregularly shaped cells. The cells of both parts of the mesophyll are very loosely packed and have many intercellular spaces. These spaces are interconnected and communicate with the atmosphere outside the leaf by way of holes in the epidermis called *stomata*. The size of the

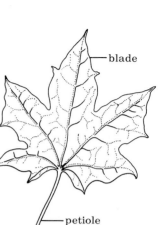

4.8. A dicot leaf. Note the many veins that radiate into the blade from the petiole.

4.9. Section of a privet leaf.

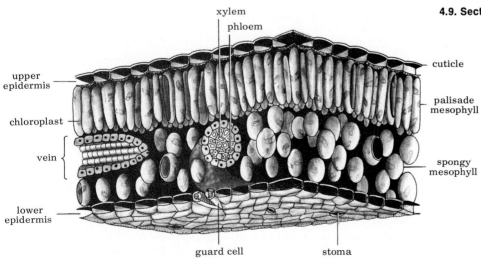

xylem
phloem
cuticle
upper epidermis
palisade mesophyll
chloroplast
vein
spongy mesophyll
lower epidermis
guard cell
stoma

stomatal openings is regulated by a pair of modified epidermal cells called *guard cells.*

A conspicuous system of veins branches into the leaf blade from the petiole (Fig. 4.8). The veins form a structural framework for the blade and also act as transport pathways connecting the leaf with the transport system of the rest of the plant. Each vein contains cells of each of the two principal vascular tissues, xylem and phloem. In most cases, the veins branch profusely within the mesophyll tissues, so that no mesophyll cell is far removed from a veinlet.

The chloroplasts are located primarily in the cells of the mesophyll layers. Chlorophyll is not simply scattered randomly within the chloroplasts, but is arranged in a very precise and orderly fashion within lamellae, which are embedded in a colorless protein matrix called the stroma (see Fig. 3.21, p. 53). In many parts of the chloroplasts of higher plants, numerous lamellae lie very close to each other, forming structures called grana; the individual granum has the appearance of a stack of coins (Fig. 4.10). Each lamella is a specialized membrane composed of a series of layers of protein, lipid, and chlorophyll. A single chloroplast contains all the electron-carrier molecules and enzymes necessary for photosynthesis, and chloroplasts isolated from the rest of the cell can still carry out the complete photosynthetic process.

4.10. Electron micrograph of a granum from a chloroplast of tobacco. The lamellae in the granum are arranged like stacks of coins. Outside the granum are some stroma lamellae. The small dark spots in the stroma are ribosomes that were present inside the chloroplast. × 200,000. [Courtesy Herbert W. Israel, Cornell University.]

CELLULAR RESPIRATION

Photosynthesis binds energy from solar radiation into complex organic compounds such as glucose and compounds synthesized from glucose. Before this potential energy can be utilized in the doing of work, either by the green plant itself or by some organism that has eaten the green plant, the large energy-rich molecules must be broken down chemically and the energy released. This oxidative breakdown is called *respiration.* In living things it is characteristically combined with processes that incorporate a large portion of the released energy into "universal energy currency"—ATP.

As you might anticipate from our examination of photosynthesis, the respiratory breakdown does not occur as a single gross reaction, but rather as a series of smaller step-by-step reactions, each catalyzed by an enzyme; and the release of packets of energy is coupled with phosphorylation reactions that synthesize ATP from ADP and inorganic phosphate.

The complete respiratory degradation of a compound such as glucose to carbon dioxide and water involves a very large number of reactions, of which only the more important will be examined here. There are two stages in this process: (1) that of *anaerobic respiration,* which does not utilize oxygen and can take place whether or not it is present; and (2) that of *aerobic respiration,* which is dependent on oxygen.

Anaerobic respiration

Our examination of cellular respiration will concentrate first on the breakdown of carbohydrate—specifically, glucose. Glucose, as we have already seen, is a stable compound; i.e. it has little tendency to break down to simpler products. If the energy locked in its molecular configuration is to be released, it must first be made more reactive. Such a change requires energy expenditure by the organism. Earlier in this chapter, we mentioned the potential energy of an automobile parked on a steep hill, and we saw that this potential energy could be transformed into the energy of motion if the brake were released. But the releasing of the brake requires an expenditure of energy by someone. Thus an initial energy investment precedes the major energy release that occurs when the car rolls down the hill. Such is also the case with the respiratory release of energy from glucose; a small amount of *activation energy* must be invested to initiate the process. Most exergonic chemical reactions in living things require activation energy, and the principle involved is an important one to understand.

As you might expect, it is ATP that provides the activation energy for initiating cellular respiration (Fig. 4.11). The initial reactions, like the succeeding ones, are facilitated by enzymes that reduce the amount of activation energy required. Two molecules of ATP donate their terminal phosphate groups to the glucose:

$$\underset{\text{glucose}}{\text{C–C–C–C–C–C}} + 2\text{ ATP} \xrightarrow{\text{enzymes}} \text{P}–\text{C–C–C–C–C–C}–\text{P} + 2\text{ ADP}$$

(The simplified summary equations given here show only the carbon skeleton, and you should bear in mind that oxygen and hydrogen are attached to the carbons; see Fig. 2.11, p. 24. Similarly, the symbol P is here used to represent an entire phosphate group.) When the phosphate groups are transferred to glucose, some of the energy in the terminal phosphate bonds of ATP is lost as heat; the bonds attaching the phosphate groups to the glucose are therefore not energy-rich.

The diphosphorylated six-carbon compound produced now splits between the third and fourth carbons, two essentially similar three-carbon molecules being formed:

$$\text{P}–\text{C–C–C–C–C–C}–\text{P} \xrightarrow{\text{enzyme}} \text{P}–\text{C–C–C} + \text{C–C–C}–\text{P}$$

The three-carbon compound is one we have encountered before; it is PGAL, which you will recall as a major product of photosynthesis. The photosynthetic process involved synthesis of glucose from PGAL. So far, then, respiration looks like photosynthesis in reverse. And so far, instead of releasing energy from the carbohydrate and forming new ATP molecules, respiration has actually resulted in the loss of two molecules of ATP used as a source of activation energy.

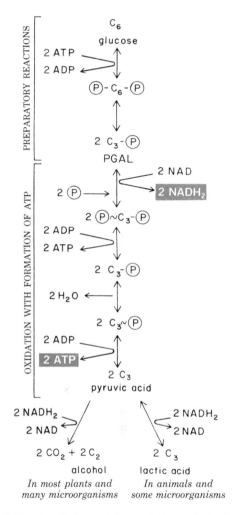

4.11. Glycolysis and fermentation. Activation energy for oxidation of glucose is supplied by two molecules of ATP. The resulting compound is then split into two molecules of PGAL. This completes the preparatory reactions. Next, the PGAL is oxidized by removal of hydrogen, which is picked up by NAD to form two molecules of NADH₂; in the same reaction inorganic phosphate is added to each of the three-carbon molecules. A series of reactions then result in synthesis of four new molecules of ATP, for a net gain of two. The pyruvic acid produced by this anaerobic breakdown can be further oxidized if O₂ is present (by reactions not shown here). But in the absence of sufficient O₂, the pyruvic acid may accept hydrogen from NADH₂ (reactions at bottom) to form CO₂ and ethyl alcohol in some kinds of organisms and lactic acid in others.

In the next reaction, which is rather complicated, the free ends of the two PGAL molecules are phosphorylated with inorganic phosphate; the result is two molecules of a three-carbon compound with a phosphate group at each end (a total of four phosphate groups for the two molecules). In the process, four hydrogen atoms are removed and are picked up by acceptor molecules of nicotinamide adenine dinucleotide, commonly abbreviated NAD; two molecules of $NADH_2$ are formed in this way. (NAD, as its name indicates, is extremely similar to NADP, a hydrogen acceptor encountered in photosynthesis.) The removal of the hydrogens has the effect of concentrating energy in two of the four phosphate groups.

The two energy-rich phosphate groups are then transferred to ADP. By the removal of two molecules of water, energy is concentrated in the two remaining phosphate groups. When these are likewise transferred to ADP, two more ATP molecules are formed, and two molecules of a three-carbon compound, *pyruvic acid*, remain.

The oxidation of carbohydrate to pyruvic acid by the pathway just described proceeds without the use of molecular oxygen and is the anaerobic portion of cellular respiration; it is often called *glycolysis.* The results of this process can be summarized as follows:

1. The breakdown of a molecule of glucose (a six-carbon compound) to two molecules of pyruvic acid (a three-carbon compound).
2. The use of two molecules of ATP as activation energy and the synthesis of four new molecules of ATP, for a *net* gain of two molecules of ATP.
3. The production of two molecules of $NADH_2$.

Glycolysis is not only the first, anaerobic, stage of respiration; it can also be considered the first stage in the process of *fermentation,* which occurs in the absence of oxygen. Under anaerobic conditions, pyruvic acid formed by glycolysis will accept the hydrogen from $NADH_2$; NAD is thus freed for further use as a hydrogen acceptor in the glycolytic pathway (Fig. 4.11). The addition of hydrogen to pyruvic acid results in the formation of *lactic acid* (a three-carbon compound) in animal cells and some unicellular organisms or in the formation of *ethyl alcohol* (a two-carbon compound) and carbon dioxide in plant cells and in many unicellular organisms:

$$\text{pyruvic acid} + NADH_2 \xrightarrow{\text{enzyme}} \text{lactic acid} + NAD$$

or

$$\text{pyruvic acid} + NADH_2 \xrightarrow{\text{enzyme}} \text{alcohol} + CO_2 + NAD$$

Depending on the end product, we can thus speak of alcohol fermentation or of lactic acid fermentation. Fermentation by yeast cells and other microorganisms is, of course, the basis for the extensive and economically important fermentation industry.

Aerobic respiration

In anaerobic respiration, as we saw, ATP is produced when a substrate molecule undergoing oxidation gains an energy-rich phosphate group—a process known as substrate-level phosphorylation—and transfers that phosphate group to ADP. Most of the ATP formed in aerobic respiration, on the other hand, is produced by a process known as *oxidative phosphorylation,* in which ATP is synthesized from ADP and inorganic phosphate by means of energy released as electrons (from hydrogen removed from a variety of compounds in the course of oxidation) are lowered step by step down an energy gradient. The process is very similar to photophosphorylation, and like it makes use of cytochromes for electron transport. But the energy that powers photophosphorylation comes from light-energized electrons in chlorophyll; the energy that powers oxidative phosphorylation comes from high-energy organic compounds.

We have seen that NAD functions as a hydrogen acceptor in anaerobic respiration, forming $NADH_2$. It is also one of the prime hydrogen acceptors for the various intermediate products of aerobic respiration. Under anaerobic conditions, however, the high potential energy of $NADH_2$ is hardly tapped, for the hydrogen is passed to pyruvic acid to form lactic acid (or alcohol), in which it remains at a high energy level. When molecular oxygen is present, on the other hand, the hydrogen can ultimately combine with it to form water, in which hydrogen is at a low energy level. The reaction can be summarized as follows:

$$\frac{1}{2}O_2 + NADH_2 \rightarrow H_2O + NAD$$

The $NADH_2$ is not oxidized directly by oxygen, as this summary equation might seem to indicate. Instead, it is oxidized by another hydrogen acceptor. When this acceptor molecule is oxidized in its turn, the electrons from the hydrogen are fed into a *cytochrome electron-transport system,* where they take part in a series of oxidation-reduction reactions on the model outlined on p. 76. The last cytochrome in the chain contributes the electrons to the formation of water:

$$2H^+ + 2e^- + \frac{1}{2}O_2 \rightarrow H_2O$$

As the hydrogen is passed from its high energy level in $NADH_2$ to its low energy level in H_2O, enough energy is released at several points along the respiratory chain to synthesize ATP from ADP and inorganic phosphate.

The hydrogen electrons entering the cytochrome transport system via NAD and other acceptor molecules (e.g. NADP) have three sources, briefly described below. Since the electrons are not all fed into the respiratory chain at the same point, their transport yields different amounts of ATP (Fig. 4.12).

1. We have seen that two molecules of $NADH_2$ are formed during glycolysis. When enough oxygen is present, the four hydrogens are not used for the reduction of pyruvic acid to lactic acid, but are donated to the respiratory chain; the energy yield from electron transport is four molecules of ATP. If lactic acid has already been formed, pyruvic acid and $NADH_2$ can be regenerated, since when sufficient oxygen is available the reaction between pyruvic and lactic acid is reversible. Note that glycolysis, which produces only two molecules of ATP under anaerobic conditions, produces a total of six when oxygen is present.

2. Under aerobic conditions, the two molecules of pyruvic acid produced for every molecule of glucose via the glycolytic pathway are

broken down to carbon dioxide and an activated form of *acetic acid,* a two-carbon compound. The complicated series of reactions can be summarized as follows:

$$2 \text{ pyruvic acid} + 2 \text{ NAD} \rightarrow 2 \text{ acetic acid} + 2CO_2 + 2 \text{ NADH}_2$$

The two $NADH_2$ molecules donate their hydrogen to the respiratory chain. As the hydrogen is lowered down the energy gradient, six molecules of ATP are synthesized.

3. The acetic acid formed by the breakdown of pyruvic acid (see equation above) is fed into a complex circular system of reactions called the *Krebs citric acid cycle* (after the British scientist Sir Hans Krebs, who was awarded a Nobel Prize for his elucidation of this system). The cycle is outlined in Fig. 4.13. It begins when a molecule of acetic acid combines with a four-carbon compound to form *citric acid,* a six-carbon compound. A succession of intermediate compounds then form before the original four-carbon compound is reconstituted, eight hydrogens being removed in the process. Since a single molecule of glucose yields two acetic acid molecules, the cycle turns twice for every molecule of glucose and 16 hydrogens are removed. These are picked up by NAD and other similar compounds and donated to the respiratory chain for

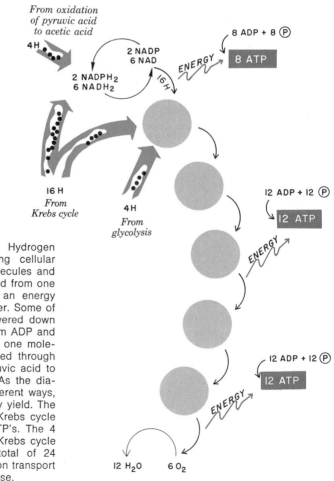

4.12. The cytochrome electron-transport system. Hydrogen removed from the substances being oxidized during cellular respiration is picked up by NAD or other acceptor molecules and transported to the cytochrome system, where it is passed from one acceptor substance to the next, step by step down an energy gradient, until it is finally combined with O_2 to form water. Some of the energy released as the hydrogen electrons are lowered down the energy gradient is used in the synthesis of ATP from ADP and inorganic phosphate. During the complete oxidation of one molecule of glucose, a total of 24 hydrogens are transported through this system (4 from glycolysis, 4 from oxidation of pyruvic acid to acetic acid, and 16 from the Krebs citric acid cycle). As the diagram indicates, the hydrogens enter the system in different ways, and this difference is reflected in a difference in energy yield. The 4 hydrogens from glycolysis and 4 of those from the Krebs cycle yield one ATP for every hydrogen, for a total of 8 ATP's. The 4 hydrogens from acetic acid and 12 of those from the Krebs cycle yield three ATP's for every two hydrogens, for a total of 24 ($16/2 \times 3$). A grand total of 32 ATP's result from electron transport during the complete oxidation of one molecule of glucose.

an energy yield of 22 ATP molecules. Two more ATP molecules are synthesized within the Krebs cycle itself by way of substrate-level phosphorylation, for a grand total of 24.

Summary of ATP yield of the respiration of glucose

We can summarize the yield of ATP molecules from the complete cellular respiration of one molecule of glucose to carbon dioxide and water as follows (Fig. 4.14):

1. A grand total of 36 ATP molecules are formed. This means that about 38 percent of the energy initially present in the glucose has been trapped; the rest has been lost, much of it as heat.

2. Only two of the 36 ATP molecules are synthesized anaerobically; the other 34 are the product of aerobic respiration. We now see why oxygen is essential to the life of human beings and most other organisms. The importance of the anaerobic processes of glycolysis and fermentation should not be overlooked, however. Many microorganisms rely exclusively on fermentation for their energy. Our own tissues, particularly our muscles during violent activity, often need so much energy so fast that the oxygen supplied from breathing is insufficient. Under such circumstances, glycolysis provides the needed energy. Later, the oxygen debt is paid back by deep breathing or panting, and the lactic acid, which accumulated in the muscles as a result of fermentation, and which in excess contributes to muscle fatigue, is removed.

3. Of the 34 aerobically synthesized ATP molecules, 32 result from hydrogen transport via NAD (or similar electron acceptors) and the cytochrome system. It is thus evident that this system is of critical importance, and it is easy to understand why cyanide and certain other poisons that block the cytochrome system are lethal.

Respiration of fats and proteins

Metabolism of fats begins with their hydrolysis to glycerol and fatty acids. The glycerol (a three-carbon compound) is then converted into PGAL and fed into the glycolytic pathway at the point where PGAL normally appears. The fatty acids are broken down into a number of two-carbon fragments, which are converted into acetic acid and fed into the respiratory pathway at the appropriate point. Since fats are more completely reduced compounds than carbohydrates, their complete oxidation yields more energy per unit weight; one gram of fat yields slightly more than twice as much energy as one gram of carbohydrate.

The amino acids produced by hydrolysis of proteins are metabolized in a variety of ways. After the amino group is removed (deamination), some amino acids are converted into pyruvic acid, some into

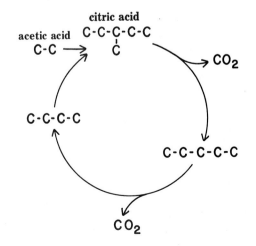

4.13. Simplified version of the Krebs citric acid cycle. The two carbons of the acetic acid combine with a four-carbon compound to form citric acid, a six-carbon compound. Then one carbon is removed as CO_2, leaving a five-carbon compound. Then another carbon is removed as CO_2, leaving a four-carbon compound, which can combine with another molecule of acetic acid and start the cycle over again. Since one molecule of glucose gives rise to two molecules of acetic acid, two turns of the cycle occur for each molecule of glucose oxidized. This diagram, which shows only the essential features of what is really a complicated series of more than ten reactions, focuses on the fate of the carbon atoms. Attached to these are hydrogen atoms, which, as they are removed in the reactions, are picked up by NAD (or other similar compounds) and transported to the cytochrome system.

acetic acid, and some into various compounds of the citric acid cycle. Complete oxidation of a gram of protein yields roughly the same amount of energy as one gram of carbohydrate.

Mitochondria

It has been demonstrated that aerobic respiration is confined to the mitochondria—cellular structures already mentioned in the last chapter. The electron microscope, you will recall, shows that each mitochondrion has a wall composed of an outer and an inner membrane and that the inner membrane has a series of folds, or cristae, which project into the matrix (Fig. 4.15).

It has been established that some of the reactions of the Krebs citric acid cycle take place in the matrix and some in the inner membrane. The enzymes in the matrix can easily be isolated and studied, but the ones in the membrane, which appear to be integral components of the membrane, are difficult to extract.

The precision with which the reactions of the cytochrome electron-transport system intermesh suggests that the components of that system are arranged in precise order. Recent research has in fact indicated that these components are built into the inner mitochondrial membrane and that they are arranged in clusters. Each cluster, it has been shown, is a self-contained functional unit comprising a complete set of the com-

4.14. Summary of ATP yield during complete oxidation of one molecule of glucose. If conditions are completely anaerobic, there is a net gain of only two ATP molecules, which are formed at substrate level. Under such conditions, the pyruvic acid must accept hydrogen from the $NADH_2$ synthesized during glycolysis, and either alcoholic fermentation or lactic acid fermentation results. But if O_2 is present, the $NADH_2$ may pass its hydrogen to the cytochrome system and ultimately to O_2 to form water. Thus the pyruvic acid may be further oxidized aerobically, yielding much more ATP. Of the grand total of 36 new ATP molecules formed by complete oxidation of glucose, 32 are formed via hydrogen-electron transport through the cytochrome system. The other four are formed at substrate level, two during the anaerobic reactions of glycolysis and two during the reactions of the Krebs citric acid cycle.

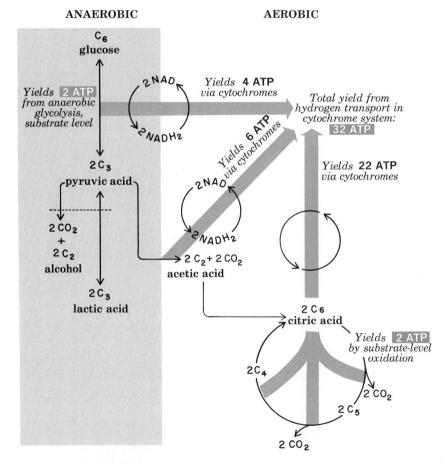

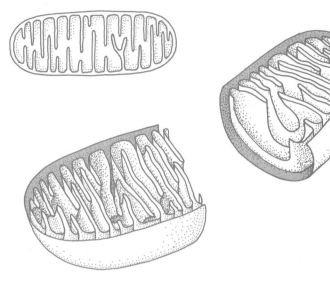

4.15. Structure of a mitochondrion. The inner membrane is folded into cristae, which project into the stroma. The mitochondria of metabolically very active cells have more cristae than those of less active cells.

ponents of the electron-acceptor chain. This functional unit is called a *respiratory assembly.*

Since 32 of the 36 new ATP molecules synthesized as a result of the complete oxidation of a molecule of glucose come from electron transport through the cytochrome system, it follows that the number of respiratory assemblies in a cell is one determinant of the capacity of that cell to synthesize ATP. It is not surprising, therefore, that the cells of tissues requiring large amounts of energy have more mitochondria than cells with low energy demands, and that these mitochondria tend to have more cristae and hence more respiratory units.

BODY TEMPERATURE AND METABOLIC RATE

We have seen that cellular respiration captures some of the energy released by the oxidation of carbohydrates, fats, and proteins and converts it into the energy of high-energy phosphate bonds in ATP. But this process fails to capture roughly 60 percent of the released energy. This unharnessed energy is largely in the form of heat, most of which is promptly lost to the environment by all plants and the vast majority of animals. The body temperature of such animals, usually designated as cold-blooded—a more accurate term is *poikilothermic,* i.e. of variable temperature—is determined by the environment; when they are at rest, their temperature is very nearly the same as that of the surrounding medium, particularly if the medium is water.

The metabolism[5] of an organism is very closely tied to temperature. Within the narrow range of temperatures to which the active organism is tolerant, the metabolic rate increases with increasing temperature and decreases with decreasing temperature in a very regular fashion.

As would be expected, the activity of poikilothermic animals is radically affected by temperature changes in their environment. As the temperature rises (within narrow limits), they become more active; as

[5] Metabolism means the chemical reactions of the body, both those, like respiration, that break down materials and those that build up complex molecules and structures.

it falls, they become sluggish and lethargic. Such animals, then, are restricted as to the habitats they can effectively occupy, because they are at the mercy of the temperatures in those habitats.

A few animals, notably mammals and birds, can make use of the heat produced during the exergonic reactions of their metabolism because they have evolved mechanisms, often including insulation by fat, hair, feathers, etc., whereby heat loss to the environment is retarded. Such animals are commonly called warm-blooded; biologists use the term *homeothermic* (homoiothermic), i.e. of uniform temperature. These animals have a body temperature that is fairly high—usually higher than the environmental one—and relatively constant even when the environmental temperature fluctuates widely. Their metabolic rate can accordingly be maintained at a uniformly high level, and they remain very active. They are thus less dependent on environmental temperatures than poikilothermic animals and are freed for successful exploitation of more varied habitats.

REFERENCES

BALDWIN, E., 1967. *Dynamic Aspects of Biochemistry,* 5th ed. Cambridge University Press, New York.

LEHNINGER, A. L., 1970. *Biochemistry.* Worth, New York.

RACKER, E., 1965. *Mechanisms in Bioenergetics.* Academic Press, New York.

SUGGESTED READING

ARNON, D. I., 1960. "The Role of Light in Photosynthesis," *Scientific American,* November. (Offprint 75.)

BASSHAM, J. A., 1962. "The Path of Carbon in Photosynthesis," *Scientific American,* June. (Offprint 122.)

LEHNINGER, A. L., 1961. "How Cells Transform Energy," *Scientific American,* September. (Offprint 91.)

————, 1971. *Bioenergetics,* 2nd ed. Benjamin, New York.

LEVINE, R. P., 1969. "The Mechanism of Photosynthesis," *Scientific American,* December. (Offprint 1163.)

MARGARITA, R., 1972. "The Sources of Muscular Energy," *Scientific American,* March. (Offprint 1244.)

RABINOWITCH, E. I., 1948. "Photosynthesis," *Scientific American,* August. (Offprint 34.)

————, and GOVINDJEE, 1965. "The Role of Chlorophyll in Photosynthesis," *Scientific American,* July. (Offprint 1016.)

RACKER, E., 1968. "The Membrane of the Mitochondrion," *Scientific American,* February. (Offprint 1101.)

STUMPF, P. K., 1953. "ATP," *Scientific American,* April. (Offprint 41.)

PART II

The biology of organisms

CHAPTER 5

Nutrient procurement and processing

To keep alive, organisms need materials for the synthesis of new proto-plasm for both growth and repair; and they need materials that will serve them as sources of energy. They obtain these materials, or *nutrients,* from the environment. How they procure and process them is the subject of the present chapter.

Two classes of organisms can be recognized on the basis of their methods of nutrition. Fully *autotrophic* organisms can subsist in an exclusively inorganic environment because they can manufacture their own organic compounds from inorganic raw materials taken from the surrounding media. Since the molecules of these raw materials are small enough and soluble enough to pass through cell membranes, autotrophic organisms do not need to pretreat, or digest, their nutrients before taking them into their cells. As you would guess, most autotrophs are photo-synthetic, although a few are chemosynthetic. The green plants are by far the most important of the earth's autotrophic organisms.

Heterotrophic organisms (most animals and all those plants, such as fungi, that lack chlorophyll) are incapable of manufacturing their own complex organic compounds from simple inorganic nutrients. Hence they must obtain prefabricated organic molecules from the en-vironment. Many of the organic molecules found in nature are too large and not sufficiently soluble to be absorbed unaltered through cell mem-branes, and they must first be broken down into smaller, more soluble molecular units; i.e. they must be digested.

It is clear, then, that autotrophic and heterotrophic organisms differ both in their nutrient requirements and in the problems associated with nutrient procurement. And it is not surprising that they should have evolved radically different adaptations in response to the different selec-tion pressures acting upon them. We shall therefore discuss these two great groups of organisms separately in this chapter, but shall indicate similarities between them where appropriate.

NUTRIENT REQUIREMENTS OF GREEN PLANTS

Raw materials for photosynthesis

The raw materials needed in greatest quantity by higher photosynthetic organisms are, of course, carbon dioxide and water. These two compounds supply the carbon, oxygen, and hydrogen that are the predominant elements in organic molecules. Carbon dioxide, one of the constituent gases of our atmosphere, is obtained directly from the air by the leaves of terrestrial plants; submerged aquatic plants absorb the dissolved gas from the surrounding water. Terrestrial plants obtain the other raw material, water, from the substrate in which they grow; most higher plants absorb water from the soil by roots.

A very high percentage of the total dry body weight of a large tree is carbohydrate, and much of the rest of it was synthesized from carbohydrate. This fact has some startling implications. The formula for glucose, which may be taken as the central carbohydrate in protoplasm, is $C_6H_{12}O_6$. Now, the combined weight of the six atoms of carbon and six atoms of oxygen in glucose is about 93 percent of the total weight of glucose. Since both the carbon and the oxygen come from carbon dioxide, it follows that about 93 percent of the weight of a large, immensely heavy tree comes initially from the air. The hydrogen in glucose comes from water, and hydrogen constitutes roughly 7 percent of the weight of glucose; hence about 7 percent of the dry weight of the tree comes initially from water. Even with a modern knowledge of photosynthesis, it is hard to comprehend that most of the mass of a plant's body comes from air, not from the solid earth in which it grows.

Mineral nutrition

Clearly, carbon dioxide and water cannot be the only nutrient materials needed by a green plant. These two compounds provide only three elements: carbon, oxygen, and hydrogen; yet we know that some vital components of the plant contain other elements as well. Protein, for example, is an essential component of protoplasm, and the amino acids of which proteins are composed always contain nitrogen; several very important proteins also contain sulfur. ATP, the "universal energy currency of life," contains phosphorus; phosphorus is also present in nucleic acids and many other critically important compounds. Chlorophyll, the essential mediator of photosynthesis, contains magnesium. Where does the green plant obtain the nitrogen, sulfur, phosphorus, magnesium, and other elements it needs? Obviously not from carbon dioxide or water. Here, finally, we see the role of the soil as a source of plant nutrients. It is from the soil that the plant derives the minerals essential to its life.

During the nineteenth century, there was much interest in Europe in determining the mineral needs of crop plants and in devising ways of supplementing the amounts of essential mineral elements in the soil. By 1900, seven of these were known: nitrogen, phosphorus, sulfur, potassium, calcium, magnesium, and iron. Three of them—nitrogen, phosphorus, and potassium—were stressed particularly, as they are to this day in the manufacture of fertilizer. Modern commercial fertilizers are often designated by their N-P-K percentages; e.g. the widely used garden fertilizer called 5-10-5 contains 5 percent nitrogen, 10 percent phosphoric acid, and 5 percent soluble potash by weight. These three are the elements most rapidly removed from the soil; consequently they must be replenished if crops are to continue to flourish. Many modern fertilizers are also fortified by small amounts of some of the other essential minerals.

It was not until about 1920 that it became apparent that other elements, in addition to the seven already known, were essential to plants. These additional minerals (e.g. boron, manganese, copper, zinc, sodium, chlorine) are required in such small amounts that the traces present as contaminants in the water or salts used in the early experiments were sufficient to meet the needs of the plants. Only with very elaborate purification procedures could their presence be controlled and their effects determined. Such elements, essential in minute amounts but sometimes toxic in excess, are now called trace elements or micronutrients.

Trace elements are needed in only slight amounts because most of them are components of enzymes or coenzymes. You will remember that enzymes can be used over and over and that a very small quantity of each is sufficient. Therefore only a small amount of mineral is required to synthesize the enzyme or coenzyme initially and to replenish the supply as the enzyme molecules are slowly broken down.

NUTRIENT PROCUREMENT BY GREEN PLANTS

We have seen that three classes of nutrients are needed by green plants: carbon dioxide, water, and minerals. Carbon dioxide, together with the other components of air, moves into the internal spaces of the leaf through openings called stomata. Inside the leaf, the air circulates throughout the numerous intercellular spaces (see Fig. 6.1, p. 118); carbon dioxide dissolves in the film of water on the surfaces of the leaf cells and diffuses into the cells, where it is used as a raw material for photosynthesis. In the higher land plants, water and minerals enter the plant primarily through the roots. In such aquatic plants as the algae, minerals are absorbed from the surrounding water, in which they are in solution; these plants usually lack specialized procurement organs and absorb their nutrients through the general body surface.

Roots as organs of procurement

Root structure.   The first root formed by the young seedling is called the *primary root.* Later, *secondary roots* branch from the primary root, and a root system is formed. If the branching results in a system of numerous slender roots, with no single root predominating, as in grass or clover, the plant is said to have a *fibrous root system* (Fig. 5.1A). If, however, the primary root remains dominant, with smaller secondary roots branching from it, the arrangement is called a *taproot system* (Fig. 5.1B). Dandelions, beets, and carrots, among others, are plants with taproots. As these examples suggest, taproots are frequently specialized as storage organs for the products of photosynthesis. Storage is a function of all roots, but particularly of taproots. Obviously, procurement of water and minerals and storage of high-energy organic compounds are

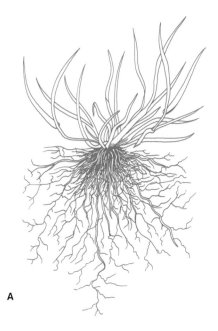

A

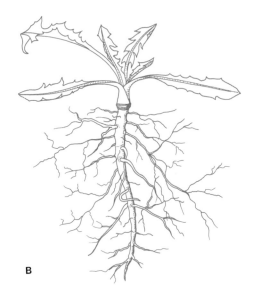

B

5.1. Two types of root systems.   (A) Fibrous root system of grass.   (B) Taproot system of dandelion.

5.2. Root of radish seedling with many prominent root hairs. [Courtesy W. E. Loomis, Iowa State University.]

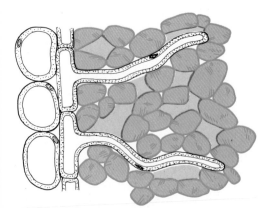

5.3. Root hairs penetrating soil. Each root hair, which is an extension of a single epidermal cell, is in contact with many soil particles and with soil spaces, some of which contain air, some water (gray).

not the only functions of roots; they also serve to anchor the plant to the substrate.

The root system of a plant is normally very extensive, far more extensive than is ordinarily realized. When we pull up a plant, we seldom get anything even approaching the entire root system, since most of the smaller roots are so firmly embedded in the soil that they break off and are lost. The amplitude of the system is of course important both in anchoring the plant to the substrate and in providing sufficient absorptive surface.

As a cell or an organism gets bigger, its volume increases much faster than its surface area (volume increases as a cube function, surface area as a square function). A large multicellular organism thus faces a serious problem; it must have an absorptive surface extensive enough to enable it to obtain the nutrients it needs to support its large volume, particularly if most of the absorption is restricted to a limited region of the body, in this case the roots. As an adaptation solving this problem, many organisms have evolved absorptive surfaces that are extensively subdivided, giving a total absorptive area far greater than that of an undivided system of the same volume. The manifold branching of a typical root system is an example of this sort of adaptation. Approximately 14 million branch roots were found in the root system of a rye plant only 2 feet tall; their combined length was about 380 miles, and their total surface area exceeded 2,500 square feet, by contrast with a surface area for the shoots of only 51 square feet.

But calculations of surface area based simply on the number of branch roots and their length and diameter fall far short of giving a true picture of the total absorptive area. Just behind the growing tip of each rootlet, there is usually an area bearing a dense cluster of tiny hairlike extensions of the epidermal cells (Figs. 5.2, 5.3). The zone of these *root hairs* on each rootlet may be only a fraction of an inch long, but it contains such vast numbers of root hairs that the total absorptive surface of the root is enormously increased.

When viewed in cross section, a root of a young dicot plant can be seen to consist of a series of different tissue layers. On its outer surface is a layer of *epidermis* one cell thick (Fig. 5.4a). Unlike the epidermis of the aerial parts of the plant, that of the root usually has no waxy cuticle on its surface; the explanation is obvious: the epidermis of a root functions in water absorption, while that of the aerial parts functions as a barrier against diffusion of water. As we saw, the root hairs extend each from a single epidermal cell (Fig. 5.3).

Beneath the epidermis is the *cortex,* a wide area composed primarily of parenchyma tissue, with numerous intercellular spaces (Fig. 5.4b). Large quantities of starch are often stored in the cells of the cortex. This tissue, so prominent and important in young roots, tends to be much reduced or even lost in older roots, where both cortex and epidermis may be replaced by a corky periderm.

Next interior to the cortex is the *endodermis,* a layer one cell thick (Fig. 5.4c). Endodermal cells are characterized by a waterproof band, the *Casparian strip,* which runs through their radial (side) and end walls, i.e. all walls where one endodermal cell abuts another; there is no Casparian strip in the walls that abut cortex or pericycle. The walls of mature endodermal cells are frequently very thick and lignified. A well-

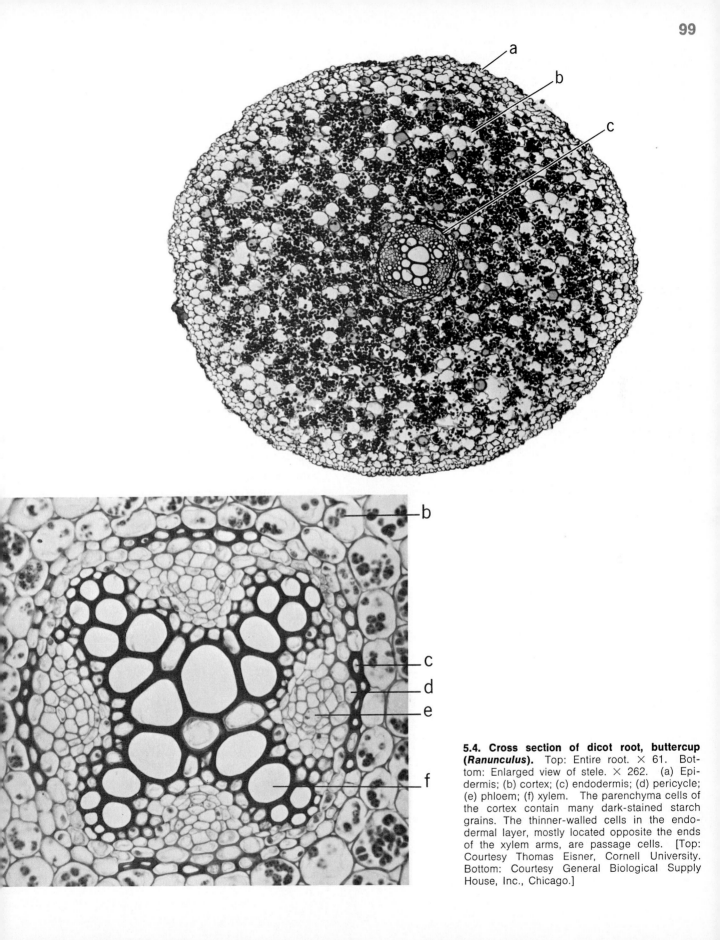

5.4. Cross section of dicot root, buttercup (*Ranunculus*). Top: Entire root. × 61. Bottom: Enlarged view of stele. × 262. (a) Epidermis; (b) cortex; (c) endodermis; (d) pericycle; (e) phloem; (f) xylem. The parenchyma cells of the cortex contain many dark-stained starch grains. The thinner-walled cells in the endodermal layer, mostly located opposite the ends of the xylem arms, are passage cells. [Top: Courtesy Thomas Eisner, Cornell University. Bottom: Courtesy General Biological Supply House, Inc., Chicago.]

differentiated endodermis is always present in roots, but occurs less regularly in stems. The endodermis forms the outer boundary of a central core of the root that contains the vascular cylinder; this core is called the *stele*. Just inside the endodermis is a layer, often only one cell thick, of thin-walled parenchymatous cells. The cells of this layer, called the *pericycle* (Fig. 5.4d), readily take on meristematic activity and may give rise to lateral roots (Pl. II-6A).

The central portion of the dicot stele, surrounded by endodermis and pericycle, is filled with the two vascular tissues, *xylem* and *phloem*. The thick-walled xylem cells frequently form a cross- or star-shaped figure (Fig. 5.4f; Pl. II-6B). Bundles of phloem cells are located between the arms of the xylem. Consequently the phloem does not form a continuous cylinder like the epidermis, cortex, endodermis, and pericycle; instead, xylem and phloem alternate in this portion of the stele.

Large roots of monocots commonly have a region of parenchyma tissue, called *pith*, located at the very center of the stele. The xylem therefore does not form the star-shaped figure described as characteristic of dicots, but even in such roots the bundles of xylem and phloem alternate.

Absorption of nutrients. Some water from the soil moves into epidermal cells by osmosis, and then moves from cell to cell across the cortex, either by osmosis or by diffusion along the plasmodesmata that interconnect the cytoplasm of the cells (Fig. 5.5). Since the concentration of dissolved substances such as sugar and other organic compounds in the epidermal cells is normally higher than the concentration of dissolved substances in the soil water, and since the cell membranes are permeable to water but not to the sugar or to the other organic compounds, a simple osmotic system is established; water is in higher con-

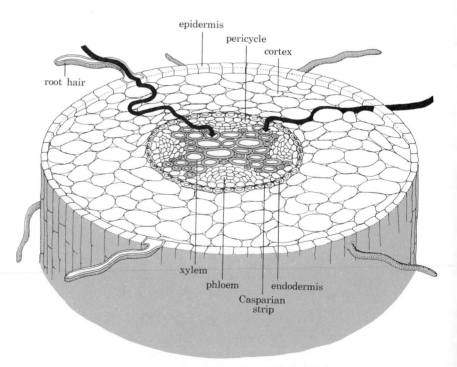

5.5. Movement of water from soil to xylem in root. Some water (arrow at left) is absorbed by the epidermal cells, particularly the root hairs, and moves from cell to cell, either by osmosis or by diffusion through plasmodesmata. Some water (arrow at right) flows along cell walls and does not cross membranes of living cells until it reaches the endodermis. The Casparian strip (color) of the endodermal cells prevents flow of water along their radial and end walls; hence all water entering the stele must move through the living cells of the endodermis.

epidermis
pericycle
cortex
root hair
xylem
phloem
endodermis
Casparian strip

centration outside the cell than inside, and it therefore moves across the membrane into the cell, obeying the normal laws of diffusion. But once water has entered an epidermal cell, it dilutes the contents of that cell, and the total concentration of dissolved substances will now be lower in the epidermal cell than in the adjacent cell of the cortex; or, to word it another way, the osmotic pressure of the epidermal cell will now be lower than that of the cortex cell. Consequently water will tend to move from the epidermal cell to the adjacent cortex cell. But now a new concentration gradient is established as water moving into the outermost cell of the cortex dilutes the contents of that cell and lowers its osmotic pressure to a point below that of the next cell of the cortex. As a result, water moves from the first cortex cell to the second cortex cell, following the concentration gradient. Again, dilution of the recipient cell occurs, a new gradient is established, and water moves on to the next cell. In this way, water can move fairly easily from the soil across the epidermis and cortex to the stele. Once inside the xylem of the vascular cylinder, the water can rise to other parts of the plant body.

There is another important way in which water can move across the epidermis and cortex of a root. The cellulose of plant cell walls has a strong tendency to imbibe water, and this water can flow along the cell walls from cell to cell, crossing the epidermis and the entire cortex without ever actually penetrating a membrane or entering a cell (Fig. 5.5). But the water cannot flow across the endodermis in this manner. Because the Casparian strips act as a barrier, all water entering the vascular cylinder must cross through the living cells of the endodermis. This gives the plant an opportunity for controlling the movement into the stele of substances dissolved in the water.

Minerals are usually absorbed in ionic form. The ions are in solution in the soil water, their concentration varying according to the fertility and the acidity of the soil and other factors. Often the minerals are not dissolved in quantity in the soil water, but are bound by ionic bonds to soil particles and are thus not available to plants. Agricultural soil management frequently involves changing the soil acidity to free more such bound minerals, so that they can be absorbed by roots. For example, the addition of lime to very acid soil may increase the availability of phosphorus, potassium, and molybdenum, but an excess of lime may decrease the available iron, copper, manganese, and zinc. Obviously, a carefully planned balance appropriate to the particular crop to be grown must be achieved for maximum yield.

Numerous experiments have shown that simple diffusion alone cannot account for all the absorption of mineral nutrients by roots. Plants can sometimes take in a mineral that is in higher concentration inside the root cells than in the soil solution and that therefore should, according to the rules of simple diffusion, move the other way. Furthermore, even when the concentration gradient favors inward movement, the rate of absorption is often greater than would be possible by passive diffusion alone. Active transport is clearly involved. The plant expends energy in the process of procuring the mineral nutrients essential to its continued existence. As will be evident throughout this book, active transport is the rule rather than the exception in most kinds of organisms, both plant and animal. Only rarely is simple diffusion the whole story when substances are moving across the membranes of living cells.

NUTRIENT REQUIREMENTS OF HETEROTROPHIC ORGANISMS

Heterotrophic organisms cannot manufacture their own high-energy compounds from low-energy inorganic raw materials. Yet they, like all living things, must extract from chemical bonds the energy necessary for both maintenance and growth. They must therefore obtain prefabricated high-energy organic nutrients. Heterotrophic plants may be *saprophytic* (living on dead organic matter) or *parasitic* (living on or in other organisms). Animals may be *herbivores* and eat green plants, thereby obtaining high-energy compounds directly from the organisms that first made them. Or they may be *carnivores* and eat the animals that ate the plants. Or they may be *omnivores,* eating both plant and animal material. Whether a heterotrophic organism is saprophytic or parasitic, or herbivorous, carnivorous, or omnivorous, it is clear that its energy-yielding nutrients came originally from green plants, which used radiant energy from the sun to make them.

Nutrients required in bulk

Carbohydrates, fats, and proteins are the main classes of compounds serving as energy sources for heterotrophic organisms. Of these, carbohydrates alone might seem at first glance to be sufficient. And, indeed, they alone would suffice if organic nutrients functioned only as an energy source. But organic nutrients also function in another very important way; they provide carbon skeletons and functional groups essential for the synthesis of new organic compounds. Carbohydrates alone cannot fulfill this second function; the various organic food materials are not fully interconvertible as raw materials for synthesis.

Proteins, or the amino acids of which they are composed, must be included in the diet of most heterotrophic organisms. A diet including only carbohydrates and fats is usually soon fatal. One reason immediately comes to mind. Carbohydrates and fats lack the nitrogen necessary to form the amino acids that the organisms require. It is true that if heterotrophic organisms had the ability of green plants to combine inorganic nitrogen (usually nitrate) with carbon skeletons from carbohydrates to make amino acids, there might be no need for them to include protein in their diet. And, indeed, some kinds of heterotrophs have this ability; many bacteria and a few heterotrophic flagellated Protozoa can survive with ammonia (NH_3) as their only nitrogen source. Higher animals, however, must obtain dietary amino acids.

Suppose an animal were fed a diet containing only one kind of amino acid. Would this single source of organic nitrogen suffice? For most animals, the answer is no, because they have apparently lost the ability to synthesize certain amino acids and must get them in their diet. These are called *essential amino acids*—a somewhat misleading term, since it seems to imply that the other amino acids commonly occurring in proteins are not essential; what is meant, of course, is that the designated amino acids are essential *in the diet,* whereas the others, which are also necessary for life, can be synthesized by the organism itself from other amino acids or organic nitrogen compounds.

Since all the essential amino acids must be included in the diet of an animal, several different proteins should be eaten, because a single protein may not contain them all. For example, zein, the main protein in corn, is deficient in tryptophan, lysine, and cysteine, while egg albumin is low in leucine. It has been recommended that an average adult human being include at least 70 grams (about 1.85 ounces) of protein in his diet each day, of which at least half should be of animal origin. The proportions of the various amino acids in plant proteins are often quite different from those in animal proteins; hence plant proteins are not as reliable a source of essential amino acids for human beings. Kwashiorkor, a disease resulting from protein deficiency, is particularly common in children in countries where the diet consists primarily of plant material, as in Indonesia, where rice forms much of the diet, or parts of Africa, where corn is the principal staple. The disease is characterized by degeneration of the liver, severe anemia, and inflammation of the skin.

Vitamins and minerals

Vitamins are organic compounds necessary in small quantities to given organisms that cannot synthesize them and must therefore obtain them prefabricated in the diet. Note that a compound which is a vitamin for species A may not be a vitamin for species B, because B may not have lost the ability to synthesize it. Vitamins are necessary in only very small quantities because they ordinarily function as coenzymes or as parts of coenzymes; you will recall that enzymes and coenzymes are catalysts that can be re-used many times and hence are not needed in large amounts.

It is often exceedingly hard to demonstrate conclusively that a particular chemical compound is a vitamin. The quantities needed are so minute that it is difficult to be sure a diet supposedly free of a compound being tested does not contain trace amounts that would be sufficient to prevent deficiency symptoms from developing in an experimental animal. Elaborate purification techniques must be employed, but even these are not always successful. A list of compounds that may be vitamins for human beings is under investigation today, but it may be years before any certainty about them is reached. It is even more difficult to establish reliable minimum daily requirements for known vitamins. Such minimum requirements as have supposedly been established are still very much open to question. Little is known, for example, about how requirements alter with age or with changing health. Much research remains to be done. One thing, however, can be asserted with reasonable confidence: If a healthy person eats a varied diet including meats, fruits, and vegetables, he will probably get all the vitamins he needs, numerous advertisements to the contrary notwithstanding.

Like the green plants, heterotrophs require certain minerals, which are usually absorbed as ions. Some of these are needed in relatively large quantity, e.g. sodium, chlorine, potassium, phosphorus, magnesium, and calcium. Other minerals are needed in much smaller amounts, e.g. iron, manganese, and iodine. And still other minerals, though essential to life, are needed only in trace amounts, e.g. copper, zinc, molybdenum, and cobalt.

The function of some of the minerals is obvious. Calcium is a major constituent of bones and teeth in vertebrates and plays a variety of other roles in most organisms. Phosphorus is a component of many high-energy organic compounds of critical importance (e.g. ADP and ATP). Iron is a constituent of the cytochromes and of hemoglobin. Sodium, chlorine, and potassium are important components of the body fluids, playing a role in osmotic phenomena and in such processes as nerve and muscle action. But the function of some of the minerals, particularly those needed only in trace amounts, is less obvious. Apparently, most of them act as components of coenzymes, or perhaps as cofactors helping to catalyze reactions without being actually incorporated into enzymes or coenzymes. Such minerals are comparable to vitamins, functioning in the same way. The only distinction, and an arbitrary one at that, is that vitamins are organic compounds and minerals are inorganic.

NUTRIENT PROCUREMENT
BY HETEROTROPHIC PLANTS

It is a common error to think of all plants as autotrophic. There are many plants, in fact, that lack chlorophyll and are heterotrophic (see Pls. I-4, 5). Their nutritional requirements resemble those of animals in that prefabricated organic compounds, necessary both as energy sources and as raw materials for synthesis, must be obtained from the surrounding environment.

The fungi constitute a large group of heterotrophic plants. Consider bread mold as an example. The bread on which the mold grows is largely starch. But starch is a polysaccharide, whose very large and insoluble molecules cannot move across cell membranes. How can the mold obtain nourishment from such a source? Clearly, something must be done to the starch before absorption can take place. The starch must be broken down to its constituent building-block compounds, the simple sugars, and these can be absorbed. In short, the starch must be digested. *Digestion* is nothing more than enzymatic hydrolysis, which, you will recall, involves the addition of water (see p. 25). In bread mold, the hydrolysis takes place outside the cell. Digestive enzymes synthesized inside the cells of the mold are released from the cells and hydrolyze the starch. The simple sugars that are the products of this digestion are then absorbed by the cells. Such digestion outside of cells is called *extracellular digestion.*

Since most green plants do not procure organic compounds from the environment, it is not necessary for them to carry out extracellular digestion. But they, like all other living things, must on occasion hydrolyze such compounds as polysaccharides, fats, and proteins into their constituent building-block compounds. If, for example, starch stored in one cell of a plant is to be moved to some other cell, it must first be hydrolyzed, since the starch itself cannot move across the intervening membranes. Such hydrolysis reactions taking place within cells constitute *intracellular digestion.* This kind of digestion is characteristic of almost all cells, both plant and animal.

NUTRIENT PROCUREMENT BY ANIMALS

Like the fungi already examined, most animals must digest their food before it can cross the membranes of their cells. Only rarely can they obtain as food such comparatively simple and diffusible compounds as glucose, glycerol, fatty acids, and amino acids; usually food material is in the form of polysaccharides, fats, proteins, etc., and must be hydrolyzed. Sometimes digestion is extracellular. In other instances, bulk food is taken into a cell by phagocytosis, or some similar process, and then digested in a food vacuole. Though the latter process is classified as intracellular digestion, it should be noted that the food material is separated from the rest of the cellular material by a membrane that it cannot cross until after digestion has occurred. Thus extracellular and intracellular digestion are alike in that digestion always precedes the actual absorption of complex foods across a membrane.

Although both the nutritional requirements and the basic processes of digestion are essentially similar in all animals, from single-celled Protozoa to man, the body plans of animals vary so greatly that the structures involved in food processing and the details of that processing are often different. In the following sections we shall briefly examine the digestive mechanisms of a variety of different animals.

Nutrient procurement by Protozoa

Since Protozoa, as single-celled organisms, have a body plan obviously very different from that of other animals, we would expect their adaptations for food procurement to be likewise markedly different from those of multicellular animals. And the differences are, in fact, considerable. But a more interesting point, one with important biological implications, is that the similarities are often more striking than the differences.

Let us look first at *Amoeba,* an organism that is constantly changing shape as its protoplasm flows along, pushing out new armlike projections and withdrawing others. When an amoeba is stimulated by nearby food, some of these armlike extensions, called *pseudopodia,* may flow around the food until they have completely surrounded it. This is the process we have called phagocytosis. The food is completely engulfed by the cytoplasm and is enclosed in a *food vacuole,* where it will be digested (see Fig. 3.9, p. 45). *Amoeba* is an example of a protozoan without specialized permanent digestive structures, though its transitory food vacuoles are functional analogues of the digestive systems of higher animals.

The ciliates, another important group of Protozoa, of which *Paramecium* is an example, are characterized by innumerable cilia covering the surface of their bodies. Like all Protozoa, they are commonly regarded as unicellular, but it is a mistake to think of them as simple (see Pl. I-1). Some of them, in fact, are among the most incredibly complex cells known. Though they lack actual subdivision into recognizable cellular units, they show much of the internal specialization usually associated with multicellularity. Unlike *Amoeba, Paramecium* has a permanent structure, an organelle, that functions in feeding. Food particles are swept into an *oral groove,* a ciliated channel on one side of the cell (Fig. 5.6A), by water currents produced by the beating cilia, and are carried into a *cytopharynx,* at the lower end of which a vacuole forms around the food (Fig. 5.6B). Eventually the vacuole breaks off and begins to move toward the anterior (front) end of the cell. Digestive enzymes are secreted into the vacuole and digestion begins. As digestion

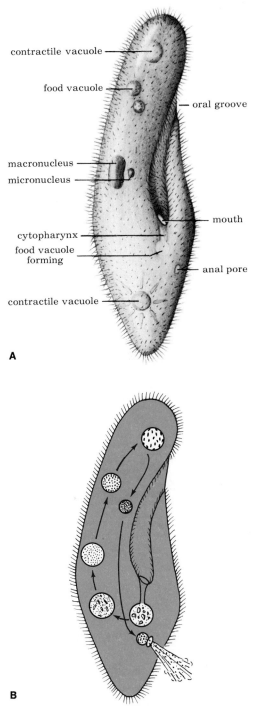

A

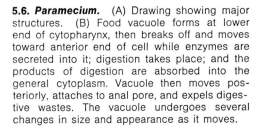

B

5.6. *Paramecium.* (A) Drawing showing major structures. (B) Food vacuole forms at lower end of cytopharynx, then breaks off and moves toward anterior end of cell while enzymes are secreted into it; digestion takes place; and the products of digestion are absorbed into the general cytoplasm. Vacuole then moves posteriorly, attaches to anal pore, and expels digestive wastes. The vacuole undergoes several changes in size and appearance as it moves.

contractile vacuole

food vacuole

oral groove

macronucleus

micronucleus

mouth

cytopharynx

food vacuole forming

anal pore

contractile vacuole

proceeds, the products (simple sugars, amino acids, etc.) diffuse across the membrane of the vacuole into the cytoplasm, and the vacuole begins to move back toward the posterior end of the cell. When the vacuole reaches a tiny specialized region of the cell surface called the anal pore, it becomes attached to the surface and then ruptures, expelling to the outside any remaining bits of indigestible material. Not only does the food vacuole function as a digestive chamber, but by its movement it serves to distribute the products of digestion to all parts of the cell.

We have said that digestive enzymes are secreted into the food vacuoles of both *Amoeba* and *Paramecium*. But if these powerful enzymes are capable of hydrolyzing such compounds as polysaccharides, fats, and proteins, and if the cell itself is composed of these kinds of compounds, how can the cell contain the digestive enzymes without being destroyed by them? A partial answer was given in the chapter on cells: Digestive enzymes are packaged in lysosomes, vesicles whose membrane is apparently both impermeable to the enzymes and capable of resisting their hydrolytic action. The digestive enzymes are presumably synthesized on the ribosomes, move through the endoplasmic reticulum to the Golgi apparatus, and there become surrounded by a membrane to form the lysosome. When a food vacuole is formed, a lysosome soon fuses with it (Fig. 5.7). Food materials and the digestive enzymes are mixed in the resulting digestive vacuole. As already described, this vacuole circulates in the cytoplasm, the products of digestion are absorbed, and indigestible materials are eventually expelled from the cell.

Though the above description of lysosome activity pertains to digestion in Protozoa, it applies equally well to intracellular digestion in any animal cell. Lysosomes, you will recall, were in fact first discovered in rat liver cells. Here is only one example of the similarity between the basic processes of digestion in Protozoa and in higher animals.

Nutrient procurement by coelenterates

With the evolution of multicellularity came a corresponding evolution of cellular specialization resulting in a division of labor among cells. The coelenterates provide a comparatively simple example of this phenomenon. These radially symmetrical animals have a saclike body composed of two principal layers of cells, with a jellylike layer, called mesoglea, between them (Fig. 5.8). The central cavity of this saclike body functions as a digestive cavity. It has only one opening to the outside, which is surrounded by movable tentacles. A digestive cavity of this sort, having only one opening that functions as both mouth and anus, is called a *gastrovascular cavity*.

Coelenterates are strictly carnivorous. Embedded in their tentacles are numerous stinging structures called *nematocysts.* Each nematocyst consists of a slender thread coiled within a capsule, with a tiny hairlike trigger penetrating to the outside. When appropriate prey comes into contact with the trigger, the nematocyst fires, the thread turns inside out, spines on its surface unfold, and it either penetrates the body of the prey or entangles the prey in sticky loops. The nematocysts also eject poisons, which have a paralyzing action on the prey. The tentacle then grasps the prey and draws it toward the mouth, which opens wide to receive it. Once the food is inside the gastrovascular cavity, digestive enzymes are secreted into the cavity and extracellular digestion begins. This

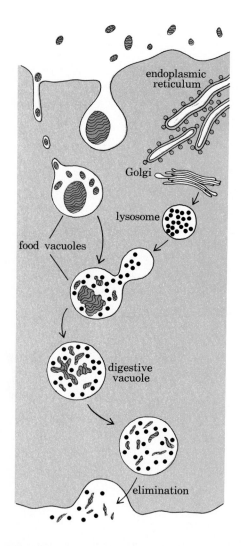

5.7. The role of lysosomes in intracellular digestion. A lysosome containing digestive enzymes fuses with a food vacuole. Digestion takes place within the composite structure thus formed (digestive vacuole), and the products of digestion are absorbed across the vacuolar membrane. The vacuole eventually fuses with the cell membrane and then ruptures, expelling indigestible wastes to the outside.

extracellular digestion, largely limited to proteins in coelenterates, does not break down these substances completely to their constituent amino acids. As soon as the food has been reduced to small fragments, the cells lining the gastrovascular cavity engulf them by phagocytosis, and digestion is completed intracellularly in food vacuoles. Indigestible remains of the food are expelled from the gastrovascular cavity via the mouth.

If phagocytosis and intracellular digestion are going to take place anyway, we can ask what adaptive advantage the evolution of the additional process of extracellular digestion might have. Why should not coelenterates rely exclusively on intracellular digestion as the Protozoa do? The answer is obvious. Intracellular digestion severely limits the size of the food the organism can handle. Extracellular digestion enables it to utilize much larger pieces of food; even whole multicellular animals become possible prey. Extracellular digestion is the rule rather than the exception in multicellular animals.

Nutrient procurement by flatworms

Unlike the radially symmetrical coelenterates, the flatworms are bilaterally symmetrical; they have distinct anterior (front) and posterior (rear) ends, and also distinct dorsal (upper) and ventral (lower) surfaces. Their bodies are composed of three well-formed tissue layers. Many flatworms are parasitic on other animals, but some are free-living, and it is to these that we shall first direct attention, using planaria as an example (Fig. 5.9).

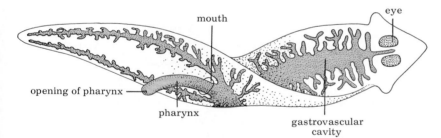

5.9. **Planaria, showing much-branched gastrovascular cavity and extruded pharynx.**

The mouth of planaria is located on the ventral surface near the middle of the animal. It opens into a muscular tubular *pharynx,* which can be protruded through the mouth directly onto prey. The pharynx leads into a gastrovascular cavity (i.e. a cavity with only one opening to the outside). This cavity, though functionally similar to that of the coelenterates, is far more branched, ramifying throughout the animal's body. The extensive branching greatly increases the total absorptive surface of the cavity. We saw with plants that as organisms increase in size, and particularly as their volume increases, the problem of sufficient absorptive surface becomes more acute. Highly subdivided absorptive surfaces, compacting much total surface area into relatively little space, have evolved in many kinds of organisms. The root hairs of plants are one example, and the branched gastrovascular cavity of planaria is another; we shall encounter many more in this and later chapters.

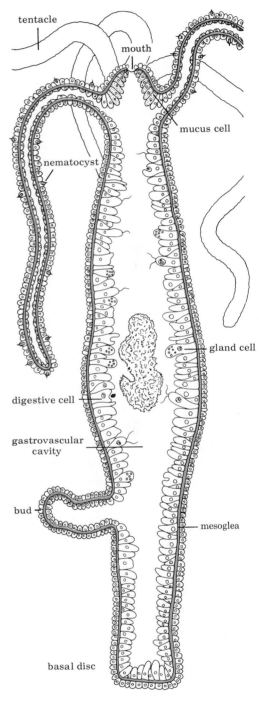

5.8. **Hydra, showing gastrovascular cavity with food material in it.**

The members of one class of flatworms have become so highly specialized as parasites living in the digestive tracts of other animals that in the course of their evolution they have lost their own digestive systems. These are the tapeworms. They are constantly bathed by the products of the host's digestion and can absorb them without having to carry out any digestion themselves. Evolutionary adaptation can involve the loss of structures as well as their acquisition.

Animals with complete digestive tracts

Animals above the level of coelenterates and flatworms have a complete digestive tract, i.e. one with two openings, a **mouth** and an **anus.** The advantages of such a system over a gastrovascular cavity are obvious. No longer must incoming food material and outgoing wastes pass through the same opening. Instead, food can be passed in one direction through a tubular system, which can be divided into a series of distinct sections or chambers, each specialized for a different function. As the food passes along this assembly line, it is acted upon in a different way in each section. The sections may be variously specialized for mechanical breakup of bulk food, temporary storage, enzymatic digestion, absorption of the products of digestion, reabsorption of water, storage of wastes, etc. The overall result is a much more efficient digestive system, as well as the potential for special evolutionary modifications fitting different animals for different modes of existence.

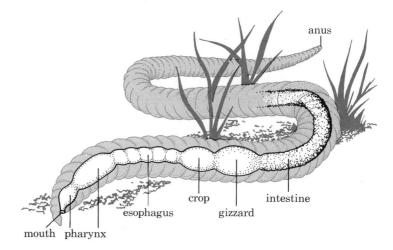

5.10. Digestive system of an earthworm.

The digestive system of an earthworm is a good example of subdivision into specialized compartments (Fig. 5.10). Food, in the form of decaying organic matter mixed with soil, is drawn into the mouth by the sucking action of a muscular chamber called the **pharynx.** It passes from the mouth through a short passageway into the pharynx and then through a connecting passage called the **esophagus,** after which it enters a relatively thin-walled **crop** that functions as a storage chamber. Next, it enters a compartment with thick muscular walls, the **gizzard,** where it is ground up by a churning action; the grinding is often facilitated by the presence of small stones. The pulverized food, suspended in water, now passes into the long **intestine,** where enzymatic digestion and absorption take place. Finally, in the rear of the intestine, some of the water involved in the digestive process is reabsorbed, and the indigestible residue is eliminated from the body through the anus.

Notice that earthworms utilize extracellular digestion. Glandular cells in the epithelial lining of the intestine secrete hydrolytic enzymes into the intestinal cavity, and the end products of digestion, the simple building-block compounds, are absorbed.

We have already seen that extracellular digestion is an adaptation for eating sizable pieces of food; the gizzard is obviously another such adaptation in earthworms. Mechanical breakup of bulk food is common among animals, and a variety of structures that serve this function have evolved. In our own case, food is torn and ground by the teeth. The grinding or chewing device need not be in the first section of the digestive tract, however; in earthworms, the grinding chamber comes after the crop, which corresponds functionally to our stomach, and mechanical breakup follows temporary storage instead of preceding it.

Not all large animals eat large pieces of food and have special masticating devices. For example, many animals are *filter feeders,* straining small particles of organic matter from water. Clams and many other molluscs filter water through tiny pores in their gills; microscopic food particles are trapped in streams of mucus that flow along the gills and enter the mouth, kept moving by beating cilia.

But let us return to the earthworm and examine the implications of another of the specialized compartments of its digestive tract, the crop. We have already said that this chamber functions in food storage. And the functional significance of a storage chamber should be clear after a moment's thought. It enables the animal to take in large amounts of food in a short time, when it is available, and then to utilize this food over a stretch of time. Such discontinuous feeding frees the animal for activities other than feeding, such as searching for a mate, mating, egg laying, and, in some animals, care of young. Our own stomachs function as storage organs analogous to the earthworm's crop; they enable us to live well on only three or four meals a day and to devote the rest of our time to other pursuits.

The digestive system of man and other vertebrates

Although an examination of the structure of man's digestive tract reveals little in the way of general principles that could not as easily be seen in an earthworm, natural interest in ourselves and our own species prompts a more detailed examination of human systems.

The oral cavity.　The first chamber of the digestive tract is, of course, the oral cavity. Located here are the teeth, which function in the mechanical breakup of food by both biting and chewing. Human teeth are of several different types, each adapted to a different function (Fig. 5.11). In front are the chisel-shaped *incisors,* four in the upper jaw and four in the lower, which are used for biting. Then come the more pointed *canines,* one on each side in each jaw, which are specialized for tearing food. Behind each canine are two *premolars* and three *molars* in adults; these have flattened, ridged surfaces, and function in grinding, pounding, and crushing food.

The teeth of different species of vertebrates are specialized in a variety of ways and may be quite unlike those of man in number, struc-

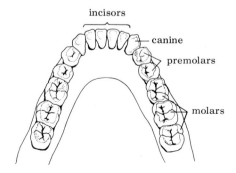

5.11. Human teeth.　Lower jaw of adult.

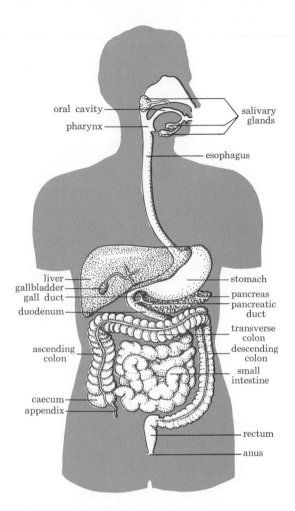

oral cavity
pharynx
salivary glands
esophagus
liver
gallbladder
gall duct
duodenum
stomach
pancreas
pancreatic duct
transverse colon
descending colon
small intestine
ascending colon
caecum
appendix
rectum
anus

5.12. Human digestive system. (The organs are slightly displaced, and the small intestine is greatly shortened.)

ture, arrangement, and function. For example, the teeth of snakes are very thin and sharp and usually curve backward. They function in capturing prey, but not in mechanical breakup; snakes do not chew their food, but swallow it whole. The teeth of carnivorous mammals, such as cats and dogs, are more pointed than those of man; the canines are long, and the premolars lack flat grinding surfaces, being more adapted to cutting and shearing. On the other hand, such herbivores as cows and horses have very large flat premolars, and molars with complex ridges and cusps; the canines are often totally absent in such animals. Notice that sharp pointed teeth poorly adapted for chewing seem to characterize meat eaters like snakes, dogs, and cats whereas broad flat teeth, well adapted for chewing, seem to characterize vegetarians. How can this difference be explained? Remember that plant cells are enclosed in a cellulose cell wall. Very few animals can digest cellulose; most of them must break up the cell walls of the plants they eat if the cell contents are to be exposed to the action of digestive enzymes. Animal cells, like those in meat, do not have any such nondigestible armor and can be acted upon directly by digestive enzymes. Therefore chewing is not as essential for carnivores as for herbivores. You have doubtless seen how dogs gulp down their food, while cows and horses spend much time chewing. But carnivores have other problems. They must capture and kill their prey, and for this, sharp teeth capable of piercing, cutting, and tearing are well adapted. Man, being an omnivore, has teeth that belong, functionally and structurally, somewhere between the extremes of specialization attained by the teeth of carnivores and herbivores.

The oral cavity has other functions besides those associated with the teeth. It is here that food is tasted and smelled, activities of great importance in food selection. And it is here that food is mixed with saliva secreted by several sets of salivary glands. The saliva dissolves some of the food and acts as a lubricant, facilitating passage through the next portions of the digestive tract. The saliva of man also contains a starch-digesting enzyme, which initiates the process of enzymatic hydrolysis.

The muscular tongue manipulates the food during chewing and forms it into a mass, called a bolus, in preparation for swallowing; it then pushes the bolus backward through a cavity called the *pharynx* and into the *esophagus* (Fig. 5.12; see also Fig. 6.8, p. 122). The pharynx functions also as part of the respiratory passageway; the air and food passages cross here, in fact. Swallowing accordingly involves a complex set of reflexes that close off the opening into the nasal passages and trachea (windpipe), thereby forcing the food to move into the esophagus. As you know, these reflexes occasionally fail to occur in proper sequence and the food enters the wrong passageway, causing you to choke.

The esophagus and the stomach. The esophagus is a long tube running downward through the throat and thorax and connecting to the stomach in the upper portion of the abdominal cavity (Fig. 5.12). Food moves quickly through the esophagus, pushed along by waves of muscular contraction in a process called *peristalsis.* Circular muscles in the wall of the esophagus just behind the food bolus contract, squeezing the food forward. As the food moves, the muscles it passes also contract, so that a region of contraction follows the bolus and constantly pushes it forward, much as though you were to keep a ball moving through a soft rubber tube by giving the tube a series of squeezes, with your hand always just behind the ball.

At the junction between the esophagus and the stomach is a special ring of muscle called a *sphincter,* which, when it is contracted, closes the entrance to the stomach. It is normally closed, thus preventing the contents of the stomach from moving back into the esophagus when the stomach moves during digestion. It opens when a wave of peristaltic contraction coming down the esophagus reaches it.

The stomach lies slightly to the left side in the upper portion of the abdomen, just below the lower ribs. It is a large muscular sac, which, as we have already seen, functions as a storage organ, making discontinuous feeding possible. It has other functions too. When it contains food, it is swept by powerful waves of contraction, which churn the food, mixing it and breaking the larger pieces. In this manner, the stomach supplements the action of the teeth in mechanical breakup of food. The glands of the stomach lining are of several types. Some secrete mucus, which covers the stomach lining—hence the name *"mucosa"* or "mucus membrane" for the inner layer of the stomach wall; others secrete *gastric*[1] *juice,* a mixture of hydrochloric acid and digestive enzymes. Enzymatic digestion, then, is a third important function of the stomach in man.

[1] The adjective "gastric" and the prefix "gastro-" always refer to the stomach.

The small intestine. The food leaves the stomach as a soupy mixture. It passes through the pyloric sphincter into the small intestine, which is the portion of the digestive tract where most of the digestion and absorption takes place. The first section of the small intestine, attached to the stomach, is called the *duodenum* (Fig. 5.12). It leads into a very long coiled section lying lower in the abdominal cavity. The entire small intestine of an adult man is about 23 feet long.

The length of the small intestine shows interesting variations in different animals. The intestine is usually very long and much coiled in herbivores, much shorter in carnivores, and of medium length in omnivores, like man. These differences, like those of the teeth, are correlated with the difficulty of digesting plant material because of the cellulose cell walls. Even if the cellulose has been well broken up, it remains mixed with the digestible portions of the cells and tends to mask them from the digestive enzymes. This interference makes digestion and absorption of plant material much less efficient than that of animal material, with the result that a longer intestine is an adaptive advantage in extracting a maximum amount of nutrients from a herbivorous diet.

Since the small intestine is the place where absorption of the products of digestion occurs, we would expect it to have special structural adaptations that increase its absorptive surface area. Clearly its great length plays a role here. But examination of its internal surface reveals other modifications that vastly increase the surface area over what it would be if the intestine were a smooth-walled tube. First, the mucosa lining the intestine is thrown into numerous folds and ridges (Figs. 5.13, 5.14). Second, small fingerlike outgrowths, called *villi*, cover the surface of the mucosa. And third, the individual epithelial cells covering the folds and villi have a border of countless, closely packed, cylindrical processes called *microvilli* (Fig. 5.15), revealed by the electron microscope. Thus the total internal surface of the small intestine, including folds, villi, and microvilli, is incredibly large.

The large intestine. In man, the junction between the small intestine and the large intestine (or colon) that follows it is usually in the lower right portion of the abdominal cavity. A blind sac, the *caecum,* projects from the large intestine near the point of juncture (Fig. 5.12). In man, there is a small fingerlike process, the *appendix,* at the tip of the caecum. As you know, the appendix frequently becomes infected and must be surgically removed.

The caecum of man is small and functionally unimportant, but in some mammals, particularly herbivorous ones, it is large and contains many microorganisms (bacteria and Protozoa) capable of digesting cellulose. Since the mammal cannot itself digest cellulose, it benefits from the action of the microbes.

From the caecum, the large intestine of man ascends on the right side to the mid-region of the abdominal cavity, then crosses to the left side, and descends again (Fig. 5.12). One of its chief functions is reabsorption of much of the water used in the digestive process. If all the water in which enzymes are secreted into the digestive tract were lost with the feces, man would have a severe problem of desiccation. A second function of the colon is excretion of certain salts, such as those of calcium and iron, when their concentration in the blood is too high.

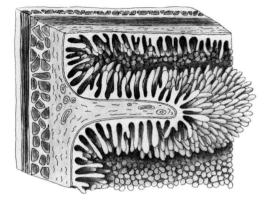

5.13. Section of a fold of human intestine, showing many villi.

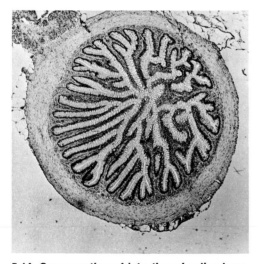

5.14. Cross section of intestine of calico bass, showing extensive folding. [Courtesy Warren Andrew, *Textbook of Comparative Histology,* Oxford University Press, 1959.]

The salts are excreted into the colon and eliminated from the body in the feces. The last portion of the large intestine, the *rectum,* functions as a storage chamber for the feces until defecation. The feces are eliminated from the rectum through an opening called the anus.

Enzymatic digestion in man

Digestion by saliva. Having traced the human digestive tract from mouth to anus, let us next consider the chemical changes that occur in a meal as it passes through this complex tubular system. We have said that enzymatic digestion starts in the mouth. The saliva contains an enzyme called *amylase*[2] (also known as ptyalin) that begins the hydrolysis of starch. Amylase, however, does not completely hydrolyze starch to glucose. What it does is split the starch into units of maltose, a double sugar (Fig. 5.16), which must be further digested in the intestine. Why, we may well ask, can't the amylase split all the bonds between sugar units instead of splitting only every other one? There is no known chemical difference between these bonds, and it would seem logical that an enzyme capable of splitting one could also split the others. Biologists do not as yet have any satisfactory answer to this question. Perhaps the bonds between the simple-sugar building blocks are not really exactly alike. Or perhaps the enzyme and its substrate fit together in such a way that only every other linkage is in the proper spatial configuration to be broken. Much remains to be explained about the amazing specificity exhibited by many enzymes.

Digestion in the stomach. Once in the stomach, food is exposed to the action of gastric juice secreted by numerous gastric glands of the stomach wall. This juice contains much hydrochloric acid and several enzymes. The acid makes the contents of the stomach very acidic (pH of about 1.5–2.5). Note that, advertisements for many patent medicines

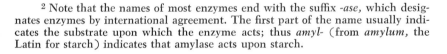

5.15. Electron micrograph of portions of two epithelial cells from intestine of a hamster, showing numerous closely packed microvilli. × 23,300. [Courtesy E. W. Strauss, University of Colorado.]

[2] Note that the names of most enzymes end with the suffix *-ase,* which designates enzymes by international agreement. The first part of the name usually indicates the substrate upon which the enzyme acts; thus *amyl-* (from *amylum,* the Latin for starch) indicates that amylase acts upon starch.

5.16. Digestion of starch. Amylase in the saliva and in the pancreatic juice hydrolyzes the bonds between every other pair of glucose units, producing the double sugar maltose. Maltose is digested to glucose by maltase, secreted by intestinal glands.

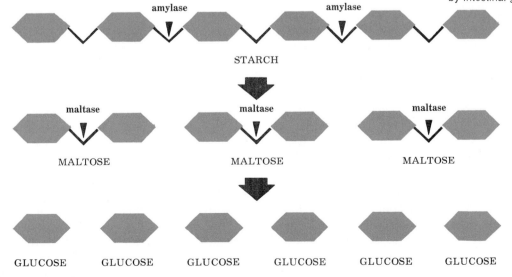

to the contrary, an acid stomach is both normal and necessary for proper function. The principal enzyme of gastric juice is *pepsin,* which digests protein.

Pepsin does not hydrolyze protein all the way to its amino acid components. It splits the peptide bonds adjacent to only a few amino acids. The specificity of proteolytic enzymes is more understandable than that of amylase, because proteins are composed of a variety of building-block compounds, not just of one, like starch. This means that the structural configuration around the various peptide bonds in a protein varies, depending on which two amino acids the bond joins. Consequently some bonds may fit on the active site of the enzyme and others may not.

Digestion in the small intestine. It is in the next section of the digestive tract, the small intestine, that by far the most digestion takes place. When partially digested food passes from the stomach into the duodenum, its acidity stimulates the release of a large number of different digestive enzymes into the lumen of the intestine. These enzymes are secreted from two principal sources, the *pancreas* and the *intestinal glands.* The pancreas, a large glandular organ lying just below the stomach (Fig. 5.12), has a connection to the duodenum called the *pancreatic duct.* When food enters the duodenum, the pancreas secretes a mixture of enzymes that flows through the pancreatic duct into the duodenum. Included in this mixture are enzymes that digest all three principal classes of foods—carbohydrates, fats, and proteins—as well as some that digest nucleic acids.

One of the pancreatic enzymes is *pancreatic amylase,* which, as its name implies, acts like salivary amylase, splitting starch into the double sugar maltose. It is far more important than salivary amylase, for it carries out most of the starch digestion.

Lipase is the fat-digesting enzyme of the pancreas. It splits molecules of fat into glycerol and fatty acids. Almost all fat digestion is catalyzed by this pancreatic enzyme.

Trypsin and *chymotrypsin* are two of the proteolytic enzymes of the pancreas. Like pepsin, they are incapable of splitting all peptide linkages in a protein molecule; each cleaves only the linkages adjacent to certain specific amino acids. The three enzymes thus effect a splitting of proteins into fragments of varying lengths, but do not produce many free amino acids. Another class of enzymes complete the digestive process. There is a great variety of such enzymes, each highly specific in its action. Most of them are secreted by intestinal glands, but some are produced in the pancreas.

Just as certain intestinal enzymes complete the digestion of protein, other intestinal enzymes complete the digestion of carbohydrate begun by salivary and pancreatic amylase. These enzymes split double sugars into simple sugars. For example, maltase splits maltose, sucrase splits sucrose, and lactase splits lactose.

Bile. One more secretion should be mentioned in this discussion of human digestion. The *liver,* a critically important organ about which much will be said in later chapters, produces a fluid called bile, which aids in fat digestion. The liver is a very large organ occupying much of the space in the upper part of the abdomen. On its surface is a small storage organ, the *gallbladder* (Fig. 5.12). Bile, produced throughout the liver, is collected by a series of branching ducts and emptied into the gallbladder. When food enters the duodenum, the muscular wall of the gallbladder is stimulated to contract, and the bile is forced down the gall duct into the duodenum. Bile is not a digestive enzyme. It is not even a protein. It is a complex solution of bile salts, bile pigments, and cholesterol. The bile salts act as emulsifying agents, causing large fat droplets to be broken up into many tiny droplets suspended in water. This action is much like that of a good detergent. The many small fat droplets expose much more surface area to the digestive action of lipase than would a few large droplets.

REFERENCES

Anonymous, 1963. *Vitamin Manual,* rev. ed. Upjohn, Kalamazoo, Mich.

BEST, C. H., and N. B. TAYLOR, 1966. *The Physiological Basis of Medical Practice,* 8th ed. Williams & Wilkins, Baltimore. (See esp. Chapters 56–63, 71–73.)

PROSSER, C. L., and F. A. BROWN, 1961. *Comparative Animal Physiology,* 2nd ed. Saunders, Philadelphia. (See esp. Chapters 4–5.)

STEWARD, F. C., ed., 1959–1963. *Plant Physiology:* vol. 2, *Plants in Relation to Water and Solutes;* vol. 3, *Inorganic Nutrition of Plants.* Academic Press, New York.

STILES, W., 1961. *Trace Elements in Plants,* 3rd ed. Cambridge University Press, New York.

WINTON and BAYLISS, 1969. *Human Physiology,* 6th ed., rev. and ed. by O. C. J. Lippold and F. R. Winton. Williams & Wilkins, Baltimore. (See esp. Chapters 27–31.)

SUGGESTED READING

CARLSON, A. J., V. JOHNSON, and H. M. CAVERT, 1961. *The Machinery of the Body,* 5th ed. University of Chicago Press, Chicago. (See esp. Chapters 7–8.)

D'AMOUR, F. E., 1961. *Basic Physiology.* University of Chicago Press, Chicago. (See esp. Chapters 11–13.)

DEDUVE, C., 1963. "The Lysosome," *Scientific American,* May. (Offprint 156.)

NEURATH, H., 1964. "Protein-Digesting Enzymes," *Scientific American,* December. (Offprint 198.)

RAY, P. M., 1971. *The Living Plant,* 2nd ed. Holt, Rinehart & Winston, New York. (See esp. Chapter 6.)

SCHMIDT-NIELSEN, K., 1970. *Animal Physiology,* 3rd ed. Prentice-Hall, Englewood Cliffs, N.J. (See esp. Chapter 1.)

SINNOTT, E. W., and K. S. WILSON, 1963. *Botany: Principles and Problems,* 6th ed. McGraw-Hill, New York. (See esp. Chapter 5.)

STEWARD, F. C., 1964. *Plants at Work.* Addison-Wesley, Reading, Mass. (See esp. Chapters 4, 9.)

CHAPTER

Gas exchange

We have seen that when nutrient compounds are broken down in the process of respiration, over 90 percent of the energy yield depends on the presence of oxygen, which makes possible the complete oxidation of the compounds to carbon dioxide and water. Thus a basic problem for the great majority of living organisms is the procurement of oxygen and the elimination of carbon dioxide.

It is a common misconception that oxygen procurement is a problem faced only by animals, and that gas exchange in green plants consists exclusively of intake of carbon dioxide and release of oxygen. The latter is the exchange that takes place in association with photosynthesis, but the carbohydrate products of photosynthesis are of little value to the plant if they cannot be respired to provide usable metabolic energy. Thus plants, like animals, are constantly taking in oxygen and releasing carbon dioxide as they carry out the process of cellular respiration. Both photosynthetic and respiratory gas exchange are usually taking place when a green plant is exposed to bright light; since the rate of photosynthesis then greatly exceeds the rate of respiration, the *net* effect is one of uptake of carbon dioxide and release of oxygen. The reverse is true, of course, when the green plant is in the dark or when it has no leaves in winter.

THE PROBLEM

Gas exchange between a living cell and its environment always takes place by diffusion across the moist cell membrane. The gases that move across the membrane are in solution. In unicellular organisms and many small multicellular ones, particularly those that are aquatic, each cell is either in direct contact with the surrounding medium or only a few cells removed from that medium. Thus exchange of oxygen and carbon dioxide is no serious problem, and these organisms have usually not evolved special respiratory devices.

Large body size, however, greatly complicates the gas-exchange process; and the evolution of specialized respiratory arrangements has been a prerequisite to the evolution of large size. Admittedly, some brown algae, the kelps, which may grow to a length of 200–300 feet, have no special gas-exchange mechanism other than direct diffusion between each cell and the surrounding water, but the large size of these plants is mostly in two dimensions. The blades of even the longest kelps remain very thin; as a result, no cell is far from the surface, and the total gas-exchange area is fairly large in relation to the volume of the plant.

When increase in body size involves three dimensions, as it generally does, the maintenance of a respiratory surface of adequate dimensions relative to the volume becomes a more complex problem, because area (a square function) increases much more slowly than volume (a cube function). The problem is most acute for the more active animals, whose rapid utilization of energy demands a large amount of oxygen per unit of volume. An additional complicating factor is that many organisms have evolved relatively impermeable skin and outer body coverings such as scales, feathers, and hair, which function as protective barriers between the fragile internal tissues and organs and the often hostile outer environment. The presence of these protective devices means that the gas-exchange surface must be confined to a restricted region of the body surface, making the problem of adequate exchange area even more critical.

Another complication brought on by large three-dimensional size in animals is that many cells are deep within the body, far from the gas-exchange surface. Diffusion alone is incapable of moving gases in adequate concentrations across the immense number of cells that may intervene between these more internal cells and the body surface. Some mechanism for conveying gases to every individual cell of the organism becomes essential.

The need for direct contact between the moist membranes across which gas exchange occurs and the environmental medium also poses serious difficulties, especially for terrestrial organisms. The moist membranes must be exposed to the environment, but they must be exposed in such a way that the chances of desiccation are minimized. Further, a large, thin, moist surface is often very fragile and easily suffers mechanical damage. In general, therefore, there has been a tendency toward the evolution of protective devices, particularly when the respiratory surfaces have been evaginated ones.

Broadly speaking, specialized respiratory surfaces belong to two categories: inward-oriented and outward-oriented extensions of the body surface (see Fig. 6.3). Each category embraces a diversity of form and detail, but the diversities become less bewildering if one bears in mind that each type of respiratory system represents merely one way of meeting the basic needs discussed above: (1) a respiratory surface of adequate dimensions; (2) for many organisms, methods of transporting gases between the area of exchange with the environment and the more internal cells; (3) means of protecting the fragile respiratory surface from mechanical injury; and (4) means of keeping the surface moist.

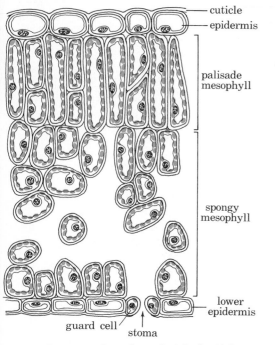

cuticle
epidermis

palisade mesophyll

spongy mesophyll

lower epidermis

guard cell — stoma

6.1. Cross section of a privet leaf. Not generally found in epidermal cells, chloroplasts (color) are present in guard cells.

6.2. Lenticel on stem of elderberry (*Sambucus*). × 55. [Courtesy Thomas Eisner, Cornell University.]

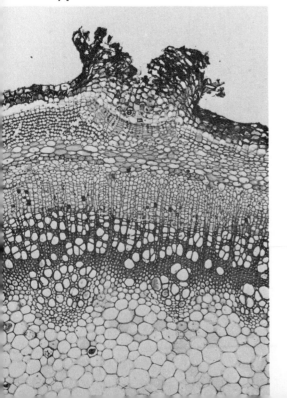

SOLUTIONS IN TERRESTRIAL PLANTS

Leaves

Gas exchange associated with both photosynthesis and cellular respiration takes place at a particularly high rate in green leaves, organs strikingly adapted for this process.

Being covered by a waxy cuticle, most of the visible outer surface of a leaf is more or less dry and impermeable, and hence ill suited for diffusion of gases. Exchange must therefore take place elsewhere. You will remember that the mesophyll parenchyma in a leaf is packed very loosely, leaving large intercellular spaces (Fig. 6.1; also Fig. 3.26, p. 59). A high percentage of the total surface of each mesophyll cell is exposed to the air in these spaces, which are interconnected and continuous with the external atmosphere by way of openings in the epidermis: the *stomata.* Each stoma is bounded by two highly specialized epidermal cells called *guard cells,* which, unlike most epidermal cells, contain chloroplasts. When the guard cells contain much water (are turgid), the stoma is open; when the guard cells are flaccid, they buckle inward and close the stoma. While the stomata are open, gases can move easily between the surrounding atmosphere and the internal spaces of the leaf. The actual gas exchange, by which we mean the diffusion of gases into and out of living cells, takes place across the thin moist membranes of the cells inside the leaf.

Let us briefly consider how the structures of the leaf help it meet the four above-mentioned requirements for respiratory systems.

1. The surface area available for gas exchange in the leaf is very large; if the total area of cell membrane exposed to the intercellular spaces could be measured accurately, it would be found enormous by comparison with the total outer area of the leaf. The principle involved is a very elementary one: A chamber irregularly shaped and greatly subdivided by partial partitions will have far more wall space than a round or square one of equal volume.

2. Internal transport of gases occurs in leaves without any special adaptations. Gases can reach each individual cell directly via the intercellular spaces.

3. The danger of mechanical injury is a relatively minor one for an internal exchange surface. The epidermis, often equipped with hairs or spines, does, of course, function as a protective covering for the entire leaf.

4. The exchange surfaces remain moist because they are exposed to air only in intercellular spaces. The humidity within those spaces being nearly 100 percent, the membranes of the mesophyll cells always retain a thin film of water on their surfaces. Gases dissolve in this water before moving into the cells. The protective epidermal tissues and the layers of waxy cuticle on their outer surfaces act as barriers between the dry outside air and the moist inside air.

Stems and roots

Gas exchange is not confined to the leaves, though these organs have particularly sophisticated adaptations for the process. In stems, procurement of oxygen and release of carbon dioxide generally take place by diffusion across the outer layer of epidermal cells. In many old

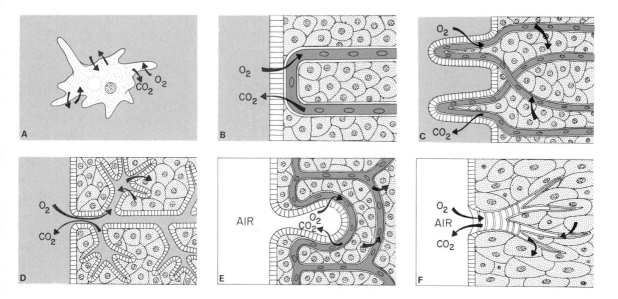

stems, however, the relatively impervious layer of bark would effectively cut off most of the oxygen supply were it not for the development of numerous small areas of loosely arranged cells with many intercellular spaces between them through which gases can move freely. Each such loose group of cells is called a *lenticel* (Fig. 6.2).

Roots also carry out gas exchange, though they usually possess no special structures for this function. Gases can diffuse readily across the moist membranes of root hairs and other epidermal cells. For roots to obtain sufficient oxygen, however, the soil in which they grow must be well aerated. One of the benefits of hoeing, raking, plowing, or otherwise cultivating the soil is increased air circulation.

There seems to be no need for special gas-transporting mechanisms in plants such as exist in animals, even though the most internal cells of a large stem or root are far removed from the surface of the plant. Most of the intercellular spaces in the tissues of land plants are filled with air, by contrast with those in animal tissues, which are filled with fluid. These air-filled spaces are interconnected to form an intercellular air-space system that opens to the outside through stomata and lenticels and penetrates to the innermost parts of the plant body. Thus incoming gases can move directly to the internal parts of the plant from the environmental atmosphere without having to cross membranous barriers, and they do not have to diffuse long distances through water or cell fluids, because they do not go into solution until they reach the film of water on the surfaces of the individual cells. Since oxygen can diffuse some 300,000 times faster through air than through fluids, the intercellular air-space system ensures that cells in the center of large stems and roots are adequately supplied.

SOLUTIONS IN AQUATIC ANIMALS

As we have already indicated, unicellular animals have no special gas-exchange devices, simple diffusion across their cell membranes being sufficient (Fig. 6.3A). Some of the smaller and simpler multicellular

6.3. Types of gas-exchange systems in animals. (A) Unicellular organisms exchange gases with the surrounding water directly across the general cell membrane. (B) Some multicellular aquatic animals use the general body surface as an exchange surface; the blood transports gases to and from the surface. (C) Many multicellular aquatic animals have specialized evaginated gas-exchange structures (gills). (D) A few aquatic animals, such as the sea cucumber, utilize invaginated exchange areas. (E) Most true air breathers have lungs, specialized invaginated areas that depend upon a blood transport system. (F) Land arthropods have tracheal systems, invaginated tubes that carry air directly to the tissues without the intervention of a blood transport system.

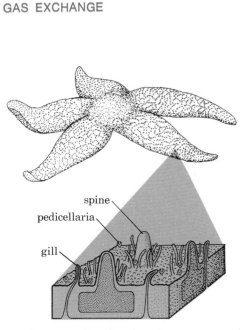

6.4. Sea star. The tiny skin gills are protected from damage by the spines and pedicellariae.

animals like jellyfish, hydra, and planaria show little further development, although their multipurpose gastrovascular cavities do facilitate the exposure of the more internal cells to environmental water (containing dissolved oxygen) drawn in through the mouth; no cell in these animals is far from the water medium. A few larger aquatic animals, particularly some of the marine segmented worms, lack special respiratory systems and utilize the skin of the general body surface, which is usually richly supplied with blood vessels (Fig. 6.3B). Most larger multicellular animals, however, have evolved true respiratory systems.

Gills

With a few exceptions, the respiratory systems of multicellular aquatic animals involve evaginated exchange surfaces, usually known as gills (Fig. 6.3C). Gills vary in complexity all the way from the simple bumplike skin gills of some sea stars (Fig. 6.4), or the flaplike parapodia of many marine segmented worms (Fig. 6.5), to the minutely subdivided gills of fish (Fig. 6.6).

Most gills, particularly those of very active animals, have such finely dissected surfaces that a few small gills may expose an immense total exchange surface to the water. Thus, although the gas-exchange surface takes up a very limited part of the animal, making possible relatively impermeable protective coverings for the rest of the body, the surface-to-volume ratio of the exchange surface remains high.

Another characteristic of most gills is that they contain a rich supply of blood vessels. Often the blood in these vessels is separated from the external water by only two cells: the single cell of the wall of the vessel and a cell of the gill surface. Oxygen moves by diffusion from the water, across the intervening cells, and into the blood, where it is ordinarily picked up by a carrier pigment (transport by the blood will be discussed in the next chapter). The blood then carries the oxygen throughout the body to the individual cells. Carbon dioxide produced by cellular metabolism moves in the opposite direction, being transported to the gills and discharged into the surrounding water.

The fragile gills are easily damaged, and a variety of protective devices have evolved for them. Frequently, these devices are coverings, such as the operculum of fish (Fig. 6.6A), but sometimes they take other forms, as in the spines and pincers that surround the skin gills of sea stars (Fig. 6.4).

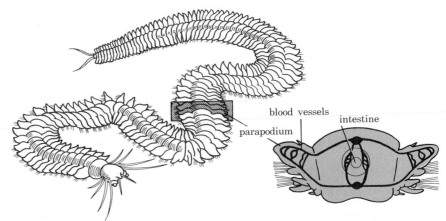

6.5. Gills of marine segmented worm (*Nereis*). The cross section shows that the flaplike parapodia on each segment are richly supplied with blood vessels. [Modified from Ralph Buchsbaum, *Animals Without Backbones,* University of Chicago Press, 1948.]

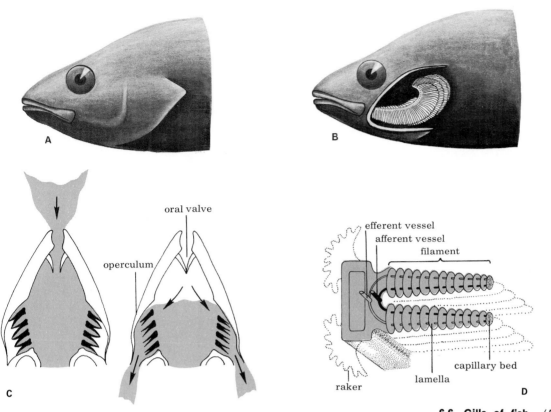

6.6. Gills of fish. (A) Head with operculum covering the gills. (B) Head with operculum cut away and gills exposed. (C) Water (gray) is drawn through the mouth into the pharynx while the opercula are closed. Then the oral valve closes as the oral cavity contracts; the water is forced across the gills (black structures) and out behind the now open opercula. (D) Structure of a gill. Each gill is composed of many filaments, each subdivided into numerous lamellae.

SOLUTIONS IN TERRESTRIAL ANIMALS

A few land animals have evolved highly modified gill-like respiratory structures that function in air (e.g. the book lungs of spiders). But the hazards of desiccation for most such evaginated surfaces are considerable, and major structural problems are associated with an array of filaments or a branched structure both sufficiently strong to maintain its shape against surface tension and gravity and sufficiently thin-walled to allow easy passage of gases. It is not surprising, therefore, that most terrestrial animals have evolved invaginated respiratory systems. These invaginated systems are of two principal types, *lungs* and *tracheae* (Fig. 6.3E, F). In both, the air inside the system is kept moist, and the cells of the exchange surface are covered by a film of water in which gases can dissolve. Thus the process of gas exchange has remained essentially aquatic in land animals, as it has in the leaf.

Lungs

Lungs, which are invaginated gas-exchange organs limited to a particular region of the animal and dependent upon a blood transport system, are most typical of two unrelated animal groups, the land snails and the higher vertebrates, including some fish, most amphibians, and all reptiles, birds, and mammals. In their simplest form, lungs are little more than chambers with slightly increased vascularization in their walls and with some sort of passageway leading to the outside. From such a rudimentary beginning, the evolution of the lung has tended

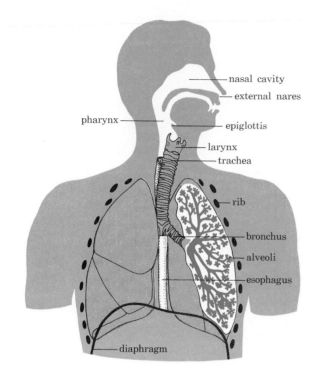

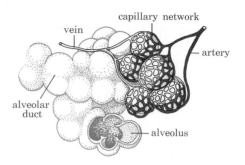

6.7. Respiratory system of man. The enlargement (above) shows a few sectioned alveoli and others surrounded by blood vessels. Actually, all alveoli, including those lying along alveolar ducts, are surrounded by networks of capillaries.

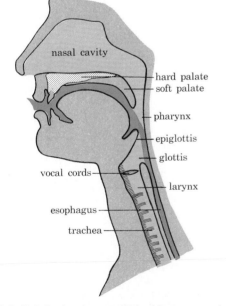

6.8. Detail of upper portion of human respiratory and digestive systems. Light color, path of air. Dark color, path of food. The two paths cross in the pharynx.

toward a greatly increased surface area, by subdivision of its inner surface into many small pockets or folds, and toward increased vascularization of its exchange surface.

Let us look at the respiratory system of man (Fig. 6.7) in some detail, as an example of the mammalian type. Air is drawn in through the *external nares,* or nostrils, and enters the *nasal cavities,* which function in warming and moistening the air, filtering out dust particles, and smelling. It then passes into the *pharynx* (Fig. 6.8), which it enters dorsally during inhalation and leaves via a ventral opening, the *glottis.* (We are here using the terms "dorsal" and "ventral" as though the human were standing on four legs like other mammals.) Since air enters the pharynx dorsally and exits ventrally, and since food enters the pharynx ventrally and exits dorsally into the esophagus, it follows that the air and food passages not only join but actually cross in the pharynx.

After leaving the pharynx through the glottis, the air enters the *larynx,* a chamber surrounded by a complex of cartilages (commonly called the Adam's apple). In many animals, including man, the larynx functions as a voice box. It contains a pair of vocal cords—elastic ridges stretched across the laryngeal cavity that vibrate when air currents pass between them; changes in the tension of the cords result in changes in the pitch of the sounds emitted.

Next, the air enters the *trachea,* a duct leading from the larynx into the thoracic cavity. Foreign particles and mucus are swept up the trachea away from the lungs by the waving cilia of its epithelial lining. A series of C-shaped rings of cartilage are embedded in the walls of the trachea and prevent it from collapsing upon inhalation. At its lower end, it divides into two *bronchi,* tubes that lead toward the two lungs.

Each bronchus branches and rebranches, and the bronchioles thus formed branch repeatedly in their turn, forming smaller and smaller ducts that terminate ultimately in tiny air pockets, each of which has a series of small chamberlike bulges in its walls termed *alveoli.* The total alveolar surface is enormous: about 1,000 square feet—an area many times greater than the total area of the skin. The walls of the alveoli are exceedingly thin, usually only one cell thick, and each alveolus is surrounded by a dense bed of blood capillaries. The alveoli are the site of the actual gas exchange and may therefore be regarded as the primary functional units of the lungs. Oxygen entering an alveolus dissolves in the film of water on its wall and then moves by diffusion across the intervening cells to the blood.

Air is drawn into and expelled from the lungs by the mechanical process called *breathing.* In mammals this process generally involves muscular contractions of two regions, the *rib cage* and the *diaphragm.* The latter is a muscular partition separating the thoracic and abdominal cavities. Inhalation, or inspiration, occurs whenever the volume of the thoracic cavity, in which the lungs lie, is increased; such an increase reduces the air pressure within the chest below the atmospheric pressure and draws air into the lungs. The increase in thoracic volume is accomplished by contractions of the rib muscles that draw the rib cage up and out and by contraction, or downward pull, of the normally upward-arched diaphragm (Fig. 6.9); the first mechanism is popularly called "chest breathing," while the second is called "abdominal breathing." Normal exhalation (or expiration) is a passive process; the muscles relax, allowing the rib cage to fall back to its resting position and the diaphragm to arch upward. This reduction of thoracic volume, combined with the elastic recoil of the lungs themselves, causes a rise in the pressure inside the lungs to a level above that of the outside atmosphere and drives out the air.

In contrast to mammalian lungs, which always retain some residual air after exhalation, the lungs of birds (Fig. 6.10) can be ventilated almost completely at each breath. This is because there are usually numerous extensions beyond the lungs, called *air sacs,* that may reach into almost every part of the body, even replacing some of the bone marrow within the bones. On inspiration, the air moves completely through the lungs and into the air sacs; expiration drives the air back through the lungs. The air sacs, of course, are important adaptations for flight, allowing a low weight-to-size ratio, and, predictably, they are usually best developed in strong-flying species and hardly developed in nonflying ones.

The mammalian and avian method of breathing is known as *negative-pressure breathing,* by contrast with *positive-pressure breathing,* where air is forced into the lungs rather than drawn in. Both these methods are used by adult frogs. With the mouth closed and nostrils open, the frog lowers the floor of the mouth, thereby sucking air into the mouth cavity (negative-pressure method). Then it closes the nostrils and raises the mouth floor; this reduction in the volume of the mouth cavity exerts pressure on the imprisoned air and forces it into the lungs (positive-pressure method). (We should note in passing that a frog is an excellent example of an animal that utilizes a variety of gas-exchange mechanisms. The lungs are only occasionally filled, much exchange surface being provided by the thin membrane of the mouth cavity and by the soft moist skin.)

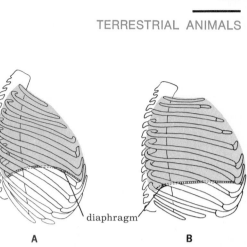

6.9. The mechanics of human breathing. (A) Resting position. (B) Inhalation: The rib cage is raised up and out, and the diaphragm is pulled downward. Both of these motions increase the volume (indicated by color shading) of the thoracic cavity, and the consequently reduced air pressure in the cavity causes more air to be drawn into the lungs.

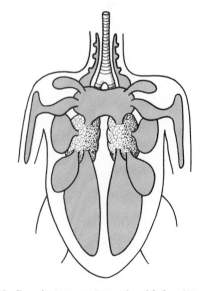

6.10. Respiratory system of a bird. Attached to the lungs are many air sacs (color), some of which even penetrate into the marrow cavities of the wing bones.

Tracheal systems

The second principal type of invaginated respiratory system evolved for air breathing is the tracheal system. It is typical of most terrestrial arthropods, in which it has evolved independently many times. Here we find no localized respiratory organ and little or no significant transport of gases by the blood. Instead, the system is composed of many small tubes that ramify throughout the body (Pl. II-5D). These tubes, called *tracheae,* carry air directly to the individual cells, where diffusion across the cell membranes takes place. Air enters the tracheae by way of *spiracles,* apertures in the body wall that can usually be opened and closed by valves. Some of the larger insects actively ventilate their tracheal systems by muscular contraction, but most small insects and some fairly large ones apparently do not. Calculations have shown that the rate of diffusion of oxygen in air is rapid enough to maintain at the tracheal endings an oxygen concentration only slightly below that of the external atmosphere.

REFERENCES

BEST, C. H., and N. B. TAYLOR, 1966. *The Physiological Basis of Medical Practice,* 8th ed. Williams & Wilkins, Baltimore. (See esp. Chapters 50–54.)

PROSSER, C. L., and F. A. BROWN, 1961. *Comparative Animal Physiology,* 2nd ed. Saunders, Philadelphia. (See esp. Chapter 7.)

WINTON and BAYLISS, 1969. *Human Physiology,* 6th ed., rev. and ed. by O. C. J. Lippold and F. R. Winton. Williams & Wilkins, Baltimore. (See esp. Chapters 12–15.)

SUGGESTED READING

CARLSON, A. J., V. JOHNSON, and H. M. CAVERT, 1961. *The Machinery of the Body,* 5th ed. University of Chicago Press, Chicago. (See esp. Chapter 6.)

CLEMENTS, J. A., 1962. "Surface Tension in the Lungs," *Scientific American,* December. (Offprint 142.)

D'AMOUR, F. E., 1961. *Basic Physiology.* University of Chicago Press, Chicago. (See esp. Chapter 10.)

RAMSAY, J. A., 1968. *Physiological Approach to the Lower Animals,* 2nd ed. Cambridge University Press, New York. (See esp. Chapter 3.)

SCHMIDT-NIELSEN, K., 1970. *Animal Physiology,* 3rd ed. Prentice-Hall, Englewood Cliffs, N.J. (See esp. Chapter 2.)

————, 1971. "How Birds Breathe," *Scientific American,* December. (Offprint 1238.)

WAGGONER, P. E., and I. ZELITCH, 1965. "Transpiration and the Stomata of Leaves," *Science,* vol. 150, pp. 1413–1420.

7

Internal transport

Every living cell, whether it exists alone as a single-celled organism or is a component of a multicellular one, must perform its own metabolic activities. It must synthesize its own ATP by cellular respiration (and/or photosynthesis) and carry out for itself those activities necessary for its growth and maintenance. It follows, then, that every cell must obtain the necessary raw materials to support its metabolism. It must obtain nutrients, and, if it utilizes aerobic respiration, it must obtain oxygen. Likewise, it must rid itself of metabolic wastes such as carbon dioxide and, in animals, nitrogenous compounds. In short, every cell must be exposed to a medium from which it can extract raw materials and into which it can dump wastes. In unicellular organisms and some of the structurally simpler multicellular ones, each cell is either in direct contact with the environmental medium or only a short distance from it. But in the larger and structurally more complex multicellular plants and animals, the more internal cells are far from the body surface and from the general environmental medium. We have already seen that in such organisms nutrient procurement, gas exchange, and waste expulsion take place in certain restricted regions of the body specialized for those functions. Obviously, some mechanism is needed for transporting substances between the specialized systems of procurement, synthesis, or elimination and the individual living cells throughout the body.

ORGANISMS WITHOUT SPECIAL
TRANSPORT SYSTEMS

In bacteria, Protozoa, and unicellular algae (and within single living cells in general), diffusion plays an important role in the movement of materials. It is also important in movement of materials from

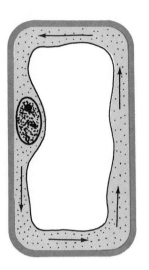

7.1. Cytoplasmic streaming in a plant cell. The cytoplasm flows around the large central vacuole.

cell to cell within the body of a multicellular organism; we have examined its part in the procurement of water by plant roots, for example. Such intercellular diffusion may be facilitated in plants by the plasmodesmata, tiny strands of protoplasm that penetrate the cell walls and interconnect the cytoplasmic contents of adjacent cells. We have also seen the importance in plants of diffusion of water along cell walls and of diffusion of gases through intercellular spaces.

But diffusion is a very slow process. If only diffusion were involved, it would take a long time for a substance to move from one cell to another, or even from one end of a single large cell to the other end. It is not surprising, therefore, that even in unicellular and small multicellular organisms diffusion is supplemented by other transport mechanisms. We have seen, for example, that food vacuoles commonly move along a fairly precise path within the cell, thereby distributing the products of digestion to all parts of the cytoplasm. We have also seen that the endoplasmic reticulum may provide a specialized pathway for intracellular movement of some substances. And the cytoplasm itself is seldom motionless; it frequently exhibits rapid massive flow within the cell. The flowing cytoplasm of an active amoeba is an example. The cytoplasm of many plant cells undergoes a characteristic movement called cytoplasmic streaming, in which the cytoplasm flows in definite currents along the surface of the cell vacuole (Fig. 7.1). Such mass flow can transport substances from one part of a cell to another many times faster than simple diffusion.

Among multicellular plants, it is not only the very tiny ones that lack a specialized internal-transport system. Many algae, particularly the brown and red algae, have large multicellular bodies, yet usually lack vascular tissue. As we have already seen, the cells of such plants are seldom far from the surrounding water or from water in intercellular spaces continuous with the external medium. Furthermore, nutrient and gas procurement is not limited to specialized restricted regions of the body, and photosynthesis is seldom localized in specific structures. Consequently each cell gets ample supplies locally, and long-distance transport is rarely necessary.

Unlike plants, animals are usually adapted for active locomotion. The more rapid metabolism required for an active life makes them less able to rely on such a slow process as diffusion, even when it is supplemented by the other intracellular processes mentioned above. Furthermore, because of their way of life, animals are much less likely than plants to have bodies large in one or two dimensions but flat and thin in the third.

In general, only very small animals lack a circulatory system; and even these frequently exhibit some adaptations for transport. Consider hydra (Fig. 5.8, p. 107). Its body wall is basically only two cells thick, but the cells of the inner as well as of the outer layer are exposed directly to water containing dissolved oxygen, because such water is drawn into the gastrovascular cavity. In planaria (Fig. 5.9) the profusely branching gastrovascular cavity ramifies into all parts of the body, functioning as a primitive transport system. In short, though animals like hydra and planaria lack a true blood circulatory system, they do have compensatory adaptations that free them from complete dependence upon diffusion and intracellular transport.

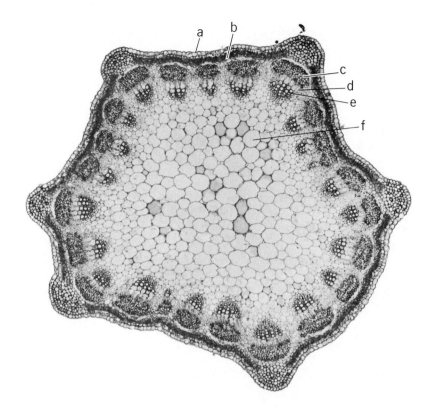

7.2. Cross section of herbaceous dicot stem, alfalfa (*Medicago*). (a) Epidermis; (b) cortex; (c) phloem fibers; (d) vascular phloem; (e) xylem; (f) pith. The cambium is difficult to distinguish and is not labeled here. × 49. [Courtesy Thomas Eisner, Cornell University.]

VASCULAR PLANTS

The vascular plants, as their name indicates, incorporate the two principal types of plant vascular tissue, xylem and phloem. Thanks to these specialized internal-transport tissues, they have been free to evolve bodies large in all dimensions and to develop far greater specialization of parts and more complete integration of function than other plant groups. Thus water and mineral uptake can be restricted primarily to the roots, while photosynthesis can be restricted largely to the leaves. In some very tall forest trees, the distance between the roots and the leaves may be enormous; yet the xylem and phloem form continuous pathways between them, and they can exchange materials with relative ease. Clearly, the successful exploitation of the land environment by plants was dependent upon the evolution of such a transport system.

Structure of stems

Stems of plants serve many functions. Some contain chlorophyll and carry out photosynthesis. Others are highly specialized as storage organs; potato tubers, which are underground stems, are an example. Here, however, we shall concentrate on stems as organs of transport and support, and examine in some detail the structural adaptations associated with these two functions. Keep in mind that, although our discussion of transport is based on stems, the vascular tissue of the stem is continuous with that in the roots and leaves, and that internal transport is no less important in those organs than it is in stems.

Gross anatomy. Let us first examine in cross section the stem of a herbaceous dicot such as alfalfa (Fig. 7.2). ("Herbaceous" usually

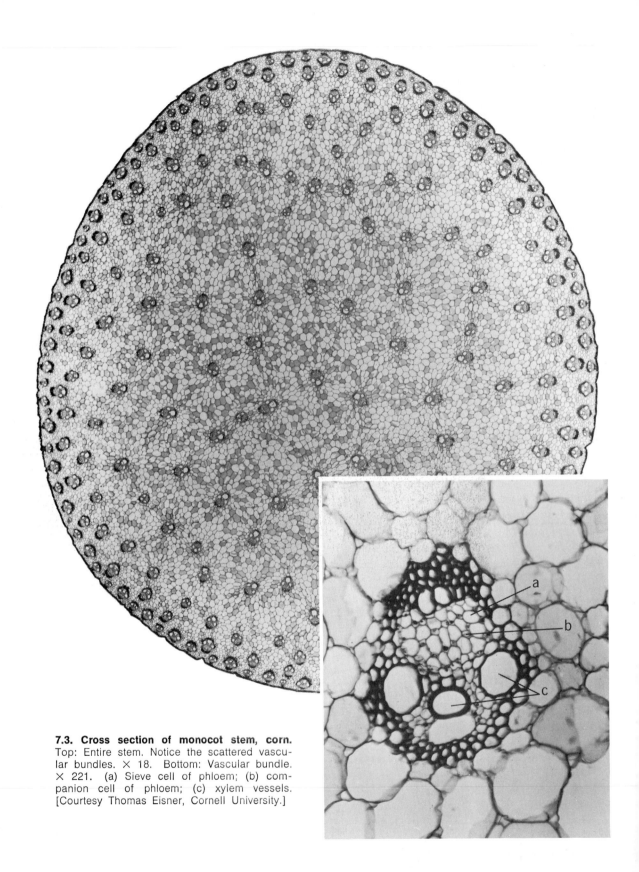

7.3. Cross section of monocot stem, corn.
Top: Entire stem. Notice the scattered vascular bundles. × 18. Bottom: Vascular bundle. × 221. (a) Sieve cell of phloem; (b) companion cell of phloem; (c) xylem vessels. [Courtesy Thomas Eisner, Cornell University.]

refers to plants whose stems remain soft and succulent; the contrasting term is "woody.") The outer tissue layer of a herbaceous stem is epidermis. Next comes the cortex, which is frequently divided into an area of collenchyma just beneath the epidermis and an area of parenchyma more internally. Internal to the cortex lies the vascular tissue, which may be arranged in a continuous hollow cylinder or in a series of discrete bundles, as in alfalfa (see also Pl. II-7B). In either case, the phloem lies outside the xylem, with a layer of meristematic tissue, called *vascular cambium,* between them. The center of the stem is filled with pith, which is parenchyma tissue and functions as a storage area.

Notice that the arrangement of the tissues in the stele of a dicot stem differs in two obvious ways from that in a typical young dicot root. (1) The phloem and xylem form separate rings in the stem, one outside the other, whereas the two tissues alternate in the young root. (2) Dicot stems characteristically have pith, whereas most dicot roots do not.

Stems of monocots are similar to those of herbaceous dicots. Their vascular tissue, however, always forms discrete bundles, never a continuous cylindrical shell, and the bundles are usually not arranged in a definite circle, as in dicots, but tend to be scattered through the stem (Fig. 7.3). Most monocots lack cambium (palms and some lilies are exceptions).

The cambium of many herbaceous dicots never becomes active and never produces additional phloem or xylem cells. In such plants, all the vascular tissue is said to be *primary tissue*—tissue derived originally from the apical meristem as the stem (or root) grew in length. The apical meristem of a stem is, of course, in the bud, while that of the root is near the root tip. The new cells produced in the apical meristem soon begin to differentiate, some forming epidermis, some forming the fundamental tissues of the cortex and pith, and some forming the primary phloem, primary xylem, and cambium.

In some species of herbaceous dicots, the cambium does become active, however. As the cambial cells divide, they give rise to new cells both to the inside and to the outside. The new cells formed on the outer side of the cambium differentiate as *secondary phloem;* those formed on the inner side of the cambium differentiate as *secondary xylem* (Fig. 7.4). Secondary vascular tissue, then, is tissue derived from the lateral meristem, the cambium, and is a result of growth in diameter rather than growth in length. As secondary phloem is produced by the cambium, it pushes the older, primary phloem farther and farther away from the cambium toward the outside of the stem. Similarly, as secondary xylem is produced, the cambium becomes increasingly distant from the primary xylem, which is left in the inner portion of the vascular cylinder. In a stem that has undergone secondary growth, therefore, the sequence of tissues in the stele (moving from the outside toward the center) is: primary phloem, secondary phloem, cambium, secondary xylem, primary xylem, pith.

A cross section of a "woody" stem made early in its first year would not appear very different from that of a herbaceous stem; in both, the primary vascular tissue is arranged in a continuous ring in some species and in discrete bundles in others. Secondary growth, however, soon makes the rings continuous in all woody stems, and as this growth pro-

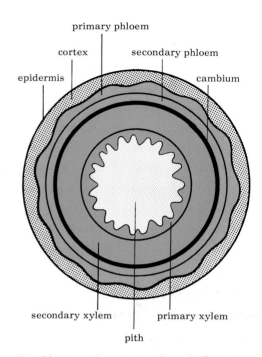

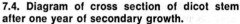

7.4. Diagram of cross section of dicot stem after one year of secondary growth.

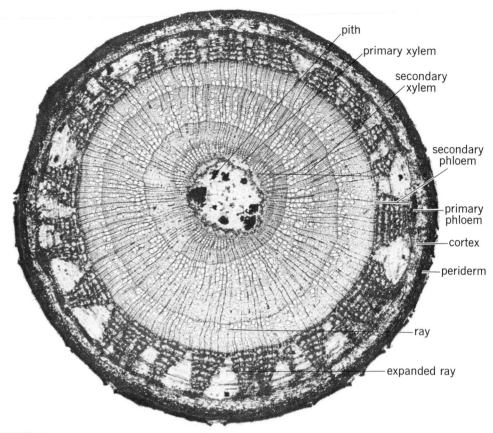

7.5. Cross section of basswood (*Tilia*) stem at the end of three years of growth. The annual growth rings in the secondary xylem are easily seen. Note that the rays, which are only one or a few cells thick in the xylem, are expanded in the secondary phloem. When an old layer of phloem is pushed farther out and hence must expand, the only part that can do so is the parenchymatous ray. × 26. [Courtesy Thomas Eisner, Cornell University.]

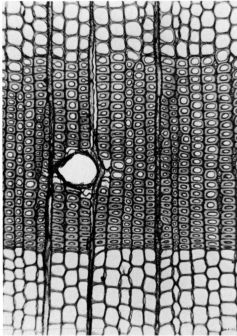

7.6. Cross section of part of secondary xylem (wood) of pine. The smaller, thicker-walled cells are summer wood and the larger, thinner-walled cells are spring wood. A large resin duct can be seen in the summer wood. × 92. [Courtesy U.S. Forest Products Laboratory.]

ceeds, the stem looks less and less like a herbaceous one (Fig. 7.5). The secondary xylem becomes thicker and thicker until almost the entire stem of an older plant is xylem tissue—commonly called wood.

Since new xylem cells produced early in the growing season, when conditions are best, grow larger than cells produced later in the season, a series of concentric annual rings, clearly visible in cross sections of the stem, are formed. Each ring is made up of an inner area of spring wood with large cells and an outer area of summer wood with smaller cells (Fig. 7.6). A fairly accurate estimate of the age of a tree can be made by counting the annual rings.

As woody stems (or roots) grow in diameter, a layer of cells outside the phloem takes on meristematic activity and becomes the **cork cambium.** As growth continues, the original epidermis and cortex flake off and are replaced by **cork cells** produced by cell division in the cork cambium. The layer of dead cork cells (periderm) constitutes the outer bark of the older stem or root. The inner bark is the phloem tissue. Since the older phloem is pushed to the outside where it is periodically sloughed off, the phloem layer never becomes thick like the xylem, and annual growth rings are very difficult, if not impossible, to detect in it.

In summary, then, the old woody stem of a tree has no epidermis or cortex. Its surface is covered by an outer bark of cork tissue. Beneath the cork cambium is the thin layer of phloem, or inner bark, and beneath this is the vascular cambium, which is usually only one cell thick. The rest of the stem is mostly secondary xylem, or wood, of which only the outer annual rings, or *sapwood*, still function in transport. The older xylem, known as *heartwood*, no longer functions in transport, but it remains an important supportive component of the tree.

The xylem. Now let us examine the cellular makeup of the xylem. Xylem is a complex tissue containing several different types of cells. Two of these, the tracheids and the vessel cells, are important as conductive elements after they have matured. Actually, the cell walls are all that remains of a tracheid or vessel cell functioning in transport; the cellular contents, both cytoplasm and nucleus, have disintegrated. The main transport in the xylem occurs, then, in tubular remnants of cells, not in the living cells themselves.

Tracheids are elongate, tapering cells with heavily lignified secondary cell walls; the walls are particularly thick in summer wood and are important as supportive elements. The walls of tracheids in secondary xylem are interrupted by numerous *pits* (Fig. 7.7). The pits may occur anywhere on the cell wall, but they are often particularly numerous on the tapered ends of the cell, where it abuts upon the next cell beyond it. Water and dissolved substances move from tracheid to tracheid through the pits.

Vessel cells are more highly specialized conductive elements than tracheids, from which they probably evolved. They are characteristic of the flowering plants and do not occur in most gymnosperms (conifers etc.). In general, vessel cells are shorter and wider than tracheids (Fig. 7.8). They have pits along their sides, through which some

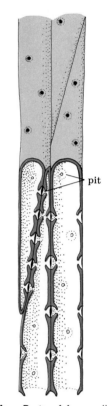

7.7. Tracheids. Parts of four cells are shown, three of them with portions of the wall cut away to expose the lumen and give a clearer view of the junction between cells. Notice the pits, which are particularly abundant along the tapering ends of the cells. [Modified from V. A. Greulach and J. E. Adams, *Plants: An Introduction to Modern Botany,* Wiley, 1962.]

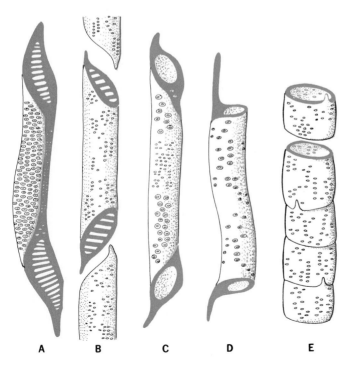

A   B   C   D   E

7.8. Vessel cells. Five different types of vessel cells are shown—those thought to be the more primitive on the left, those thought to be the more advanced on the right. (E) shows a single vessel cell on top, four cells linked in sequence to form a vessel below. The evolutionary trend seems to have been toward shorter and wider cells, larger perforations in the end walls until no end walls remained, and less oblique, more nearly horizontal ends.

lateral movement of substances may take place, but materials move chiefly through their ends, which are more extensively perforated and may even lack a wall altogether. Since the perforations lack both secondary and primary walls, material moving from one vessel cell to the next in a vertical sequence forms a continuous column. A vertical series of vessel cells is called a *vessel.*

In addition to tracheids and vessel cells, xylem contains fiber and parenchyma cells. The fibers are elongate, very thick-walled cells that function as supportive elements. Some of the parenchyma cells are scattered among the other cells of the xylem, but many are grouped together to form *rays* that run through the xylem in a radial direction (Fig. 7.5) and function as pathways for lateral movement of materials and as storage areas.

The phloem.   Like xylem, phloem is a complex tissue. It contains supportive fibers and also parenchyma; the phloem rays are continuous with the xylem rays. The principal vertical conductive elements in phloem are the *sieve elements,* which, in vertical series, form a *sieve tube.* In their most advanced form, sieve elements are elongate cells with specialized areas on their end walls, called *sieve plates* (Fig. 7.9). As the name implies, a sieve plate is an area with numerous perforations or pores, through which strands of protoplasm connect the contents of one cell with those of the next. Unlike the tracheids and vessels of xylem, the sieve elements retain their cytoplasm at maturity, but their nucleus disintegrates.

The sieve elements of most flowering plants usually have closely associated with them one or more specialized, elongate, parenchymatous *companion cells.* Mature companion cells retain both their cytoplasm and their nucleus. Some biologists have suggested that the nucleus of the companion cell controls both its own cytoplasm and the cytoplasm of the adjoining sieve element after the nucleus of the latter has disintegrated. Such an association, they think, would help explain why the mature sieve element can continue to carry out many of the normal activities of a living cell even though it has no nucleus of its own.

The ascent of sap

Water, absorbed by the roots, moves upward through the plant body in the mature tracheids and vessels of the xylem. That the upward movement of water is primarily in the xylem can easily be demonstrated by ringing experiments; if a ring of cork, phloem, and cambium is removed from the trunk of a tree, the leaves will still remain turgid, even though they are connected to the roots only by xylem. But a complete explanation of this upward movement has eluded botanists for centuries. We shall here examine briefly some of the possible explanations, together with evidence both for and against them.

Any general explanation for the ascent of sap (water plus dissolved materials) in xylem must identify the forces capable of raising water to the tops of the tallest trees, which may be 300–400 feet high. A pressure of one atmosphere can support a column of water approximately 34 feet high at sea level. It follows that a pressure of about 12 atmospheres would be needed to support a column 400 feet high. But the column must be more than supported; the water must be moved

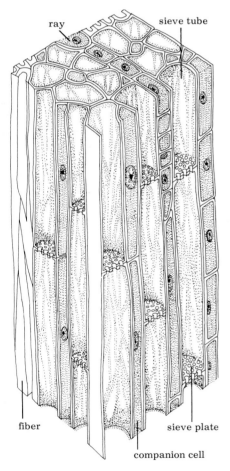

ray   sieve tube

fiber   sieve plate

companion cell

7.9. Phloem. [Adapted from S. and O. Biddulph, "The Circulatory System of Plants," *Sci. Am.,* February, 1959. Copyright © 1959 by Scientific American, Inc. All rights reserved.]

upward at a rate that may sometimes be as fast as 40 or more inches per minute, and this movement must take place in a system that offers frictional resistance to it. It has been calculated that an additional pressure of at least 18 atmospheres is necessary to achieve the observed rate. Therefore, a total force of at least 30 atmospheres is necessary in the tallest trees, and any general theory of water movement in the xylem must account for forces of this magnitude.

The driving force might be at the base of the plant and push the water upward, or it might be at the top of the plant and pull the water up, or forces at both positions might be jointly involved. Each of these possibilities has had its advocates, and each has some evidence in its favor.

Root pressure. Let us first consider the possibility of a force applied as a push from below. When the stems of certain species of plants are cut, sap flows from the surface of the stump for some time, and if a tube is attached to the stump, a column of water several feet high may rise in it. Something similar happens when conditions are optimal for water absorption by the roots but the humidity is so high that little water is lost by *transpiration,* as evaporation from living plant tissue, chiefly through the stomata, is termed. Water under pressure may then be forced out at the ends of the leaf veins, forming droplets along the edges of the leaves in a process called *guttation* (Fig. 7.10). When the water in the xylem is under pressure, as in these instances of bleeding and guttation, the pushing force involved is apparently in the roots, and is called *root pressure.*

The manner in which root pressure is built up is not fully understood. We have already seen that water moves from the soil, through the epidermis, cortex, endodermis, and pericycle of the root to the xylem, in which it then flows upward to the rest of the plant. Much of this movement across the tissues of the root is simple diffusion along a concentration gradient. But this cannot be the whole story. Water may be in much higher concentration in the xylem than in the protoplasm of the endodermal cells and yet continue to move from the endodermis into the xylem; if it were undergoing simple diffusion, it would go the other way. Activity by living cells is immediately surmised in a situation like this. And, indeed, if the roots are killed, all root pressure disappears. Or if the roots are simply deprived of oxygen, the root pressure ceases, an indication that respiratory production of ATP is necessary to provide the energy for the active inward secretion of water. Such secretion is perhaps one of the important functions of the living cells of the endodermis; they may function as an active barrier between the stele and the cortex of the root, secreting water into the stele against the osmotic gradient.

Next, we must ask if root pressure can reach the magnitude we have said is required. The answer appears to be no. Some plants, particularly the conifers and their relatives, are incapable of developing much root pressure at all. Attempts to measure the root pressure in species in which it does occur have rarely yielded values exceeding 1 or 2 atmospheres, though occasionally pressures as high as 6–10 atmospheres have been found. Another reason to doubt that root pressure is the principal motive force for the ascent of sap is that, when a punc-

7.10. Guttation by a strawberry leaf. [Courtesy J. Arthur Herrick, Kent State University.]

ture is made in a xylem vessel during the summer, it is uncommon to find water under pressure; i.e. water is seldom forced out of the wound. Instead, one can often hear a short hissing sound as air is drawn into the vessel.

In short, root pressure is not the explanation we are seeking, though it may be involved in the ascent of sap in some plants some of the time. It may be particularly important in very young plants, and perhaps in a few species of trees in early spring.

The cohesion theory.   What about the alternative possibility, that the water is pulled up from above? The mechanism has been explained as follows: As water in the walls of parenchyma cells in the leaves (or other parts of the shoot) is lost by transpiration, it is replaced by water from the cell contents. The osmotic concentration in the cells is thus raised, and they take up water from adjacent cells, which, in turn, withdraw water from cells adjacent to them. In this way, a gradient extends to the xylem in the veins of the leaf, and the parenchyma cells next to the xylem withdraw water from the column in the xylem. This removal of water from the top of the column pulls the column upward. Notice that this is not a matter of pulling the column up by air pressure or vacuum; the mechanism is not strictly analogous to sucking up a liquid through a straw. Air pressure could raise water only about 34 feet, but we are dealing with a mechanism presumed to move a water column that may be hundreds of feet high. What is assumed here is a continuity between the water on the evaporating surfaces of the cell wall and the water in the xylem, and a continuity between the water at the top of the xylem and that in the roots.

This theory of pull from above as a result primarily of transpiration depends on proof that water has certain physical properties. Water molecules moving out of the top of the xylem must pull other water molecules behind them; there can be no break, no separation, between water molecules. In short, the validity of this theory depends on the existence of great cohesive forces between the individual water molecules. And water does indeed exhibit great cohesive strength. It is pertinent to recall that water molecules do not ordinarily exist as separate entities in the liquid phase. Measurements of surface-tension relationships, internal pressure, heats of vaporization, etc. lead to a predicted cohesive strength for water as high as 15,000 atmospheres.

Although the theory has not been tested under conditions duplicating those in very tall trees, some interesting experiments have been made on a smaller scale. For example, if water (previously boiled to remove all dissolved air) is evaporated from the top of a thin tube whose walls are made of a material to which water molecules can adhere and whose lower end is immersed in mercury, a column of mercury can be pulled up the tube to a far greater height than it could be pulled by a vacuum (Fig. 7.11). If the base of a cut branch is inserted tightly in the upper end of the tube and the leaves of the branch become the site of evaporation, similar results are obtained; the tension developed pulls the mercury to a height above the barometric height. The water molecules have adhered to each other (and to the walls of the tube) tightly enough to lift a heavy column of mercury. Such experiments support the *cohesion theory* (or transpiration theory).

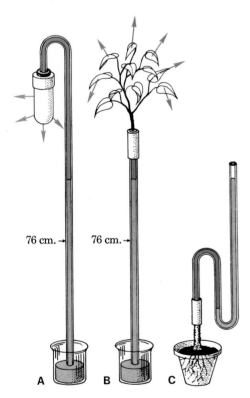

7.11. Demonstrations of rise of water by pull from above (A, B) and root-pressure push from below (C).   (A) Water (color) is evaporated from a clay pot attached to the top of a thin tube whose lower end is in a beaker of mercury. The water in the tube rises and pulls a column of mercury to a point well above the 76 cm. to which a vacuum could pull it.   (B) The same results are obtained when transpiration from the leaves of a shoot is substituted for evaporation from a clay pot.   (C) In some plants, root pressure can raise a column of mercury.

Nevertheless, the cohesion theory has failed to gain complete acceptance. The majority of botanists favor some form of it, while admitting that modifications are needed. A few scientists reject it outright and suggest that a totally new hypothesis must be advanced. Clearly, the rise of sap remains an intriguing problem for research.

Translocation of solutes

Translocation of organic solutes. Two principal classes of solutes are transported—or translocated, as plant physiologists generally call it —within the plant body: organic solutes and inorganic solutes. Let us consider the organic solutes first. We can conveniently divide these into two principal types, carbohydrates (usually transported as sucrose) and organic nitrogen compounds. (So little is known about the translocation of fats and related compounds in plants that we shall disregard them in this discussion.)

The classical picture of the translocation of solutes was that all upward movement was through the xylem and all downward movement through the phloem. About 1920, however, this view had to be revised. It became apparent that most of the movement of carbohydrates, whether up or down, was through the phloem. Most of the early work on the path of movement involved ringing experiments. It could be demonstrated that if all the bark (which includes, of course, the phloem) was removed as a ring from the trunk of a tree, the supply of carbohydrates to all parts of the plant below the ring was cut off, and those parts eventually died when they had depleted their stored reserves. Downward movement of carbohydrates was clearly through the phloem, not through the xylem, which had been left intact in these experiments. But it could also be demonstrated that if a branch was ringed a short distance behind the growing bud, the supply of carbohydrates moving to the bud was cut off. Again, the movement must have been in the phloem, but in this case the movement was upward. From numerous such experiments as these, most botanists came to regard it as a valid generalization that almost all carbohydrate movement is through the phloem.

Some botanists, however, objected that far too much carbohydrate moves within the plant body, and moves too rapidly, for the phloem to be the exclusive channel for this movement. After all, the total amount of functional phloem tissue in the trunk of a large tree is rather small. Surely, the argument ran, it is physically inconceivable that so much material should pass through so few sieve tubes, particularly since these tubes are not open pathways like xylem vessels. Numerous workers, however, showed by careful ringing experiments and by radioactive-tracer studies that, hard to conceive as it is, the phloem is indeed the pathway of sugar movement, and that this movement is amazingly rapid.

The situation is less clear with organic nitrogen compounds. It was formerly thought that nitrogen, absorbed by the roots primarily as nitrate, was carried upward in inorganic form through the xylem to the leaves, there to be used in synthesis of organic compounds, which were then transported through the phloem. This sequence probably holds true for some plants. But there is now good evidence that many species promptly incorporate incoming nitrogen into organic compounds in the roots. Opinion is divided on whether these organic nitrogen compounds move upward in the xylem or in the phloem. Probably they move to some extent in both, perhaps primarily in the xylem in some species and primarily in the phloem in other species. Physiological traits often vary widely from one species to another, and generalizations are risky.

Translocation of inorganic solutes. Inorganic ions such as those of calcium, sulfur, and phosphorus are translocated upward from the roots to the leaves primarily through the xylem. Use of radioactive forms of these minerals, however, indicates that some are quite mobile in the plant, traveling rapidly back down the plant in the phloem, or moving out of the older leaves through the phloem and being transferred to the newer, more actively growing leaves. Phosphorus, for example, easily moves upward in the xylem and downward in the phloem, often circulating rapidly throughout the plant in this manner. If a plant is grown for a short time in a solution containing radioactive phosphorus, and the plant is then placed against a photographic plate, the younger leaves will be found to contain the greatest concentrations of radioactive phosphorus. If the plant is then moved into a normal solution (one without the radioactive tracer), allowed to grow for a day or so, and again placed against a photographic plate, the resulting pictures will show that the radioactive phosphorus has moved from the leaves in which it was first concentrated to the new leaves just beginning to develop. Calcium, on the other hand, is not mobile in the phloem, and thus cannot move from old leaves to newer ones. Consequently plants must obtain a steady supply of new calcium from the soil, whereas they can easily survive with only intermittent feedings of phosphorus, since this element can shift from place to place within the plant and be re-used many times. Well-designed fertilization programs take into account such differences in the properties of the different mineral nutrients.

Hypotheses of phloem function. We have seen that most transport of organic solutes, both up and down, is through the phloem, and that most downward transport of minerals is also through the phloem. How phloem functions in this transport is a problem that has been under investigation for a very long time. Several hypotheses have been put forward, but it must be admitted that none is fully convincing.

One hypothesis is that materials are carried the length of each sieve cell by cytoplasmic streaming. The suggestion is that materials diffusing into one end of a sieve cell through the sieve plate are picked up by the streaming cytoplasm and carried to the other end of the cell, where they diffuse across the sieve plates at that end and, upon entering the next cell in the tube, are again picked up by streaming cytoplasm. In this way, by alternately streaming within cells and diffusing between cells, the materials would move long distances through the sieve tubes of the phloem. It has been objected, however, that there is little, if any, evidence that cytoplasmic streaming occurs in mature sieve-tube cells, and, further, that measurements of the velocities of streaming in other cells, where the process does occur, yield values much lower than the known rates of solute movement through sieve tubes. At the present time, not many botanists accept the streaming hypothesis.

A second hypothesis rests on the fact that substances which lower the surface tension at interfaces spread rapidly along these interfaces.

According to this hypothesis, substances move through the sieve cells by flowing along the intracellular membranous interfaces. The evidence, however, makes it doubtful that there is sufficient interface surface area in the sieve cells to account for the quantities of material known to be transported.

A third hypothesis, which invokes *pressure flow* or mass flow, is probably the one most widely supported by botanists today. According to this hypothesis, there is a mass flow of water and solutes through the sieve tubes along a turgor-pressure gradient. Cells like those of the leaf contain high concentrations of such osmotically active substances as sugar. Much water therefore tends to diffuse into them, and their turgor pressure rises. This pressure, impinging upon the next cell, tends to force substances from the first cell into the second. Thus, under pressure, substances are forced en masse into the sieve-tube cells in the upper parts of the plant. This means that the upper portions of the sieve tubes are under pressure. But in storage organs or actively growing tissues, sugars are being used up; as sugars are removed from the sieve tubes in these regions, the osmotic concentrations in the tubes are lowered. They therefore tend to lose water, which results in a drop in their turgor pressure. We have, then, a sieve-tube system in which the contents in some portions of the plant are under considerable turgor pressure and the contents in other portions of the plant are under lower turgor pressure. The result is a mass flow of the contents of the sieve tubes from the regions under high pressure (usually in the leaves, but sometimes in storage organs when reserves are being mobilized for use, as in early spring) to the regions under lower pressure (usually actively growing regions or storage depots). The whole process is dependent upon massive uptake of water by cells at the one end, because of their high osmotic concentrations, and massive loss of water by cells at the other end, because their osmotic concentrations are lowered by their loss of sugar.

The chief objection to the mass-flow hypothesis is an obvious one. It seems to assume that material can flow with relative freedom from one sieve-tube cell to the next. But the openings in the sieve plates between successive sieve-tube cells are very tiny indeed. Furthermore, the cytoplasm of sieve-tube cells, particularly that in the vicinity of the sieve plates, seems to be rather viscous and should offer great resistance to mass flow. The most that can be said at present is that the mass-flow hypothesis is considered by many botanists to be the best so far proposed, but that there are major weaknesses in it.

CIRCULATION IN HIGHER ANIMALS

Animal circulatory systems usually include some sort of pumping device called a *heart.* There may be only one heart, as in our own case, or a number of separate hearts, as in earthworms, where five blood vessels on each side of the animal pulsate, pumping blood from the main dorsal longitudinal vessel into the main ventral longitudinal vessel (Fig. 7.12). Many insects have both a large general heart and a series of smaller accessory hearts at the bases of their legs and wings.

The one-way pumping action of the heart, usually combined with a system of one-way valves, moves the blood in a regular fashion through the circuit. This circuit may be rigidly encompassed in well-defined

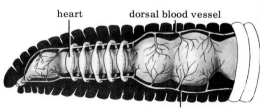

7.12. Circulatory system of earthworm, showing five of the ten hearts.

vessels, in which case it is called a *closed circulatory system.* Or the circuit may have some sections where definite vessels are absent and the blood flows through large open spaces known as sinuses; such a system is called an *open circulatory system.* Closed circulatory systems are characteristic of a great variety of animals, including our old friends the earthworms and all vertebrates. Open circulatory systems are characteristic of most molluscs (snails, oysters, clams, etc.) and all arthropods (insects, spiders, crabs, crayfish, millipeds, etc.).

The insect circulatory system

Since movement of the blood through an open system is not as fast, orderly, or efficient as through a closed system, it may seem surprising that such active animals as insects, which must have relatively high metabolic rates and precise internal regulation, should have open circulatory systems. But you will recall that insects do not rely on the blood to carry oxygen to their tissues, this function being fulfilled by the much-branched tracheal system; consequently it is not vital for insects that their blood flow very fast and in a precise pathway. This is a good example of the complex interrelationship between the various systems of a living creature.

The circulatory systems of insects are even more reduced than those of most other arthropods. Ordinarily, the only definite blood vessel in an insect is a longitudinal vessel running through the dorsal portion of the animal's thorax and abdomen (Fig. 7.13). The posterior portion of this vessel is pierced by a series of openings, or ostia, each regulated by a valve that will allow movement of blood only into the vessel. This vessel functions as a heart and is often so designated. When it contracts it forces blood out of its open anterior end into the head region. When it relaxes again, blood is drawn in through the ostia. Once outside the heart, the blood is no longer in vessels; there are no veins, capillaries, or arteries, other than the heart itself with its valve segment and the short so-called artery that forms its anterior end. The blood simply fills the spaces between the internal organs of the insect, so that each organ is bathed directly by blood.

The action of the heart causes the blood to move sluggishly through the body spaces from the anterior end where it was released to the posterior end where it will again enter the heart. The movement of the blood is accelerated by the stirring and mixing action of the muscles of the body wall and gut during activity. Thus, when the animal is most active, as in running or flying, and its organs are in most need of rapid delivery of nutrients and removal of wastes, the blood moves with relative rapidity because of the activity itself. That insects are as successful as they are is proof enough that their open circulatory systems are sufficient for their needs.

The human circulatory system

Man, like all vertebrates, has a closed circulatory system, which consists basically of a heart and numerous arteries, capillaries, and veins. An *artery* is a blood vessel carrying blood away from the heart, while a *vein* is a vessel carrying blood back toward the heart. Note that, contrary to a common impression, the definitions of these two types of vessels are not based on the condition of the blood carried. Although it

7.13. The dorsal heart of a grasshopper. Blood enters the heart through the ostia and is pumped forward and out at its open anterior end.

heart

ostium

is true that the majority of arteries carry oxygenated blood and the majority of veins deoxygenated blood, oxygen content is not always a reliable way to distinguish them. *Capillaries* are tiny blood vessels that interconnect the arteries with the veins. It is across the thin walls of the capillaries that exchange of materials between the blood and the other tissues takes place.

The circuit.   Let us trace the movement of blood through the human circulatory system, beginning with that returning to the heart from the legs or arms. Such blood enters the upper right chamber of the heart, called the *right atrium* (or auricle) (Figs. 7.14, 7.15). This chamber then contracts, forcing the blood through a valve (the tricuspid valve) into the *right ventricle,* the lower right chamber of the heart. Now, this blood, having just returned to the heart from its circulation through tissues, contains little oxygen and much carbon dioxide. It would be of little value to the body simply to pump this deoxygenated blood back out to the general body tissues. Instead, contraction of the right ventricle sends the blood through a valve (the pulmonary semilunar valve) into the *pulmonary artery,* which soon divides into two branches, one going to each lung. In the lungs, the pulmonary arteries branch into many small arteries, called arterioles, which connect with dense beds of capillaries lying in the walls of the alveoli. Here gas exchange takes place, carbon dioxide being discharged from the blood into the air in the alveoli and oxygen being picked up by the hemoglobin in the red cells of the blood. From the capillaries, the blood passes into small veins, which soon join to form large *pulmonary veins* running back toward the heart from the lungs. The four pulmonary veins (two from each lung) empty into the upper left chamber of the heart, called the *left atrium* (or auricle). When the left atrium contracts, it forces the blood through a valve (the bicuspid or mitral valve) into the *left ventricle,* which is the lower left chamber of the heart. The left ventricle, then, is a pump for recently oxygenated blood. When it contracts, it pushes the blood through a valve (the aortic semilunar valve) into a very large artery called the *aorta.*

After the aorta emerges from the anterior portion of the heart (the upper portion, in humans standing erect), it forms a prominent arch and runs posteriorly along the middorsal wall of the thorax and abdomen (Fig. 7.15). Numerous branch arteries arise from the aorta along its length, and these arteries carry blood to all parts of the body. Each of these arteries, in turn, branches into smaller arteries, until eventually the smallest arterioles connect with the numerous tiny capillaries embedded in the tissues.

Very little, if any, exchange of materials occurs across the walls of the arteries or veins, which are apparently impermeable to the substances in the blood and tissue fluid. In the capillaries, oxygen, nutrients, hormones, and other substances move out of the blood into the tissues; such waste products as carbon dioxide and nitrogenous wastes are picked up by the blood, and substances to be transported, such as hormones secreted by the tissues, or nutrients from the intestine and liver, are also picked up.

From the capillary beds the blood runs into tiny veins, which fuse to form larger and larger veins, until eventually one or more large

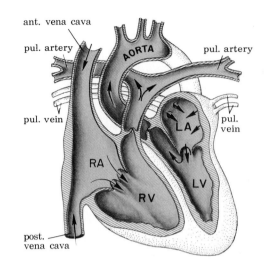

7.14. The human heart. The arrows show direction of blood flow. RA, right atrium; RV, right ventricle; LA, left atrium; LV, left ventricle. [Modified from N. D. Millard and B. G. King, *Human Anatomy and Physiology,* Saunders, 1951.]

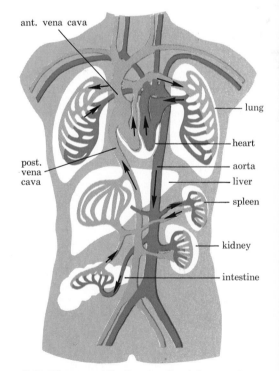

7.15. Diagram of the human circulatory system. Dark vessels contain oxygenated blood; light vessels contain deoxygenated blood. Only a very few of the vast number of arteries that branch off the aorta are shown here.

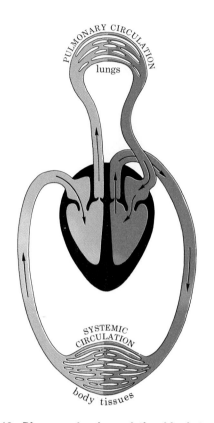

7.16. Diagram showing relationship between the pulmonary and systemic circulations. Dark vessels contain oxygenated blood; light vessels contain deoxygenated blood.

veins exit from the organ in question. These veins, in turn, empty into one of two very large veins that empty into the right atrium of the heart: the *anterior vena cava* (sometimes called the superior vena cava), which drains the head, neck, and arms, and the *posterior vena cava* (or inferior vena cava), which drains the rest of the body.

Let us now retrace the complete circuit traveled by blood. First, it entered the right side of the heart and was pumped to the lungs, where it picked up oxygen and gave up carbon dioxide, and then it returned to the left side of the heart. This portion of the circulatory system is called the *pulmonary circulation* (Fig. 7.16). Note that in the pulmonary circuit the arteries carry deoxygenated blood and the veins carry oxygenated blood. From the left side of the heart, the blood was pumped into the aorta and its numerous branches, from which it moved into capillaries, and then into veins, and finally back in the anterior or posterior vena cava to the right side of the heart. This portion of the circulatory system is called the *systemic circulation.* The arteries of the systemic circulation carry oxygenated blood and the veins carry deoxygenated blood—a reversal of their roles in the pulmonary circulation.

The heart. Let us look more closely at the pumping device, the heart. We have seen that the human heart is, in effect, two hearts in one, since blood in the left side of a normal heart is completely separated from blood in the right side. This type of heart—four-chambered, with complete separation of sides—is characteristic of mammals and birds, the two groups of vertebrates commonly termed "warm-blooded." These animals, which maintain relatively constant high body temperatures, regardless of fluctuations in the environmental temperature, have high metabolic rates and very precise internal control mechanisms. Constant perfusion of the tissues with blood rich in oxygen is clearly essential to them. It would be highly disadvantageous to such animals if the oxygen-rich blood returning to the heart from the pulmonary circulation were mixed with the oxygen-poor blood returning from the systemic circulation. Most "cold-blooded" vertebrates do not have complete separation of the two sides of the heart.

Even though the human heart is double, the two halves beat essentially in unison. The beating is inherent in the heart itself and not dependent upon stimulation from the central nervous system. This can easily be demonstrated. If all nerve connections to the heart are cut, the heart will continue to beat in a normal manner, although the rate of beat may change slightly. As you probably know, the heart of a frog or turtle can continue to beat even after its complete removal from the animal's body, if it is placed in a solution with the proper osmotic concentration. While the initiation of the beat and the beat itself are intrinsic properties of the heart, the rate of beat is partly regulated by stimulation from two sets of nerves.

The initiation of the heartbeat normally comes from a small mass of tissue on the wall of the right atrium near the point where the anterior vena cava empties into it. This mass of tissue, called the sino-atrial node, or *S-A node,* is very unusual and very important. A second mass of *nodal tissue* called the atrio-ventricular node, or *A-V node,* is located in the lower part of the partition between the two atria. A bundle of nodal-tissue fibers runs from the A-V node into the walls of the two ventricles, branching to penetrate into all parts of the ventricular musculature. Nodal tissue, which is unique to the heart, has some of the properties of muscle and some of the properties of nerve; it can contract like muscle and it can transmit impulses like nerve. At regular intervals, a wave of contraction spreads from the S-A node across the walls of the atria. When this wave of contraction reaches the A-V node, the node is stimulated and excitatory impulses are rapidly transmitted from it to the ventricles, which contract. Notice that the atria contract a fraction of a second before the ventricles.

The alternation of systole (contraction) and diastole (relaxation) occurs at an average rate (pulse rate) of about 70 times per minute in a normal human being at rest; there is much individual variation. In the course of the beat, the heart emits several characteristic sounds, which can be heard easily through a stethoscope placed against the chest. Changes in these sounds often indicate to a physician that the heart is defective. For example, if a valve has been damaged and cannot shut completely, a hissing or murmuring sound can be heard as blood leaks backward through the damaged valve. This condition is called a diastolic heart murmur; it is a common result of rheumatic fever and some other diseases.

In the course of contraction, the heart muscle undergoes a series of electrical changes. These changes can be detected by electrodes attached to the skin and can be graphed by a device called an electrocardiograph. Abnormalities in the heart's action alter the pattern of the graph, or electrocardiogram.

Blood pressure. When the left ventricle contracts, it forces blood under high pressure into the aorta, and blood surges forward in each of the arteries. The walls of the arteries are elastic and the pulse wave stretches them. During diastole, the relaxation phase of the heart cycle, the heart is not exerting pressure on the blood in the arteries and the pressure in them falls, but elastic recoil of the previously stretched artery walls maintains some pressure on the blood. There is thus a regular cycle of pressure in the larger arteries, the pressure reaching its high point during systole and its low point during diastole.

Both the systolic and the diastolic pressure are important diagnostic indicators to the physician, as you know. Ordinarily these pressures are measured in the artery of the upper arm. In a normal young adult at rest, the systolic pressure thus obtained averages about 120 mm. of mercury; the diastolic pressure averages 80 mm. The values would not be the same for the lower arm or for the leg or for any other part of the body. The average blood pressure decreases continuously as the blood moves farther and farther away from the heart. Greatest in the part of the aorta close to the heart, it falls off steadily in the more distant parts of the aorta and its branches, falls even more rapidly in the arterioles and capillaries, then declines more slowly in the veins, reaching its lowest point in the veins nearest the heart, where its value may be as low as one atmosphere or even less (Fig. 7.17). The gradual decline of the blood pressure in successive parts of the circuit is the result of friction between the flowing blood and the walls of the vessels. Such a gradient of pressure

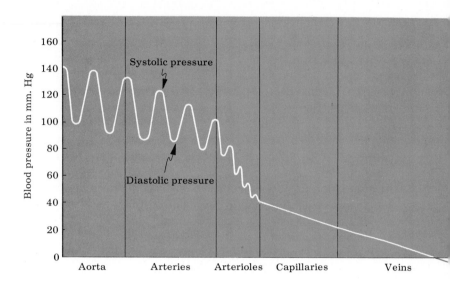

7.17. Graph of blood pressure in different parts of the human circulatory system. In the arteries, there is considerable fluctuation between the systolic pressure and the diastolic pressure. This fluctuation diminishes in the arterioles and no longer occurs in the capillaries and veins. The most rapid fall in pressure is in the arterioles. Pressure in the veins near the heart may fall below zero.

is essential, of course, if the blood is to continue to flow; the fluid can only move from a region of higher pressure toward a region of lower pressure.

The hydrostatic pressure is so low by the time the blood reaches the veins that some other mechanisms besides pressure from the beating heart must be at work in moving the blood. The walls of veins are relatively thin and easily collapsible. When nearby muscles contract as the body moves, they put pressure on the veins, compressing their walls and forcing the fluid in them forward. The fluid can move only toward the heart, because veins are equipped with numerous one-way valves.

Capillary function. The capillaries are so numerous that they penetrate into all parts of every tissue; no cell is far removed from at least one capillary. One worker estimates that there are more than 1,500,000 capillaries per square inch in muscle tissue. Their diameters are very small, being seldom much larger than those of the blood cells that must pass through them. The extensive branching and small diameters of individual capillaries are functionally important in several respects. They ensure not only that all portions of the tissues will be supplied with capillaries, but also that a very great capillary surface area will be available for the exchange process. It has been estimated that every cubic centimeter of blood contacts about 8 square feet of capillary surface each time it passes through a capillary bed! The branching also increases the total cross-sectional area of the system and thus makes blood flow more slowly in the capillaries than in the arteries or veins. This slower flow allows more time for the exchange process. Furthermore, the very small bore of the capillaries results in high frictional resistance to blood flow and causes a considerable drop in blood pressure in the capillary bed. This drop in blood pressure, which plays an extremely important role in the exchange process, deserves a more detailed examination here.

At the arteriole end of a representative capillary, the hydrostatic blood pressure averages about 35 mm. of mercury (Fig. 7.18). The pressure has fallen to about 15 by the time the blood reaches the venule end of the capillary. The hydrostatic pressure tends to force materials out of the capillaries into the surrounding tissue fluid. If this were the only force involved, there would be a steady loss from the blood of both water and those dissolved substances that can readily cross the capillary walls. It can be demonstrated, however, that normally there is relatively little net loss of water from the blood in the capillaries. Clearly, some other force must act in opposition to the hydrostatic force. This other force derives from the difference in osmotic concentration between the blood and the tissue fluid. The blood contains a relatively high concentration of proteins, and these large molecules cannot easily pass through the capillary walls. The same kinds of proteins occur in the tissue fluids, but in much lower concentration. Because of the difference in protein concentration on the two sides of the capillary wall, the blood and tissue fluids will have different osmotic pressures. Normally, the osmotic pressure of the blood is about 25 mm. of mercury higher than that of the tissue fluid, with the result that water tends to move into the capillaries from the tissue fluid by osmosis. We have, then, a system in which hydrostatic pressure developed by the heart tends to force water out of the capillaries and osmotic pressure reflecting differences in protein concentration tends to force water into the capillaries. Obviously, the net movement of water will be determined by the relative magnitudes of these two opposing forces. Notice that at the arteriole end of our representative capillary the hydrostatic blood pressure is 35 and the osmotic pressure is 25. Subtracting one from the other, we find that there is a net pressure of 10 tending to force water out of the capillary. At approximately the midpoint of the capillary, the hydrostatic and osmotic pressures are equal (both being nearly 25) and there is no net movement of water. At the venule end of the capillary, the hydrostatic pressure has fallen to 15, while the osmotic pressure has not changed greatly. Therefore, there is now a net pressure of at least 10 tending to force water into the capillary. In summary, the balance between hydrostatic blood pressure and osmotic pressure is such as to force water out of the capillaries at the arteriole end and into the capillaries at the venule end.

The capillary walls are freely permeable to most of the smaller molecules dissolved in the blood plasma and tissue fluid, and these molecules tend to move with the water in which they are dissolved. Apparently, much of this movement is by bulk flow or filtration rather than by normal diffusion across cell membranes. There is evidence suggesting that much of this filtration is through the intercellular spaces between the cells rather than across the cells themselves. Electron-microscope studies show that transport also takes place in pinocytic vesicles that pinch off from one side of a capillary-wall cell, migrate through the cytoplasm to the other side of the cell, and there release their contents (Fig. 7.19).

There is, then, a net movement out of the capillaries at the arteriole end not just of water but of dissolved ions and nutrients as well. Similarly, there is a net movement into the capillaries at the venule end not only of water but also of waste materials from the tissues and of any special products synthesized by the tissues, such as hormones from the endocrine glands. In short, the blood in the capillaries first unloads materials for the tissues at the arteriole end and then picks up materials for transport at the venule end. In the process, there is normally very little net loss of water, and the blood volume is not appreciably altered.

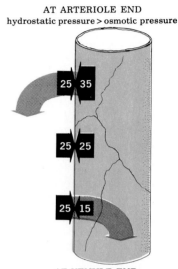

AT ARTERIOLE END
hydrostatic pressure > osmotic pressure

25 35

25 25

25 15

AT VENULE END
osmotic pressure > hydrostatic pressure

7.18. Diagram of forces involved in exchange of materials across capillary walls. In this hypothetical capillary, the hydrostatic pressure of the blood (35) at the arteriole end exceeds the osmotic-pressure difference (25) and therefore materials move out of the capillary. At the venule end, the osmotic-pressure difference (25) exceeds the hydrostatic pressure of the blood (15) and materials enter the capillary.

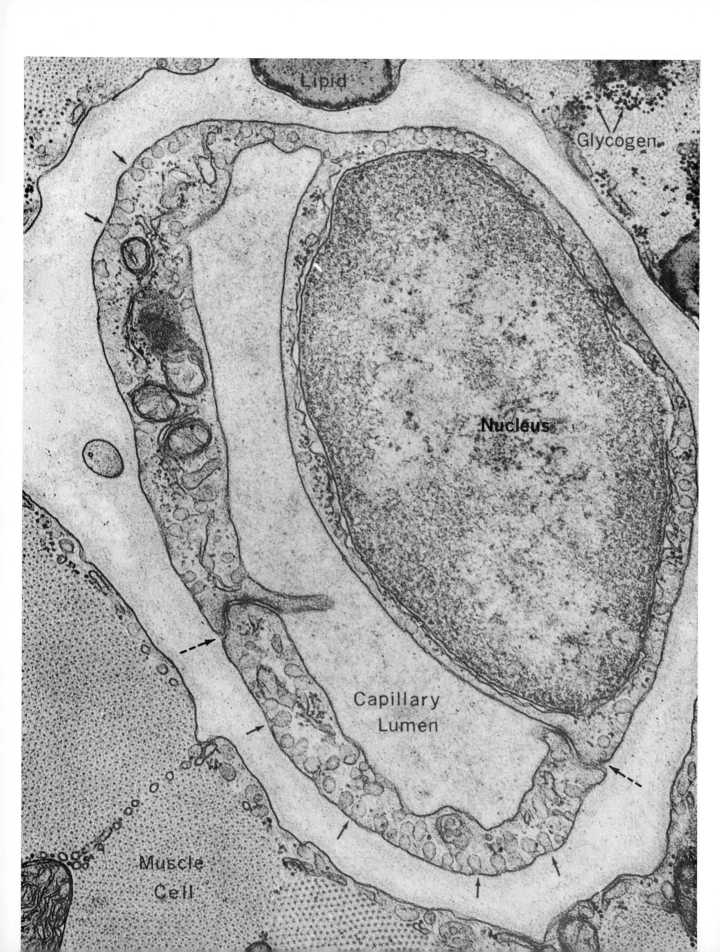

The balance of hydrostatic and osmotic pressures in the capillaries is a very delicate one. Since it plays such an important role in the exchange of materials between the blood and the tissue fluid, any disturbance of it may have profound effects on the condition of the organism. For example, an increase in blood pressure would tend to increase loss of fluid from the blood, while a decrease in blood pressure would have the opposite effect. Such changes in blood pressure could be produced by any one or more of a variety of factors such as changes in rate or strength of heart action, increase or decrease in total blood volume, changes in the elasticity of the walls of the arteries, or increased dilation or constriction of capillaries.

Changes in the relative concentrations of proteins in the blood and in the tissue fluids can also severely alter the balance of forces operating in the capillaries. Numerous experiments have been performed in which the protein concentration in the blood supply to a limb of a frog, cat, or dog was artificially regulated. As was predicted, increasing the protein concentration in the blood decreased loss of fluids from the blood and increased absorption from the tissue fluid. Conversely, decreasing the protein concentration in the blood increased loss of fluids from the blood and decreased reabsorption of fluid from the tissues, the result being an abnormal accumulation of fluid in the tissues.

The lymphatic system. Let us suppose that in a particular tissue of a healthy person there is a slight net loss of fluid from the blood to the tissue and also a slight leakage of proteins from the blood. Can the excess tissue fluid and the protein return to the blood by any other means than direct reabsorption into the blood capillaries? The answer is yes. Vertebrates have a special system of vessels that function in returning materials from the tissues to the blood. These vessels are called lymph vessels, and together they constitute the lymphatic system, which includes lymph veins and lymph capillaries, but no arteries. The lymph capillaries, which like the blood capillaries are distributed throughout most of the body, are closed at one end. They absorb tissue fluid, henceforth known as *lymph,* which slowly flows through them into small lymph veins that unite to form larger and larger veins until finally two very large lymph ducts empty into veins of the blood circulatory system in the upper portion of the thorax near the heart. Besides returning excess tissue fluid and proteins to the blood, the lymph vessels pick up much of the fat absorbed from the intestine. Absorption of fats thus differs from absorption of sugars and amino acids, which are picked up by blood capillaries.

Lymph nodes are present in the lymphatic systems of mammals and some birds, but are absent in most lower vertebrates. Located along major lymph vessels and composed of a meshwork of connective tissue harboring many phagocytic cells, they act as filters and are sites of for-

7.19. Electron micrograph of cross section of capillary. The capillary wall is composed of two cells; the section shows the large nucleus of one of them. The spaces at the junctions between the two cells are quite distinct (dashed arrows). Numerous pinocytic vesicles (solid arrows) can be seen in the cytoplasm of the cells. These vesicles may play a very important role in movement of materials across the wall of the capillary. [Courtesy D. W. Fawcett, Harvard University.]

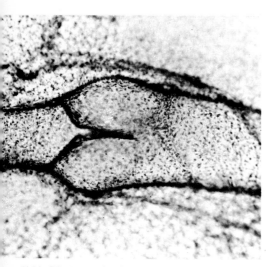

7.20. Photograph of a valve in a lymph vessel. [Courtesy General Biological Supply House, Inc., Chicago.]

A

B

7.21. Erythrocytes. The cells have a biconcave disclike shape. (A) Surface view. (B) Side view.

mation of certain types of white blood cells. As the lymph trickles through the nodes, it is filtered and such particles as dead cells, cell fragments, and invading bacteria are destroyed by the phagocytic cells.

Since the lymphatic system is not connected to the arterial portion of the blood circulatory system, it is obvious that lymph is not moved by hydrostatic pressure developed by the heart. Its movement, like that of blood in the veins, is due to the contractions of muscles that press on the lymph vessels and push the lymph forward past one-way valves (Fig. 7.20).

Blood

We have discussed the routes followed by the circulating blood, the mechanism of circulation, and the process of exchange of certain materials with the tissue fluid. We should now examine the blood itself in more detail. It is one of the most important and unusual tissues in the animal body.

Composition of the blood. In Chapter 3, we classified blood as a type of connective tissue with a liquid matrix. The extracellular liquid matrix of blood is called *plasma.* Suspended in the plasma are the formed elements, which are of three major types in vertebrates: (1) the red blood cells or *erythrocytes,* (2) the white blood cells or *leukocytes,* and (3) the *platelets,* which are small disc-shaped bodies that arise as cell fragments. Normally, the formed elements constitute about 40–50 percent of the volume of whole blood, while the plasma constitutes the other 50–60 percent.

The basic solvent of the plasma is, of course, water, which constitutes roughly 90 percent of the plasma. A great variety of substances are dissolved in the water; the relative concentrations of these vary with the condition of the organism and with the portion of the system under examination. The solutes can be divided into six categories: (1) inorganic ions and salts, which make up about 0.9 percent by weight of the plasma of mammals; (2) plasma proteins, which constitute 7–9 percent by weight of the plasma; (3) organic nutrients, including glucose, fats, phospholipids, amino acids, and lactic acid; (4) nitrogenous waste products; (5) special products being transported (e.g. hormones); and (6) dissolved gases, including oxygen and carbon dioxide, and also nitrogen, which diffuses into the blood in the lungs but seems to be physiologically inert.

Erythrocytes and their function. Human erythrocytes or red blood cells are small, biconcave, disc-shaped cells (Fig. 7.21) that lack nuclei. Normally, there are roughly five million of them per cubic millimeter of blood. Though the number of red cells remains amazingly constant from day to day, there is continual destruction of some cells and formation of new ones; the normal survival time of an erythrocyte is 120 days.

The erythrocytes of adults are formed in the red bone marrow. Though the mature erythrocytes of mammals are devoid of nuclei, mitochondria, Golgi apparatus, etc., and therefore lack many of the characteristics of living cells, they arise from normally nucleated, rapidly dividing connective-tissue cells of the bone marrow. Toward the end of their development, they lose their nuclei and acquire the red oxygen-carrying pigment called *hemoglobin.* They then enter the circulating blood. The developmental sequence in vertebrates other than mammals is somewhat different in that their mature erythrocytes retain nuclei. It has been suggested that the evolutionary loss of the nuclei in mammals has the adaptive advantage of leaving room for more hemoglobin in each cell. Extra hemoglobin might, in turn, be correlated with the high metabolic rates—and therefore high oxygen demands—of the tissues of homeothermic (warm-blooded) animals.

A hemoglobin molecule is a protein enfolding four prosthetic groups, each with an iron atom at its center. Each of these iron atoms is capable, by virtue of its structural relationships within the hemoglobin molecule, of combining loosely with one molecule of oxygen (O_2). Oxygenation makes the hemoglobin a brighter red; hence the typical difference in color between systemic arterial and venous blood. Though many invertebrates have hemoglobin, many others have different oxygen-transporting pigments, which are like hemoglobin in that they all contain a metal combined with protein. For example, many molluscs and arthropods have a pigment called hemocyanin, which contains copper instead of iron; when oxygenated, it is blue instead of red. Hemocyanin never occurs in cells; it is dissolved in the plasma of the animals that have it.

Of the invertebrates that have hemoglobin, some have it in cells, like vertebrates, but many simply have it dissolved in the plasma. Though a hemoglobin molecule can function just as well in the plasma as in an erythrocyte, there is a decided adaptive advantage, in animals with high metabolic rates, for the pigment to be in cells; more pigment molecules can then be carried per unit volume of blood, and the oxygen-transporting capacity is correspondingly increased. If all the hemoglobin in the erythrocytes of a human being were in his plasma instead, the high concentration of protein would not only have a profound effect on the osmotic balance, but would make it impossible for the heart to force the blood through the vessels, for it would be as thick as syrup. Erythrocytes, then, are a convenient method of packaging large amounts of hemoglobin with relatively little disturbance of the viscosity and the osmotic concentration of the blood.

There are other gases besides oxygen that will also bind loosely to hemoglobin. One that binds even more readily than oxygen is carbon monoxide (CO). This gas, common in coal gas used for heating and cooking, in the exhaust from automobiles, and in tobacco smoke, is dangerous because even when its concentration in the air is relatively low such a high percentage of the hemoglobin may bind with it that not enough is left to carry sufficient oxygen to the tissues. Severe symptoms of asphyxiation (impairment of vision, hearing, and thought) or even death may thus result from exposure to carbon monoxide.

The blood not only transports oxygen from the lungs to the tissues, but it also has the very important function of transporting carbon dioxide in the reverse direction, from the tissues to the lungs. Some of the CO_2 is carried in the plasma, some in loose combination with hemoglobin in the red cells. It is at lower pressure in the lungs than in the blood, and the gradient therefore favors its release.

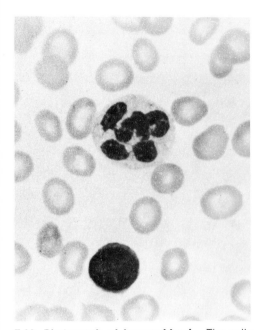

7.22. Photograph of human blood. The cells without nuclei are erythrocytes. The two large nucleated cells are leukocytes. × 1,000. [Courtesy M. S. Greenberg, Tufts University.]

Leukocytes and their functions. Human leukocytes (white blood cells) are larger than the erythrocytes and have large, often irregularly shaped nuclei (Fig. 7.22). They are not restricted to the blood, being even more abundant in the lymphatic system. And they are also found wandering free in loose connective tissue and occasionally in other tissues. Capable of amoeboid movement, they can escape from the blood and lymph vessels by squeezing through the vessel walls at the points of contact between cells. In essence, then, all connective tissues, including blood and lymph, form one continuous system so far as the leukocytes are concerned.

Leukocytes play a very important role in the body's defenses against disease and infection. Apparently, both damaged tissues and invading bacteria release chemicals that attract leukocytes, some of which act as phagocytes engulfing and destroying bacteria and remnants of damaged tissue cells. In a severe infection, the leukocyte count in the blood and lymph increases enormously, and vast numbers of them may invade the infection area.

Certain nonphagocytic leukocytes known as *lymphocytes* give rise to specialized cells called *plasma cells,* which play a central role in immunologic reactions. Plasma cells respond to the presence of certain kinds of foreign substances called *antigens,* which are usually (but not always) proteins, by synthesizing special types of proteins called *antibodies* that destroy or inactivate the antigens. Each type of antibody is usually very specific and will inactivate only the antigen that stimulated its synthesis. Whereas phagocytosis is particularly important as a first defense against acute infections, production of antibodies confers a degree of active immunity and is critical both in fighting long-term chronic infections and in building resistance against further infection.

REFERENCES

BEST, C. H., and N. B. TAYLOR, 1966. *The Physiological Basis of Medical Practice,* 8th ed. Williams & Wilkins, Baltimore. (See esp. Chapters 22–49.)

BOLLARD, E. G., 1960. "Transport in the Xylem," *Annual Review of Plant Physiology,* vol. 11. Annual Reviews, Inc., Palo Alto, Calif.

PROSSER, C. L., and F. A. BROWN, 1961. *Comparative Animal Physiology,* 2nd ed. Saunders, Philadelphia. (See esp. Chapters 8, 13.)

WINTON and BAYLISS, 1969. *Human Physiology,* 6th ed., rev. and ed. by O. C. J. Lippold and F. R. Winton. Williams & Wilkins, Baltimore. (See esp. Chapters 3, 5–11.)

ZIMMERMANN, M. H., 1960. "Transport in the Phloem," *Annual Review of Plant Physiology,* vol. 11. Annual Reviews, Inc., Palo Alto, Calif.

SUGGESTED READING

ADOLPH, E. F., 1967. "The Heart's Pacemaker," *Scientific American,* March. (Offprint 1067.)

BIDDULPH, O., and S. BIDDULPH, 1959. "The Circulatory System of Plants," *Scientific American,* February. (Offprint 53.)

CARLSON, A. J., V. JOHNSON, and H. M. CAVERT, 1961. *The Machinery of the Body,* 5th ed. University of Chicago Press, Chicago. (See esp. Chapters 3–5.)

CRAFTS, A. S., 1961. *Translocation in Plants.* Holt, Rinehart & Winston, New York.

KILGOUR, F. G., 1952. "William Harvey," *Scientific American,* June.

MAYERSON, H. S., 1963. "The Lymphatic System," *Scientific American,* June. (Offprint 158.)

RAY, P. M., 1971. *The Living Plant,* 2nd ed. Holt, Rinehart & Winston, New York. (See esp. Chapters 5, 7.)

STEWARD, F. C., 1964. *Plants at Work.* Addison-Wesley, Reading, Mass. (See esp. Chapter 10.)

WOOD, J. E., 1968. "The Venous System," *Scientific American,* January. (Offprint 1093.)

ZIMMERMANN, M. H., 1963. "How Sap Moves in Trees," *Scientific American,* March. (Offprint 154.)

ZWEIFACH, B. W., 1959. "The Microcirculation of the Blood," *Scientific American,* January. (Offprint 64.)

Regulation of body fluids

Evidence of many types has led biologists to the conclusion that life had its origin in the ancient seas. Of the major environmental media of the earth—sea water, fresh water, air—sea water exhibits by far the greatest stability. In such crucial characteristics as temperature, acidity, and salt concentration, the seas fluctuate remarkably little over immense spans of time, their vast bulk making any change very gradual and slow.

We have already seen that a living cell interacts constantly with its surrounding environmental medium. Such critical functions as nutrient procurement, gas exchange, metabolism—indeed life itself—are closely dependent upon the properties of the surrounding medium. It is not surprising, therefore, that the protoplasm of the early cells had many characteristics in common with the sea water that bathed them, and that the life processes evolved a close dependence on the stable conditions existing in sea water. Similarly, it is not surprising that the evolution of complex multicellular marine animals involved the development of body fluids—tissue fluid, blood, etc.—that could provide even the innermost body cells with a relatively nonfluctuating aquatic environment, and that the internal body fluids of those primitive marine animals resembled in many important ways the sea water that had been the cradle of life.

As the ages passed and evolution continued, the body fluids of different organisms, like their other characteristics, evolved in different ways. Comparison of the chemical makeup of the body fluids of a variety of present-day marine animals reveals many differences; even more noticeable differences are found if the comparison is extended to fresh-water and terrestrial animals, and very great differences indeed if it is extended to plants. Thus we should not make the mistake of exaggerating the similarities of these fluids. Nonetheless, it remains true that all of them have much in common, and that, as Ernest Baldwin of Cambridge

University has said, "The conditions under which cell life is possible are very restricted indeed and have not changed substantially since life first began." The evolutionary development of the immense diversity now seen among living organisms has necessarily involved the concomitant evolution of mechanisms for maintaining within each organism a fluid environment with the properties requisite for the continued life of its cells.

THE EXTRACELLULAR FLUIDS OF PLANTS

Multicellular marine algae differ greatly from multicellular marine animals in the sort of fluid environment to which their cells are exposed. Roughly 50 percent of the water in the body of a complex animal is extracellular, being in the form of tissue fluid, lymph, or blood plasma. This extracellular body fluid, which bathes most of the cells, is separated from the environmental water by cellular barriers and has a characteristic composition, differing from both that of the intracellular fluid and that of the surrounding water. By contrast, most of the fluid content of a multicellular alga is intracellular. The fluid filling its intercellular spaces is essentially continuous with the environmental water and cannot be regarded as separate or distinct. The alga thus has no fluid that is fully analogous to the tissue fluid and blood of an animal. Hence, unlike the animal, which must regulate the composition of both intracellular and extracellular fluids, the alga must regulate the composition only of its intracellular fluids.

A similar contrast appears between an animal and a large vascular land plant. Such a plant obviously contains much extracellular fluid in the form of xylem sap and the water imbibed in its cell walls. But this fluid is not as fully distinct from the environmental water as the tissue fluid and blood of an animal. You will recall that water can penetrate far into the cortex of a root by flowing along cell walls without having to cross any membranous barrier. Thus, much of the fluid that directly bathes the plant cells, even those far inside the plant body, is essentially continuous with the environmental water and is therefore not fully analogous to animal tissue fluid, which is separated from the environmental medium by a membranous barrier. This means, of course, that the composition of much of the extracellular fluid of the plant cannot be as well regulated as the tissue fluid and blood of animals.

We can easily understand why the inability of marine algal cells to regulate the composition of the fluid that bathes them poses no serious problem for the life of the cells; that fluid, after all, is essentially the same as sea water, the nonfluctuating medium in which life arose. (It is true that modern sea water has a composition quite different from that of the ancient seas, but the change was so very gradual that the organisms living in the seas had ample time to evolve with their evolving environment.) But what about a plant living in fresh water or on land? The fluids to which the internal cells of these plants are exposed will fluctuate much more than the tissue fluids of animals, and even their normal composition is one that would be quickly fatal to the internal cells of most higher animals. The reasons why plant cells seem able to withstand much greater fluctuations in the makeup of the fluids bathing them than animal cells are complex, and can be explained only in part.

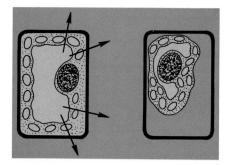

8.1. Plasmolysis. A plant cell in a hyperosmotic medium will lose so much water (left) that, as it shrinks, it will pull away from its more rigid wall (right).

Animal cells are seriously affected by changes in the osmotic concentration of the extracellular body fluids, which, under normal conditions, are approximately isosmotic with the cells; osmotic shifts often severely alter the physiology of the cells or even kill them. Unlike the animal cell, the cell of a land or fresh-water plant almost always exists in a medium much more dilute than the cell's contents. In other words, the plant cell is decidedly hyperosmotic relative to the fluid that bathes it. In such a situation, an animal cell would take in so much water by osmosis that it would burst, unless it had some special mechanism for expelling the excess water. But the plant cell is surrounded by its cell wall, and as the cell takes in more water and becomes more turgid, the wall pressure becomes greater and resists further expansion. Eventually the wall pressure becomes as great as the opposing osmotic pressure, and then no further net gain of water by the cell is possible.

The plant cell, then, can withstand rather pronounced changes in the osmotic concentration of the surrounding fluids as long as those fluids remain more dilute than the cell's contents, i.e. as long as the fluids remain appreciably hypoosmotic relative to the cell. If the external fluids become decidedly hyperosmotic relative to the cell, the cell may lose so much water and shrink so grievously that it pulls away from its more rigid wall, a process known as *plasmolysis* (Fig. 8.1). The presence of the cell wall in plants and its absence in animals thus make the problem of salt and water balance quite different in these two types of cells. Although changes in the osmotic concentration of the surrounding medium affect plant cells far less than animal cells, changes in the concentration of individual ions may have pronounced effects on their health and growth. These effects are usually attributable to an alteration in the chemical makeup of the plant.

THE VERTEBRATE LIVER

As a first example of the problems involved in keeping the internal fluids of complex animals relatively constant in composition, consider the blood leaving the intestinal capillaries of your body shortly after you have eaten a meal. Digestion is taking place in the small intestine, and the products of digestion are moving in large quantities into the capillaries of the intestinal villi. This means that the blood leaving these capillaries contains high concentrations of such compounds as simple sugars and amino acids—concentrations considerably greater than normally found in the blood in most parts of the circulatory system. But wholesale addition of these materials to the blood, if not controlled, would drastically alter the composition of the blood and other body fluids and make impossible the maintenance of a relatively nonfluctuating fluid environment for the cells.

Your body and the bodies of other vertebrates meet this difficulty with the help of a very important organ, the liver. Blood from the intestine and stomach is collected in the *portal vein,* which does not empty into the vena cava as might be expected, but goes to the liver, where it breaks up into a network of capillaries in the liver tissue (Fig. 8.2). The liver is one of only three places in your body where blood passes through a second set of capillaries before returning to the heart; other blood circuits involve only a single capillary bed.

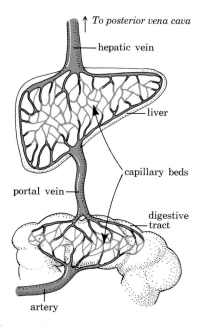

8.2. The hepatic portal circulation. Blood from the intestinal capillaries is carried by the portal vein to a second bed of capillaries in the liver.

The liver's role in regulation of the blood-sugar level. After a meal, the blood coming to the liver via the portal vein has a higher than normal concentration of sugar—or, more specifically, glucose. Under these conditions, the liver removes most of the excess, converting it into the insoluble polysaccharide glycogen, which is the principal storage form of carbohydrate in animal cells. Thus the blood leaving the liver via the *hepatic vein* (which leads into the posterior vena cava) contains a concentration of glucose only slightly higher than that normally found in the arteries. After this blood is mixed in the vena cava with blood from other parts of the body—blood that contains a lower concentration of sugar because it has given up glucose to the tissues through which it has passed—the blood entering the heart has a glucose concentration within the normal tolerance range. If the incoming supply of glucose exceeds all the body's immediate needs, and the liver has stored its full capacity of glycogen, the liver begins converting glucose into fat, which can then be stored in the various regions of adipose tissue throughout the body. Thus, in spite of the great quantities of sugar absorbed by the intestine, the blood-sugar level in most of the circulatory system is not greatly raised.

The whole process is reversed if some time has elapsed since the last meal and no sugar is being absorbed from the intestine. At such times, blood in the intestinal capillaries gives up glucose to the intestinal cells, just as blood in any capillary bed will give up nourishment to the surrounding tissues. This means that the blood reaching the liver via the portal vein is poor in glucose. Under these conditions, the liver converts some of its stored glycogen into glucose and adds it to the blood. The result is that blood leaving the liver in the hepatic vein has its normal glucose concentration, and that the blood entering the heart likewise has the normal blood-sugar level. We see, then, that the liver functions in helping maintain the blood-sugar concentration in a steady state.

The human liver is capable of storing enough glycogen to supply glucose to the blood for a period of about 24 hours. What happens if at the end of this period no new glucose has come to the liver from the intestine? A drop in the blood-sugar concentration to a level much below normal would soon be fatal; the brain cells are particularly sensitive to such a drop. Under such conditions, the liver begins converting other substances, such as amino acids, into glucose, and in this way maintains the normal blood-sugar level.

The liver's activity in carbohydrate metabolism is regulated in a complex fashion by several hormones, as will be described in Chapter 9. An abnormal balance of these hormones may result in an unusually high or unusually low blood-sugar level. Either condition can be dangerous.

The liver's role in the metabolism of amino acids. Like the sugar, the amino acids absorbed by the villi of the intestine pass into the portal vein and thence to the liver. The liver removes many of these amino acids from the blood, temporarily storing small quantities and later gradually returning them to the blood, which carries them to other tissues for use in the synthesis of enzymes, hormones, or new protoplasm. But the usual diet contains far more amino acid than can be utilized in such syntheses. Since the animal body, unlike the plant, is capable of very little long-term storage of amino acids, the excess must be converted into other substances such as glucose, glycogen, or fat. Such conversions take place in the liver.

You will recall that amino acids differ from carbohydrates and fats in always containing nitrogen in the form of an amino group ($-NH_2$). It is not surprising, therefore, that the first step in converting amino acids into these other substances is *deamination,* or the removal of the amino group. In the deamination reaction, the amino group is converted into *ammonia* (NH_3). The livers of some animals simply release this ammonia, which is a waste material, into the blood, and it is soon removed from the blood and from the body of the organism by excretory mechanisms. The livers of many other animals, including man, first combine the ammonia with carbon dioxide to form a more complex but less toxic nitrogenous compound called *urea,* and then release the urea into the blood. The livers of still other animals convert the waste ammonia into a compound more complex than urea, called *uric acid,* and release this into the blood. In short, whether the nitrogenous waste product is ammonia, urea, uric acid, or some other compound, the liver dumps it into the blood, and it becomes necessary for another system of the body to prevent the wastes from reaching too high a concentration in the body fluids.

Notice that animals differ markedly from green plants in being unable to re-use much of the nitrogen from the amino acids they metabolize. Green plants can shift nitrogen-containing groups from one organic compound to another more freely than animals, and they can also utilize inorganic nitrogen to synthesize organic nitrogen-containing compounds. Hence excretion of nitrogenous wastes is essentially an animal activity.

THE PROBLEM OF EXCRETION AND SALT AND WATER BALANCE IN ANIMALS

We have seen that animals need mechanisms for ridding their bodies of metabolic wastes—particularly nitrogenous ones, but many others as well. The process of releasing such useless substances is called excretion. In general, excretory mechanisms also serve a second very important function: They help regulate the water and salt balance of the organism. Our examination of excretion will be focused on both these aspects, which in most cases are inextricably intertwined.

The problem in aquatic animals

As we saw, the first nitrogenous waste formed by deamination of amino acids is ammonia. Now, ammonia is an exceedingly poisonous compound, and no organism can survive if its concentration in the body fluids gets very high. But the small, highly soluble molecules of ammonia readily diffuse across cell membranes, and there is no great difficulty in getting rid of them if an adequate supply of water is available. The water keeps the solution dilute while the ammonia is in the body, acts as a vehicle for the expulsion of the ammonia from the body, and flushes the ammonia rapidly away from the vicinity of the animal. In view of the plentiful supply of water available to aquatic animals, it is not strange that for many of these the characteristic nitrogenous excretory product is ammonia.

Marine invertebrates. Many marine invertebrates lack special excretory systems, relying instead on release of wastes across the general surface membranes. Such organisms seldom have any problem with water balance, because they are essentially isosmotic with the surrounding sea water, and hence neither take in much excess water nor lose too much.

Maintenance of the proper nonfluctuating internal fluid environment is relatively simple for marine invertebrates as long as they remain in the sea; it is quite a different matter when they move into hypoosmotic media such as the brackish water of estuaries or the fresh water of rivers and lakes. Many marine animals are incapable of moving into such habitats. Since their body fluids always lose salts until they have about the same salinity and osmotic concentration as the external fluids, and since their cells generally cannot tolerate much change in the makeup of the fluids bathing them, these animals soon die when they are put into brackish or fresh water.

Some marine animals, however, have evolved adaptations that enable them to move into hypoosmotic media. The adaptations may be of an evasive character, as in oysters and clams, which simply close their shells and thereby exclude the external water during those parts of the tidal cycle when the water in the estuaries is very dilute. But by far the most important adaptations for survival in dilute media—and the ones that have played the principal role in the evolutionary movement of animals into fresh water—are those that enable animals to regulate the osmotic concentrations of their body fluids and keep them constant despite changes in the external medium. Such organisms are said to have the power of *osmoregulation.* They generally have mechanisms for secreting salt into their blood from the surrounding medium and for bailing out excess water.

Fresh-water animals. Once the ancestors of the modern fresh-water animals had made the transition to the fresh-water environment, natural selection seems to have favored a reduction of the osmotic concentration of the body fluids within the bounds possible for the continuance of the life of the tissues. Modern fresh-water animals, both invertebrate and vertebrate, have osmotic concentrations decidedly lower than sea water. It seems incompatible with cellular existence, however, for the body fluids to be as dilute as fresh water, for no organisms are actually isosmotic with their fresh-water medium.

Because fresh-water animals are hyperosmotic relative to the surrounding environmental medium, they have a strong tendency to gain water and to lose salts. In compensation, they usually have excretory organs that can pump out water as fast as it floods in—preferably through the production of urine more dilute than the body fluids—and/or special secretory cells somewhere on the body that can absorb salts from the environment and release them into the body. Both corrective measures—production of dilute urine and absorption of salts—entail movement of materials against concentration gradients and therefore necessitate expenditure of energy in the doing of osmotic work.

REGULATION OF BODY FLUIDS

An examination of the water and salt regulation typical of modern fresh-water bony fishes will provide a good example of the above-mentioned processes. The blood and tissue fluids of the fish are more concentrated than the environmental water. Although the fish almost never drink, there is a constant osmotic intake of water across the membranes of the gills and of the mouth, and a constant loss of salts across the same membranes. Osmoregulatory correction is utilized in two ways: The excess water is eliminated in the form of very dilute and copious urine produced by the kidneys, and salts are actively absorbed by specialized cells in the gills (Fig. 8.3).

Marine bony fishes. Curiously enough, bony fishes living in the sea have the reverse problem: They live in water, yet they steadily lose water to their environment and are in constant danger of dehydration. The explanation is that the ancestors of the bony fishes apparently lived in fresh water, not in the sea, and that when some of their descendants moved to the marine environment they retained their dilute body fluids. Thus marine bony fishes are hypoosmotic relative to the surrounding water, and they have the problem of excessive water loss and excessive salt intake. Their solution is to drink almost continuously and actively excrete salt in very concentrated form by means of specialized cells in the gills (Fig. 8.3). Most of the nitrogenous wastes are excreted as ammonia through the gills; hence only a small quantity of urine is produced by the kidneys, and little water need be lost in this manner. Apparently, fish kidneys have not evolved the capacity to produce concentrated urine, and they are consequently of no help in salt elimination.

The problem in terrestrial animals

On land, the greatest threat to life is desiccation. Water is lost by evaporation from the respiratory surfaces (lungs, tracheae, etc.), by evaporation from the general body surface, by elimination in the feces, and by excretion in the urine. The lost water must obviously be replaced if life is to continue. It is replaced by drinking, by eating foods containing water, and by the oxidation of nutrients (remember that water is one of the products of cellular respiration).

We saw that ammonia is a satisfactory nitrogenous excretory product for aquatic animals. It is far from satisfactory for terrestrial ones, because of the difficulty of getting rid of this highly toxic substance on land, where an unlimited water supply is not available. Amphibians and mammals rapidly convert ammonia into urea, a compound that, though very soluble, is relatively nontoxic. Urea can remain in the body for some time before being excreted, and we can regard its production as an adaptation to the conditions of water shortage characteristic of terrestrial existence.

Although urea is a far more satisfactory excretory product than ammonia for land animals, it has the disadvantage of draining away some of the critically needed water, for this highly soluble compound must be released from the body in an aqueous solution. If, however, uric acid, a very insoluble compound, is excreted instead of urea, almost no water need be lost. It is not surprising, therefore, that many terrestrial animals—most reptiles, birds, insects, and land snails—excrete uric acid

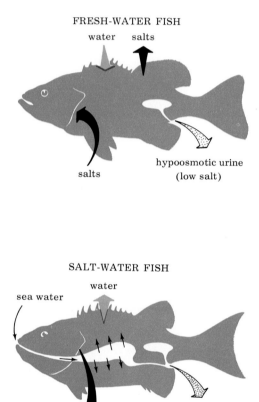

FRESH-WATER FISH

water   salts

salts

hypoosmotic urine
(low salt)

SALT-WATER FISH

water

sea water

salts

isosmotic urine

8.3. Osmoregulation in bony fishes. Fresh-water fishes, which are hyperosmotic relative to the water in which they live, tend to take in excessive amounts of water and to lose too much salt. They compensate by seldom drinking, by actively absorbing salts through specialized cells on their gills, and by excreting copious dilute urine. Salt-water fishes, which are hypoosmotic relative to sea water, tend to lose too much water and to take in too much salt. They compensate by drinking constantly and by actively excreting salts across their gills. They cannot produce hyperosmotic urine; hence the kidneys are of little aid to marine fishes in osmoregulation.

or its salts. The excretion of this substance not only allows them to conserve water, but has another advantage perhaps even more important in the evolution of uric acid metabolism. All these animals lay eggs enclosed within a relatively impermeable shell or membrane. If the embryos excreted ammonia, they would rapidly be poisoned, and if they produced urea, the concentration in the egg by the latter part of development would become decidedly harmful. Uric acid, on the other hand, is so insoluble that it can be precipitated in almost solid form and stored in the egg without harmful toxic or osmotic effects. In the nitrogen metabolism of fully terrestrial animals, uric acid excretion is correlated with egg laying, while urea excretion is correlated with viviparity (giving birth to living young).

EXCRETORY MECHANISMS IN ANIMALS

Contractile vacuoles

Special excretory structures are absent in many unicellular and simple multicellular animals. Nitrogenous wastes are simply excreted across the general cell membranes into the surrounding water. Some Protozoa do, however, have a special excretory organelle, the contractile vacuole (Fig. 8.4). Each vacuole goes through a regular cycle, filling with liquid, becoming larger and larger, and finally contracting and ejecting its contents from the cell. Although there is now evidence that contractile vacuoles excrete some nitrogenous wastes, it seems clear that their primary function is elimination of excess water.

Flame-cell systems

The beginnings of a tubular excretory system can be seen in the flatworms (planaria, flukes, tapeworms, etc.). These animals are relatively small and lack a functional body cavity; i.e. there is no major break in the tissue mass between the outer epithelium of the body and the gastrovascular cavity. There is no circulatory system.

Flatworm excretory systems usually consist of two or more longitudinal branching tubules running the length of the body. In planaria and its relatives, the tubules open to the body surface through a number of tiny pores (Fig. 8.5). The critical portions of the system are many small bulblike structures located at the ends of side branches of the tubules. Each bulb has a hollow center into which a tuft of long cilia projects. The hollow centers of the bulbs are continuous with the cavities of the tubules. Water and some waste materials move from the tissue fluids into the bulbs. The constant undulating movement of the cilia creates a current that moves the collected liquid through the tubules to the excretory pores, where it leaves the body. The motion of the tuft of cilia resembles the flickering of a flame, and for this reason this type of

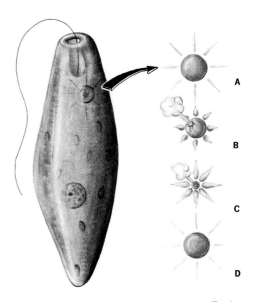

8.4. Contractile vacuole of *Euglena*. *Euglena* is a unicellular organism with some animal-like and some plantlike characteristics: It swims by means of a long flagellum, and it has chlorophyll. Its contractile vacuole removes the excess water that tends to move into the cell. The vacuole fills with water led into it by a system of radiating canals (A). When full, the vacuole contracts, expelling the water from the cell (B, C).

8.5. Flame-cell system of planaria. Each excretory canal, or tubule, forms a longitudinal network with numerous offshoots, of which some end in flame cells (two are shown enlarged at right) and others in excretory pores. The cilia in the flame cells create currents that move water and waste materials through the canals and out through the pores.

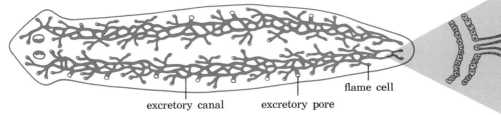

flame cell

excretory canal   excretory pore

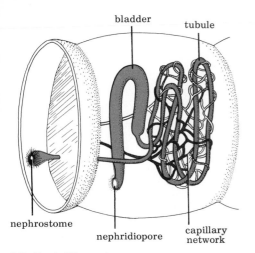

8.6. Nephridium of an earthworm. The open nephrostome of each nephridium is located in the segment ahead of the one containing the rest of that nephridium. The tubule penetrates through the membranous partition between the two segments and is then thrown into a series of coils, with which a network of blood capillaries is closely associated. The tubule empties into a storage bladder that opens to the outside through a nephridiopore.

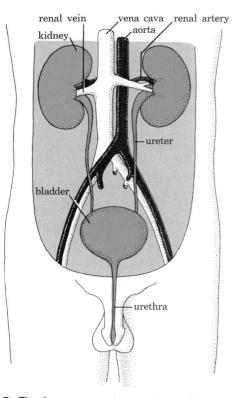

8.7. The human excretory system. (The organs and vessels are shown larger relative to the body than they actually are.)

excretory system is often called a flame-cell system. Like the contractile vacuoles discussed earlier, flame-cell systems seem to function primarily in the regulation of water balance; most metabolic wastes of flatworms are excreted from the tissues into the gastrovascular cavity and eliminated from the body through the mouth.

Nephridia of earthworms

Note that flame-cell systems are found in animals without circulatory systems, and that, as a result, they pick up substances only from the tissue fluids. In animals that have evolved closed circulatory systems, the blood vessels have become intimately associated with the excretory organs, making possible direct exchange of materials between the blood and the excretory system.

The earthworm is an example of an animal in which the circulatory system plays a critical role in excretion. Its body is composed of a series of segments internally partitioned from each other by membranes. In general, each of the compartments thus formed has its own pair of excretory organs, called nephridia, which open independently to the outside. A typical nephridium (Fig. 8.6) consists of an open ciliated funnel, or nephrostome (which corresponds functionally to the bulb of a flame-cell system), a coiled tubule running from the nephrostome, an enlarged bladder into which the tubule empties, and a nephridiopore through which materials are expelled from the bladder to the outside. Blood capillaries form a network around the coiled tubule. Materials move from the body fluids into the nephridium through the open nephrostome, but some materials are also picked up by the coiled tubule directly from the blood in the capillaries. There is probably also some reabsorption of materials from the tubule into the blood capillaries. The principal advance of this type of excretory system over the flame cell, then, is the association of blood vessels with the coiled tubule.

The vertebrate kidney

Structure of the kidney. Like the nephridial system of earthworms, the excretory systems of vertebrates are closely associated with a closed circulatory system. When an efficient circulatory system can bring wastes to the excretory organs, the functional excretory units no longer have to be scattered throughout the body tissues, as in planaria. And the absence of internal segmentation of the body obviates the need for a series of individual excretory organs, as in earthworms. Higher vertebrates have typically evolved compact discrete organs, the kidneys, in which the functional units are massed. In man, the kidneys are located in the back of the abdominal cavity (Fig. 8.7).

The functional units of the kidneys of higher vertebrates are called *nephrons* (Fig. 8.8). Each nephron consists of a closed bulb called a *Bowman's capsule* and a fairly long coiled tubule. The tubules of the various nephrons empty into collecting tubules, which in turn empty into the central cavity of the kidney, the pelvis. From the pelvis, a large duct leaves each kidney and runs posteriorly. In mammals, these ducts, called *ureters,* empty into the *urinary bladder.* This storage organ drains to the outside via another duct, the *urethra.*

Blood capillaries and the capsules and tubules of the nephrons are intimately associated in the modern vertebrate kidney. No longer are materials picked up from the general body fluids; exchange of substances takes place almost exclusively between blood capillaries and nephrons. Blood reaches each kidney via a *renal artery,* a short vessel leading directly from the aorta to the kidney. The renal artery enters the kidney at its median depression and then breaks up into many tiny branch arterioles, each of which penetrates into a cuplike depression in the wall of a Bowman's capsule. Within the capsule, the arteriole breaks up into a tuft of capillaries called the *glomerulus* (Fig. 8.8). Blood leaves the glomerulus via an arteriole formed by the rejoining of the glomerular capillaries. After emerging from the capsule, the arteriole promptly divides again into many small capillaries that form a second dense network around the tubule of the nephron. Finally, these capillaries unite once more to form a small vein. The veins from the many nephrons then fuse to form the *renal vein,* which leads from the kidney to the posterior vena cava. The kidney, you will note, is the second place where we have encountered blood circuits that involve two sets of capillaries.

The formation of urine.   With the structural relationships in mind, we are now in a position to examine the mechanism of urine formation in the human kidney. In 1844 the German physiologist Carl Ludwig suggested that a glomerulus acts as a simple mechanical filter— that molecules small enough to pass through the capillary walls and through the thin membranous walls of the Bowman's capsule filter from the blood into the nephron as a result of the high hydrostatic pressure in the glomerulus. In confirmation of this theory, it has been shown that the liquid entering the lumen of the nephron has basically the same percentage composition as blood, lacking only the formed elements and the plasma proteins, both of which are too large to filter through the membranes to any appreciable extent. It has been demonstrated, further, that the cells of the glomerular capillaries and of the Bowman's capsules do not carry out active transport in the movement of materials from the glomeruli into the capsules; the work involved is performed by the beating heart as it drives the blood under high hydrostatic pressure into the glomeruli.

If the filtrate were expelled from the body without modification, many very valuable, indeed essential, substances would be lost and the process would be wasteful in the extreme. If we consider water alone, it is overpowering to contemplate the drinking that would be necessary to replace the 180 quarts of filtrate formed every day in the average person's kidneys! Selective reabsorption of most of the water and many of the dissolved materials is one of the functions of the tubules of the nephrons. In man, the filtrate passes first through the *proximal convoluted tubule,* then through the long *loop of Henle,* then through the *distal convoluted tubule,* and finally into the *collecting tubule* (Fig. 8.8). As the filtrate moves through the tubules, as much as 99 percent of the water may be reabsorbed by the cells of the tubule walls and returned to the blood in the capillary network. Thus the kidneys can produce concentrated urine, i.e. urine that is hyperosmotic relative to the blood plasma even though the initial filtrate was isosmotic. Whether the collecting tubule finally releases dilute or concentrated urine depends on whether there is a deficiency or an excess of water in the body at the moment; the pituitary gland responds to changes in the

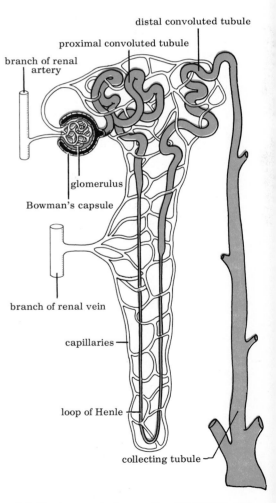

distal convoluted tubule

proximal convoluted tubule

branch of renal artery

glomerulus

Bowman's capsule

branch of renal vein

capillaries

loop of Henle

collecting tubule

8.8. The human nephron.   For description, see text. [Modified from H. W. Smith, *The Kidney,* Oxford University Press, 1951.]

amount of water in the body by releasing a hormone, vasopressin, that regulates the permeability of the collecting tubule to water.

Water is not the only substance reabsorbed by the tubules of the nephrons. In a normal healthy person, all the glucose, almost all the amino acid, and much of the salt are also reabsorbed and returned to the blood. Much of this reabsorption involves active transport, and thus energy expenditure by the tubule cells.

In spite of the extensive reabsorption that may take place, urine is more than a concentrated solution of urea. Most substances have what is called a kidney threshold level. If the concentration of such a substance in the blood exceeds its kidney threshold level, the excess is not reabsorbed from the filtrate by the tubules but instead appears in the urine. Glucose is an example of a substance with a high threshold value; ordinarily all glucose in the filtrate is reabsorbed because the threshold level for glucose is higher than the normal blood-glucose level. If, however, the blood-sugar level is abnormally high, as in diabetes, sugar appears in the urine. This elimination of excess sugar by the kidneys points up once again that excretory organs do far more than just remove nitrogenous wastes; they play a critical role in maintaining the relatively nonfluctuating internal fluid environment of the organism. In this case, when the liver and/or the peripheral tissues are not functioning properly and the blood-sugar level rises, the kidneys act as a second line of defense. The kidneys likewise help regulate the composition of the blood by keeping the relative concentrations of such inorganic ions as sodium, potassium, and chloride in the blood plasma at a nearly constant level.

Malpighian tubules

Proceeding from flame-cell systems to earthworm nephridia to vertebrate kidneys, we have noted an increasingly close interrelationship between the excretory structures and closed circulatory systems. In the flame-cell system, no circulatory system is involved. In the earthworm nephridium, blood capillaries are associated with the tubule but not with the nephrostome. Finally, in the advanced vertebrate kidney, there are both tubule capillaries and glomerular capillaries, the glomerulus and Bowman's capsule forming a compact interacting unit. We do not mean to imply that this sequence represents a true evolutionary progression; indeed the evidence indicates that the evolution of earthworms and of vertebrates had little to do with each other, and that the excretory systems of the two animal groups almost certainly evolved independently. Nonetheless, these systems illustrate the trend, seen in many animal groups, of increasing dependence of the excretory process on blood-capillary beds.

But we have said before and must say again that in biology almost all generalizations have exceptions. Insects are a case in point. The phylum to which this immense class of animals belongs probably evolved from an ancestral form similar to the ancestor of segmented worms like the earthworm. The evidence indicates that this ancestor had nephridia. Yet insects do not have nephridia, nor have their excretory organs evolved from nephridia. The evolution of an open circulatory system in insects (the ancestral system was most likely a closed one) and the consequent lack of blood capillaries probably account for the evolutionary loss of nephridia, which are dependent upon capillaries.

Insects and many of their relatives evolved an entirely new excretory system, one that functions well in association with an open circulatory system.

The excretory organs of insects are called Malpighian tubules. They are diverticula of the digestive tract located at the junction between the midgut and the hindgut (Fig. 8.9). These blind sacs, variable in number, are bathed directly by the blood in the open sinuses of the animal's body. Fluid is secreted from the blood into the blind distal end of the Malpighian tubules. As the fluid moves through the proximal portion of the tubules, the nitrogenous material is precipitated as uric acid and much of the water and various salts are reabsorbed. The concentrated, but still fluid, urine next passes into the hindgut and then into the rectum. The rectum has very powerful water-reabsorptive capacities, and the urine and feces leave the rectum as very dry material.

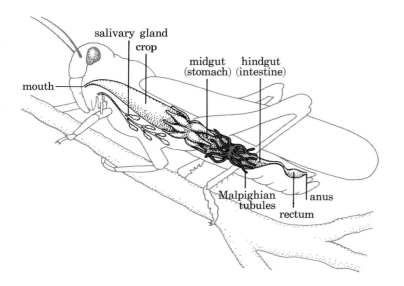

8.9. Malpighian tubules of an insect. These excretory organs arise as diverticula of the digestive system at the junction between the midgut and the hindgut.

The cellular basis of active transport of salt

In our discussion of excretion and osmoregulation so far, we have paid little attention to events at the cellular level. We have indicated that active transport of some substances, particularly salts, is carried out by the cells in the walls of the kidney tubules and Malpighian tubules, and by the salt-secreting cells of the gills of bony fishes. What is known about the active-transport process?

The sodium-potassium pump. We have stated repeatedly that animals tend to maintain a nonfluctuating fluid environment for their cells, and that this fluid has approximately the same osmotic concentration as the cells, but these statements should not be understood to imply that the extracellular and intracellular fluids have the same ionic composition. On the contrary, their compositions are very different. All cells, plant and animal, tend to accumulate certain ions in much higher concentrations than are found in the surrounding fluids and to stabilize intracellular concentrations of other ions at levels far below those in the extracellular fluids. For example, the vast majority of cells maintain an internal concentration of sodium ions (Na^+) far below that in the fluids

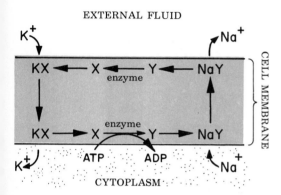

EXTERNAL FLUID

8.10. Model of sodium-potassium pump. See text for explanation.

bathing them, while at the same time accumulating potassium ions (K^+) to a concentration many times that in the extracellular fluid.

It can be demonstrated that the unequal concentrations of most small ions, such as sodium and potassium, on the two sides of the cellular membrane cannot be maintained under conditions that prevent the cell from carrying out energy-yielding chemical reactions such as cellular respiration. It may be inferred, therefore, that the cell must do work to accumulate some ions and expel others; in short, active transport across the membrane must be involved. Evidently, active transport of ions is not a phenomenon restricted to the cells of excretory and osmoregulatory organs, but is a general property of cells. Excretory and osmoregulatory cells have simply become highly specialized for this activity, regulating not only their own intracellular composition but also the composition of the extracellular fluids for the whole animal.

Unfortunately, we have very little knowledge about the actual mechanism whereby sodium and potassium are actively transported across the cell membrane. One widely accepted model for the so-called sodium-potassium pump assumes that a carrier compound in the membrane is involved (Fig. 8.10). Let us look briefly at this model. It is hypothesized that near the outer surface of the membrane a carrier compound exists in a form, X, in which it readily unites with potassium ions to form a new compound, KX. Since KX is in higher concentration near the outer surface of the membrane than near the inner surface, the molecules of this compound tend to diffuse passively across the membrane toward the inner surface. Once a molecule of KX reaches the inner surface of the membrane, it is assumed to dissociate, releasing the potassium ion into the interior of the cell. According to the model, the carrier molecule, X, is then immediately converted into another form, Y, by an energy-requiring reaction driven by ATP. In its new form, Y, the carrier molecule readily unites with sodium ions at the inner surface of the membrane to form a new compound, NaY. But NaY is in much higher concentration near the inner surface of the membrane where it is being synthesized than near the outer surface. Hence the molecules of NaY tend to diffuse passively across the membrane toward the outer surface. When such a molecule reaches the outer surface, it is assumed to dissociate, releasing the sodium ion into the fluid outside the cell. Y is then immediately reconverted into X by an enzyme-catalyzed reaction. The carrier molecule can now pick up another potassium ion to form KX and start the cycle over again. Thus the carrier molecule in its two forms, X and Y, constantly shuttles back and forth across the membrane, bringing potassium ions into the cell and taking sodium ions out. The energy that keeps the pump operating is provided by the ATP-driven conversion of X into Y at the inner surface of the membrane.

We must emphasize that this is only a hypothetical model of what happens in the membrane, and should not be taken as fact. Ion transport may actually proceed quite differently (e.g., instead of a diffusible carrier, there may be a fixed carrier that undergoes conformational changes; see p. 44), and the mechanism by which the conversion of ATP into ADP is coupled with the transport system may be completely unlike that proposed by the model. Much more research is needed on this subject.

Let us return to the epithelial cells involved in active transport in excretory and osmoregulatory organs. Most investigators believe that these cells utilize, at least in part, the same basic type of pump described above as characteristic of cells in general. But these specialized cells perform a particularly complex task. Consider the osmoregulatory cells in the gills of a marine fish. These cells must remove sodium from the blood and tissue fluid and actively secrete it into the surrounding sea water (chloride will follow passively to balance the electric charge). The cells do more than simply expel sodium ions from their cytoplasm: They pick up sodium ions on one side and expel them on the other side. In other words, sodium ions are moved completely across the cell barrier that separates the tissue fluids of the fish from the sea water. Apparently the membranes on the two sides of the cell function differently.

It has been hypothesized that sodium ions diffuse passively into the cell on one side and are then actively expelled from the cell on the other side. How would this hypothesis apply to the osmoregulatory cells in the gill of a fresh-water fish? These cells must take in salt from the surrounding water and extrude it into the blood-derived tissue fluid. The pump would be active only at the tissue-fluid side of the cell, pumping sodium ions from the cell contents into the tissue fluid. This removal of sodium would lower the sodium concentration in the cell to a point below that in the environmental water, and if the membrane on the environmental side of the cell were permeable to sodium, sodium would tend to diffuse passively into the cell from the environmental water. In other words, sodium would diffuse passively into the cell from the environment on one side and then be actively secreted by the pump from the cell into the tissue fluid on the other side. In a marine fish, the situation would be reversed: The pump would be in the membrane on the side of the cell exposed to the sea water, and sodium would diffuse passively into the cell on the tissue-fluid side.

REFERENCES

BEADLE, L. C., 1957. "Comparative Physiology: Osmotic and Ionic Regulation in Aquatic Animals," *Annual Review of Physiology,* vol. 19, pp. 329–358.

BEST, C. H., and N. B. TAYLOR, 1966. *The Physiological Basis of Medical Practice,* 8th ed. Williams & Wilkins, Baltimore. (See esp. Chapters 79–80.)

BLACK, V. S., 1951. "Osmotic Regulation in Teleost Fishes," *University of Toronto Studies: Biological Series,* vol. 59, pp. 53–89.

EDNEY, E. B., 1957. *Water Relations of Terrestrial Arthropods.* Cambridge University Press, New York.

LOCKWOOD, A. P. M., 1964. *Animal Body Fluids and Their Regulation.* Harvard University Press, Cambridge, Mass.

MAXIMOV, N. A., 1929. *The Plant in Relation to Water.* Allen & Unwin, London.

PITTS, R. F., 1968. *Physiology of the Kidney and Body Fluids,* 2nd ed. Year Book Medical Publishers, Chicago.

PROSSER, C. L., and F. A. BROWN, 1961. *Comparative Animal Physiology,* 2nd ed. Saunders, Philadelphia. (See esp. Chapters 2–3, 6.)

SCHMIDT-NIELSEN, B., 1958. "Urea Excretion in Mammals," *Physiological Reviews,* vol. 38, pp. 139–168.

WINTON and BAYLISS, 1969. *Human Physiology,* 6th ed., rev. and ed. by O. C. J. Lippold and F. R. Winton. Williams & Wilkins, Baltimore. (See esp. Chapter 38.)

SUGGESTED READING

D'AMOUR, F. E., 1961. *Basic Physiology.* University of Chicago Press, Chicago. (See esp. Chapters 12, 14.)

BALDWIN, E., 1964. *An Introduction to Comparative Biochemistry,* 4th ed. Cambridge University Press, New York. (See esp. Chapters 1–5.)

CARLSON, A. J., V. JOHNSON, and H. M. CAVERT, 1961. *The Machinery of the Body,* 5th ed. University of Chicago Press, Chicago. (See esp. Chapters 8–9.)

RAMSAY, J. A., 1968. *Physiological Approach to the Lower Animals,* 2nd ed. Cambridge University Press, New York. (See esp. Chapter 4.)

SCHMIDT-NIELSEN, B., 1965. "Comparative Morphology and Physiology of Excretion," *Ideas in Modern Biology,* ed. by J. A. Moore. Natural History Press, Garden City, N.Y.

————, 1970. *Animal Physiology,* 3rd ed. Prentice-Hall, Englewood Cliffs, N.J. (See esp. Chapter 4.)

————, and B. SCHMIDT-NIELSEN, 1953. "The Desert Rat," *Scientific American,* July.

SMITH, H. W., 1953. "The Kidney," *Scientific American,* January. (Offprint 37.)

————, 1953. *From Fish to Philosopher.* Little, Brown, Boston. (Paperback edition by Doubleday Anchor Books, 1961.)

SOLOMON, A. K., 1962. "Pumps in the Living Cell," *Scientific American,* August. (Offprint 131.)

CHAPTER 9

Chemical control

That living things function in an orderly fashion despite their immense complexity shows clearly that control mechanisms are at work, coordinating the innumerable activities of their cells, tissues, organs, and systems. Two principal types of control mechanisms can be recognized: chemical control mechanisms, which are found in all organisms, and nervous control mechanisms, which, in the strict sense, are found only in multicellular animals. This chapter will be concerned with the first of these mechanisms—chemical control.

Control chemicals produced in a regular fashion by tissues or organs specialized for that function and exerting their highly specific effects on other tissues and organs are usually called *hormones.* They are effective in very low concentrations, as would be expected if, as recent investigations indicate, they act by influencing the synthesis or the activity of enzymes or coenzymes.

PLANT HORMONES

Plant hormones, at least those so far known, are produced most abundantly in the actively growing parts of the plant body, such as the apical meristems of the shoot and the root, young growing leaves, or developing seeds. The tissues in which these hormones are produced, frequently the meristematic tissues themselves, are specialized for hormone production, but they are not so highly specialized as to be concerned with little else, as is frequently the case with the most highly specialized hormone-producing tissues in animals. There are no separate hormone-producing organs in plants analogous to the endocrine glands of higher animals. Furthermore, plant hormones are almost exclusively involved in regulating growth patterns and are often called growth

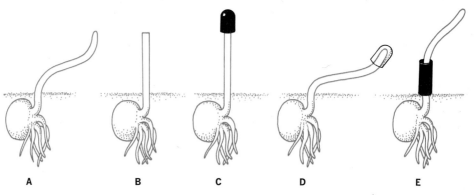

A   B   C   D   E

9.1. The Darwins' experiments on phototropism. (A) Coleoptile of canary grass bends toward the light. (B–C) The coleoptile does not bend if its tip is removed or is covered by an opaque cap. (D) The coleoptile does bend if its tip is covered by a transparent cap. (E) It also bends if its base is covered by an opaque tube.

regulators, while animal hormones mediate a great variety of functions in addition to growth.

Work on plant hormones is one of the most active areas of modern botanical research, and our knowledge of this subject is growing rapidly. The following sections do not pretend to be complete in their coverage; they are intended, rather, as an introduction providing the background necessary for understanding advances as they occur.

Auxins

One of the earliest-investigated and best-known groups of plant hormones, or growth regulators, includes those hormones collectively known as auxins. Of the amazing variety of effects that auxins have on different parts of the plant, probably the most extensively studied is their control of cell elongation in stems. Let us begin with an examination of this function of auxins.

Auxins and phototropism of shoots. Everyone is familiar with the strong tendency of many plants to turn toward the light. A potted plant in the living room bends toward a window; you turn the plant so that it will look nicer to people in the room, but discover that in a disconcertingly short time the shoot is again oriented toward the light of the window. This phenomenon of responding to light by turning is called phototropism, from the Greek words for light and turning. (Other tropisms involve turning responses to other stimuli. Geotropism is a turning response to gravity, hydrotropism a turning response to water.) In plant shoots, the phototropism is positive, a turning toward the stimulus; roots, on the other hand, exhibit negative phototropism, a turning away from a light stimulus.

One of the first to investigate the phototropism of plants was the incredibly wide-ranging Charles Darwin, who worked on this problem with his son Francis, about 1880. They, like many who followed them, performed their experiments on the hollow cylindrical sheath that encloses the first leaves of seedlings of grasses and their relatives. This sheath, called the *coleoptile*, grows principally by cell elongation, and it exhibits a very strong positive phototropic response. The Darwins showed that if the tip of the coleoptile is covered by a tiny black cap, it fails to bend toward light coming to it from one side, while control coleoptiles with their tips exposed or covered with transparent caps bend, as expected, toward the light (Fig. 9.1). They observed that a black tube placed over the base of the coleoptile, but not covering the tip, fails to prevent bending. It seemed to be the tip of the coleoptile, therefore, that played the key role in the phototropic response. The Darwins concluded that light is detected by the tip of the coleoptile and that "some influence is transmitted from the upper to the lower part, causing the latter to bend."

About thirty years later, P. Boysen-Jensen in Denmark obtained the first clear evidence that the "influence" postulated by the Darwins was probably material rather than electrical or nervous. He removed the tips of oat coleoptiles (which made the coleoptiles stop growing), placed a thin layer of gelatin on the cut end of the stump, and then placed the tip on the gelatin. Thus the tip was separated from the rest of the coleoptile by a thin layer of gelatin (Fig. 9.2). The coleoptiles resumed growing. If the tip was then illuminated with a light from the side, the coleoptile base bent toward the light. The tip had received the light stimulus, and a message from the tip had moved across the gelatin barrier and induced bending in the base. Although this experiment did not completely rule out an electrical or nervous message, it made such a possibility appear highly unlikely and strongly indicated that a diffusible chemical was involved.

That the tip could cause the base of the coleoptile to bend even in the dark was demonstrated by A. Paál in Hungary in 1918. He cut off the tip and then replaced it off center on the stump (Fig. 9.3). If he put the tip on the left side of the stump in the dark, the coleoptile bent to the right; if he put the tip on the right side, the coleoptile bent to the left. Apparently the part of the coleoptile directly under the replaced tip grew much faster than the part not under the tip. This asymmetric elongation of the coleoptile caused it to bend away from the side undergoing the greatest elongation.

Experiments conclusively demonstrating that the growth stimulus moving downward from the tip is a chemical were reported in 1926 by Frits Went in Holland. He removed the tips from coleoptiles and placed these isolated tips, base down, on blocks of agar for about an hour (Fig. 9.4). (Agar, a gelatinlike material made from seaweeds, is often used as the base for laboratory culture media.) He then put the blocks of agar, minus the tips, on the cut ends of the coleoptile stumps. The stumps behaved as though their tips had been replaced; they resumed growth, responded to lateral light by bending toward it, and, if the agar blocks were put on off center, could be made to bend even in darkness. Plain agar blocks used as controls produced none of these effects. Apparently a growth-stimulating substance had diffused out of the tips and into the blocks of agar while the tips were sitting on the blocks. When the blocks containing the chemical were placed on the stumps, the chemical moved down into the stumps and stimulated elongation. Went named the diffusible hormone presumably involved "auxin" (from a Greek word meaning to grow).

Many chemicals, some found naturally in plants and some synthesized only in the laboratory, have passed Went's test and are commonly called auxins. The one most often encountered and most thor-

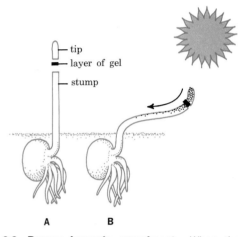

9.2. Boysen-Jensen's experiment. When the tip of an oat coleoptile is cut off, a layer of gelatin put on the end of the stump, and the tip replaced (A), the coleoptile will grow and turn toward the light (B).

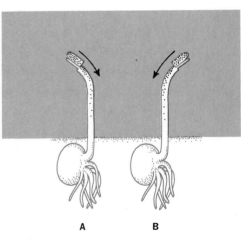

9.3. Paál's experiment. This experiment is performed in the dark. If the tip of a coleoptile is cut off and then replaced right of center, the coleoptile will bend to the left (A); if the tip is placed left of center, it bends to the right (B).

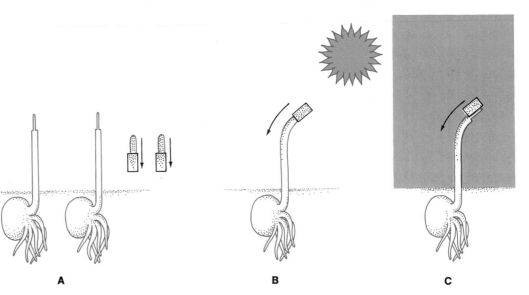

A B C

9.4. Went's experiment. When the tips of coleoptiles are cut off and placed on blocks of agar for about an hour (A), and one of the blocks alone is then put on a stump (B), the stump will resume growing and will respond to light; if a block is placed off center on a stump in the dark (C), the stump will grow and will bend away from the side on which the block rests. Apparently a hormone that has diffused from the tips into the blocks then diffuses from the blocks into the stumps.

oughly investigated is **indoleacetic acid,** which has been isolated from numerous natural sources.

The experiments we have discussed have shown, then, that the tip of the coleoptile releases auxin, which moves downward and stimulates cell elongation in the coleoptile. But how do these facts help explain the phototropic response? According to recent evidence, light induces lateral movement of auxin from the illuminated side of the tip to the shaded side. The mechanism of this movement is unclear, but it seems plain that the effect of light is to reduce the amount of auxin on the illuminated side and to increase it on the shaded side. Paál's and Went's results suggest that there is normally little lateral movement once the auxin is released from the tip; apparently the hormone reaches and stimulates only the cells directly under the point of release. Hence the shaded side grows faster, and this asymmetrical growth produces bending toward the illuminated side.

Auxins and geotropism of shoots. Auxin is involved in another plant tropism besides phototropism. If you lay a potted plant on its side and leave it for a few hours, you will find that the shoot has begun to bend upward (Fig. 9.5). This is a negative geotropic response; the shoot turns away from the pull of gravity. (How could you prove that the shoot is responding primarily to gravity and not to some other stimulus such as light?) Herman Dolk in Holland showed that the concentration of auxin in the lower side of a horizontally placed shoot increases while the concentration in the upper side decreases. This unequal distribution of auxin stimulates the cells in the lower side to elongate faster than the cells in the upper side, and the shoot thus turns upward as it grows. Again, the external stimulus, in this case gravity, is apparently detected by the meristematic tissue in the shoot tip. It has been suggested that the meristematic cells sense the pull of gravity by its effect on the distribution of cellular inclusions such as starch grains and other small bodies. Responding to the pull of gravity, these particles would tend to accumulate in the lower parts of the cell and would, by some unknown mechanism, bring about an asymmetric auxin distribution.

9.5. When a growing plant is left lying on its side, the shoot will bend upward and the roots will bend downward.

Auxins and geotropism of roots. Roots, unlike shoots, exhibit positive geotropism; they turn toward the pull of gravity. The gravitational stimulus is apparently detected by cells in the root cap. While the effects of auxin discussed previously are consistent with each other —in both phototropic and geotropic turning by shoots, the side receiving the more auxin elongates faster—something different seems to happen in roots. Examination of the auxin distribution in the roots of a plant placed on its side reveals a greater quantity of auxin in their lower than in their upper sides; in other words, auxin distribution in the roots seems to duplicate that in the shoot. Why, then, does the root turn toward the pull of gravity while the shoot turns away from it? An obvious possibility is that auxin stimulates elongation in the shoot but inhibits it in the root. Further experiments have shown, however, that auxins do not always inhibit root elongation; in very low concentration, they stimulate it. The basic difference between the reactions of root and shoot cells to auxin seems to be one of sensitivity. Root cells must be more sensitive to auxin, being stimulated to elongate at very low concentrations of the hormone and inhibited at higher concentrations (Fig. 9.6). An increase in auxin concentration above the normal moderate level thus inhibits root elongation but stimulates shoot elongation (provided the increase is not too great).

Auxins and inhibition of lateral buds. The fact that the response curves to auxins of different plant organs may differ—i.e. that a stimulating concentration for one organ may be an inhibiting concentration for another—explains another important function of auxin: its inhibiting effect on lateral buds in many plants. Auxin produced in the terminal bud moves downward in the shoot and inhibits development of the lateral buds, while at the same time stimulating elongation of the main stem. An examination of the graph in Fig. 9.6 shows how these two apparently opposite effects can occur at the same time; an auxin concentration that stimulates stem elongation is high enough to inhibit the more sensitive buds. The terminal bud thus exerts apical dominance over the rest of the shoot, ensuring that the plant's energy for growth will be funneled into the main stem and produce a tall plant with relatively short lateral branches. Longer branches usually develop only from buds far enough below the terminal bud to be partly free of the apical dominance. If the terminal bud is removed, however, apical dominance is temporarily destroyed, and several of the upper lateral buds will begin to grow, producing branches whose terminal buds soon exert dominance over any buds below them (Fig. 9.7). Flower and shrub growers frequently pinch out the terminal buds of their plants to produce bushy well-branched plants with many flowering points instead of tall spindly ones with sparse flowers.

Other functions of auxins. We cannot discuss all the other known functions of auxins here. Suffice it to say that they play a role in stimulating growth of fruit, in controlling the time of fruit and leaf drop, in stimulating renewed activity of the cambium in spring, and in stimulating the cell division in the pericycle that leads to production of new lateral roots.

Commercial compounds similar to auxins are often used as weed

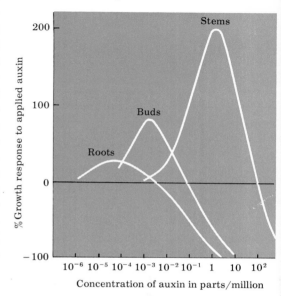

9.6. Graph showing different sensitivities of roots, buds, and stems to auxins. The percent growth is here calculated on the basis of normal growth. [Redrawn from L. J. Audus, *Plant Growth Substances*, Leonard Hill, 1959.]

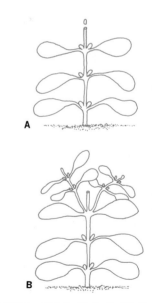

9.7. Inhibition of lateral buds by the terminal bud. When the terminal bud, which inhibits the lateral buds, is removed (A), several upper lateral buds begin to grow (B), and then the terminal buds of the new branches inhibit lateral buds below them.

killers. They apparently achieve their effect by stimulating uncoordinated and distorted growth of some body parts, while inhibiting proper function of other parts.

Gibberellins

The Japanese have long been familiar with a disease of rice that they call "foolish-seedling disease." Afflicted plants grow unusually tall but seldom live to maturity. In 1926 a Japanese botanist, E. Kurosawa, found that all such plants are infected with a fungus named *Gibberella fujikuroi*. He showed that when the fungus was moved to healthy seedlings they developed the typical disease symptom of rapid stem elongation. He could also produce the symptoms with an extract made from the fungus, and even with an extract made from culture media on which the fungus had grown. Clearly, some chemical was involved.

Japanese scientists, working on the problem of foolish-seedling disease during the 1930's, succeeded in isolating and crystallizing a substance from *Gibberella* that produced typical disease symptoms when applied to rice plants. Western scientists did not become aware of the Japanese investigations until about 1950, but since that time work on *gibberellins* has become widespread. More than twenty different substances that can be classed as gibberellins have been isolated from fungi and from higher plants; the one most often used in experimental work is called gibberellic acid. The most dramatic effect of gibberellins is their stimulation of rapid stem elongation in dwarf plants and other plants that normally undergo little stem elongation (Fig. 9.8). They have much less effect on most normally tall plants.

If both auxins and gibberellins stimulate stem elongation, why must gibberellins be considered a separate class of hormones?

9.8. Effect of gibberellic acid on cabbage. The plant at left is normal. The one at right was treated with gibberellic acid. [Courtesy S. H. Wittwer, Michigan State University.]

1. Gibberellins have a drastic effect on intact stems of dwarf plants, whereas auxins applied to an intact stem have no effect; auxins have an effect on elongation only after removal of the terminal bud. Gibberellins probably function in conjunction with auxins, not as replacement for them.
2. Gibberellins produce no effect on decapitated coleoptiles when given the Went test for auxins, apparently because they cannot move freely out of the plant tips into blocks of gelatin or agar.
3. Even more important, gibberellins cannot produce the bending movements of shoot and root characteristic of auxin-induced tropistic responses.
4. Gibberellins do not inhibit growth of lateral buds.

Other developmental hormones

Development has three basic aspects: cell division, cell enlargement, and differentiation. Both auxins and gibberellins, as we have seen, play important roles in controlling cell enlargement in plants. Although auxins likewise play a role in cell division and differentiation, that role is accessory. The main promoters of cell division in plants are the *cytokinins;* auxins enhance their action. The ratio of cytokinin to auxin appears to be of fundamental importance in determining how the new cells formed by cell division will differentiate. Thus, when there is more cytokinin than auxin in tissue cultures, buds tend to form; when there is more auxin, root growth is initiated.

Relatively little is known about growth *inhibitors*, which have effects opposite to those of auxins, gibberellins, and cytokinins. A few have been isolated and identified, but the existence of many others has simply been inferred. The role of inhibitors in maintaining dormancy in the buds and seeds of some plants has attracted particular interest. It is believed that inhibitors block the activity of some buds and seeds in autumn, thus ensuring that they will not begin to grow during a few warm days, only to be killed by the rigors of the winter climate. The inhibitors either break down gradually with time or are destroyed by cold, so that the buds and seeds are free to become active in the next growing season. In some cases, inhibitors in seeds must be leached out by water before the seeds can germinate.

The old saying "One rotten apple spoils the lot" rests on a common observation: When one apple in a barrel goes bad, most of the other apples in that barrel soon go bad too. We now know that the bad apple affects other fruit by the production of a very volatile compound called *ethylene* (C_2H_4), which plays a variety of other roles as well in the life of a normal plant.

One of the best-studied effects of ethylene is stimulation of fruit ripening. Once the ovary (the part of the flower that forms the fruit) has passed through an early cell-division stage followed by a cell-growth stage, and has attained its maximum size, a host of chemical changes begin that cause the fruit to ripen. The ripening process starts with a sudden sharp increase in CO_2 output, followed soon after by a sharp decline. It can be shown that this burst of metabolic activity, called the climacteric, is triggered by an approximately hundredfold increase in the concentration of ethylene. Inhibition of ethylene production, or removal of the ethylene as fast as the fruit produces it, prevents the climacteric, and no ripening occurs.

Ethylene has also been shown to contribute to leaf drop and to lateral bud inhibition. In general, most of the effects of this hormone can be characterized as contributing to the senescence, or aging, of the plant. Other such compounds are thought to be important in plants, but they have not yet been isolated.

The following picture emerges from our discussion of chemical control of plant growth and development so far:

1. Cell division, which is the first phase of growth, is stimulated by cytokinins and by other factors that enhance their activity.
2. Cell division is inhibited by a variety of substances, which are only poorly known.
3. It is the balance between the cytokinins and the inhibitors that determines whether a cell will divide or not.
4. Control of cell enlargement, which is the second phase of growth, involves auxins, gibberellins, and other substances less well studied.
5. The aging process, leading to death, is brought on by various senescence-inducing substances, of which ethylene is one of the most important.

Chemical control of flowering

As you are doubtless aware, flowering is not a random process. Some plants flower early in the spring, others flower in midsummer, and still others, like chrysanthemums, flower in the fall. These simple facts have been known for centuries. But only since 1920 has anything been known of the control mechanisms involved, and the subject is still not well understood.

Photoperiodism and flowering. The intense modern interest in the flowering process dates from the investigations of W. W. Garner and H. A. Allard of the U.S. Department of Agriculture. They were working with a new mutant variety of tobacco called Maryland Mammoth, which grew unusually large (as much as 10 feet tall) but would not flower, thus making it unusable in breeding experiments; however, when they propagated the new variety by cuttings, they found that it would flower in the greenhouse in winter. Though flowering was not the subject they were originally investigating, they became interested in the question why Maryland Mammoth would flower in the greenhouse in winter but not in the fields in summer, and began a series of experiments that were to open the way to a whole new area of botanical re-

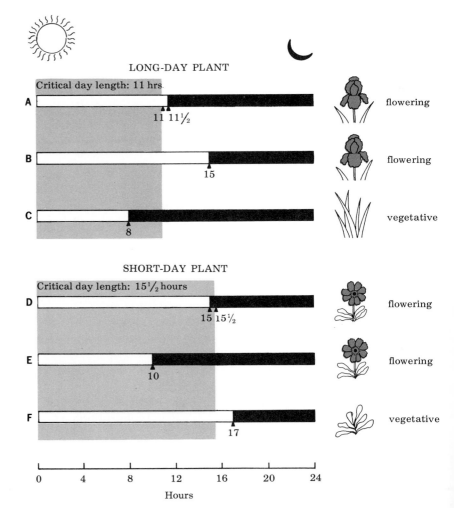

9.9. Comparison of long-day and short-day plant. White bars indicate days and black bars nights. The hypothetical long-day plant has a rather short critical day length of 11 hours, and the hypothetical short-day plant has a rather long critical day length of 15½ hours. In other words, in this example, the critical day length for the long-day plant is shorter than that for the short-day plant. The point to remember is that the critical day length is a *minimum* value for the long-day plant and a *maximum* value for the short-day plant. Thus the long-day plant will flower when the day length is slightly *above* the critical value (A) or when it is much above the critical value (B), but will not flower when it is below the critical value (C). Conversely, the short-day plant will flower when the day length is slightly *below* the critical value (D) or when it is much below the critical value (E), but will not flower when it is above the critical value (F).

search. Here is but one example of an important lead uncovered almost accidentally through research originally devoted to a different subject. It is the mark of the good scientist that he does not overlook such leads, but recognizes them when he finds them and pursues them, even though he might have to change the course of his research.

Garner and Allard realized that winter greenhouses and summer fields differ in temperature, moisture, light intensity, day length, etc. They began experiments that painstakingly eliminated one after another of these environmental factors until only one was left as the probable controlling factor in flowering—day length. They concluded that the short days of late autumn and early winter induced flowering in Maryland Mammoth tobacco. They could get the plants to bloom in summer if they shielded them from the light for a part of each day. Conversely, they could prevent blooming in the greenhouse in winter by extending the day length with electric lights.

Garner and Allard also experimented with Biloxi soybeans. They planted soybeans at two-week intervals from early May through July and found that all the plants flowered at the same time in September, even though their growing periods had differed by as much as 60 days. It was as though they were waiting for some signal from the environment. Garner and Allard were sure that the signal was short days.

Experiments with other species revealed that most plants can be placed in one of three groups: (1) short-day plants, which flower when the day length is below some critical value, usually in spring or fall (examples are chrysanthemum, poinsettia, dahlia, aster, cocklebur, goldenrod, ragweed, Maryland Mammoth tobacco, and Biloxi soybean); (2) long-day plants, which bloom when the day length exceeds some critical value, usually in summer (beet, clover, gladiolus, larkspur, black-eyed Susan); and (3) day-neutral plants, which are independent of day length and can bloom under conditions of either long or short days (dandelion, sunflower, carnation, pansy, tomato, corn, string bean).

The difference between long-day and short-day plants does not depend upon the actual day length at the time of flowering. The difference is that a long-day plant will flower only when the day length is *longer than* a critical value, whereas a short-day plant will flower only when the day length is *shorter than* a critical value (Fig. 9.9). The critical day length is thus a minimum value for flowering by long-day plants and a maximum value for flowering by short-day plants. Garner and Allard called the phenomenon they had discovered—response by an organism to the duration and timing of the light and dark conditions —*photoperiodism.*

Flowering hormone. The question now became: How does day length exert its effect on flowering? It was suggested long ago that hormones might be involved, but the first convincing evidence for a floral hormone did not come until 1936, from the experiments of M. H. Chailakhian in Russia. Chailakhian removed the leaves from the upper half of chrysanthemums (which are short-day plants), but left the leaves on the lower half (Fig. 9.10). He then exposed the lower half to short days while simultaneously exposing the defoliated upper half to long days; the plants flowered. Next, he reversed the procedure, exposing the lower half to long days and the defoliated upper half to short

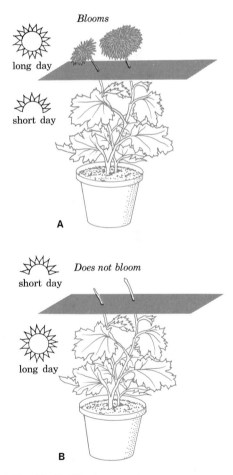

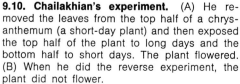

9.10. Chailakhian's experiment. (A) He removed the leaves from the top half of a chrysanthemum (a short-day plant) and then exposed the top half of the plant to long days and the bottom half to short days. The plant flowered. (B) When he did the reverse experiment, the plant did not flower.

days; the plants did not flower. He concluded that day length does not exert its effect directly on the flower buds, but causes the leaves to manufacture a hormone that moves from the leaves to the buds and induces flowering. This hypothetical hormone has been named *florigen.*

The explanation of flowering that emerges from numerous experiments is that an inducing photoperiod (short days for short-day plants and long days for long-day plants) causes the leaves to produce florigen, which moves to the buds and stimulates development of the flower. A noninducing photoperiod (long days for short-day plants and short days for long-day plants) causes the leaves of many, but not all, plants to destroy florigen or to inhibit its action in some way. Under natural conditions, flower induction is triggered when, as the season changes, the day length passes a critical value and production of florigen exceeds its destruction.

This is a tidy explanation, but it still awaits proof. All attempts to isolate a flower-inducing hormone have failed. The florigen hypothesis will not rest on firm foundations until a flower-inducing hormone is actually isolated and identified. Perhaps this will have been accomplished by the time you read this book.

Detection of the photoperiod.   In our discussion so far, we have avoided asking one fundamental question: How do plants detect the photoperiod, how do they measure the relative lengths of the light and dark periods? Before we discuss current ideas about the detection process itself, let us find out more precisely what is the critical element in the photoperiod. If the critical element is day length, as the terms "long day" and "short day" imply, then we should be able to prevent a long-day plant from flowering at the proper season by shielding it from light for an hour or so during the middle of the day. But if this is done, nothing happens; the plant flowers normally. If, however, a short-day plant is illuminated by a bright light for a few minutes, or even seconds, in the middle of the night during the normal flowering season, it will not bloom. The same sort of experiment will induce flowering at the wrong season by a long-day plant (Fig. 9.11). It is clear, then, that the critical element of the photoperiod is actually the length of the night, not the length of the day. Our terminology would be more accurate if, instead of speaking of long-day and short-day plants, we spoke of short-night and long-night plants. Having established this, we must rephrase our question and ask: How does the plant detect and measure the dark period?

Since interrupting the dark period prevents flowering by a short-day (long-night) plant and induces flowering by a long-day (short-night) plant, the light itself must be detected by the plant. What wavelengths of light are involved? H. A. Borthwick, S. B. Hendricks, and their associates of the U.S. Department of Agriculture, Beltsville, Maryland, began investigating this question in 1944. They exposed Biloxi soybeans to light of different wavelengths, and found that red light (wavelength about 6,600 angstroms) is by far the most effective in inhibiting flowering in these short-day plants; the same red light is very effective in inducing flowering by long-day plants. Later it was found that far-red light (wavelength about 7,300 angstroms), which is invisible to the human eye, has effects exactly contrary to those of

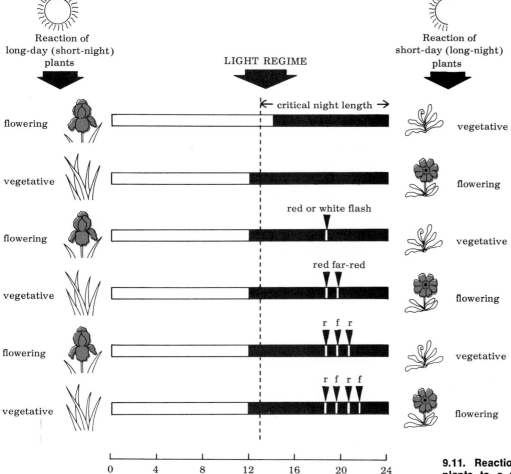

9.11. Reactions of long-day and short-day plants to a variety of light regimes. White bars are days; black bars are nights. Long-day (short-night) plants flower when the night is shorter than the critical value, or when a longer night is interrupted by an intense red or white flash or by a series of flashes of which the last is red or white. Short-day (long-night) plants give the reverse responses; they flower when the night is longer than the critical value, or when a night is interrupted by a far-red flash or by a series of flashes of which the last is far-red.

red light; it induces flowering in short-day plants and inhibits flowering in long-day plants. Not only do red and far-red light have opposite effects, but each reverses the effect of prior exposure to the other (Fig. 9.11). A short-day (long-night) plant will not flower if its long night is interrupted by a bright flash of red light; if, however, the red flash is followed immediately by a far-red flash, the plant flowers normally. Almost any number of successive flashes can be used, the final effect depending solely on whether the last flash was red or far-red.

The fact that red and far-red light can reverse each other led Borthwick and Hendricks to conclude that a single receptor pigment is involved and that this pigment exists in two forms: one that absorbs red light (R form) and one that absorbs far-red light (F form). They called this pigment *phytochrome.* It has since been shown to be a protein. When R-phytochrome absorbs red light, it is rapidly converted into F-phytochrome. Conversely, absorption of far-red light by F-phytochrome rapidly converts it into R-phytochrome. The R form is apparently the more stable of the two; in darkness, F-phytochrome is slowly converted metabolically into R-phytochrome. We can summarize these conversions as follows:

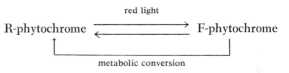

When phytochrome is exposed to both red and far-red light simultaneously, the red light dominates and the pigment is converted into the F form. Sunlight or light from ordinary electric lamps contains both red and far-red wavelengths; hence, during the day, the phytochrome exists predominantly in the F form. During the night, the F-phytochrome is slowly converted metabolically into R-phytochrome. The pigment thus gives the plant a way of sensing whether it is day or night.

We began this part of our discussion with the question: How does the plant detect and measure the dark period? We have answered the first half of the question: The plant possesses a sensitive pigment, phytochrome, that responds to presence or absence of light. But what about the crucial second half of the question? It is, after all, the measuring of the dark period that is fundamental to control of flowering. The most obvious hypothesis is that the metabolic conversion of F-phytochrome into R-phytochrome in the dark proceeds so slowly that the amount of conversion occurring between two light periods provides a measure of the length of the intervening dark period. In other words, the system would work like an hourglass. Light would convert all the pigment into F-phytochrome. Then during the following dark period the amount of F-phytochrome converted into R-phytochrome before the next light period would tell the plant how long the dark period had lasted. This would explain why a burst of bright red or white light in the middle of the night would inhibit flowering in short-day (long-night) plants; the burst of light would reset the hourglass by rapidly converting all R-phytochrome into F-phytochrome, with the result that the plant would measure two short nights, one before and one after the light burst, instead of the single long night necessary for flowering. The hourglass hypothesis would also explain why this effect of a burst of red or white light would be canceled if followed immediately by a burst of far-red light; the far-red light would rapidly reconvert F-phytochrome into R-phytochrome (light conversion is much more rapid than metabolic conversion), so that at the end of the night a high proportion of the phytochrome would be in the R form, indicating to the plant that the night had been long.

This is an attractive and simple hypothesis, but unfortunately the evidence is against it. The rate of metabolic conversion would be temperature-dependent, but all the indications are that the plant's measure of time is not influenced by temperature.

The mechanism whereby the plant measures the length of the dark period is apparently tied to a phenomenon, now believed to occur in all living cells, that involves persistent and regular rhythms in function, rhythms dependent on some internal time-measuring system, or "internal clock." Phytochrome enables the plant to sense whether it is in light or in darkness, but the actual measuring of the time lapse between the moment the plant senses onset of darkness and the moment

II-1 GETTING THERE

"Fowl of the air" and "thing that creepeth upon the earth," they travel by means as different as the different environments they have made their own. Yet the Cardinal, which displaces the air with its beating wings, and the water snake, whose flexible body gains traction from winding among the pebbles, have a common reptilian ancestor. The wings of the bird are modified reptilian forelimbs, and the feathers modified scales. The limbs of the reptilian ancestor have been lost in the snake—which nonetheless outdistances many a legged creature.

A, Arthur A. Allen. B, Roman Vishniac.

II-2A

II-2B

II-2, II-3 · TOOLS FOR GRASPING

Animals have evolved a varied assortment of structures for capturing or manipulating food. Often these structures come in pairs, e.g. the functionally analogous forelimbs of the red squirrel (2A) and pincerlike front legs of the praying mantis (2B). And often, too, they are multiple. The octopus (literally, "eight-footed one") has eight tentacles, lined with suckers (3B). The sea anemones (3C)—Large Pink Sea Roses and Brown Anemones are among those seen here—have numerous stinging tentacles, with which they can paralyze their prey. Even some plants, such as the Venus'-flytrap (3A), have prey-capturing adaptations; an insect touching the red surface of one of the hinged leaves stimulates the leaf to snap shut, and is imprisoned.

2A, David G. Allen. 2B, 3A, courtesy Carolina Biological Supply Co. 3B, Marineland of Florida. 3C, Woody Williams from National Audubon Society.

II-3A

II-3B

II-3C

II-4

II-4, II-5 ADAPTATIONS OF A TRIUMPHANT GROUP

If numbers are taken as the criterion of successful adaptation to life on earth, then insects, which outnumber all other animals combined, must be accounted the most successful of all animals. They are indeed adapted to an incredible array of habitats and ways of life. One of the prime attributes of most species is the ability to fly. The grasshopper (above), like most flying insects, has four wings. The front pair are leathery and, when the animal is at rest, provide a protective covering for the fragile pleated, often colorful, rear wings, which are the ones important for flight. The compound eye of a mosquito (5A) and the antenna of a male polyphemus moth (5C) are characteristic of the highly developed sense organs of insects; the antennae bear extremely sensitive smell receptors. The sting of a bee (5B) is but one of the vast arsenal of defensive adaptations among insects. Though air-breathing, insects do not have lungs, nor does their blood function as a transport medium for oxygen. Air enters their bodies through holes called spiracles (5D, top center) and moves through numerous branching tubes—the tracheae—directly to the individual cells.

5C, courtesy Carolina Biological Supply Co. All others, Roman Vishniac.

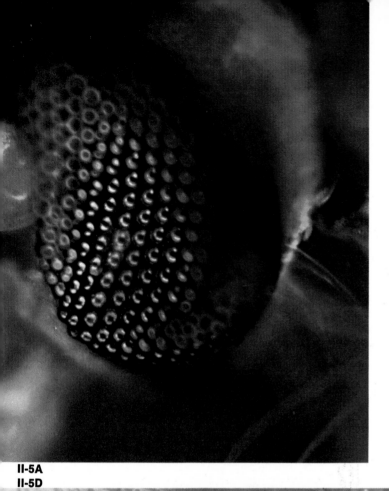

II-5B

II-5C

II-5A
II-5D

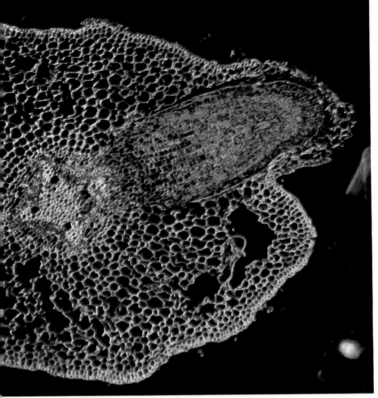

II-6, II-7   ANATOMY OR ART?

Revealed in these photographs of unstained sections are the striking geometry and unexpected color of plant tissues. The cross sections are from (6A) a willow root in which a developing lateral root juts forth like a tongue; (6B) a buttercup root; (7A) a pine leaf (needle); (7B) a buttercup stem. In all four sections note the xylem—a vascular tissue composed of several kinds of cells that serves as a conduit for fluids; in the two roots (6A and B) it forms a cross inscribed in a circle; in the pine needle (7A) it is the central portion with many small white cells; in the stem (7B) it is seen as a so-called vascular bundle, one of many located around the periphery of the stem. The thick-walled epidermal cells (outermost layer) of the pine needle show many stomata, particularly along the lower right edge; six large resin ducts are just outside the prominent endodermis that bounds the vascular core of the needle.

All photos, Roman Vishniac.

-6A

II-6B

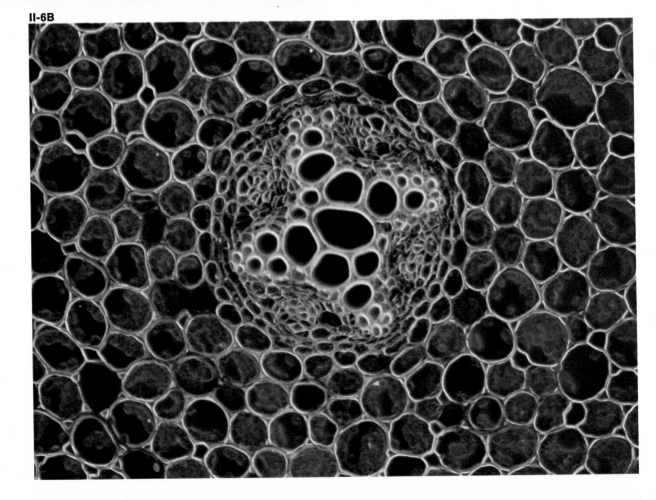

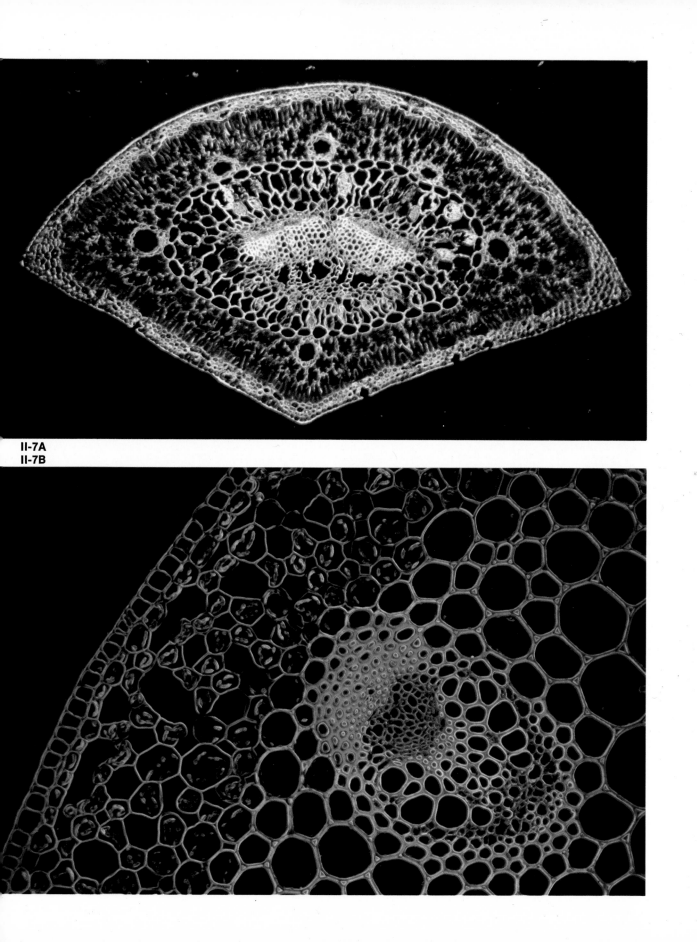

A

II-8 METAMORPHOSIS

The green larva (A) of the polyphemus moth, *Telea,* feeds on leaves. After growing in size through several molts, it spins a silken cocoon around itself, with dead leaves attached (B), and goes into the pupal stage. The pupa, seen in its protective coat inside the sectioned cocoon (C), undergoes extensive tissue reorganization and development of new organs. The newly formed adult (D) is a male, as shown by the large feathery antennae.

A–D, courtesy Carolina Biological Supply Co.

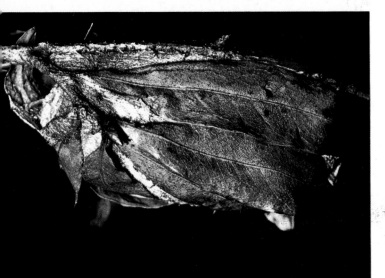

B

C

D

it senses the next exposure to light must depend on its internal clock. Once the phytochrome mechanism and the internal-clock mechanism have together indicated to the plant that the photoperiod is appropriate to flowering, the leaves must begin synthesizing flowering hormone.

HORMONES IN VERTEBRATE ANIMALS

It is not our purpose here to enumerate all known vertebrate hormones and their attributes. Instead, as in our discussion of plant hormones, we shall try to give you some insight into the way the diverse functions of a complex organism are regulated and indicate what is currently known of chemical control in vertebrates. Since far more is known about hormones in mammals, especially man, than about those of any other group of animals, it is the mammalian hormonal system that we shall emphasize here (Fig. 9.12).

We have already said that hormones are specific chemical messengers that exert effects at points some distance removed from their sites of production. Hormones may, of course, diffuse from one place to another, but, as would be expected in animals with well-developed circulatory systems, most of their transport in mammals is by the blood. The tissues and organs that produce and release hormones are termed *endocrine* tissues and endocrine organs. The use of the word "endocrine" —i.e. secreting internally—is meant to convey that the hormones are secreted directly into the blood in the capillaries supplying the endocrine tissues and that no special ducts or tubes are involved. In fact, endocrine glands are frequently called the ductless glands.

The pancreas as an endocrine organ

Diabetes (or, more precisely, diabetes mellitus), a disease in which much sugar is excreted in the urine, has been known for centuries, but its causes did not begin to be understood until the latter half of the nineteenth century.

In 1889 two German physicians, Johann von Mering and Oscar Minkowski, who were interested in the role of the pancreas as a producer of digestive enzymes, surgically removed the pancreas of a dog. A short time later, it was noticed that the dog's urine was attracting an unusual number of ants. Analysis showed that the urine contained a high concentration of sugar. Furthermore, the dog soon developed other symptoms strikingly like those of human diabetes. Von Mering and Minkowski removed the pancreas of other dogs, and diabetes invariably developed. To eliminate the possibility that the extensive damage resulting from so severe an operation might be the causal factor, they performed operations in which all the damage usually associated with the operation was produced, but the pancreas was not actually removed. These dogs did not develop symptoms of diabetes. Clearly, the onset of diabetes was directly correlated with extirpation of the pancreas. But operations in which the pancreatic duct was destroyed without producing the disease made it clear also that diabetes was not correlated with absence of the pancreatic digestive enzymes. The unavoidable inference was that the pancreas functioned not only in digestion, but in some other way as well.

Mounting evidence pointed to secretion by the pancreas of some

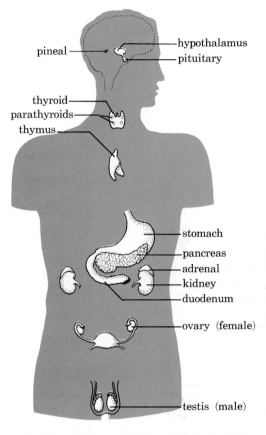

9.12. The major endocrine organs of man. Whereas certain organs, such as the pituitary, the thyroid, and the adrenals, are specialized as endocrine glands, hormone production is only one of the functions of organs such as the stomach and duodenum, whose hormones help regulate digestion, and the kidneys, which produce a hormone that helps regulate blood flow.

substance that prevents diabetes in the normal animal, but all attempts at proof failed. Feeding bits of pancreas to diabetic dogs had no effect; if the pancreas contained a control chemical—a hormone—it was destroyed by digestive enzymes. Repeated efforts by numerous investigators to show that injection of an extract made from the pancreas would alter diabetic symptoms failed. The reason for these failures was soon realized: Grinding pancreatic tissue to produce the extracts mixed the hormone with the pancreatic digestive enzymes, which destroyed the hormone. But how could this be avoided?

It was known that the pancreas is a compound organ, i.e. that it contains several types of cells, which were thought to function independently. There are the cells involved in production and release of digestive enzymes, and there are other quite different cells, called *islet cells* or islets of Langerhans. It seemed likely that the hormone so many people were searching for was produced by the islet cells. In a critical experiment that supported this hypothesis and opened the way for isolation of the hormone, it was shown that tying off the pancreatic duct resulted in atrophy of most of the pancreas but not in development of diabetes. Examination of the atrophied pancreas revealed that it was the enzyme-producing portion that had atrophied, while the islet cells had remained essentially intact. The hormone that prevented diabetes must have come from this portion of the pancreas.

The hormone, *insulin,* was finally isolated in 1922 by F. G. Banting and C. H. Best, working in the laboratory of J. J. R. MacLeod at the University of Toronto. They tied off the pancreatic ducts of a number of dogs, waited until the enzyme-producing tissue had atrophied, removed the degenerated pancreas and froze and macerated it in an isosmotic medium (freezing prevents any remaining digestive enzymes from acting), filtered the solution, and quickly injected the filtered material into diabetic dogs. The dogs showed marked improvement. Banting and Best also obtained good results with extracts prepared from the pancreases of embryonic animals; since the islet cells develop in the embryo before the enzyme-producing cells, there are no enzymes to destroy the insulin during the extraction procedure.

Banting and Best followed a procedure considered standard for demonstrating that a particular organ or tissue has an endocrine function. Let us outline the essential criteria of that procedure:

1. Removal or destruction of the organ in question should result in predictable symptoms presumed to be associated with absence of the hormone.
2. Administration of material prepared from the organ in question should relieve the symptoms.
3. It should be demonstrated that the hormone is present in both the organ and the blood, and the hormone should be extractable from each.

Fortunately, administration of extracts of suspected organs has not always been as difficult as with the pancreas before Banting and Best solved the problem.

High concentration of sugar in the urine is familiar as a major symptom of diabetes. How is insulin related to this symptom? Before attempting an answer, we must examine the symptom further. The presence of sugar in the urine of a diabetic generally indicates, not that the kidneys are functioning improperly, but that the blood-sugar concentration is higher than normal and that the kidneys are removing part of the excess. You will recall that, when blood coming to the liver via the portal vein from the intestines contains a higher than normal concentration of sugar, the liver removes much of the excess and stores it as glycogen. Conversely, when blood coming to the liver is low in sugar, the liver converts some of its stored glycogen into glucose and adds this to the blood. Other parts of the body, particularly the muscles, are also important elements in this regulatory system; e.g. when the blood-sugar concentration rises after a carbohydrate meal, part of the excess glucose is stored as glycogen in the muscles, and the rate of oxidation of carbohydrate in the muscles may also increase under these conditions.

As this brief outline of the interplay between liver, blood, and muscles suggests, insulin probably functions in at least four ways to reduce the concentration of glucose in the blood:

1. It stimulates (probably by altering membrane permeabilities) the muscles to remove more glucose from the blood.
2. It stimulates both the muscles and the liver to convert more glucose into glycogen for storage.
3. It inhibits the liver from producing glucose from glycogen or other stored materials.
4. It stimulates the muscles and the liver to oxidize carbohydrates at a more rapid rate.

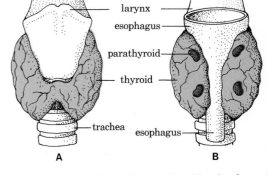

9.13. The thyroid and parathyroid glands. (A) Anterior view of thyroid. (B) Posterior view of thyroid, with parathyroids.

Too much insulin in the system, as from an overactive pancreas or from administration of too large a dose to a diabetic, can produce a severe reaction called insulin shock. The blood-sugar level falls so low that the brain, which has few stored food reserves of its own, becomes over-irritable; convulsions may result, followed by unconsciousness and often death. A naturally occurring excess of insulin is, however, extremely rare. Far more common is a deficiency of insulin, and it is this that we call diabetes.

The pancreas secretes another hormone besides insulin. This hormone, called *glucagon,* has effects opposite to those of insulin; it causes an increase in blood-glucose concentration. Several hormones produced by the adrenal glands also cause a rise in the blood-sugar level, and hormones from the pituitary may do so also. We see, then, that normal function depends on a delicate balance between opposing control systems; if either of the opposing factors is disturbed, the proper balance is destroyed and abnormalities result.

The thyroid

Most vertebrates have two thyroid glands located in the neck; in man the two have fused to form a single gland (Figs. 9.12, 9.13). Years ago, a condition known as goiter, in which the thyroid may become so enlarged that the whole neck looks swollen and deformed (Fig. 9.14), was very common in some areas of the world, such as the Swiss Alps and the Great Lakes region of the United States. Goiter is often associated with a group of other symptoms, including dry and puffy skin, loss of hair, obesity, a slower than normal heartbeat, physical lethargy, and mental dullness. No cause for this condition was known. Then in

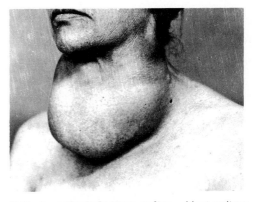

9.14. A particularly large goiter. Most goiters are much smaller. [Courtesy A. J. Carlson, V. Johnson, and H. M. Cavert, *The Machinery of the Body,* University of Chicago Press, 1961.]

1883 a Swiss surgeon, who believed that the thyroid had no important function, removed the gland from a number of his patients. Most of these patients developed all the symptoms usually associated with goiter. The results suggested that the normal thyroid must secrete some chemical that prevents these symptoms. The curious fact that patients with no thyroid and patients with the excessively large thyroid of a goiter showed the same complex of symptoms could be explained if the malformed gland of the goiter was, despite its large size, secreting too little hormone. By the 1890's, patients with goiters or the other symptoms of hypothyroidism[1] were being successfully treated with injections of thyroid extract.

In 1905 David Marine of Western Reserve University noticed that many people in Cleveland had goiters. A high percentage of the dogs also had goiters, and so did many of the trout in the streams. Marine came to suspect that the goiters might be caused by an insufficiency of iodine in the food and water. When he administered tiny traces of iodine in water to his experimental animals, their goiters and other symptoms disappeared. In 1916 he tried his treatment on approximately 2,500 schoolchildren in Akron, Ohio. He fed these children iodized salt. Another 2,500 children used as controls were fed uniodized salt. At the end of a specified period, he found only two cases of goiter among the children who had eaten iodized salt, as against 250 among the controls. Though it took years to convince a skeptical public, use of iodized salt finally became widespread, and hypothyroidism caused by insufficient iodine in the soil and water now seldom occurs. Hypothyroidism caused by malfunction of the thyroid gland itself is treated by administration of thyroid hormone.

The various symptoms of hypothyroidism all result from a slowing down of the oxidative energy-releasing reactions of the body. Thyroid hormone must therefore speed up the metabolism in some manner. Hyperthyroidism, the opposite of hypothyroidism, produces many symptoms that you might predict: a higher than normal body temperature, profuse perspiration, high blood pressure, loss of weight, irritability, and muscular weakness.

A thyroid hormone, now known as *thyroxin,* was isolated in 1914. It proved to be an amino acid containing four atoms of iodine. More recently, another thyroid hormone, identical to thyroxin except that it contains only three atoms of iodine, has been found. This hormone, called *triiodothyronine,* is three to five times more active than thyroxin but is secreted in smaller amounts. Our understanding of how thyroxin and triiodothyronine speed up oxidative metabolism is as yet only rudimentary.

The parathyroids

The parathyroid glands in man are small pealike organs, usually four in number, located on the surface of the thyroid (Fig. 9.13). They were long thought to be part of the thyroid or to be functionally associated with it. Now, however, we know that their close proximity to the thyroid is misleading; both developmentally and functionally, they are totally separate from the thyroid.

[1] The prefix *hypo-* means less than normal, the prefix *hyper-* more than normal. Thus hypothyroidism means less than normal thyroid activity, and hyperthyroidism means more than normal thyroid activity.

The parathyroid hormone, often called *parathormone,* functions in regulating the calcium-phosphate balance between the blood and the other tissues. Consequently it is an important element in maintaining the relative constancy of the internal fluid environment of the body, a subject discussed at length in Chapter 8. Parathormone increases the concentration of calcium in the blood and decreases the concentration of phosphate by acting on at least three different organs: the kidneys, the intestines, and the bones. It inhibits excretion of calcium by the kidneys and intestines, and it stimulates release of calcium into the blood from the bones (which contain more than 98 percent of the body's calcium and 66 percent of its phosphate). But calcium in bone is bonded with phosphate, and breakdown of bone releases phosphate as well as calcium. Parathormone compensates for this release of phosphate into the blood by stimulating excretion of this material by the kidneys. Actually, it overcompensates, causing more phosphate to be excreted than is added to the blood from bone; the result is that the concentration of phosphate in the blood drops as the secretion of parathormone increases.

Naturally occurring hypoparathyroidism is very rare, but the parathyroids are sometimes accidentally removed during surgery on the thyroid. The result is a rise in the phosphate concentration in the blood and a drop in the calcium concentration (as more calcium is excreted by the kidneys and intestines and more is incorporated into bone). This change in the fluid environment of the cells produces serious disturbances, particularly of muscles and nerves. These tissues become very irritable, responding even to very minor stimuli with tremors, cramps, and convulsions. Complete absence of parathormone is usually soon fatal unless very large quantities of calcium are included in the diet.

Hyperparathyroidism sometimes occurs naturally when the glands become enlarged or develop tumors. The most obvious symptom of this condition is weak bones, easily bent or fractured, because of excessive withdrawal of calcium from the bones.

The adrenals

The two adrenal glands lie close to the kidneys (Fig. 9.12). Each adrenal in mammals is actually a double gland, composed of an inner corelike *medulla* and an outer barklike *cortex.* The medulla and cortex arise in the embryo from different tissues, and their mature functions are unrelated; we shall discuss them separately.

The adrenal medulla. The adrenal medulla secretes two hormones, *adrenalin* (also known as epinephrine) and *noradrenalin* (norepinephrine), whose functions are very similar but not identical. Adrenalin causes rise in blood pressure, acceleration of heartbeat, decreased secretion of insulin by the pancreas, increased conversion of glycogen into glucose and release of glucose into the blood by the liver, increased oxygen consumption, release of reserve erythrocytes into the blood from the spleen, vasodilation and increased blood flow in skeletal and heart muscle, resistance to fatigue, vasoconstriction and decreased blood flow in the smooth muscle of the digestive tract, inhibition of intestinal peristalsis, erection of hairs, production of "gooseflesh," and dilation of the pupils. At first glance, this list (which could be repeated, with some variations, for noradrenalin) may look like a curious assortment of seemingly unrelated effects, but a more careful examination shows that these reactions occur together in response to stress, variously caused by physical exertion, pain, fear, anger, or other heightened emotional states; they have sometimes been called fight-or-flight reactions. It has been suggested, therefore, that adrenalin and noradrenalin help mobilize the resources of the body in emergencies by stimulating reactions that increase the supply of glucose and oxygen carried by the blood to the skeletal and heart muscles and, at the same time, help inhibit functions not immediately important during the emergency, such as digestion, which might otherwise compete with the skeletal muscles for oxygen.

The adrenal cortex. A person can live normally without the adrenal medullae, but not without the cortices. These are essential for life, and their removal is soon fatal. Death is preceded by a severe fall in the concentration of sodium and chloride in the blood and tissue fluids, a rise in the concentration of potassium in these fluids, loss of water from the blood resulting in diminished total blood volume and lowered blood pressure, impairment of kidney function with accompanying rise in concentration of certain metabolic wastes in the blood, impairment of carbohydrate metabolism with a marked decrease in both blood-glucose concentration and stored glycogen, loss of weight, general muscular weakness, and a peculiar browning of the skin. These same symptoms are seen in varying degrees in individuals whose adrenal cortices are insufficiently active.

The numerous symptoms of adrenal cortical insufficiency listed above are not related to a single hormone. The adrenal cortex is, in fact, an amazing endocrine factory, producing so many different hormones that we still have no idea of the total number. All the cortical hormones are steroids, often differing from each other by only one or two atoms of hydrogen or oxygen; yet these differences, minor as they may appear to us, give the various hormones strikingly different properties. In mammals, only the hormones of the adrenal cortex and those of the gonads and other reproductive structures are steroids. All the hormones of other endocrine organs, as far as we know, are proteins, peptides, or amino acids (or derived from these). Our understanding of the role of steroids in the chemistry of life is still very limited, but what we have learned so far strongly suggests that for years to come steroid research will be a highly rewarding field. On the basis of function, the cortical hormones may be grouped into three categories: (1) those that act primarily in regulating carbohydrate and protein metabolism, called glucocorticoids; (2) those that act primarily in regulating salt and water balance, called mineralocorticoids; and (3) those that function primarily as sex hormones.

Hormones in the first category (e.g. cortisone and, more important to man, hydrocortisone) inhibit the uptake of amino acids and their incorporation into proteins in muscles, and stimulate conversion of amino acids into carbohydrate and formation of glycogen by the liver.

Hormones in the second category (e.g. aldosterone, the most important in man) stimulate the cells of the convoluted tubules of the kidneys to decrease reabsorption of potassium and increase reabsorption of sodium, which leads also to increased reabsorption of chloride and water. The reabsorption of these substances, in turn, causes a rise in blood volume and blood pressure. It is clear, then, that these cortical hormones, together with insulin, glucagon, parathormone, and adrenalin, are important elements in the intricate regulation of the body's internal fluid environment.

Hormones in the third category (e.g. adrenosterone) are very similar both chemically and functionally to the sex hormones produced by the gonads. They stimulate development of such secondary sexual characteristics in the male as growth of the beard, deepening of the voice, and maturation of the genital organs. Although there are both male and female cortical sex hormones, the male hormones greatly predominate. Consequently, when tumors of the adrenal cortex increase secretion of these hormones to a level far above normal in a female, she begins to develop masculine characteristics such as hair on chest and face, a deeper voice, and more masculine musculature, while feminine characteristics such as breast development and menstruation are suppressed.

The pituitary

The pituitary is a small gland lying just below the brain. Like the adrenals, the pituitary is a double gland (Fig. 9.15). It consists of an anterior lobe, which develops in the embryo as an outgrowth from the roof of the mouth, and a posterior lobe, which develops as an outgrowth from the lower part of the brain. The two lobes eventually contact each other as they grow, and the anterior lobe partly wraps itself around the posterior lobe. In time, the anterior lobe loses its original connection with the mouth, but the posterior lobe retains its stalklike connection with a part of the brain called the *hypothalamus.* Despite their intimate spatial relationship, the two lobes remain fully distinct functionally, and we shall consider them separately here.

The anterior pituitary. The anterior pituitary is an immensely important organ that produces a number of different hormones of far-reaching effect. There are at least seven of them in man. One, called lactogenic hormone or *prolactin,* stimulates milk production by the female mammary glands shortly after birth of a baby. Another, *growth hormone,* plays a critical role in the normal processes of growth. If the supply of this hormone is seriously deficient in a child, growth will be stunted and the child will be a midget. Oversupply of the hormone in a child results in a giant. Both pituitary midgets and pituitary giants have relatively normal body proportions and features. If, however, oversecretion of growth hormone begins during adult life, only certain bones, such as those of the face, fingers, and toes, can resume growth. The result is a condition known as acromegaly, characterized by disproportionately large hands and feet and distorted features—a greatly enlarged and protruding jaw, enlarged cheekbones and eyebrow ridges, and a thickened nose.

The anterior pituitary also secretes a number of very important hormones that exert controlling action on other endocrine organs. These hormones are *thyrotrophic hormone,* which stimulates the thyroid; *adrenocorticotrophic hormone* (abbreviated ACTH), which stimulates the adrenal cortex; and at least two *gonadotrophic hormones* (follicle-stimulating hormone, abbreviated FSH, and luteinizing hormone, abbre-

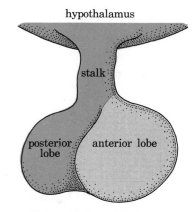

9.15. The pituitary gland.

viated LH), which act on the gonads. Proper growth and development of these endocrine glands depend on adequate secretion of the appropriate trophic (i.e. stimulatory) hormone from the pituitary; if the pituitary is removed or becomes inactive, these organs atrophy and function at very low levels. It is easy to understand why the pituitary is often called the master gland of the endocrine system.

The interaction between the anterior pituitary and the other endocrine glands over which it exerts control is an example of *feedback,* a type of interaction very common in living systems. Thyrotrophic hormone, released by the pituitary when the concentration of thyroxin in the blood is low, stimulates increased production of thyroxin by the thyroid, but the resulting rise in concentration of thyroxin in the blood inhibits secretion of more thyrotrophic hormone by the pituitary. In other words, the pituitary responds to a low thyroxin level in the blood by sending a chemical messenger that stimulates increased activity by the thyroid. But once the thyroid becomes more active, the increased amount of thyroxin produced tells the pituitary that release of thyrotrophic hormone can now be reduced. There is thus a feedback of information from the thyroid to the pituitary. The pituitary exerts control over the thyroid, and the thyroid, in turn, exerts some control over the pituitary. The pituitary tends to speed up the system, and the thyroid tends to slow it down. The interaction between the two opposing forces produces a delicately balanced system.

The interaction between the pituitary and the adrenal cortex or the gonads is similar to that outlined for the thyroid. The pituitary responds to low levels of cortical hormones by secreting more ACTH and to low levels of sex hormones by secreting more gonadotrophic hormone. The resulting rise in concentration of cortical hormones or of sex hormones inhibits further secretion by the pituitary.

The control function of the hypothalamus.   Suppose an animal faces a situation that puts it under stress and its adrenals release more hydrocortisone. Or suppose it encounters an attractive potential mate and its gonads secrete more sex hormone. How did its perception of the stressful stimulus or of the sexual stimulus affect its endocrine glands? Its perception of the stimulus involved its nervous system, but there are no nervous connections to the anterior pituitary or to the endocrine cells in the adrenal cortex or the gonads. Somehow information must move from the nervous system to the endocrine system, but it cannot do so by way of nervous connections. Might the nervous system itself secrete chemical messengers that stimulate the endocrines? Most biologists believe that it does.

Located just above the pituitary is a portion of the brain called the hypothalamus (Figs. 9.12, 9.15). Now, even though there is no direct physical connection between it and the anterior pituitary, there is an unusual connection between their blood supplies. Arteries to the hypothalamus break up into capillaries, and these capillaries eventually join to form several veins leading away from the hypothalamus. But unlike most veins, these do not run directly into a larger branch of the venous system; instead, they pass downward into the anterior pituitary and there break up into a second capillary bed. We have encountered two other places in the body where the circulation depends on two beds of capillaries arranged in sequence: the kidney nephrons and the hepatic portal system. In both places, the special type of circulation reflects an important functional arrangement. In the hepatic portal system, for example, many substances picked up by the blood in the first capillary bed are removed from the blood in the second capillary bed. Might the portal system linking the hypothalamus and the anterior pituitary function in a similar fashion? The answer seems to be yes. The hypothalamus, when appropriately stimulated, apparently releases hormonelike substances into the blood, substances that, carried directly to the anterior pituitary, stimulate it to greater activity. The relationship between the body's two principal control systems—nervous and endocrine—thus becomes much easier to understand. It is apparently by sending such substances, called *releasing factors,* of which a number have been discovered, from the hypothalamus to the anterior pituitary that the nervous system can influence the endocrine system. Thus our hypothetical animal would detect a sexual object with its sense organs; a nervous impulse would be transmitted from the sense organs to the hypothalamus of the brain; the hypothalamus would secrete appropriate releasing factors into the blood; the releasing factors would stimulate the anterior pituitary to increase its secretion of gonadotrophic hormones; the gonadotrophic hormones would stimulate the gonads to secrete more sex hormone; and the sex hormones would help prepare the animal physiologically to react appropriately to the stimulus.

The posterior pituitary.   Two hormones are released by the posterior pituitary, *oxytocin* and *vasopressin.* Oxytocin, which acts mainly on the muscles of the uterus, causing them to contract, is probably involved in childbirth. Vasopressin causes constriction of the arterioles, with a consequent marked rise in blood pressure. It also stimulates the kidney tubules to reabsorb more water.

We said earlier that the posterior pituitary originates as an outgrowth of the hypothalamus of the brain. Even in the adult it retains a stalklike connection with the hypothalamus (Fig. 9.15). There is now good evidence that oxytocin and vasopressin do not originate in the posterior pituitary, but are produced in the hypothalamus and flow along nerves in the stalk to the posterior pituitary, where they are stored. The storage organ releases the hormones when stimulated to do so by nervous impulses from the hypothalamus.

The pineal

The pineal, a lobe of the roof of the rear portion of the forebrain, has long intrigued investigators by its glandular appearance, but only recently has it been demonstrated to have an endocrine function. In some primitive vertebrates, the pineal is eyelike and responds to light both by generating nervous impulses and by secreting a hormone called *melatonin,* which lightens the skin by concentrating the pigment granules in specialized cells called melanophores. Although the mammalian pineal, too, secretes melatonin, it lacks light-sensitive cells and is without direct nerve connections to the rest of the brain; its only innervation originates outside the skull cavity in a cervical (neck) ganglion. What's more, mammals have no melanophores.

What function, then, might the pineal and its hormone have in mammals? It has recently been proposed that this baffling organ might act as a neuroendocrine transducer, converting neural information about light conditions into hormonal output. Apparently, impulses from the eyes reach the pineal via sympathetic nervous pathways from the cervical ganglia, and the pineal responds by secreting more or less melatonin. The melatonin no longer affects skin pigmentation, but appears to have become a regulator of the anterior pituitary, particularly of the secretion of gonadotrophic hormones. The likelihood is that this regulation of the pituitary is indirect, and that the direct effect of melatonin is on hypothalamic releasing centers, but the evidence on this point is still meager. Melatonin may also exert effects on other parts of the brain, particularly those concerned with locomotor rhythms, feeding rhythms, and other biological rhythms. But until more research is done, all proposals concerning the functioning and effects of the pineal and its hormone remain highly speculative.

HORMONAL CONTROL OF VERTEBRATE REPRODUCTION

Reproduction is the central theme of life. All the other aspects of living discussed in this book—nutrient procurement, gas exchange, internal transport, waste excretion, osmoregulation, growth, hormonal and nervous control, and behavior—can be viewed, in a sense, as processes that enable organisms to survive to reproduce. It has been said that "the hen is the egg's way of producing another egg," and the idea is equally applicable to man; we are, in a way, elaborate devices for producing eggs and sperms, for bringing them together in the process of fertilization, and for giving birth to young. We shall not attempt here to discuss all aspects of animal reproduction; such fundamental topics as genetics, differentiation, and growth will be the subjects of later chapters. This section will be concerned only with the physiology of vertebrate (particularly mammalian) reproduction as an example of the complex interplay between a variety of different control mechanisms.

The process of sexual reproduction

Sexual reproduction depends on the bringing together of two *gamete* cells, an egg cell and a sperm cell, which then unite in the process of fertilization to form the first cell of the new individual. There are two basic ways in which egg cells and sperm cells are brought together: external fertilization, where both types of gametes are shed into the surrounding medium and the sperms swim or are carried by water currents to the eggs; and internal fertilization, where the egg cells are retained within the reproductive tract of the female until after they have been fertilized by sperms inserted into the female by the male.

External fertilization is limited essentially to animals living in aquatic environments, because the flagellated sperm cells must have fluid in which to swim and the egg cells, in the absence of a protective coat or shell, which would prevent the sperms from penetrating and fertilizing them, would become desiccated on land. Almost all aquatic invertebrates, most fishes (but not sharks), and many amphibians utilize external fertilization. As you would expect, shedding eggs and sperms into the water of a lake or stream is an uncertain method of fertilization; many of the sperms never locate an egg and many eggs are never fertilized, even if both types of gametes are shed at the same time and in the same place, as is usually the case. Consequently animals utilizing external fertilization generally release vast numbers of eggs and sperms at one time. And they often go through elaborate behavioral sequences (in which hormonal control is very important) that ensure concurrence in both time and space in the release of gametes by the two sexes.

Most land animals, both invertebrate and vertebrate, utilize internal fertilization. In effect, the sperm cells are provided with the sort of fluid environment that is no longer available to them outside the animals' bodies. Thus the sperms remain aquatic and swim through the film of fluid always present on the walls of the female reproductive tract. Once fertilized, the egg is either enclosed in a protective shell and released by the female or held within the female's body until the embryonic stages of development have been completed. Internal fertilization requires, of course, very close physiological and behavioral synchronization of the sexes, and this synchronization involves extensive hormonal control.

Among land vertebrates, reptiles and birds use internal fertilization and lay eggs with an outer shell and four membranes—the amnion, the allantois, the yolk sac, and the chorion (Fig. 9.16). The *amnion* encloses a fluid-filled chamber housing the embryo, which can thus develop in an aquatic medium even though the egg as a whole may be laid on dry land. The *allantois* functions as a receptacle for the urinary wastes of the developing embryo, and its blood vessels, which lie near the shell, function in gas exchange. The *yolk sac,* as its name indicates, encloses the yolk, which is food material used by the developing embryo. The *chorion* is an outer membrane surrounding the embryo and the other membranes.

Like reptiles and birds, mammals utilize internal fertilization, but (with a few rare exceptions) no shell is deposited around the fertilized egg and it is not laid. Instead, the early embryo with its membranes becomes implanted in a specialized chamber of the female genital tract, and there embryonic development is completed. The young animal is then born alive. The remainder of our discussion here will be concerned with mammalian, and in particular human, reproduction.

The human reproductive system

The genital system of the human male. The male gonads, or sex organs, are the *testes,* oval glandular structures that form in the dorsal portion of the abdominal cavity from the same embryonic tissue that gives rise to the ovaries in females. In the human male, the testes descend about the time of birth from their points of origin into the *scrotal sac* (or scrotum), a pouch whose cavity is initially continuous with the abdominal cavity via a passageway called the *inguinal canal.* After the testes have descended through the inguinal canal into the scrotum, the canal is slowly plugged by growth of connective tissue, so that the scrotal and abdominal cavities are no longer continuous.

Each testis is composed of two functional components: the *seminiferous tubules,* in which the sperm cells are produced, and the *interstitial cells,* which secrete male sex hormone. The seminiferous tubules

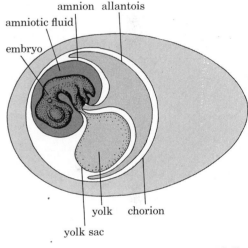

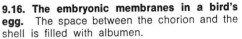

9.16. The embryonic membranes in a bird's egg. The space between the chorion and the shell is filled with albumen.

of the human are functional at temperatures below that of the abdominal cavity; if the testes fail to descend, the germinal epithelium of the tubules eventually degenerates. If, however, the testes descend normally into the scrotal sac, where the temperature is about three degrees cooler, the germinal epithelium becomes functional at the time of puberty. Mature sperm cells pass from the seminiferous tubules via many tiny ducts into a much-coiled tube, the *epididymis,* which lies on the surface of the testis (Fig. 9.17). The sperms are stored in the epididymis until they are released during copulation.

A long sperm duct, or **vas deferens,** runs from each epididymis through the inguinal canal and into the abdominal cavity, where it loops over the bladder and joins with the **urethra** just beyond the point where the urethra arises from the bladder. The urethra, in turn, passes through the *penis* and empties to the outside. Notice, then, that the urethra in the mammalian male is a common passageway used by both the excretory and reproductive systems; urine passes through it during excretion and semen passes through it during sexual activity. In more primitive vertebrates, the relationship between the excretory and reproductive systems is even closer. In frogs, for example, sperm cells pass from the testes into the kidneys and down the excretory ducts to the cloaca; the reproductive system has no separate vasa deferentia. In the vertebrates, there has been an evolutionary trend toward increasing liberation of the reproductive system from its ancestral dependence on the excretory system. There is far more separation in mammalian males than in fish or frogs, but the two systems still share the urethra and thus do not have separate openings to the outside. As we shall see, only in mammalian females has complete separation arisen.

As sperms pass through the vasa deferentia and urethra, seminal fluid is added to them to form **semen,** which is a mixture of seminal fluid and sperm cells. The seminal fluid is secreted by three sets of glands: the **seminal vesicles,** which empty into the vasa deferentia just before these join with the urethra; the **prostate,** which empties into the urethra near its junction with the vasa deferentia; and the **Cowper's glands,** which empty into the urethra at the base of the penis (Fig. 9.17). Seminal fluid has a variety of functions: (1) It serves as a vehicle for transport of sperms; (2) it lubricates the passages through which the sperms must travel; (3) as an effectively buffered fluid, it helps protect the sperms from the harmful effects of the acids in the female genital tract; (4) it contains much sugar, which the active sperms can use as a source of energy. The tiny sperm cells can store very little food themselves and hence depend on an external source of nutrients for use in the cellular respiration that provides them with the ATP necessary to keep their flagella active.

During sexual excitement, the arteries leading into the penis dilate and the veins from the penis constrict in response to stimulation by nerves of the autonomic system. Much blood is pumped under considerable pressure through the arteries into the spaces in the spongy erectile tissue of which the penis is largely composed. The engorgement of the penis by blood under high arterial pressure causes the penis to increase greatly in size and to become hard and erect, thus preparing it for insertion into the female vagina during copulation. Note that erection of the penis does not involve activity of skeletal muscles but is entirely a vasomotor phenomenon.

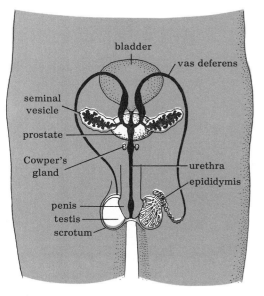

9.17. Reproductive system of the human male.

When the penis is sufficiently stimulated by friction during copulation, nervous reflexes involving pathways of the sympathetic system cause waves of contraction in the smooth muscles of the walls of the epididymides, vasa deferentia, seminal glands, and urethra. These contractions move sperms from the epididymides through the vasa deferentia, combine seminal fluid from the various glands with the sperms, and expel the semen from the urethra. An average of about 100 million sperm cells in about 3.5 ml. of semen are released during one ejaculation by a human male.

Hormonal control of sexual development in the male. No sperms and very little male sex hormone are produced by the testes before puberty. The onset of puberty is apparently triggered by increased release of gonadotrophic hormones by the anterior pituitary, and this activity of the pituitary is presumably itself stimulated by releasing factors secreted by the hypothalamus and carried to the pituitary by blood in the portal system that links these two organs. As we have seen, two gonadotrophic hormones are secreted by the anterior pituitary: *FSH*, which stimulates maturation of the seminiferous tubules, and *LH*, which stimulates maturation of the interstitial cells and induces them to begin secretion of the male sex hormone, *testosterone* (Fig. 9.21).

Once testosterone appears in appreciable quantity in the system, it stimulates development of the secondary sexual characteristics normally associated with puberty: growth of the beard, growth of pubic hair, deepening of the voice, maturation of the seminal vesicles and the prostate gland, development of larger and stronger muscles, etc. If the testes are removed (castration) before puberty, these changes never occur. If castration is performed after puberty, there is some retrogression of the adult sexual characteristics, but they do not disappear entirely. Castration after puberty abolishes the sex urge in many animals, but not in man, where psychological factors are far more important than in other animals. Cutting the vasa deferentia (vasectomy), an operation that prevents movement of sperms into the urethra and is sometimes performed as a birth-control measure, causes no retrogression of sexual characteristics, because there is no alteration of hormone levels.

The genital system of the human female. The female gonads are the *ovaries,* which are located in the lower part of the abdominal cavity, where they are held in place by large ligaments. Like the testes, the ovaries have the two main functions of producing gametes (in this case, egg cells) and secreting sex hormones. At the time of birth, the ovaries already contain a huge number of primordial egg cells, estimates of which range from 100,000 to 1,000,000. Since, during the approximately 30 years of her reproductive life, a woman ovulates about 13 times per year, producing one mature egg cell, or ovum, each time, it follows that only about 390 primordial egg cells ever mature and leave the ovaries. The rest eventually degenerate, and none can be found in the ovaries of women past the age of about 50.

Each primordial egg cell is enclosed within a cellular jacket called a *follicle.* The egg cell fills most of the space in the small immature follicle. In the process of maturation, however, the follicle grows bigger relative to the egg cell and develops a large cavity filled with a granular fluid (Fig. 9.18); the egg cell, embedded in a mass of follicular epithe-

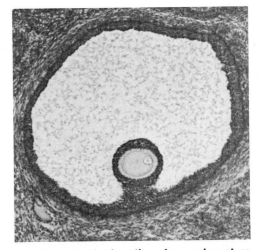

9.18. Photograph of section of a nearly mature follicle of a cat. The follicle has a large cavity. The egg cell is embedded in a pedestal of epithelial cells that projects into the cavity. [Courtesy Thomas Eisner, Cornell University.]

lial cells, protrudes into the cavity. A ripe follicle bulges from the surface of the ovary; when ovulation occurs, its outer wall ruptures and both the liquid and the detached ovum are expelled. In the human, only one ovum is normally released at each ovulation. There is no apparent regularity as to which ovary will ovulate at any given period.

A ripe egg is released from the ovary into the abdominal cavity. From there, it is usually promptly drawn into the large funnel-shaped end of one of the *oviducts* (or Fallopian tubes), which partly surround the ovaries but are not continuous with them (Fig. 9.19). Cilia lining the funnel of the oviduct produce currents that help move the egg into the oviduct. If sperms are present, the egg is fertilized while it is still in the upper third of the oviduct.

Each oviduct empties directly into the upper end of the *uterus* (or womb). This organ, which is about the size of a fist, lies in the lower portion of the abdominal cavity just behind the bladder. It has very thick muscular walls and a mucous lining containing many blood vessels. If an egg is fertilized as it moves down the oviduct, it becomes implanted in the wall of the uterus, and there the embryo develops until the time of birth. One method of birth control involves insertion of a plastic ring or spiral into the uterus. Such intra-uterine devices (IUD) seem to be very effective in preventing pregnancy, probably by preventing implantation in the uterus.

At its lower end, the uterus connects with a muscular tube, the *vagina,* which leads to the outside. The vagina acts as the receptacle for the male penis during copulation. The great elasticity of its walls makes possible not only the reception of the penis but the passage of the baby during childbirth.

The uterus and vagina do not lie in a straight line, as Fig. 9.19A might seem to indicate. Instead, the uterus projects forward nearly at a right angle to the vagina, as shown in Fig. 9.19B. The *cervix,* a muscular ring of tissue at the mouth of the uterus, protrudes into the vagina. Devices that block the mouth of the uterus by covering the cervix are widely used in birth control. One such device, the diaphragm, is a shallow rubber cup with a spring around its rim. It is inserted into the vagina and positioned so that it covers the entire cervical region. It is very effective in preventing sperms from entering the uterus, particularly if used in conjunction with spermicidal jellies or creams.

The opening of the vagina in young human females is partly closed by a thin membrane called the *hymen.* Traditionally, the hymen has been regarded as the symbol of virginity, to be destroyed the first time sexual intercourse takes place. Frequently, however, the membrane is ruptured during childhood, by disease or by a fall or as a result of strenuous exercise.

The external female genitalia are collectively termed the *vulva.* The vulvar region is bounded by two folds of skin that enclose the vestibule. The vagina opens into the rear portion of the vestibule, and the urethra opens into the midportion of the vestibule. Note, then, that in the adult mammalian female there is no interconnection between the excretory and reproductive systems, and that the urethra carries only excretory materials.

In the anterior portion of the vestibule, in front of the opening of the urethra, is a small erectile organ, the *clitoris,* which forms from the same embryonic tissue that gives rise to the penis in the male. Like the

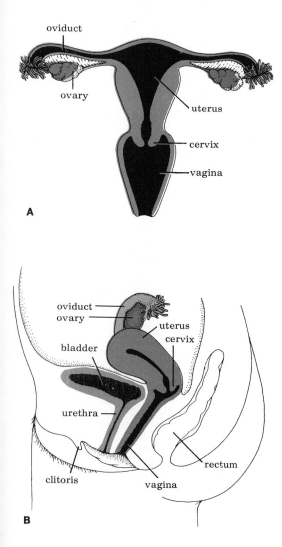

9.19. Reproductive tract of human female.
(A) Anterior view.   (B) Lateral view.

penis, it becomes engorged with blood during sexual excitement and is a major site of stimulation during copulation.

Hormonal control of the female reproductive cycle. As in the male, puberty in the female is thought to begin when the hypothalamus stimulates the anterior pituitary to release increased amounts of FSH and LH. These gonadotrophic hormones cause maturation of the ovaries, which then begin secreting the female sex hormones, *estrogen* and *progesterone.* The sex hormones, particularly estrogen, stimulate development of the female secondary sexual characteristics: growth of pubic hair, broadening of the pelvis, development of the breasts, increase in the size of the uterus and vagina, some change in voice quality, and the onset of *menstrual cycles.* We shall be particularly concerned here with the menstrual cycles as an example of the complex interplay between several hormones and between the endocrine and nervous systems.

Rhythmic variations in the secretion of gonadotrophic hormones in the females of most species of mammals lead to what are known as *estrous cycles*—rhythmic variations in the condition of the reproductive tract and in the sex urge. The females of most species will accept the male in copulation only during those brief periods of the cycle near the time of ovulation when the uterine lining is thickest and the sex urge is at its height. During such periods, the female is said to be "in heat" or in estrus. Many mammals have only one or a few estrous periods each year, but some, like rats, mice, and their relatives, may have them as often as every five days. If fertilization does not occur during the heat period, the thickened lining of the uterus is gradually reabsorbed by the female's body; ordinarily, no bleeding is associated with this process.

The reproductive cycle in humans and other higher primates differs in several important ways from that of other mammals. There is not so pronounced a heat period, the female being to some degree receptive to the male during most parts of the cycle. And the thickened lining of the uterus is not completely reabsorbed if no fertilization occurs; instead, part of the lining is sloughed off during a period of bleeding known as menstruation. Human menstrual cycles average about 28 days; there are consequently about 13 of them each year. This is an extremely rough average, however, since there is much variation from person to person and from period to period in the same person.

Let us trace the sequence of events in the menstrual cycle. It is customary in medical practice to consider the first day of menstruation as the first one of the cycle. From a biological point of view, however, it is more appropriate to regard the end of the period of bleeding as the beginning of the cycle. At this point, the uterine lining is thin and there are no ripe follicles in the ovaries. The first event in the new cycle is an increase in secretion of FSH by the anterior pituitary as a result of stimulation by the hypothalamus (Fig. 9.20). FSH (the initials, you will recall, stand for follicle-stimulating hormone) stimulates growth of follicles in the ovaries. One of the follicles soon gains ascendancy, and the other follicles cease growing. If there is also a small amount of LH in the system to act synergistically with the FSH, the growing follicles begin secreting the first of the two female sex hormones, estrogen. The estrogen, in turn, stimulates the lining of the uterus to thicken. This follicular or growth phase of the cycle lasts, on the average, about nine to ten days after cessation of the previous menstrual flow.

As the follicles grow under the influence of FSH from the pituitary, they produce more and more estrogen. This increase in the level of estrogen in the blood exerts an inhibitory effect on the FSH-stimulating center in the hypothalamus, the result of which is a drop in the secretion of FSH by the anterior pituitary. Here we have an example of the principle of feedback discussed earlier. But the increasing concentration of estrogen in the blood apparently stimulates the LH-stimulating center in the hypothalamus, with the result that secretion of LH by the pituitary rises abruptly about eight or nine days after cessation of the previous menstrual flow. When the concentration of LH in the system has reached a critical level, the ascendant follicle—by this time mature and bulging from the surface of the ovary—ovulates. Ovulation marks the end of the follicular or growth phase of the menstrual cycle.

Following ovulation, LH induces changes in the follicular cells that convert the old follicle into a yellowish mass of cells rich in blood vessels. The new structure formed from the ruptured follicle under the influence of LH—luteinizing hormone—is called the *corpus luteum* (Latin for yellow body). The corpus luteum continues secreting estrogen, though not as much as was secreted by the follicle just prior to ovulation. But the corpus luteum also secretes a second female sex hormone, progesterone (Fig. 9.21).

Progesterone functions in preparing the uterus to receive the embryo. Acting on the uterine lining, which has already become much thicker under the stimulation of estrogen during the follicular phase, it causes maturation of the complex system of glands in the lining. The luteal phase of the menstrual cycle is, in fact, sometimes called the secretory phase—somewhat misleadingly, since there is already some glandular activity in the uterine lining during the latter part of the follicular phase. Repeated experiments have shown that implantation of a fertilized ovum in the uterus cannot occur in the absence of the changes in the uterine lining produced by progesterone. Progesterone is, in a very real sense, the hormone of pregnancy.

9.20. The sequence of events in the human menstrual cycle. See text for description.

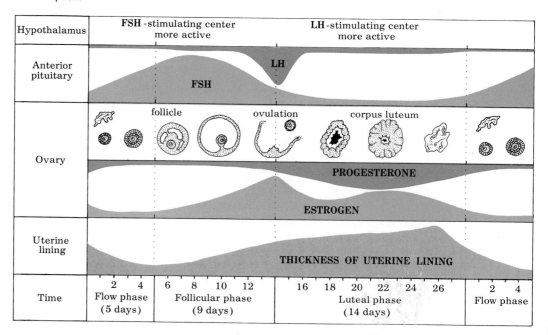

In addition, progesterone inhibits the FSH-stimulating center in the hypothalamus, and thus prevents a rise in FSH secretion by the pituitary, which would trigger the start of a new cycle. As long as progesterone is present in quantity in the system, then, there is little follicular growth. This FSH-inhibiting effect of progesterone is, of course, an important element in regulating the duration of the menstrual cycle. It is also the basis for the action of birth-control pills. These pills contain synthetic compounds similar to progesterone and estrogen. Taken daily, they inhibit secretion of FSH (and LH) and thus prevent follicular growth and ovulation and, consequently, conception.

If no fertilization occurs during a normal cycle, the continued high levels of progesterone begin to exert negative feedback action on the LH-stimulating center in the hypothalamus, with the result that secretion of LH by the pituitary falls. But LH is necessary for maintenance of the functioning corpus luteum. Hence, when the level of LH in the system has fallen appreciably, the corpus luteum begins to atrophy and ceases secreting progesterone. When this happens, the thickened lining of the uterus can no longer be maintained, and reabsorption of part of the lining begins. Unlike most mammals, humans and other higher primates cannot reabsorb all the extra tissue laid down during the follicular and luteal phases of the cycle; part of it must be sloughed off during the flow phase, which lasts about four or five days. Absence of sex hormone, resulting from atrophy of the corpus luteum, frees the FSH-stimulating center in the hypothalamus from inhibition and allows it to stimulate the pituitary to increase secretion of FSH, which triggers the follicular growth marking the beginning of a new cycle. The sequence of events in a normal menstrual cycle is depicted diagrammatically in Fig. 9.20. The hormonal interactions are shown in Fig. 9.22.

During the flow phase, particularly its first few days, a woman's body must function temporarily in the near absence of estrogen and progesterone. The result may be irritability, depression, and nausea, among other signs of physiological and psychological disturbance. In

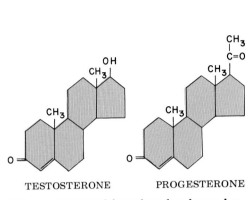

TESTOSTERONE   PROGESTERONE

9.21. The structural formulas of male sex hormone (testosterone) and female sex hormone (progesterone). Differing in only one side group, these two steroids have amazingly different effects on the body.

9.22. Diagram of hormonal interactions during human menstrual cycle. Black arrows indicate stimulation and white arrows inhibition.

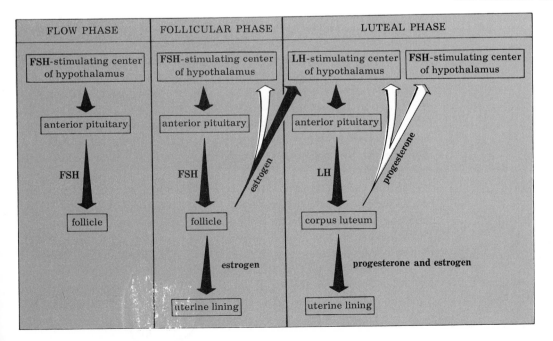

addition, strong contractions of the uterus often cause abdominal cramps. Emotional stress sometimes also accompanies the *menopause,* a period lasting a year or two at the end of a woman's reproductive life. The menopause usually comes between the ages of 45 and 50. It is apparently attributable both to cessation of secretion of LH by the pituitary as a result of declining activity of the LH-stimulating center in the hypothalamus, and to declining sensitivity of the ovaries to the stimulatory activity of gonadotrophic hormones. In the absence of LH, secretory activity in the ovaries falls off; neither estrogen nor progesterone can be produced in significant quantities. Consequently, no cyclic thickening of the uterine lining occurs, and hence no menstruation. The drop in concentration of sex hormones during menopause may cause physiological and psychological disturbances until a new physiological balance has been established.

We have seen that ovulation in human beings occurs roughly midway in the menstrual cycle. This ovulation is spontaneous; it does not depend on a copulatory stimulus. But in some mammals, such as rabbits and cats, ovulation is reflex-controlled; the nervous stimulation of copulation is necessary for the hypothalamus to trigger the release of LH by the pituitary that will lead to ovulation. In such reflex ovulators, it is possible to predict with great precision just when ovulation will occur; in the rabbit, for example, it occurs about 10½ hours after copulation. Such precision is not possible with spontaneous ovulators like humans. Yet predictions of the time of ovulation are important in the practice of the so-called rhythm method of birth control.

The rhythm method is based on the premise that fertilization can take place only during a very short period in each menstrual cycle. Copulation without risk of pregnancy should be possible during all other parts of the cycle. Present evidence indicates that human egg cells begin to deteriorate about 12 hours (maximum 24) after ovulation and can no longer be fertilized after that time. In other words, fertilization can occur only if fertile sperms are in the upper third of the oviduct during the 12 to 24 hours immediately following ovulation; conception is not possible during the other 27 days of a 28-day cycle.

Immediately, the question of the fertile life of sperms in the female reproductive tract becomes pertinent. We have said that one ejaculation releases about 100 million sperm cells into the vagina. Conditions in the vagina are very inhospitable to sperms, and vast numbers of sperm cells are killed before they have a chance to pass the cervix. Millions of others die or become infertile in the uterus and oviducts, and millions more go up the wrong oviduct or never find their way into an oviduct at all. Current evidence indicates that, in the female genital tract, human sperm cells remain fertile only about 48 hours or less after their release.

Since the fertile life of the egg cell lasts at most one day and that of the sperm cell at most two days, there is a period of somewhat less than three days during which copulation can result in conception (the day when the egg is fertile and the two preceding days). But which three days? Ovulation, as we have seen, comes at the end of the follicular phase, but this is a very variable phase even in women with fairly regular cycles; it may be prolonged when sickness or emotional upset alters the hypothalamic control of LH secretion, with whose sudden increase ovulation is correlated. It is our inability to say precisely when ovulation will occur that makes the rhythm method of birth control unreliable.

Hormonal control of pregnancy. Our discussion so far has assumed that conception did not occur and that each cycle was terminated by a menstrual flow. Let us now assume that the egg cell is fertilized at some time during the 12 hours after ovulation. Only one of the millions of sperm cells released into the vagina actually penetrates the egg cell and fertilizes it. As soon as that one cell has fertilized the egg, the outer membrane of the egg changes in consistency and becomes impenetrable to the other sperm cells, which soon die. The fertilized egg, or *zygote,* goes down the oviduct, probably carried by fluid moved by contractions of the circular muscles in the walls of the oviduct. During the days of transit, cell divisions begin and an embryo is formed.

The human embryo becomes implanted in the wall of the uterus eight to ten days after fertilization. During the interval between fertilization and implantation, the embryo is nourished by its limited supply of yolk and by materials secreted by the glands of the female genital tract. After implantation in the uterine lining, the embryonic membranes form the *umbilical cord,* through which blood vessels contributed by the allantois run to a large structure, the *placenta,* formed from the embryonic membranes (primarily the chorion) and from the adjacent uterine tissue (Fig. 9.23). Within the placenta, the blood vessels of the embryo and those of the mother lie very close together, but they are not joined and there is no mixing of maternal and fetal blood. Exchange of materials takes place in the placenta by diffusion between the blood of the mother and that of the embryo; nutritive substances and oxygen move from the mother to the embryo, and urinary wastes and carbon dioxide move from the embryo to the mother.

We saw earlier that progesterone is essential for maintenance of the uterine lining during implantation and pregnancy. But we saw also that in a normal menstrual cycle the high level of progesterone exerts negative feedback action on the hypothalamus, with a consequent decrease in LH secretion by the pituitary and eventual atrophy of the corpus luteum. The atrophy of the corpus luteum cuts off the supply of progesterone, with the result that menstruation occurs. Clearly, this sequence of events cannot take place after conception, or the uterine lining with the implanted embryo would be sloughed off and lost. It can be shown that when conception occurs the corpus luteum does not atrophy but lasts through most of the term of pregnancy. How is this possible? Apparently the chorionic portion of the placenta secretes a hormone very similar to LH. This hormone takes the place of LH from the pituitary and preserves the corpus luteum, which continues to secrete progesterone and thus sustains the pregnancy.

So much chorionic hormone is produced in a pregnant woman that much of it is excreted in the urine. Many commonly used tests for pregnancy are based on this phenomenon. Urine from the subject is injected into a test animal such as a rat, rabbit, or frog. If chorionic hormone is in the urine, it induces development of corpora lutea and changes in the vagina of female rats 72 to 96 hours after injection. The rabbit test is faster; pregnancy urine induces corpus luteum formation

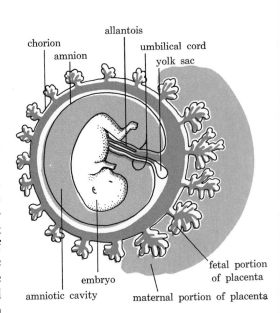

9.23. **Diagram of human embryonic membranes and placenta.**

within 24 hours after injection. And frog tests are even faster; pregnancy urine induces ovulation by females and release of sperms by males within 6–8 hours after injection.

Hormonal control of parturition and lactation. Much research is needed to clarify the complex interactions of hormones that control the birth process (parturition). One important factor in parturition seems to be a shift in the balance of estrogen and progesterone. It is known that estrogen stimulates contractions of the uterine muscles and that progesterone severely inhibits muscular contraction. It is also known that secretion of estrogen by the placenta rises sharply just prior to parturition. It seems reasonable to assume, therefore, that during most of pregnancy progesterone blocks contraction of the uterine muscles, but that the rise of estrogen concentration late in pregnancy overcomes this progesterone block and stimulates the onset of the contractions that eventually expel the fetus from the uterus. Since oxytocin, secreted by the posterior pituitary (and perhaps also by the placenta), is known to have a powerful stimulatory effect on uterine contractility, it seems likely that this hormone acts in conjunction with estrogen in inducing labor.

Like the hormonal control of parturition, that of milk secretion is complicated and not completely understood. Growth and development of the mammary glands seem to be controlled by a complex interaction between estrogen, progesterone, thyroxin, insulin, growth hormone, prolactin, and hydrocortisone. Initiation and maintenance of lactation by mature mammary glands following parturition seem to be controlled primarily by prolactin and growth hormone; the process may be aided by hydrocortisone and thyroxin. These hormones apparently become effective in inducing lactation when the high levels of sex hormones, which inhibit lactation, disappear at the time of parturition.

The actual release of milk from the mammary glands involves both neural and hormonal mechanisms. The stimulus of suckling, or, in conditioned cows, of seeing the calf or hearing rattling milk pails, causes nervous stimulation of that part of the hypothalamus that stimulates release of oxytocin stored in the posterior pituitary. The oxytocin, in turn, induces constriction of the many tiny chambers in which the milk is stored in the mammary glands. The constriction forces the milk into ducts that lead to the nipple. Adrenalin inhibits this milk-ejection process.

THE MECHANISM OF HORMONAL ACTION

The 1960's saw some exciting progress in the attempt to learn how hormones act on their target cells. One of the great breakthroughs was accomplished by E. W. Sutherland, who was awarded a Nobel Prize in 1971, and T. W. Rall, in the course of investigations into the mechanism by which glucagon and adrenalin stimulate liver cells to release more glucose into the blood. These hormones, they discovered, stimulate an increase in the intracellular concentration of a compound called cyclic adenosine monophosphate—*cyclic AMP* or cAMP for short—which, in turn, leads to activation of an enzyme necessary for breakdown of glycogen to glucose. Subsequent research has demonstrated that a large number of other hormones act on their target cells either to increase or to decrease the concentration of cAMP. At this writing, cAMP has been firmly implicated in the action of 14 different mammalian hormones, and there is suggestive evidence for more.

As you have doubtless deduced from its name, cAMP is a compound related to ATP (adenosine triphosphate). Very widely distributed in nature, it has been found in almost all animal tissues studied (vertebrate and invertebrate), in bacteria, and in plants. It is synthesized from ATP in living cells by a reaction catalyzed by an enzyme called *adenyl cyclase,* which appears to be built into the cell membrane.

According to a recently proposed *two-messenger model,* hormonal control involves an extracellular first messenger, which is the hormone itself, and an intracellular second messenger, which is often cAMP. The hormone is thought to react with a specific receptor site on the outer surface of the membrane of the target cell. If the reaction stimulates rather than inhibits adenyl cyclase on the inner surface of the membrane, the enzyme will catalyze production of cAMP (Fig. 9.24). The increased cAMP then interacts with various cellular components, which may be at some distance from the site of cAMP production, and thus initiates the cell's characteristic responses to the hormonal stimulation. In other words, the initial extracellular signal (the hormone or first messenger) is converted into an intracellular signal (cAMP or second messenger) that the cell's chemical machinery can more readily understand. There are indications that certain drugs (e.g. caffeine) and some hormones may increase the intracellular cAMP concentration in another way—by inhibiting an enzyme (phosphodiesterase) that breaks down cAMP.

This model of hormonal action is pleasing in that it proposes a unified basis for endocrinology, a single mode of operation for a multitude of different hormones (and many drugs). However, it obviously raises some important questions. What is the basis for hormonal specificity if most hormones simply act to regulate the adenyl cyclase system of their target cells? Why do different target cells respond so differently to increases or decreases in their cAMP content?

The first question is relatively easy to answer. We need only assume that different types of cells have different hormone-specific receptor sites. Thus cells in the thyroid might have receptors on which only thyrotrophic hormone could react, while cells in the adrenal cortex might have receptors on which only ACTH could react. Such an arrangement would ensure that the thyroid would be stimulated by thyrotrophic hormone and not by ACTH, and that the adrenal cortex would be stimulated by ACTH and not by thyrotrophic hormone, despite the fact that both thyrotrophic hormone and ACTH act to stimulate cAMP production by the adenyl cyclase system once they have reacted with the receptor sites specific for them. Presumably cells that respond to many different hormones have a variety of receptor sites on their membranes. For example, liver cells might have receptor sites for insulin, glucagon, adrenalin, noradrenalin, thyroxin, and perhaps for a number of other hormones as well; some of these sites would be stimulatory for adenyl cyclase (e.g. glucagon sites), some would be inhibitory for adenyl cyclase (e.g. insulin sites), and still others might be inhibitory for phosphodiesterase, the enzyme that breaks down cAMP. Clearly, the characteristics of the

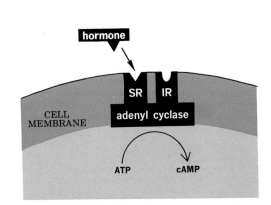

9.24. Model for hormonal activity via cyclic AMP. Stimulatory receptors (SR) and inhibitory receptors (IR) are built into the outer surface of the membrane of the target cell. When a hormone for which one of these receptors is specific complexes with the receptor, the activity of the adenyl cyclase system, which is built into the inner surface of the membrane, is altered. Adenyl cyclase catalyzes synthesis of cAMP from ATP. The cAMP influences chemical reactions within the cell to bring about the cell's characteristic response to the hormone.

various receptor sites should receive much attention in future research.

The second question—why different cells respond so differently to changes in their cAMP content—is much more difficult to answer. It is by no means clear why an increase in cAMP should cause increased thyroxin production by thyroid cells, increased hydrocortisone production by adrenal cortex cells, and increased glucose production by liver cells. We must assume, first, that a vast array of different chemical processes in cells can be regulated by cAMP. Second, we must assume that the response of a given cell to a change in cAMP concentration depends on its own chemical makeup, which in turn depends on its previous developmental history. Since thyroid, adrenal, and liver cells have followed quite different developmental pathways, they must have quite different chemical profiles. Thus a hormonally induced rise in cAMP will take place against vastly different enzymatic backgrounds in each of these three cell types, and the effects of the rise may be correspondingly different.

The role of cyclic AMP as a second messenger in the action of hormones so impressed some researchers that they envisioned cAMP as a possible universal mediator of hormonal action. More recent evidence, however, suggests that cAMP is primarily a mediator of the action of hormones classifiable as proteins, peptides, or amino acids (or derived from these). The steroid hormones seem to behave in a different way. Rather than react with a receptor site on the outer surface of the target-cell membrane, they actually appear to penetrate into the cytoplasm of the target cell. One recent model postulates that the steroid (S) binds to a cytoplasmic receptor (R), and that the complex (S-R) then moves into the nucleus, where, interacting with the genetic material, it influences the instructions for protein synthesis sent from the nucleus to the ribosomes.

Research on cyclic AMP is still so new that this account of the mechanism of hormone–target cell interactions must be viewed merely as a provisional interpretation of what occurs. New data might well force extensive revision of the proposed model.

REFERENCES

AUDUS, L. J., 1959. *Plant Growth Substances,* 2nd ed. Leonard Hill, London.

BEST, C. H., and N. B. TAYLOR, 1966. *The Physiological Basis of Medical Practice,* 8th ed. Williams & Wilkins, Baltimore. (See esp. Chapters 74–78.)

O'MALLEY, B. W., 1971. "Mechanisms of Action of Steroid Hormones," *The New England Journal of Medicine,* vol. 284, pp. 370–377.

PINCUS, G., and K. V. THIMANN, eds., 1948–1964. *The Hormones* (5 vols.). Academic Press, New York.

ROBISON, G. A., R. W. BUTCHER, and E. W. SUTHERLAND, 1970. "On the Relation of Hormone Receptors to Adenyl Cyclase," in *Fundamental Concepts in Drug-Receptor Interactions,* ed. by J. F. Danielli, J. F. Moran, and D. J. Triggle. Academic Press, New York.

SAWIN, C. T., 1969. *The Hormones: Endocrine Physiology.* Little, Brown, Boston.

SWYER, G. I. M., ed., 1970. Issue devoted to "Control of Human Fertility," *British Medical Bulletin,* vol. 26, no. 1.

TURNER, C. D., and J. T. BAGNARA, 1971. *General Endocrinology,* 5th ed. Saunders, Philadelphia.

VAN TIENHOVEN, A., 1968. *Reproductive Physiology of Vertebrates.* Saunders, Philadelphia.

WURTMAN, R. J., and F. ANTON-TAY, 1969. "The Mammalian Pineal as a Neuroendocrine Transducer," *Recent Progress in Hormone Research,* vol. 25. Academic Press, New York.

YOUNG, W. C., ed., 1961. *Sex and Internal Secretions* (2 vols.), 3rd ed. Williams & Wilkins, Baltimore.

SUGGESTED READING

BORTHWICK, H. A., and S. B. HENDRICKS, 1960. "Photoperiodism in Plants," *Science,* vol. 132, pp. 1223–1228.

BUTLER, W. L., and R. J. DOWNS, 1960. "Light and Plant Development," *Scientific American,* December. (Offprint 107.)

CSAPO, A., 1958. "Progesterone," *Scientific American,* April. (Offprint 163.)

GALSTON, A. W., and P. J. DAVIES, 1970. *Control Mechanisms in Plant Development.* Prentice-Hall, Englewood Cliffs, N.J.

GRAY, G. W., 1950. "Cortisone and ACTH," *Scientific American,* March. (Offprint 14.)

HILLMAN, W. S., 1962. *The Physiology of Flowering.* Holt, Rinehart & Winston, New York.

LEVINE, S., 1971. "Stress and Behavior," *Scientific American,* January. (Offprint 532.)

LI, C. H., 1963. "The ACTH Molecule," *Scientific American,* July. (Offprint 160.)

PASTAN, I., 1972. "Cyclic AMP," *Scientific American,* August. (Offprint 1256.)

RASMUSSEN, H., 1961. "The Parathyroid Hormone," *Scientific American,* April. (Offprint 86.)

SALISBURY, F. B., 1963. *The Flowering Process.* Pergamon Press, Oxford.

———, and R. V. PARKE, 1970. *Vascular Plants: Form and Function,* 2nd ed. Wadsworth, Belmont, Calif. (See esp. Chapters 8, 14–18.)

STEWARD, F. C., 1963. "The Control of Growth in Plant Cells," *Scientific American,* October. (Offprint 167.)

VAN OVERBEEK, J., 1968. "The Control of Plant Growth," *Scientific American,* July. (Offprint 1111.)

WENT, F. W., 1962. "Plant Growth and Plant Hormones," in *This Is Life,* ed. by W. H. Johnson and W. C. Steere. Holt, Rinehart & Winston, New York.

WURTMAN, R. J., and J. AXELROD, 1965. "The Pineal Gland," *Scientific American,* July. (Offprint 1015.)

Nervous control

As we saw in the last chapter, there is an intimate relationship between the endocrine and nervous control systems of multicellular animals. Together, these systems make possible the integrated control so typical of animal behavior. Although plants are complex, dynamic organisms that grow, change, react to external stimuli, and move—indeed it is no exaggeration to say that plants behave—their behavior is fundamentally different from that of animals. This difference is an outgrowth of the early evolutionary divergence of the two groups toward different ways of life—sedentary in plants, mobile in animals. Animals probably differ most obviously from plants in the greater speed of their movements and other behavior patterns. Much of the behavior of plants depends on variations in growth rates or changes in the turgidity of cells, both rather slow ways of bringing about movement. Animals do not rely on such processes, having evolved tissues specialized for production of rapid movement, notably the muscles. Correlated with the difference in the speed of movement—a result of the very different ways the movement is produced—are basic differences in the control systems involved.

Hormonal control is a relatively slow process. Even when a hormone is transported via phloem or bloodstream, there is an appreciable delay between the release of the hormone and its arrival at the target organ. Response to the stimulus that induced secretion of the hormone is therefore not immediate; there is a lag of seconds or often of minutes. Given the inherent slowness of the mechanisms of plant movement, however, the delay involved in chemical control is insignificant. Slow chemical control is also sufficient for animals when instantaneous response is not needed, as in control of digestion, salt and water balance, metabolism, and growth. But when rapid response is required, as in the movements produced by skeletal muscles, chemical control is not sufficient.

It is here that nervous control is essential. A nerve impulse can move several hundred feet per second, thus reducing the interval between stimulus and response to milliseconds. Evolution of nerve and muscle tissues, then, was basic to the evolution of active multicellular animals as we know them today.

EVOLUTION OF NERVOUS SYSTEMS

Irritability. Irritability—the capacity to respond to stimuli—is a universal characteristic of protoplasm. Any manifestation of irritability—any reaction to stimulus—ordinarily involves three principal components: (1) reception of a stimulus, (2) conduction of a signal, and (3) response by an effector.

A stimulus is an environmental change of some sort; hence it always involves energy. Any change in the environment is potentially a stimulus, but whether it actually becomes a stimulus depends, first, on the presence of protoplasm; second, on the capacity of the protoplasm to detect the change. Many environmental changes never function directly as stimuli, so far as we know, because no protoplasm can detect them. Examples are radio and TV radiations. Human beings can use these in communication because they have learned to convert undetectable energy changes into detectable ones—sound vibrations and light—which function as the actual stimuli.

Conduction of a signal, the second component in a reaction to stimulus, is a capacity inherent in the nature of protoplasm itself. If a stimulus can produce a change in protoplasm at some point (i.e. if it can be received), neighboring regions of protoplasm will almost surely be influenced to some extent; the initial change will spread. This spread may be limited and slow, or it may be extensive and rapid. It is this tendency of protoplasmic changes to spread from the point of origin, of changes at one point to induce changes at neighboring points, that was the raw material for evolution of nervous conduction.

The response, the third component, is the action prompted by receipt of a stimulus. What different organisms do upon receiving the same stimulus may be very different. And what a single organism does upon receiving the same stimulus on two different occasions may also be different. Thus the response depends more on the characteristics of the organism than on the characteristics of the stimulus. The response is not produced directly by the stimulus—the way the breaking of a vase is by hitting it with a hammer.

Simple nervous pathways. All animal groups above the level of the sponges have some form of nervous system, though in some groups it is very primitive. Most nervous pathways comprise at least three separate cells, which may be adapted for the three functions involved in reaction to stimulus: a receptor cell specialized for reception of a particular kind of stimulus, a conductor cell specialized for conducting impulses over long distances, and an effector cell (frequently a muscle cell) specialized for giving a response (Fig. 10.1B). More complex pathways may involve any number of additional conductor cells interposed between the receptor cell and the effector cell (Fig. 10.1C, D). Once the pathways include several conductor cells, response can be more flexible, because more than one route is usually open to the impulse coming from the receptor; any one of several alternative effectors, or all of the possible effectors, may be activated. In general, the more conductor cells in the circuitry, the more flexible the response.

The typical nerve cell, or *neuron,* consists of an enlarged region, the *cell body,* which contains the nucleus, and one or more long processes, or *nerve fibers,* which extend from the cell body and may measure as much as 6 feet or more (Fig. 3.34, p. 66). Neurons leading from receptor cells are called *sensory neurons;* those leading to effector cells, *motor neurons;* and those lying between the sensory and motor neurons, *association neurons.* Notice that the neurons in a nervous pathway do not actually contact each other (Fig. 10.1); their fibers come very close to each other but a tiny gap remains between them. A junction of this sort between adjacent neurons is called a *synapse.*

Nerve nets and radial systems. The simplest form of organized nervous system is seen in coelenterates of the hydra type, which have reached the level of separate receptor, conductor, and effector cells. The conductor cells do not, however, form definite pathways, but interlace to form a diffuse *nerve net* running throughout the body (Fig. 10.2). There is apparently no central control of any sort. Conduction is slow, and the impulse can move in either direction along the fibers. An im-

10.1. Representative nervous pathways. (A) Pathway in which the receptor cell (which also functions as a conductor cell) is in direct contact with the effector cell. Such short pathways are extremely rare. (B) Pathway in which separate cells perform the three roles of receptor, conductor, and effector. Pathways with only one conductor cell (neuron) are rare. (C) Pathway with separate sensory and motor neurons. (D) The more usual type of pathway, in which association neurons are interposed between the sensory and motor neurons and in which the impulse may follow alternative routes.

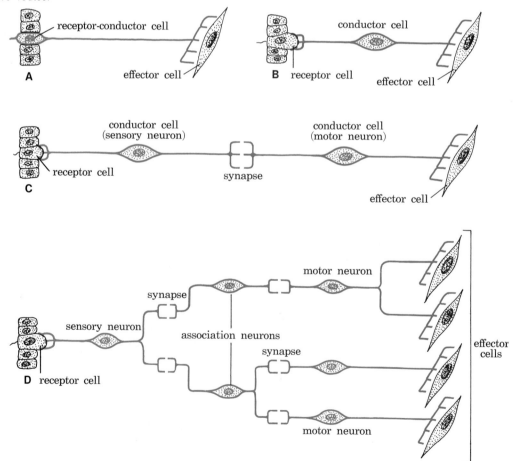

pulse simply spreads from the region of initial stimulation to adjacent regions, becoming less intense as it spreads. The stronger the initial stimulus, the farther the impulses will spread. Reactions are mostly limited to local contractions. Such a system, lacking the potential for central coordination of complex reactions, can produce only a limited behavioral repertoire.

Apparently, radial symmetry, like that of coelenterates, severely restricts the evolutionary potential for extensive centralization of nervous systems. The behavior of even relatively advanced radial animals such as echinoderms is simple compared with that of many bilaterally symmetrical animals. It is to these that we must turn if we are to understand the major trends in the evolution of nervous systems.

Evolutionary trends in bilateral nervous systems. The major trends in the evolution of nervous systems in bilaterally symmetrical animals can be detected even in the lowly flatworms. Let us summarize these trends briefly:

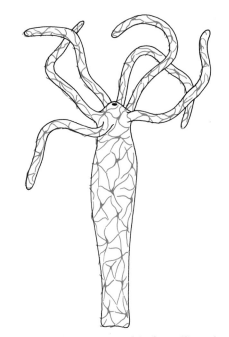

10.2. Nerve-net system of hydra. There is no central nervous system. (The gaps between the fibers of adjacent nerve cells are not as wide as shown here.)

1. Increasing centralization of the nervous system by formation of major longitudinal nerve cords (the *central nervous system*). The central nervous system developed as more and more pathways between receptors and effectors had to pass through the longitudinal cords and the cell bodies of most neurons came to lie in or near them.
2. Increasing complexity of nervous pathways within the central nervous system by interpolation of large numbers of association neurons, with a concomitantly increased flexibility of response.
3. Increasing segregation within the central nervous system of cells performing different functions, with eventual formation of distinct functional areas and structures.
4. Increasing dominance of the front end of the longitudinal cords, leading to formation of an ever more dominant *brain.*
5. Limitation of conduction along nervous pathways to one direction only. A distinction thus developed between sensory fibers leading toward the central nervous system (afferent fibers) and motor fibers leading away from the central nervous system (efferent fibers).
6. Increasing number and complexity of sense organs.

These trends are not yet very distinct in the most primitive flatworms (those thought to be most like the ancient ancestral forms); such flatworms have only a nerve net much like that of hydra. Some slightly more advanced flatworms (less like the ancestral forms) show the beginnings of a condensation of major longitudinal cords within their nerve nets, but very little evidence of any special structure at the anterior end that could be called a brain; biologists have, however, charitably labeled "brain" the tiny swellings present there. Flatworms at more advanced stages show a much better developed brain (Fig. 10.3), though even this brain exerts only limited dominance over the rest of the central nervous system.

The most primitive version of the brain was probably almost exclusively concerned with funneling impulses from the sense organs into the cords. Then, because of the adaptive advantage of shortening

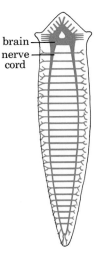

brain
nerve
cord

10.3. Nervous system in a flatworm. Planaria has two cords and a moderately developed brain.

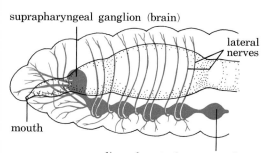

suprapharyngeal ganglion (brain)

lateral
nerves

mouth

ganglion of ventral nerve cord

10.4. Anterior portion of earthworm nervous system. The two parts of the double ventral nerve cord separate at the anterior end and encircle the pharynx. The ring of nervous tissue thus formed consists of the paired suprapharyngeal ganglia, the paired subpharyngeal ganglia, and the cords connecting them. The suprapharyngeal ganglia are customarily regarded as the brain, but they are only slightly larger than the segmental ganglia of the ventral nerve cord. [Modified from T. I. Storer and R. L. Usinger, *General Zoology*, McGraw-Hill Book Co., 1957. Used by permission.]

the pathway these impulses have to follow before reaching the main coordination areas of the central nervous system, natural selection must have favored grouping those areas toward the anterior ends of the cords. Thus the brain came to be more than a sensory funneling area; as coordination increasingly became concentrated in it, it became more and more dominant over the rest of the central nervous system. This dominance has its greatest development in mammals, especially man.

The evolutionary trends whose beginnings can be seen so clearly in flatworms have their most extensive development in the vertebrates and, among invertebrates, in the annelids and arthropods. In all these animals, there is a high degree of centralization. In annelids and arthropods, the central nervous system is a pair of longitudinal cords in which the cell bodies of the neurons form masses called *ganglia* and the fibers, gathered into huge bundles, function as through-conducting systems. Thus in primitive annelids and arthropods, prominent ganglionic masses, a pair in each body segment, are connected by bundles of fibers running between the segments (Fig. 10.4); almost all the cell bodies are located in ganglia. The brain is simply another ganglion located in the animal's head. It is little if any larger than the segmental ganglia; its dominance over the other ganglia is noticeable, but limited in comparison to that of the vertebrate brain.

The central nervous system (spinal cord plus brain) of vertebrates differs in several important ways from those of annelids and arthropods:

1. The vertebrate spinal cord is single; it is located dorsally; and it forms in the embryo as a tube with a hollow central canal, a remnant of which survives in the adult (see Fig. 10.7). The cords of annelids and arthropods, on the other hand, are double (two cords lying side by side and often partly fused); they are located ventrally; and they are always solid.
2. The vertebrate spinal cord is not so obviously organized into a series of alternating ganglia and connecting tracts.
3. Although many coordinating functions in vertebrates are still performed by the spinal cord, there has been extensive development of a brain, which exerts far more dominance over the entire nervous system than that of any annelid or arthropod.

NERVOUS PATHWAYS IN VERTEBRATES

Far more is known about the structure and general function of the nervous system of vertebrates than about those of any other group of animals. For this reason, and because of our natural interest in the system that more than any other makes man man, we shall use vertebrates as our models in discussing basic nervous pathways.

Neurons of vertebrates

Vertebrate neurons may have one, two, or more fibers. When a neuron has more than one fiber, those fibers are usually of two types: *dendrites,* which receive excitation from other cells and conduct impulses toward the cell body; and *axons,* which conduct impulses away from the cell body (Fig. 10.5). A neuron often has many dendrites, but generally only one axon. The dendrites tend to be short; the axon is usually longer. But the most fundamental distinction between dendrites and axons is that dendrites receive excitation from other cells whereas axons generally do not, and that axons can stimulate other cells whereas dendrites cannot.

Vertebrate axons are usually enveloped in a sheath formed by special cells, the **Schwann cells,** that almost completely encircle the axons. The Schwann cells play a role in the nutrition of the nerve fibers, and they provide a conduit within which damaged fibers can grow from the cell body back to their proper target tissues. Many axons, though not all, are also enveloped in a **myelin sheath** lying between most of the Schwann-cell cytoplasm and the axon. The myelin sheath is interrupted at regular intervals by **nodes** (Fig. 10.5). The myelin, which contains

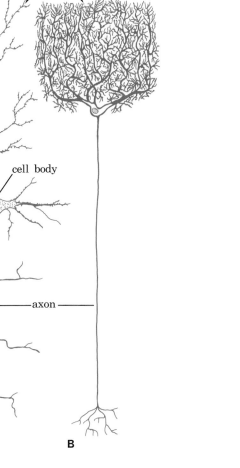

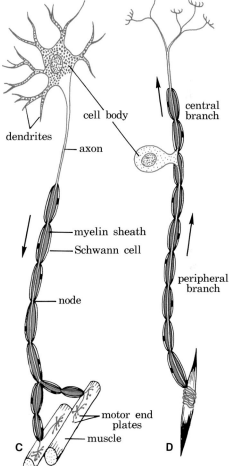

10.5. A variety of neuron types in human beings. (A) The dendrites, unlike the axon, often have a spiny look. (B) The dendrites of certain brain cells branch profusely, giving the cell a treelike appearance. (C) Motor neurons have long axons that run from the central nervous system to the effector (in this case muscle); these axons are frequently, but not always, myelinated. Note the presence of many dark granules in the cell body and dendrites. (D) Sensory neurons generally have only one fiber, which branches a short distance from the cell body, one branch (peripheral) running between the receptor site and the dorsal-root ganglion in which the cell body is located, and the other branch (central) running from the ganglion into the spinal cord or brain. Except for its terminal portions, the entire fiber is structurally and functionally of the axon type, even though the peripheral branch conducts impulses toward the cell body. A sensory neuron thus has no true dendrites, although the peripheral branch is often called a dendrite because of the direction in which it conducts impulses.

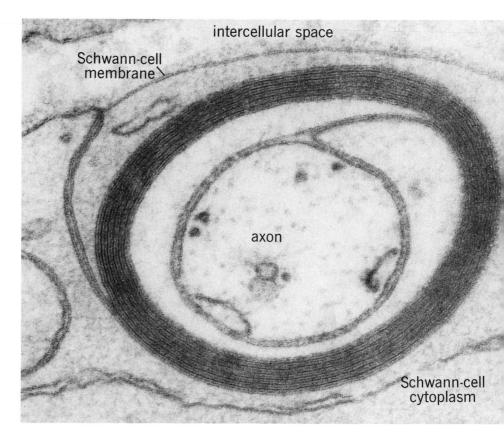

intercellular space

Schwann-cell
membrane

axon

Schwann-cell
cytoplasm

10.6. Myelin sheath. This electron micrograph shows conclusively that the myelin sheath is a coil of the Schwann-cell membrane. × 124,800. [Courtesy J. David Robertson, *Ann. N.Y. Acad. Sci.,* vol. 94, 1961.]

much fatty material, functions in speeding up the conduction of impulses in the axons it envelops. A long-held belief that the sheath is secreted by either the Schwann cells or the axons themselves was disproved when electron microscopy showed that the sheath is a tightly packed spiral of the cell membrane of the Schwann cells (Fig. 10.6).

Reflex arcs

A reflex arc is a simple neural pathway linking a receptor and an effector. Such arcs are important functional units of the nervous system. The reflexes they produce, which are responses to specific stimuli, are usually rapid and relatively automatic.

The simplest reflex arcs in vertebrates involve only two neurons, sensory and motor. A classic example is the knee-jerk reflex arc, commonly checked by physicians during physical examinations (Fig. 10.7). In this reflex, the receptors are stretch receptors. The sensory neuron is a very long one, running all the way from the stretch receptor in the knee to the spinal cord. The cell body of the sensory neuron is located in a *dorsal-root ganglion,* which lies just outside the spinal cord near its dorsal[1] surface. The axon of the sensory neuron enters the cord dorsally and synapses with the dendrites or cell body of a motor neuron within the spinal cord. The axon of the motor neuron then exits ventrally from

[1] The terms "dorsal" and "ventral" as used here refer to the usual vertebrate standing on four legs. In man the dorsal becomes the posterior side, and the ventral becomes the anterior side.

the spinal cord and runs all the way to the effector cells, which in this case are muscle fibers in the leg. Thus, when the physician taps the knee and a stretch receptor is stimulated, impulses travel up a sensory neuron to the spinal cord and back down a motor neuron to the leg, where they stimulate muscle fibers, which contract, causing the leg to jerk.

This very simple reflex illustrates certain features common to all spinal reflex arcs:

1. There is never more than one sensory neuron, which, however, may be exceedingly long.
2. The cell body of a sensory neuron is always outside the spinal cord in a dorsal-root ganglion.
3. The axons of sensory neurons always enter the spinal cord dorsally.
4. The axons of motor neurons always leave the spinal cord ventrally.

So far, we have not used the term "nerve" in our discussions of neural pathways. A **nerve** is a compound structure consisting of a number of neuron fibers bound together. Although there may be thousands of fibers in a single nerve, each is insulated from all the others and conducts impulses independently of the others. A nerve, therefore, is much like a telephone cable containing many functionally separate telephone wires; the many independent communications pathways have been packaged together for structural convenience. Thus the sensory and motor neurons of the knee-jerk reflex run through the same nerve, even though they carry impulses in opposite directions.

Very few reflexes involve only two neurons. At least one association neuron is usually interposed between the axon of the sensory neuron and the dendrites of the motor neuron, and it is common for many association neurons to be involved even in relatively simple reflex arcs. Another factor that makes reflex arcs much more complicated than we have so far indicated is that they always interconnect with other neural pathways. For example, there are always interconnections with pathways leading to the brain, and the brain can send impulses that modify the reflexes.

Let us suppose you decide to try to inhibit the knee-jerk reflex. What do you do? You send many excitatory impulses to those leg muscles that oppose, or antagonize, the extensor muscles. Thus, when your knee is struck and impulses are sent by way of the knee-jerk reflex arc to the extensor muscles, those muscles cannot bring about much jerk (extension) of the leg; there are too many other muscles pulling against them. But you can do more than this. You can send inhibitory impulses to the motor neurons of the extensor muscles themselves. In other words, you can make it more difficult to stimulate the motor neurons to those muscles. This is an important point: Impulses are not all excitatory. In the nervous system as in the endocrine system, it is the interaction between excitatory and inhibitory impulses that makes possible precisely integrated control.

We have mentioned here only a few of the many pathways activated by impulses coming to the cord from the receptors stimulated in the knee-jerk reflex. Evidently, a "simple reflex" is not really simple; even apparently automatic responses to stimuli may involve intricate and

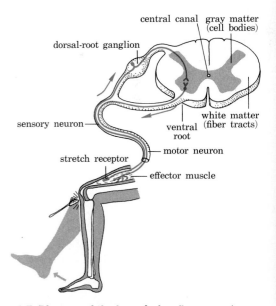

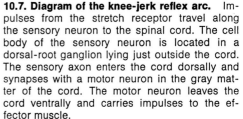

10.7. Diagram of the knee-jerk reflex arc. Impulses from the stretch receptor travel along the sensory neuron to the spinal cord. The cell body of the sensory neuron is located in a dorsal-root ganglion lying just outside the cord. The sensory axon enters the cord dorsally and synapses with a motor neuron in the gray matter of the cord. The motor neuron leaves the cord ventrally and carries impulses to the effector muscle.

complex coordinating mechanisms. Our discussion may suggest how the linking together of sufficient numbers of neural pathways can result in the many complex behavior patterns exhibited by animals.

The autonomic nervous system

The central nervous system serves as a coordinating system for two kinds of pathways: somatic and autonomic. *Somatic* pathways, exemplified by the reflex arcs discussed above, usually innervate skeletal muscle and include one sensory and one motor neuron, both lying largely outside the central nervous system. They involve, at least potentially, some conscious control of the reflex or an awareness that the reflex has occurred. *Autonomic* pathways, by contrast, are not ordinarily under the control of the will and usually function without our being aware of them. They innervate the heart, some glands, and the smooth muscle in the walls of the digestive tract, respiratory system, excretory system, reproductive system, and blood vessels. The pathways of the autonomic nervous system differ structurally from somatic pathways in having two motor neurons instead of one.

The autonomic nervous system is separated into two parts, both structurally and functionally. These are called the *sympathetic* and the *parasympathetic* systems. Most internal organs are innervated by both sympathetic and parasympathetic fibers, with the two systems functioning in opposition to each other. Thus, if the sympathetic system excites a particular organ, the parasympathetic system usually inhibits that organ, and vice versa. In general, the sympathetic system produces the same effects as the hormones of the adrenal medulla, i.e. the effects we have termed the fight-or-flight responses. Present evidence suggests that this nervous mechanism is far more important than the endocrine mechanism in preparing an animal for emergency situations. You can easily figure out for yourself the effects produced by the parasympathetic system by simply reversing the fight-or-flight responses. The condition of an organ innervated by the autonomic nervous system is determined at any given moment by the relative amounts of stimulation coming to it via each of the two parts of the autonomic nervous system.

TRANSMISSION OF NERVOUS IMPULSES

The transmission of impulses along neural pathways may be as fast as 100–300 feet per second. In higher animals, as we have seen, impulses generally move in a specified direction along the dendrites and axon of a neuron; transmission across synapses is also directional, being from the terminal portion of an axon to the dendrites or cell body of the next neuron in the sequence. Two components can be recognized in the propagation of an impulse: conduction along nerve cells and transmission across the synaptic gaps between cells.

Conduction along neurons

General features of the nerve impulse. Nerves will respond to a great variety of stimuli, but electrical stimuli are the most frequently used in laboratory experiments. Let us suppose that we are working with an isolated nerve fiber. We have touched two electrodes to the surface of the fiber at points several centimeters apart. These electrodes are connected to recording equipment, which enables us to detect any electrical changes that may occur at the points on the nerve fiber with which the electrodes are in contact. Now we apply an extremely mild electrical stimulus to the nerve fiber. Nothing happens; our recording equipment shows no change. We increase the intensity of the stimulus and try again. This time our equipment tells us that an electrical change occurred at the point in contact with the first electrode and that a fraction of a second later a similar electrical change occurred at the point in contact with the second electrode. We have succeeded in stimulating the nerve fiber, and a wave of electrical change has moved down the fiber from the point of stimulation, passing first one electrode and then the other. Let us suppose that our recording equipment has enabled us to measure the intensity of this electrical change. We next apply a still more intense stimulus. Again we record a wave of electrical change moving down the fiber, but the intensity and speed of this electrical change are the same as those recorded from the previous stimulation. Again we increase the intensity of the stimulus, but again the recorded change shows the same intensity and speed.

We have learned several important facts from this experiment: (1) A nervous impulse can be detected as a wave of electrical change moving along a nerve fiber. (2) A potential stimulus must be above a critical intensity (and duration) if it is actually to stimulate a nerve fiber; this critical intensity is known as the *threshold* value, and it differs for different nerve fibers. (3) Increasing the intensity of the stimulus above the threshold value does not alter the intensity or speed of the nervous impulse produced; i.e. the nerve fiber fires maximally or not at all, a type of reaction commonly called an *all-or-none response.*

Immediately, an important question comes to mind. If a nerve fiber exhibits the all-or-none property with respect to intensity of impulse and speed of conduction, how do animals normally detect the intensity of a stimulus? They do this in several ways. First, a nerve fiber does not exhibit an all-or-none response with respect to frequency. The more intense the stimulus, the more frequent are the impulses moving along the fiber (up to a maximum value, of course). Second, because different fibers have different thresholds, a more intense stimulus ordinarily stimulates more nerve fibers; individual fibers exhibit all-or-none properties, but nerves (which are composed of many fibers) do not. Apparently the brain interprets both a greater frequency of impulses coming to it via individual fibers and a greater number of stimulated fibers as indicating greater intensity of the stimulus.

The nature of the impulse. When it was discovered over a century ago that a nerve impulse involves electrical changes, scientists assumed that the impulse was a simple electric current flowing through a nerve, just as other currents flow through wires. It was soon shown, however, that the speed of electricity is far greater than the speed of a nerve impulse and, further, that the cytoplasmic core of nerve fibers offers so much resistance to simple electric currents that they die out after moving only a few millimeters through a nerve. This and other evidence led to the conclusion that impulse conduction depends on activity by the living cell, activity that almost certainly involves chemical processes. According to this view, the impulse is not an electric current but an electrochemical change propagated along the nerve fiber.

The now accepted theory of propagation of the nerve impulse is a modified version of one proposed in 1902 by the German physiologist Julius Bernstein. The condition that makes propagation by electrochemical change possible, according to this theory, is the difference in the concentration of certain ions inside nerve cells and in the surrounding fluids. The concentration of sodium ions inside the cell is very low and that of potassium and negative organic ions very high (we saw in Chapter 8 that this is true to some extent of most cells). This unequal distribution of ions results in an electric potential difference across the cell membrane in the resting state, the inside being negatively charged relative to the outside (Fig. 10.8A). In the resting state, the membrane is relatively impermeable to sodium ions. Stimulation, however, radically alters its permeability characteristics. At the point stimulated, it undergoes an initial great increase in permeability to sodium ions, and these rush into the cell (Fig. 10.8B). The inward flux of sodium is so great, in fact, that for a moment the inside actually becomes positively charged relative to the outside (Fig. 10.8C). A fraction of a second later, the membrane becomes highly permeable to potassium ions, which rush out of the cell. This exit of positively charged potassium ions restores the electric charge inside the cell to its original negativity (Fig. 10.8D). In short, the inside surface of the membrane is initially negative; it becomes positive when sodium ions flood inward and then

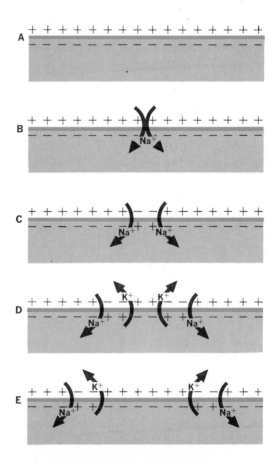

10.8. Model of propagation of a nerve impulse. (A) The interior of a resting nerve fiber is negative relative to the exterior because the ratio of negative to positive ions is higher inside the cell than outside. The interior has a high concentration of potassium ions (K^+), the exterior a high concentration of sodium ions (Na^+). (B) When a fiber is stimulated, the membrane, previously relatively impermeable to Na^+ ions, becomes highly permeable to them at the point of stimulation, and a large number rush into the cell (arrows). (C) The result is a reversal of polarization at that point, the inside of the fiber becoming positive relative to the outside. Meanwhile, because the change in membrane permeability at the point initially stimulated has altered the permeability at adjacent points, Na^+ ions begin rushing inward at those points. (D) An instant later, the membrane at the initial point of stimulation becomes highly permeable to K^+ ions; a large number of these rush out of the cell, and the inside of the fiber once again becomes negative relative to the outside. (E) The cycle of changes at each point alters the permeability of the membrane at adjacent points and initiates the same cycles of changes there; Na^+ ions rush into the cell, and K^+ ions rush out a moment later. As the impulse is propagated along the length of the fiber, active transport restores the original ion distribution (high concentration of K^+ inside, of Na^+ outside), and the fiber is ready to conduct another impulse.

negative again when potassium ions rush outward. Each such cycle of electrical changes is known as an *action potential.*

The nerve impulse is propagated along the neuron because the action potential at each point alters the permeability of the membrane at the adjacent point and initiates an action potential there. The effect is much like that of falling tenpins; when the bowling ball hits the first pin, that pin falls and strikes the next pin, which falls and hits the pin behind it, etc. In a nerve fiber, stimulation at one point produces an action potential at that point and initiates a wave of rapid polarization reversals along the fiber.

The sodium-potassium pump. At this point, we encounter the really perplexing aspect of nerve activity. If impulse conduction involves inward flow of sodium followed by outward flow of potassium, how does the neuron re-establish its original ionic balance? In other words, how does it get rid of the sodium and regain the potassium, so that the initial low concentration of sodium and high concentration of potassium inside the cell will be restored?

Since expelling the sodium means making it move against its concentration gradient and against the electrostatic gradient, and since regaining the lost potassium means making it move against its concentration gradient, some form of active transport across the membrane must be involved. It has been hypothesized that there exists in the membrane of the neuron a so-called sodium-potassium exchange pump of the same basic sort as that discussed in Chapter 8 (see Fig. 8.10, p. 162). We saw there that many cells actively extrude sodium and take in potassium; the neuron, then, is a type of cell that has evolved an extraordinarily effective version of the ionic pump so widely found in other cells. The development of such a pump, combined with the evolution of extreme susceptibility to induction of membrane-permeability changes by external stimuli, has been the basis for the neuron's high degree of specialization for impulse conduction.

The pumping model explains how the cell can maintain its low concentration of sodium ions and its high concentration of potassium ions. But an important question remains: If the pump simply exchanges positive sodium ions for positive potassium ions, i.e. if it is electrically neutral, how does it give rise to a potential difference across the membrane? At first glance, the model seems to provide no way to establish a separation of charge, positive to the outside and negative to the inside. Two facts must be kept in mind, however. First, inside the cell there is a high concentration of negative organic ions, which cannot cross the membrane. Second, the membrane of the resting cell is more permeable to potassium than to sodium. Therefore, even if there were always an exact one-to-one exchange of sodium for potassium—and there is good evidence that this need not be true—potassium would leak back out of the cell faster than sodium would leak back into the cell. This excess of outward-leaking positive potassium ions over inward-leaking positive sodium ions would result in a net outward movement of positively charged ions. There could be no correspondingly large outward movement of negatively charged ions, since the organic ions, which account for most of the negative charge inside the cell, cannot cross the membrane. Thus a separation of charge is established.

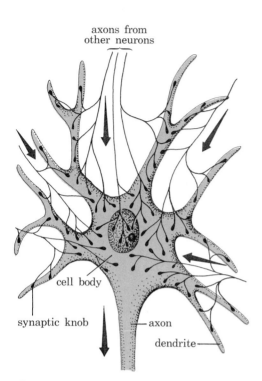

axons from
other neurons

cell body

synaptic knob

axon

dendrite

10.9. Synapses on a motor neuron. Many different axons, each of which branches repeatedly, synapse on the dendrites and cell body of a single motor neuron. Each branch of an axon terminates in a swelling called a synaptic knob.

Transmission across synapses

The nature of synaptic transmission. We have said that the axon of one neuron synapses with the dendrites or cell body of other neurons. Since the terminal portion of an axon usually branches repeatedly, a single axon may synapse with many other neurons, and it usually synapses at numerous points with each of these neurons (Fig. 10.9). Each tiny branch of an axon ends in a small swelling called a *synaptic knob* (Fig. 10.10), whose membrane is separated from the *postsynaptic membrane* of the adjoining cell by a space of about 20 millimicrons. Many scientists once thought that transmission across this gap was electrical, but the evidence now strongly supports the view that most synaptic transmission is by a diffusible chemical.

The electron microscope has revealed that each synaptic knob contains numerous tiny *synaptic vesicles* (Fig. 10.11). Apparently these vesicles contain molecules of the transmitter chemical, which were probably synthesized in the axonal cytoplasm and then taken into the vesicles. It seems probable that an impulse traveling along an axon reaches the synaptic knob and somehow causes vesicles lying near the membrane to discharge their contents into the synaptic cleft. (There is recent evidence that this process may involve cyclic AMP, a substance we have already seen to be of major importance in hormonal activity.) The transmitter molecules released into the cleft diffuse across it and alter the membrane potential of the adjoining nerve cell.

It is the synapses that make transmission of impulses along the neural pathways of higher animals one-way. A neuron can conduct impulses in both directions. If, for example, we stimulate an axon at a point between its base and its terminus, an impulse will move in both directions along the axon from the point of stimulation, as shown in Fig. 10.8. But the impulse moving back toward the cell body and dendrites will die when it reaches the end of the cell; it cannot bridge the gap to the next cell. Only the ends of axons can secrete transmitter substances.

For synapses between neurons outside the central nervous system, the transmitter chemical is *acetylcholine.* Once this chemical is released by the vesicles in the synaptic knobs of axons, it diffuses across the synaptic cleft, exerts its effect at receptor sites on the postsynaptic membrane of the dendrite or cell body of the next cell, and is then promptly destroyed by an enzyme called *cholinesterase.* This destruction is of critical importance. If the acetylcholine were not destroyed, it would continue its stimulatory action indefinitely and all control would be lost. In fact, many insecticides, such as the organophosphates (also known as nerve gases), are cholinesterase inhibitors. They block destruction of acetylcholine, with the result that the animal's nervous system soon runs wild; the insect goes into uncontrollable tremors and spasms, and death ensues.

Acetylcholine has also been implicated as one of a number of transmitter chemicals inside the central nervous system. Others include *noradrenalin* (a substance also produced as a hormone by the adrenal medulla), serotonin, dopamine, glutamate, and gamma-aminobutyric acid (GABA); there are probably still other transmitter chemicals not yet discovered.

The action of transmitter substances. Let us now examine more closely the effects of transmitter substances on postsynaptic membranes. When such a substance has diffused across the synaptic cleft, how does it affect the polarization of the postsynaptic membrane of the dendrite or cell body of the next neuron? If the transmitter substance is an excitatory one, it apparently slightly increases the permeability of the postsynaptic membrane to sodium ions. The resulting increased inward flow of sodium ions along their concentration gradient slightly decreases the polarization of the neuron; i.e. the inside becomes less negative relative to the outside. If the decrease is sufficiently great, it may spread to the base of the cell's axon and there trigger a nerve impulse, which will move down the axon to the next synapse. In effect, excitatory transmitter substances produce a short circuit in the postsynaptic membrane potential, and this triggers the nerve impulse.

Synapses are points of resistance in the nervous pathways, and some impulses reaching the synapses are not transmitted to the next neurons. In such a case, not enough transmitter substance is released to depolarize the postsynaptic membrane sufficiently; hence no impulse is triggered. Ordinarily, excitatory impulses must arrive at more than one synapse on the cell within a short space of time if the cell is to be sufficiently depolarized to trigger an impulse.

Neurons differ in their thresholds, i.e. in the levels of depolarization at which they will fire. Depending on its threshold, a neuron may fire an impulse when stimulated by only a few incoming axons or it may not fire unless stimulated by many incoming axons. The difference in thresholds at different synapses plays an extremely important role in determining the routes impulses will follow through the nervous system. Synapses, in effect, are the regulatory valves of the nervous system. It has been suggested that learning may involve a reduction of the threshold at certain synapses as a result of their use. Later impulses would thus encounter less synaptic resistance if they followed the same pathways as earlier impulses.

The excitatory transmitter substances we have discussed so far reduce the polarization of the postsynaptic membrane. Inhibitory transmitter substances, on the other hand, increase the polarization of the postsynaptic membrane and thus make the neuron harder to fire.

Because synapses act as control valves in the nervous system, and because their proper function depends on a very delicate balance between transmitter substance, deactivating enzyme, and membrane sensitivity, it is not surprising that synaptic malfunctions have been implicated in several mental disorders, among them schizophrenia. Nor is it surprising that many neurological drugs exert their effects at synapses. Used with caution and under medical supervision, some of these drugs can give relief from anxiety and tension or from neurological diseases involving biochemical lesions of synapses. But used improperly, the same agents can induce symptoms strikingly reminiscent of those seen in certain mental disorders, and in some cases the symptoms may be long-lasting or even permanent.

Neurological drugs can alter synaptic function in a variety of ways. On the one hand, they may turn off certain synapses by (1) interfering with synthesis of the appropriate transmitter substance; (2) blocking

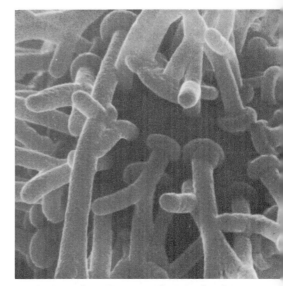

10.10. Scanning electron micrograph of synaptic knobs from a sea hare (*Aplysia californica*). × 8,750. [Courtesy E. R. Lewis, T. E. Everhart, and Y. Y. Zeevi, *Science*, vol. 165, 1969. Copyright 1969 by the American Association for the Advancement of Science.]

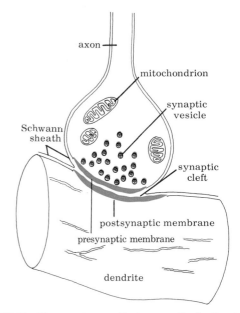

10.11. The synapse. Each synaptic knob at the end of an axon encloses numerous synaptic vesicles containing transmitter substance. When vesicles release this substance into the synaptic cleft, the substance diffuses across the cleft and alters the polarization of the postsynaptic membrane of the dendrite or cell body of the next cell.

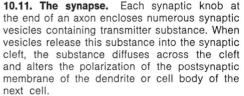

uptake of the transmitter into the synaptic vesicles; (3) preventing release of the transmitter from the vesicles; or (4) blocking the receptor sites on the postsynaptic membranes, so that the transmitter has no effect even if released. On the other hand, drugs may induce excessive and uncontrolled firing of postsynaptic cells by (1) stimulating massive release of transmitter substance from the vesicles; (2) mimicking the effect of the transmitter; or (3) inhibiting destruction of the transmitter once it has done its job, like the insecticides mentioned above.

The physiological mode of action of a drug may help explain the behavioral symptoms it induces. Consider amphetamine (of which a well-known form is Dexidrin) and reserpine. Amphetamine, a stimulant, causes increased release of noradrenalin in the brain. Reserpine, a tranquilizer, blocks uptake of noradrenalin into synaptic vesicles and hence prevents its release. Thus the contrasting behavioral symptoms induced by these drugs are correlated with their opposite effects on some of the same synapses.

Recent research has provided partial explanations for the action of some other important neurological drugs. Nicotine acts as a stimulant because it mimics the effect of acetylcholine. Chlorpromazine, a commonly used tranquilizer, inhibits transmission of impulses at both acetylcholine- and noradrenalin-mediated synapses by combining with receptor sites on the postsynaptic membranes and thus blocking the transmitter substances. LSD (lysergic acid diethylamide) produces its characteristic derangement of certain mental functions by combining indiscriminately with receptor sites for serotonin. The mode of action of marihuana (*Cannabis*) remains unknown.

Transmission between the motor axon and the effector

Just as there is a gap at the synapses between successive neurons in a neural pathway, so is there a gap between the terminus of an axon and the effector it innervates.[2] When the effector is skeletal muscle, the gap is usually contained within a specialized structure, the *motor end plate* (or neuromuscular junction), formed from the end of the axon and the adjacent portion of the muscle surface (Fig. 10.12). Transmission across this gap is via transmitter chemicals. In somatic endings in vertebrates, the transmitter substance is acetylcholine, which produces a similar effect on the membrane of muscle cells as on the postsynaptic membrane in synapses between neurons. Much less is known about the transmitter substances at the motor end plates of invertebrates; it is clear, however, that other substances are important besides acetylcholine (which is not the transmitter substance at the neuromuscular junctions of insects, for example).

Parasympathetic motor fibers of vertebrates release acetylcholine at their junctions with effectors, just as somatic motor fibers do. But many sympathetic motor fibers release noradrenalin. The fact that parasympathetic and sympathetic neuro-effector junctions employ different transmitter chemicals helps explain why these two components of the autonomic nervous system produce different responses by effectors. We have said previously that the sympathetic nervous system and the hor-

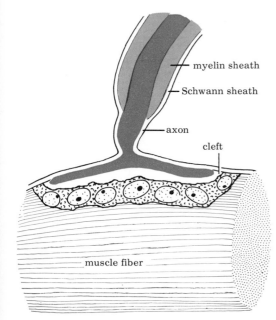

10.12. Motor end plate. The end of the axon and the specialized adjacent portion of the muscle fiber together constitute the motor end plate. As in a synapse between two neurons, there is a cleft between the two cells.

---

[2] Some biologists apply the term "synapse" to both the gap between two neurons and the gap between the end of an axon and an effector cell. Other biologists restrict "synapse" to gaps between neurons; we shall follow this usage.

mones of the adrenal medulla have amazingly similar effects on the body, both eliciting the so-called fight-or-flight syndrome of responses. It now becomes clear why their effects are so similar. Both sympathetic fibers and the adrenal medulla release noradrenalin.

SENSORY RECEPTION

Virtually the entire body of a unicellular organism is directly exposed to environmental stimuli. Even in such an organism, however, certain regions of the cell may be specialized for stimulus reception. As animals became larger and more complex, and as many of their cells lost all direct contact with the outside world, natural selection apparently placed a premium on increasing specialization of certain cells for stimulus reception. For higher animals, these specialized receptor cells came to be the body's principal means of gaining information about the surrounding environment.

In general, each type of receptor is responsive to a particular kind of stimulus: stretching, heat or cold, certain kinds of chemicals, vibrations, light. Most receptors will not respond to stimuli other than those for which they are specialized. Each type of receptor functions as a transducer, converting the energy that constitutes the particular stimulus to which it is attuned into the electrochemical energy of the nerve impulse.

The traditional view is that man has five senses: touch, taste, smell, vision, and hearing. That some persons are endowed with "a sixth sense," as an instrument of special insight, is debatable, but all of us literally have a sixth sense—and a seventh, eighth, ninth, and many more. In short, the traditional five senses are only a few of many senses with which all human beings are endowed. We shall examine some of them here.

Sensory receptors of the skin, skeletal muscles, and viscera

Types of receptors.   There are numerous types of sensory receptors in the skin concerned with at least five different senses, although the traditional classification mentioned above recognizes only one—touch. These five senses are touch, pressure, heat, cold, and pain. Some of the skin receptors, particularly those for pain, are simply the unmyelinated terminal branches of neurons. Others are nets of nerve fibers surrounding the bases of hairs; these fibers, which are particularly important in the sense of touch, are stimulated by the slightest displacement of the tiny hairs present on most parts of the body. Other skin receptors are more complex, consisting of nerve endings surrounded by a capsule of specialized connective-tissue cells.

Unlike the skin receptors, which function in receiving information from the outside environment, some other receptors widely dispersed over the body function primarily in receiving information about the condition of the body itself. Though the senses these receptors mediate are not included in the traditional classification of five, they are of immense importance in the life of the individual. Among them are the stretch receptors (proprioceptors) in the muscles and tendons

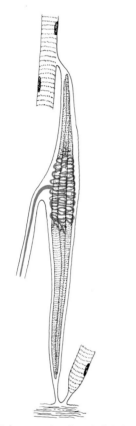

10.13. A stretch receptor in skeletal muscle. The branches of a sensory nerve fiber are intimately associated with several specialized muscle fibers in a region where the muscle fibers have few cross striations. The entire apparatus is called a neuromuscular spindle.

(Fig. 10.13), which we mentioned earlier when discussing the knee-jerk reflex. They send impulses to the central nervous system informing it of the position and movements of the various parts of the body. Other dispersed receptors include those of the so-called visceral senses, located in the internal organs; examples are those in the neck artery sensitive to carbon dioxide concentrations in the blood and those in the aorta sensitive to blood pressure. The firing of such visceral receptors seldom results in sensation (i.e. we are not aware of their action); the responses to their stimulation are usually mediated by the autonomic system. Sometimes, however, stimulation of visceral receptors produces conscious sensations, such as thirst, hunger, and nausea.

Sensations. The sensations you experience when touch receptors are stimulated are very different from those you experience when cold receptors are stimulated. Yet both types of receptors send the same sort of message to the central nervous system; in each case the message consists of waves of electrochemical change moving along neurons, the intensity of the stimulus being signaled by the frequency of impulses along individual neurons and by the number of parallel neurons stimulated. How, then, do these messages, which qualitatively are essentially identical, give rise to the entirely different sensations of touch and cold? The difference lies neither in the receptors nor in the messages they send, but in the destinations of the messages in the brain. The sensation is not an inherent property of the stimulus or of the neural message; it is a creation of the brain.

Each type of receptor sends impulses to a particular part of the brain; cold receptors send impulses to cold centers in the brain, touch receptors to touch centers, and pain receptors to pain centers. You experience the sensation of cold whenever excitatory impulses arrive at a cold center, the sensation of pain whenever excitatory impulses arrive at a pain center. It doesn't matter where the impulses originate. It only matters what part of the brain is stimulated.

Normally, of course, impulses from a given receptor go to the appropriate part of the brain because of the way the nerve circuitry is organized. But the circuitry can be changed experimentally. Thus if the fibers leading from one of your touch receptors and from one of your cold receptors were rerouted so that impulses from the touch receptor would go to a cold center in the brain and vice versa, every time you were touched and a touch receptor fired, you would experience the sensation of cold, and every time a cold object stimulated a cold receptor to fire, you would experience the sensation of touch. In other words, you would experience a sensation "inappropriate" to the stimulus, because the impulses would be going to the wrong part of the brain. But note that this inappropriateness would not change the quality of the sensations; these would be the normal ones of touch and cold, even though they were in response to the wrong stimuli. The sensation is simply your brain's interpretation of incoming stimuli, and depends on the part of the brain stimulated. Someone has dramatically emphasized this point by suggesting that if nerve fibers from your eyes could be crossed with nerve fibers from your ears, you would hear lightning and see thunder.

It is also the brain that is responsible for the localization of the sensation. Each part of the body has its own sensory area in the brain.

Thus fibers from your big toe run to the big-toe center in your brain, fibers from your ankle to the ankle center, fibers from the thumb to the thumb center, etc. Again, it is the part of the brain to which the impulses go, not the stimulus or the receptor or the message itself, that determines localization of the sensation. If the fibers from the pain receptors in your big toe were crossed with those from your thumb, and if your big toe were then pricked with a needle, you would experience a sensation of pain in your thumb and promptly examine it for the cause of the trouble. In other words, the pain sensation is a creation of your brain, and it exists only in your brain, but it is referred by the brain to some other part of the body, where it then seems to you to exist.

The senses of taste and smell

The receptors of taste and smell are chemoreceptors; i.e. they are sensitive to solutions of certain types of chemicals. The two senses are much alike, and when we speak of a taste sensation we are often referring to a compound sensation produced by stimulation of both taste and smell receptors. One reason why hot foods often have more "taste" than cold foods is that they vaporize more, the vapors passing from the mouth upward into the nasal passages and there stimulating smell receptors. And one reason why we cannot "taste" foods well with a cold is that, with nasal passages inflamed and coated with mucus, the smell receptors are essentially nonfunctional. In other words, much of what we call taste is really smell. Conversely, some vapors entering our nostrils pass across the smell receptors and down into the mouth, where they stimulate taste receptors. In each case, taste and smell, chemicals must go into solution in the film of liquid coating the membranes of receptor cells before they can be detected. The major functional difference between the two kinds of receptor is that taste receptors are specialized for detection of chemicals present in quantity in the mouth itself, while smell receptors are more specialized for detecting vapors coming to the organism from distant sources; they are thus much more sensitive than taste receptors, as much as 3,000 times more in some cases.

Taste. The receptor cells for taste are located in *taste buds* on the upper surface of the tongue and, to a lesser extent, on the surface of the pharynx and larynx. The receptor cells themselves are not neurons but specialized cells with hairlike processes on their outer ends (Fig. 10.14). The ends of nerve fibers lie very close to these receptor cells, and when a receptor cell is stimulated, it generates impulses in the fibers.

There are apparently four basic taste senses: sweet, sour, salt, and bitter. The receptors for these four basic tastes have their areas of greatest concentration on different parts of the tongue—sweet and salt on the front, bitter on the back, and sour on the sides (Fig. 10.14). A few substances stimulate only one of the four types of receptors, but most stimulate two, three, or four types in varying degrees. The sensations we experience are thus produced by a blending of the four basic sensations in different relative intensities.

Smell. The receptor cells for the sense of smell (olfaction) in man are located in two clefts in the upper part of the nasal passages.

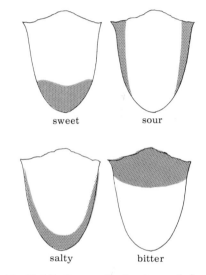

sweet   sour

salty   bitter

10.14. Distribution and structure of human taste buds. The receptors for the four taste senses are on different parts of the tongue. Each taste bud contains two specialized types of cells: supporting cells and receptor cells; the latter bear sensory hairs that are exposed on the tongue surface. The ends of sensory neurons are closely associated with the receptor cells.

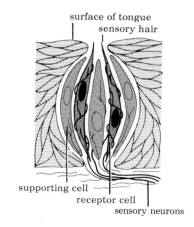

surface of tongue
sensory hair

supporting cell
receptor cell
sensory neurons

Many attempts have been made to identify a group of primary odors from which all more complicated odors can be derived. But the olfactory sense has not proved as easily analyzable as taste, the four basic components of which were identified years ago. As far back as 1895, a system was devised in which all odors were explained in terms of nine basic odors. It soon became clear, however, that this system did not reduce odors to their fundamental components. For example, although most sense receptors exposed to a constant and unchanging stimulus for a long period of time become adapted to the stimulus (i.e. they cease responding to it or respond only weakly), it was shown that when the olfactory receptors became adapted to one odor and were then suddenly exposed to another odor presumed to be in the same one of the nine classes, they frequently responded well. The inference was that different receptors must be involved and that the odors were not really made up of the same basic components. Many other classifications of fundamental odors have been proposed, but most have proved just as unsatisfactory.

The olfactory sense, like all senses, may well differ in different species of animals. It is common knowledge that many animals have a far more acute sense of smell than man, who depends far less on olfaction than most mammals, for example. Some moths have incredibly sensitive smell receptors, the males being capable of detecting the females at a distance of several miles. If the sense of smell depends on a limited number of primary odors, it seems entirely possible that these are not the same in all animals. Mammals and moths, for instance, are so different in their other characteristics that one would expect their smell receptors to function differently also.

The sense of vision

Light receptors of animals. Almost all animals respond to light stimuli. Even Protozoa react quickly to changes in light intensity, often moving away from brightly lit areas. In fact, the single cells of many Protozoa have a special region that serves as a sensitive detector of light. This region contains a pigment that, when exposed to light energy, undergoes chemical changes, which "tell" the protozoan that light is present. Most multicellular animals have evolved specialized light receptor cells, but the basic mechanism of detection is the same as in protozoans: Light energy produces changes in a light-sensitive pigment.

The light receptors of many invertebrates do not function as eyes, in the usual sense of that word. They do not form images, but simply indicate to the animal whether or not light is present and, frequently, whether the intensity of the light is increasing or decreasing. Some of these receptors give the animal no clue to the direction in which the light source lies, and the animal responds by essentially random movements. However, many light receptors are structurally arranged in such fashion that direction becomes an additional type of information detectable by the animal. The eyespots of planaria (see Fig. 5.9, p. 107) are an example; the sensory cells in these organs are stimulated primarily by light coming from above and slightly to the front. More complex eyes commonly include a lens capable of concentrating light on the receptor cells, thereby increasing the sensitivity of the eye to light of weak intensity. Lenses made possible the later evolution of image-forming eyes in some molluscs, most arthropods, and most vertebrates.

The independent evolution of image-forming eyes by different groups of animals has resulted in two quite different basic types of eyes: camera-type eyes, such as those of molluscs and vertebrates, and compound eyes, such as those of insects and crustaceans.

A *camera-type eye* utilizes a single-lens system to focus light on a surface containing many receptor cells packed close together; the receptor surface, called the *retina,* thus functions in a manner analogous to a piece of film. The light pattern focused on the retina produces differential stimulation of different receptor cells, just as a light pattern focused on a piece of film produces different amounts of chemical reaction at different points on the film.

A *compound eye,* on the other hand, utilizes many closely packed lenses, each associated with only a few sensory cells (Pl. II-5A). Each lens with its associated receptor cells forms a functional unit called an ommatidium. Image formation depends on the light pattern falling on the surface of the compound eye; this light pattern determines which ommatidia will be stimulated and at what intensity. Thus there is no structure strictly analogous to the retina of a camera eye, the critical surface being the outer surface of the compound eye itself, composed of the closely packed individual lenses. Since each ommatidium points in a slightly different direction, each is stimulated by light coming from different points in the surrounding area. The insect's brain must integrate the messages coming to it from many ommatidia to produce (presumably) an image that represents the sum of many separate much smaller images.

Structure of the human eye. The adult human eye is globe-shaped and has a diameter of approximately one inch (Fig. 10.15). It is encased in a tough but elastic coat of connective tissue, the *sclera.* The anterior portion of the sclera, called the *cornea,* is transparent and more strongly curved, and functions as the first element in the light-focusing system of the eye. Just inside the sclera is a layer of darkly pigmented tissue, the *choroid,* through which many blood vessels run; the choroid is important both as the structure that provides a blood supply to the rest of the eye and as a light-absorbing layer that (like the black inner surface of a camera) helps prevent internally reflected light from blurring the image. Just behind the junction between the main part of the sclera and the cornea, the choroid becomes thicker and has smooth muscles embedded in it; this portion of the choroid is called the *ciliary body.* At a point anterior to the ciliary body, the choroid leaves the surface of the eyeball and extends into the cavity of the eye as a ring of pigmented tissue, the *iris.* The iris contains smooth muscle fibers arranged in both circular and radial directions; when the circular muscle fibers contract, the opening in the center of the iris, called the *pupil,* is reduced; when the radial muscles contract, the pupil is dilated. The iris thus regulates the size of the opening admitting light (the pupil) in about the same way as the diaphragm of a camera regulates the lens aperture.

The *lens,* which functions as the second element in the light-focusing system, is suspended just behind the pupil by a *suspensory ligament* attached to the ciliary body. The lens and its suspensory ligament thus divide the cavity of the eyeball into two chambers, each filled

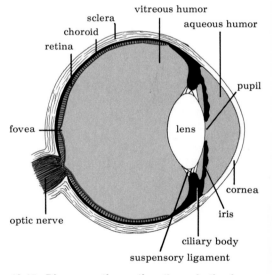

10.15. Diagrammatic section through the human eye.

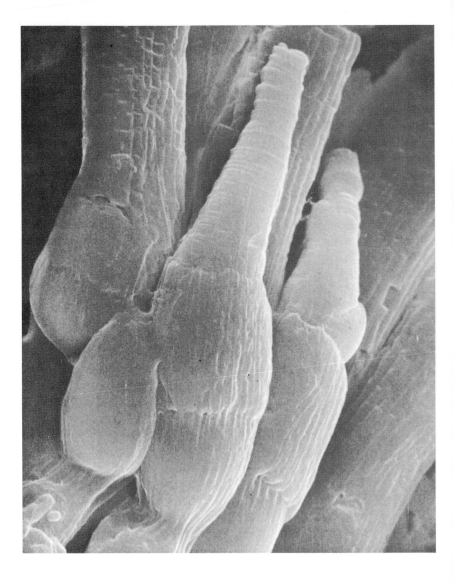

10.16. Scanning electron micrograph of two cones from the retina of a mudpuppy. Some rods can be seen in the background. × 4,060. [Courtesy E. R. Lewis, Y. Y. Zeevi, and F. S. Werblin, *Brain Research,* vol. 15, 1969.]

with transparent fluid or semifluid material (aqueous and vitreous humor).

The *retina,* which contains the receptor cells for the sense of vision, is a thin tissue covering the inner surface of the choroid. It is composed of several layers of cells. The receptor cells are of two types: rod cells and cone cells. The *rod cells* are more abundant toward the periphery of the retina. They are exceedingly sensitive and enable us to see in light too dim to stimulate cone cells. They cannot detect colors, however, and the images to which they give rise are coarse and poorly defined. The *cone cells* (Fig. 10.16), which are more abundant in the central portion of the retina, are used for vision in bright light. They give rise to detailed well-defined images, and they enable us to detect color.

Function of rods and cones. Both rods and cones contain light-sensitive pigments. The pigment in rods (called rhodopsin) is converted into a slightly different form when struck by photons of light and is regenerated to its original form in the dark. Apparently it takes one photon to convert one molecule of rhodopsin. The rod-cell membrane seems to be exceedingly sensitive to the photochemical reaction; available evidence indicates that conversion of one molecule of rhodopsin by one photon is sufficient to cause impulse induction. The impulses going to the brain from the rods signal that light is present, but they convey no information concerning color.

The mechanism of cone vision is much more complex, and our understanding of color perception is still elementary. The experimental evidence indicates that there are three functional types of cones, each containing a different pigment. Each pigment is sensitive to wavelengths of light covering a broad band of the visible spectrum, but has its maximum absorption in a different portion of that spectrum. Thus the three pigments can be designated as blue-absorbing, green-absorbing, and red-absorbing. This evidence for a three-color, three-receptor mechanism of cone reception agrees with a theory of color vision long supported by psychological experiments.

Human beings are so accustomed to their own color vision that they tend to assume other animals see colors in the same way. Yet man and other primates are rather unusual among mammals in possessing color vision; most mammals see only in shades of gray. Among vertebrates that apparently do see in color are many fishes and reptiles and most birds. It is curious that in this characteristic man bears a greater resemblance to these animals, to which he is only distantly related, than to the other mammals, to which he is more closely related.

Insects, too, often have color vision—a very important attribute for the ones that feed on flowers. Many of them, however, do not have the same visible spectrum as man. Their eyes cannot always detect light of the longest wavelengths seen by man; hence a room in which there is only pure red light will be in total darkness to many insects. But these insects can see light of wavelengths in the near end of the ultraviolet band, which human beings cannot see.

Refraction and accommodation.   Since the pattern of illumination of retinal receptor cells is fundamental to image formation, high-resolution image vision depends on precise focusing of incoming light beams on the retina, just as clear high-resolution photography depends on precise focusing of incoming light on the film. If focusing is not good, the image is blurred. The object of focusing is to bring together at one point on the receptor surface all rays of light originating from a single point source. Suppose, for example, that you are looking at the face of another person. If you are to experience a clear image of his face, all light rays reflected from each point on the face must be brought together at a single point on your retina; thus all rays reflected from the point of the chin must be brought together at one point on your retina, all rays from the tip of the nose at another point on your retina, all rays from the center of the forehead at still another point on your retina, etc. In short, the projection of a true image of the observed object onto the retina requires a lens system capable of bending incoming rays of light and focusing them on the retina (Fig. 10.17), just as the lens system of a movie projector focuses onto the screen light that has come from the film.

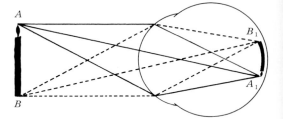

10.17. Image formation on the retina. Incoming light rays from each point on the object being viewed are bent by the cornea in such a manner that they come together at a single point on the retina. Thus an image of the object is projected on the retina. Note, however, that the image is inverted.

Because light rays reflected from a given point travel away from that point in all directions, the rays that strike the eye from a near source will be both more numerous and more divergent than the rays that strike it from a distant source (Fig. 10.18). If all the divergent rays from a near object are to be brought into focus on the retina, they must be strongly bent by the lens system; much less bending is necessary to bring into focus the essentially parallel rays coming from a distant object.

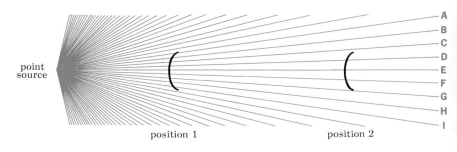

10.18. Difference in degree of divergence of incoming light rays from near and far sources. Light rays from a point source travel outward in all directions. A cornea near the point source (position 1) will thus be struck by many strongly divergent light rays (rays A through I). A cornea farther away (position 2) will be missed by the most divergent rays (A–C and G–I) and will be struck by rays traveling at much smaller angles to each other (D–F). If the cornea is 20 or more feet from the point source, the light rays reaching it will be traveling almost parallel to each other.

The human lens system has two components, the cornea and the lens. Most bending, or *refraction,* of light from distant objects is performed by the cornea. But the cornea cannot refract the strongly divergent light from near objects sufficiently to bring the image into clear focus on the retina. It is the lens, which is the alterable part of the system, that can produce enough additional refraction to bring the image into focus.

The lens is an elastic biconvex structure. It is attached, you will recall, to the ciliary body by the suspensory ligament. When the eye is viewing distant objects (i.e. objects more than 20 feet away), a considerable tension on the suspensory ligament stretches the lens, which is thus flattened and becomes less convex. Now, the less convex a lens is, the less refractive power it has. Hence, when the lens is maximally stretched, it exerts very little influence on incoming light rays; under these conditions, refraction is primarily a function of the cornea. When, however, the eye is viewing a near object, the tension on the suspensory ligament is partly relaxed, and the front surface of the lens bulges outward as a result of its natural elasticity. In other words, the reduced tension on the lens allows it to become more convex, which means that its refractive power increases. Tension on the lens is relaxed until its refractive power adequately supplements the refractive power of the cornea and the image is brought into clear focus on the retina. The nearer the viewed object is to the eye, the more the tension on the lens is relaxed. This process of correcting the focus of images of near objects by changes in the shape of the lens is called *accommodation.*

Structural defects in the shape of the eye are quite common in human beings. Figure 10.19 shows how deviations from normal shape can make eyes far- or nearsighted, and how lenses can compensate for these deviations. Another common defect, astigmatism, stems from unequal curvature of the cornea; it can be corrected with a lens ground unequally to compensate for the irregularities of the cornea.

The sense of hearing

Receptors of the sense of hearing are specialized for detection of vibrations. The human ear is sensitive to vibrations of an amazing range of frequencies—from about 16 to 20,000 cycles per second in young people. Some other animals can hear much higher frequencies; dogs respond to whistles at 30,000 cycles, which few human beings can hear, and bats and some moths can hear frequencies of 100,000 cycles or higher. Like smell and vision, hearing enables an animal to gain information from distant parts of its environment.

Structure of the human ear. The human ear (Fig. 10.20A) is divided into three parts: the *outer ear,* the *middle ear,* and the *inner ear.* The outer ear consists of the ear flap, or pinna, and the auditory canal. At the inner end of the auditory canal is the *tympanic membrane,* more commonly called the eardrum.

On the other side of the tympanic membrane is the chamber of the middle ear. This chamber is connected to the pharynx via the *Eustachian tube,* which functions as a duct making possible the equalization of air pressure between the outer and middle ear. When you ascend a high hill, the reduced pressure at the higher altitude results in a lower pressure in the auditory canal than in the middle-ear chamber, and the tympanic membrane is stretched outward. The pressure is

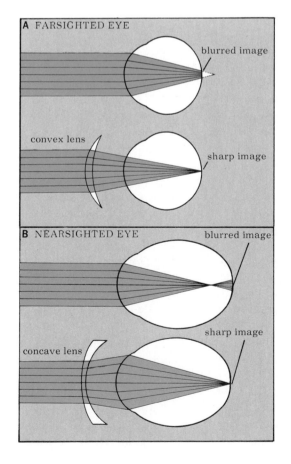

10.19. Farsighted and nearsighted eyes. (A) The farsighted eye is so short that the focal point would be behind the retina; i.e. the cornea and lens cannot bend the light rays enough to bring them together on the retina. Since rays from a distant source are almost parallel, they need less bending than the divergent rays from a near source; hence the person can see distant objects more clearly. Farsightedness is corrected by a convex lens (one that is thicker in the center than at the edges). Such a lens bends the incoming light rays before they reach the cornea, thus aiding the cornea in bringing them together on the retina. (B) The nearsighted eye is so long that the focal point is in front of the retina and the rays have started to diverge again by the time they reach the retina, thus producing a blurred image. The strongly divergent rays from a near source require more bending; the focal point for them is therefore nearer the retina than the focal point for the almost parallel rays from a distant source. Hence the person can see near objects more clearly. Nearsightedness is corrected by a concave lens (one that is thinner in the center than at the edges). Such a lens spreads the incoming parallel rays from a distant source, thus making them strike the cornea at divergent angles just as the rays from a near source would do.

equalized when air escapes from the middle ear through the Eustachian tube into the pharynx. When you descend quickly from a high altitude, the reverse process occurs. As you know, passage of air through the Eustachian tube is facilitated by swallowing, yawning, or coughing. Three small bones (the malleus, incus, and stapes) arranged in sequence extend across the chamber of the middle ear from the tympanic membrane to a membrane called the *oval window*. Another membrane, the *round window*, lies just below the oval window in the wall of the middle-ear chamber.

On the inner side of the oval and round windows is the inner ear, a complicated labyrinth of interconnected fluid-filled chambers and canals. The upper group of chambers and canals is concerned with the sense of equilibrium and will be discussed later. The lower portion of the inner ear consists of a long tube coiled like a snail shell. This is the *cochlea,* which is the organ of hearing. Inside the cochlea are three canals (Fig. 10.20B–D): the vestibular canal, which begins at the oval window; the tympanic canal, which connects with the vestibular canal and ends at the round window; and the cochlear canal, which lies between the other two. All three canals are filled with fluid. The sensory portion of the cochlea, called the *organ of Corti,* projects into the cochlear canal from the *basilar membrane,* which forms the lower boundary of the cochlear canal. The organ of Corti consists of a layer of epithelium on which lie rows of specialized receptor cells bearing sensory hairs at their apexes. Dendrites of sensory neurons terminate on the surfaces of the hair cells. Overhanging the hair cells is a gelatinous structure, the *tectorial membrane,* into which the hairs project. When vibrations of the basilar membrane cause the sensory hairs to move up and down against the less movable tectorial membrane, the deformations of the hairs thus produced apparently give rise to a generator potential in the hair cells, and these cells, in turn, stimulate the sensory neurons.

Reception of vibratory stimuli. Let us now trace briefly the steps involved in the reception of vibratory stimuli by the ear. Vibrations in the air pass down the auditory canal of the outer ear and strike the tympanic membrane, causing it to vibrate. These vibrations are transmitted across the cavity of the middle ear to the oval window by the three small middle-ear bones, which are arranged so as to constitute a lever system that diminishes the amplitude of the vibrations but increases their force. The resultant movements of the oval window in their turn produce movements of the fluid in the canals of the cochlea— movements at the same frequencies as those of the air that entered the outer ear. The pressure waves in the fluid of the cochlea cause the basilar membrane to move up and down and rub the hairs of the hair cells against the tectorial membrane. The stimulus to the hair cells is passed on to their associated sensory neurons, which carry impulses to the auditory centers in the brain.

The characteristics of sounds. We can ordinarily distinguish three characteristics of the sounds we hear: pitch, volume (intensity), and tone quality. A satisfactory theory of hearing must explain all three.

Pitch is a function of frequency; low-frequency vibrations stimulate a sensation of low pitch, and high-frequency vibrations stimulate a sensation of high pitch. Apparently, low-frequency vibrations stimulate hair cells near the apex of the cochlea; high-frequency vibrations, hair cells near its base; and intermediate frequencies, hair cells of correspondingly intermediate regions of the cochlea. Thus the hair cells, like the keys of a piano graduated from low pitch to high pitch, are arranged in sequence, from those stimulated by low frequencies at the

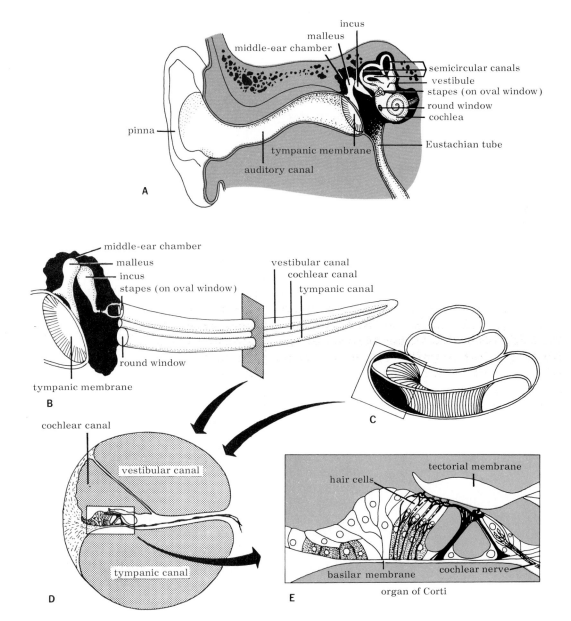

10.20. The human ear. (A) Diagram showing the major parts of the outer, middle, and inner ear (see text for description). (B) Diagram of the relationship between the middle ear and the cochlea, which has here been uncoiled to show its canal system more clearly. (C) A section through the cochlea in its normal coiled state. (D) Enlarged cross section through one unit of the coil, showing the relationship between the vestibular, cochlear, and tympanic canals and the location of the organ of Corti. (E) Enlarged diagram of the organ of Corti, which rests on the basilar membrane separating the cochlear and tympanic canals. When the basilar membrane vibrates and moves the sensory hair cells up and down, the hairs rub against the tectorial membrane overhanging them. The resulting deformation of the hairs produces a generator potential in the hair cells, which triggers impulses in sensory neurons running from the organ of Corti to auditory centers in the brain.

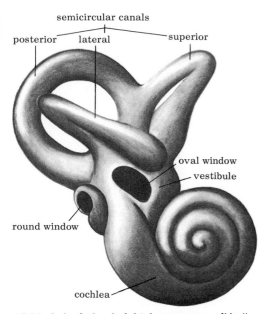

10.21. Labyrinth of right human ear. [Modified from *Sobotta-Figge: Atlas of Human Anatomy,* 8th English ed., Hafner Publishing Co., New York.]

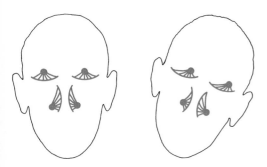

10.22. Function of the otolithic organs. Crystals of calcium carbonate (otoliths) rest on sensory hair cells in the utriculi and sacculi. Changes in the position of the head relative to gravity displace the crystals, thereby altering the pattern of push or pull they exert on the hair cells. This, in turn, alters the pattern of impulses sent to the brain.

apex to those stimulated by high frequencies at the base. The neurons from each region along the length of the cochlea lead to slightly different areas in the brain. The pitch sensation we experience depends on which of these areas of the brain is stimulated.

Volume is a function of the amplitude of vibrations. Thus very intense vibrations cause the fluid of the cochlea to oscillate at greater amplitude, and the correspondingly greater amplitude of oscillation of the basilar membrane produces more intense stimulation of the hair cells. The result is the transmission of more impulses to the brain per unit time. The brain interprets this increased stimulation as loudness.

If a violin, a piano, and a clarinet all play a note at the same pitch and volume, each will sound different. We call this difference tone quality. It is apparently the result of stimulation of hair cells in other regions of the cochlea besides the main region stimulated. The secondary vibrations that produce such stimulation are known as harmonics or overtones. Different instruments (and different voices) produce different patterns of harmonics. Tone quality is thus the interpretation put by the brain on the pattern of hair cells stimulated.

Not all animals with an acute sense of hearing detect all three of the characteristics of sound important to man. Many insects, for example, have well-developed sound receptors, but all available evidence indicates that they cannot detect pitch or tone quality. On the other hand, they are more responsive than human beings to changes in the intensity and in the rhythm (duration and pattern) of bursts of sound.

The senses of equilibrium and acceleration

The upper portion of the labyrinth of the inner ear is composed of three *semicircular canals* and a large vestibule that connects them to the cochlea (Fig. 10.21). Inside the vestibule are two chambers, the *utriculus* and the *sacculus.* Each contains a bed of sensory hair cells. upon which rest crystals of calcium carbonate. Changes in the position of the head cause these crystals, called otoliths, to exert more pull on some hair cells than on others (Fig. 10.22), thereby stimulating them more. The relative strength of the pulls signals to the cerebellum of the brain what the position of the head is at any given moment.

The semicircular canals are concerned with sensing acceleration (changes in the speed or direction of motion). Each of the three canals is oriented in a different plane of three-dimensional space (Fig. 10.21). At the base of each canal is a small chamber containing a tuft of sensory hair cells. When the head is moved or rotated in any direction, the fluid in the canals lags behind because of its inertia, and thus exerts increased pressure on the hair cells. This pressure stimulates the hair cells to initiate impulses to the cerebellum of the brain. The brain, by integrating the different amounts of stimulation coming to it from each of the three canals, can then determine very precisely the direction and the speed of the movement.

THE BRAIN

As we saw earlier (p. 203), the brains of invertebrate animals are much smaller in relation to the size of their bodies than those of vertebrates, and their dominance over the rest of the central nervous system is usually less pronounced. The brains of the most primitive vertebrates are not much more dominant than those of invertebrates, but they do show the beginnings of the evolutionary trends that have made extensive brain development one of the most prominent characteristics of vertebrates. The immense importance of the vertebrate brain makes it worthy of more extensive treatment than we can give it here. The following is a brief introduction to a few of the most important aspects of this subject.

Evolution of the vertebrate brain

The most primitive vertebrate brains and the partly developed brains of embryos consist of three irregular swellings at the anterior end of the longitudinal nerve cord. These three regions undergo much modification in the course of the development of more advanced vertebrates, showing specially thickened areas in their walls and distinctive outgrowths in other places. Despite these changes, however, the original three divisions of the brain can still be recognized even in the most advanced vertebrates, including man. The three divisions are the *forebrain,* the *midbrain,* and the *hindbrain* (Fig. 10.23).

Very early in its evolution, the brain underwent modifications that set the stage for later evolutionary trends. Briefly, the modifications were these:

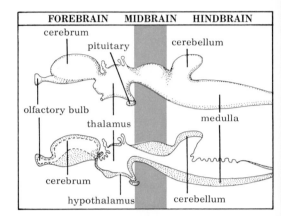

10.23. Diagram of the principal divisions of the vertebrate brain. Top: Lateral view. Bottom: Longitudinal section. [Modified from A. S. Romer, *The Vertebrate Body,* Saunders, 1962.]

1. The ventral portion of the hindbrain, or *medulla oblongata,* became specialized as a control center for some autonomic and somatic pathways concerned with visceral functions (e.g. breathing and heartbeat) and as a connecting tract between the spinal cord and the more anterior parts of the brain, while the anterior dorsal portion of the hindbrain became much enlarged as the *cerebellum,* a structure concerned with balance, equilibrium, and muscular coordination.
2. The dorsal part of the midbrain became specialized as the *optic lobes,* visual centers associated with the optic nerves.
3. The forebrain became divided into an anterior portion consisting of the *cerebrum,* with its prominent olfactory bulbs, and a posterior portion consisting of the *thalamus,* which plays a role in sensory integration and arousal, and the *hypothalamus.*

Later evolution made few changes in the hindbrain, though the cerebellum became larger and more complex in many animals. The really major evolutionary change was the steady increase in size and importance of the cerebrum, with a corresponding decrease in relative size and importance of the midbrain (Fig. 10.24).

The ancestral cerebrum was only a pair of small smooth swellings concerned almost exclusively with the sense of smell. As in the spinal cord, the gray matter (cell bodies and synapses) was mostly internal. The synapses functioned only as relays between the olfactory bulbs and more posterior parts of the brain; little if any correlation of information from different sources occurred in the cerebrum. The cerebrums of modern fishes are still little more than this, although the areas of gray matter are more massive. In amphibians, which evolved from ancestral fish, there was an expansion of the gray matter and a multiplication of synapses between neurons. No longer was the cerebrum merely a relay

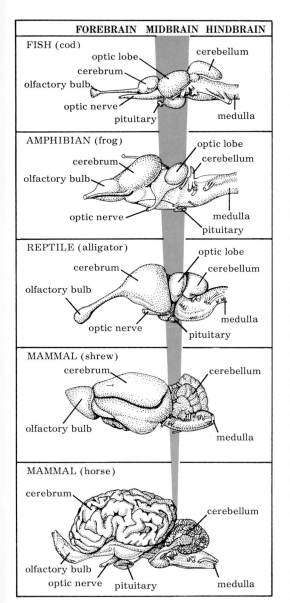

FOREBRAIN MIDBRAIN HINDBRAIN

FISH (cod)
optic lobe
cerebellum
cerebrum
olfactory bulb
optic nerve
pituitary
medulla

AMPHIBIAN (frog)
optic lobe
cerebrum
cerebellum
olfactory bulb
optic nerve
medulla
pituitary

REPTILE (alligator)
optic lobe
cerebrum
cerebellum
olfactory bulb
optic nerve
medulla
pituitary

MAMMAL (shrew)
cerebrum
cerebellum
olfactory bulb
medulla

MAMMAL (horse)
cerebrum
cerebellum
olfactory bulb
optic nerve
pituitary
medulla

10.24. Evolutionary change in relative size of midbrain and forebrain in vertebrates. In the sequence from fish through amphibian, reptile, and primitive mammal (shrew) to advanced mammal (horse), the relative size of the midbrain markedly decreases, while the forebrain expands enormously. [Modified in part from A. S. Romer, *The Vertebrate Body,* Saunders, 1962; and in part from G. G. Simpson, C. S. Pittendrigh, and L. H. Tiffany, *Life: An Introduction to Biology,* copyright, © 1957, by Harcourt Brace Jovanovich, Inc. Used by permission of the publishers.]

station; it now functioned as a correlation center between impulses coming to it from the smell receptors and impulses sent to it by other sensory areas of the brain. Slowly much of the gray matter moved outward from its initially internal position, until it came to lie on the surface of the cerebrum. This surface layer is known as the *cerebral cortex* (cortex means bark). In amphibians and many reptiles, it is concerned largely—but not exclusively—with the sense of smell.

But in certain advanced reptiles, a new component of the cortex, called the *neocortex,* arose at a point on the anterior surface of the cerebrum. Mammals, which evolved from reptiles of this type, show the greatest development of the neocortex. Even in primitive mammals, the neocortex has expanded to form a surface layer covering most of the forebrain. This does not mean that the old cortex of the ancestral cerebrum predominantly concerned with smell has been reduced; as you know, the sense of smell remains of prime importance in most mammals. Furthermore, the so-called smell brain of higher vertebrates performs many important functions that have nothing to do with smell; e.g. it plays a role in control of emotions. The old cortex has simply been pushed to an internal position by the immense increase in relative size of the neocortex, which is a major coordinating center for sensory and motor functions involving all senses and all parts of the body.

We have repeatedly said that the neurons from different sensory receptors lead to different parts of the brain. These parts can be located and mapped by carefully following the paths of nerve fibers in dissections and by sectioning and staining. The results thus obtained can be checked by electrical techniques; sensory receptors are stimulated, and recordings made from the brain tell which area of the brain receives the incoming impulses. Similar research approaches allow mapping of the motor areas. It has been shown by mapping the brains of various mammals that the proportion of the total area of the cerebrum devoted to sensory and motor functions differs greatly from one species to another. In general, the larger and the more convoluted the cerebral cortex, the smaller the proportion devoted exclusively to sensory and motor activities. Man represents the extreme example of this trend; the so-called associative areas constitute by far the largest proportion of his cerebral cortex. It is, of course, precisely this characteristic, with its behavioral consequences, that most clearly distinguishes man from other animals.

The mammalian forebrain

Our information about the function of the various parts of the brain, acquired through experiments utilizing electrical stimulation and recording, has reference particularly to such animals as rats, cats, monkeys, and chimpanzees, on which the experiments were performed. But about the human brain, too, a considerable body of information has accumulated, most of it derived from electrical stimulation during brain surgery and from observation of the effects of tumors and of accidental damage or destruction of parts of the brain. We cannot hope to summarize the whole fund of current knowledge here, but a few comments must be made about the forebrain, which plays so vital a part in all our lives.

The hypothalamus. Extensive research on the hypothalamus (Fig. 10.23) has led to the conclusion that this part of the forebrain is the most important control center for the visceral functions of the body. Stimulation of the hypothalamus with microelectrodes has made it possible to locate centers that control hunger, thirst, body temperature, water balance, blood pressure, reproductive behavior, pleasure, hostility, pain, etc. It is possible to fit experimental animals with electrodes inserted into various of these control centers and then, by turning the electricity on or off, make the animal feel hungry or sated, cold or hot, angry or benign. Cats wired in this way may be friendly one moment and in a rage, with fur erect, eyes wide, and claws out, the next moment, depending on whether their rage center is being stimulated or not. Rats with electrodes in their pleasure centers will spend much of the time pressing levers that turn on the current; the sensation is apparently one they cannot resist. Cats that have just eaten a large meal will resume gulping food as soon as stimulation to their hunger centers is turned on. Sometimes these centers are only a millimeter or less apart; e.g. stimulation can sometimes be shifted from extreme pleasure to extreme pain or fright by moving the electrode only 0.02 inch.

The cerebral cortex. Because the cerebral cortex has been identified with intellectual capacity, we tend to think of it as synonymous with the brain. But we have seen that other parts of the brain play a critical role in almost all our activities. The cortex has, in fact, been viewed by some workers as an organ of elaboration and refinement of functions that, in its absence, could be performed to some extent by other parts of the brain. This certainly seems to be true in lower vertebrates. A frog whose entire cerebrum has been removed shows almost no behavioral changes and can see as well as before. A decorticated rat shows no obvious motor defects, and, though its ability to distinguish complex visual patterns is impaired, it can tell light from dark and can respond to movement. A cat without its cerebral cortex can move around, albeit sluggishly, swallow, react appropriately to pain stimuli, say miaow, and even purr, but it has the appearance of an automaton, seeming essentially unconscious. Human beings are almost completely disabled by loss of their cerebral cortices; they become totally blind and extensively paralyzed and, though they can carry out such vegetative functions as breathing and swallowing, they usually soon die. It seems, then, that the cortex does not exert exclusive control over any of the bodily processes vital to life, but that, in the course of its evolution from the simple olfactory forebrain of the primitive vertebrate, it took over more and more of the function of some of the older parts of the brain until, as in human vision, its role became predominant or even essential.

We said earlier that the percentage of the cortex taken up by purely motor and sensory areas is smaller in man than in other animals. Probing with electrodes shows, however, that within these limited areas each part of the body is represented by its own control center. These centers are not arranged in random fashion, but form a regular pattern. Thus a map of the surface of the somatic sensory area of the cortex (Fig. 10.25) gives, on the right side, a picture of the entire left side of the body and, on the left side, a picture of the entire right side of the body (the fiber tracts running between the brain and the spinal cord cross to the opposite side; hence the left brain controls the right side of the body and the right brain the left side). Similar pictures are obtained by mapping the surface of the motor area of the cortex. These pictures, however, are not faithful reproductions of the bodily proportions; they are distorted and grotesque because the area of the cortex devoted to each part of the body is proportional not to the size of the part but to its sensory or motor capabilities.

If human sensory and motor areas can be mapped, is it likewise possible to map the association areas so prominent in the human cortex? This is more difficult, but progress is being made. Wilder Penfield of the Montreal Neurological Institute has located three speech areas on the cortex. Curiously enough, these are almost always restricted to the left hemisphere in the normal brain; most functions are symmetrically represented in the two hemispheres.

Brain surgery, particularly on patients suffering from severe epilepsy or other types of brain damage, has not only allowed mapping of motor, sensory, and speech areas, but has also made possible investigations of memory storage. When Penfield stimulated certain regions of the temporal lobes of the cerebral cortex, he sometimes caused patients to experience remarkably detailed recollections of past events, events that often had not been remembered for many years. One patient relived an episode from her childhood. Another heard her small son playing in the yard, accompanied by sounds of automobiles, barking dogs, and all the other usual neighborhood noises. Another watched a scene from a play she had not seen in years. Another heard Christmas carols being sung in the church in her home town in Holland. About these recollections Penfield says, "The patients have never looked upon an experiential response as a remembering. Instead of that it is a hearing-again and seeing-again—a living-through of past time."

What is memory? We cannot say as yet. We assume that it is a composite of many separate sensory "memory traces," each corresponding to a neuronal circuit facilitated by prior use—probably as a result of some change in the synapses that interconnect the neurons participating in that memory trace. Presumably, the more impulses are sent along a given memory trace, the easier it becomes for later impulses to travel the same circuit. Facilitation induced by repeated use of a circuit would help explain why practice improves performance.

10.25. Function maps of the surface of the somatic sensory area (left) and motor area (right) of the cerebral cortex of man. Note that the area of cortex devoted to each body part is proportional to the importance of the sensory or motor activities of that part, not to its size; hence the face and hand are especially prominent. [Modified from W. Penfield and T. Rasmussen, *The Cerebral Cortex of Man*, Macmillan, 1950.]

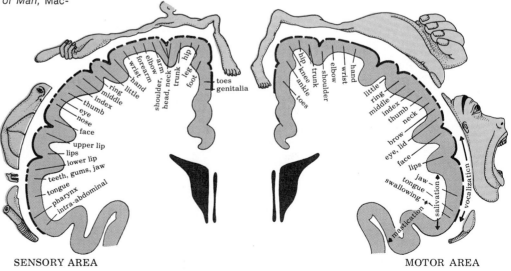

SENSORY AREA   MOTOR AREA

It has been found significant that nerve cells have a high RNA (ribonucleic acid) content and, moreover, that their RNA content increases during neuronal activity. RNA is a nucleic acid, and it is in the nucleic acids that all the information of heredity is stored. The intriguing suggestion has been made that the establishment of memory traces involves encoding information in the RNA of the nerve cells. A more likely possibility is that RNA, which plays a major role in directing protein synthesis, is only indirectly concerned with memory storage and that it is the altered protein content of the cells that in some way facilitates the circuits.

Recent evidence suggests that experience does more than facilitate circuits already present anatomically; it may actually influence the growth of circuits in young animals. Thus whether an infant experiences a stimulus-rich or a stimulus-poor environment may permanently influence his later mental potential.

Our knowledge of how the brain functions is still very rudimentary. But progress is being made at an ever accelerating pace, and the day can be foreseen when the "higher faculties" that set man apart can be explained in physicochemical terms. And with this understanding will come, inevitably, increased power to control the functioning of men's minds (power that is already much further developed in its potential than most people realize). The use man makes of this knowledge may well determine, more than any other factor, what the future of our species will be. The day may come when the problems associated with the ability to control men's brains will outweigh the problems associated with the harnessing of nuclear energy.

REFERENCES

Best, C. H., and N. B. Taylor, 1966. *The Physiological Basis of Medical Practice,* 8th ed. Williams & Wilkins, Baltimore. (See esp. Chapters 2–20.)

Bloom, W., and D. W. Fawcett, 1968. *A Textbook of Histology,* 9th ed. Saunders, Philadelphia. (See esp. Chapters 12, 34–35.)

Eccles, J. C., 1957. *The Physiology of Nerve Cells.* Johns Hopkins Press, Baltimore.

———, 1964. "Ionic Mechanism of Postsynaptic Inhibition," *Science,* vol. 145, pp. 1140–1147.

Gardner, E., 1968. *Fundamentals of Neurology,* 5th ed. Saunders, Philadelphia.

Huxley, A. F., 1964. "Excitation and Conduction in Nerve: Quantitative Analysis," *Science,* vol. 145, pp. 1154–1159.

Penfield, W., and L. Roberts, 1959. *Speech and Brain Mechanisms.* Princeton University Press, Princeton, N.J.

Prosser, C. L., and F. A. Brown, 1961. *Comparative Animal Physiology,* 2nd ed. Saunders, Philadelphia. (See esp. Chapters 10–12, 21.)

Quarton, G. C., T. Melnechuk, and F. O. Schmitt, eds., 1967. *The Neurosciences: A Study Program.* Rockefeller University Press, New York.

Rall, T. W., and A. G. Gilman, 1970. "The Role of Cyclic AMP in the Nervous System," *Neurosciences Research Program Bulletin,* vol. 8, no. 3.

Ruch, T. C., H. D. Patton, J. W. Woodbury, and A. L. Towe, 1965. *Neurophysiology,* 2nd ed. Saunders, Philadelphia.

Walls, G. L., 1942. *The Vertebrate Eye and Its Adaptive Radiation.* Cranbrook Institute of Science, Bloomfield Hills, Mich.

SUGGESTED READING

Amoore, J. E., J. W. Johnston, and M. Rubin, 1964. "The Stereochemical Theory of Odor," *Scientific American,* February. (Offprint 297.)

Carlson, A. J., V. Johnson, and H. M. Cavert, 1961. *The Machinery of the Body,* 5th ed. University of Chicago Press, Chicago. (See esp. Chapters 10–12.)

Delgado, J. M. R., 1969. *Physical Control of the Mind.* Harper & Row, New York.

DiCara, L. V., 1970. "Learning in the Autonomic Nervous System," *Scientific American,* January. (Offprint 525.)

Eccles, J. C., 1958. "The Physiology of Imagination," *Scientific American,* September. (Offprint 65.)

————, 1965. "The Synapse," *Scientific American,* January. (Offprint 1001.)

Fisher, A. E., 1964. "Chemical Stimulation of the Brain," *Scientific American,* June. (Offprint 485.)

French, J. D., 1957. "The Reticular Formation," *Scientific American,* May. (Offprint 66.)

Gazzaniga, M. S., 1967. "The Split Brain in Man," *Scientific American,* August. (Offprint 508.)

Gray, G. W., 1948. "The Great Ravelled Knot," *Scientific American,* October. (Offprint 13.)

Haagen-Smit, A. J., 1952. "Smell and Taste," *Scientific American,* March. (Offprint 404.)

Hodgkin, A. L., 1964. "The Ionic Basis of Nervous Conduction," *Science,* vol. 145, pp. 1148–1154.

Hubel, D. H., 1963. "The Visual Cortex of the Brain," *Scientific American,* November. (Offprint 168.)

Kandel, E. R., 1970. "Nerve Cells and Behavior," *Scientific American,* July. (Offprint 1182.)

Katz, B., 1966. *Nerve, Muscle, and Synapse.* McGraw-Hill, New York.

Kennedy, D., 1967. "Small Systems of Nerve Cells," *Scientific American,* May. (Offprint 1073.)

Keynes, R. D., 1958. "The Nerve Impulse and the Squid," *Scientific American,* December. (Offprint 58.)

Land, E. H., 1959. "Experiments in Color Vision," *Scientific American,* May. (Offprint 223.)

Loewenstein, W. R., 1960. "Biological Transducers," *Scientific American,* August. (Offprint 70.)

MacNichol, E. F., 1964. "Three-Pigment Color Vision," *Scientific American,* December. (Offprint 197.)

Melzack, R., 1961. "The Perception of Pain," *Scientific American,* February. (Offprint 457.)

Michael, C. R., 1969. "Retinal Processing of Visual Images," *Scientific American,* May. (Offprint 1143.)

Miller, W. H., F. Ratliff, and H. K. Hartline, 1961. "How Cells Receive Stimuli," *Scientific American,* September. (Offprint 99.)

Olds, J., 1956. "Pleasure Centers in the Brain," *Scientific American,* October. (Offprint 30.)

Rosenzweig, M. R., E. L. Bennett, and M. C. Diamond, 1972. "Brain Changes in Response to Experience," *Scientific American,* February. (Offprint 541.)

Snider, R. S., 1958. "The Cerebellum," *Scientific American,* August. (Offprint 38.)

Sperry, R. W., 1959. "The Growth of Nerve Circuits," *Scientific American,* November. (Offprint 72.)

———, 1964. "The Great Cerebral Commissure," *Scientific American,* January. (Offprint 174.)

Von Békésy, G., 1957. "The Ear," *Scientific American,* August. (Offprint 44.)

Wald, G., 1950. "Eye and Camera," *Scientific American,* August. (Offprint 46.)

Wooldridge, D. E., 1963. *The Machinery of the Brain.* McGraw-Hill, New York.

CHAPTER 11

Effectors

We discussed the sensory and conductor components of reflex arcs at some length in the preceding chapter. The last components in those arcs, the effectors, will be our subject here. Unlike sensory receptors and conductor cells, effectors are not themselves components of the nervous system; they are the parts of the organism that do things, that carry out the organism's response to stimuli. Their activity may be controlled by the nervous system, but this is not always so; there are numerous effector systems not under nervous control, among them the effector systems of plants, many (though not all) glands of animals, the nematocysts of coelenterates, and the pigment cells of many animals.

The response of an organism to stimulation may not involve movement of the organism in the usual sense. Effector actions that can occur without gross movement of the organism include secretion from glands; changes in the size of pigment cells, or in the distribution of pigment within the cells, with resulting changes in an animal's color (see Pl. IV-3C); and light production by fireflies and by other luminescent organisms. But the responses usually most obvious to us involve motion. It is these that constitute most of what we call behavior. And it is with the effectors that produce motion that we shall be especially concerned in this chapter.

EFFECTORS OF NONMUSCULAR MOVEMENT

Movements produced by differential growth or by turgor changes in plants. We do not ordinarily think of plants as actively moving organisms. Yet their phototropic and geotropic responses are instances of active movement. No specialized effector cells are involved in these responses, the slow movements being produced by differential growth rates under the control of hormones.

But plants are frequently capable of other movements besides growth responses. Some leaves droop or fold at night and expand again in the morning. The flowers of many plants open and close in a regular fashion at different times of day. The leaves of the Venus'-flytrap rapidly close around insects that have landed on them (see Pl. II-3A). All these movements are far too rapid to depend on differential-growth changes. Another mechanism, turgor-pressure change, is involved. Leaves droop when certain of their cells lose so much water that they are no longer turgid enough to give rigidity to the leaf. Flowers fold when specially sensitive cells arranged in rows along the petals lose their turgidity, and they open again when these cells regain their turgidity. Similarly, rapid changes in turgidity in special effector cells located along the hinge of the leaf of the Venus'-flytrap are responsible for that plant's curious behavior.

Cytoplasmic streaming.   Another type of movement in plant cells, as yet only poorly understood, is cytoplasmic streaming. A process of this kind is likewise responsible for the amoeboid movement characteristic of many protozoans, slime molds, human leukocytes, etc. As we saw in Chapter 5, an amoeba moves as its cytoplasm flows into new armlike projections of the cell called *pseudopodia.* For years scientists have sought explanations for this movement in the hope of gaining insight into some fundamental properties of protoplasm, particularly contractile properties. But a definitive explanation has not yet emerged.

Movement by cilia and flagella.   Another type of nonmuscular movement exhibited by some plant and animal cells is that produced by the beating of cilia or flagella. Many protozoans and primitive algae move in this way, and the sperm cells of all animals and of some multicellular plants (large algae, mosses, ferns, etc.) swim by means of cilia or flagella. Even some small multicellular animals use cilia in their movements, and most multicellular animals possess ciliated epithelia that function in moving small particles along the epithelial surface, as in the human trachea.

THE EVOLUTION AND STRUCTURAL ARRANGEMENTS OF MUSCULAR EFFECTORS

The most obvious effectors in all multicellular animals except sponges are the muscles—tissues composed of specialized contractile cells that evolved as multicellular animals became larger and more complex, and as division of labor increased among their cells and tissues.

Animals with hydrostatic skeletons

The first multicellular animals (disregarding the sponges) were doubtless small, perhaps on the order of one millimeter in length. They probably swam by means of cilia. Even today, the smallest flatworms and the tiny larvae of many coelenterates depend primarily on cilia as

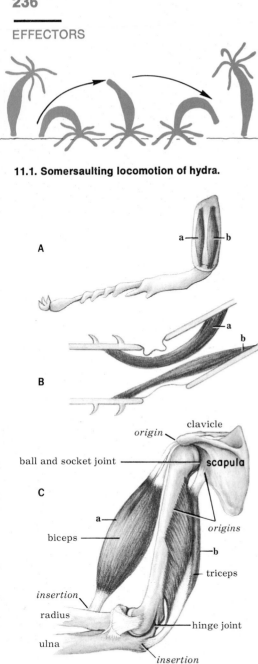

11.1. Somersaulting locomotion of hydra.

11.2. Mechanical arrangement of muscle and skeleton in human arm and insect leg. When the biceps (muscle **a**) of the human arm (B) contracts, the arm is flexed (bent) at the elbow. The triceps (muscle **b**) has the opposite action; when it contracts, the lower arm is extended. The comparable muscles (**a** and **b**) in the insect leg (A) have reverse actions: **a** is an extensor and **b** is a flexor. Thus, although the muscles in both (A) and (B) span the joint, the actions of those in (A) are the reverse of the ones in the same position in (B) because the muscles of (A) are inside the skeleton while those of (B) are outside.

their locomotory effectors. But cilia are practical only in very small organisms. With larger size, animals evolved contractile tissues that first supplemented and then supplanted cilia as the chief effectors of locomotion. Though coelenterates have only very primitive contractile fibers, which do not constitute distinct tissues, rhythmic contractions of such fibers in the bell of a jellyfish or other medusa (Pl. I-7C) enable it to swim weakly, and contractions of other fibers enable a jellyfish or a hydra to move its tentacles. The hydra is even able to move by turning somersaults (Fig. 11.1), a type of movement that is rather surprising in an animal with such primitive nerve and muscle cells.

The musculature of the body wall in the most primitive flatworms is only feebly developed and can produce only minor changes of shape. But in more advanced flatworms, like planaria, the muscle fibers are organized in longitudinal and circular layers. The fibers in these two layers are antagonistic to each other; i.e. they produce opposite actions. Contraction of the longitudinal muscles shortens the animal, contraction of the circular muscles lengthens it. Because the semifluid body contents resist compression, and thus function as a *hydrostatic skeleton,* the body volume remains constant and the shortening is accompanied by a compensating increase in diameter, the lengthening by a compensating decrease in diameter. Notice that in such an animal locomotion is possible only because the force of the muscular contractions can be applied against the noncompressible body contents.

The most complete exploitation of the potentialities of hydrostatic skeletons is seen in certain annelid worms, such as earthworms. Here the body cavity is partitioned into a series of separate fluid-filled chambers. Correlated with this *segmentation* of the body cavity is a similar segmentation of the musculature; each segment of the body has its own circular and longitudinal muscles. It is thus possible for the animal to elongate one part of the body while simultaneously shortening another part. For a worm with an unsegmented body cavity, it would not be so easy to perform a variety of localized movements, because changes in the fluid pressure would be freely transmitted to all parts of the body.

Animals with hard jointed skeletons

Exoskeletons versus endoskeletons. The arthropods and the vertebrates are much the most mobile of the multicellular animals. Both groups possess paired locomotory appendages—legs and sometimes wings. Neither depends on a hydrostatic skeleton as the mechanical resistance against which their muscles act; each has evolved, instead, a hard jointed skeleton, with most of the skeletal muscles so arranged that one end is attached to one section of the skeleton and the other end to a different section (Fig. 11.2). Thus, when the muscle contracts, it causes the skeletal joint between its two points of attachment to bend. In many ways, then, the skeletal and muscular systems of arthropods and vertebrates show striking functional similarities. But these two great groups of animals evolved (if we read the evidence correctly) from entirely different ancestral stocks; they represent the highly successful products of two very different evolutionary lines. It is not surprising, therefore, that an examination of the bases of their striking similarities reveals equally significant differences. The two groups of animals have evolved many similar adaptations to the same functional problems, but they have arrived at those adaptations in entirely different ways.

The most obvious difference between the skeletal systems of arthropods and vertebrates is that the arthropods have an *exoskeleton*— a hard body covering with all muscles and organs located inside it— whereas vertebrates have an *endoskeleton*—a framework embedded within the organism, with the muscles outside. In addition to functioning as structures against which muscles can pull, both types of skeleton are important in providing shape and structural support for large animals, particularly animals living on land, where the buoyancy of water is not available for support; in this respect they are analogous to the rigid xylem, which is a critical factor in enabling land plants to attain large size. Exoskeletons, which are composed of noncellular material secreted by the epidermis, function also as a protective armor for the softer body parts and as a waxy barrier preventing excessive water loss by terrestrial arthropods. The rib cage of the vertebrate endoskeleton protects the organs of the thorax, and the skull and vertebral column protect the brain and spinal cord.

Exoskeletons obviously impose difficulties in overall growth, and periodic molting of the exoskeleton and deposition of a new one are necessary to permit size increase. Further, the mechanics of an exoskeletal system are such as to impose limitations on the possible size of the animal.

The vertebrate skeletal and muscular systems.   Vertebrate skeletons are composed primarily of bone and/or cartilage, two types of connective tissue mentioned earlier (see p. 65).

Cartilage is firm, but not as hard or as brittle as bone. The skeletons of some adult vertebrates, such as sharks and rays, are composed almost entirely of cartilage. Cartilage is the primary component of the skeletons of embryos of all vertebrates; it is progressively replaced by bone as development proceeds. In the adult skeletons of most higher vertebrates, cartilage is retained wherever firmness combined with flexibility is needed, as at the ends of ribs, on the articulating surfaces in skeletal joints, in the walls of the larynx and trachea, in the external ear, and in the nose.

Some bones are partly "spongy," consisting of a network of hardened bars with the spaces between them filled with marrow. Other bones are more compact, their hard parts appearing as an almost continuous mass with only microscopic cavities in them. The shafts of typical long bones, like those of the upper arm and thigh, consist of compact bone surrounding a large central marrow cavity.

Compact bone is composed of structural units called *Haversian systems* (Fig. 11.3). Each such unit is irregularly cylindrical and is composed of concentrically arranged layers of hard inorganic matrix surrounding a microscopic central Haversian canal. Blood vessels and nerves pass through this canal. The scattered irregularly shaped bone cells lie in small cavities located along the interfaces between adjoining concentric layers of the hard matrix. Exchange of materials between the bone cells and the blood vessels in the Haversian canals is by way of radiating canalicules that penetrate and cross the layers of hard matrix.

11.3. Photograph of cross section of bone, showing Haversian systems. Each Haversian system is seen as a nearly round area. The light circular core of each system is the Haversian canal, through which blood vessels pass. Around the Haversian canal is a series of concentrically arranged hard lamellae. The elongate dark areas between the lamellae are cavities, called lacunae, in which the bone cells are located. The numerous very thin dark lines running radially from the central canal across the lamellae to the lacunae are canalicules through which tissue fluid can diffuse. × 300. [Courtesy Thomas Eisner, Cornell University.]

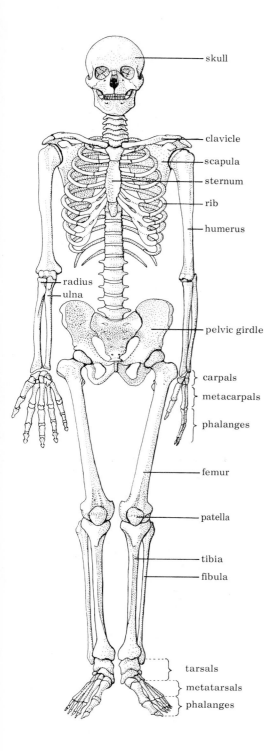

skull

clavicle

scapula

sternum

rib

humerus

radius
ulna

pelvic girdle

carpals

metacarpals

phalanges

femur

patella

tibia

fibula

tarsals

metatarsals

phalanges

11.4. Human skeleton.

Vertebrate skeletons are customarily divided into two components: the axial skeleton, which is the main longitudinal portion, composed of the skull and the vertebral column with its associated rib cage, and the appendicular skeleton, which includes the bones of the paired appendages (fins, legs, wings) and their associated pectoral and pelvic girdles (Fig. 11.4). Some bones are joined together by immovable joints or sutures, as in the case of the numerous small bones that together constitute the skull. But many others are held together at movable joints by *ligaments.* Skeletal muscles, attached to the bones by means of *tendons,* produce their effects by bending the skeleton at these movable joints. The force causing the bending is always exerted as a pull by contracting muscles; muscles cannot actively push. Reversal of the direction in which a joint is bent must be accomplished by contraction of a different set of muscles.

If a given muscle is attached to two bones with one or more joints between them, contraction of the muscle generally causes movement of only one of the two bones, the other being held relatively rigid by other muscles. The end of muscle attached to the essentially stationary bone— generally the proximal end in limb muscles—is called the *origin,* and the end of the muscle attached to the bone that moves—generally the distal end in limb muscles—is called the *insertion* (see Fig. 11.2B). The movable bones behave like a lever system with the fulcrum at the joint. A single muscle sometimes has multiple origins and/or insertions, which may be on the same or on different bones. The action resulting from contraction of any specific muscle depends primarily on the exact positions of its origins and insertions and on the type of joint between them.

Actually, under normal circumstances, muscles do not contract singly. The nervous system does not send impulses to one muscle without sending impulses to other nearby muscles. Thus the various muscles operate in antagonistic groups; if one group of muscles is strongly contracted, an antagonistic group is exerting an opposing pull, and these stretched muscles are ready to reverse the direction of the movement. In addition, other muscles (synergists) serve to guide and limit the principal movement. To understand fully the action of a muscle, therefore, it is necessary to know, in addition to its own origins and insertions, the positions, actions, and relations of its antagonists and synergists.

The types of muscle. Three different types of muscle tissue are recognized in vertebrates: smooth muscle, skeletal muscle, and heart or cardiac muscle.

Smooth muscle (also called visceral muscle) forms the muscle layers in the walls of the digestive tract, bladder, various ducts, and other internal organs. It is also the muscle present in the walls of arteries and veins. The individual smooth-muscle cells—or fibers, as they are commonly called—are thin, elongate, and usually pointed at their ends (Fig. 11.5A). Each has a single nucleus. The fibers are not striated. They interlace to form sheets of muscle tissue rather than bundles. Smooth muscle is innervated by the autonomic nervous system.

Skeletal muscle (also called voluntary or striated muscle) produces the movements of the limbs, trunk, face, jaws, eyeballs, etc. It is by far the most abundant tissue in the vertebrate body. Most of what we commonly call "meat" is skeletal muscle. Each skeletal-muscle fiber is roughly cylindrical, contains many nuclei, and is crossed by alternating light and dark bands called *striations* (Fig. 11.5B). The fibers are usually bound together by connective tissue into bundles rather than sheets; these bundles, in turn, are bound together by connective tissue to form muscles. A muscle, then, is a composite structure made up of many bundles of muscle fibers, just as a nerve is made up of many nerve fibers bound together. Skeletal muscle is innervated by the somatic nervous system.

Cardiac muscle, the tissue of which the heart is composed, shows some characteristics of skeletal muscle and some of smooth muscle. Its fibers, like those of skeletal muscle, are striated and contain numerous nuclei. But like smooth muscle, it is innervated by the autonomic nervous system. Until only a few years ago, it was thought that cardiac muscle differed from both skeletal and smooth muscle in lacking distinct fibers (cells). Ordinary microscopes showed no separation into individual fibers, and it was assumed that the entire heart was a single mass of cytoplasm containing many nuclei. Studies with the electron microscope, however, have revealed that there are, in fact, separate fibers in cardiac muscle, but that the adjacent surfaces of these fibers are so tightly appressed against each other and so complexly interdigitated that they had not previously been recognized as cellular junctions. The sites of these junctions had been visible under light microscopes as dark-colored discs (Fig. 11.5C), but they had been variously interpreted as special contraction bands, as nutritive structures, and as structures for intracellular conduction.

The descriptions of the muscle types given above do not apply in all respects to the muscles of invertebrate animals. For example, all the muscles of insects are striated, even those in the walls of their internal organs; many other invertebrates possess only smooth muscles.

THE PHYSIOLOGY OF MUSCLE ACTIVITY

In vertebrates, skeletal muscle is primarily concerned with effecting adjustments to the organism's external environment, while smooth muscle is responsible for movements in response to internal changes. These differences in action are reflected in differences in their physiological characteristics. Cells of skeletal muscle are innervated by only one nerve fiber; they contract when stimulated by nerve impulses and relax when no such impulses are reaching them. Smooth-muscle cells, by contrast, are usually innervated by two nerve fibers, one from the sympathetic and one from the parasympathetic system; they contract in response to impulses from one of the fibers and relax in response to impulses from the other. Skeletal muscles cannot function normally in the absence of nervous connections and actually degenerate when deprived of their innervation, but smooth muscle (like cardiac muscle) can often contract without any nervous stimulation, as is commonly the case in peristaltic contractions of the intestine, for example. The action of skeletal muscle is more rapid, but smooth muscle can remain contracted longer. Skeletal muscle is more sensitive to electrical stimuli than smooth muscle, but the latter is more sensitive to chemical stimuli. Skeletal muscle has a definite resting length, whereas smooth muscle

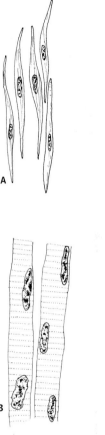

11.5. Fibers of vertebrate muscle. (A) Smooth-muscle fibers. (B) Portions of two skeletal-muscle fibers. Each has many nuclei and is crossed by alternating light and dark bands, or striations. (C) Cardiac muscle. The dark black lines, which were called intercalated discs, are now known to be the places where one cell ends and the next begins.

does not; yet smooth muscle contracts more readily in response to stretching.

More is known at present about the processes involved in contraction of skeletal muscle than of smooth muscle. Consequently most of the discussion below will be restricted to skeletal muscle.

The general features of muscle contraction

Individual muscle fibers resemble individual nerve cells in firing only if an impinging stimulus is of threshold intensity, duration, and rate. In vertebrates, muscle fibers also seem to exhibit the all-or-none property. If an excised vertebrate muscle fiber is administered a stimulus above the threshold value, the same degree of contraction is obtained whatever the value of the stimulus, provided that it is not so strong as to damage the cell.

You know, of course, that you can use the same muscles to perform tasks as different as lifting a pencil and lifting a 20-pound weight. Clearly, an individual muscle can give graded responses depending on the strength of the stimulation reaching it. We can demonstrate this in the laboratory by removing a leg muscle from a frog and attaching it to a device that will measure the extent of contraction of the muscle when it is stimulated. If we administer a stimulus barely above threshold intensity, the muscle gives a very weak twitch (Fig. 11.6). If this is followed after a few seconds' delay by a slightly stronger stimulus, the muscle gives a slightly stronger twitch. We can keep increasing the strength of the stimulus and getting a stronger contraction from the muscle until we reach a point where further increases in the stimulus do not increase the strength of the response. The muscle has reached its maximal response.

How can these results be explained if muscle fibers give all-or-none responses? One possible explanation is that the threshold values of the different muscle fibers of which a muscle is composed are not the same. Furthermore, different muscle fibers may be innervated by different nerve fibers, and these nerve fibers may not all fire at the same time. Thus, although single fibers give an all-or-none response to stimuli, an increase in the strength of the stimulus above the threshold level may elicit a greater response from the whole muscle by stimulating more muscle fibers. Ultimately, however, all the fibers will be stimulated to respond and the muscle will thus have reached a maximal response; increasing the intensity of the stimulus to supramaximal levels will not increase the response, there being no more fibers to stimulate. It must be stressed that the description given here applies only to vertebrate skeletal muscles. The striated-muscle fibers of invertebrates seldom exhibit the all-or-none property; the strength of their contraction is proportional to the frequency of stimulation.

In the experiment described above, we were careful to allow an appreciable time delay between stimuli. Thus each response was induced by a single brief stimulus, and the muscle had sufficient time to relax fully before it was stimulated again. Such a muscular response is called a *simple twitch.* Let us examine its characteristics, again using an isolated frog muscle. If a single adequate stimulus is administered to the muscle, there is a brief interval during which no contraction occurs; this is the *latent period,* the interval between stimulation of the muscle

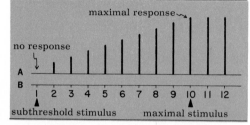

11.6. Response of a muscle to stimuli of various intensities. Line B indicates the intensities at which stimuli are administered. The height of the black bars on line A shows the strength of muscle response. Stimulus 1 is very weak and elicits no response; i.e. it is subthreshold. Stimulus 2 is somewhat stronger and proves to be above threshold, for the muscle contracts. Each stimulus from 3 to 10 is slightly stronger than the preceding, and each elicits a correspondingly stronger muscle contraction. Stimuli 11 and 12 are stronger than 10, but the muscle gives no greater response, an indication that 10 elicits a maximal response.

and the commencement of the shortening process (Fig. 11.7). The latent period is followed by the *contraction period,* and this is followed immediately by a *relaxation period.* These three periods comprise a single simple twitch of the muscle.

Now let us suppose that a series of very frequent stimuli is applied to the muscle. In this case, the muscle will not have completely relaxed after contracting in response to one stimulus when the next stimulus arrives. When this happens, a contraction is elicited that is greater than either stimulus alone would produce (Fig. 11.8). There has been a *summation* of contractions, the second adding to the first. If the initial stimulus was submaximal, the summation may be due in part to recruitment of additional muscle fibers by the second stimulus. But summation can occur even if the individual stimuli are of maximal intensity, i.e. even if all fibers in the muscle are activated by each stimulus. Some change in the physiological condition of the muscle fibers during activity must account for the increased strength of contraction. Our earlier statement that individual muscle fibers give an all-or-none response must therefore be slightly qualified. Each fiber will respond maximally to a single isolated stimulus if that stimulus is above the threshold level, but the fiber may respond a little more strongly if it is given a rapid series of stimuli, because its initial contraction produces chemical changes that make it more irritable. These chemical changes may be in part due to the increase in temperature resulting from the work of contraction.

When stimuli are given with extreme rapidity, the muscle cannot relax at all between successive stimuli. Consequently the individual contractions become indistinguishable and fuse into a single sustained contraction known as *tetanus* (do not confuse this normal muscular response with the bacterial infection known also as lockjaw). For the reasons given in the explanation of summation, of which tetanus is a form, a tetanic contraction is greater than a maximal simple twitch of the same muscle. Normally, a high percentage of our actions involve tetanic contractions rather than simple twitches, because a volley of nerve impulses is sent to the muscle. If, however, a tetanic contraction is maintained too long, the muscle will begin to fatigue, and the strength of its contraction will fall even though the stimuli continue at the same intensity. Fatigue is probably due to an accumulation of lactic acid, depletion of stored energy reserves, and other chemical changes.

Some muscles are never completely relaxed, but are kept in a state of partial contraction called muscle tone or *tonus.* Tonus is maintained by alternate contraction of different groups of muscle fibers, so that no single fiber has a chance to fatigue.

The molecular basis of contraction

It has been known for many years that when a muscle contracts it releases heat. Now, if a contracting muscle can perform work and if it releases heat in the process, an energy supply must be involved. This energy, in the form of ATP, comes from oxidation of food materials such as glycogen, glucose, and fatty acids. In fact, a high percentage of the oxidative phosphorylation in an animal's body occurs in the muscles. Much of the early work on cellular respiration was performed on muscle tissue.

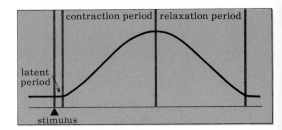

11.7. Record of a simple twitch.

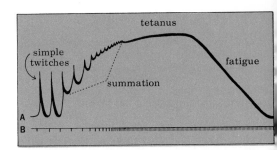

11.8. Record of summation and tetanus. When the stimuli (line B) are widely spaced, the muscle has time to relax fully before the next stimulus arrives, and simple twitches result. As the frequency of the stimuli increases, the muscle does not have time to relax fully from one contraction before the next stimulus arrives and causes it to contract again. The result is summation—contractions that are stronger (and hence produce taller spikes in the trace) than any single simple twitch. If the stimuli are very frequent, the muscle may not relax at all between successive stimulations; the resulting strong sustained contraction is called tetanus. If the very frequent stimulation continues, however, the muscle may fatigue and be unable to maintain the contraction.

During violent muscular activity, as when you are exercising strenuously or lifting a very heavy object, the energy demands of the muscles may be greater than can be met by complete respiration alone, because sufficient oxygen cannot be gotten to the tissues fast enough. Under these circumstances, lactic acid fermentation occurs. The muscles obtain the extra energy they need from anaerobic processes, and thus incur what physiologists call an *oxygen debt*. Some of the lactic acid accumulates in the muscles, but much of it is promptly transported by the blood to the liver. When the violent activity is over, you continue to breathe hard or to pant for some time, thus supplying the liver with the large quantities of oxygen it requires as it reconverts the lactic acid into pyruvic acid and oxidizes some of it, utilizing the energy thus obtained to resynthesize glycogen from the rest of the lactic acid. In this manner, the oxygen debt is paid off.

If ATP is the immediate source of energy for muscle contraction, how is the ATP coupled to the contraction process and what is that process? Analysis of muscle shows that the major components of its contractile parts are two proteins, *actin* and *myosin*. It has been shown that myosin can function as an enzyme that catalyzes removal of the terminal phosphate group from ATP. This is, of course, precisely the energy-liberating reaction we would expect to find coupled to the contraction process. Might myosin function in the intact muscle both as one of the contractile elements and as the enzyme that makes energy available for contraction? The answer seems to be yes. Here is an excellent example of the dual function of some (and perhaps all) proteins; they can act simultaneously as structural elements and as enzymes.

Evidence for the dual function of myosin in muscle contractions comes from experiments performed by Albert Szent-Györgyi at Woods

11.9. Electron micrograph of skeletal muscle from a rabbit. The myofibrils run diagonally across the micrograph from upper left to lower right; each looks like a ribbon crossed by alternating light and dark bands. The wide light bands are called I-bands; there is a narrow dark Z-line in the middle of each I-band. The wide dark bands are A-bands, each of which has a lighter H-zone across its middle. × 24,360. [Courtesy H. E. Huxley, Cambridge University.]

Hole, Massachusetts. Szent-Györgyi showed that if actin and myosin are separately extracted from muscle, purified, and put into solution together, they will combine spontaneously to form a loose complex known as actomyosin. If the actomyosin complex is then precipitated and artificial fibers are prepared from it, the fibers will contract when exposed to ATP. Neither actin nor myosin alone contracts. Clearly, then, the actomyosin complex is a contractile material. And clearly, the complex (actually its myosin component) can itself liberate from ATP the energy necessary for its own contraction, since in this experiment no other possible enzyme is present.

If actomyosin really is the contractile component of muscle, by what process does it contract? One way a protein complex of this sort could shorten would be for it to fold more tightly. This was a hypothesis once supported by many workers. More recently, however, anatomical studies have shown that it is incorrect. Let us turn, then, from the physiological to the anatomical evidence.

We have seen that a skeletal muscle is composed of numerous muscle fibers (cells) bound together by connective tissue. Examination of these fibers under very high magnification reveals that they, in turn, are composed of numerous long thin *myofibrils,* with large numbers of mitochondria in the cytoplasm between them. The myofibrils show the same pattern of cross striations as the fibers of which they are a part (Fig. 11.9). There is an alternation of fairly wide light and dark bands, which have been called *I-bands* and *A-bands* respectively. In the middle of each dark A-band is a region that is lighter than the rest of the A-band but darker than the I-bands—the *H-zone.* In the middle of the light I-band is a very dark thin line called the *Z-line.* The entire region of a myofibril from one Z-line to the next is called a *sarcomere.*

Might the striations so characteristic of skeletal muscle be a structural reflection of the functional contractile units? According to a variety of evidence, yes. Chemical analysis shows that myosin is concentrated in the A-bands and actin in the I-bands. Furthermore, A. F. Huxley of Cambridge University showed in 1954 that the relative widths of the bands change as the fiber contracts; the I-bands and H-zones become narrower but the A-bands remain the same, with the result that the A-bands are moved closer together.

At about the time A. F. Huxley was performing his experiments under a high-power light microscope, H. E. Huxley (no kin to A. F.) of University College, London, developed a new technique for viewing muscle with the electron microscope. He found that within each myofibril there are two types of filaments, thick ones and thin ones (Fig. 11.10F), arranged in a very precise pattern. The two are interdigitated, with the thick ones located exclusively in the A-bands and the thin ones primarily in the I-bands but extending some distance into the A-bands. This explains the different appearances of the A-bands, I-bands, and H-zones. Each dark A-band is precisely the length of one region of thick filaments; it is darkest near its borders, where the thick and thin filaments overlap, and lighter in its mid-region or H-zone, where only the thick filaments are present (Fig. 11.10E). Each light I-band corresponds to a region where only the thin filaments are present. The Z-line is interpreted as a structure to which the thin filaments are anchored at their midpoints; it may function to hold the filaments in proper register,

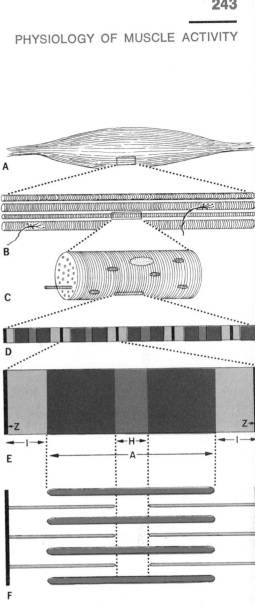

11.10. The component parts of skeletal muscle. (A) A whole muscle. (B) A small part of the muscle magnified to show the muscle cells, or fibers. (C) Part of a fiber much magnified. (D) A myofibril removed from a fiber. (E) A sarcomere of the myofibril much magnified to show the pattern of striations. (F) The myosin and actin filaments that give rise to the pattern of light bands (gray) and dark bands (here shown in color for better contrast). The A-band corresponds to the length of the thick myosin filaments; the H-zone is the region where only the thick filaments occur, while the darker ends of the A-band are regions where thick and thin filaments overlap. The I-band corresponds to regions where only thin actin filaments occur. [Adapted from H. E. Huxley, "The Contraction of Muscle," *Sci. Am.,* November, 1958. Copyright © 1958 by Scientific American, Inc. All rights reserved.]

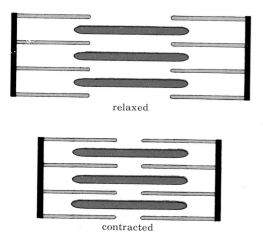

11.11. Arrangement of actin and myosin filaments in a sarcomere in relaxed and contracted states.

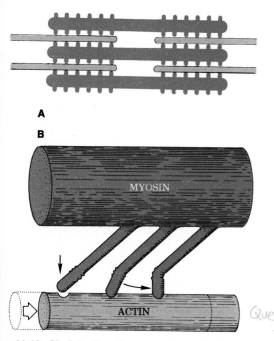

11.12. Model of action of cross bridges between myosin and actin filaments. (A) The cross bridges are thought to be part of the thick myosin filaments. (B) Possible mode of action: The distal portion of each cross bridge may bend, hook onto an active site, and, bending further, pull the actin inward.

and it may also function as the structure against which the filaments exert their pull during contraction.

The observations of the two Huxleys led each of them independently to propose a new theory of muscle contraction—that instead of folding, the filaments telescope together by sliding past each other (Fig. 11.11). If the filaments slide together, the zone of overlap between thick and thin filaments would increase until the thin filaments from the I-bands on the two sides of an A-band might actually meet; this sliding together would reduce the width of the H-zone and even obliterate it entirely if the thin filaments met. The sliding together would also pull the Z-lines closer together and greatly reduce the width of the I-bands. But it would not change the width of the A-bands, since these correspond to the full length of the thick filaments, which remains the same. Thus the sliding theory accounts for the changes that have been observed to occur.

Numerous electron micrographs of muscle fibers in various degrees of contraction have shown that the more contracted the muscle is, the more telescoped the filaments are—a confirmation of the sliding-filament theory. The older folding hypothesis has been largely discredited by the observation that the lengths of the individual filaments do not change appreciably unless the contraction is so great that filaments meet and crumple.

If the sliding-filament theory is the one most widely accepted today, what explanation is given for the sliding? What mechanism is supposed to bring it about? Analysis shows that the thick filaments are composed of myosin and the thin filaments of actin. We have already seen that myosin and actin must unite to form the actomyosin complex before they possess contractile properties. Some sort of connection must exist, then, between the thick myosin filaments and the thin actin filaments. Electron micrographs do indeed show what appear to be small cross bridges between the filaments (Fig. 11.12A). The evidence suggests that they are portions of the myosin molecules. H. E. Huxley has proposed that they act as hooks or levers that enable the myosin filaments to pull the actin filaments (Fig. 11.12B). His hypothesis is that the cross bridges are movable and that they bend toward the actin, hook onto it at specialized receptor sites, and then bend in the other direction, pulling the actin with them; they would then let go, bend in the first direction again, hook onto the actin at a new active site, and again pull. In other words, the sliding together of the filaments would be effected by a ratchet mechanism. Huxley has assumed that each oscillation of a cross bridge would require the energy of one ATP molecule.

The stimulus for contraction

Like the membrane of a resting neuron, that of a resting muscle fiber is polarized, with the outer surface positively charged in relation to the inner one. Stimulatory transmitter substance released by a nerve axon at a neuromuscular junction (see Fig. 10.12, p. 214) causes a momentary reduction of this polarization. If the reduction reaches the threshold level, an impulse, or action potential, is triggered and propagated over the surface of the fiber.

In nerve cells, you will recall, stimulation results in an action potential marked by an inward flow of sodium ions; in muscle cells, it results in an inflow of calcium ions (Ca^{++}) instead. Now, calcium can be shown to stimulate muscle contraction. It is reasonable to suppose, therefore, that the calcium ions, which flow into the cell as the wave of polarization changes moves across its surface, induce the myosin cross bridges to become active. When this idea was first put forward, however, there were at least two major objections to it. First, contraction of a vertebrate muscle fiber requires the essentially simultaneous shortening of all its many myofibrils, but the myofibrils in the center of a fiber are so far from the surface that calcium ions from outside could not possibly diffuse fast enough to reach them in the short interval between stimulation of the fiber and its contraction. Second, there was evidence that not enough calcium enters the cell to account for the sustained contraction resulting from rapid volleys of nerve impulses.

The problem of the coupling of excitation and contraction was solved in 1955, when attention was drawn to an extensive network of tubules in muscle fibers. These tubules comprise two separate but functionally related systems: the *sarcoplasmic reticulum,* which does not open to the exterior, and the *T-system* (for transverse system), which does. The sarcoplasmic reticulum is the muscle cell's highly specialized version of the ubiquitous endoplasmic reticulum (Fig. 11.13). Its membranous canals form a cufflike network around each of the sarcomeres of the myofibrils. At the ends of each sarcomere, in the I-bands, the reticulum forms a series of sacs called *terminal cisternae.* Lying next to these, at the level of the Z-line, is usually a tubule of the T-system. Though the terminal cisternae and the T-tubules are in direct contact, there is no interconnection between their lumina (cavities) and hence no mixing of their contents.

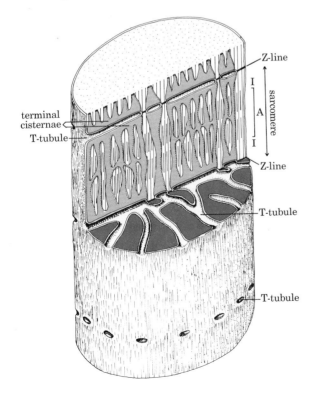

11.13. Portion of a muscle fiber, showing the relationship between the sarcoplasmic reticulum (color) and the T-system. The longitudinal section reveals the intimate association between the terminal cisternae of the sarcoplasmic reticulum and the T-tubules. The cross section at the level of a Z-line shows that the tubules of the T-system are invaginations of the plasma membrane.

The tubules of the T-system have been shown to be deep invaginations of the cell membrane (Fig. 11.13). Therefore when an action potential is propagated across the cell surface, it also penetrates into the interior of the fiber via the membranes of the T-tubules. The action potential moves much faster than diffusing ions, fast enough so that the stimulus for contraction can reach all the myofibrils at nearly the same instant and the myofibrils near the surface and those in the center of the fiber can contract together.

The intimate association discovered between the T-tubules and the terminal cisternae of the sarcoplasmic reticulum suggested that an action potential moving along the membrane of a T-tubule might alter the properties of the adjacent cisternal membranes. It was soon found that this is indeed what happens. The cisternae contain very large amounts of calcium. The action potential induces a sharp increase in the permeability of the cisternal membranes to the calcium ions, allowing them to escape in large numbers (Fig. 11.14). It is this suddenly released intracellular calcium that is the direct stimulant for contraction. Relaxation occurs when a calcium pump in the cisternal membrane causes the calcium ions to move back into the cisternae.

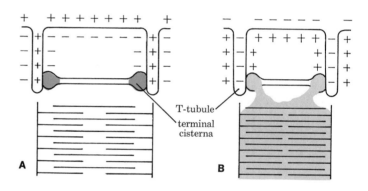

11.14. The role of calcium in stimulation of muscle contraction. (A) A sarcomere in the resting (relaxed) condition. Calcium ions (color) are stored in high concentration in the terminal cisternae of the sarcoplasmic reticulum. (B) The polarization of the membranes of the T-tubules is momentarily reversed during an action potential (impulse), and this reversal of polarization induces release of the calcium ions, which spread over the sarcomere and stimulate contraction.

T-tubule
terminal cisterna

A

B

REFERENCES

BENDALL, J. R., 1969. *Muscles, Molecules, and Movement.* Heinemann, London.

BLOOM, W., and D. W. FAWCETT, 1968. *A Textbook of Histology,* 9th ed. Saunders, Philadelphia. (See esp. Chapters 9–11.)

GIBBONS, I. R., 1968. "The Biochemistry of Motility," *Annual Review of Biochemistry,* vol. 37, pp. 521–546.

GRAY, J., 1968. *Animal Locomotion.* Norton, New York.

HOLWILL, N., 1967. "Contractile Mechanisms in Cilia and Flagella," in *Current Topics in Bioenergetics,* vol. 2, ed. by D. R. Sanadi. Academic Press, New York.

HOWELL, A. B., 1944. *Speed in Animals: Their Specialization for Running and Leaping.* University of Chicago Press, Chicago.

RUCH, T. C., and H. D. PATTON, eds., 1965. *Physiology and Biophysics,* 19th ed. Saunders, Philadelphia. (See esp. Chapter 5.)

ZENKEVICH, L. A., 1945. "The Evolution of Animal Locomotion," *Journal of Morphology,* vol. 77, pp. 1–52.

SUGGESTED READING

ALLEN, R. D., 1962. "Amoeboid Movement," *Scientific American,* February. (Offprint 182.)

CARLSON, A. J., V. JOHNSON, and H. M. CAVERT, 1961. *The Machinery of the Body,* 5th ed. University of Chicago Press, Chicago. (See esp. Chapter 10.)

HOYLE, G., 1970. "How Is Muscle Turned On and Off?" *Scientific American,* April. (Offprint 1175.)

HUXLEY, H. E., 1958. "The Contraction of Muscle," *Scientific American,* November. (Offprint 19.)

————, 1965. "The Mechanism of Muscular Contraction," *Scientific American,* December. (Offprint 1026.)

————, 1969. "The Mechanism of Muscular Contraction," *Science,* vol. 164, pp. 1356–1366.

McELROY, W. D., and H. H. SELIGER, 1962. "Biological Luminescence," *Scientific American,* December. (Offprint 141.)

PORTER, K. R., and C. FANZINI-ARMSTRONG, 1965. "The Sarcoplasmic Reticulum," *Scientific American,* March. (Offprint 1007.)

ROSENBLUTH, J., 1965. "Smooth Muscle: An Ultrastructural Basis for the Dynamics of Its Contraction," *Science,* vol. 148, pp. 1337–1339.

SATIR, P., 1961. "Cilia," *Scientific American,* February. (Offprint 79.)

SMITH, D. S., 1965. "The Flight Muscles of Insects," *Scientific American,* June. (Offprint 1014.)

WILKIE, D. R., 1968. *Muscle.* St. Martin's Press, New York.

CHAPTER 12

Behavior

What animals do and how they do it constitutes one of their outstanding attributes—behavior. The study of behavior, in our view, is no less valid a part of biology than, say, anatomy or physiology. Biology is the study of life, and surely behavior is one of the most fundamental characteristics of animal life.

Admittedly, many aspects of behavior are customarily classified as psychology, which, in turn, is often grouped with the social rather than the biological sciences. But demarcations between allied disciplines are bound to be arbitrary. One of the most invigorating and productive developments in the recent history of science has been the increased blending of the traditional physical and biological sciences, so that it is now often difficult to classify a scientist as a chemist or a biologist. A similar interfusion of the biological and social sciences has begun and will almost certainly gather momentum—surely to the benefit of research in both areas.

THE NATURE OF BEHAVIOR

Behavior is a highly complex subject that demands the application of the most refined techniques of neurophysiology, endocrinology, effector physiology, anatomy, ecology, evolutionary biology, mathematical analysis, etc. It is, in short, a synthetic study, which depends for its own advances on advances in a great variety of other disciplines. It is thus not surprising that rigorous scientific investigation of behavior is a relatively recent development and that our knowledge of this crucial subject is still in its infancy.

Even the mere detailed description of the behavior of an animal is a difficult task—not only because it requires incredibly careful observation over a long period of time, but also because our language lacks adequate words to convey accurately what has been seen. Almost all words have some sort of human connotation, imply some type of human motivation and purpose. But such motivation and purpose may have no relevance to the behavior of other animals, and we must constantly guard against anthropomorphism—against unwarranted attribution of human characteristics to other species.

For centuries men have observed the behavior of animals and have explained it in terms of their own experience. Observing insects fly away at the approach of a man, they have said that the insects are frightened by the man and fly away in order to avoid him. Observing earthworms squirm when pierced by fishhooks, they have said that the hooks hurt the worms, causing them to writhe in pain. Observing adult birds feeding their young, they have said that the birds love their babies and want to feed and protect them. But such descriptions are unacceptable; we have no evidence that being afraid, feeling pain, loving, wanting to do something, as those descriptions apply to human beings, are meaningful when applied to insects, earthworms, or birds. Such interpretations are projections to animals of the sensations human beings would experience in similar circumstances, but as was emphasized in our discussion of the nervous system, sensations are products of the brain and depend upon conscious awareness. Insects and earthworms have brains so different from man's that extrapolations from one to the other require the utmost caution. It is more than unlikely that insects and earthworms experience conscious awareness in the human sense, if, indeed, they experience it in any sense. The same arguments apply, though with slightly less force, to comparisons between birds and men.

What, then, can we say about the insect that flies away, the earthworm that squirms on the fishhook, and the bird that feeds her young? Restricting ourselves to what is observable and provable, we may say that the insect receives visual stimuli from an approaching man, whereupon impulses travel along reflex circuits and stimulate its muscles of flight. About the earthworm, we may say that the hook stimulates receptors that initiate impulses along nerve circuits and that these impulses result in squirming. Clearly, squirming, as a reflex response to stimulation of sensory receptors by fishhooks or other objects, has an adaptive advantage; it might help the worm avoid further physical damage. No conscious awareness or volition need be associated with such an adaptive response. Even the more complex behavior of a mother bird feeding her young can be explained in terms of responses to stimuli within the context of the physiological condition of the bird at the time; no "love" or "desire" need be assumed.

Describing behavior in terms of the simplest neural mechanisms that can account for the observed actions rules out conscious thought, deliberate decision, purposive determination, foresight, as explanations of animal behavior. In short, it is a much-needed safeguard against the unfounded attribution of human characteristics to other animals.

Taxes and reflexes

Numerous experiments have shown that in some cases animals orient so that the critical stimulus registers equally on their left and on their right side. Consider grayling butterflies, which escape from predators by flying toward the sun, thereby causing the predators to be partly

12.1. Phototaxis of a planarian. Two equally bright lights are located at equal distances from the worm. The animal moves toward a point midway between the two lights.

blinded. These butterflies orient to the sun by turning to that position in which both eyes are equally stimulated. If one eye is experimentally blinded, the butterfly flies in circles; it cannot orient toward the sun because it cannot achieve equal stimulation of the two eyes. Similarly, planarians will move toward a light by orienting in such a manner that both eyes are equally stimulated. If two equally bright lights a short distance apart are placed near a planarian, the animal will orient toward a point midway between them, thus attaining equal stimulation of the two eyes (Fig. 12.1). A simple continuously oriented movement of this kind is called *taxis.*

Taxes do not necessarily depend on a comparison within the animal's central nervous system of the incoming stimuli from two sides. Although orientation of the grayling butterfly toward the sun depends on both eyes, male graylings can orient toward a female and follow her in flight even if one eye has been blinded. Similarly, a dragonfly with only one eye can orient toward and pursue its prey.

Responses of animals to simple stimuli need not involve any orientation relative to the stimulus. Paramecians tend to congregate at a fairly precise distance from a bubble of carbon dioxide, but there is no evidence that the animals actually swim toward the bubble. What happens is that when their random swimming movements bring them into a region where the CO_2 has made the water mildly acidic, they swim more slowly. As a result, they gradually collect in the region of mild acidity, even though no oriented response is involved.

Taxes are extremely interesting behavior patterns, of great importance in the lives of lower animals. But since such simple behavior as a paramecian's response to a bubble of carbon dioxide is not dependent on taxes and since, moreover, most taxes involve several reflexes—the whole body being usually oriented in response to the stimulus—we cannot accept taxes as the fundamental units of behavior. Should we then regard reflexes as the basic units of behavior? In a sense, yes. Reflexes are certainly more general and more fundamental than taxes. And it is true that there is no difference in kind between simple reflexes and more complex responses; every possible intermediate stage exists between the simplest reflex pathway and the most complicated neural pathway. In a sense, even the most complex behavior could be viewed as the result of an intricate interaction among many enormously complex reflexes. But such a use of the term "reflex" is rather unproductive; we cannot study complex behavior in the same way as simpler reflexes. And applying the term "reflex" so broadly simply makes it synonymous with "behavior," which doesn't help us at all. It is therefore customary to define *reflex* more restrictively, confining the term to relatively simple and essentially automatic responses to stimuli and to designate more complicated behavior patterns by other terms.

The more complicated behavior patterns, which cannot profitably be interpreted simply as reflexes or taxes, predominate in higher animals. When these behavior patterns are rather rigid, stereotyped, and little modifiable by learning, as in such invertebrates as worms and insects, and in many vertebrates, they are known as *instincts.* In the higher vertebrates, especially the mammals, behavior becomes increasingly modifiable by learning. Taxes, as they are usually understood, are almost nonexistent in the higher mammals, and simple reflexes, though still important, constitute only a very small portion of the total behavioral repertoire.

Inheritance and learning in behavior

Nature versus nurture. The relative importance of inheritance and learning in animal behavior was a subject of fierce controversy during much of the first half of this century. Some psychologists, particularly some members of the "white rat" school in America, went so far as to deny that inheritance plays any significant role in behavior. Their opponents, many of them biologists, pointed out that the developmental potentials for nervous pathways and effectors are inherited, and that an animal can exhibit only those behavior patterns for which it has the appropriate neural and effector mechanisms. They argued further that learning itself depends on inherited neural pathways; if the necessary connections are lacking, no amount of experience can establish a given behavior pattern.

Fortunately, much of the furor of the old "nature versus nurture" controversy has now subsided. Psychologists and biologists alike increasingly recognize that inheritance and learning are both fundamental in determining the behavior of higher animals, and that the contributions of these two elements are inextricably intertwined in most behavior patterns. As in so many instances in the history of science, insistence on an either-or approach proved unproductive.

One way of viewing the interaction of inheritance and learning in animal behavior is to regard inheritance as determining the limits within which a particular type of behavior can be modified and to regard learning as determining, within those limits, the precise nature of the behavior. In some cases, as in the simplest reflexes, the limits imposed by inheritance leave little room for modification by learning; the available neural pathways and effectors rather rigidly determine the response to a given stimulus. In other cases, the inherited limits may be so wide that learning plays the major role in determining the behavior elicited by a given stimulus.

A dramatic example of the interaction between inheritance and learning is that of the song of the European Chaffinch. W. H. Thorpe of Cambridge University raised Chaffinches in isolation and found that such birds gave recognizable Chaffinch calls but were unable to sing a normal Chaffinch song. One might immediately conclude that the call is inherited but that the song is not. The matter is more complicated, however. Thorpe demonstrated that young Chaffinches raised in isolation and permitted to hear a recording of a Chaffinch song when about six months old would quickly learn to sing properly. But young Chaffinches permitted to hear recordings of songs of other species that sing similar songs did not ordinarily learn to sing those other songs. Apparently Chaffinches must learn to sing by hearing other Chaffinches, but they inherit ability to recognize and respond to the songs of their own species. This conclusion was reinforced by experiments in which Thorpe played Chaffinch songs backward to his young birds; they responded to these, even though to our ears such a recording sounds completely unlike a typical Chaffinch song. Here, then, is an example where the inherited limits ordinarily preclude an animal's learning something uncharacteristic of its species.

Clearly, to ask whether Chaffinch song is inherited or learned is to ask a meaningless question. The song is neither wholly inherited nor wholly learned; it is both inherited and learned. Chaffinches inherit the neural and muscular mechanisms requisite for Chaffinch song, and they inherit, apparently, the ability to recognize a Chaffinch song when they hear it, and they inherit severe limits on the type of song they can learn, but the experience of hearing another Chaffinch sing is necessary to trigger their inherited singing abilities into action, and in this sense their song is learned.

Human speech obviously involves a much greater learning component than Chaffinch songs. Human languages differ from one another much more than they do, and the inherited limits for human language must be far less restrictive. But this does not mean that inheritance plays no role in human speech. The fact that all efforts to train apes to speak have failed is evidence enough that human speech depends on human genes.[1]

In general, the inherited limits within which behavior patterns can be modified by learning are much narrower in the invertebrates than in the vertebrates, and narrower in the so-called lower vertebrates (fish, amphibians, reptiles, and birds) than in the mammals. These differences are responsible, in part, for the old nature versus nurture controversy. Psychologists, particularly those who adopted the most extreme views in favor of learning, worked largely on mammals, while zoologists concerned with animal behavior—ethologists, to use the technical term —worked mostly on insects, fishes, and birds. It is not surprising that the psychologists tended to overgeneralize from the learning abilities that impressed them in their mammalian subjects, while the ethologists tended to overemphasize the numerous apparently innate behavior patterns they observed in their insect, fish, and bird subjects.

To say that the inherited limits of behavior patterns in insects or birds tend to be narrower than those of higher mammals is a valid generalization, but it does not mean that the limits of *all* insect or avian behavior are narrower than the limits of *all* mammalian behavior. In any animal, the limits for different behavior patterns are different; each animal, whether insect, bird, or mammal, has some behavioral traits that are rather rigidly determined by inheritance, with very little possibility of modification by learning, and other behavioral traits capable of much modification. Such a difference can be observed in Herring Gulls. The adult birds nesting in colonies learn to recognize their own young about five days after they hatch; thereafter, if young of the same age from another nest are substituted for their own, the adults will not accept them and will neglect or even kill them. Yet these same gulls show amazingly little aptitude for learning to recognize their own eggs. They can be given substitute eggs quite different in color, pattern, shape, or size and will accept them without hesitation. These gulls, then, exhibit great aptitude for learning to recognize individual young so alike in appearance that human beings can tell them apart only with great

[1] The failure to teach apes to speak more than a few simple words is probably due in part to their having very little tendency to attempt to imitate sounds. It may also be due in some degree to anatomical limitations; their larynx is not in the same position, relative to the pharynx, as that of adult human beings. There are recent reports that chimpanzees can be taught to communicate some simple ideas in sign language.

difficulty, if at all, but they show very little aptitude (at least in the experimental context) for learning to recognize eggs so different that human beings can distinguish them at a glance. From an evolutionary point of view, this difference is readily understandable; there must have been far more selection pressure for evolution of recognition of young, which might stray from their own nest under normal circumstances, than for recognition of individual eggs, which in nature seldom wander from nest to nest.

Difficulties in studying learning. It is probably impossible to separate completely what is inherited and what is learned in any behavior pattern. Behavior is not a simple combination of these two elements, but is the outcome of a fusion between them. So far as the part played by learning can be distinguished, however, the learning process is one well worth studying. Before we examine some commonly recognized categories of learning, we should mention several factors that complicate its study.

First, it is often difficult to determine whether improvement in the performance of a behavior pattern is due to experience or simply to greater maturity or to a different physiological condition. For example, observations that young birds just leaving the nest cannot fly well, but improve rapidly over the next few days, have led to the widespread belief that the birds must learn to fly and that they improve with practice. But, as repeated experiments have demonstrated, when young birds are reared in narrow tubes or other devices that prevent them from flapping their wings and are released at an age when they normally would already have "learned" to fly, they are able to fly as well as control birds raised under normal conditions. In other words, it is not practice that causes the flight of a newly fledged bird to improve; it is greater maturity. Numerous other examples could be cited of improvement that appears to be a result of learning but is really a result of maturation. Furthermore, injections of hormones have sometimes caused behavior patterns to change in ways previously thought to be produced only by learning experiences; hence one must be exceedingly careful to rule out physiological changes as a possible cause before asserting that a particular behavioral change is due to learning.

A second complication in studying the learning ability of animals is that an animal may readily learn something in one context and be completely incapable of learning it in some other context. Thus one may erroneously conclude that the animal cannot learn something when, in fact, the negative results are simply a product of the experimental situation used.

Another difficulty in determining what an animal can learn is that a particular behavior can often be learned only during a rather limited critical period in its life. If the animal does not encounter the necessary learning situation during the critical period, it may never learn the behavior. Exposure to the learning situation before or after the critical period may be ineffective in producing learning. For example, during his studies of the development of Chaffinch song, Thorpe demonstrated that, unless young Chaffinches hear a Chaffinch song during a certain period in their development, they never learn to sing properly, despite frequent later exposure to singing Chaffinches. Critical periods are seldom so rigid in human beings, but there is abundant evidence that various types of learning ability are greatest at certain ages. For example, children between the ages of two and ten can learn languages far more easily than adults.

Still another difficulty is that one cannot always tell immediately whether or not learning has occurred. There may be considerable delay between exposure to the learning situation and the performance of a behavior pattern that shows effects of learning. For example, if young Chaffinches only a few weeks old are allowed to hear a tape recording of a singing adult for a few days and are then raised in isolation, they will sing a nearly normal Chaffinch song when they first begin to sing the following spring. Exposure to the song during their first summer, long before they themselves are old enough to sing, results in learning, but the proof does not come until months later.

Much caution is needed in any attempt to compare the learning capabilities of different species. As might be deduced from the differences between their nervous systems, superficially similar learning in different species may actually involve different underlying mechanisms and fulfill entirely different functions in the lives of the animals. Thus, as shown by T. C. Schnierla of Columbia University, rats and ants can learn to run the same maze, but they do so in very different ways; rats appear to learn a "map" of the maze as a whole, whereas ants appear to learn the maze as a series of separate problems, one at each choice point. Mastery of the maze tends to improve the performance of rats when they are subsequently placed in new mazes, but it actually seems to hinder the performance of ants in new mazes. In other words, rats not only learn the particular maze but can also generalize to some extent from this experience and thus develop increased competence at running other mazes, whereas ants learn only the particular maze and this achievement makes their behavior in new mazes less flexible.

Types of learning. There are many different classifications of learning. We shall restrict our discussion here to some of the categories recognized by one of the most commonly used systems. All forms of learning involve, by definition, relatively enduring changes in behavior due to experience rather than maturation. More transient changes, such as those due to sensory adaptation, fatigue, fluctuations in physiological condition, and differences in motivation, are not considered to be learning.

1. One of the simplest types of learning is *habituation,* a gradual decline in response to "insignificant" stimuli upon repeated exposure to them without any positive *reinforcement* (reward). In effect, it is a learning to ignore stimuli that are unimportant in the life of the animal. Its relative durableness distinguishes it from sensory adaptation or fatigue.

2. *Conditioning* is the associating, as a result of reinforcement, of a response with a stimulus with which it was not previously associated. The simplest form of conditioning is seen in conditioned reflexes, first studied scientifically by the great Russian physiologist Ivan Pavlov. Pavlov rang a bell each time he fed meat to a group of dogs, and eventually the salivary reflex of the dogs became conditioned to the auditory stimulus of the ringing bell. Pavlov could then ring the bell, and the dogs

would salivate even if they could not see, smell, or taste meat. A new reflex had been established, presumably by facilitation of neural pathways previously unused; a stimulus elicited a reflex response that it had never elicited before the training. The new stimulus (sound of ringing bell) had apparently been associated in the dogs' nervous systems with the original stimulus (sight, smell, or taste of meat), and the same response was now given to both.

Conditioning is not restricted to behavior patterns as simple as reflexes. Animals may be conditioned to perform such complex activities as running, pushing levers, opening doors, and performing complicated tricks. For example, conditioning is the basis for much of the training of domesticated animals such as cats, dogs, and horses; the animals learn to associate stimuli such as whistles or spoken commands with responses not normally elicited by such stimuli.

3. In *trial-and-error learning,* an animal does something and, if the result is rewarding, may do the same thing again; if the result is not rewarding, or is disagreeable, it may, after several trials, learn not to do the same thing anymore. This is, of course, an extremely common type of learning with which we are all personally familiar. We learned as children not to touch hot stoves because we tried it and were burned. We learned to play games by trying the various actions involved and profiting from our mistakes and successes. Much psychological research has been devoted to studying the sorts of influences that speed up or slow down trial-and-error learning, as in the running of mazes by experimental animals.

4. *Imprinting* is characterized by an extremely short critical period, which, in all cases so far known, occurs early in the animal's life. The concept of imprinting was first formulated in 1935 by the great Austrian zoologist Konrad Lorenz, who is one of the fathers of modern ethology. He was studying, at the time, birds such as geese, chickens, and partridges whose young, being precocial, are able to move around and feed themselves soon after hatching. He found that the young of such species will follow the first moving object they see and form a strong and lasting attachment to it, particularly if the object emits a sound. In effect, they adopt this object as their parent. Ordinarily the first moving vocal object such a young bird sees is its mother, and imprinting on her has obvious survival value. But under experimental conditions, the young bird may be imprinted on a toy train or a box pulled around by a string (particularly if the box contains a loudly ticking clock or some other sound-producing device), or even on a dog, cat, or human being. Once the critical period, usually only about 36 hours, has passed and the young birds have been imprinted on such surrogate mothers, they cannot be imprinted on any other object, including their true mother. This kind of imprinting is called social imprinting; it has obvious importance not only in establishing a bond between the young and their mother under natural conditions, but also in establishing proper species recognition and interaction.

Imprinting is an interesting example of the interaction of inheritance and learning. Inheritance determines the critical period, the class of objects to which the response may be directed, the tendency to respond promptly and strongly to the first object in that class to which the animal is exposed, and the near irrevocability of the attachment once it is

12.2. Lack of insight in a raccoon. Tied to stake A, the animal cannot quite reach the food dish as long as its leash is looped around stake B. In this situation, a man or a chimpanzee would immediately turn, walk around stake B, and go to the food; they would do this without any previous trial-and-error experience of such a situation. But the raccoon will not perform the task correctly at first. It must find the solution by trial and error, though once it has done so it will learn very quickly. [Modified from *High School Biology*, Rand McNally, 1963. Used by permission of the Biological Sciences Curriculum Study.]

formed, while learning establishes the tie between the animal and the particular object upon which the imprinting occurs.

5. *Insight learning* is most prevalent in the higher primates, particularly man. Some workers prefer to call it reasoning, and to distinguish it from learning as such. Essentially, insight is the ability to respond correctly the first time to a situation different from any previously encountered. Through insight, the animal is able to apply its prior learning in other situations to the new situation and, in effect, solve the new problem mentally without the necessity of overt trial and error (Fig. 12.2).

It is important to distinguish insight from simple *generalization,* which is a characteristic feature of most learning. A dog conditioned to salivate upon hearing a sound of 1,200 cycles per second will also salivate to some extent if it hears sounds of 1,000 or 1,400 cycles. Or a pigeon trained to peck a red button to receive food may peck an orange one if a red one is not available. These are examples of generalization, the ability of an animal conditioned to one stimulus to respond in the same way to other similar stimuli. In simple generalizations, the new stimuli differ very little from previously experienced ones. In insight, the new stimuli may be qualitatively as well as quantitatively different from previous ones. Insight often requires the mental putting together of several elements originally learned separately.

Motivation

Motivation can be defined as the complex of internal factors that determine an animal's behavior. When the behavior (or sequence of behaviors) fulfills some basic need of the animal, the motivation is often referred to as a drive. Students of behavior recognize hunger drives, thirst drives, sex drives, attack drives, escape drives, etc. The numerous interacting factors that determine any given behavior are hard to analyze, but certain factors can be singled out as playing major roles in most of the behavior of higher animals: the general health of the animal, hormones, integrative activity of the central nervous system, sensory stimuli, and previous experience that led to learning.

Clearly, the relative contributions of these different factors vary with the behavior pattern. Thus hormones and learning are of little significance in simple behaviors like the knee-jerk reflex, where sensory stimuli are the predominating causal factors, but hormones are of enormous importance in the reproductive behavior of cows, which will not receive the bull in copulation when the level of their sex hormones is low. Similarly, the relative contributions of the different motivational factors vary in different animal species, even when they are behaving in analogous ways. Thus the relative contribution of hormones to the sex drive is considerably smaller in primates, particularly man, than in rats, dogs, or cows, while the relative contributions of sensory stimuli, activity of the cerebral cortex, and learning are greater.

Available evidence points to the hypothalamus as the part of the brain that acts as the principal controller of behavioral drives in higher vertebrates. Located here are the excitatory and inhibitory centers for the various drives. Arousal of a particular drive must result from activation of its hypothalamic excitatory centers through the action of hormones, incoming impulses from sensory receptors, and impulses from other parts of the brain. Similarly, reduction or satiation of a drive must result from activation of its hypothalamic inhibitory centers and from reduced stimulation of its excitatory centers.

Motivated behavior is goal-directed, not in the sense that animals consciously decide on a goal and then strive to attain it, but in that such behavior tends to lead to the fulfillment of biological needs of the animal, that it tends to be biologically functional and adaptive. This is true of both innate and learned behavior patterns. Let us examine feeding behavior as an example of motivated behavior directed toward meeting certain biological needs.

When the hypothalamic excitatory centers for hunger become activated, usually as a result of such influences as a low glucose concentration in the blood, the strength of the animal's hunger drive rises, or, to put it another way, behavior of a sort that may result in feeding becomes more highly motivated. This behavior may include both learned and innate components, and it may involve many sorts of activities. Thus the animal may first become restless; it may begin to move around in what appears to be a rather random manner. But then its activities may become more obviously coordinated into some form of "searching" behavior. If this behavior results in the finding of food, the food may then be eaten, which may entail a variety of motor patterns such as tearing, chewing, and swallowing. Once the food has been eaten, the animal's behavior may well undergo radical changes; searching activities may cease, and some completely different behavior, such as courtship, or nest building, or preening, or sleeping, may begin. We can say, then, that if the behavior caused by one drive results in attainment of its biological goal, the strength of that drive is diminished, and the animal's next behavior pattern will be determined by whatever other drives are now strongest. Attainment of a goal tends to satiate the corresponding drive and leave the animal free to respond to other drives. But attainment of the goal also acts as positive reinforcement to the successful behavior patterns. Thus, if the behavior is of a sort that can be modified by learning, the animal will be more likely to perform similar activities when the drive is again high than to perform activities that did not lead to goal attainment.

Sometimes the goal of motivated behavior is less obvious than in feeding. Feeding or drinking supplies materials necessary to the continued existence of the organism; hence such behavior can be said to contribute to maintenance of stability in the organism. Similarly, escape from pain plainly contributes to maintenance of stability, because pain functions as a warning to the animal that the stimuli involved may well be harmful. But what about mating, or nest building, or care of the young? Such behavior patterns involve strong drives, and they obviously contribute to continuation of the species. But can they be said to contribute to maintenance of stability in the individual organism? We must apparently answer in the affirmative. Such behavior seems to fulfill biological needs built into the animal's nervous system; if those needs are not met, they apparently constitute a source of internal instability. Behavior, like the regulatory mechanisms of organisms already studied, seems to function, then, in maintaining *homeostasis,* an equilibrium among the animal's various physiological functions and between the animal and its environment. Under natural conditions, anything that contributes to homeostasis constitutes a positive reinforcement, and

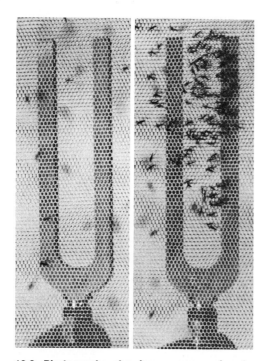

12.3. Photographs showing response of male mosquitoes to a tuning fork. Left: The fork is silent. Right: The fork is vibrating and its sound is attracting males. [Courtesy E. R. Willis, Illinois State University.]

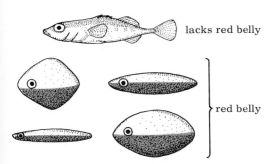

lacks red belly

red belly

12.4. Models of male stickleback. The realistically shaped model lacking the red belly was attacked by male sticklebacks much less often than the oddly shaped models with red bellies. [Modified from N. Tinbergen, *The Study of Instinct*, Oxford University Press, 1951.]

anything that contributes to instability constitutes a negative reinforcement.

The concept of releasers

Let us now examine the role of the stimuli that trigger behavioral patterns, restricting ourselves here to behavior patterns that are primarily innate, since these are generally easier to analyze than learned behavior patterns and have been studied extensively by ethologists.

Male *Aedes* mosquitoes are attracted to a tuning fork producing a sound similar in pitch to that of a female mosquito in flight (Fig. 12.3). The males can surely detect by sight and smell that a tuning fork is different from a female mosquito; yet the sound continues to attract them. Similarly, males of some species of insects will attempt to copulate with bits of paper on which female scent has been placed, even though they can certainly see and feel the difference between the paper and a female insect. Observations of this type lead to the conclusion that an animal responds at any given time only to a limited number of the stimuli its receptors are detecting. In a somewhat less rigid way, the same is true of human beings. A man driving an automobile along a highway sees hundreds, or perhaps thousands, of objects every minute—trees, telephone poles, houses, grass, clouds, etc.—but (one devoutly hopes) he does not respond to all of them. His driving behavior is determined largely by certain limited classes of stimuli, including the visual stimuli that indicate the width and curvature of the road, the visual stimuli from oncoming automobiles, and the auditory stimuli from automobile horns. Furthermore, a careful analysis would doubtless show that a driver responds to only a few of the clues that could provide him with information about road conditions; in other words, he ignores not only irrelevant stimuli but many relevant ones as well.

From an evolutionary standpoint, it is readily understandable that animals should be prompted to action by only a few of the many stimuli they encounter. An animal cannot possibly respond to all the stimuli impinging on its receptors. It must be selective. In the case of instinctive behavior, natural selection seems to have led to the evolution of special behavioral sensitivity to a few stimuli that under natural conditions would be reliable clues to the situation in which an animal found itself. Thus tuning forks and bits of paper specially impregnated with the scent of a female insect are not common in the insect cosmos. In short, the animal's seemingly blind, rigid response to certain stimuli to the exclusion of others is biologically functional and adaptive most of the time; it is only when an unpredictable factor like man intervenes that things go wrong.

Stimuli particularly effective in triggering behavior are termed *releasing stimuli,* and the structures, actions, sounds, etc., that give rise to the releasing stimuli are termed *releasers.*[2] The lower an animal's motivation to perform a given behavior, the greater the intensity of releasing stimulation necessary to trigger that behavior. The motivation,

[2] When Konrad Lorenz introduced the term "releaser" into ethology, he applied it only to characteristics of animals that give rise to releasing stimuli to which other animals respond. According to a more recent usage, followed here, the term is used more broadly to denote any object, sound, or action that releases complex stereotyped behavior in an animal; Lorenz's restricted type of releaser is called a social releaser.

as we have already seen, is determined by such factors as hormones, influences from various neural centers like the cerebral cortex, previous learning experience, and recent sensory stimulation. The animal must possess neural mechanisms that are selectively sensitive to the releasing stimuli; it is these *releasing mechanisms* that initiate the behavior when they are activated by the releasing stimuli appropriate to them. It should be emphasized that the term "releasing mechanism," as currently used, is not meant to apply to any particular part of the nervous system or to any particular type of neural function; indeed, the releasing mechanisms for different behavior patterns may well have entirely different neurological bases. Let us examine some examples of releaser-induced behavior.

N. Tinbergen and his associates at Oxford University studied fighting between male stickleback fish in spring. In the spring, the throat and belly of the males become intensely red. It seemed probable, therefore, that the red color was an important stimulus. The investigators presented their subjects with a series of models, some quite like actual male sticklebacks except that they lacked the red coloration, and some showing little resemblance to actual sticklebacks except that they were red on the lower surface (Fig. 12.4). The male fish attacked the red-bellied models, despite their un-fishlike appearance, much more vigorously than they did the fishlike ones that lacked red. Surely the sticklebacks could see the other characteristics of the models, but they reacted essentially only to the releasing stimuli from the red belly.

G. H. Brückner of the University of Rostock, Germany, studied a situation in which a sound acts as a releaser while associated visual stimuli have little if any effect. He showed that a hen reacts to her chick's distress calls, but not to its distress actions (Fig. 12.5). A chick fastened to a peg under a soundproof glass dome can struggle agitatedly in full view of the hen without eliciting any reaction from her, but if she hears the chick's call, she reacts vigorously, even if the chick is hidden from view.

It has been shown repeatedly that once the releasing stimuli for a particular behavior pattern have been carefully analyzed, it is often possible to design releasers that are even more effective than the natural one. For example, N. Tinbergen showed that the size of Oystercatcher eggs is important in determining the releasing properties of the eggs for the adult birds. An adult Oystercatcher provided with normal Oystercatcher eggs and with larger eggs of other species will usually react preferentially to the largest egg, even if she cannot possibly hatch it (Fig. 12.6).

ANIMAL COMMUNICATION

Communication between members of a single species (and, less commonly, between members of different species) is an important aspect of animal behavior. The ability to communicate is not restricted to species that live in societies, such as bees, ants, termites, and human beings; animals that live in less complex social groupings also communicate, and even the least social of animals must communicate with other individuals at certain critical times, such as the time of mating. The highly varied methods of communication animals have evolved utilize

12.5. Difference in response by a hen to her chick's visual and vocal distress signals. Top: The hen ignores the chick if she cannot hear its calls, even though its actions are clearly visible. Bottom: Distress calls elicit vigorous reaction from the hen even when she cannot see the chick. [Modified from N. Tinbergen, *The Study of Instinct,* Oxford University Press, 1951.]

12.6. Oystercatcher reacting to giant egg. She chooses it instead of her own egg (foreground) or a Herring Gull's egg (left). [Redrawn from N. Tinbergen, *The Study of Instinct,* Oxford University Press, 1951.]

particularly the senses of hearing, sight, and smell. A brief examination of some of these methods will not only aid in understanding a subject that is in itself fascinating, but will also illuminate some of the most fundamental principles of animal behavior.

Communication by sound

Being vocal animals ourselves, we are very familiar with the use of sound as a medium of communication. No other species has a sound language that even approaches the complexity and refinement of human spoken languages. But many other species can communicate an amazing amount of information via sound, information on which both the life of the individual and the continued existence of the species may depend.

Sound communication in insects. Let us return to the male *Aedes* mosquitoes attracted by the buzzing sound produced by the female's wings during flight (or by devices such as tuning forks that emit sounds of a similar pitch). The head of a male mosquito bears two antennae, each covered with long hairs. When sound waves of certain frequencies strike the antennae, these are caused to vibrate in unison. The vibrations stimulate sensory cells packed tightly into a small segment at the base of each antenna. The male responds to such stimulation by homing in on the source of the sound, thus locating the female and copulating with her. A striking demonstration of the adaptiveness of this communication system is that during the first 24 hours of the adult life of the male, when he is not yet sexually competent, the antennal hairs lie close to the shaft and thus make him nearly deaf. Only after he becomes fully developed sexually do the hairs stand erect, enabling the antennae to receive sound stimuli from the female. Thus the male does not waste energy responding to females before he is sexually competent, but once he becomes competent he has a built-in system for locating a mate without random searching. Furthermore, his built-in receptor system is species-specific; it is stimulated by sounds of the frequency characteristic of females of his own species, not by the frequencies characteristic of other species of mosquitoes. Hence the sound produced by the female's wings functions as both mating call and species recognition signal.

The sounds made by many other insects function in a similar way. For example, the calls of male crickets, produced by rasping together specialized parts of their wings, function in species recognition, in attracting females and stimulating their reproductive behavior, and in warning away other males. So species-specific are the calls of crickets that in several cases closely related species can best be told apart by human beings on the basis of the calls; the species may be almost indistinguishable on an anatomical basis, but have distinctly different calls.

Sound communication in frogs. The calls of frogs serve functions similar to those of cricket calls, and like these they are very species-specific. The male frogs attract females to their territory by calling. In one experiment, C. M. Bogert of the American Museum of Natural History recorded the call of male toads. He then captured 24 female toads and released them in the dark in the vicinity of a loudspeaker over which he was playing the recorded male call. Thirty minutes later, he turned on the lights and determined the position of each female.

Nineteen of the females had moved nearer the loudspeaker, four had moved farther away, and one had escaped. Eighteen of the nineteen females that had moved toward the speaker were physiologically ready to lay eggs; of the four females that moved away from the speaker, three had already laid their eggs and the other was not yet of reproductive age. In a control experiment, Bogert showed that 24 females released in the dark near a silent speaker had scattered randomly in all directions by the time the lights were turned on 30 minutes later. This demonstration that females ready for mating are strongly attracted by the male's vocalizations while other females are not serves to emphasize that the effectiveness of a releasing stimulus depends, among other things, on the condition of the recipient of the stimulus; the call of the male frog is an effective releaser for movement by the female toward the male only if the female's reproductive drive is high, largely as a result of a high level of sex hormones.

Robert R. Capranica, of Cornell University, and his colleagues have recently shown that female cricket frogs will respond to the mating calls of males from their own or nearby populations, but not to those of males of the same species from more distant populations. This curious parochialism has a physiological basis. The ears of the female frogs are sensitive to only a very narrow band of frequencies, and the frequency at which they are maximally sensitive differs for frogs from different localities. The calls of the males also vary geographically in the frequency of their maximal intensity; i.e. the males have different local dialects. A female's lack of response to the call of a male from a distant locality may be due to a mismatch between her ears and his call; she may be deaf to his calling frequency. Because the continuation of the species depends on the females' responding to the males, it is clear that the females' auditory apparatus and the males' vocal apparatus must always evolve in tandem. Thus, in any given locality, the males call at the frequency to which the females in that locality are most sensitive. This is a particularly dramatic example of a general phenomenon: the coupled evolution of social releasers and of the sensory and releasing mechanisms responsive to them.

Bird songs. Of all the familiar animal sounds—the buzzing of mosquitoes, the calling of crickets and frogs, the barking, roaring, purring, grunting, etc. of various mammals—perhaps none, with the exception of human speech, has received so much attention as the singing of birds. The popular explanation for bird song is simple: The bird is happy and sings with joy, welcoming the morning and the spring and expressing love for his mate. Unfortunately for these pleasing anthropomorphic fancies, objective investigation has demonstrated that bird song functions primarily as a species recognition signal, as a display that attracts females to the male and contributes to the synchronization of their reproductive drives (increasing sexual motivation and decreasing attack and escape motivations), and as a display important in defense of territory. In its defensive function, a bird's singing is certainly no indication of happiness or joy; if such human-oriented concepts could properly be applied to birds, which they cannot, the singing would more accurately be taken as an indication of combativeness.

The role of singing in the establishment and defense of territories is an especially interesting one. A *territory* may be defined as an area defended by one member of a species against intrusion by other members of the same species (and occasionally against members of other species). A male bird chooses an unoccupied area and begins to sing vigorously within it, thus warning away other males. The boundaries between the territories of two males are regularly patrolled, and the two may sing loudly at each other across the border. Although during early spring there is often much shifting of boundaries as more and more males arrive and begin competing for territories, later in the season the boundaries usually become fairly well stabilized and each male knows where they are. During the period when the boundaries are being established, it is often the males that can sing loudest and most vigorously that successfully retain large territories or even expand their territories at the expense of other males that sing less loudly and vigorously.

Experiments performed on thrushes by William C. Dilger of Cornell University illustrate especially well the role of singing in territorial defense. Dilger placed a stuffed thrush near a loudspeaker in the territory of a male Wood Thrush. Wires led from the loudspeaker to a tape player in a blind in which Dilger could sit. When Dilger played a recording of a singing male Wood Thrush, the male bird in whose territory the loudspeaker was located responded as though another male had entered its territory. If the volume at which the recording was played was very low, the defending male attacked the stuffed bird. If the volume was high, the defending bird retreated. By alternately turning the volume up and down, Dilger could make the defending male move alternately toward or away from the stuffed bird and loudspeaker, almost as one might work a yo-yo. So precisely was he able to control the defending bird's movements by this method that he could make it teeter on one leg, its conflicting attack and escape drives almost exactly balancing each other.

Notice that agonistic (hostile) encounters between individuals of the same species may often be resolved by vocal and/or visual displays without any physical combat. It is, in fact, rather rare that individuals of the same species engage in combat serious enough to cause significant damage. The adaptive importance of this is obvious. Physical damage to one or both of the combatants is biologically deleterious. Agonistic encounters between individuals of the same species occur frequently, and if they often led to serious physical damage, many otherwise fit individuals might be eliminated from the population. It is not surprising, therefore, that most animals have evolved other methods of resolving conflicts, methods usually involving displays by which the combatants convey to each other the intensity of their attack motivation (see Pls. III-2, 3). The individual showing the higher attack motivation is ordinarily the winner. The animals can, in effect, tell without fighting which would be the probable winner if they were to fight; hence an actual fight is unnecessary. In our example, the vigor and loudness with which male Wood Thrushes sing convey to other Wood Thrushes very precise information concerning the motivational states of the singers. A singing duel can, therefore, resolve a boundary dispute without fighting.

Dilger's experiments also illustrate the importance of song in species recognition. Several other species of thrushes breed in the area where the experiments on Wood Thrushes were conducted. Some of these species look much like Wood Thrushes; they have brown backs and spotted breasts. Dilger showed that a male Wood Thrush would attack models of any of these species set up in its territory, provided the models were silent. It would even attack models of other species unrelated to thrushes if these happened to be brown with spotted breasts. If, however, the models were set up on a loudspeaker and Dilger played a recording of the song appropriate to the model, the Wood Thrush would pay the models of other species little attention. Apparently it could not distinguish visually (in this context) between the various brown thrushlike birds with spotted breasts, and responded to all, if they remained silent, as though they were invading members of its own species. But it could distinguish between the songs of the other species and that of its own species, and would not attack a model "singing" the song of some other species, even if it was brown and had a spotted breast. Auditory stimuli were thus shown to be far more important than visual stimuli in species recognition by thrushes. This is frequently the case among birds living in dense woods and thickets, where vision is obstructed; visual displays are often more important for birds living in more open habitats.

Echolocation. Echolocation by bats is probably the most widely known example of sonar orientation in animals. Flying bats produce a great variety of vocalizations, some at frequencies as high as 100,000 cycles per second. The extremely high frequencies and correspondingly very short wavelengths of these sounds make them particularly well suited for echolocation, because they can be reflected from smaller objects and they spread less widely and are less diffuse than sounds of longer wavelengths. They thus permit more sensitive detection and more precise localization of objects. For example, some bats can avoid wires that are only 0.2 mm. in diameter. Similarly, they can locate and catch very small insects. Bats have a remarkable ability to detect the echoes of their own clicks, even in the midst of an enormous amount of noise. All attempts to confuse them by exposing them to intense vibrations while they are flying have failed. It is biologically important, of course, that they should be able to analyze sounds and eliminate noise. They often fly in flocks of thousands, and it is essential that each bat be able to distinguish the echoes of its own sounds from those of all the other bats.

Echolocation is not, strictly speaking, a form of communication, since only one individual is involved. The sounds used by bats in echolocation have, however, come to have some interesting interspecies communication properties, to the disadvantage of the bats themselves. The bats that use echolocation feed on insects, which they capture in flight. Among these insects are a number of species of moths, some of which have evolved simple ears capable of detecting the high-pitched clicks of the bats. When a moth of such a species hears an approaching bat, it begins evasive tactics and often escapes. It has recently been discovered that some moths produce high-pitched sounds of their own, which, in some manner as yet not understood, seem to interfere with the bat's ability to locate the moth. We see, then, that there is constant interplay in the evolution of interacting species. In this example, the evolution of more effective echolocation by bats has resulted in greater selection pressure for improved ears and evasive tactics in moths, and these adaptive changes, in turn, may be expected to affect the selection pressure acting on the bats' echolocation system.

Communication by chemicals

Since most communication between human beings involves auditory or visual signals, we tend to think of these as inherently the most appropriate methods of communication in animals. And biologists studying behavior have understandably focused much of their attention on animals such as birds that, like man, rely strongly on their auditory and visual senses. Such animals are generally easier for us to study than those more dependent on olfaction, because we can more readily detect the stimuli to which they respond. But we should not let the limitations of our own senses prevent us from recognizing that the olfactory sense is immensely important in the lives of many animals and constitutes a basis for effective communication.

Many animals secrete substances that influence the behavior of other members of the same species. Such substances are called *pheromones.* Most pheromones can be classified in one or the other of two groups: those that act as releasers, triggering a more or less immediate and reversible behavioral change in the recipient, and those that act as primers, initiating more profound physiological changes in the recipient but not necessarily triggering any immediate behavioral reaction.

Releaser pheromones. Many insects, and also other types of animals, utilize chemical sex attractants. Female silkworm moths release a sex attractant so powerful that males are drawn to a single female from distances of two miles or more, even though each female releases less than 0.01 microgram of the attractant chemical. The chemical acts as a releaser to which the male moth responds by flying upwind, thus moving toward the female. Only in the immediate vicinity of the female is the concentration of pheromone sufficient to establish a gradient; when the male comes into this region, he stops flying upwind and follows the gradient instead, thus locating the female. It can be shown that in this behavior the male responds only to the chemical releaser, not to visual stimuli. He will be attracted to a female in a gauze cage even if he cannot see her, but he will not be attracted to a female clearly visible in a tightly sealed glass cage from which none of the pheromone can escape. Sex-attractant pheromones occur also in cockroaches, queen honeybees, gypsy moths, and many other insects. A currently very active field of research is the use of sex attractants in nontoxic species-specific insect control, in place of the toxic and relatively nonspecific insecticides now in use.

The trail substances of ants constitute another class of releaser pheromones (Fig. 12.7). A foraging ant returning to the nest from a food source intermittently touches the tip of her abdomen to the ground and secretes a tiny amount of trail substance. Other worker ants can follow the trail to the food source; these ants, too, will lay trail substance as they return with food to the nest. Workers that do not find food do not lay trail; hence when the food has been consumed no more trail is laid and, since the trail pheromone is very volatile, the trail disappears within a few minutes. It has been demonstrated that trail substances

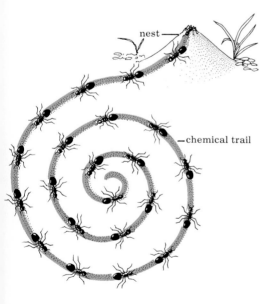

nest

—chemical trail

12.7. Ants following a spiral path of trail substance laid down by an experimenter.

are species-specific; no two species have been found to secrete the same substance. This specificity is, of course, biologically adaptive, because it ensures that workers will not mistakenly follow trails of other ant species that may cross their own.

Other releaser pheromones of ants include alarm substances and death substances. The latter show vividly to what extent much insect behavior is stimulus-bound and rigidly stereotyped. Certain substances released from a dead and decomposing ant act as pheromones that stimulate worker ants to pick up the carcass and carry it to a refuse pile outside the nest. When living ants are experimentally painted with these substances, workers pick them up and dump them on the refuse pile. The hapless victims of the experimenter promptly return to the nest, only to be thrown back on the refuse pile, again and again. The workers can surely see and feel that the object they are carrying is struggling in a most undead manner, but they disregard the evidence from all their other senses, respond only to the death pheromone, and continue to treat the painted ant as a carcass to be taken to the refuse pile.

Releaser pheromones are also common in mammals, most of which, as you know, rely much more on olfaction than man. Male mammals can often tell by smell when the female is in heat, because she secretes certain pheromones at that time. Besides playing a part in sexual recognition and reproduction, mammalian pheromones are often important in marking territories and home ranges. You are familiar with the way dogs and many other mammals use their urine as a marking substance.

Primer pheromones. Primer pheromones produce relatively long-term alterations in the physiological condition of the recipient and thus change the effects that later stimuli will have on the recipient's behavior; they do not necessarily produce any immediate behavioral change.

Some experiments with mice have indicated various effects of primer pheromones. In one, it was shown that the estrous cycles of female mice in a laboratory colony can be initiated and synchronized by the odor of a male mouse, even if the male cannot be seen or heard, and, in another, that the pregnancy of a newly impregnated female mouse can be blocked by the odor of a strange male mouse; the blockage will not occur if the olfactory bulbs of the female mouse's brain are removed. In a third experiment, it was found that crowding of female mice results in disturbance and even blockage of estrous cycles, but that removal of the olfactory bulbs restores the cycles to normal. This experiment and others suggest that pheromones help regulate population density in some species.

Another type of primer pheromone is seen in social insects such as ants, bees, and termites. These pheromones are ingested rather than simply smelled, and they play an important role in caste determination. For example, termite queens and kings secrete substances that prevent the workers from developing reproductive capabilities. The number of soldiers in a termite colony is regulated by similar pheromones secreted by the fully developed soldiers.

Communication by visual displays

Displays in reproductive behavior. A display may be defined as a behavior that has evolved specifically as a signal. According to this definition, a song or a call is a display. Many animals have also evolved a variety of often complex actions, frequently including vocal elements, that function as signals when seen by other individuals. You may have seen male pigeons strutting, with tail spread and dragging on the ground, neck fluffed, and wings lowered, or courting songbirds in spring going through odd and seemingly senseless antics. If you have the opportunity, watch a flock of ducks on a pond in early spring. You may see a male give a loud whistle and raise both his head and his tail as high in the air as he can while also raising his wings (Fig. 12.8B), or he may raise his stern in the air, dip his head in the water, and then abruptly raise it and whistle (Fig. 12.8C), or he may put his bill in the water and then quickly flick his head to the side, toss an arc of droplets into the air while arching his body upward, and follow this acrobatic feat with a whistle and a grunt (Fig. 12.8D). Such carryings-on may seem senseless to a casual observer in the park on a Sunday afternoon, but biologically they are far from senseless. They function in synchronizing the sexual physiology of the male and female and in making the female more receptive to the male. They also assure that the female will choose a male of her own species as her mate; the males of each species—there may be more than one on the same pond—give somewhat different displays. Special structures or bright patches of color, which clearly evolved in association with the displays as part of the signal system, are often elaborately exposed (see Pls. III-4, 5).

12.8. Courtship displays of a male Mallard. (A) Normal swimming posture. (B) The "Head-Up-Tail-Up" display. (C) Part of the "Down-Up" display. (D) The "Grunt-Whistle" display. [Adapted from K. Z. Lorenz, "The Evolution of Behavior," *Sci. Am.*, December, 1958. Copyright © 1958 by Scientific American, Inc. All rights reserved.]

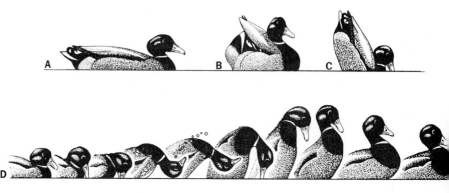

Complex behavior patterns are often composed of several separate activities occurring in sequence, with each activity serving as the initiator of the next. The mating behavior of the three-spined stickleback provides a good example. In spring, the male fish has bright red underparts, and the female's abdomen is swollen by the large number of eggs it contains. Tinbergen and his associates have shown that the female is attracted by the male's red belly and that the male is stimulated by the sight of the female's swollen abdomen. When a female swims into the territory of a male in reproductive condition, she often adopts an unusual head-up posture (Fig. 12.9). The combination of her swollen abdomen and courting posture acts as a releaser for the male to swim toward her in a curious zigzag fashion. The combination of the male's red belly and zigzag dance acts as a releaser for the female to swim toward the male in the head-up posture. Her approach acts as a releaser for the male to turn and swim rapidly toward the nest that he has already constructed. His swimming away, in turn, stimulates the female to follow him, which stimulates the male to make a series of rapid thrusts with his snout into the nest entrance and then to turn on his side and raise his dorsal spines. This "showing-of-the-nest-entrance" behavior of the male acts as a releaser for the female to enter the nest. Her occupation of the nest, in turn, stimulates the male to thrust his snout against her rump in a series of quick rhythmic trembling movements, which induce the female to spawn. It can be shown that without the stimulus of the male's tremble-thrusts, the female is incapable of spawning; the stimulus can be effectively duplicated, however, by prodding her with a glass rod or other hard object. Once the female has spawned, the fresh eggs stimulate the male to fertilize them.

Though the sequence of actions described here is the usual one, it is not absolutely rigid and variations do occur. If, however, the sequence or the way in which the displays are performed is altered too much, the final acts—spawning and fertilization—will not occur. Since the mating behavior of each species of stickleback differs in certain critical elements, it is unlikely that a male of one species and a female of another will get through enough of the ritual for spawning and fertilization to occur. This elaborate behavior functions, therefore, both in bringing together and synchronizing the two sexes in the mating act and in avoiding mating errors.

Displays in agonistic behavior. Visual displays are important in many aspects of an animal's life. You are familiar with the way a dog wags its tail as a greeting display and with the way it tucks its tail between its legs as a display of appeasement. And you have surely seen two dogs or cats displaying in an antagonistic encounter, with hackles raised, teeth bared, ears laid back, body raised as high off the ground as possible, and movements stiff-legged and exaggerated. Or you have seen the loser of such an encounter giving appeasement displays—fur sleeked, tail tucked under, head down and often turned away from the antagonist, legs bent. Such visual displays function like vocalizations in similar situations: They communicate the current balance between the individual's attack and escape motivations.

Analogous agonistic displays can be observed in many animals (see Pls. III-2, 3). A high attack motivation is often conveyed, as it is in dogs, by directing the face straight at the antagonist and spreading and raising the body, making it look as large as possible. Appeasement displays usually involve making the body appear as small as possible and turning the face away from the antagonist or exposing to the antagonist the appeaser's most vulnerable spot; such appeasement displays tend to inhibit further attack by the antagonist.

Communication in honeybees

A last example of communication should be mentioned. This is the amazing ability of scout honeybees to inform the workers in the hive of the quality of a food source and its direction and distance from the hive. This communication depends on displays that utilize auditory, visual, chemical, and tactile elements.

The work of Karl von Frisch of the University of Munich, Germany, on the language of bees is a biological classic. Von Frisch had long

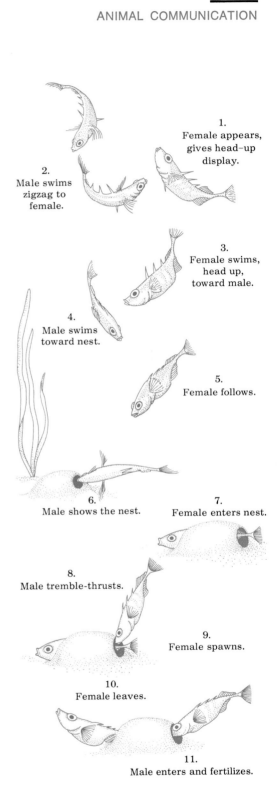

1. Female appears, gives head–up display.

2. Male swims zigzag to female.

3. Female swims, head up, toward male.

4. Male swims toward nest.

5. Female follows.

6. Male shows the nest.

7. Female enters nest.

8. Male tremble-thrusts.

9. Female spawns.

10. Female leaves.

11. Male enters and fertilizes.

12.9. Courtship behavior in the three-spined stickleback. [Modified from N. Tinbergen, *The Study of Instinct*, Oxford University Press, 1951.]

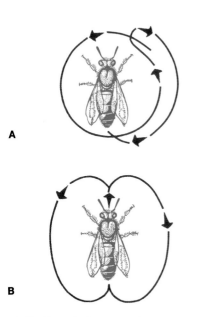

A

B

12.10. Round dance and wagging dance of scout honeybee. (A) In the round dance, the bee circles first one way and then the other, over and over again. The dance tells other bees that there is a source of food near the hive. (B) In the wagging dance, the scout runs forward in a straight line while wagging her abdomen, circles, runs forward again, circles in the other direction, and runs forward again. The orientation of the run indicates the direction from the hive of the food source, and the number of turns per unit time indicates the distance. [Redrawn from K. von Frisch, *Bees: Their Vision, Chemical Senses, and Language.* Copyright 1950, Cornell University. Used by permission of Cornell University Press.]

been interested in the ability of bees to distinguish between different colors and scents. In the course of his experiments, he would set up in the vicinity of a hive a table with sheets of paper on which he had smeared honey. He would then have to wait—sometimes for several hours—for the bees to find the honey. He noticed that when one bee finally discovered the feeding place, many others appeared at the table within a short time. It seemed likely that the first bee had somehow informed the others of the existence of the new feeding place. In order to see what happens in the hive when a scout bee returns from a new food supply, von Frisch set up an observation hive with glass sides. When a bee landed at the new feeding place and began to feed, von Frisch daubed a spot of paint on her thorax so that he could recognize her when she returned to the hive. He discovered that the returning bee first feeds several other bees and then performs a dance on the surface of the honeycomb. The dance consists of circling first to the right, then to the left, and repeating this pattern over and over with great vigor (Fig. 12.10A). Von Frisch named this the round dance. The dance excites other bees in the vicinity of the dancer, and they begin to follow her, with their antennae held close to her. Suddenly, however, they turn away one by one and leave the hive; a short time later they appear at the feeding place. Apparently the round dance is a display that informs the other bees of the existence of the food supply.

Von Frisch wanted to know exactly what sorts of information the round dance conveys. He fed several bees at a dish containing sugar water scented with honey that he had located 10 meters west of the hive. He also put out dishes of sugar water to the north, south, and east of the hive. Other bees began appearing in approximately equal numbers at all four dishes a few minutes after the bees that had been fed at the west dish began performing a round dance in the hive. There was no evidence that the round dance indicated direction, and other similar experiments brought no evidence that it indicated distance. It seemed simply to say, "Fly out and seek in the neighborhood of the hive." Von Frisch did find, however, that if each of the dishes of sugar water was scented with a different flower, the other bees came in significantly greater numbers to the dish that the dancer had visited. He showed that these bees determined what scent to search for in two ways: They smelled the body of the dancer by holding their antennae near her, and they detected the odor in the droplets of material she fed to them.

Even though von Frisch had satisfied himself that the round dance indicates neither direction nor distance, he began to suspect that at times bees are able to communicate this type of information, presumably in some other way. In 1944 he performed the following experiment. He set up two dishes of sugar water, one at 10 meters from the hive and the other at 300 meters. Each was scented with lavender oil. He then fed a few bees at the dish 10 meters from the hive; shortly thereafter, numerous bees appeared at this dish but only a few appeared at the distant dish. When he reversed the procedure and fed forager bees at the dish 300 meters from the hive, other bees appeared in large numbers at this dish but only a few appeared at the nearer dish. Distance was clearly being communicated in some manner. When von Frisch observed the dances of the forager bees returning from the two dishes, he saw immediately that they were entirely different. The foragers from the dish 10 meters from the hive danced the familiar round dance, but the foragers from the dish 300 meters away danced a different dance, one that von Frisch named the wagging dance (Fig. 12.10B). He found an inverse relationship between the distance of the food source and the number of turns per unit time in this wagging dance, and concluded that the number of turns tells the other bees the distance to the food. Other investigators have suggested that sounds made by the bee during the dance may also be important in communicating distance.

Von Frisch found that bees can also communicate the direction of a food source. The location of the food relative to the position of the sun is indicated by the direction of the straight portion of the wagging dance. In the dark hive, a run straight up the vertical comb means that the food lies in the direction of the sun; a run straight down the vertical comb means the food lies in the opposite direction from the sun; a run at an angle indicates that the food is to be found at that angle to the sun (e.g. a run 30 degrees to the right of vertical indicates that the feeding place is 30 degrees to the right of the sun). In other words, bees use the force of gravity as a symbol for the sun when inside the dark hive. When, however, they perform the wagging dance on a horizontal surface outside, they abandon the symbolism and orient the dance relative to the sun itself.

BIOLOGICAL CLOCKS

The leaves of many plants show regular movements in a cycle of approximately 24 hours, even if kept under constant conditions. Many flowers open and close in a similar 24-hour cycle. Most animals show activity rhythms that vary with a period of approximately 24 hours, even if they are kept under constant conditions and have no known external indication of the actual daily environmental cycle. Indeed, *circadian rhythms* (from the Latin *circa*, about, and *dies*, day) seem to be characteristic of all living things, whether organisms or individual cells. Cellular phenomena that vary in approximately 24-hour cycles include enzyme activity, respiration rate, sensitivity to light and temperature, and reactions to various drugs. Physicians are becoming increasingly aware that the proper dosage of a drug may be very different at different times of day; in some cases, what constitutes a beneficial dose at one time may actually be lethal at another.

Though the clock is probably innate, it is not unresponsive to environmental conditions; indeed, it is strongly influenced by them. Under normal conditions, the clock is constantly being reset by the environmental cycle. If an organism is kept under constant conditions for a long time, its clock will slowly get more and more out of phase with the environmental diurnal cycle. Thus, if an insect whose activity rhythm has an innate period of 23 hours and 45 minutes (the innate period often deviates slightly from 24 hours) is kept for 20 days under constant conditions, at the end of that time its activity cycle will be five hours out of phase with the actual diurnal cycle.

The fact that environmental conditions can reset the biological clock means that it can be experimentally manipulated. Suppose an animal with a pronounced circadian activity rhythm is put in a light-proof room for five days and exposed to an artificial day that begins and ends six hours later than the natural day. If the animal is then exposed to constant light, its activity rhythm of roughly 24 hours continues, despite the absence of external time cues, but this activity rhythm is now in phase with the artificial day, not with the true day—an indication that the animal's internal clock has been successfully reset. A similar internal displacement is experienced by a person who flies from Chicago to Paris. At first he finds himself disconcertingly out of phase with the people around him; he feels like sleeping when they are wakeful, he is hungry when their minds are on higher things. After three or four days, however, his internal clock has shifted to Paris time, and his problems are over.

The clock is a crucial element in the navigation of migratory birds that use the sun as a compass. As you know, such birds may fly thousands of miles, over precise routes, from their summer to their winter ranges and back again. They could not tell direction from the sun unless they were able to correlate its position with the time of day. Numerous experiments have confirmed the working of an internal clock of some sort in such birds. In one series of experiments, it was found that starlings trained to eat in the north compartment of a circular cage will go to the north compartment whatever the position of the cage, at any time of day, whether they see a sun low in the east in the morning, a higher sun nearer noon, or a sun in the west in the afternoon; they evidently identify north by combining the time of day with the position of the sun. In another experiment, birds trained to orient toward the north in a circular cage, using the sun as their cue, were put in a light-proof room for four to six days and exposed to an artificial day that began and ended six hours later than the natural day. When they were then again placed in the circular cage and exposed to the sun, they oriented toward the east instead of the north. Because their clock had been shifted six hours out of phase with the natural day, they interpreted the sun's position erroneously.

It has been shown that when an organism is exposed to new conditions that shift the setting of its biological clock, the clocks of its various cells or organs do not all necessarily shift together. Some may adjust to the new conditions in only two or three days, while others may take a week or more. Thus it is possible for the different organs of an individual to be thrown out of phase with each other. For example, an endocrine gland might be in the phase of its maximal secretion of hormone while the target organ was in a phase of relative unresponsiveness to the hormone. Or an enzyme system might be potentially most active at a time when its substrate was not available. Such uncoordinated phase shifts of the various biological clocks in an individual may well lead to serious physiological disturbances and perhaps to disease. It has even been suggested that the onset of cancer may be triggered by unbalanced phase shifts of this type.

THE EVOLUTION OF BEHAVIOR

Implicit throughout this chapter has been the assumption that behavior is a biological attribute, one to be investigated like anatomy or physiology and subject to the same kinds of evolutionary processes. The corollary assumptions—that behavior is adaptive and that natural selection brings about an increase in well-adapted and a decrease in poorly adapted behavior patterns in the population—are basic to the elucidation of many types of behavior.

If behavior patterns evolve and can be studied in terms of the selection pressures that produce them, it follows that we should be able to make reasonable conjectures concerning the ancestral behavior patterns from which newer behavior patterns have arisen, just as we can make inferences about the ancestral structures from which our hands or other structures have evolved. And we should be able to study the genetics of behavior just as we study the genetics of other characters. Both of these approaches to behavior have been much used in recent years.

The derivation of behavior patterns

Evidence from comparison of species. One fruitful way of studying the evolutionary derivation of behavior is to compare the behavior patterns of a number of related species. Because these patterns often represent different stages of development of the same basic behavior, they may give a clue to the ancestral condition. Let us look at several examples.

There are certain species of flies (family Empididae) in which the males always present the females with a silken balloon before mating. This is a curious bit of behavior, and we could hardly guess what selection pressures brought it about, and from what ancestral beginnings, were it not that there are other species of empidid flies still living today that exhibit various stages of development of this courtship pattern. In many species of empidid flies where the male does not give anything to the female when courting her, she sometimes captures and devours him. In other slightly more advanced species, the male captures prey, and presents this to the female, and then mates with her while she is occupied with eating the prey. The selection pressure for this behavior seems easy to explain. Males that divert the females' attention by giving them prey succeed in mating and escaping more often than those that do not. The male of still more advanced species captures prey, wraps it in a ball of silk, and presents this to the female. Presumably the fact that the female is occupied longer in opening the balloon and eating the prey gives the male more time to accomplish copulation. In still more advanced species, the male encloses only a tiny prey or fragments of prey in the balloon, and the female does not actually eat the prey. In other words, at this stage in the evolution of the courtship behavior, the balloon has replaced the prey as the important element, becoming part of a display that functions in making the female more receptive to the male. We can understand, therefore, how in the most advanced species an empty balloon or even some other bright object such as a rose petal can suffice. By examining a whole group of species in this way, we can get a reasonably good idea of the derivation of behavior that would otherwise seem odd and enigmatic, and we can identify the probable selection pressures involved.

Another good example is the evolution of nest building in the parrot genus *Agapornis,* which has been studied by W. C. Dilger, whose experiments with Wood Thrushes we mentioned earlier. These small African parrots are unusual in building nests; most parrots simply lay their eggs on the bare floor of a cavity in a tree. The various species of *Agapornis* exhibit a variety of stages in the evolution of nest-building behavior. Females of the most primitive living species use their very sharp bills to cut small irregular bits of bark, leaf, wood fiber, or paper and thrust these amid their feathers at any point on the body. When a number of pieces are lodged among their feathers, they fly to the nest cavity and unload them. The nest they make from this material is only a simple pad. Females of a somewhat more advanced species (see Pl. III-6) cut long regular strips of material and tuck the ends of these into the feathers of their rumps. Flying to the nest cavity with three or four such strips dangling in the air behind them, they construct more elaborate nests with a deep cup for reception of the eggs. Females of the most advanced species no longer tuck nesting material amid their feathers, but carry it to the cavity one piece at a time in their bills. This means that they can carry sticks and other stronger material and hence can construct very elaborate roofed nests with two chambers and a passageway. We have, then, what appears to be an evolutionary trend (1) from carrying material tucked all over the body, to carrying it only in the rump feathers, and finally to carrying it in the bill; (2) from cutting small irregular pieces, to cutting long regular ones, and finally to using twigs in addition to the strips; (3) toward increasing complexity of the nests from a simple pad, to a well-formed cup, and finally to an elaborate roofed structure. If the more advanced species were the only ones still living today, we would have little hope of understanding the evolution of nest-building behavior in these parrots. But the fact that more primitive species are still extant enables us to make comparative studies that help us not only to understand how this particular behavior evolved but also to get new insight into the ways animal behavior in general evolves.

Dilger has also used the carrying behavior of *Agapornis* to demonstrate the importance of genetics in behavior. He has succeeded in hybridizing the species that carries strips tucked in the rump feathers with the species that carries material in the bill. The hybrids clearly show effects produced by the genes of both parents. They cut strips and try time and again to tuck them, but without success. Sometimes they fail to let go of the strip; after a lengthy bout of tucking, when their heads come forward again, the strip is still held tight in their bills. Sometimes they let go of the strip, only to see it fall from the rump to the floor. It is as though the genes of one parent made them try to tuck but the genes of the other parent prevented them from doing it correctly. Eventually the hybrids learn that they cannot tuck, and begin carrying nesting material in their bills, but even the oldest and most experienced still gives at least a perfunctory flick of her head over her shoulder before flying to the nest with the material in her bill.

Evidence from behavioral analysis.   Through detailed analysis of the motor patterns involved in a particular behavior and of the precise context in which each component of the behavior occurs, and through comparison of the results with those from similar analyses of other behaviors, a biologist can often convincingly identify the actions from which the behavior in question is derived.

A

B

III-1　SOCIAL ORGANIZATION

Honeybees and other social insects live in complex societies based on extensive division of labor. A queen honeybee, such as the one seen here surrounded by the smaller workers (B), is the sole fertile female in the hive, hence the sole egg producer of a community typically several thousand strong. The males, or drones—stingless and subject to extermination when food is scarce—have the one function of inseminating the queen. The workers, all sterile females, gather nectar, build and stock the combs, nurse the larvae—and, in short, labor so unremittingly that they survive only a few months, whereas a queen may flourish for several years. The interdependence in bird societies is much less complex than in insect societies. But there is evidence of leadership and social cooperation in the characteristic V formation of the Canada geese (A), which is thought to provide favorable aerodynamic conditions for the birds. Note the coordination of wing movement; the leader has just begun a downbeat, and successive birds along the arm of the V are at correspondingly successive stages of the beat.

A, David G. Allen. B, courtesy Carolina Biological Supply Co.

III-2A
III-2C

III-2B

III-2, III-3 THE LANGUAGE OF AGGRESSION

Agonistic (hostile) encounters between individuals are common in many animal species, and a variety of displays are used in such situations to communicate the motivational state of the respective parties. (2A) A female Goshawk spreads her "pantaloons" (under tail coverts) in a display that indicates she is aggressively alert. (2B) Two Common Egrets "challenge" each other. (2C) A frilled lizard of Australia spreads its specialized neck membrane, gapes, and hisses. (3A) A male bison gives a threat display, with tail "flag" raised. (3B) Three males, all with tail flags, charge. (3C) After the three clash head-on, one (right) lowers and tucks his tail, indicating diminished aggressive motivation.

2C, Australian News and Information Bureau. All others, David G. Allen.

III-3A

III-3B

III-3C

III-4

III-4, III-5 THE LANGUAGE OF COURTSHIP

(Above) A male Ruffed Grouse, with tail and neck ruff spread, mounts his drumming log and beats his wings rapidly and forcefully, producing a loud sound that attracts females. (5A) The bright coloration of the male Wood Duck, contrasting with the drabness of the female, is typical of many species of birds. In most species it is the female that does the choosing of the mate, and the distinctive plumage of the male is an adaptation that helps avoid mating errors. (5B) A male Cowbird displays to a female. (5C) A male Superb Lyrebird of Australia spreads his enormous tail, holding it forward like an umbrella; the bird itself is completely concealed in this pose, which climaxes a long and elaborate courtship dance performed atop a mound built by the bird in its territory. (5D) A male American Goldeneye performs a display in which he throws his head back and simultaneously kicks up a spray of water.

4, 5A, 5D, David G. Allen. 5B, Arthur A. Allen. 5C, Australian News and Information Bureau.

III-6A

III-6B

III-6C

III-6, III-7   CARING FOR YOUNG

Unlike the young of most animal groups, which can fend for themselves as soon as they are born, the offspring of many birds and mammals are helpless at birth—inadequately clad, unable to feed themselves—and thus wholly dependent on parental care for survival. Many birds, such as the lovebird *Agapornis roseicollis,* have evolved elaborate nest-building behaviors, peculiar to their species. The captive lovebird in Pl. 6 cuts strips of nesting paper by punching a line of holes with its supersharp beak (A), tucks them amid specialized rump feathers (B), and flies to its nest with the strips dangling behind (C); in the wild the birds cut leaves and bark, and use these as nesting material. Many baby birds, like the Cedar Waxwings about to snap up a berry (7A), have brightly colored mouth linings, which act as releasers of the parental feeding response. Cedar Waxwings are born nearly featherless (7B) and may die of cold if not brooded.

All photos, David G. Allen.

III-7A

III-7B

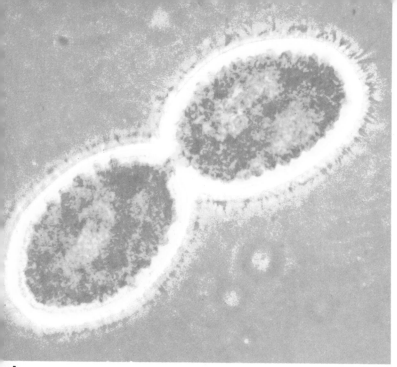

III-8 SYMBIOSIS

(A) A ciliate protozoan containing mutualistic green algae has just undergone fission, producing two daughter cells. The ciliates benefit from photosynthetic products of the algae, and the algae benefit from the favorable environment provided by the protozoans. (B) A hermit crab inhabits the abandoned shell of a gastropod mollusc; two sea anemones are attached to the shell. The anemones, with their stinging tentacles, provide the crab with protection and camouflage; the crab provides, in return, transportation to new feeding areas and scraps from its own meals.

A, Roman Vishniac. B, Marineland of Florida.

A
B

Consider, for example, the origin of courtship or agonistic displays. A courting animal has simultaneously high attack, escape, and mating drives, and an animal in an agonistic encounter experiences conflict between his attack and escape drives. The conflict often thwarts the complete performance of the behavior most appropriate to each of the separate drives. Hence the animal may perform acts that seem out of context in the given situation, or perform an appropriate act but direct it at the wrong object, or start to do something and not finish doing it. Because such apparently irrelevant or redirected or incomplete actions occur so often in conflict situations, they have commonly been the raw material from which natural selection has molded new displays that communicate the motivations of the animals and thus help resolve the conflict situation. Detailed analysis can help reveal what actions were the raw material for any particular display.

Let us look more closely at the types of actions mentioned above as raw materials for displays. The first type, irrelevant action—often called *displacement activity*—is an action performed in what appears to be the wrong context. A male Wood Thrush responding to Dilger's loudspeaker may be in such a state of conflict between attack and escape that he suddenly starts to scratch his head or to preen his wings or to eat, or he may even go to sleep. He has to do something, it seems, and if he cannot do something appropriate he does something inappropriate. In a similar way, when you are nervous over making some decision, you may drum your fingers on the desk or crack your knuckles or doodle. Such actions, performed by the male thrush or by you, may indicate to a perceptive observer something about the performer's current motivational state. Natural selection tends to exaggerate and ritualize such actions, increasing their information content and thus giving them a true display function. In other words, actions that were irrelevant in their original context come to have great relevance, and become appropriate and integral parts of other kinds of situation. Displacement activities are most often actions that ordinarily function in the daily maintenance of the individual, such as preening, scratching, yawning, stretching, and eating.

A second type of action that may serve as raw material for evolution of displays is one performed in the appropriate context but directed at the wrong objects; such an action is often called *redirected activity*. A bird thwarted in attacking an intruder on his territory may peck vigorously at the branch on which he stands instead. A dog prevented from attacking an antagonist may bite viciously at a nearby object. A man angry at his boss may come home and yell at his wife. Like displacement activities, redirected activities may indicate to a perceptive observer the motivational state of the performer, and natural selection can ritualize them and cause their incorporation into displays.

A third source of motor patterns for displays includes actions called *intention movements*, which are the incomplete initial stages of other actions. A bird crouches slightly before it flies. A man clenches his fist before striking an opponent. Even if the bird does not actually fly or if the man does not actually strike his opponent, the crouch or the clenched fist may serve as a signal to another individual, and may therefore come to have a ritualized display function quite apart from its initial significance as simply a movement preparatory to further action.

We see, then, that careful study can reveal to us the evolutionary source of the various components of behavior patterns such as displays, just as similar study reveals the evolutionary history of anatomical characters. In short, the usual methods of biological research can rewardingly be applied to animal behavior.

REFERENCES

ASCHOFF, J., ed., 1965. *Circadian Clocks.* North-Holland Publishing Co., Amsterdam.

CARTHY, J. D., 1958. *An Introduction to the Behavior of Invertebrates.* Macmillan, New York.

Cold Spring Harbor Symposium on Quantitative Biology: Biological Clocks, 1960. Biological Laboratory, Cold Spring Harbor, N.Y.

DILGER, W. C., 1956. "Hostile Behavior and Reproductive Isolating Mechanisms in the Avian Genera *Catharus* and *Hylocichla,*" *Auk,* vol. 73, pp. 313–353.

———, 1960. "The Comparative Ethology of the African Parrot Genus *Agapornis,*" *Zeitschrift für Tierpsychologie,* vol. 17, pp. 649–685.

EIBL-EIBESFELDT, I., 1970. *Ethology: The Biology of Behavior.* Holt, Rinehart & Winston, New York.

ETKIN, W., ed., 1964. *Social Behavior and Organization Among Vertebrates.* University of Chicago Press, Chicago.

HINDE, R. A., 1970. *Animal Behavior: A Synthesis of Ethology and Comparative Psychology,* 2nd ed. McGraw-Hill, New York.

IERSEL, J. J. A. VAN, 1953. "An Analysis of the Parental Behaviour of the Male Three-Spined Stickleback," *Behaviour,* sup. 3, pp. 1–153.

LINDAUER, M., 1971. *Communication Among Social Bees,* 3rd printing with appendices. Harvard University Press, Cambridge, Mass.

MARLER, P. R., and W. J. HAMILTON III, 1966. *Mechanisms of Animal Behavior.* Wiley, New York.

MOYNIHAN, M., 1955. "Remarks on the Original Sources of Displays," *Auk,* vol. 72, pp. 240–246.

THORPE, W. H., 1961. *Bird-Song: The Biology of Vocal Communication and Expression in Birds.* Cambridge University Press, New York.

———, 1963. *Learning and Instinct in Animals,* 2nd ed. Methuen, London.

TINBERGEN, N., 1951. *The Study of Instinct.* Oxford University Press, New York.

———, 1960. *The Herring Gull's World,* rev. ed. Basic Books, New York.

———, 1965. *Social Behaviour in Animals,* 2nd ed. Methuen, London.

WILSON, E. O., 1971. *Insect Societies.* Belknap Press-Harvard University Press, Cambridge, Mass.

SUGGESTED READING

Animal Behavior, 1972. National Geographic Society, Washington, D.C.

DETHIER, V. G., and E. STELLAR, 1970. *Animal Behavior,* 3rd ed. Prentice-Hall, Englewood Cliffs, N.J.

DILGER, W. C., 1962. "The Behavior of Lovebirds," *Scientific American,* January. (Offprint 1049.)

FRISCH, K. VON, 1962. "Dialects in the Language of the Bees," *Scientific American,* August. (Offprint 130.)

———, 1971. *Bees: Their Vision, Chemical Senses, and Language,* rev. ed. Cornell University Press, Ithaca, N.Y.

GRIFFIN, D. R., 1959. *Echoes of Bats and Men.* Doubleday Anchor Books, New York.

HESS, E. H., 1958. "Imprinting in Animals," *Scientific American,* March. (Offprint 416.)

HOCKETT, C. F., 1960. "The Origin of Speech," *Scientific American,* September. (Offprint 603.)

JACOBSON, M., and M. BEROZA, 1964. "Insect Attractants," *Scientific American,* August. (Offprint 189.)

JOHNSGARD, P. A., 1971. *Animal Behavior,* 2nd ed. W. C. Brown, Dubuque, Iowa.

LEHRMAN, D. S., 1964. "The Reproductive Behavior of Ring Doves," *Scientific American,* November. (Offprint 488.)

LORENZ, K. Z., 1952. *King Solomon's Ring.* Crowell, New York.

———, 1958. "The Evolution of Behavior," *Scientific American,* December. (Offprint 412.)

———, 1965. *Man Meets Dog.* Penguin, Baltimore.

———, 1966. *On Aggression.* Harcourt, Brace & World, New York.

MENAKER, M., 1972. "Nonvisual Light Reception," *Scientific American,* March. (Offprint 1243.)

MORRIS, D., 1967. *The Naked Ape.* McGraw-Hill, New York.

ROEDER, K. D., 1965. "Moths and Ultrasound," *Scientific American,* April. (Offprint 1009.)

SMITH, N. G., 1967. "Visual Isolation in Gulls," *Scientific American,* October. (Offprint 1084.)

TAVOLGA, W. N., 1969. *Principles of Animal Behavior.* Harper & Row, New York.

THORPE, W. H., 1956. "The Language of Birds," *Scientific American,* October. (Offprint 145.)

TINBERGEN, N., 1952. "The Curious Behavior of the Stickleback," *Scientific American,* December. (Offprint 414.)

———, 1960. "The Evolution of Behavior in Gulls," *Scientific American,* December. (Offprint 456.)

———, and the Editors of *Life,* 1965. *Animal Behavior.* Time Inc., New York.

VAN DER KLOOT, W. G., 1968. *Behavior.* Holt, Rinehart & Winston, New York.

WILSON, E. O., 1963. "Pheromones," *Scientific American,* May. (Offprint 157.)

PART III

The perpetuation of life

13

Cellular reproduction

All living things are composed of cells, and all cells arise from previously existing ones. This statement of the cell theory, so simple and familiar as to seem almost commonplace, acquires another dimension when one considers the far from simple process by which old cells give rise to new ones, a process generally called cell division.

As you know, the nucleus is the control center of the cell. It contains the chromosomes, which bear the genes—the units of information that, passed down from generation to generation, determine the characteristics of each new organism and direct its myriad activities. Now, if the nucleus carries the information or blueprint for the development of the new individual, it follows that when the cell divides the nuclear information must be transmitted in orderly fashion to both of the new cells. The division cannot be a simple splitting of the information into two halves, because this would give neither of the new cells a satisfactory blueprint. Just as you cannot have two buildings erected by cutting one blueprint in two and giving half to each of two contractors, so two new cells cannot develop if each receives only half the necessary information from the parental cell. If you wanted two contractors to erect identical buildings simultaneously, you would logically duplicate the necessary blueprint and give a complete copy to each. The same applies to the cell. If a parental cell is to divide and produce two viable new cells, it must first make a complete copy of the genetic information in its nucleus and then, as it divides, give one complete copy to each daughter cell. In other words, division of the nucleus is not simply a process of halving; it is a process of duplicating genetic information and distributing the duplicates.

MITOTIC CELL DIVISION

Cell division in eucaryotic cells involves two fairly distinct processes that often but not always occur together: division of the nucleus and division of the cytoplasm. The process whereby the nucleus divides to produce two new nuclei, each with the same number of chromosomes as the parental nucleus, is called *mitosis.* The process of division of the cytoplasm is called *cytokinesis.* We shall discuss mitosis and cytokinesis separately below.

Mitosis

Nuclear division entails, first, precise duplication of the genetic material and, second, distribution of a complete set of the material to each daughter cell. All available evidence indicates that the duplication step occurs in the nucleus of the nondividing cell prior to the start of division proper. Mitosis, therefore, is largely concerned with distribution of the genetic material in an orderly fashion.

Evidence of many different sorts shows conclusively that the units of genetic information, the genes, are located on the chromosomes. The number of chromosomes is usually constant for normal somatic[1] cells of all individuals of the same species. For example, the nuclei of human somatic cells contain 46 chromosomes, those of the fruit fly *Drosophila melanogaster* contain 8, and those of onion 16. Thus, if mitosis is to distribute a complete set of genetic instructions to each daughter nucleus, it must ensure that each receives a full set of chromosomes exactly like the set initially present in the parental nucleus; only in this way can a basic constancy of chromosome number and gene content in the somatic cells be maintained.

The duration of mitosis varies from about five minutes to several hours in different types of cells and under different environmental conditions. For convenience, it is customary to divide the mitotic process into a series of stages, each designated by a special name. Although each stage will be discussed separately here, it should be kept in mind that the entire process is a continuum, not a series of discrete occurrences.

Interphase. The nondividing cell is said to be in the interphase state. In past years, such a cell was commonly called a resting cell, but this terminology has been rejected as grossly inappropriate. The interphase cell is definitely not resting; it is carrying out all the innumerable activities of a living, functioning cell—respiration, protein synthesis, growth, differentiation, etc. Furthermore, as already noted, it is during interphase that the genetic material is replicated in preparation for the next division sequence.

During interphase, the nucleus is clearly visible as a distinct membrane-bounded organelle, and one or more nucleoli are usually prominent. But chromosomes, as ordinarily pictured, are not visible in the nucleus; there are no distinct rodlike bodies such as are seen so easily in a dividing cell (see Fig. 3.15, p. 48). The only manifestation of the chromosomes is an irregular granular-appearing mass of chromatin material. Formerly, some workers, in fact, insisted that the chromosomes

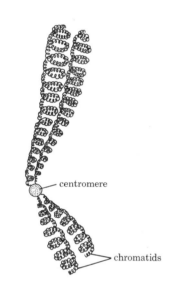

13.1. A late-prophase chromosome. It consists of two identical chromatids united at some point along their length by a single centromere. The chromatin of each chromatid is coiled.

---

[1] "Somatic," as used here, refers to all cells in the body except reproductive cells—eggs and sperms.

exist as distinct structures only during cell division and break up during interphase. More recent chemical and genetic evidence indicates that they continue to exist during interphase, but that they become very long, thin, and intertwined and hence cannot be recognized by present techniques.

In interphase animal cells, but not in most plant cells, there is a special region of cytoplasm just outside the nucleus that contains two small cylindrical bodies oriented at right angles to each other. These are the *centrioles* (see Fig. 3.24, p. 56), which will move apart and organize the mitotic apparatus of the dividing cell. In many animal cells, this separation of the centrioles occurs just before the onset of mitosis. No centrioles have yet been detected in the cells of most seed plants, but they do occur in some algae, fungi, bryophytes (mosses, liverworts, etc.), and ferns in association with the production of motile sperms.

Prophase. The first stage of mitosis is called the prophase. It is, in a sense, a preparatory stage that readies the nucleus for the crucial event of mitosis—the separation of two complete sets of chromosomes into two daughter nuclei. As the two centrioles of an animal cell move toward opposite sides of the nucleus, the initially indistinct chromosomes begin to condense into visible threads, which become progressively shorter and thicker and more easily stainable with dyes (see Fig. 13.3: 2–4). When the chromosomes first become visible during early prophase, they appear as long, thin, intertwined filaments, but by late prophase the individual chromosomes can be clearly discerned as much shorter rodlike structures. As the chromosomes become more distinct, the nucleoli become less distinct, often disappearing altogether by the end of prophase. It seems probable that the condensing of the chromosomes is largely a matter of their becoming coiled.

If an individual chromosome from a late-prophase nucleus is examined under very high magnification, it can be seen to consist of two separate strands called *chromatids* (Fig. 13.1). The two chromatids are united by a small body called a *centromere*. (Note that, in determining the number of chromosomes present in a nucleus, we count the number of separate centromeres, not the number of chromatids.) The replication process that occurred during interphase produced one of the chromatids by using the other as a model, in a manner to be discussed at length in Chapter 15. Thus the two chromatids of a prophase chromosome are identical. It must be emphasized that, contrary to a common impression, the interphase process that produced the second chromatid was not a splitting, but a duplicating.

As the two centrioles of an animal cell move apart, a system of thin fibrils, which radiate in all directions (Fig. 13.3: 2), appears near each centriole. Some fibrils from one of the centrioles apparently link up with fibrils from the other centriole, so that by the end of prophase the fibrils are divided into two groups, the *spindle fibrils,* which form continuous links between the two centrioles, and the blindly ending astral fibrils, collectively called an *aster,* which radiate in other directions from each centriole (Fig. 13.2). Even though the cells of most higher plants lack centrioles, they develop a spindle much like that in animal cells, but no asters are formed.

During late prophase, the double-stranded chromosomes, at first

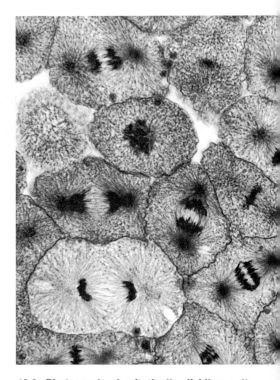

13.2. Photograph of mitotically dividing cells of whitefish blastula. The asters around each centriole are prominent. Most of the cells shown here are in either metaphase or anaphase. [Courtesy General Biological Supply House, Inc., Chicago.]

distributed essentially at random within the nucleus, begin to move toward the middle, or equator, of the biconical spindle formed by the spindle fibrils. Prophase ends when the nuclear membrane has disappeared and each chromosome has moved to the midpoint of one of the spindle fibrils.

Metaphase. During the brief stage termed metaphase, the chromosomes are arranged on the equatorial plane of the spindle, and in side view appear to form a line across the middle of the spindle (Fig. 13.3: 5). Each double-stranded chromosome is attached by its centromere to a fibril of the spindle, and it is actually the centromeres that are lined up precisely along the equatorial plane. Metaphase ends when each chromosomal centromere divides and each of the former chromatids thus becomes a separate single-stranded chromosome—in other words, when the total number of independent chromosomes in the nucleus is doubled.

Anaphase. Now begins the separation of two complete sets of chromosomes, the critical event for which the previous stages were the preparation. At the beginning of anaphase, the two new single-stranded chromosomes derived from each original double-stranded chromosome begin to move away from each other, one going toward one pole (centriole) of the spindle and the other going toward the opposite pole. A chromosome cannot move during anaphase unless its centromere is attached to a spindle fibril; it can be demonstrated that both the centromere and the fibril are necessary for the movement. A cell in early anaphase can be recognized by the fact that the chromosomes are in two equal groups a short distance apart. A cell in late anaphase contains two groups of chromosomes that are more widely separated, the two clusters having almost reached their respective poles of the spindle (Fig. 13.3: 7). Cytokinesis often begins during late anaphase.

Telophase. Telophase is essentially a reverse of prophase. The two sets of chromosomes, having reached their respective poles, become enclosed in new nuclear membranes as the spindle disappears. These new nuclear membranes are assembled from membranous vesicles derived from the endoplasmic reticulum. Then the chromosomes begin to uncoil and to resume their interphase form, while the nucleoli slowly reappear (Fig. 13.3: 8). The centriole of each nucleus is usually replicated during late telophase. Cytokinesis is often completed during telophase. Telophase ends when the new nuclei have fully assumed the characteristics of interphase, thus bringing to a close the complete mitotic cycle.

Note that when the individual chromosomes fade from view at the end of telophase they are single-stranded, but when they reappear during prophase in the next division sequence they are double-stranded—the replication, which cannot be observed by light microscopy, having occurred during interphase. Thus, if prophase begins with a nucleus containing, say, six double-stranded chromosomes, telophase ends with two new nuclei each containing six single-stranded chromosomes. Mitosis separates the strands of the initially double-stranded chromosomes. Thus each new daughter cell has the same number and kinds of

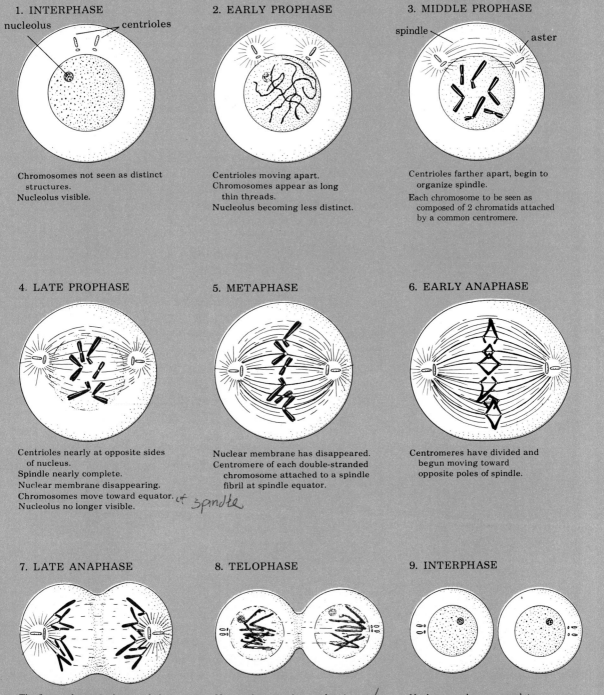

1. INTERPHASE

nucleolus — centrioles

Chromosomes not seen as distinct structures.
Nucleolus visible.

2. EARLY PROPHASE

Centrioles moving apart.
Chromosomes appear as long thin threads.
Nucleolus becoming less distinct.

3. MIDDLE PROPHASE

spindle — aster

Centrioles farther apart, begin to organize spindle.

Each chromosome to be seen as composed of 2 chromatids attached by a common centromere.

4. LATE PROPHASE

Centrioles nearly at opposite sides of nucleus.
Spindle nearly complete.
Nuclear membrane disappearing.
Chromosomes move toward equator. of spindle
Nucleolus no longer visible.

5. METAPHASE

Nuclear membrane has disappeared.
Centromere of each double-stranded chromosome attached to a spindle fibril at spindle equator.

6. EARLY ANAPHASE

Centromeres have divided and begun moving toward opposite poles of spindle.

7. LATE ANAPHASE

The 2 sets of new single-stranded chromosomes nearing respective poles.
Cytokinesis beginning.

8. TELOPHASE

New nuclear membranes forming.
Chromosomes become longer, thinner, and less distinct.
Nucleolus reappearing.
Centrioles replicated.
Cytokinesis nearly complete.

9. INTERPHASE

Nuclear membranes complete.
Chromosomes no longer visible.
Cytokinesis complete.

13.3. Mitosis and cytokinesis of an animal cell.

chromosomes as the parental cell, and hence each has the complete set of genetic information necessary to determine its characteristics and to guide its activities.

Cytokinesis

We said earlier that division of the cytoplasm frequently accompanies division of the nucleus, often beginning in late anaphase and reaching completion during telophase. But this is not always the case. Mitosis without cytokinesis is common in some algae and fungi, producing plant bodies with many nuclei but with no, or few, cellular partitions. It regularly occurs during certain phases of reproduction in seed plants and certain other vascular plants. It is also common in a few lower invertebrate animals. And it produces the multinucleate cells of vertebrate skeletal and cardiac muscle.

Cytokinesis in animal cells. Division of an animal cell normally begins with the formation of a *cleavage furrow* running around the cell. When cytokinesis occurs during mitosis, the location of the furrow is ordinarily determined by the orientation of the spindle, in whose equatorial region the furrow forms (Fig. 13.3: 7). The furrow becomes progressively deeper, until it cuts completely through the cell (and its spindle), producing two new cells. The recent discovery of a ring of contractile microfilaments in the cytoplasm at the site of the furrow makes it seem likely that it is formed by the same sort of microfilament activity known to be involved in other types of cell movement (see Fig. 16.10, p. 341).

Cytokinesis in plant cells. Plant cells possess relatively rigid cell walls, which cannot develop cleavage furrows. It is therefore not surprising that cytokinesis in most plant cells is very different from cytokinesis in animal cells. A special membrane, called the *cell plate*, forms halfway between the two nuclei (at the equator of the spindle if cytokinesis accompanies mitosis; Fig. 13.4). The cell plate begins to form in the center of the cytoplasm and slowly becomes larger until its edges reach the outer surface of the cell and the cell's contents are cut in two. In plant cells, then, cytokinesis progresses from the middle to the periphery, whereas in animal cells it progresses from the periphery to the middle.

The cell plate forms from membranous vesicles that first line up and then unite. The endoplasmic reticulum apparently makes major contributions to the developing cell plate. Even though the cell plate develops from membranes, it probably does not form the plasma membranes of the new cells. Instead, it seems to become impregnated with pectin and to form the middle lamella upon which the cellulose walls of the two new cells are deposited by the plasma membranes that soon form.

MEIOTIC CELL DIVISION

The nuclei of the cells shown in Fig. 13.3, belonging to a hypothetical organism, contain six chromosomes. Notice that those six chromosomes are of only three different types, there being two of each type. Cells of this sort, with two of each type of chromosome, are said to be *diploid.* The somatic cells of the human body, as well as those of most higher plants and animals, are diploid. We have said that each somatic cell of a human being has 46 chromosomes and that each somatic cell of a fruit fly has eight; since these cells are diploid, it follows that the chromosomes in a human somatic cell are of 23 types and those in a fruit-fly somatic cell are of four types. Mitosis, as we have seen, produces new cells with exactly the same chromosomal endowment as the parental cell. Hence, whenever a human somatic cell divides mitotically, the new cells thus produced have 46 chromosomes of 23 different types. But we also know that, in sexual reproduction, two cells (the gametes: an egg and a sperm) unite to form the first cell (zygote) of the new individual. If the gametes were produced by normal mitosis in human beings or fruit flies or our hypothetical organism, the zygote produced by their union would have double the normal number of chromosomes, and at each successive generation the number would again double, until the total chromosome number per cell would approach infinity. This does not happen; the chromosome number normally remains constant within a species. At some point, therefore, a different kind of cell division must take place, a division that reduces the number of chromosomes by half so that when the egg and sperm unite in fertilization the normal diploid number is restored. This special process of reduction division is called *meiosis.*

In all multicellular animals, meiosis occurs at the time of gamete production. Consequently each gamete possesses only half the species-typical number of chromosomes. It is important to note that in the reduction division of meiosis the chromosomes of the parental cell are not simply separated into two random halves; the diploid nucleus contains two of each type of chromosome, and meiosis partitions these chromosome pairs so that each gamete contains one of each type of chromosome. Such a cell, with only one of each type of chromosome, is said to be *haploid.* When two haploid gametes unite in fertilization, the resulting zygote is diploid, having received one of each chromosome type from the sperm of the male parent and one of each type from the egg of the female parent.

13.4. Photograph of mitotically dividing cells in onion root tip. A cell plate (CP) can be seen as a faint dark line in the equatorial plane between two telophase nuclei. [Courtesy General Biological Supply House, Inc., Chicago.]

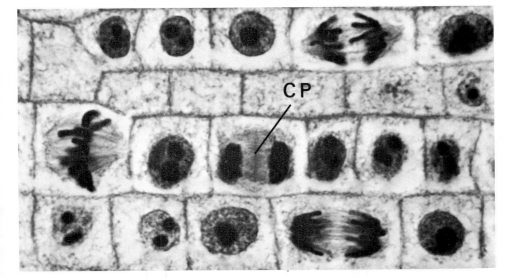

CP

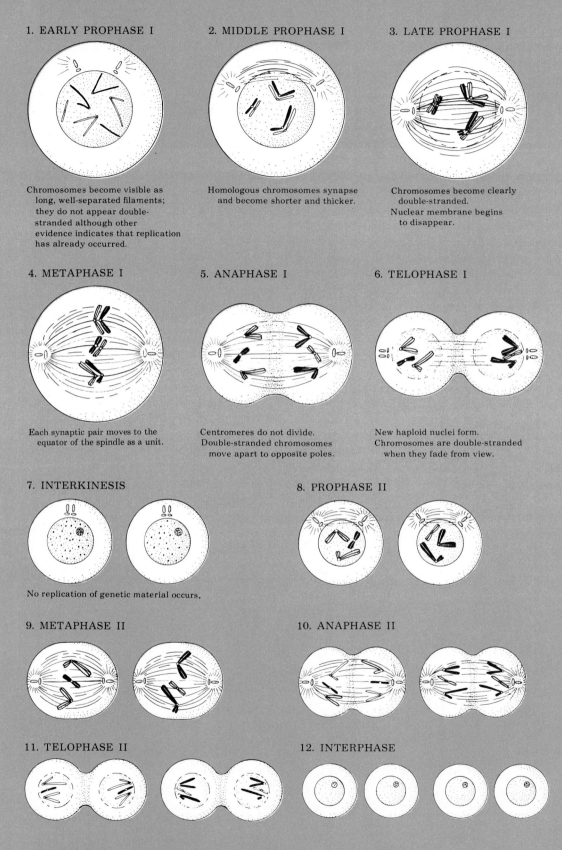

1. EARLY PROPHASE I

Chromosomes become visible as long, well-separated filaments; they do not appear double-stranded although other evidence indicates that replication has already occurred.

2. MIDDLE PROPHASE I

Homologous chromosomes synapse and become shorter and thicker.

3. LATE PROPHASE I

Chromosomes become clearly double-stranded.
Nuclear membrane begins to disappear.

4. METAPHASE I

Each synaptic pair moves to the equator of the spindle as a unit.

5. ANAPHASE I

Centromeres do not divide.
Double-stranded chromosomes move apart to opposite poles.

6. TELOPHASE I

New haploid nuclei form.
Chromosomes are double-stranded when they fade from view.

7. INTERKINESIS

No replication of genetic material occurs.

8. PROPHASE II

9. METAPHASE II

10. ANAPHASE II

11. TELOPHASE II

12. INTERPHASE

13.5. Meiosis in an animal cell.

The process of meiosis

Complete meiosis involves two successive division sequences, which result in four new haploid cells. It is the first division sequence that accomplishes the reduction in the number of chromosomes; the second division sequence is essentially a mitotic one. Let us examine the nuclear events characteristic of meiosis, recognizing four stages, as in mitosis—namely, prophase, metaphase, anaphase, and telophase.

First prophase. Many of the events in the first prophase of meiosis resemble those in the prophase of mitosis. The individual chromosomes come slowly into view as they coil and become shorter, thicker, and more easily stainable. The nucleoli slowly fade from view, and, finally, the nuclear membrane disappears and the spindle is organized. Radioactive-tracer studies show that the replication of the genetic material occurs during the interphase that precedes prophase I, as in mitosis.

The chief difference between prophase of meiosis and mitosis is that in meiosis the members of each pair of chromosomes (homologous chromosomes) move together and come to lie side by side in an intimate association (Fig. 13.5: 2). They do not fuse together, but they do often intertwine. This pairing process is known as *synapsis.* Each chromosome being visibly double-stranded by late prophase, a synaptic pair can be seen to consist of two identical double-stranded chromosomes lying next to each other. Toward the end of prophase, the synaptic pair moves as a unit to the equator of the spindle.

First metaphase. In mitosis, the two chromosomes of each type are completely independent of each other in their movements. They do not synapse, and each individual chromosome moves on its own to a separate fibril of the spindle. Hence, at metaphase of mitosis in our hypothetical organism of Fig. 13.3, each of the six chromosomes occupies a different fibril; similarly, each of the 46 chromosomes in a mitotically dividing human cell occupies a different fibril. But in meiosis, the two chromosomes of each type synapse, as we have seen, and move onto the spindle as a single unit. Hence, at metaphase I of meiosis in our hypothetical organism (Fig. 13.5: 4), only three of the spindle fibrils are occupied, not six. An entire synaptic pair, consisting of two chromosomes and four chromatids, is attached to each of these three fibrils. Now, at the start of metaphase in mitosis, there is only one chromosome on each fibril and thus only one centromere attached to it (before division of the centromere occurs). But at the start of metaphase I in meiosis, there are two chromosomes on each occupied fibril and thus two separate centromeres attached to it.

First anaphase. In mitosis, metaphase ends and anaphase begins when the single centromere of each double-stranded chromosome divides and the two separate single-stranded chromosomes thus formed move away from each other toward opposite poles of the spindle. But in the first division sequence of meiosis, two chromosomes with separate centromeres are attached to each occupied fibril from the beginning of metaphase. Hence there is no division of centromeres. The two double-stranded chromosomes of each synaptic pair move away from each other toward opposite poles during anaphase. This means that, in our hypothetical organism, only three chromosomes move to each pole (Fig. 13.5: 5), in contrast to the six that move to each pole in mitosis. Notice, however, that because the synaptic pairing was not random but involved the two homologous chromosomes of each type, the two daughter nuclei get not just any three chromosomes but, rather, one of each of the three types.

First telophase. Telophase of mitosis and meiosis are essentially the same, except that each of the two new nuclei formed in mitosis has the same number of chromosomes as the parental nucleus, whereas each of the new nuclei formed in meiosis has half the chromosomes that were present in the parental nucleus (Fig. 13.5: 6). At the end of telophase of mitosis, the chromosomes are single-stranded when they fade from view; at the end of telophase I of meiosis, the chromosomes are double-stranded when they fade from view.

Interkinesis. Following telophase I of meiosis, there is a short period called interkinesis, which is similar to an interphase between two mitotic division sequences except that no replication of the genetic material occurs and hence no new chromatids are formed (replication is unnecessary, since the chromosomes are already double-stranded when interkinesis begins).

Table 13.1 summarizes the differences between mitosis and the first division sequence of meiosis.

Second division sequence of meiosis. The second division sequence of meiosis, which follows interkinesis, is essentially a mitotic one (Fig. 13.5: 8–11). The chromosomes do not synapse; they cannot, since the nucleus is haploid and there are no homologous chromosomes. Each double-stranded chromosome moves onto the spindle independently, and its centromere attaches to a fibril. At the end of metaphase II

Table 13.1. **Comparison of mitosis and first division sequence of meiosis**

Phase	Mitosis	Meiosis
Prophase	No synapsis; chromosomes move to spindle individually	Synapsis; chromosomes move to spindle in pairs
Metaphase	Each chromosome attached to separate fibril; centromeres divide	The two chromosomes of each pair attached to the same fibril; centromeres do not divide
Anaphase	Separation of new single-stranded chromosomes derived from one original double-stranded chromosome	Separation of old double-stranded chromosomes of each synaptic pair
Telophase	Formation of two new nuclei, each with same number of chromosomes as parental nucleus; chromosomes are single-stranded when they fade from view	Formation of two new nuclei, each with half the chromosomes present in parental nucleus; chromosomes are double-stranded when they fade from view
Interphase, Interkinesis	Replication of genetic material, with formation of new chromatids	No replication of genetic material and hence no new chromatids

the centromeres divide, and during anaphase II the new single-stranded chromosomes thus formed move away from each other toward opposite poles of the spindle. The new nuclei formed during telophase II are haploid like the nuclei formed during telophase I, but their chromosomes are single-stranded instead of double-stranded.

In summary, then, the first meiotic division produces two haploid cells containing double-stranded chromosomes. Each of these cells divides in the second meiotic division; thus a total of four new haploid cells containing single-stranded chromosomes are produced.

The timing of meiosis in the life cycle

Meiosis in the life cycle of animals. With rare exceptions, higher animals exist as diploid multicellular organisms through most of their life cycle. At the time of reproduction, meiosis produces haploid gametes, which, when their nuclei unite in fertilization, give rise to the diploid zygote. The zygote then divides mitotically to produce the new diploid multicellular individual. The gametes—sperms and eggs—are thus the only haploid stage in the animal life cycle (see Fig. 13.7B).

In male animals, sperm cells (spermatozoa) are produced by the germinal epithelium lining the seminiferous tubules of the testes. When one of the epithelial cells undergoes meiosis, the four haploid cells that result are all quite small but approximately equal in size. All four soon differentiate into sperm cells with long flagella, but with very little cytoplasm in the head, which consists primarily of the nucleus.

In female animals, the egg cells are produced within the follicles of the ovaries. When a cell in the ovary undergoes meiosis, the haploid cells that result are very unequal in size. The first meiotic division produces one relatively large cell and a tiny one called a first *polar body* (Fig. 13.6). The second meiotic division of the larger of these two cells produces a tiny second polar body and a large cell that soon differentiates into the egg cell. Thus, when a diploid cell in the ovary undergoes complete meiosis, only one mature egg cell (or ovum) is produced; the polar bodies are essentially nonfunctional. By contrast, a diploid cell undergoing complete meiosis in the testis gives rise to four functional sperm cells.

The advantage of the unequal cytokinesis in the production of egg cells is obvious. By this mechanism, an unusually large supply of cytoplasm and stored food is allotted to the nonmotile ovum for use by the embryo that will develop from it. In fact, the ovum provides almost all the cytoplasm and initial food supply for the embryo. The tiny, highly motile sperm cell contributes, essentially, only its genetic material.

Meiosis in the life cycle of plants. That meiosis produces gametes in animals does not mean that it must do so in all organisms. There is no inherent reason why the cells resulting from meiosis must be specialized for sexual reproduction. And, indeed, they are not specialized in this way in most plants. Meiosis in plants usually produces haploid reproductive cells called *spores,* and these spores often divide mitotically to develop into haploid multicellular plant bodies.

Let us briefly examine the life cycle of a hypothetical plant in which meiosis produces spores rather than gametes (Fig. 13.7A). While this plant is in the diploid portion of its life cycle, certain cells in its

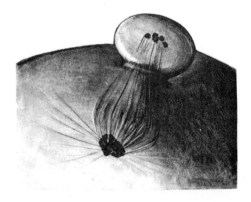

13.6. Formation of a tiny first polar body on the surface of the much larger developing egg cell of a whitefish. Before meiosis began, the nucleus migrated to the periphery of the cell. The meiotic spindle is oriented at right angles to the surface of the larger cell.

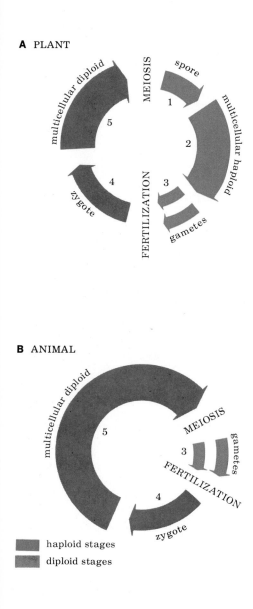

A PLANT

B ANIMAL

haploid stages
diploid stages

13.7. Two types of life cycles. (A) In most multicellular plants, there are two multicellular stages, one haploid and one diploid (stages 2 and 5); the relative importance of these two stages varies greatly from one plant group to another. (B) Animals have a life cycle in which meiosis produces gametes directly—the spore stage (stage 1) and the multicellular haploid stage (stage 2) being absent.

reproductive organs divide by meiosis to produce haploid spores (stage 1). These spores divide mitotically and develop into haploid multicellular plants (stage 2). The haploid multicellular plant eventually produces cells specialized as gametes (stage 3). Notice, however, that the gametes of the plant are produced by mitosis, not meiosis, because the cells that divide to produce the gametes are already haploid. Two of these gametes unite to form the diploid zygote (stage 4), which divides mitotically and develops into a diploid multicellular plant (stage 5). In time, this plant produces spores, and the cycle starts over again. The various groups of plants vary greatly as to the relative importance of the diploid and haploid phases in their life cycles.

The adaptive significance of sexual reproduction

Sexual reproduction is so widespread, being characteristic of the vast majority of both plants and animals, that we tend to regard it as a universal attribute of life. But it is not. Many plants and animals reproduce asexually, i.e. give rise to new individuals by mitotic cell division. Such reproduction is undeniably simpler than sexual reproduction, where the complicated processes of meiosis and fertilization must alternate with each other. We can legitimately ask, therefore, why asexual reproduction is not adequate for most organisms, why natural selection has so often favored sexual reproduction, with all the complex problems it entails.

We have seen that mitosis produces new cells with a genetic endowment identical to that of the parental cell; each new cell gets a set of chromosomes copied from the parental set. Hence the characteristics of offspring produced by mitosis will be essentially the same as those of the parent. In other words, asexual reproduction does not give rise to genetic variation, for mitosis provides no way in which the genes of one individual can be recombined with those of another.

But now let us look at the chromosomes of a diploid individual produced by sexual reproduction. The first cell (zygote) of this individual was formed by the union of two gametes, an egg cell from one parent and a sperm cell from a second parent. Each of these gametes contributed one chromosome of each type characteristic of the species. Hence some of the genes of one parent have recombined with genes from the other parent, and this new combination may produce progeny with characteristics different from those of either parent. Unlike mitotic asexual reproduction, sexual reproduction with its meiosis and fertilization augments variation in the population by recombining chromosomes and the genes they bear.

Such variation is clearly advantageous to the species. Without genetic variation, there could be no evolution, no response to inevitable environmental changes. Genetic variation is the evolutionary raw material upon which natural selection acts. If a species had no genetic variation, and thus could not evolve, it would soon be doomed to extinction. The environmental forces impinging upon any species are constantly changing, and the survival of the species depends, ultimately, on its ability to respond to those forces with evolutionary changes that maintain or increase its fitness. Sexual reproduction is one important factor in providing the raw material for such changes.

REFERENCES

AUSTIN, C. R., 1965. *Fertilization.* Prentice-Hall, Englewood Cliffs, N.J.

———, 1968. *Ultrastructure of Fertilization.* Holt, Rinehart & Winston, New York.

DEROBERTIS, E. D. P., W. W. NOWINSKI, and F. A. SAEZ, 1970. *Cell Biology,* 5th ed. Saunders, Philadelphia. (See esp. Chapters 3, 13.)

DUPRAW, E. J., 1968. *Cell and Molecular Biology.* Academic Press, New York. (See esp. Chapter 19.)

SUGGESTED READING

ALSTON, R. E., 1967. *Cellular Continuity and Development.* Scott, Foresman, Glenview, Ill.

DUPRAW, E. J., 1970. *DNA and Chromosomes.* Holt, Rinehart & Winston, New York. (See esp. Chapters 4–6, 15.)

MAZIA, D., 1953. "Cell Division," *Scientific American,* August. (Offprint 27.)

———, 1961. "How Cells Divide," *Scientific American,* September. (Offprint 93.)

CHAPTER

Patterns of inheritance

It is time now that we offer evidence for our oft-repeated, but so far un-
supported, assertion that there are units called genes located on the
chromosomes in the nucleus, and that these units, passed on from
generation to generation, exert control over the characteristics of organisms. The earliest evidence on which the modern concepts of chromosomal inheritance are founded came from breeding experiments, and
this is the evidence we shall examine first.

MONOHYBRID INHERITANCE

Experiments by Mendel

Let us begin our examination of the chromosomal theory of inheritance by considering a series of experiments performed on ordinary
garden peas during the years 1856 to 1868 by a modest Austrian monk,
Gregor Mendel. Mendel published the results of these experiments in
1866.

Of the numerous characteristics of garden peas that Mendel
studied, seven were particularly interesting to him. He noticed that each
of these seven characteristics occurred in two contrasting forms (Table
14.1). Thus the seeds were either round or wrinkled, the flowers either
red or white, the pods either green or yellow, etc. When Mendel cross-
pollinated plants with contrasting forms of one of these characteristics,
all the offspring were alike and resembled one of the two parents. When
these offspring were crossed among themselves, however, some of their
offspring showed one of the original contrasting traits and some showed
the other. In other words, a trait present in the grandparental generation but not in the parental generation reappeared.

Let us examine more closely some of Mendel's crosses. When

plants with red flowers were crossed with plants with white flowers, all the offspring—the F₁ generation[1]—had red flowers (Fig. 14.1). Similarly, when plants characterized by round seeds were crossed with plants characterized by wrinkled seeds, all the offspring had round seeds. Apparently, one form of each characteristic had taken precedence over the other; i.e. red color had taken precedence over white in the flowers, and round form had taken precedence over wrinkled in the seeds. Mendel termed the traits that appear in the F₁ offspring of such crosses (in these examples, red flowers or round seeds) *dominant characters,* and the traits that become latent (in these examples, white flowers or wrinkled seeds) *recessive characters.*

When Mendel allowed the F₁ peas from the cross involving flower color, all of which were red, to breed freely among themselves, their offspring (the F₂ generation) were of two types; there were 705 plants with red flowers and 224 plants with white flowers. The recessive character had reappeared in about one fourth of the F₂ plants. Similarly, when F₁ peas from the cross involving seed form, all of which had round seeds, were allowed to breed freely among themselves, the F₂ offspring were of two types; 5,474 had round seeds and 1,850 had wrinkled seeds. Again, the recessive character had reappeared in about one fourth of the F₂ plants. The same thing was true of crosses involving the other five characters that Mendel studied (Table 14.1); in each case, the recessive character disappeared in the F₁ generation but reappeared in about one fourth of the plants in the F₂ generation. We can summarize the results of the experiment involving flower color as follows:

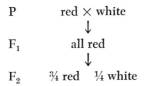

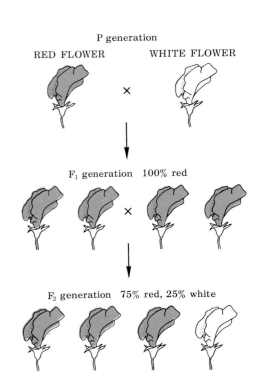

14.1. Results of Mendel's cross of red-flowered and white-flowered peas.

[1] The parental generation in such a cross is customarily designated as the P generation. The offspring are designated as the F₁ generation (meaning first filial generation), and the offspring of the F₁ generation as the F₂ generation (meaning second filial generation). Mendel did not use these terms, but we shall use them throughout our discussion of his experiments.

Table 14.1. Mendel's results from crosses involving single character differences

P characters	F₁	F₂	F₂ ratio
1. Round × wrinkled seeds	All round	5,474 round : 1,850 wrinkled	2.96 : 1
2. Yellow × green seeds	All yellow	6,022 yellow : 2,001 green	3.01 : 1
3. Red × white flowers	All red	705 red : 224 white	3.15 : 1
4. Inflated × constricted pods	All inflated	882 inflated : 299 constricted	2.95 : 1
5. Green × yellow pods	All green	428 green : 152 yellow	2.82 : 1
6. Axial × terminal flowers	All axial	651 axial : 207 terminal	3.14 : 1
7. Long × short stems	All long	787 long : 277 short	2.84 : 1

Mendel's conclusions. From experiments of this type, Mendel drew the important conclusion that each pea plant possesses two hereditary factors for each character, and that when gametes are formed the two factors segregate and pass into separate gametes, so that each gamete possesses only one factor for each character. Each new plant thus receives one factor for each character from its male parent and one for each character from its female parent. The fact that two contrasting parental traits, such as red and white flowers, can both appear in normal form in the F_2 offspring indicates that the hereditary factors must exist as separate particulate entities in the cell; they do not blend or fuse with each other. Thus the cells of an F_1 pea plant from the cross involving flower color contain, according to Mendel, one factor for red color and one for white color, the factor for red being dominant and the factor for white recessive. But their existence together in the same nucleus does not change the factors; they do not alter each other, remaining distinct and segregating unchanged when gametes are formed.

You will have noticed that Mendel's conclusions are consistent with what we know about the chromosomes and their behavior in meiosis. The diploid nucleus contains two of each type of chromosome. Presumably each of the two chromosomes of any given pair bears genes for the same characters; hence the diploid cell contains two doses, which may or may not be identical, of each type of gene; these are the two hereditary factors for each character that Mendel described.[2] Since the members of each pair of chromosomes segregate during meiosis, gametes contain only one chromosome of each type and hence only one dose of each gene, just as Mendel deduced. In short, Mendel's theories seem rather obvious in the light of the events of cell division. But it should be remembered that Mendel did his work before the details of cell division had been learned, before, in fact, the significance of chromosomes for heredity had been discovered. He arrived at his conclusions purely by reasoning from the patterns of inheritance he detected in his experiments, without reference to the structural components of the cell or its nucleus. The chromosomal theory of inheritance, therefore, rests today on two independent lines of evidence, one from breeding experiments, the other from microscopic examination of the nucleus. That these two lines of evidence should agree amazingly well is testimony to the strength of the theory.

The publication of Mendel's classic paper in 1866 did not all at once change the prevalent ideas about inheritance, giving birth to a flourishing new branch of biology. Far from it. The scientific community of his day was unprepared for so radical a view of heredity, and it paid little heed to Mendel's results or theories. His paper was soon forgotten. It was not until 1900, after the details of cell division had been worked out and the scientific community was in a more receptive mood, that Mendel's paper was rediscovered.

A modern interpretation of Mendel's experiments. Let us now interpret in modern terms the results of Mendel's cross involving garden

---

[2] It will be necessary to consider other definitions of the gene in the next chapter. For the moment, however, we can define a gene (Mendel's "factor") as a hereditary unit, located at a specific place or locus on a chromosome, that determines a particular character of the organism.

peas of different colors. Mendel was working with two forms of the gene for flower color, one that produced red flowers and another that produced white flowers. When a gene exists in more than one form, in this way, the different forms are called *alleles.* In the gene for flower color in peas, the allele for red is dominant, the allele for white recessive. It is customary to designate genes by letters, using capitals for dominant alleles and small letters for recessive alleles. We may thus designate the allele for red flowers in peas as C and the allele for white flowers as c.

Now, a diploid cell contains two doses of each gene, one on each of two homologous chromosomes. It may have two doses of the same allele or one dose of one allele and one dose of another allele. Thus cells of a pea plant may contain two doses of the allele for red flowers (CC), or two doses of the allele for white flowers (cc), or one dose of the allele for red and one dose of the allele for white (Cc). Cells with two doses of the same allele (CC or cc) are said to be *homozygous.* Those with one each of two different alleles (Cc) are said to be *heterozygous.*

Note that one cannot tell by visual inspection whether a given pea plant is homozygous dominant (CC) or heterozygous (Cc), because the two types of plants will look alike; both will have red flowers. In other words, where one allele is dominant over another, the dominant allele takes full precedence over the recessive allele, and a heterozygous organism exhibits the trait determined by that dominant allele; one dose of the dominant allele is as effective as two doses in determining the character trait. This means that there is often no one-to-one correspondence between the different possible genic combinations—*genotypes*—and the possible appearances—*phenotypes*—of the organisms. Thus, in the example of flower color in peas discussed here, there are three possible genotypes, CC, Cc, and cc, but only two possible phenotypes, red and white.

We can now apply this understanding of genes to Mendel's pea cross and rewrite the summary of p. 293 as follows:

$$P \qquad CC \quad \times \quad cc$$
$$\text{red} \qquad \text{white}$$
$$\downarrow$$
$$F_1 \qquad Cc \quad \times \quad Cc$$
$$\text{red} \qquad \text{red}$$
$$\downarrow$$
$$F_2 \qquad CC \quad Cc \quad cC \quad cc$$
$$\text{red} \quad \text{red} \quad \text{red} \quad \text{white}$$

Here we have shown both the genotypes and the phenotypes of the plants in the three generations. Mendel began with a cross in the parental generation between a plant with a homozygous dominant genotype (red phenotype) and a plant with a homozygous recessive genotype (white phenotype). All of the F_1 progeny had red phenotypes, because all of them were heterozygous, having received a dominant allele for red (C) from the homozygous dominant parent and a recessive allele for white (c) from the homozygous recessive parent. But when the F_1 individuals were allowed to cross freely among themselves, the F_2 progeny they produced were of three genotypes and two phenotypes; one fourth were homozygous dominant and showed red phenotypes, two fourths were heterozygous and showed red phenotypes, and one fourth were homo-

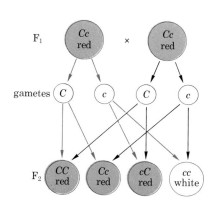

14.2. Gametes formed by F₁ individuals in Mendel's cross for flower color, and their possible combinations in the F₂.

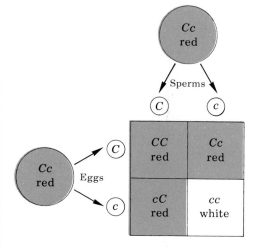

14.3. Punnett-square representation of the same information as shown in Fig. 14.2.

zygous recessive and showed white phenotypes. Thus the ratio of genotypes in the F₂ was 1 : 2 : 1, and the ratio of the phenotypes was 3 : 1.

How do we figure out the possible genotypic combinations in the F₂? This is an easy matter in a monohybrid cross (a cross involving only one character) such as this. All individuals in the F₁ generation are heterozygous (Cc); i.e. they have one of each of the two types of alleles. Each of these two alleles is located on a different one of the two chromosomes of a homologous pair. In meiosis, these two chromosomes synapse, move onto the spindle as a unit, and then separate, moving to opposite poles, so that the chromosome bearing the C allele is incorporated into one new haploid nucleus and the chromosome bearing the c allele is incorporated into the other new haploid nucleus. This means that half the gametes produced by such a heterozygous individual will contain the C allele and half the c allele. When two such individuals are crossed (Fig. 14.2), there are four possible combinations of their gametes:

C from male parent, C from female parent
C from male parent, c from female parent
c from male parent, C from female parent
c from male parent, c from female parent

The first of these four possible combinations produces homozygous dominant offspring (red). The second and third possible combinations produce heterozygous offspring (also red). And the fourth possible combination produces homozygous recessive offspring (white). Since each of these four combinations is equally probable, we would expect, if a large number of F₂ progeny are produced, a genotypic ratio close to 1 : 2 : 1 and a phenotypic ratio close to 3 : 1, just as Mendel found.

An easy way to figure out the possible genotypes produced in the F₂ is to construct a so-called Punnett square. To do this, write along a horizontal line all the possible kinds of gametes the male parent can produce; in a vertical column to the left, write all the possible kinds of gametes the female parent can produce; then draw squares for each possible combination of these, as follows:

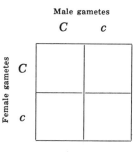

Next, write in each box, first, the symbol for the female gamete and, second, the symbol for the male gamete. Each box will then contain the symbols for the genotype of one possible zygote combination from the cross in question. A glance at the completed Punnett square in Fig. 14.3 shows that the cross yields the expected 1 : 2 : 1 genotypic ratio and, since dominance is present, the expected 3 : 1 phenotypic ratio.

Extensive investigation of a vast array of plant and animal species by thousands of scientists has demonstrated conclusively that the results Mendel obtained from his monohybrid crosses, and the interpretations he placed on them, are not limited to garden peas but are of general validity. Whenever a monohybrid cross is made between two contrasting homozygous individuals, and whatever character is involved, the expected genotypic ratio in the F_2 is 1 : 2 : 1. And whenever dominance is involved, the expected phenotypic ratio is 3 : 1. If these ratios are not obtained in large samples, it can be assumed that some complicating condition is present, which must be discovered.

The test cross

We have seen that in a monohybrid cross involving dominance the homozygous dominant progeny and the heterozygous progeny have the same phenotype and cannot be distinguished by inspection. But it is often of practical importance—notably in breeding—to distinguish between individuals with these two different genotypes. Homozygous individuals breed true; i.e. matings between two such individuals produce offspring all of a single genotype and phenotype like their parents. But heterozygous individuals do not breed true; as we have seen, a cross between two such individuals may produce offspring of three different genotypes and two different phenotypes. Consequently the identification of homozygous individuals is of the utmost importance to an animal or plant breeder who is trying to establish true-breeding strains of animals or plants. This is no problem when the breeder is interested in a recessive character, because homozygous recessive organisms can readily be recognized by their phenotype. But if the breeder is interested in establishing a strain that is true-breeding for a dominant character, he needs a test that will tell him whether a given individual is homozygous dominant or heterozygous.

The test used for this purpose is a cross between the individual of unknown genotype and a homozygous recessive individual, recognizable by its phenotype. Suppose, for example, we want to know whether a particular red-flowered pea plant has a genotype of CC or Cc. The most we can say from simple inspection is that it is red and hence must have at least one C allele; we write what we know about its genotype as C—; in other words, half of its genotype for this character is known and half is unknown and is designated by a dash. We now cross this plant with a white-flowering plant, which can only have the genotype cc. The results should give us the information we seek. If the plant in question has CC as its genotype, then the cross will turn out as follows:

$$
\begin{array}{ccc}
\text{C}\underline{\text{C}} & \times & cc \\
\text{red} & & \text{white}
\end{array}
$$

$$\downarrow$$

$$
\begin{array}{cccc}
Cc & Cc & Cc & Cc \\
\text{red} & \text{red} & \text{red} & \text{red}
\end{array}
$$

If, however, the plant has the other possible genotype, C$\underline{c}$, then the cross will turn out as follows:

$$
\begin{array}{ccc}
C\underline{c} & \times & cc \\
\text{red} & & \text{white}
\end{array}
$$

$$\downarrow$$

$$
\begin{array}{cccc}
Cc & Cc & cc & cc \\
\text{red} & \text{red} & \text{white} & \text{white}
\end{array}
$$

If we obtain a large number of progeny from the cross $C- \times cc$ and all of them show the dominant phenotype (in this case, red flowers), the chances are great that the test plant's genotype is CC (Fig. 14.4, top). If, however, we obtain progeny of which some show the dominant phenotype (red) and some the recessive phenotype (white), we know that the test plant's genotype is Cc (Fig. 14.4, bottom). We expect a 1 : 1 ratio in this case, but if our results happened to depart considerably from this expected ratio, we should still feel certain that the test plant's genotype is Cc. For if any of the progeny show the recessive phenotype, their genotype must be cc, which means that they received a c allele from each parent; hence the test plant must have a c allele in its genotype. Since we already know that it has at least one C allele, we combine these two pieces of information to write its genotype as Cc.

Intermediate inheritance

The seven characters of peas cited in Mendel's published experiments are all of the kind in which one allele shows complete dominance over the other. Many characteristics in a variety of organisms show this mode of inheritance. But many others do not. In fact, there is evidence that Mendel himself studied some characters that do not exhibit dominance, even though he did not take them into account in establishing his theory. Inheritance in which heterozygous individuals clearly show effects of both alleles is termed intermediate inheritance.

In many cases of intermediate inheritance, heterozygous individuals have a phenotype that is actually intermediate between the phenotype of individuals homozygous for one allele and the phenotype of individuals homozygous for the other allele. For example, crosses between homozygous red snapdragons and homozygous white snapdragons yield pink snapdragons. When these pink plants are crossed among themselves, they yield red, pink, and white offspring in a ratio of 1 : 2 : 1, as follows:

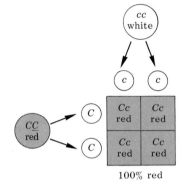

100% red

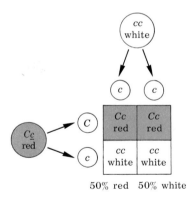

50% red   50% white

$$
\begin{array}{lcc}
P & RR & \times \quad R'R' \\
 & \text{red} & \text{white} \\
 & & \downarrow \\
F_1 & RR' & \times \quad RR' \\
 & \text{pink} & \text{pink} \\
 & & \downarrow \\
F_2 & RR \quad RR' & R'R \quad R'R' \\
 & \text{red} \quad \text{pink} & \text{pink} \quad \text{white}
\end{array}
$$

14.4. A test cross. Homozygous dominant ($C\underline{C}$) and heterozygous ($C\underline{c}$) red-flowered individuals are to be distinguished by crossing them with a homozygous recessive (cc) white-flowered individual. Top: If the red-flowered plant is homozygous, all the progeny of the test cross will be heterozygous (Cc) and will have red flowers. Bottom: If the red-flowered plant is heterozygous, 50 percent of the progeny will be heterozygous (Cc) and will have red flowers, and 50 percent will be homozygous recessive (cc) and will have white flowers.

Notice that when dominance is lacking both alleles are designated by a capital letter, the one being distinguished from the other by a prime, as here, or by a superscript (e.g. C^r for red, C^w for white).

The inheritance pattern of a system involving incomplete dominance differs from that of a system involving complete dominance in two ways: (1) The F_1 offspring of a monohybrid cross between parents each of which is homozygous for a different allele have a phenotype different from both parents. (2) The F_2 phenotypic ratio is 1 : 2 : 1 (just like the genotypic ratio) rather than 3 : 1.

DIHYBRID AND TRIHYBRID INHERITANCE

We have limited our discussion so far to crosses involving a single phenotypic character. Or, rather, we have so far chosen to ignore all but one character. The latter is a more accurate statement because all crosses involve many more than one character. Organisms contain thousands of genes, and it would be impossible to set up a cross in which only one character was allowed to vary. Let us now turn to crosses involving two characters (dihybrid crosses), three characters (trihybrid crosses), or more.

The basic dihybrid ratio

Mendel's experiments on garden peas were not limited to single characters, but sometimes involved two or more of the characters listed in Table 14.1. For example, he crossed plants characterized by round yellow seeds with plants characterized by wrinkled green seeds. The F_1 plants all had round yellow seeds. When these plants were crossed among themselves, the resulting F_2 progeny showed four different phenotypes:

315 had round yellow seeds
101 had wrinkled yellow seeds
108 had round green seeds
 32 had wrinkled green seeds

These numbers represent a ratio of about 9 : 3 : 3 : 1 for the four phenotypes.

This experiment demonstrated that a dihybrid cross can produce some new plants phenotypically unlike either of the original parental plants; in this particular case, the new phenotypes were wrinkled yellow and round green. It demonstrated, in other words, that the genes for seed color and the genes for seed form do not necessarily stay together in the combinations in which they occurred in the parental generation. In modern terms, it demonstrated that the genes for seed color are on the chromosomes of one homologous pair and the genes for seed form on the chromosomes of a different pair, and hence that the genes for the two characters segregate independently during meiosis.

The 9 : 3 : 3 : 1 phenotypic ratio is characteristic of the F_2 generation of a dihybrid cross (with dominance) in which the genes for the two characters are *independent* (i.e. are located on nonhomologous chromosomes). Each independent gene behaves in a dihybrid cross in exactly the same way as in a monohybrid cross. If we examine Mendel's F_2 results given above, and think of the experiment as a monohybrid cross for seed color (ignoring seed form), we find that there were 416 yellow seeds (315 + 101) and 140 green seeds (108 + 32), which closely approximates the 3 : 1 F_2 ratio expected in a monohybrid cross. Similarly, if we treat the experiment as a monohybrid cross for seed form and ignore seed color, the F_2 results also show a phenotypic ratio of approximately 3 : 1. The dihybrid F_2 ratio of 9 : 3 : 3 : 1 is thus simply the product of two separate and independent 3 : 1 ratios.

Let us examine in somewhat more detail a cross of this type, using the symbols R for the allele for round seed and r for the allele for wrinkled seed, and the symbols G for the allele for yellow seed and g

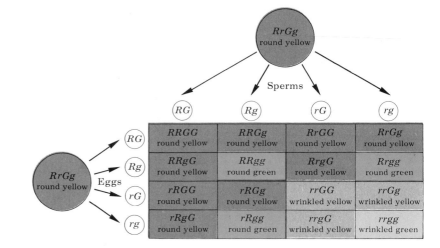

14.5. Punnett-square representation of a mating of two F₁ individuals in one of Mendel's dihybrid crosses in peas.

for the allele for green seed. In the summary of Mendel's cross below, the dash means that it does not matter phenotypically whether the dominant or the recessive allele occurs in the spot indicated.

P $RRGG$ × $rrgg$
            round yellow   wrinkled green

                       ↓

F₁ $RrGg$ × $RrGg$
            round yellow   round yellow

                       ↓

F₂	9 R–G–	3 rrG–	3 R–gg	1 $rrgg$
	round yellow	wrinkled yellow	round green	wrinkled green

The round yellow parent could produce gametes of only one genotype, RG. The wrinkled green parent could produce only rg gametes. When RG gametes from the one parent united with rg gametes from the other parent in the process of fertilization, all the resulting F₁ offspring were heterozygous for both characters ($RrGg$) and showed the phenotype of the dominant parent (round yellow). Each of these F₁ individuals could produce four different types of gametes, RG, Rg, rG, and rg. When two such individuals were crossed, there were 16 possible combinations of gametes (4 × 4), as shown in Fig. 14.5. These 16 combinations included nine different genotypes ($RRGG$, $RRGg$, $RRgg$, $RrGG$, $RrGg$, $Rrgg$, $rrGG$, $rrGg$, $rrgg$), which determined four different phenotypes in the ratio of 9 : 3 : 3 : 1. What would the phenotypic ratio have been if both characters had exhibited intermediate inheritance rather than dominance-recessiveness? What would have been the ratio if one character had exhibited dominance-recessiveness and the other intermediate inheritance?

One way to calculate the genotypic and phenotypic ratios in a dihybrid cross is to construct a Punnett square and then count the number of boxes representing each genotype and phenotype. This method, satisfactory in a monohybrid cross and only moderately laborious in a dihybrid cross, becomes prohibitively tedious in a trihybrid cross or a cross involving even more than three characters. There is an alternative procedure that is much easier. It is based on the principle that *the chance that a number of independent events will occur together is equal to the product of the chances that each event will occur separately.* Suppose we wanted to know how many of the 16 combinations in Mendel's cross would produce the wrinkled yellow phenotype. We know that wrinkled is recessive; hence it would be expected in $\frac{1}{4}$ of the F_2 individuals in a monohybrid cross. We know that yellow is dominant; hence it would be expected in $\frac{3}{4}$ of the F_2 individuals. Multiplying these two separate values ($\frac{1}{4} \times \frac{3}{4}$) gives us $\frac{3}{16}$; three of the 16 possible combinations will produce a wrinkled yellow phenotype. Similarly, if we had wanted to know how many of the combinations would produce a round yellow phenotype, we could have multiplied the separate expectancies for two dominant characters ($\frac{3}{4} \times \frac{3}{4}$) to get $\frac{9}{16}$. Or, to use a more complex example, suppose we were interested in a trihybrid cross involving the two seed characters and flower color, and we wanted to know what fraction of the F_2 individuals (produced by allowing $CcRrGg$ individuals to cross among themselves) would exhibit a phenotype combining red flowers, wrinkled seeds, and yellow seeds. The separate probability for red flowers in a monohybrid cross is $\frac{3}{4}$, that for wrinkled seeds is $\frac{1}{4}$, and that for yellow seeds is $\frac{3}{4}$. Multiplying these three values ($\frac{3}{4} \times \frac{1}{4} \times \frac{3}{4}$) gives us $\frac{9}{64}$. This tells us that of the 64 possible combinations in a trihybrid cross, nine would produce the phenotype here specified. What proportion would show a phenotype combining white flowers and round green seeds?

Gene interactions

Some dihybrid crosses involve two genes that are inherited independently but exert their phenotypic effect on the same character. When two or more genes interact in determining a single character, the ratios obtained from crosses in which they are involved are sometimes different from the basic ratio.

An example is provided by the two genes that together determine the color of guinea pigs: (1) a gene that controls production of a dark pigment, melanin, and (2) a gene that controls the deposition of the pigment. The first gene has two alleles: C, which causes the pigment to be produced, and c, which causes no pigment to be produced; hence a homozygous recessive individual, cc, is an albino. The second gene has an allele B that causes deposition of much melanin, which gives the guinea pig a black coat, and an allele b that causes deposition of only a moderate amount of melanin, which gives the guinea pig a brown coat. Neither B nor b can cause deposition of melanin if C is not present to make the melanin. We can summarize a cross involving these two genes as follows:

P		$CCBB$ black	$\times$	$ccbb$ albino
			$\downarrow$	
F_1		$CcBb$ black	$\times$	$CcBb$ black
			$\downarrow$	
F_2	9 C–B– black	3 C–bb brown	3 $ccBb$ albino	1 $ccbb$ albino

Instead of an F_2 phenotypic ratio of 9 : 3 : 3 : 1, this cross has yielded a ratio of 9 : 3 : 4. The last two phenotypic classes of the normal 9 : 3 : 3 : 1 ratio have been combined.

Probably no inherited characteristic is controlled exclusively by one gene pair. Even when only one principal gene is involved, its expression is influenced to some extent by countless other genes with individual effects often so slight that they are very difficult to locate and analyze. An example is eye color in human beings.

Human eye color can be regarded as controlled by one gene with two alleles—a dominant allele, B, for brown eyes, and a recessive allele, b, for blue eyes. Brown-eyed people (BB or Bb) have branching pigment cells containing melanin in the front layer of the iris. Blue-eyed people (bb) lack melanin in the front layer; the blue is an effect of the black pigment on the back of the iris as faintly seen through the semi-opaque front layer. Even though most people can be classified as either brown-eyed or blue-eyed, it is common knowledge that eyes exhibit endless variations in hue, some of which are recognized in everyday terminology; we describe eyes as gray (genetically a form of blue) or black (genetically a form of brown). It is obvious, then, that an explanation of eye color in terms of a single-gene system is an oversimplification. Many *modifier genes* are also involved, some affecting the amount of pigment in the iris, some the tone of the pigment (which may be light yellow, dark brown, etc.), and some its distribution (even over the whole iris, in scattered spots, in a ring around the outer edge of the iris, etc.).

PENETRANCE AND EXPRESSIVITY

It is not always possible to assume, as we did in the first few sections of this chapter, that whenever an individual has a dominant gene he will show the phenotypic effect of that gene. As we have just seen, the phenotypic expression of a dominant gene may depend on the presence of some other gene, or it may be affected by a large number of modifier genes. And when a gene is expressed, there are many possible degrees of intensity of expression. We can, therefore, speak of the penetrance and expressivity of a gene. *Penetrance* is the percentage of individuals that, when carrying a given gene in proper combination for its expression, actually express that gene's phenotype. *Expressivity* denotes the manner in which the phenotype is expressed.

The concepts of incomplete penetrance and variable expressivity are illustrated by a gene in human beings that causes a condition known as blue sclera, in which the whites of the eyes appear bluish. This gene usually behaves as a simple dominant, and anyone who possesses the gene, whether in heterozygous or homozygous form, might be expected to show the blue-sclera phenotype. But it has been found that only about nine out of ten people who have the gene actually show the phenotype. We can say, therefore, that the penetrance of this gene is about 90 percent (or 0.9). Among those showing the phenotype, the expressivity is variable, the intensity of the bluish coloration ranging from very pale whitish blue to very dark blackish blue.

As we have indicated, incomplete penetrance and variable expressivity may be due to factors in the genetic background against which the gene in question must act. This explanation points up a general truth: *The action of any gene can be fully understood only in terms of the overall genetic makeup of the individual organism in which it occurs.*

Penetrance and expressivity may also be affected by environmental influences. For example, if *Drosophila* (fruit flies) homozygous for the gene for vestigial wings are reared at normal room temperatures (about 72°F), their wings remain tiny stumps. But if they are reared at temperatures as high as 88°F, their wings grow almost as long as normal ones. Himalayan rabbits are normally white with black ears, nose, feet, and tail (Fig. 14.6), but if the fur on a patch on the back is plucked and an ice pack is kept on the patch, the new fur that grows there will be black; the gene for black color can express itself only if the temperature is low, which it normally is only at the body extremities. A human being with genes for great height will not grow tall if raised on a starvation diet. Numerous other examples could be cited.

We see, then, that the expression of a gene depends both on the other genes present (the genetic environment) and on the physical environment (temperature, sunlight, humidity, diet, etc.). We don't inherit characters. We inherit only genes, only potentialities; other factors govern whether or not the potentialities are realized. All organisms are products of both their inheritance and their environment. The familiar either-or question is meaningless here.

MULTIPLE ALLELES

From our discussion so far, you may have gotten the impression that genes can have only two allelic forms, which may exhibit either a dominant-recessive or an intermediate relationship to each other. This is not so. Genes may exist in any number of different allelic forms. Under normal circumstances, of course, the maximum number of different alleles for each gene that any individual can possess is two, because he has only two doses of each gene. But many other alleles may be present in the population to which he belongs.

Human A-B-O blood types. A well-known example of multiple alleles in man—and a relatively simple one, since only a few alleles are involved—is that of the A-B-O blood series, in which four blood types are recognized: A, B, AB, and O. The red cells in type-A blood bear antigen A; in type B, antigen B; in type O, neither A nor B; in type AB, both A and B. These two kinds of cellular antigen, A and B, react with certain antibodies, anti-A and anti-B, that may be present in the blood plasma. If red cells bearing a particular antigen and plasma containing the corresponding antibody come to be mixed in a person's blood, the cells agglutinate, or clump, and he may die. Normally, a person with antigen A on his red cells does *not* have anti-A in his plasma, but he does have anti-B; a person with antigen B on his red cells has only anti-A in his plasma; a person lacking both antigens A and B (type-O blood) has both anti-A and anti-B; and a person with both antigens A and B (type-AB blood) lacks both types of antibodies (see Table 14.2). Normally, in short, a person's plasma contains antibodies corresponding to any antigens his own cells do not bear; this is one of the few cases where the body normally synthesizes antibody against an antigen to which it has not actually been exposed.

14.6. Effect of temperature on the expression of a gene for coat color in the Himalayan rabbit. (A) Normally, only the feet, tail, ears, and nose are black. (B) Fur is plucked from a patch on the back, and an ice pack is applied to the area. (C) The new fur grown under the artificially low temperatures is black. Himalayan rabbits are normally homozygous for the gene that controls synthesis of the black pigment, but the gene is active only at low temperatures (below about 92°F). [Modified from A. M. Winchester, *Genetics*, Houghton Mifflin, 1972.]

Table 14.2. Antigen and antibody content of the blood types of the A-B-O series

Blood type	Blood contains	
	Cellular antigens	*Plasma antibodies*
O	None	anti-A and anti-B
A	A	anti-B
B	B	anti-A
AB	A and B	None

The presence of these antigens and antibodies in the blood has important implications for blood transfusions. Because the antibodies present in the plasma of blood of one type will tend to react with the antigens on the red cells of other blood types and cause clumping, it is always best, when transfusions are to be given, to obtain a donor who has the same blood type as the patient. It has been found, however, that when such a donor is not available blood of another type may be used, provided that the *plasma of the patient* and the *red cells of the donor* are compatible; in other words, one can usually ignore the red cells of the patient and the plasma of the donor. The reason is that, unless the transfusion is to be a massive one or is to be made very rapidly, the donor's plasma is sufficiently diluted during transfusion so that little or no agglutination occurs. This means that type-O blood can be given to anyone because its red cells have no antigens and thus are obviously compatible with the plasma of any patient; type-O blood is sometimes called the universal donor. But type-O patients can receive transfusions only from type-O donors because their plasma contains both anti-A and anti-B and thus is obviously not compatible with the red cells of any other class of donor. Conversely, people with type-AB blood, whose plasma contains no antibodies, are universal recipients but cannot act as donors for any except type-AB patients. Table 14.3 summarizes these transfusion relationships.

Table 14.3. Transfusion relationships of the A-B-O blood groups

Blood group	*Can act as donor to*	*Can receive blood from*
O	O, A, B, AB	O
A	A, AB	O, A
B	B, AB	O, B
AB	AB	O, A, B, AB

At first glance, you might suppose that two independent genes are involved in the A-B-O system, one determining whether the A antigen is present and another determining whether the B antigen is present.

But this is not the case. It has been shown that the inheritance of the A-B-O groups is best described by a theory of three alleles,[3] I^A, I^B, and i. Both I^A and I^B are dominant over i, but neither I^A nor I^B is dominant over the other. Accordingly, the four blood-type phenotypes correspond to the genotypes indicated in Table 14.4.

Table 14.4. Genotypes of the A-B-O blood types

Blood type	Genotype
O	ii
A	$I^A I^A$ or $I^A i$
B	$I^B I^B$ or $I^B i$
AB	$I^A I^B$

Human Rh blood factors. Another series of blood antigens is designated as Rh (which stands for rhesus monkey, the animal in which the antigens were first discovered). The Rh series includes at least nine different antigens, and some authorities believe that the inheritance of these antigens is best explained in terms of multiple alleles; if this view is correct, the Rh alleles would constitute one of the largest series of multiple alleles known in man (much larger series are known in some other species). Other authorities prefer to explain Rh inheritance in terms of three or more separate genes that lie close together on the same chromosome and have very similar functions. Which, if either, of these two views is correct has some importance for our understanding of the nature of the gene, but from a medical standpoint the matter is far simpler.

The clinician ordinarily divides the subjects he tests into two phenotypic classes designated Rh-positive and Rh-negative. Rh-positive individuals have the Rh antigen on their red blood cells, while Rh-negative individuals do not (more precisely, their antigen is so weak that it can be disregarded). About 85 percent of the white population in the United States is Rh-positive, and 15 percent is Rh-negative. Rh-negative is much rarer in people of Mongoloid or Negroid extraction.

For convenience, let us assume that Rh-positive is the phenotype produced by a dominant gene Rh, and that Rh-negative is the phenotype produced by its recessive allele rh.[4] This assumption allows us to treat the genetics of Rh blood types like a simple monohybrid cross, even though we actually know that the Rh-positive phenotype can be produced by any one of a number of slightly different alleles that produce

[3] Actually, four alleles are now known. What we here designate as the I^A allele is really two different but very similar alleles, I^{A1} and I^{A2}. This means that there are actually six blood types (O, A_1, A_2, B, A_1B, A_2B), rather than four. For our purposes, however, this complication can be ignored.

[4] Notice that we are here using a two-letter symbol for a single gene. Thus Rh is the symbol for a single dominant allele, and rh is the symbol for a single recessive allele. Do not make the mistake of thinking that the r and the h stand for different things.

slightly different antigens. On this assumption, the possible genotypes and their phenotypes would be as follows:

Phenotypes	Genotypes
Rh$^+$	*RhRh* or *Rhrh*
Rh$^-$	*rhrh*

Whereas a person's blood plasma may contain anti-A or anti-B antibody without stimulation by the corresponding antigen, it will not contain any antibody against the Rh antigen unless the body is sensitized to the antigen by exposure to it. Thus the blood of a normal unsensitized Rh-negative person contains neither Rh antigen on its red cells nor Rh antibody in its plasma. But if an Rh-negative person is mistakenly given a transfusion of Rh-positive blood, this exposure to the antigen will sensitize him and stimulate his plasma cells to begin synthesizing antibodies. If the same patient is later given a second transfusion of Rh-positive blood, the antibodies now present in his plasma will react with the antigens in the donated blood, and the patient may die. It is important, therefore, that patients and donors be carefully typed for both the A-B-O blood group and the Rh factor.

The Rh factor also has medical importance in certain cases of pregnancy. If an Rh-negative woman (genotype *rhrh*) marries an Rh-positive man (genotype *RhRh* or *Rhrh*), some of their children may be heterozygous and have the Rh-positive phenotype. Often this causes no trouble; there is no direct connection between the circulatory systems of the mother and the fetus in the normal placenta, and hence there can be no mixing of blood. But in the late stages of some pregnancies, there may be some seepage of blood between the two circulatory systems. If the blood of an Rh-positive fetus seeps into the circulatory system of an Rh-negative mother, the effect is the same as though she had been given a transfusion of Rh-positive blood. Her plasma cells are stimulated to begin synthesizing Rh antibody. Ordinarily there is no immediate harm, because the baby is born before the mother's blood contains appreciable quantities of antibody. But if this sensitized mother later bears a second or third Rh-positive fetus, and if seepage again develops, antibodies from the maternal blood may enter the fetal circulation and react with the red cells of the fetus, causing a disease known as erythroblastosis fetalis (or simply Rh disease), which, unless treated promptly, is often fatal to the baby.

It must be emphasized that erythroblastosis fetalis does not occur in all second or later pregnancies of Rh-negative women married to Rh-positive men. Furthermore, a method has recently been developed for immunizing Rh-negative mothers so that they will not form Rh antibodies. If an Rh-negative woman bears an Rh-positive child, she is injected immediately after delivery with anti-Rh, which kills any fetal red cells that may have gotten into her circulation before those cells can stimulate her to begin producing antibodies. The injected anti-Rh soon disappears, and the woman is left in the unsensitized condition and thus able to bear a second Rh-positive child without danger. Even if a fetus should develop erythroblastosis fetalis, physicians have learned that if the newborn infant is promptly given a massive transfusion that replaces all its blood (contaminated with antibodies from the mother) with blood free of antibodies, the chances are good that it will live and develop normally.

MUTATIONS AND DELETERIOUS GENES

A variety of influences can cause slight changes in the chemical structure of a gene. Such changes are called mutations. The rate at which any particular gene undergoes mutation is ordinarily extremely low. But every individual organism has a very large number of different genes, and the total number of genes in all the individuals of a species is vast indeed. Hence mutations are constantly occurring within a species, pure chance determining in which individual any given mutation will occur. Now, every living organism is the product of billions of years of evolution and is a finely tuned, smoothly running, astoundingly intricate mechanism, in which the function of every part in some way influences the function of every other part. By comparison, the best Swiss watch is simple indeed. If you were to take such a watch, remove its back, and make some random change in its parts, the chances are very great that you would make it run worse rather than better. A random change in any delicate and intricate mechanism is far more likely to damage it than to improve it. Mutational changes in genes being random, it is easy to understand why in the vast majority of cases new mutations are deleterious. Only very rarely is a new mutation beneficial.

When a new deleterious allele arises by mutation, natural selection can act against it only if it causes some change in the organism's characteristics. Selection acts directly on phenotypes and only indirectly on genotypes. Because dominant deleterious mutations will be expressed phenotypically, they can be eliminated from the population rapidly by natural selection. But many new mutations are recessive to the normal alleles. And since the probability that the same mutation will occur twice in the same individual is vanishingly slight, most new alleles occur in combination with the normal allele in diploid cells; i.e. the diploid cell is heterozygous, containing one normal allele and one new allele produced by mutation. If a new allele is recessive, and if it occurs in heterozygous condition, it can have little immediate phenotypic effect on the organisms that possess it—its deleterious effects cannot be fully expressed—and therefore natural selection cannot eliminate it from the population very rapidly. Deleterious alleles that are not dominant may be retained in the population in heterozygous condition for a long time.

When two individuals both carrying the same deleterious recessive allele in heterozygous condition mate, about one fourth of their progeny will be homozygous for the deleterious allele, and these homozygous offspring will have the harmful phenotype. In some cases, the phenotype may even kill the organism. An allele whose phenotype, when expressed, results in the death of the organism is called a *lethal*.

In many instances, alleles harmful or even lethal when homozygous are actually beneficial when heterozygous. An example in human beings is the gene for sickle-cell anemia. When homozygous, this gene results in a serious abnormality of the red blood cells, which are curved like a sickle and bear long filamentous processes (Fig. 14.7). These abnormal cells tend to form clumps and to clog the smaller blood ves-

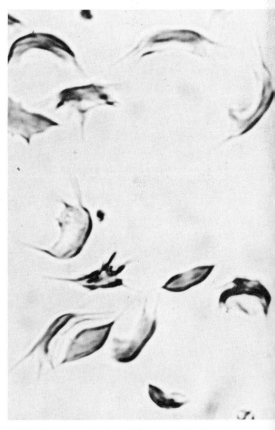

14.7. Photograph of red blood cells of a person with sickle-cell anemia. The cells are curved and bear long filamentous processes. [Courtesy J. W. Harris, Cleveland Metropolitan General Hospital.]

sels. The resulting impairment of the circulation leads to severe pains in the abdomen, back, head, and extremities, and to enlargement of the heart and atrophy of brain cells. In addition, the tendency of the deformed red blood cells to rupture easily brings about severe anemia. There is no cure for sickle-cell anemia, and its victims usually suffer an early death. Individuals heterozygous for the sickle-cell gene sometimes show mild symptoms of the disease, but the condition is usually not serious. Since sickle-cell anemia occurs only when the gene is present in homozygous form, the obvious preventive measure is avoidance of marriage between heterozygous bearers of the gene, for which there is a relatively simple test.

It might be supposed that natural selection would operate against the propagation of any gene so obviously harmful as the sickle-cell gene and that it would be held at very low frequency in the population. This seems to be true among blacks in the United States, in whom sickle-cell anemia is relatively rare. But the gene is surprisingly common in many parts of Africa, being carried by as much as 20 percent of the black population. What is the explanation? A. C. Allison of Oxford, England, found that individuals heterozygous for this gene have a much higher than normal resistance to malaria. Since malaria is very common in many parts of Africa, the gene must be regarded as beneficial when heterozygous. Thus, in Africa, there is selection for the gene because of its heterozygous effect on malarial resistance and selection against it because of its homozygous production of sickle-cell anemia. The balance between these two opposing selection pressures determines the frequency of the gene in the population.

The gene for sickle-cell anemia is a dramatic example of a gene that has more than one effect. Such a gene is said to be **pleiotropic.** Pleiotropy is, in fact, the rule rather than the exception. All genes probably have many effects on the organism. Even when a gene produces only one noticeable phenotypic effect, it doubtless has numerous physiological effects more difficult for us to detect.

The conditions under which genes cause deleterious phenotypes explain the danger in marriages between closely related human beings. Everyone probably carries in heterozygous combination many genes that would cause harmful effects if present in homozygous combination, including some lethals. But because most of these deleterious genes originated as rare mutations, and are limited to a tiny percentage of the population, the chances are slight that two unrelated persons who marry will be carrying the same deleterious recessive genes and produce homozygous offspring that show the harmful phenotype. The chances are much greater that two closely related persons will be carrying the same harmful recessives, having received them from common ancestors, and that, if they marry, they will have children homozygous for the deleterious traits.

SEX AND INHERITANCE

Sex determination

We have said repeatedly that a diploid individual has two of each type of chromosome, identical in size and shape, and hence two doses of each gene. But we must now qualify that statement somewhat. In most higher organisms where the sexes are separate (i.e. where males and females are separate individuals), the chromosomal endowments of males and females are different, and one or the other of the two sexes has one chromosomal pair consisting of two chromosomes that differ markedly from each other in size and shape. These are the **sex chromosomes,** which play a fundamental role in determining the sex of the individual. All other chromosomes are called **autosomes.**

Let us look first at the chromosomes in *Drosophila* and in human beings. In each case, the sex chromosomes are of two sorts: one bearing many genes, conventionally designated the **X chromosome,** and one of a different shape and bearing only a few genes, designated the **Y chromosome.** Females characteristically have two X chromosomes and males have one X and one Y. The diploid number in *Drosophila* is eight (four pairs); a female therefore has three pairs of autosomes and one pair of X chromosomes, and a male has three pairs of autosomes and a pair of sex chromosomes consisting of one X and one Y (Fig. 14.8). The diploid number in human beings is 46 (23 pairs); a female therefore has 22 pairs of autosomes and one pair of X chromosomes, and a male has 22 pairs of autosomes plus one X and one Y. This means that when a female produces eggs by meiosis, all the eggs receive one of each type of autosome plus one X chromosome. When a male produces sperm cells by meiosis, half the sperm cells receive one of each type of autosome plus one X chromosome and half receive one of each autosome plus one Y chromosome. In short, all the egg cells are alike in chromosomal content, but the sperm cells are of two different types occurring in equal numbers (Fig. 14.8). When fertilization takes place, the chances are approximately equal that the egg will be fertilized by a sperm carrying an X chromosome or by a sperm carrying a Y chromosome. If fertilization is by an X-bearing sperm, the resulting zygote will be XX and will develop into a female. If fertilization is by a Y-bearing sperm, the resulting zygote will be XY and will develop into a male.[5] We see, therefore, that the sex of an individual is normally determined at the moment of fertilization and depends on which of the two types of sperm fertilizes the egg.

Sex-linked characters

There are many genes that occur on the X chromosome and not on the Y chromosome. Such genes are said to be sex-linked. The inheritance patterns for the characteristics controlled by such genes are quite different from those for characteristics controlled by autosomal genes, for obvious reasons. The females have two doses of each sex-linked gene, one from each parent, but the males have only one dose of each sex-linked gene, and that one dose always comes from the mother since the father contributes a Y chromosome instead of an X. Hence, in the male,

[5] The XY system, where XX is female and XY is male, is characteristic of many animals including all mammals. It is also found in many plants with separate sexes. Birds, butterflies and moths, and a few other animals have just the opposite system, where XX is male and XY is female (to distinguish this system from the usual XY system, the symbols Z and W are often substituted—ZZ being male and ZW being female). A completely different mechanism of sex determination exists in the Hymenoptera (bees, wasps, ants, etc.), where the males hatch from unfertilized eggs and are haploid while the females hatch from fertilized eggs and are diploid.

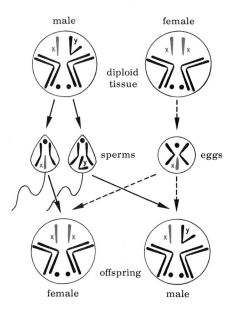

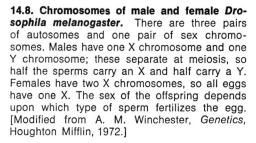

14.8. Chromosomes of male and female *Drosophila melanogaster.* There are three pairs of autosomes and one pair of sex chromosomes. Males have one X chromosome and one Y chromosome; these separate at meiosis, so half the sperms carry an X and half carry a Y. Females have two X chromosomes, so all eggs have one X. The sex of the offspring depends upon which type of sperm fertilizes the egg. [Modified from A. M. Winchester, *Genetics,* Houghton Mifflin, 1972.]

all sex-linked characteristics are inherited from the mother only. And since the male has only one dose of each sex-linked gene, recessive genes cannot be masked; consequently, recessive sex-linked phenotypes occur much more frequently in males than in females.

Sex linkage was discovered in 1910 by the great American geneticist Thomas Hunt Morgan of Columbia University. The first sex-linked trait he observed was white eye color in *Drosophila*. This trait is controlled by a recessive allele *r*. The normal red eye color is controlled by a dominant allele *R*. If a homozygous red-eyed female is crossed with a white-eyed male, all the F_1 offspring, regardless of their sex, have red eyes, since they receive from their mother an X chromosome bearing an allele for red. In addition, the F_1 females receive from their father an X chromosome bearing an allele for white eyes, but the allele for red, being dominant, masks its presence. The F_1 males, like the females, receive from their mother an X chromosome bearing an allele for red eyes. But unlike the females, they receive no gene for eye color from their father, who contributes a Y chromosome instead of an X (in writing the genotype of a male for a sex-linked character, the Y is customarily shown in order to indicate clearly that no second X chromosome is present and hence that there is no second dose of the sex-linked gene). We can summarize this cross as follows (♀ denotes females, ♂ males):

P $\qquad$ *RR* $\quad\times\quad$ *rY*
red-eyed ♀ $\quad$ white-eyed ♂
↓
F_1 $\qquad$ *Rr* $\quad\times\quad$ *RY*
red-eyed ♀ $\quad$ red-eyed ♂
↓
F_2 $\quad$ *RR* $\qquad$ *rR* $\qquad$ *RY* $\qquad$ *rY*
red- $\qquad$ red- $\qquad$ red- $\qquad$ white-
eyed ♀ $\quad$ eyed ♀ $\quad$ eyed ♂ $\quad$ eyed ♂

Notice that when the F_1 flies of this cross are allowed to mate among themselves, the F_2 flies thus produced show the customary 3 : 1 phenotypic ratio of a monohybrid cross where dominance is present. But notice also that this 3 : 1 ratio is rather different from the 3 : 1 ratio obtained in a cross involving autosomal genes. In an autosomal cross, there is no correlation of phenotype with sex, but in this cross all F_2 individuals showing the recessive phenotype are males. In other words, an autosomal cross gives a 3 : 1 F_2 ratio for both females and males, but this cross yielded females of only one phenotype and males with a 1 : 1 phenotypic ratio.

Now let us examine the reciprocal cross, where the parental generation consists of homozygous white-eyed females and red-eyed males. We can summarize this cross as follows:

P $\qquad$ *rr* $\quad\times\quad$ *RY*
white-eyed ♀ $\quad$ red-eyed ♂
↓
F_1 $\qquad$ *rR* $\quad\times\quad$ *rY*
red-eyed ♀ $\quad$ white-eyed ♂
↓
F_2 $\quad$ *rr* $\qquad$ *Rr* $\qquad$ *rY* $\qquad$ *RY*
white- $\qquad$ red- $\qquad$ white- $\qquad$ red-
eyed ♀ $\quad$ eyed ♀ $\quad$ eyed ♂ $\quad$ eyed ♂

Notice that the phenotypic makeup of both the F_1 and the F_2 generations differs both from a normal autosomal cross and from the reciprocal cross for this same sex-linked trait. In the F_1, the dominant phenotype appears only in females, rather than in all individuals regardless of sex; all males show the recessive phenotype. In the F_2, instead of a 3 : 1 ratio, there is a 1 : 1 ratio in each sex. Comparison of the two reciprocal crosses shown above makes it clear that when a sex-linked trait is involved in a cross, the results depend on which parent shows the trait (or carries the gene for the trait). By contrast, in crosses involving autosomal genes, it makes no difference which parent possesses the gene in question; the results of reciprocal crosses are identical.

Two well-known examples of recessive sex-linked traits in man are red-green color blindness and hemophilia ("bleeder's disease"). Color blindness occurs in about 8 percent of white men in the United States and in about 4 percent of black men. It occurs in only about one percent of white women and about 0.8 percent of black women. It is expected, of course, that more men than women will show such a trait, because a man needs only one dose of the gene to show the phenotype, and he can inherit this one dose from a heterozygous mother who is not herself color-blind. But for a woman to be color-blind, she must have two doses of the gene (i.e. be homozygous), which means not only that her father must be color-blind but also that her mother must be either color-blind or a heterozygous carrier of the gene. Since the gene is not very common in the population, it is not likely that two such people will marry; hence the low number of color-blind women.

Although females have two X chromosomes in every cell of their bodies, it has recently been discovered that in any given cell only one of the X chromosomes has active genes; the other condenses into a tiny dark object called a **Barr body,** whose genes are inactive. Which of the two X chromosomes is active in any particular cell, and which becomes a Barr body, seems to be a matter of chance. The presence of Barr bodies is easily verified, and they thus provide a simple technique for determining the genetic sex of fetuses and of individuals suffering from certain hormonal imbalances.

The suppression of one or the other X chromosome throughout the female body does not seem to affect the expression of most sex-linked traits. In a female heterozygous for red-green color blindness, for example, the X chromosome carrying the allele for color blindness will be active in about half her cells, but the fact that the X chromosome carrying the normal allele is active in her remaining cells apparently suffices for normal color perception.

In our discussion of sex linkage, we have referred only to genes on the X chromosome and ignored the Y, on which there are apparently very few genes. Genes that are on the Y and not on the X are termed *holandric.* The phenotypic traits they control appear, of course, only in males.

Sex-influenced characters

The fact that some genes are sex-linked should not lead you to assume that all genes for characters commonly associated with sex are sex-linked. They are not. As we saw above, sex-linked genes may control characters not customarily regarded as "sexual." And many genes that do control "sexual" characters are located on the autosomes. For example, a number of genes that control growth and development of the sexual organs, such as the penis, the vagina, the uterus, or the oviducts, or that control distribution of body hair, size of breasts, pitch of voice, or other secondary sexual characteristics, are autosomal and are present in individuals of both sexes. That their phenotypic expression is different in the two sexes indicates that they are sex-limited, not that they are sex-linked. Apparently, the sex hormones influence the activity of the genes, either inhibiting or stimulating them.

LINKAGE

Observed patterns of inheritance amply support Mendel's so-called Law of Segregation—that in each individual the genes occur in pairs,[6] and that in the formation of gametes the members of each pair separate and pass into different gametes, so that each gamete has only one of each type of gene. And all the dihybrid and trihybrid crosses we have discussed so far agree with Mendel's Law of Independent Segregation, which states that when two or more pairs of genes are involved in a cross, the members of one pair segregate independently of the members of all other pairs. But, as numerous experiments have shown, many crosses fail to obey this second law. Why?

The chromosomal basis of linkage. Mendel knew nothing about chromosomes; he derived his principles exclusively from data from crosses, and either he was lucky enough to have worked only with dihybrid crosses that yielded 9 : 3 : 3 : 1 phenotypic ratios in the F_2 generation or he chose to ignore crosses that did not fit this ratio (or obvious modifications of it). Shortly after the rediscovery of Mendel's paper in 1900, W. S. Sutton of Columbia University pointed out that Mendel's conclusion that hereditary factors (genes) occur in pairs in somatic cells and separate in gametogenesis was in striking accord with the recent cytological evidence that somatic cells contain two of each kind of chromosome and that these chromosomes segregate in meiosis. Sutton suggested that the chromosomes are the bearers of the genes and that this explained the correspondence between Mendel's results and the cytologists' discoveries. Our discussion in most of this chapter has followed Sutton's theory.

But a moment's thought will convince you that Sutton's theory that the genes are located on the chromosomes is incompatible with the unmodified form of Mendel's second law. *Drosophila* have only four pairs of chromosomes, garden peas have only seven pairs, and human beings have only 23 pairs. Since each species has thousands of different genes, it follows that there must be many different genes on each chromosome. And since it is whole chromosomes that segregate independently in meiosis, it follows that only genes that are located on different chromosomes can segregate independently of each other; genes located on the same chromosome cannot separate and hence must move together during meiosis. Such genes are said to be **linked.** We must therefore modify Mendel's second law, restricting its application to pairs of genes located on different chromosomes, i.e. to genes that are not linked.

---

[6] To be more precise, we should say that in each *diploid* individual the genes occur in pairs, since the haploid individuals of lower plants obviously have only one dose of each gene.

One of the first examples of linkage was reported in 1906 by William Bateson and R. C. Punnett of Cambridge University. They crossed sweet peas that had purple flowers and long pollen with ones that had red flowers and round pollen. All the F_1 plants had purple flowers and long pollen, as expected (it was already known that purple was dominant over red and that long was dominant over round). The F_2 plants from this cross did not show the expected 9 : 3 : 3 : 1 ratio, however, but a highly anomalous one. Next, Bateson and Punnett tried a test cross, crossing the F_1 plants back to homozygous recessive plants (with red flowers and round pollen). According to the Law of Independent Segregation, such a test cross should have yielded a ratio of 1 : 1 : 1 : 1. But it did not; it yielded a ratio of approximately 7 : 1 : 1 : 7. Using the symbols B for purple, b for red, L for long, and l for round, we can summarize this test cross as follows:

BbLl	×	*bbll*	
purple-long		red-round	

↓

7 *BbLl*	1 *Bbll*	1 *bbLl*	7 *bbll*
purple-long	purple-round	red-long	red-round

You will recall that, in a test cross, the phenotypic ratio of the offspring depends on the genotype of the parent showing the dominant phenotype, since the recessive parent produces only one kind of gamete. In this example, the homozygous recessive red-round parent can produce only bl gametes. Hence it is the gametes of the heterozygous purple-long parent that determine the phenotype of the offspring. According to the Law of Independent Segregation, this parent should produce four kinds of gametes (BL, Bl, bL, and bl) in equal numbers. When united with the bl gametes from the homozygous recessive parents, BL gametes should give rise to purple-long offspring, Bl gametes to purple-round, bL to red-long, and bl to red-round, and these four phenotypes should occur in equal numbers; hence the expected 1 : 1 : 1 : 1 ratio. But the results Bateson and Punnett actually obtained make it appear that the heterozygous parent produced far more BL and bl gametes than Bl and bL gametes. It was not until 1910 that T. H. Morgan, who had obtained similar results from *Drosophila* crosses, provided the explanation accepted today. He postulated that the anomalous ratios were caused by linkage.

Crossing-over. If in Bateson and Punnett's cross the genes for purple and long and the genes for red and round were linked, we would expect the *BbLl* parent in the test cross to have produced only two kinds of gametes, BL and bl. This means that the test cross should have yielded offspring of only two phenotypes, purple-long and red-round, in equal numbers. Yet the cross also yielded some purple-round and red-long offspring. How could the *BbLl* parent have produced Bl and bL gametes? Morgan suggested that some mechanism occasionally breaks the original linkages between purple and long and between red and round and establishes in a few individuals new linkages between purple and round and between red and long. The mechanism whereby this recombination is presumed to occur is called crossing-over.

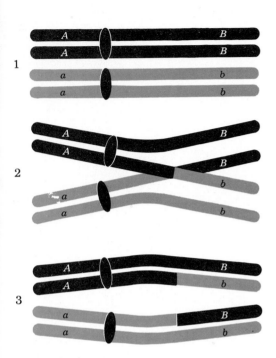

14.9. Schematic model of crossing-over. (1) The two homologous double-stranded chromosomes, one bearing alleles *A* and *B* and the other alleles *a* and *b*, lie side by side in synapsis. (2) Corresponding breaks occur in one chromatid of each chromosome, and the fragments are exchanged. (3) After crossing-over, one chromatid of the first chromosome bears alleles *A* and *b* and one chromatid of the second chromosome bears alleles *a* and *B*.

Crossing-over apparently takes place while the homologous chromosomes synapse during the first division sequence of meiosis. It is believed that chromatids of each of the two homologous chromosomes break at corresponding points and then exchange parts (Fig. 14.9). Suppose one of the chromosomes in a synaptic pair bears a gene *A* near one end and a gene *B* near the other end, and that the other chromosome bears genes *a* and *b* at corresponding points. If one chromatid of the first chromosome and one chromatid of the second chromosome break at corresponding points between the two genes, and if the two chromatids exchange parts before repair of the break occurs, then this breakage-fusion mechanism will give rise to one chromatid bearing genes *A* and *b* and to another chromatid bearing genes *a* and *B*. If this crossing-over had not occurred, two of the four gametes produced by meiosis would have carried genes *A* and *B* and two would have carried genes *a* and *b*; there would have been no *Ab* or *aB* gametes. But after crossing-over, each of the four gametes carries different genes (*AB*, *Ab*, *aB*, *ab*). Crossing-over thus increases the number of different genetic combinations that any given cross can produce, and it therefore contributes to variability in the population.

Chromosomal mapping. If we assume, as Morgan did, that breakage is about equally probable at any point along the length of a chromosome, it follows that the farther apart two linked genes are on the chromosome, the more frequently will breakage occur between them, because there are more points between them at which a break may occur. Or, to be more precise, the frequency of crossing-over between any two linked genes will be proportional to the distance between them. Consequently we can use the percentage of crossing-over as a tool for mapping the locations of genes on chromosomes.

The percentage of crossing-over gives us no information about the absolute distances between genes—we cannot state these distances in microns or millimicrons—but it does give us relative distances. By convention, one unit of map distance on a chromosome is the distance within which crossing-over occurs one percent of the time. In Bateson and Punnett's test cross, 2 out of 16 of the offspring were recombinant products of crossing-over. Two is 12.5 percent of 16; hence the genes controlling flower color and pollen shape in the sweet peas of this cross are located 12.5 map units apart.

Suppose we know that linked genes *B* and *L* are 12.5 map units apart. And suppose we find another gene, *A*, linked with these, that crosses over with gene *L* 5 percent of the time. How do we determine the order of the genes? The order could be *B–A–L*:

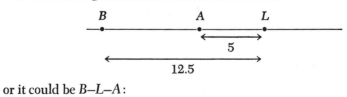

or it could be *B–L–A*:

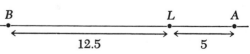

Obviously, the way to decide between these two alternatives is to determine the frequency of crossing-over between *A* and *B*. If this frequency is 7.5 percent (12.5 − 5), then we know that the first alternative is correct; if it is 17.5 percent (12.5 + 5), then we know that the second alternative is correct. In this way, by determining the frequency of crossing-over between each gene and at least two other known genes, it is possible to build up a map showing the arrangement of many different genes on a chromosome.

One important implication of the frequencies of crossing-over and the mapping they make possible should be noted. All known frequencies agree with a model of the chromosome in which the genes are sequentially arranged along the chromosome. Therefore, one possible definition of the gene is that it is a point (or locus) on a chromosome that controls one or more characteristics of the organism. If two characters are always linked, and recombination by crossing-over never occurs between them, then we assume that they are controlled by the same point on the chromosome, i.e. by the same gene. If, on the other hand, crossing-over does occur between them, even if extremely seldom, then we can say that they must be controlled by different points on the chromosome, i.e. by different genes. Crossing-over is thus the test whether two characters are controlled by one chromosomal locus (gene) or by two separate chromosomal loci (genes). According to this view, the gene is the smallest unit of recombination. We shall later examine other possible definitions of the gene.

Giant chromosomes. Another approach to chromosomal mapping is based on the study of the giant chromosomes that occur in the salivary glands of the larvae of many flies, including *Drosophila*. These chromosomes are more than 200 times larger than normal chromosomes, and they can easily be studied in detail through an ordinary microscope.

When stained appropriately, these giant chromosomes have a banded appearance (Fig. 14.10). The bands, which differ in width and the spacings between them enable a worker thoroughly familiar with them to recognize with great precision the various regions of the chromosome. It has been possible by detailed comparative studies of chromosome abnormalities to determine the location of individual genes in relation to the bands. For a while, geneticists hoped that each band represented one gene, but this did not prove to be true; each band usually contains a number of different genes. Cytological studies of this sort have provided a second way of mapping the arrangement of the genes on the chromosomes. Such mapping has fully corroborated that based on crossing-over frequencies as to the sequence of the genes, but not the spacing between them (Fig. 14.11). Apparently breaks do not occur with equal facility at all points along the chromosomes; some parts of the chromosome are more susceptible to breakage than others. However, even though maps based on cytological examination give a more accurate picture of the relative distances between the genes, maps based on crossing-over frequencies are still constructed because they are more useful than cytological maps in predicting the results of crosses involving linked genes.

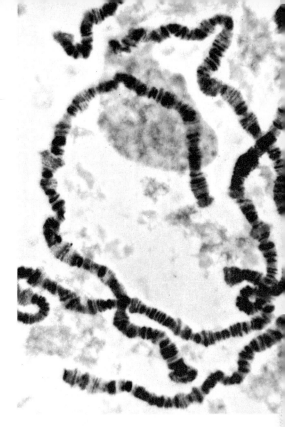

14.10. Photograph of giant chromosomes from salivary gland of *Drosophila melanogaster.* Note the pattern of banding by which different parts of the chromosomes can be identified. [Courtesy General Biological Supply House, Inc., Chicago.]

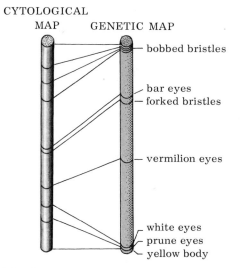

CYTOLOGICAL MAP   GENETIC MAP

— bobbed bristles

— bar eyes
— forked bristles

— vermilion eyes

— white eyes
— prune eyes
— yellow body

14.11. Comparison of cytological and genetic (cross-over) maps of a portion of the X chromosome in *Drosophila melanogaster.* The two methods of mapping yield the same sequence for the genes, but the spacing is very different. [Modified from *Biological Science*, Houghton Mifflin, 1963. Used by permission of the Biological Sciences Curriculum Study.]

GENETICS PROBLEMS

The best way to learn the principles of genetics is to apply them. They will become clearer to you, and you will grasp them more surely, if you carefully think through the following problems, which illustrate the various patterns of inheritance treated in this chapter. Additional problems will be found in the genetics textbooks listed at the end of the chapter.

1. Assume that white color is dominant over yellow color in squash. If pollen from the anthers of a heterozygous white-fruited plant is placed on the pistil of a yellow-fruited plant, show, using ratios, the genotypes and phenotypes you would expect the seeds from this cross to produce.

2. In human beings, brown eyes are usually dominant over blue eyes. Suppose a blue-eyed man marries a brown-eyed woman whose father was blue-eyed. What proportion of their children would you predict will have blue eyes?

3. If a brown-eyed man marries a blue-eyed woman and they have ten children, all brown-eyed, can you be certain that the man is homozygous? If the eleventh child has brown eyes, will that prove what the father's genotype is?

4. A brown-eyed man whose father was brown-eyed and whose mother was blue-eyed married a blue-eyed woman whose father and mother were both brown-eyed. The couple has a blue-eyed son. For which of the individuals mentioned can you be sure of the genotypes? What are their genotypes? What genotypes are possible for the others?

5. If the litter resulting from the mating of two short-tailed cats contains three kittens without tails, two with long tails, and six with short tails, what would be the simplest way of explaining the inheritance of tail length in these cats? Show genotypes.

6. In the fruit fly *Drosophila melanogaster,* vestigial wings and hairy body are produced by two recessive genes located on different chromosomes. The normal alleles, long wings and hairless body, are dominant. Suppose a vestigial-winged hairy male is crossed with a homozygous normal female. What types of progeny would be expected? If the F_1 from this cross are permitted to mate randomly among themselves, what progeny would be expected in the F_2? Show complete genotypes, phenotypes, and ratios for each generation.

7. In some breeds of dogs, a dominant gene controls the characteristic of barking while trailing. In these dogs, another independent gene produces erect ears; it is dominant over its allele for drooping ears. Suppose a dog breeder wants to produce a pure-breeding strain of droop-eared barkers, but he knows that the genes for silent trailing and erect ears are present in his kennels. How should he proceed?

8. In hogs, a gene that produces a white belt around the animal's body is dominant over its allele for a uniformly colored body. Another gene produces a fusion of the two hoofs on each foot; this gene is dominant over its allele, which produces normal hoofs. Suppose a uniformly colored hog homozygous for fused hoofs is mated with a normal-footed hog homozygous for the belted character. What would be the phenotype of the F_1? If the F_1 individuals are allowed to breed freely among themselves, what genotype and phenotype ratios would you predict for the F_2?

9. In watermelons, the genes for green color and for short shape are dominant over their alleles for striped color and for long shape. Suppose a plant with long striped fruit is crossed with a plant heterozygous for both of these characters. What phenotypes would this cross produce and in what ratios?

10. In peas, a gene for tall plants (T) is dominant over its allele for short plants (t). The gene for smooth peas (S) is dominant over its allele for wrinkled peas (s). Calculate both phenotypic and genotypic ratios for the results of each of the following crosses:

$$TtSs \times TtSs$$
$$Ttss \times ttss$$
$$ttSs \times Ttss$$
$$TTss \times ttSS$$

11. A dominant gene, A, causes yellow color in rats. The dominant allele of another independent gene, R, produces black coat color. When the two dominants occur together $(A{-}R{-})$, they interact to produce gray. Rats of the genotype $aarr$ are cream-colored. If a gray male and a yellow female, when mated, produce offspring approximately ⅜ of which are yellow, ⅜ gray, ⅛ cream, and ⅛ black, what are the genotypes of the two parents?

12. What are the genotypes of a yellow male rat and a black female that, when mated, produce 46 gray and 53 yellow offspring?

13. In Leghorn chickens, colored feathers are due to a dominant gene, C; white feathers to its recessive allele, c. Another dominant gene, I, inhibits expression of color in birds with genotypes CC or Cc. Consequently both $C{-}I{-}$ and $cc{-}{-}$ are white. A colored cock is mated with a white hen and produces many offspring, all colored. Give the genotypes of both parents and offspring.

14. If the dominant gene K is necessary for hearing, and the dominant gene M results in deafness no matter what other genes are present, what percentage of the offspring produced by the cross $kkMm \times Kkmm$ will be deaf? (Assume that there is no linkage.)

15. Suppose two $DdEeFfGgHh$ individuals are mated. What would be the predicted frequency of $ddEEFfggHh$ offspring from such a mating?

16. If a man with blood type B, one of whose parents had blood type O, marries a woman with blood type AB, what will be the theoretical percentage of their children with blood type B?

17. Both Mrs. Smith and Mrs. Jones had babies the same day in the same hospital. Mrs. Smith took home a baby girl, whom she named Shirley. Mrs. Jones took home a baby girl, whom she named Jane. Mrs. Jones began to suspect, however, that her child and the Smith baby had been accidentally switched in the nursery. Blood tests were made: Mr. Smith was type A, Mrs. Smith type B, Mr. Jones type A, Mrs. Jones type A, Shirley type O, and Jane type B. Had a mixup occurred?

18. When Mexican Hairless dogs are crossed with normally-haired dogs, about half the pups are hairless and half have hair. When, however, two Mexican Hairless dogs are mated, about a third of the pups produced have hair, about two thirds are hairless, and some deformed puppies are born dead. Explain these results.

19. Suppose a pigeon breeder finds that about one fourth of the eggs produced by one of his prize pairs do not hatch. Of the young birds produced by this pair, two thirds are males. Give a possible explanation for these results. (Remember the mechanism of sex determination in birds; see p. 309, footnote 5.)

20. Red-green color blindness is inherited as a sex-linked recessive. If a color-blind woman marries a man who has normal vision, what would be the expected phenotypes of their children with reference to this character?

21. Suppose that gene b is sex-linked, recessive, and lethal. A man marries a woman who is heterozygous for this gene. If this couple had many normal children, what would be the predicted sex ratio of these children?

22. A man and his wife both have normal color vision, but a daughter has red-green color blindness, a sex-linked recessive trait. The man sues his wife for divorce on grounds of infidelity. Can genetics provide evidence supporting his case?

23. The diagram shows three generations of the pedigree of deafness in a family. Black circles indicate deaf persons. An arrow on a circle indicates a male; a cross below a circle indicates a female. Is the condition of deafness in this pedigree inherited as (1) a dominant autosomal characteristic? (2) a recessive autosomal characteristic? (3) a sex-linked dominant characteristic? (4) a sex-linked recessive characteristic? (5) a holandric characteristic?

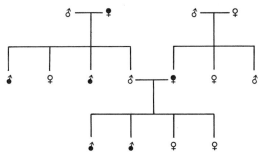

24. In *Drosophila melanogaster*, there is a dominant gene for gray body color and another dominant gene for normal wings. The recessive alleles of these two genes result in black body color and vestigial wings respectively. Flies homozygous for gray body and normal wings were crossed with flies that had black bodies and vestigial wings. The F_1 progeny were then test-crossed, with the following results:

Gray body, normal wings	236
Black body, vestigial wings	253
Gray body, vestigial wings	50
Black body, normal wings	61

Would you say that these two genes are linked? If so, how many units apart are they on the chromosome?

25. The cross-over frequency between linked genes *A* and *B* is 40%; between *B* and *C*, 20%; between *C* and *D*, 10%; between *C* and *A*, 20%; between *D* and *B*, 10%. What is the sequence of the genes on the chromosome?

REFERENCES

DeRobertis, E. D. P., W. W. Nowinski, and F. A. Saez, 1970. *Cell Biology*, 5th ed. Saunders, Philadelphia. (See esp. Chapters 14–15.)

Gardner, E. J., 1972. *Principles of Genetics*, 4th ed. Wiley, New York.

Moody, P. A., 1967. *Genetics of Man*. Norton, New York.

Peters, J. A., ed., 1959. *Classic Papers in Genetics*. Prentice-Hall, Englewood Cliffs, N.J. (A collection of 28 important papers.)

Sinnott, E. W., L. C. Dunn, and T. Dobzhansky, 1958. *Principles of Genetics*, 5th ed. McGraw-Hill, New York.

Snyder, L. H., and P. R. David, 1957. *The Principles of Heredity*, 5th ed. Heath, Boston.

Srb, A. M., R. D. Owen, and R. S. Edgar, 1965. *General Genetics*, 2nd ed. Freeman, San Francisco.

Stern, C., 1960. *Principles of Human Genetics*, 2nd ed. Freeman, San Francisco.

SUGGESTED READING

BEARN, A. G., and J. L. GERMAN III, 1961. "Chromosomes and Disease," *Scientific American,* November. (Offprint 150.)

BONNER, D. M., and S. E. MILLS, 1964. *Heredity,* 2nd ed. Prentice-Hall, Englewood Cliffs, N.J.

CLARKE, C. A., 1968. "The Prevention of 'Rhesus' Babies," *Scientific American,* November. (Offprint 1126.)

LEVINE, R. P., 1968. *Genetics,* 2nd ed. Holt, Rinehart & Winston, New York.

McKUSICK, V. A., 1971. "The Mapping of Human Chromosomes," *Scientific American,* April.

MENDEL, G., 1965. *Experiments in Plant-Hybridisation.* Harvard University Press, Cambridge, Mass. (Translation of Gregor Mendel's original paper first published in 1866.)

MITTWOCH, U., 1963. "Sex Differences in Cells," *Scientific American,* July. (Offprint 161.)

The nature of the gene and its action

In the last chapter, we postulated the existence of physical units of inheritance, the genes, which interact with environmental influences to determine the phenotypic characteristics of the organism; and we cited evidence for their arrangement in linear sequence along the chromosomes. Presenting, as we did, a model of the gene based almost entirely on deductions from inheritance patterns, we obviously left some fundamental genetic problems untouched. We said nothing about the chemical nature of the genes; we did not say how the genetic material is replicated, how genes influence the characteristics of the organism, or how they regulate the myriad activities of a living cell. In effect, we treated the genes as though they were a string of beads of unknown composition somehow exerting almost magical control over the manifest attributes of living things.

Not many years ago this was all any textbook could do. Nothing was known of the chemical makeup of genes; nor was it known how they are replicated or how they influence the phenotype. But all that is changed. The enormous advances of the last twenty years in our understanding of the control mechanisms of living cells constitute nothing short of a revolution, a revolution with such far-reaching implications that many scientists have claimed it will eventually prove more important to the future of the human species than the birth of the atomic age in physics.

THE GENETIC MATERIAL

The discovery of DNA and its function

If the genes were located on the chromosomes, an obvious first step in ascertaining the chemical nature of genes was to determine the types of compounds present in the chromosomes. Once this was done, experiments could be devised to discover which of the chromosomal compounds is the bearer of genetic information.

Most chromosomes, it was found, consist principally of a complex of protein and nucleic acid of a type called *deoxyribonucleic acid,* or *DNA* for short. The obvious deduction was that genes are composed of protein or of nucleic acid or of a complex of the two (nucleoprotein). In the early 1940's, each of these three possibilities had its advocates, with the majority favoring protein. Today, almost all biologists are convinced that DNA is the genetic material. To understand how this shift came about, we must leave our old friends *Drosophila* and peas, and turn to microorganisms such as bacteria, viruses, and molds.

In 1928 Fred Griffith, an English medical bacteriologist, showed that if live cells of a nonvirulent strain and dead cells of a virulent strain of pneumococci (the bacteria that cause pneumonia) are injected together into a mouse, the live bacteria are transformed from nonvirulent into virulent by material from the dead virulent bacteria. Later, other workers demonstrated that the transforming principle was DNA; nothing else was necessary. From our present perspective, this seems like strong evidence that DNA rather than protein or a nucleoprotein complex is the essential genetic material, but at that time many scientists remained unconvinced.

During the next ten years, however, the evidence for DNA steadily became stronger. At least 30 different examples of bacterial *transformation* by purified DNA were described. And strong evidence came from another source, studies of a special type of virus that attacks the bacterium *Escherichia coli,* which is abundant in the human digestive tract. This type of bacteria-destroying virus is called *bacteriophage*—phage, for short.

It has been established by a variety of methods that viruses are composed of two chief components, a protein coat and a nucleic acid core. In phage viruses, which the electron microscope has shown to be structurally more complex than many other types of viruses, the protein coat is divided into a head region, which contains all the DNA, and into an elongate tail region made up of a hollow core, a surrounding sheath, and six distal fibers (Fig. 15.1). Electron micrographs show that when phage attack a bacterial cell they become attached by the tip of their tails to the wall of the bacterial cell (Fig. 15.2). There is no evidence that the protein coats of the phage ever enter the cell; yet

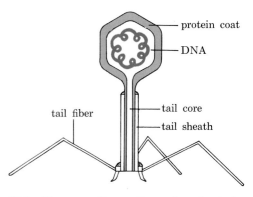

15.1. Diagrammatic section of a bacteriophage.

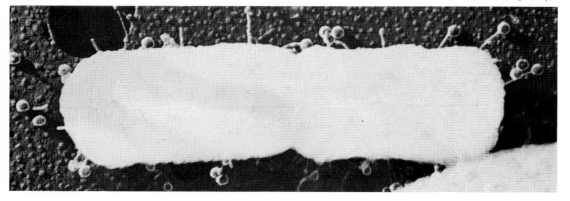

15.2. Electron micrograph of a bacterial cell with numerous phage attached to its surface. × 50,000. [Courtesy T. F. Anderson, *Ann. Inst. Pasteur* (Paris), vol. 93, 1957.]

within a few minutes after the phage become attached to the wall, new phage appear within the bacterium and the bacterial cell soon ruptures, releasing hundreds of new phage into the surrounding medium. It can be shown that these new phage are genetically identical with those that initiated the infection. Hereditary material must have been injected into the bacterial cell by the phage attached to its wall, and this hereditary material must have usurped the metabolic machinery of the bacterium and put it to work manufacturing new phage.

Alfred D. Hershey and Martha Chase of the Carnegie Laboratory of Genetics designed an experiment to determine whether the infecting phage inject into the bacterium only DNA or only protein or some of both. They made use of the fact that DNA contains phosphorus but no sulfur, whereas protein contains sulfur but no phosphorus. They cultured phage on bacteria grown on a medium containing a radioactive isotope of phosphorus (P^{32}) and a radioactive isotope of sulfur (S^{35}). The phage incorporated the S^{35} into their protein, and they incorporated the P^{32} into their DNA. Hershey and Chase then infected nonradioactive bacteria with the radioactive phage. They allowed sufficient time for the phage to become attached to the walls of the bacteria and inject hereditary material. Then they agitated the bacteria in a blendor in order to detach what remained of the phage from their surfaces. Analysis of these remains showed that they contained S^{35} but no P^{32}, an indication that only the empty protein coat had been left outside the bacterial cell. Analysis of the bacteria showed that they contained P^{32} but no S^{35}, an indication that only DNA had been injected into them by the phage. DNA alone was sufficient to transmit to the bacteria all the genetic information necessary to cause them to produce new phage. This experiment, reported in 1952, supported the earlier conclusions based on transformation experiments that nucleic acids, not proteins, constitute the genetic material.

The molecular structure of DNA

The chemical components of DNA. A single molecule of DNA is very large, containing many thousands of atoms. How could the arrangement of such a huge number of atoms be worked out in detail to give a picture of the molecular structure of DNA? Fortunately for the sake of analysis, DNA, like the other large organic molecules we have examined, is composed of a few relatively simple building-block compounds bonded together in sequence. These building-block compounds are called *nucleotides.* Each nucleotide, in turn, is composed of three still smaller constituent parts: a phosphate group, a five-carbon sugar called deoxyribose, and an organic nitrogen-containing base. Both the phosphate group and the base are bonded to the sugar (Fig. 15.3). Four different kinds of nucleotides occur in DNA; they are alike in containing the phosphate and deoxyribose, but they differ in their nitrogenous bases. These comprise the purines *adenine* and *guanine,* which have a double-ring structure, and the pyrimidines *cytosine* and *thymine,* which are single-ring structures (Fig. 15.4).

The nucleotides within a DNA molecule are bonded together in such a way that the sugar of one nucleotide is always attached to the phosphate group of the next nucleotide in the sequence (Fig. 15.5). Thus a long chain of alternating sugar and phosphate groups is estab-

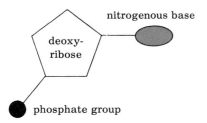

15.3. Diagram of a nucleotide from DNA. A phosphate group and a nitrogenous base are attached to deoxyribose, a five-carbon sugar.

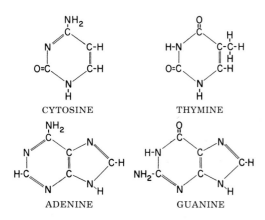

15.4. The four nitrogenous bases in DNA. The two single-ring bases, cytosine and thymine, are pyrimidines; the two double-ring bases, adenine and guanine, are purines.

lished, with the nitrogenous bases oriented as side groups off this chain. The order in which the four different nucleotides occur differs for different DNA molecules, and it is apparently this order that determines the specificity of the DNA.

The spatial configuration of the DNA molecule. By 1950 much was known about the chemical composition of DNA. But almost nothing was known about the spatial arrangement of the atoms within the DNA molecule; nor was it known how this molecule, if it really was the stuff of which genes were made, could contain within it the necessary information for replicating itself and for controlling cellular function. About this time, several workers began applying the techniques of X-ray diffraction analysis to DNA. Outstanding among them was Maurice H. F. Wilkins of King's College, London, who succeeded in obtaining much sharper X-ray diffraction patterns than had previously been obtainable. These diffraction patterns revealed three major periodicities in the crystalline DNA: one of 3.4 angstroms, one of 20 angstroms, and one of 34 angstroms.

Now began a collaboration whose results would rank among the major milestones in the history of biology. James D. Watson and Francis H. C. Crick, working at Cambridge University, decided to try to develop a model of the structure of the DNA molecule by combining what was known about the chemical content of DNA with the information gained from Wilkins' X-ray diffraction studies and with certain facts established by physical chemistry concerning the exact distances between bonded atoms in molecules, the angles between bonds, and the sizes of atoms. Watson and Crick built scale models of the component parts of DNA and then attempted to fit them together in a way that would agree with the data from all these separate sources.

They were certain that the 3.4-angstrom periodicity discovered by Wilkins corresponded to the distance between successive nucleotides in the DNA chain and that the 20-angstrom periodicity corresponded to the width of the chain. But what about the 34-angstrom periodicity? To explain it, they postulated that the chain of nucleotides was coiled in a helix. (To visualize a helix, think of the chain as wound around a long cylinder.) The 34-angstrom periodicity would thus correspond to the distance between successive turns of the helix, and it would indicate how tight the chain was wound. Since 34 is exactly ten times the 3.4-angstrom distance between successive nucleotides, it would follow that each turn of the helix must be ten nucleotides long.

Having made these essential assumptions about the meaning of the X-ray diffraction data, Watson and Crick then tried to correlate them with the information from other sources. They immediately ran into a discrepancy. They calculated that a single chain of nucleotides coiled in a helix that was 20 angstroms wide and had turns 34 angstroms long would have a density only half as great as the known density of DNA. An obvious inference was that the DNA molecule is composed of two nucleotide chains rather than one. Now they had to determine the relationship between the two chains within the double helix. They tried several arrangements of their scale model and found that the one that best fitted all the data was one in which the two nucleotide chains were wound in opposite directions around a hypo-

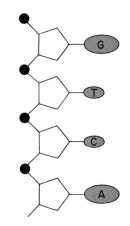

15.5. Portion of a single chain of DNA. Nucleotides are hooked together by bonds between their sugar and phosphate groups. The nitrogenous bases (G, guanine; T, thymine; C, cytosine; A, adenine) are side groups.

thetical cylinder of appropriate diameter, with the purine and pyrimidine bases oriented toward the interior of the cylinder (Fig. 15.6). With the bases oriented in this manner, hydrogen bonds between the bases of opposite chains could supply the force to hold the two chains together and to maintain the helical configuration. In other words, the DNA molecule, when unwound, would have a ladderlike structure, with the uprights of the ladder formed by the two long chains of alternating sugar and phosphate groups, and with each of the cross rungs formed by two nitrogenous bases loosely bonded to each other by hydrogen bonds (Fig. 15.7).

It soon became clear to Watson and Crick that each cross rung must be composed of one purine base and one pyrimidine base. Their scale model showed that the available space between the sugar-phosphate uprights was just sufficient to accommodate three ring structures. Hence two purines opposite each other occupied too much space, because each had two rings for a total of four, and two pyrimidines opposite each other did not come close enough to bond properly, because each had only one ring. This left four possible pairings: adenine–thymine, adenine–cytosine, guanine–thymine, and guanine–cytosine. Further examination revealed that, although adenine and cytosine were of the proper size to fit together into the available space, they could not be arranged in a way that would permit hydrogen bonding between them; the same was true of guanine and thymine. Only adenine–thymine and guanine–cytosine seemed to fulfill all requirements. It did not seem to matter in which order the bases occurred; thymine–adenine was as satisfactory as adenine–thymine, and cytosine–guanine was as satisfactory as guanine–cytosine. The essential requirement seemed to be that adenine and thymine always be paired with each other and that guanine and cytosine always be paired.

In summary, then, the Watson-Crick model of the DNA molecule shows a double helix in which the two chains, composed of alternating sugar and phosphate groups, are loosely bonded together by hydrogen bonds between adenine and thymine from opposite chains and between guanine and cytosine from opposite chains. This model, essentially in the form in which it was first proposed in 1953, has been consistently supported by later research, and it has received general acceptance.

The replication of DNA

DNA, if it is the genetic substance, must have built into it the information necessary to replicate itself and to control the cell's attributes and functions. One of the most satisfying things about the Watson-Crick model of DNA is that it immediately suggests a way in which the first of these two requirements may be met.

Since the DNA of all organisms is alike in being a polymer composed of only four different nucleotides, the characteristics that distinguish the DNA of one gene from the DNA of another gene must be the total number of nucleotides, the ratio of adenine and thymine to guanine and cytosine, and the sequence in which the four possible types of cross rungs (adenine–thymine, thymine–adenine, guanine–cytosine, and cytosine–guanine) occur. The basic question of genetic replication is, then: Assuming that an adequate supply of the four

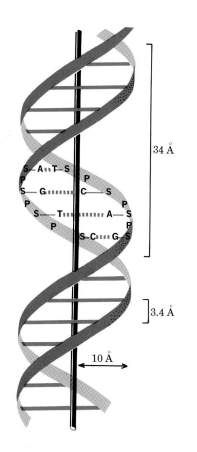

15.6. The Watson-Crick model of DNA. The molecule is composed of two polynucleotide chains held together by hydrogen bonds between their adjacent bases (S, sugar; P, phosphate; A, T, G, C, nitrogenous bases). The double-chained structure (shown here wound around a hypothetical rod) is coiled in a helix. The width of the molecule is 20 angstroms; the distance between adjacent nucleotides is 3.4 angstroms; and the length of one complete coil is 34 angstroms.

nucleotides is already synthesized in the cells, what tells the cell's biochemical machinery how to put these nucleotide building blocks together in exactly the quantities and sequences characteristic of the DNA already present in the cell?

Watson and Crick pointed out that if the two chains of a DNA molecule are separated by rupturing the hydrogen bonds between the base pairs, each chain provides all the information necessary for synthesizing a new partner. Since an adenine nucleotide must always pair with a thymine nucleotide, and since a guanine nucleotide must always pair with a cytosine nucleotide, the sequence of nucleotides in one chain tells precisely what the sequence of nucleotides in its complementary chain must be. In other words, if the cell separated the two chains in its DNA molecules, it could then line up separate nucleotides next to each of the single chains, putting each type of nucleotide next to its proper partner. Once the nucleotides had been arranged in the proper sequence, they could be bonded together to form a complete new chain. Thus, separating the two chains of a DNA molecule and then using each chain as a template or mold against which to synthesize a new partner for it would result in two complete double-chained molecules identical to the original molecule.

In the years since 1953, experiments by many workers have supported this explanation of DNA replication, and it is now generally accepted.

THE MECHANISM OF GENE ACTION

We have seen that the Watson-Crick model of DNA provides an easily understood explanation for genetic replication. Now we must ask what our knowledge of the chemistry of heredity can tell us about the way genes control cellular function. How can the DNA gene be related to phenotypically expressed inherited traits? How can nucleic acid in the nucleus control what goes on in the cytoplasm?

The one gene—one enzyme hypothesis

Alkaptonuria is a hereditary disease in human beings in which the urine is very dark-colored. More than a hundred years ago, it was shown that the darkening of the urine is caused by the presence in it of a chemical called alkapton. Normal individuals possess an enzyme that catalyzes the oxidation of alkapton to carbon dioxide and water, but those who have alkaptonuria lack this enzyme and must excrete alkapton in its undegraded form. In 1909 it was shown that alkaptonuria is inherited as a simple Mendelian recessive. Apparently the normal gene is necessary for production of the enzyme that catalyzes the oxidation of alkapton; persons homozygous for the mutant form of the gene do not produce the enzyme.

Alkaptonuria and other hereditary diseases correlated with the lack of an enzyme should have suggested to geneticists that there is a close relationship between genes and enzymes. But for many years most geneticists failed to make the connection. Then in the 1930's it was shown that synthesis of normal eye pigment in *Drosophila* involves a series of reactions that produce identifiable intermediate compounds and that each of these reactions, catalyzed by its own enzyme, is under

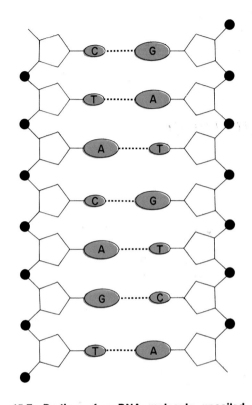

15.7. Portion of a DNA molecule uncoiled. The molecule has a ladderlike structure, with the two uprights composed of alternating sugar and phosphate groups and the cross rungs composed of paired nitrogenous bases. Note that each cross rung has one purine base (large oval) and one pyrimidine base (small oval). When the purine is guanine (G), then the pyrimidine with which it is paired is always cytosine (C); when the purine is adenine (A), then the pyrimidine is thymine (T).

genetic control. It was demonstrated that eye-color mutants arise when gene mutations result in alterations of these chemical reactions and thus of the pigment produced by them. A relationship between genes and specific enzymes was strongly suggested.

But concepts of gene action remained rather nebulous until the early 1940's, when George W. Beadle and Edward L. Tatum, then at Stanford University, performed their pioneering experiments on red bread mold, *Neurospora crassa,* which in the last three decades has taken its place alongside such organisms as *Drosophila* and corn among the genetically most extensively studied living things. They used X rays to induce mutations interfering with the mold's capacity to synthesize certain nutrients that it can normally synthesize for itself. In every case where the mutant had lost the ability to synthesize a particular nutrient, they found that the character was inherited in a pattern indicating the action of only one gene. Reasoning that the mutated gene could no longer determine an enzyme essential for the synthesis of the nutrient in question, they advanced their one gene–one enzyme hypothesis—that each enzyme in a cell is controlled by a single gene.

According to the broader implications of this hypothesis, genes exert their control over cellular function by controlling the enzymes that control the chemical reactions of the cell; these reactions, in turn, determine the phenotypic characteristics. Since enzymes are proteins, the one gene–one enzyme hypothesis is, in effect, a one gene–one protein hypothesis—a proposal that, somehow, each gene controls the synthesis of a protein. With greater precision, it would be called the one gene–one polypeptide hypothesis, for it has recently been demonstrated that, when a protein is composed of two or more chemically different polypeptide chains, each chain is determined by its own gene.

The synthesis of proteins

If it is assumed that genes control the cell's activities by controlling the synthesis of protein enzymes, the next question becomes: By what mechanism is this gene-controlled synthesis of proteins accomplished?

The problem of sequence.   Polysaccharides such as starch and cellulose are long polymers built up by serial repetition of identical units —molecules of glucose—and the order in which the building blocks are linked presents no problem. Proteins, too, are long polymers. But they are composed of some 20 different units—the amino acids—and the sequence in which those units are linked is of critical importance. It is the necessity for specifying this sequence that makes protein synthesis so much more complex than polysaccharide synthesis.

If genes really specify amino acid sequences for proteins, and if DNA really is the stuff of which genes are made, then the sequence of bases in the DNA must somehow indicate the sequence in which amino acids must be linked in protein synthesis. The problem then becomes one of translating nucleotide base sequence into amino acid sequence. One of the first points that had to be established was whether the translation is normally direct or indirect, i.e. whether protein is synthesized on the DNA template itself or whether some intermediate agent is involved. There had long been evidence that most protein synthesis takes place in the cytoplasm. Since, with a few notable exceptions, DNA is restricted largely to the nucleus, the probabilities were that the protein is not synthesized along a DNA template as new DNA is. The information, presumably coded in nucleotide sequences in the DNA, must be transmitted from the nucleus to the sites of protein synthesis in the cytoplasm. How?

RNA and the ribosomes. Several workers showed in the early 1940's that cells in tissues where protein synthesis is particularly active contain large amounts of a nucleic acid named *ribonucleic acid,* usually designated **RNA.** This nucleic acid is present in only limited quantities in cells that do not produce protein secretions. Furthermore, it had long been known that RNA, unlike DNA, occurs in the cytoplasm as well as in the nucleus. Radioactive-tracer experiments revealed that RNA is synthesized in the nucleus and moves from the nucleus into the cytoplasm. All these lines of evidence pointed to the possibility that RNA might be the chemical messenger between the DNA of the nucleus and the metabolic machinery of the cytoplasm.

Though RNA and DNA are very similar compounds, they differ in three important ways: (1) The sugar in RNA is ribose, while that in DNA is deoxyribose (it is the sugars that give the two nucleic acids their different initials). (2) RNA contains *uracil* in place of the thymine in DNA. And (3) RNA is ordinarily single-stranded, while DNA is usually double-stranded.

Despite these differences, it was obvious that DNA could easily act as a template for the synthesis of RNA. The synthesis would proceed in essentially the same way as that of new DNA. The two strands of a DNA molecule would uncoil. Then ribonucleotides (i.e. nucleotides containing ribose instead of deoxyribose) would be lined up along one of the DNA strands—a uracil ribonucleotide opposite each adenine on the DNA; an adenine ribonucleotide opposite each thymine; a cytosine ribonucleotide opposite each guanine; and a guanine ribonucleotide opposite each cytosine (Fig. 15.8). In other words, the sequence of deoxyribonucleotides in the single DNA strand would determine the sequence of ribonucleotides for the synthesis of an RNA strand. Once the ribonucleotides were arranged in the proper order, they could be bonded together, and then the new RNA molecule could separate from the DNA and travel from its place of synthesis on the chromosome of the nucleus to its functional location in the cytoplasm. According to this model, then, DNA acts as a template or mold not only for the synthesis of new DNA just before a cell undergoes division, but also for the synthesis of RNA when the cell is metabolically active but is not dividing.

The accuracy of the model has been confirmed, since it was first proposed, through the verification of many details of cellular protein synthesis. It was observed that several different types of RNA occur in actively synthesizing cells, and that there is normally a steady flow of one type from the nucleus to the *ribosomes,* where the actual synthesis of protein takes place. This type of RNA, which has been called *messenger RNA* (mRNA), is apparently responsible for carrying instructions for protein synthesis from the DNA in the nucleus to the ribosomes in the cytoplasm. The indications are that each molecule of

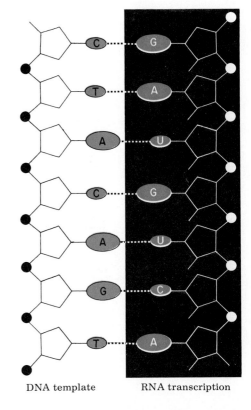

DNA template   RNA transcription

15.8. Synthesis of RNA on a DNA template. The sugar in RNA is slightly different from that in DNA, and uracil (U) takes the place of the thymine in DNA.

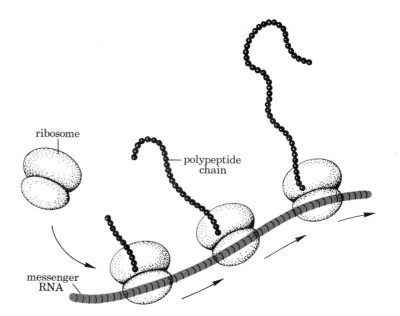

15.9. Synthesis of polypeptide chains by the ribosomes. As the ribosomes move along the messenger RNA, they "read" the information coded into its nucleotide sequence and synthesize a polypeptide chain according to that information.

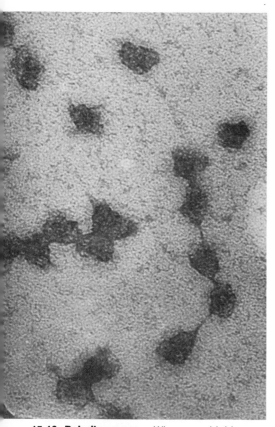

15.10. Polyribosomes. When very highly magnified (× 400,000), the ribosomes of each polyribosome are seen to be connected by a thread, which is presumed to be messenger RNA. [Courtesy Alexander Rich, Massachusetts Institute of Technology.]

mRNA is synthesized along a chromosomal gene (or genes) and then moves through a "pore" in the nuclear membrane to the cytoplasm, where it complexes with one or more ribosomes and acts as the template for synthesis of a relatively few protein molecules before it is destroyed by an enzyme that breaks it down to its constituent nucleotide building blocks.

Although the ribosomes are essential for protein synthesis in the living cell, they apparently do not play any role in determining what proteins will be produced. Ribosomes, as far as is known, function as unspecific devices capable of synthesizing any protein for which they are supplied the necessary information by messenger RNA. Each ribosome, when it is attached to mRNA, is made up of two subunits, one larger than the other (Fig. 15.9). The smaller subunit is apparently responsible for binding the mRNA. The larger subunit contains the enzymes that catalyze the peptide bonding of amino acids. The two subunits exist in the cytoplasm as separate entities when not carrying out protein synthesis, i.e. when not complexed with mRNA. When several ribosomes become associated with a molecule of mRNA, the complex they form is known as a *polyribosome* (Fig. 15.10). As each ribosome moves along the nucleotide chain of the mRNA, it translates the nucleotide sequence into an amino acid sequence, thus building a polypeptide chain (Fig. 15.9).

Transfer RNA and its role. Does the combination of amino acids into polypeptide chains entail direct interaction between the amino acids and messenger RNA or is some intermediate agent or adapter molecule involved? Strong evidence in support of the second alternative came from a demonstration that amino acids become attached to RNA *before* they arrive at the ribosomes. The RNA to which they become attached is not mRNA but another type consisting of relatively small soluble cytoplasmic molecules. It has been shown that each molecule of this RNA binds a single molecule of amino acid and transports it to the ribosome; hence the name *transfer RNA* (tRNA). At least one form of tRNA is specific for each of the 20 amino acids; the amino acid arginine, for example, will combine only with tRNA specialized to transport arginine, the amino acid leucine will combine only with tRNA specialized to transport leucine, etc.

There is evidence, which we shall examine later, that the code for each amino acid in messenger RNA is a particular sequence of three bases, often called a *codon.* Apparently the transfer RNA for each amino acid has a complementary sequence of three bases, the *anticodon,* which can couple with the appropriate codon in mRNA by base pairing of the Watson-Crick type (Fig. 15.11). To illustrate, let us assume that the messenger sequence CCG (cytosine, cytosine, guanine) codes for the amino acid proline and that GUA (guanine, uracil, adenine) codes for valine; then proline tRNA will have the anticodon GGC, which is complementary to CCG, the messenger code sequence for proline, and valine tRNA will have the anticodon CAU, which is complementary to GUA, the messenger code sequence for valine. Since each tRNA carrying an amino acid to a strand of mRNA can bond with the mRNA only at the point where its anticodon matches an mRNA codon, the scheme of complementary base triplets ensures that the tRNA's will automatically order the amino acids in the sequence coded by the mRNA.

An outline of the model. Let us now bring together the findings and hypotheses discussed in the previous sections and outline what is currently the most widely accepted model of genic control of protein synthesis (Fig. 15.12).

When the double-stranded DNA that constitutes a particular gene is activated, its two nucleotide chains uncoil and separate, and one of the chains acts as the template for synthesis of a molecule of single-stranded messenger RNA. This mRNA leaves the nucleus and moves into the cytoplasm, where it becomes associated with a cluster of ribosomes. The mRNA acts as the template for synthesis of polypeptide chains as the ribosomes move along it. Amino acids to be incorporated into the polypeptide chains are first activated by ATP and then picked up by molecules of transfer RNA, of which one type or more is specific for each of the 20 different amino acids. Each of these tRNA's has an exposed sequence of bases (the anticodon) complementary to the sequence of bases on the mRNA that codes for its particular amino acid. Each molecule of tRNA, having picked up an amino acid from the cytoplasmic pool, moves to a ribosome and attaches to the mRNA at a point where the appropriate base sequence occurs. This ordering of the tRNA's along the mRNA molecule also orders the amino acids attached to them. Once the amino acids have been moved into the proper sequence in this way, peptide linkages are formed between them, and the resulting polypeptide falls away. The tRNA's uncouple from the mRNA and move away to pick up another load of amino acids.

According to this model, then, the flow of information proceeds as follows: The DNA of the gene determines the messenger RNA, which determines protein enzymes, which control chemical reactions, which produce the characteristics of the organism.

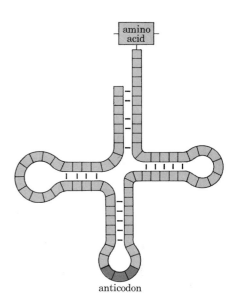

15.11. Cloverleaf model of transfer RNA for alanine in yeast. The single polynucleotide chain folds back on itself, forming four regions of complementary base pairing, three loops with unpaired bases, and an unpaired terminal portion to which the amino acid can attach. The anticodon (color) is an unpaired triplet of bases that can bond with a complementary triplet in messenger RNA. (The molecule is here shown flattened, with all coiling eliminated.)

The genetic code

The codon. In discussing the model of protein synthesis, we assumed that the nucleic acid code unit, or codon, for each amino acid is three nucleotides long. We should now examine the basis for this assumption. We are dealing with a code that has only four elements—the four different nucleotides in messenger RNA (which reflect a corresponding four nucleotides in DNA)—and a system that must be capable of coding for at least 20 different amino acids. Hence the codon cannot be only one nucleotide long, because such a system could code for only four amino acids. Nor can the codon be two nucleotides long, because only 16 combinations would be possible. If the codon is three nucleotides long, there are 64 possible combinations, which is more than enough to code for 20 amino acids. There is convincing evidence that the codon is not more than three nucleotides long.

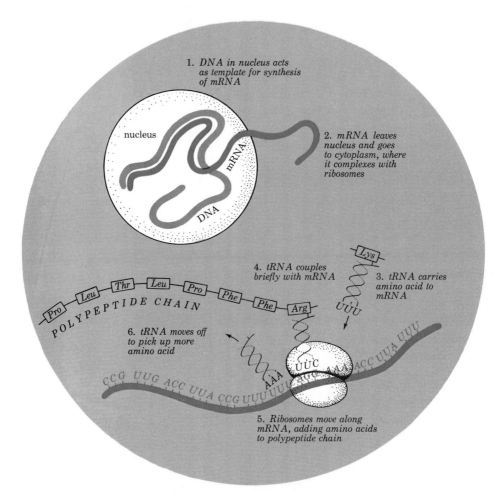

1. *DNA in nucleus acts as template for synthesis of mRNA*

2. *mRNA leaves nucleus and goes to cytoplasm, where it complexes with ribosomes*

3. *tRNA carries amino acid to mRNA*

4. *tRNA couples briefly with mRNA*

5. *Ribosomes move along mRNA, adding amino acids to polypeptide chain*

6. *tRNA moves off to pick up more amino acid*

15.12. Model for control of protein synthesis by the genes. Messenger RNA is synthesized on one of the polynucleotide chains of the DNA of the gene. This mRNA then goes into the cytoplasm and becomes associated with one or more ribosomes. The various types of tRNA in the cytoplasm pick up the amino acids for which they are specific and bring them to a ribosome as it moves along the mRNA. Each tRNA bonds to the mRNA at a point where a triplet of bases (codon) complementary to an exposed triplet (anticodon) on the tRNA occurs. This ordering of the tRNA molecules automatically orders the amino acids, which are then linked by peptide bonds. Synthesis of the polypeptide chain thus proceeds one amino acid at a time in an orderly sequence as the ribosomes move along the mRNA. As each tRNA donates its amino acid to the growing polypeptide chain, it uncouples from the mRNA and moves away into the cytoplasm, where it can be used again.

Overlapping versus nonoverlapping codes. A genetic code based on triplet codons might be either overlapping or nonoverlapping. Suppose that a particular piece of messenger RNA has the following sequence of nucleotides:

CAUCAGGUA

If the code were a completely overlapping one, this piece of mRNA would code for seven amino acids; CAU would indicate the first amino acid, AUC would indicate the second, UCA the third, CAG the fourth, etc., as follows:

CAU
AUC
UCA
CAG
AGG
GGU
GUA

If the code were a nonoverlapping one, this piece of RNA would code for only three amino acids; CAU would indicate the first, CAG the second, and GUA the third, as follows:

CAUCAGGUA

What sort of evidence would allow us to choose between these alternatives? Notice that in a completely overlapping code each nucleotide (except the first two and the last two) would be included in the codons for three amino acids; thus the first U in our hypothetical piece of messenger RNA would be in the following three codons: CAU, AUC, UCA. By contrast, each nucleotide in a nonoverlapping code would be included in only one codon; e.g. the first U would appear only in the CAU codon. It is clear that if the code is completely overlapping, a mutation substituting one nucleotide for another in a nucleic acid strand should result in a change of three adjacent amino acids in the protein determined by this nucleic acid; but if the code is nonoverlapping, such a substitution should change only one amino acid in the protein. Now, it is known that the hemoglobin of sickle-cell anemia differs from normal hemoglobin in only one amino acid in each half molecule, which would seem to indicate that a single mutation can result in an alteration of a single amino acid; it is apparently not necessary for three adjacent amino acids to be changed together. This conclusion has been supported by numerous experiments on tobacco mosaic virus, in which a number of mutations that change only one amino acid have been found.

Spacing. Let us now turn to another important question about the genetic code: What designates the ending of one gene[1] and the beginning of another? It might be easier to answer this question with reference to higher organisms if we knew how the DNA and the protein of the chromosomes are arranged. If, for example, the DNA for each gene were a separate molecule attached to a chromosomal backbone of

---

[1] We are here using a modified Beadle and Tatum definition of a gene as the length of DNA that codes for a single polypeptide chain.

protein, there would be no problem; the end of a gene would be the end of its DNA molecule, and the hundreds of DNA molecules corresponding to the hundreds of genes on each chromosome would somehow be attached in sequence to the protein backbone of the chromosome.

But there is good evidence that the DNA of each gene is not a separate molecule. The chromosomes of viruses and bacteria, which are composed exclusively of nucleic acid, clearly have no protein backbone; the entire chromosome gives every indication of being a single long DNA molecule, with no break where one gene ends and another begins. Recent electron microscopy has revealed that each DNA molecule of higher organisms also represents more than one gene; indeed there is a strong likelihood that all the DNA of a eucaryotic chromosome is a single immensely long folded and coiled molecule.

The end of the nucleotide sequence coding for a single polypeptide chain might possibly be indicated by a nonsense triplet (i.e. one that does not code for any amino acid). The nonsense triplet could cause a break either in transcribing the DNA into messenger RNA or in the ribosomal reading of the mRNA.[2] In other words, if ATT is a nonsense triplet in DNA (and there is evidence that it is), then the nucleotide sequence

CTT TTG TTT ATA ATT CAC GCC ACG

might represent parts of two genes, with the ATT triplet acting as a spacer between them:

last condon   spacer   first condon of
of one gene   another gene
↓   ↓   ↓
CTT TTG TTT ATA | ATT | CAC GCC ACG

The evidence supports this proposal.

Deciphering the code. How has it been possible to determine which nucleotide sequence codes for which amino acid? For several years, the most profitable approach was to produce synthetic messenger RNA in a test tube, using a special enzyme that does not require a DNA template but simply links nucleotides together in random order. If the only nucleotide made available to the enzyme was uracil, then a long polyuracil chain was synthesized:

UUUUUUUUUUUU

Similarly, if only adenine was made available, a polyadenine chain was formed. When poly-U was used in place of normal mRNA in a cell-free system for protein synthesis, a polypeptide chain composed of only phenylalanine resulted—a clear indication that UUU codes for phenylalanine. Similarly, AAA was shown to code for lysine.

Understandably enough, it was more difficult to assign triplets composed of two or three different nucleotides. If, for example, uracil

---

[2] Since there are only 20 different amino acids in protein, and there are 64 possible nucleotide triplets, it was at first thought that all the 44 surplus ones might be nonsense. However, present evidence indicates that most triplets do code for some amino acid (a single amino acid may be coded by several different codons). There are probably only two or three nonsense triplets.

nucleotide and guanine nucleotide were made available to the enzyme in a 2 : 1 ratio, messenger RNA in which the codons GUU, UGU, and UUG predominated was synthesized. It could be shown that when this mRNA was used as a template in polypeptide synthesis, the amino acids cysteine, valine, and leucine were the principal ones incorporated into the polypeptide, but it was impossible to determine by this technique which of the three triplets coded for which of the amino acids.

Then, in 1964, Philip Leder and Marshall W. Nirenberg of the National Institutes of Health developed a technique for forming a complex between ribosomes and RNA trinucleotides (three nucleotides bonded together in sequence) of known composition. They showed that these trinucleotides would act as though they were short pieces of messenger RNA, and that transfer RNA's would couple with them. For example, if they used a trinucleotide with the composition UUU, phenylalanine tRNA would couple with it. This technique proved to be the key to unlocking the details of the code. Since it is relatively easy to synthesize trinucleotides with a particular base sequence, each of the 64 possible triplets could be synthesized, complexed with ribosomes, and exposed to a mixture of tRNA's. By establishing which tRNA coupled with which trinucleotide, it became possible to determine the codons for each amino acid. Nirenberg and his associates performed a series of such experiments during 1965; their results made possible the construction of genetic dictionaries showing which codons determine which amino acids. It appears that all but two amino acids are coded by more than one codon; some are coded by as many as six.

NOW WHAT IS A GENE?

We have used the word "gene" in a variety of ways in this and the last chapter. Mendel did not use this term but spoke of characters and their corresponding "factors" or "elements" in the germ cells. From 1902 onward, it was generally agreed that the factors of inheritance, first called genes by Wilhelm Johannsen, a Danish botanist, in 1911, are associated with the chromosomes.

The work of T. H. Morgan with *Drosophila* led to a conception of the gene as the smallest unit of recombination. As we indicated in Chapter 14, two characters were regarded as determined by different genes if they could be recombined, and as determined by the same gene if they could not be recombined. Physically, the genes were thought of as tiny particles arranged in linear sequence on the chromosomes. Recombination between linked genes was explained by a crossing-over mechanism when the chromosomes broke at points between the particles.

Though defined essentially as the unit of recombination, the gene in classical genetics was also regarded as the unit of mutation and as the unit of function, i.e. as the smallest unit whose alteration by mutation would change the phenotype and as the smallest unit of control over the phenotype. This concept of the gene was satisfyingly unified. But modern knowledge of the structure of the genetic material and its function has destroyed this unified concept and given rise to a variety of different candidates for the designation "gene." It is now clear that the units of recombination, mutation, and function are not identical.

Seymour Benzer of Purdue University has given these units the following names:

> recon: unit of recombination
> muton: unit of mutation
> cistron: unit of function

A fourth unit—the codon, unit of coding—can be added to the list.

All available evidence indicates that recombination can take place between any two nucleotides—which would make the recon only one nucleotide long. Some mutations apparently change only one nucleotide in a nucleotide chain; hence the muton would also be only one nucleotide long and would be the same as the recon. The codon is larger than the recon and muton, being three nucleotides long. And longest of all is the cistron, the unit of function, which is usually the nucleotide chain determining one polypeptide chain. Since an average polypeptide chain contains about 300–500 amino acids, it follows that the average cistron must be 900–1,500 nucleotides long, and that some must be even longer.

According to a strict application of Morgan's definition, the gene would be synonymous with the recon. Since the recon and the muton are probably the same thing, the gene would then be the unit of both recombination and mutation. But such a gene would be a very small entity, difficult to analyze and lacking in the functional attributes traditionally assigned to the gene. A second alternative would be to regard the gene and the codon as synonymous. The gene would then be a slightly larger entity, but it would lack almost all the attributes of the traditional gene. At the other extreme, the definition of the gene implicit in the one gene–one enzyme hypothesis of Beadle and Tatum would make the gene synonymous with the cistron; it would then be the unit of function but not the unit of recombination or of mutation. Equating the gene with the cistron has the advantage of emphasizing physiological activity and of postulating a less complex relationship between genes and phenotypic characteristics. Probably the majority of biochemical geneticists today follow the one gene–one polypeptide principle and regard the gene as equivalent to the cistron. It is this definition, according to which genes are fairly large and complex entities, that we used in the present chapter in our discussions of the mechanism of gene action.

Although our discussion of the biochemical gene (= cistron) has focused on the function of determining a polypeptide chain, it must not be overlooked that some functional units of DNA, instead of determining polypeptide chains, act as templates for transfer RNA or as units exerting control over the activity of other genes. The functional definition of the gene is taken to include such units.

It must be emphasized that, while most geneticists accept the biochemical or physiological definition of the gene, they often find it inapplicable in work on higher plants or animals. When they speak of the genes for vestigial wings or forked bristles in *Drosophila,* or of the genes for tall plants or wrinkled seeds in garden peas, they have no proof that the entities in question would agree with the biochemical definition of the gene, no proof that these are cistrons. For only a very small number of "genes" is anything known about the DNA of which they are composed or about the enzymes associated with their phenotypic expression. For practical purposes, most genes of higher organisms are—and will continue to be for many years—Morgan-type genes based on analysis of recombination of phenotypic traits.

REFERENCES

CARLSON, E. A., 1966. *The Gene: A Critical History*. Saunders, Philadelphia.

DRAKE, J. W., 1970. *The Molecular Basis of Mutation*. Holden-Day, San Francisco.

HARTMAN, P. E., and S. R. SUSKIND, 1969. *Gene Action*, 2nd ed. Prentice-Hall, Englewood Cliffs, N.J.

SAGER, R., and F. J. RYAN, 1961. *Cell Heredity*. Wiley, New York.

STAHL, F. W., 1969. *The Mechanics of Inheritance*, 2nd ed. Prentice-Hall, Englewood Cliffs, N.J.

STENT, G. S., 1971. *Molecular Genetics*. Freeman, San Francisco.

SUGGESTED READING

BEADLE, G. W., 1948. "The Genes of Men and Molds," *Scientific American*, September. (Offprint 1.)

CLARK, B. F. C., and K. A. MARCKER, 1968. "How Proteins Start," *Scientific American*, January. (Offprint 1092.)

CRICK, F. H. C., 1954. "The Structure of the Hereditary Material," *Scientific American*, October. (Offprint 5.)

————, 1962. "The Genetic Code," *Scientific American*, October. (Offprint 123.)

EDGAR, R. S., and R. H. EPSTEIN, 1965 "The Genetics of a Bacterial Virus," *Scientific American*, February. (Offprint 1004.)

FRAENKEL-CONRAT, H., 1964. "The Genetic Code of a Virus," *Scientific American*, October. (Offprint 193.)

GAREN, A., 1968. "Sense and Nonsense in the Genetic Code," *Science*, vol. 160, pp. 149–159.

HANAWALT, P. C., and R. H. HAYNES, 1967. "The Repair of DNA," *Scientific American*, February. (Offprint 1061.)

HOLLEY, R. W., 1966. "The Nucleotide Sequence of a Nucleic Acid," *Scientific American*, February. (Offprint 1033.)

NIRENBERG, M. W., 1963. "The Genetic Code: II," *Scientific American*, March. (Offprint 153.)

NOMURA, M., 1969. "Ribosomes," *Scientific American*, October. (Offprint 1157.)

RICH, A., 1963. "Polyribosomes," *Scientific American*, December. (Offprint 171.)

SPIEGELMAN, S., 1964. "Hybrid Nucleic Acids," *Scientific American*, May. (Offprint 183.)

WATSON, J. D., 1968. *The Double Helix*. Atheneum, New York.

————, 1970. *Molecular Biology of the Gene*, 2nd ed. Benjamin, New York.

YANOFSKY, C., 1967. "Gene Structure and Protein Structure," *Scientific American*, May. (Offprint 1074.)

CHAPTER **16**

Development

Probably no aspect of biology is more amazing to both biologists and nonbiologists than the development of a complete new organism from one cell, a development so precisely controlled that the entire intricate organization of cells, tissues, organs, and organ systems characterizing the functioning adult comes into being with rarely a flaw. We have seen that genetic information controls development and ensures that a mouse zygote develops into a mouse, an oak zygote into an oak, and an earthworm zygote into an earthworm. Let us now briefly examine a few representative patterns in the development of plants and animals and then try to relate them to possible control mechanisms.

DEVELOPMENT OF AN ANGIOSPERM PLANT

As a first representative pattern of development, let us take the principal events in the development of an angiosperm—a flowering plant. We shall make no attempt to discuss this development in great detail or even to mention all the important events; our purpose is simply to give you some familiarity with the kinds of events that any model of developmental control must seek to explain.

The seed and its germination

The egg cell of an angiosperm plant is retained within the ovary of the maternal plant and is fertilized there by a sperm nucleus from a pollen grain. After fertilization, the zygote undergoes a series of mitotic divisions and develops into a tiny *embryo.* This embryo, together with a food-storage tissue called the *endosperm,* becomes enclosed in a tough protective *seed coat.* The resulting composite structure, made up of embryo, endosperm, and seed coat, is called a *seed.* The embryo in some seeds, such as peas and beans, absorbs all the endosperm before the seed

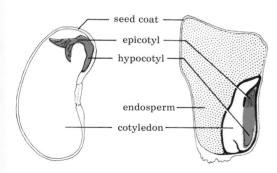

16.1. Diagram of dicot and monocot seeds. The embryo is shown in color. Left: A dicot seed (bean). Right: A monocot seed (corn).

seed coat
epicotyl
hypocotyl
endosperm
cotyledon

is released from the parent plant. In other species, such as corn, the embryo does not absorb significant quantities of the endosperm until the seed begins to germinate. The early division stages of the embryo usually do not last long, and by the time the ripe seed is released from the parent plant it is quite dry and its embryo has usually become dormant. The seed may last for months or years in this dormant state.

Germination of a seed begins with the imbibition of much water, which greatly increases the volume of the seed (sometimes as much as 200 percent). The resulting hydration of the protoplasm increases enzymatic activity, and the metabolic rate of the embryo shows a marked rise. This higher metabolic rate makes possible resumption of active cell division, synthesis of new protoplasm, and increase in cell size by uptake of water. The growing embryo soon bursts out of the seed coat and rapidly assumes a characteristic plant form, with distinguishable shoot and root.

The embryo consists of an elongate axis, to which are attached one or two *cotyledons* (commonly called seed leaves) (Fig. 16.1). In most plants, the principal function of the cotyledons is the absorption of food from the endosperm; the cotyledons thus become the immediate source of stored food on which the early growth of the embryo depends. The portion of the embryonic axis above the point at which the cotyledons are attached is called the *epicotyl;* it consists primarily of a pair of miniature leaves and a tiny bud. The epicotyl is the primordium, or rudiment, from which most of the shoot develops. The portion of the embryonic axis below the point of attachment of the cotyledons is called the *hypocotyl;* its lower end will form the primary root of the plant.

The first part of the embryo to emerge from the seed is the hypocotyl, which promptly turns downward (no matter what the orientation of the seed may be) and develops root hairs; it also soon gives rise to secondary roots. By the time the epicotyl begins its rapid development, the hypocotyl has already formed a young root system capable of anchoring the plant to the substrate and of absorbing water and minerals. In some dicots (which have, as the term "dicot" indicates, two cotyledons), the upper portion of the hypocotyl elongates and forms an arch, which pushes upward through the soil and emerges into the air (Fig. 16.2). Once the hypocotyl arch is exposed to light a phototropic response causes it to straighten, the cotyledons and the epicotyl being thus pulled out of the soil. The epicotyl then begins to elongate. In such plants, of which the garden bean is an example, the shoot of the mature plant is mostly of epicotyl origin, but a short region (usually less than an inch) at the base of the stem is derived from the hypocotyl. Other dicots, of which the garden pea is an example, show a slightly different pattern of germination (Fig. 16.3). In these plants, no hypocotyl arch forms and the cotyledons are never raised above ground. Instead, the epicotyl begins to elongate soon after the young root system has begun to form; it always grows upward and soon emerges from the soil. In such plants, the entire shoot is of epicotyl origin. A similar pattern is seen in monocots like corn (which have only one cotyledon but a large endosperm) (Fig. 16.1).

Growth and differentiation of the plant body

Growth in length. The seedling plant grows in length rather slowly at first, then enters a longer period of much more rapid growth,

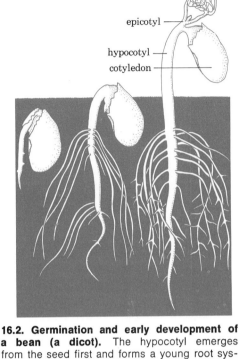

16.2. Germination and early development of a bean (a dicot). The hypocotyl emerges from the seed first and forms a young root system (left). As the upper portion of the hypocotyl elongates, it forms an arch that pushes out of the soil into the air (middle). It then straightens, pulling the cotyledons out of the ground as the epicotyl begins its development (right).

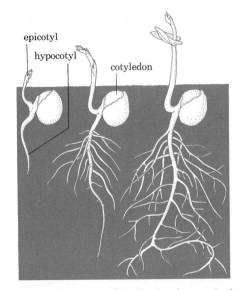

16.3. Germination and early development of a pea (a dicot). The development of a pea differs from that of a bean in that no hypocotyl arch is formed and the cotyledons remain beneath the soil.

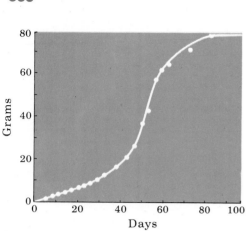

16.4. A typical S-shaped growth curve. This particular graph shows increase in weight of a young corn plant. [Redrawn from D'A. W. Thompson, *On Growth and Form,* Cambridge University Press, 1942 (from G. Backman, after Stefanowska).]

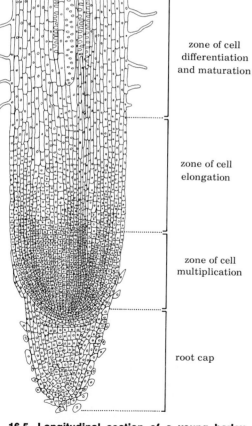

16.5. Longitudinal section of a young barley root. Each zone is only a few millimeters long. [Modified from H. J. Fuller and O. Tippo, *College Botany,* Holt, Rinehart and Winston, Inc., 1954.]

and finally slows down again or even stops as it approaches maturity. If the height (or weight) of the plant is plotted against age, the resulting growth curve is roughly S-shaped (Fig. 16.4); this same general shape also characterizes the growth curves of animals. There are, however, important differences between the growth curves of different groups of organisms, e.g. between those of perennial plants, which continue to grow to some extent throughout their lives, and those of mammals, which usually cease growing after they reach maturity.

Growth in length of the shoot and root of a young plant involves two principal processes, cell multiplication and cell elongation. Both processes are ordinarily restricted to a relatively limited region near the apex of the shoot or root. Let us look first at the root.

The extreme tip of a root is covered by a conical *root cap* consisting of a mass of nondividing parenchyma cells (Fig. 16.5). As the tip moves through the soil, some of the cells on the surface of the root cap are abraded. They are replaced by new cells added to the cap by the *apical meristem,* a zone of cell multiplication located just behind the cap. This zone, which is composed of relatively small, actively dividing cells, is usually restricted to an area a millimeter long or less. Most of the new cells produced by the meristem are laid down on the side away from the root cap. These cells are left behind as the meristem lays down additional new cells in front of them and the tip continues to move through the soil. It is these new cells, derived from the apical meristem, that will form the primary tissues of the root.

As the new cells become further removed from the meristem by the deposition of additional intervening cells, the mitotic activity of most of them slows down and eventually terminates. Once the cells have ceased dividing, they begin to absorb water rapidly and to enlarge; most of this enlargement is elongation rather than increase in width. As we saw in Chapter 9, this cell elongation is under the control of hormones, particularly auxins. Most of the absorbed water enters the rapidly expanding vacuole, which occupies most of the volume in the fully elongated cell. Note that this cell growth in plants differs markedly from cell growth in animals, where most increase in size is attributable to formation of more cytoplasm rather than to vacuolation. The zone of cell elongation, usually only a few millimeters long, is just behind the zone of cell multiplication, or meristem (Fig. 16.5).

Growth of the stem is basically similar to growth of the root. New cells are produced by an apical meristem, and these cells then elongate, pushing the apex upward. The most obvious difference between growth of the stem and of the root is the lateral production of leaves by the growing tip of the stem. At regular intervals, an increase in the rate of cell division under a localized region of the sloping surface of the apical meristem of a stem gives rise to a series of swellings that function as leaf primordia (Fig. 16.6). The point at which each leaf primordium arises from the stem is called a *node,* and the length of stem between two successive nodes is called an *internode* (see Fig. 3.29, p. 61). Most increase in length of the stem results from elongation of the cells in the young internodes.

At the tip of the stem is a series of internodes that have not yet undergone much elongation. The tiny leaf primordia that separate these internodes curve up and over the meristem, with the older, larger ones

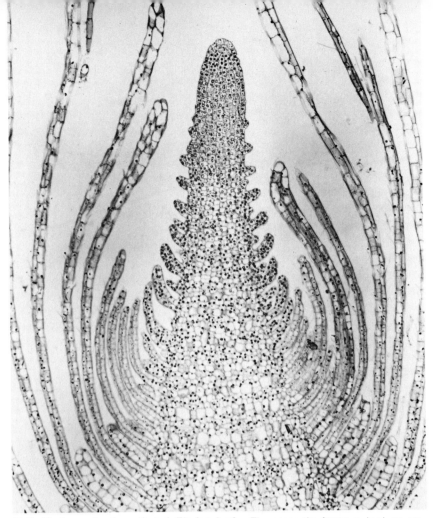

16.7. Photograph of sectioned *Elodea* bud. The stem tip forms a bud as up-curving leaf primordia enclose the apical meristem. [Courtesy Thomas Eisner, Cornell University.]

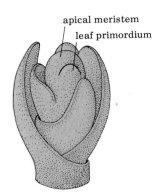

16.6. A bud. [Used by permission from J. D. Dodd, *Form and Function in Plants,* © 1962 by The Iowa State University Press.]

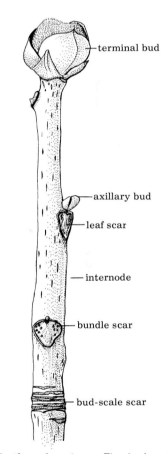

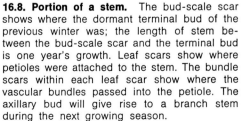

16.8. Portion of a stem. The bud-scale scar shows where the dormant terminal bud of the previous winter was; the length of stem between the bud-scale scar and the terminal bud is one year's growth. Leaf scars show where petioles were attached to the stem. The bundle scars within each leaf scar show where the vascular bundles passed into the petiole. The axillary bud will give rise to a branch stem during the next growing season.

enveloping the younger, smaller ones (Fig. 16.7). The resulting compound structure, consisting of the apical meristem and a series of unelongated internodes enclosed within the leaf primordia, is called a ***bud.*** Commonly, the bud is protected on its outer surface by overlapping scales, which are modified leaves that grow from the base of the bud (Fig. 16.6). When a dormant bud "opens" in the spring, the scales curve away from the bud and then fall off, and the internodes that were contained within the bud begin to elongate rapidly. As the nodes become farther and farther separated, mitotic activity in the leaf primordia gives rise to young leaves. Once the leaves are fully formed, the leaf primordia lose their meristematic activity. But before this happens, a small mound of meristematic tissue usually arises in the angle between the base of each leaf and the internode above it. Each of these new meristematic regions gives rise to a lateral or ***axillary bud*** with the same essential features as the terminal buds already described (Fig. 16.8). Elongation of the internodes of the lateral buds produces branch stems.

Differentiation of tissues. Cell division and cell elongation cannot alone produce all the essential features of the fully developed plant, of course. All the new cells produced by the apical meristems are fundamentally alike. Yet some of these cells will become collenchyma, some will become xylem vessels, some will become sieve cells, etc. The process whereby a cell changes from its immature form to some one mature form is called *differentiation.* In the growing root or stem, cells have begun to differentiate by the time they have finished elongating. Thus, immediately behind the zone of cell elongation in a root is a zone of cell differentiation and maturation in which the definitive tissues of the adult root take shape (Fig. 16.5).

Three concentric areas can be distinguished even in the zone of elongation of a root when it is viewed in cross section (Fig. 16.9A). These are (1) an outer layer called the *protoderm;* (2) a wide area of parenchymatous *ground tissue* located beneath the protoderm; and (3) an inner core of *provascular tissue* composed of particularly elongate cells. The protoderm rapidly matures into the epidermis. The ground tissue of the middle layer matures into the cortex and endodermis. And the provascular core differentiates into the primary tissues of the stele: primary xylem, primary phloem, pericycle, and vascular cambium (Fig. 16.9B–C). Differentiation in the growing stem follows a similar pattern except that there are usually two areas of ground tissue—one between the protoderm and the provascular cylinder, which gives rise

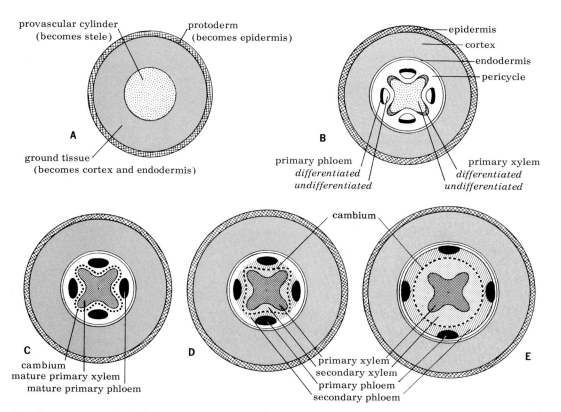

16.9. Differentiation in a young root. (A) Cross section in the zone of elongation. Three distinct concentric areas can already be detected. (B) At a slightly later stage of development, the protoderm has differentiated into epidermis; the ground tissue has differentiated into cortex and endodermis; and the provascular cylinder has begun to differentiate into primary xylem and primary phloem. (C) Differentiation in the provascular cylinder is complete, and the cambium is about to become active. (D) Divisions in the cambium have given rise to secondary xylem and secondary phloem, which are located between the primary xylem and primary phloem. (E) The areas of secondary tissue continue to thicken as more new cells are produced by the cambium.

to the cortex and endodermis, and a second inside the provascular cylinder, which becomes the pith.

As we saw in Chapter 7, increase in circumference of the root or stem depends on the formation of secondary tissues composed of cells derived from lateral meristems, particularly the vascular cambium. As cells of the vascular cambium undergo mitosis, many new cells are produced on the inner face of the cambium and these differentiate into secondary xylem, while other new cells are produced on the outer face of the cambium and differentiate into secondary phloem (Fig. 16.9D–E).

DEVELOPMENT OF A MULTICELLULAR ANIMAL

Our examination of some of the major events in the development of multicellular animals is not an attempt to describe these events fully or even to mention all the important ones. Like the discussion of plant development, it is simply meant to give you some familiarity with the kinds of events that occur, particularly in the vertebrate groups.

Embryonic development

Early-cleavage and morphogenetic stages.   In normal development, the zygote begins a rapid series of mitotic divisions immediately after fertilization has taken place. These early cleavages are not accompanied by protoplasmic growth. They produce a grapelike cluster of cells called a *morula,* which is little if any larger than the single egg cell from which it is derived (see Figs. 16.11C, 16.12C). The cytoplasm of the one large cell is simply partitioned almost equally into many new cells that are much smaller. As cleavage continues, the cells become arranged in a hollow sphere called a *blastula* (Figs. 16.11D–E, 16.12D). The cavity inside the sphere is called the *blastocoel.*

Next begins a series of complex movements that are important in establishing the definitive shape and pattern of the developing embryo. The establishment of shape and pattern in all organisms is called *morphogenesis* (meaning the genesis of form). Morphogenetic movements of cells (Fig. 16.10) in large masses always occur during the early developmental stages of animals; they are much less common in plants.

The pattern of cleavages and cell movement is greatly influenced by the amount of yolk (stored food) in the egg. We shall first examine the pattern of development in an animal whose eggs have little yolk, then the pattern of development in animals whose eggs have more yolk.

In amphioxus (Fig. 22.41, p. 525), a tiny marine chordate whose egg has very little yolk, the movements that occur after formation of the blastula convert it into a two-layered structure called a *gastrula.* The process of gastrulation begins when a small depression, or invagination, starts to form at a point on the surface of the blastula where the cells are somewhat larger than those on the opposite side (Fig. 16.11F). The differences in cell size are not very great in amphioxus embryos; they are more pronounced in many other animals. The smaller cells make up the *animal hemisphere* of the embryo. The larger cells make up the *vegetal hemisphere.* It is at the pole of the vegetal hemisphere that the invagination of gastrulation typically occurs. As gastrulation proceeds, and more and more cells move to the point of invagination

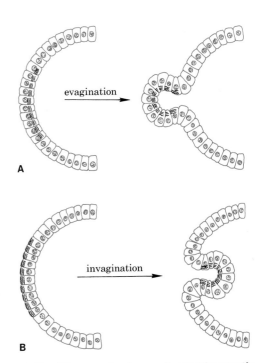

16.10. The mechanism of morphogenetic movements in cells. Contraction of microfilaments (black), asymmetrically positioned in the cells, may change the shape of the cells and produce evaginations (A), invaginations (B), or other alterations of the arrangement of cells in a developing organ. The contractile microfilaments are of the same type as those involved in amoeboid movement, mitosis, and cytokinesis.

and then fold inward, the invagination becomes larger and larger. Eventually the invaginated cell layer comes to lie almost against the outer layer, thus nearly obliterating the old blastocoel (Fig. 16.11G). The resulting gastrula is a two-layered cup, with a new cavity that opens to the outside via the **blastopore,** which is at the point where invagination first began. The new cavity, called the **archenteron,** will become the cavity of the digestive tract, and the blastopore will become the anus.

Gastrulation, as it occurs in amphioxus, first produces an embryo with two primary cell layers, an outer *ectoderm* and an inner *endoderm.* A third primary layer, the *mesoderm,* soon begins to form between the ectoderm and the endoderm. In amphioxus, the mesoderm originates as pouches pinched off the endoderm (Fig. 16.11H–J). In many other animals, it arises from inwandering cells derived primarily from the area around the blastopore where the ectoderm and endoderm meet.

In the amphioxus egg, in which the distinction between animal and vegetal hemispheres is only slight owing to the small amount of yolk in the vegetal hemisphere, the early cleavages are nearly equal (i.e. the new cells are of nearly the same size) and gastrulation can occur in a direct and uncomplicated manner. Many eggs have far more yolk in their vegetal hemisphere than the amphioxus egg, however, and this stored food imposes complications and limitations on such processes as cleavage and gastrulation. Generally, the more yolk an egg contains, the more cleavage tends to be restricted to the animal hemisphere and the more gastrulation departs from the pattern in amphioxus.

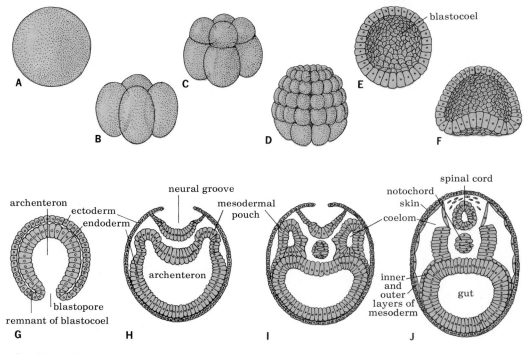

16.11. Early embryology of amphioxus. (A) Zygote. (B–D) Early cleavage stages forming a morula (C) and then a blastula (D). (E) Longitudinal section through a blastula, showing the blastocoel. (F–G) Longitudinal sections through an early and a late gastrula. Notice that the invagination is at the vegetal pole of the embryo, where the cells are largest. (H–I) Cross sections through an early and a late neurula. Invagination of the dorsal ectoderm is giving rise to the spinal cord, and pouches off the endoderm are giving rise to the mesoderm. (J) A later embryo in which both the spinal cord and the mesoderm are taking their definitive form. Notice that there is a cavity (the coelom) in the mesoderm.

Frog eggs, which contain far more yolk than those of amphioxus but much less than those of most birds, may serve as examples of eggs with an intermediate yolk mass (Fig. 16.12). Gastrulation begins early in the second day of development. Simple invagination at the vegetal pole is not mechanically feasible because of the large mass of inert yolk. Instead, portions of the cell layer of the animal hemisphere move down around the yolk mass and then fold in at the edge of the yolk (Fig. 16.12E). This infolding slowly spreads to all sides of the yolk and eventually encloses this material almost completely within the cavity of the archenteron.

Birds' eggs contain so much yolk that the small disc of cytoplasm on its surface is dwarfed by comparison (Fig. 16.13). No cleavage of the massive yolk is possible, and all cell division is restricted to the small cytoplasmic disc. (Note that the yolk and the small lighter-colored cytoplasmic disc on its surface constitute the true egg cell; the white of the egg is outside the cell.)

The fates of cells in different parts of the three primary layers of vertebrates have been determined by staining them with dyes of different colors and then following their movements. As you might expect, the ectoderm eventually gives rise to the outermost layer of the body—the epidermal portion of the skin—and to structures derived from the epidermis, such as hair, nails, the eye lens, many glands, and the epithelium of the nasal cavity, mouth, and anal canal. As you might also expect, the endoderm gives rise to the innermost layer of the body—the epithelial lining of the digestive tract and of other structures derived

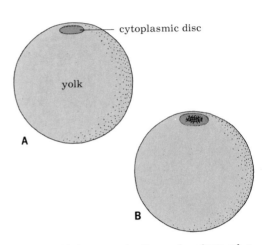

16.13. Initial stages in the embryology of a chick. (A) The zygote. A small cytoplasmic disc lies on the surface of a massive yolk. (B) Early cleavage. There is no cleavage of the yolk.

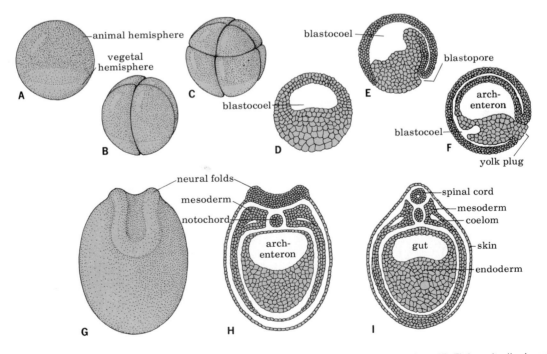

16.12. Early embryology of a frog. The large amount of yolk in the frog egg causes its pattern of gastrulation to differ from that in amphioxus. (A) Zygote. (B–C) Early cleavage stages. Note that the first horizontal cleavage is nearer the animal pole, much larger cells thus being produced at the vegetal pole. (D) Longitudinal section of a blastula. (E–F) Longitudinal sections of two gastrula stages. (G) An early neurula, showing the neural folds and neural groove. (H) Cross section of a neurula after formation of mesoderm. (I) Cross section of a later embryo, showing definitive spinal cord.

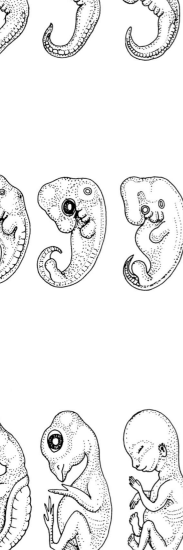

16.14. A comparison of vertebrate embryos at three stages of development. [Redrawn from G. J. Romanes, *Darwin and After Darwin*, Open Court Publishing Co., 1901.]

TORTOISE   CHICK   MAN

from the digestive tract. The mesoderm gives rise to most of the tissues in between, such as muscle and connective tissue, including blood and bone.

One major tissue located topographically between the skin and the gut does not develop from the mesoderm. This is the nervous tissue, which, curiously enough, is derived from the ectoderm. Soon after gastrulation, the ectoderm becomes divided into two components, the epidermis and the neural tube. A sheet of ectodermal cells lying along the midline of the embryo above the newly formed digestive tract bends inward and forms a long groove extending most of the length of the embryo (Figs. 16.11H–I, 16.12G–H). The dorsal folds that border this groove then move toward each other and fuse together, converting the groove into a long tube lying beneath the surface of the back. This neural tube becomes detached from the epidermis above it, and in time differentiates into the spinal cord and brain (Figs. 16.11J, 16.12I).

We see, then, that the morphogenetic movements of gastrulation and neurulation give shape and form to the embryo, and bring masses of cells into the proper position for their later differentiation into the principal tissues of the adult body. In effect, the movements mold the embryonic mass into the structural configuration upon which differentiation will superimpose the finer detail of the finished organism.

Later embryonic development.   Much must happen to convert a gastrula into a fully developed young animal ready for birth. The individual tissues and organs must be formed; in a vertebrate, the four limbs must develop; the elaborate system of nervous control must be established; and so forth. The complexity and precision characterizing these developmental changes are staggering to contemplate. For example, approximately 43 muscles, 29 bones, and many hundreds of nervous pathways must form in each human arm. If the arm is to function properly, all these components must be precisely correlated. Incredibly sensitive mechanisms of developmental control must operate if such an intricate structure can arise from a mass of initially undifferentiated cells. Yet the developmental processes that produce all these later embryonic changes are the same ones we have seen at work in the early embryo—cell division, cell growth, cell differentiation, and morphogenetic movements.

It is beyond the scope of this book to discuss in detail the many events that occur during later embryonic development. Yet it is these events that mold morphologically similar gastrulas into a fish in one instance, a rabbit in another, and a human being in still another, depending on the genetic endowment of the gastrula in question. The developmental events are programed differently for each species, and an understanding of how such different programs arise and how they are carried out is one of the important goals of developmental biologists.

The relationship between ontogeny and phylogeny.   One interesting aspect of the differences in the developmental programs of different species should be mentioned here. It has long been recognized that the early embryos of most vertebrates closely resemble one another. For example, the early human embryo has a well-developed tail and also a series of gill pouches in the pharyngeal region (Fig. 16.14). Consequently it looks very much like an early fish embryo. And it looks even more like an early tortoise embryo. Soon after Darwin formulated his theory of evolution in 1858, it was realized that such similarities in development reflect an ancestral relationship. Careful study also revealed that, in the course of its development, the embryo of a species seems to pass through a succession of stages that resemble the stages through which the evolution of that species passed. This apparent parallelism led E. H. Haeckel to formulate his "principle of recapitulation"—that ontogeny repeats phylogeny (ontogeny is the course of an individual's development, phylogeny its evolutionary history).

The modern view is that Haeckel's idea was an oversimplification. Ontogeny does not repeat phylogeny in any strict or literal sense. An individual's developmental stages do not correspond to its successive adult ancestors. What does happen is that a mammalian embryo passes through some of the same stages as an early fish embryo, and it likewise passes through stages similar to some an amphibian or reptile embryo passes through. Ontogeny does not repeat the adult stages of phylogeny, but it does repeat, in an altered form, some of the ontogeny of ancestral forms.

Postembryonic development

Growth. Though postembryonic development seldom involves any major morphogenetic movements, there is some cell multiplication and cell differentiation. But the preponderant factor by far in many animals is growth in size. As indicated earlier, growth usually begins slowly, becomes more rapid for a time, and then slows down again or stops. This pattern yields the characteristic S-shaped growth curve shown in Fig. 16.4.

Growth does not occur at the same rate and at the same time in all parts of the body. It is obvious to anyone that the differences between a baby chick and an adult hen or rooster, or between a newborn baby and an adult human being, are differences not only in overall size but also in body proportions. The head of a young child is far larger in relation to the rest of his body than that of the adult. And the child's legs are much shorter in relation to his trunk than those of the adult. If the child's body were simply to grow as large as an adult's while maintaining the same proportions, the result would be a most unadultlike individual (Fig. 16.15). Normal adult proportions arise because the various parts of the body grow at quite different rates or stop growing at different times.

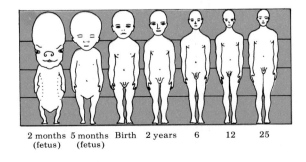

2 months 5 months Birth 2 years 6 12 25
(fetus) (fetus)

16.15. Changes in body proportions during human fetal and postnatal growth. The head grows proportionately much more slowly than the limbs. [Modified from *Morris' Human Anatomy*, ed. by C. M. Jackson, Blakiston, 1925.]

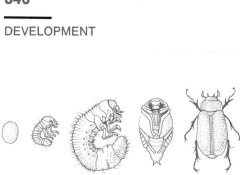

16.16. Complete metamorphosis and gradual metamorphosis. Above: The stages shown here in the complete metamorphosis of the beetle *Phyllophaga* are (from left to right): egg, young larva, full-grown larva, pupa, adult. Several intermediate larval stages are omitted. Note that as the larva grows it does not become more like the adult. The shift from larval to adult ch•racteristics occurs during the pupal stage. Below: In gradual metamorphosis, the insect (here, a grasshopper) that emerges from the egg (top) goes through several nymphal stages that bring it gradually closer to the adult form (bottom). [Above: Redrawn from H. H. Ross, *A Textbook of Entomology,* Wiley, 1956.]

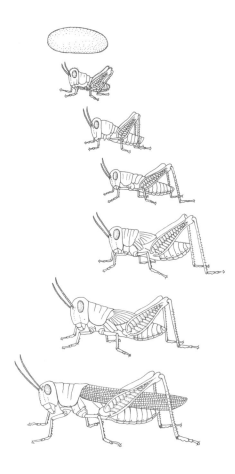

Larval development and metamorphosis. Growth in size, rather than cell division or cell differentiation, is not always the principal mechanism of postembryonic development. Many aquatic animals, particularly those leading sessile lives as adults, go through *larval* stages that bear little resemblance to the adult. The series of developmental changes (not just increase in size) that convert an immature animal into the adult form is called *metamorphosis.* The rather drastic metamorphosis of a larva into an adult often involves extensive cell division and differentiation, and sometimes even morphogenetic movement; growth alone could not accomplish so complete a change in form.

A larval stage occurs in the life history of many aquatic animals, e.g. frogs, in which the larval stage is the tadpole. But the most familiar larvae are probably those of certain groups of terrestrial insects, including flies, beetles, wasps, butterflies, and moths. The young fly, wasp, or beetle is a grub that bears no resemblance to the adult. The young butterfly or moth is a caterpillar. In the course of their larval lives, these insects molt several times and grow much larger, but this growth does not bring them any closer to adult appearance; they simply become larger larvae (Fig. 16.16, above). Finally, after completing their larval development, they enter an inactive stage called the *pupa,* during which they are usually enclosed in a case or cocoon. During the pupal stage, most of the old larval tissues are destroyed, and new tissues and organs develop from small discs of cells that were present in the larva but never underwent much development. The adult that emerges from the pupa is thus radically different from the larva; it is almost a new organism built from the raw materials of the larval body (see Pl. II-8).

Insects that have a pupal stage and undergo the type of development described above are said to have *complete metamorphosis.* Others, such as grasshoppers, cockroaches, bugs, and lice, have *gradual metamorphosis* (Fig. 16.16, below). The young of such insects, called *nymphs,* resemble the adults except for their body proportions (the wings and reproductive organs, especially, are poorly developed). They go through a series of molts during which their form gradually changes and becomes more and more like that of the adult, largely as a result of differential growth of the various body parts. They have no pupal stage, and there is no wholesale destruction of the immature tissues.

THE PROBLEM OF DIFFERENTIATION

Since mitosis gives each new daughter cell a complete set of chromosomes exactly like that in the parental cell, and since nuclear genes are located in precise sequences on the chromosomes, it follows that all the somatic cells in a multicellular organism are genotypically identical. There is no evidence that any genes are normally lost or gained in the course of a cell's development; there is, indeed, every reason to think that the genetic endowment of a nerve or muscle cell is exactly the same as that of a liver or bone cell in the same individual. If, then, all the cells in the embryo of a multicellular plant or animal have the same genetic potentials, the factors that for any given cell determine which of those potentials will be realized, and which will not, are presumably separate from the genes. One of the principal tasks of developmental biology is to identify those factors.

The effects of qualitatively unequal cleavages

If embryonic cells cannot be distinguished on the basis of genetic content, it is logical to look for differences in their cytoplasm. We have already seen that the cytoplasm of the unfertilized egg is often not homogeneous. Most animal eggs contain stored food material, or yolk, which, being usually concentrated in one part of the cell, establishes a distinction between animal and vegetal hemispheres. Other materials are similarly restricted to certain regions of the cytoplasm. Hence the daughter cells produced by cleavage of the egg cell may very well not share equally in all the cytoplasmic materials. It is reasonable to think that this difference in cytoplasmic content influences the future development of the cells.

Suppose that in a hypothetical zygote all the molecules of an important compound are distributed in a band in one part of the cell (Fig. 16.17). The orientation of the band relative to the direction of the first cleavage is obviously very important in such a cell. If the cleavage cuts across the band, each of the daughter cells will receive approximately equal amounts of the important compound. If, however, the first cleavage runs parallel to the band, one daughter cell will receive all of the compound and the other will receive none. If cells containing the compound tend to develop in one way and cells lacking the compound tend to develop in another way, a cleavage that gives all of the compound to one daughter cell determines, to some extent, the future course of development of the daughter cells.

There is good experimental evidence that the first cleavage is determinate in some animals and indeterminate in others. In annelid worms and molluscs, for example, if the two cells produced by the first cleavage are separated, each will develop into only half an embryo; the first cleavage has separated important cytoplasmic regions and thus restricted the developmental potential of the cells. The first cleavage is not determinate in most vertebrates and echinoderms (sea stars, sea urchins, etc.). As early as 1891, Hans Driesch, working in Naples, Italy, showed that if he separated the cells of a two-celled embryo of a sea urchin by shaking the embryo in a vial of sea water, each cell developed normally into a whole sea-urchin larva.

In animals in which the first few cleavages are not determinate, a determinate cleavage eventually occurs. It is a valid generalization, therefore, to say that one of the factors that first affect the developmental direction of embryonic cells is their cytoplasmic content, as determined by the extent of regional cytoplasmic differences in the zygote and the pattern of the early cleavages.

The effects of environmental influences

The cytoplasm is the immediate environment of the nucleus, and we have seen that differences in the cytoplasm can exert a profound influence on differentiation. It seems reasonable to suppose that a comparable influence is exerted by the environment external to the cytoplasm. We would expect that different cells, depending on their

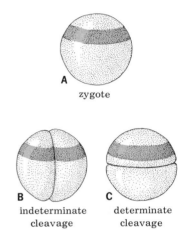

16.17. Indeterminate and determinate cleavage. (A) In this hypothetical zygote, a particularly important material is distributed in a band near one end of the cell. (B) A cleavage that cuts across this band, giving equal amounts of the important material to each daughter cell, might well be indeterminate; i.e. each cell might retain full developmental potential. (C) A cleavage running parallel to the band, giving all the material to one daughter cell, might well be determinate; the nuclei of the two daughter cells would have different cytoplasmic environments and thus different developmental potentials.

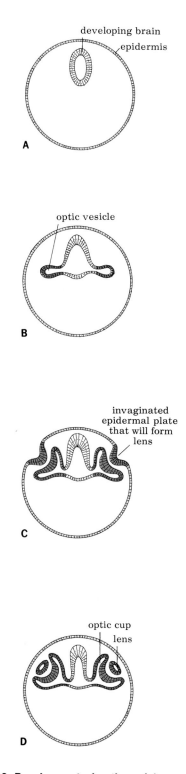

16.18. Development of optic vesicles and their induction of lenses in a frog. See text for discussion.

location in the embryo, would be exposed to a somewhat different combination of environmental factors, and that these factors might help determine the developmental direction followed by the cells.

Effects of the physical environment. A number of physical factors may affect the activity of a developing cell. Among these are temperature, light, humidity, gravity, and pressure. We have previously mentioned the influence of temperature on the development of pigmentation in Himalayan rabbits and on the vestigial-wing character in *Drosophila* (see p. 303). Numerous other examples of the influence of physical factors on development could be cited. If a seedling plant is grown exclusively in the dark, no mature chloroplasts form, the entire shoot remains colorless, and the plant eventually dies; light is an essential factor in inducing the development of chloroplasts and the synthesis of chlorophyll. And we saw in our discussion of plant hormones that light, through its effect on auxin and phytochrome, plays a critical indirect role in such developmental phenomena as increase in shoot length and initiation of flowering.

Effects of neighboring cells. One long-recognized effect of cells on one another is contact inhibition. When cultured tissue cells come into contact, both their motility and their mitotic activity are reduced, and they adhere to one another to produce a multicellular organization. Such mutual inhibition plays an important role in normal development. Significantly, one of the changes that characterize cancer cells is loss of contact inhibition, with resulting uncontrolled proliferation.

Much of the influence of one cell on another in the course of development is exerted by means of diffusible chemicals such as auxin. Auxin from a terminal bud influences the development of lateral buds and the vascular cambium. If an intact bud is grafted onto a mass of cultured cells that, taken from a stem or root, have reverted to a relatively undifferentiated state, auxin from the bud induces some of the cells to differentiate into xylem and others into phloem. Often these newly differentiated vascular tissues are arranged in a circle resembling the normal vascular cylinder.

It can be shown that the inductive action of auxin varies as a result of differences in other factors influencing the cells. If pith cells are cultured in a thin layer on a medium containing two parts per million of auxin, the cells do not divide but grow to an unusually large size. If, however, the pith tissue is molded into a cylinder and the same concentration of auxin is then introduced into one end of the cylinder, the cells undergo many divisions and some of them differentiate into xylem. The difference in response to the same inducing stimulus must be attributed to the different locations—superficial or internal—of the cells.

Some of the first definitive studies on embryonic induction in an animal were performed in 1905 by Warren H. Lewis of Johns Hopkins University. Lewis worked on the development of the eye lens in frogs. In normal development, the eyes form as lateral outpockets from the brain. When one of these outpockets, or optic vesicles as they are called, comes into contact with the epidermis on the side of the head, the contacted epidermal cells promptly undergo a series of changes and form a thick plate of cells that sinks inward, becomes detached from the epidermis, and eventually differentiates into the eye lens (Fig. 16.18). Lewis cut the connection between one of the optic vesicles and the brain before the vesicle came into contact with the epidermis. He then moved the vesicle posteriorly into the trunk region of the embryo. Despite its lack of connections to the brain, the vesicle continued to develop, and when it came into contact with the epidermis of the trunk, that epidermis differentiated into a lens. The epidermis on the head that would normally have formed a lens failed to do so. Clearly, the differentiation of epidermal tissue into lens tissue depends on some inductive stimulus from the underlying optic vesicle. Other experiments have shown that the regulation is not all one-way: The lens, once it begins to form, also influences the further development of the vesicle. If epidermis from a species with normally small eyes is transplanted to the sides of the head of a species with large eyes, the eyes that are formed do not have a large cup (formed from the vesicle) and a small lens, as might be expected. Instead, both the cup and the lens are intermediate in size and correctly proportioned to each other. Obviously, each influences the other as they develop together. Feedback is just as important in the control of embryonic development as it is in nervous or hormonal control of the fully-formed organism.

The gradualness of differentiation

Which of its potentialities a given cell will express depends, as we have seen, on a number of factors, believed to influence the activity of the genes: the intracellular environment of the nucleus, first affected by the early cleavages; the extracellular environment, which, according to the location of the cell, may subject it to different degrees of illumination, heat, pressure, pH; chemicals from other cells, which may have either inducing or inhibiting effects on its development.

As cells differentiate, their developmental potential increasingly narrows. Suppose a cell has differentiated as ecotoderm. Then it cannot ordinarily go back and form a mesodermal or endodermal structure. It may either sink inward as the neural groove forms and differentiate into nervous tissue, or it may remain on the surface and differentiate as epidermis. If it does the first, it abandons its potential as an epidermal cell but may form part of the brain or the spinal cord. Once committed as a spinal-cord cell, it can no longer become a brain cell, but it may become part of the somatic or of the autonomic nervous system. In short, differentiation is a matter of progressive determination, a gradual restricting of development to one of the many initially possible pathways (Fig. 16.19).

Regeneration of lost body parts in animals

Almost all embryonic animals have extensive regenerative capacities. Some animals retain these capacities after they reach maturity, while others lose them. An adult sea star or hydra can be chopped into many pieces and each piece can regenerate all necessary parts to become a whole individual. Half a planarian can regenerate the other half. Salamanders and lizards can regenerate new tails. But adult birds and mammals cannot regenerate whole new organs; regeneration in these animals is mostly limited to the healing of wounds.

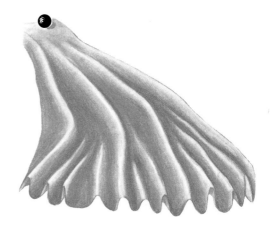

16.19. Differentiation as progressive restriction in developmental potential. C. H. Waddington has likened a differentiating cell to a ball rolling down a series of successively forking valleys. Every time the ball passes a fork, the range of points it may reach at the bottom decreases. In the same way, the range of specialized cells into which an early embryonic cell may develop narrows with increasing differentiation.

16.20. Regeneration of a salamander arm. [Adapted from M. Singer, "The Regeneration of Body Parts," *Sci. Am.,* October, 1958. Copyright © 1958 by Scientific American, Inc. All rights reserved.]

It was shown in the early 1950's that regeneration, at least in vertebrates, normally depends on the nerve supply in the regenerative region. For example, if a leg of a salamander is amputated, one of the first things that happen during the healing of the wound is penetration of the wound tissue by nerve fibers growing outward from the spinal cord. Meanwhile the scar tissue that first formed begins to disappear, and a mound of undifferentiated cells, or blastema, develops. The process of dedifferentiation that produced the blastema is succeeded by rapid cell division and redifferentiation, until the mound of cells comes to look more and more like the normal limb bud of an embryonic salamander (Fig. 16.20). This rudimentary limb continues to grow, and its cells continue to differentiate into muscle, tendon, bone, connective tissue, etc., until finally it has become a fully functional new leg. But if the nerves leading to the stump of an amputated leg are removed, no regeneration occurs, and the stump itself shrivels and disappears. Regeneration of the leg will take place only if an abundant supply of nerve fibers is present. Marcus Singer of Case Western Reserve University has shown that some animals (e.g. adult frogs) that cannot naturally regenerate lost limbs can be induced to do so if the nerve supply to the amputation stump is augmented surgically. Adequate innervation is also necessary for normal maintenance; if the nerve supply to a limb is partly destroyed (as often happens in human spinal injuries), the muscles of that limb usually begin to atrophy.

If regenerative capacity depends on an adequate nerve supply, by what means might nerve fibers bring about regeneration? It has been suggested by Singer and by other investigators that neurons secrete some inducing chemical, but repeated attempts to induce regeneration by application of nerve extract have failed. R. O. Becker of the Upstate Medical Center in Syracuse, New York, suggested another possibility— that, although chemicals secreted by the nerves might be important to regeneration after the process has been initiated, regeneration is actually triggered by electrical stimulation from the nerves. Exploring this possibility, Becker found that a very weak nonuniform current induces the same kind of mitotic activity and redifferentiation of cells at the site of injury that occurs in spontaneous regeneration. In 1972, using electrical stimulation, he was able to induce partial limb regeneration in adult rats—an indication that mammalian cells have not lost the dedifferentiation-redifferentiation potential that makes regeneration possible in lower vertebrates. Someday, perhaps, artificially induced regenerative healing will be an alternative to prosthetic devices for man.

THE REGULATION OF GENE ACTION

In our discussion so far, we have put much emphasis on the role of the extranuclear environment in regulating development. But we have also said repeatedly in this book that the genes control cellular activities and through them the developmental process. Let us now examine some possible mechanisms by which the environment might interact with the genes to regulate cellular activities.

Regulation of enzymes versus regulation of genes. There is evidence that some cells contain an inactive form of messenger RNA and that this mRNA may be triggered into protein-synthesizing activity at a later developmental stage. Here, then, is one way in which both the genes and the environment may contribute to the regulation of cellular activities. The genes synthesize the mRNA, and the environment determines when enzymes will be synthesized on the mRNA.

Alternatively, the environment may regulate the activity of the enzymes once they are synthesized. We have seen repeatedly that changes in temperature, pH, and the general chemical environment influence the rates at which enzymes catalyze reactions. A more precise type of regulation of enzyme activity involves switching the three-dimensional configuration of the molecule from one stable form to another, as in the light-induced conversion of phytochrome from the R to the F state, or the reverse. Changes in enzyme activity are sometimes brought about by specific control chemicals. Thus adrenalin stimulates glycolysis in muscle cells by triggering increased production of cyclic AMP in the cells, which in turn causes activation of a particular enzyme, which activates a second enzyme, which activates a third enzyme, which catalyzes the first reaction in the breakdown of glycogen.

Besides regulating the products of the genes—messenger RNA and the enzymes synthesized on it—might environmental factors also regulate the activity of the genes themselves? If the answer is yes, a necessary aspect of cellular differentiation must be changes in the properties of the nucleus. Such changes have, in fact, been demonstrated by nuclear-transplant experiments. When a nucleus from a cell in a frog blastula is transplanted into an enucleated frog egg, the egg develops normally—an indication that nuclei in the blastula have not undergone any stable change and still retain the capacity of directing all aspects of development. But when nuclei from cells of swimming tadpoles are transplanted into egg cells, most of the embryos show arrested development at a very early stage (though a small number may develop normally). Such experiments seem to indicate that the more differentiated a cell becomes the more likely are some relatively stable changes in its nucleus that limit its potentialities.

To the question whether the nuclear changes are irreversible, most biologists would answer no, for the following reasons: (1) A small percentage of nuclei from swimming tadpoles do direct normal development when transplanted into an egg cell. (2) Cells in regions where regeneration of lost parts is taking place resume embryonic characteristics. (3) Single differentiated cells from the body of an adult plant can revert to the embryonic state and then give rise to a complete new plant when cultured under carefully controlled conditions; thus whole carrot plants can be grown from single phloem cells cut out of the root.

That the nuclear genes do not simply churn out all their possible products in a continuous and unalterable way, but that they themselves, as well as their products, are subject to regulatory influences, has been confirmed by many lines of evidence. One of these deserves special mention. The giant chromosomes in the cells of the salivary glands of larval flies (Diptera) sometimes have curiously puffed-out regions at certain points along their length (Fig. 16.21). It has been suggested that these *chromosome puffs* indicate regions on the chromosomes where genes are especially active. The locations of puffs are different on chromosomes in different tissues, and they are different in the same

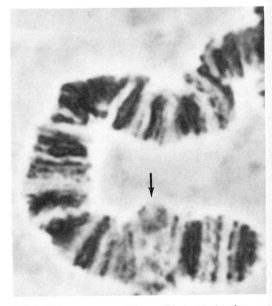

16.21. Chromosome puffing. Photograph of a giant chromosome of *Drosophila*, showing a puff (arrow). [Courtesy Edwin G. Vann, Northwestern University.]

tissue at different stages of development, although at any given time all the cells of any one type in any given tissue show the same pattern of puffing. As would be expected if each gene codes for a different messenger RNA, it can be shown that the mRNA made in one puff differs chemically from the mRNA made in a puff at a different position on the chromosome.

Puffs thus provide a way of determining visually whether or not changes in the extranuclear environment can alter the pattern of gene activity. As expected, they can. For example, if the hormone that causes molting in insects is injected into a fly larva, the chromosomes rapidly undergo a shift in their puffing pattern, taking on that characteristically found at the time of molting in normal untreated individuals. If treatment is stopped, the puffs characteristic of molting disappear. If treatment is begun again, they reappear. Or, to give another example, if chromosomes from one type of cell are transplanted into a different type of cell or into the same type of cell at a different developmental stage, they quickly lose the puffs characteristic of the donor cells and develop puffs characteristic of the type and stage of the recipient cells.

Qualitative versus quantitative regulation. Regulatory factors can evidently act directly on the genes, but by what process? As we have indicated, there is no evidence that the environmental factors influencing development normally add or delete genes. If the number of genes remains the same, then, a reasonable inference is that development involves changes in the activity of the various genes—that gene activity is susceptible to regulation. The question then arises whether the regulatory changes are qualitative or quantitative or both. In short, do environmental influences regulate what a gene makes or only how much it makes?

A classic experiment performed in 1932 by Hans Spemann and Oscar E. Schotté at the University of Freiburg suggested the answer to that question. In salamanders the ectodermal tissue of the mouth is induced to form teeth by the endodermal tissue with which it is in contact. In frogs the corresponding ectoderm forms horny jaws instead of true teeth. Spemann and Schotté transplanted a piece of ectoderm from the flank of a frog embryo to the mouth region of a salamander. The transplanted tissue developed into horny jaws. It had been induced by the salamander endoderm, but instead of forming salamander teeth it formed the corresponding structures of a frog, namely horny jaws. Apparently all the salamander inducers could do was to turn on the appropriate genes; they could not alter what those genes would make. Numerous other experiments have yielded similar results, and it is now generally accepted that environmental influences act by turning on or off the synthetic activity of the various genes; there is no evidence that they in any way determine what those genes will make. According to the current view, each gene can make one and only one kind of messenger RNA, as coded by the sequence of its nucleotides. Regulators of the genes determine if and when each gene will synthesize its particular mRNA.

Regulation of gene action in procaryotes and eucaryotes. In the course of the last decade, a now well-supported model of gene regulation in procaryotic cells (particularly bacteria) was elaborated by the French team of François Jacob and Jacques Monod. Studying enzyme synthesis in the bacterium *Escherichia coli,* Jacob and Monod found that there are other kinds of genes besides the ones that specify the structure of enzymes, that genes of several types form functional groups in procaryotic cells. Each functional group apparently controls the synthesis of the enzymes catalyzing a particular metabolic pathway. As shown in Fig. 16.22, each group is believed to include three types of genes: (1) *structural genes,* which synthesize the messenger RNA that codes for the enzymes subsequently synthesized on the ribosomes; (2) an *operator gene,* which functions as an on-off switch determining whether or not the structural genes will synthesize mRNA; and (3) a *regulator gene,* which synthesizes *repressor substance* that, together with molecules from the cytoplasm, determines whether the operator gene will be activated or inhibited. The operator and the structural genes it controls are apparently adjacent on the chromosome, forming a single complex, or *operon.*

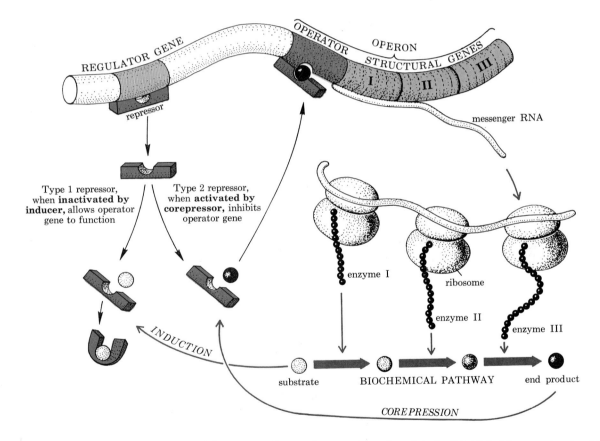

16.22. The Jacob-Monod model of gene control in procaryotic cells. An operon consists of structural genes, which synthesize mRNA, and an operator gene, which controls the activity of the structural genes something like an on-off switch. Ordinarily an entire operon is transcribed as a single molecule of mRNA. Here the mRNA carries information from three structural genes and hence directs synthesis of three different enzymes catalyzing the reactions of a particular biochemical pathway. The operator gene is itself controlled by a regulator gene, which produces one or the other of two types of repressor substance. Type 1, which the regulator synthesizes in active form, inhibits the operator unless it is itself inhibited by a molecule of inducer (often the substrate or some other molecule from the cytoplasm); when Type 1 is inhibited, the effect is induction of synthesis by the operon. Type 2, which the regulator synthesizes in inactive form, will inhibit the operator only if it is activated by a corepressor molecule (often, as here, the end product of the reactions catalyzed by the enzymes); when Type 2 is activated, the effect is inhibition of synthesis by the operon. Note that although both induction and corepression are represented here, one and the same operator gene is not ordinarily subject to both control processes.

The Jacob-Monod model has not been found to apply well to eucaryotic cells (the cells of all organisms except bacteria and blue-green algae). This is not surprising, perhaps, when one considers the differences between procaryotic and eucaryotic chromosomes. Whereas procaryotic chromosomes are naked DNA, eucaryotic chromosomes are a complex of DNA and protein; it would be strange if this protein had no effect on gene activity. In eucaryotes, the genes coding for the enzymes of a given biochemical pathway are rarely lined up in a single compact operon, as in procaryotes, and repressor substance is probably not synthesized by specific regulator genes. Repression of RNA synthesis seems to be a relatively nonspecific function of the *histones,* which are positively charged proteins unique to eucaryotic chromosomes. Histones complex readily with the negatively charged DNA, and when they do they tend to inhibit RNA synthesis.

Probably less than 10 percent of the genes in a eucaryotic cell are actively synthesizing RNA at any given moment; the rest are repressed. But the genes comprising that 10 percent may not be the same from one moment to the next. Accordingly there must be some mechanism for derepression—if histones are the repressors, some mechanism whereby the histones can selectively be uncoupled from genes, presumably through the agency of inducers. It seems likely that the coupling to DNA is loosened by a chemical alteration of the histones, which means that the inducers probably act indirectly, via enzymes.

There is obviously a long way to go before we reach an understanding of the mechanisms of induction and repression of gene activity in eucaryotic cells.

DEVELOPMENT OF IMMUNOLOGIC CAPABILITIES IN VERTEBRATES

The immune response is essentially the body's reaction against substances foreign to it, and as such it is a prime defense against disease; vertebrates have the ability to manufacture antibodies that can inactivate or destroy pathogens such as bacteria, an ability that has been utilized to create artificial immunity with vaccines and antiserums (see p. 456). But the immune response is not invariably beneficial. Occasionally it inflicts more damage on the organism than the foreign substances themselves; it is responsible for the symptoms of allergy, for example. It also accounts for the rejection of tissue transplants; the body does not distinguish between helpful and harmful substances, it seems—only between "self" and "not self."

Antibodies. In most cases, the initiation of synthesis of a particular type of antibody depends on stimulation by an *antigen,* which is frequently but not always a protein. The antigen is usually a substance foreign to the organism's own body, such as a part of a bacterium or virus. The receptor cells for the antigen are certain white blood cells called *lymphocytes.* Upon reacting with the antigen, these cells begin dividing and give rise both to more lymphocytes and to cells that rapidly differentiate into *plasma cells,* which begin secreting antibody (see p. 148).

Antibodies are highly specific for the particular antigen that stimulated their synthesis and release. They have a great affinity for the antigen and react with it, either destroying or inactivating it. In some cases, the antibody apparently combines with the antigen in such a way as to cover or mask the active sites of the antigen, much as a sheath may cover the cutting edge of a knife (Fig. 16.23); poisons released by invading bacteria are often neutralized in this manner. Other antibodies dissolve parts of the cell walls of bacteria, causing them to disintegrate. Still others cause bacteria to clump together, thus making them less active and more susceptible to being captured through the filtering action of the lymph nodes. And others seem to make the antigens more susceptible to phagocytosis by white blood cells.

According to every indication, the body can manufacture specific antibody against almost any one of the billions of different proteins that must exist, as well as against certain other compounds. How can the lymphocytes give rise to such a variety of antibodies? It has been proposed that every organism has an enormous variety of different lymphocyte types, each with the ability to recognize one specific kind of antigen. An invading antigen would stimulate the particular lymphocytes sensitive to it, and these would then give rise to plasma cells capable of producing antibody specific to the inducing antigen. The stimulated lymphocytes would also give rise to additional lymphocytes like themselves, and this multiplication would provide more potentially reactive cells to respond if antigen of the same type is encountered again. It is this proliferation of selected cells that would confer on the organism a lasting immunity to the stimulating antigen—an "immunologic memory," if you will.

Many hypotheses have been put forward to explain how a diversity of lymphocytes adequate to the great diversity of potential antigens might arise. One that has aroused much interest is that every lymphocyte may carry a million or more alternative genes for antibody, with all but one repressed. If the repression occurred randomly during development, the one gene left unrepressed would rarely be the same in different cells. Since the immune response has very great survival value in higher vertebrates, use of a sizable fraction (perhaps a tenth) of all the genetic material of the cell to code for antibody in this manner might not be as improbable as it seems at first glance.

Recognition of "self." The immune reaction responsible for the rejection of skin grafts and heart, kidney, and other transplants is apparently not mediated by the lymphocytes that give rise to antibody-producing cells, but by other lymphocytes believed to bind to the foreign tissue and secrete substances toxic to it, as well as substances that stimulate increased phagocytic activity by special white blood cells.

Tissue rejections seem to hinge on a distinction between "self" and "not self," i.e. on the body's ability to tell which substances are part of itself and which are not. Transplants between identical twins are successful presumably because such individuals develop from the same fertilized egg cell and thus have identical genes, which it is assumed control the synthesis of identical substances in the two individuals. The body of one twin therefore cannot distinguish between its own parts and parts taken from the other twin; both are identified as "self."

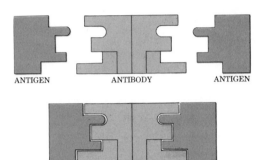

16.23. Antigen-antibody reaction. Top: There is evidence to show that an antibody molecule has two identical halves, each composed of one large component and one small component. Bottom: The two active surfaces of the antibody molecule probably fit against the active surfaces of two antigen molecules, thus masking and inactivating them.

But few people have identical twins who can come to their aid when they need a transplant. Transplantation from healthy unrelated donors or, for such organs as the heart, from the bodies of recently deceased persons or from animals has often been tried, with X-ray or chemical suppression of the tissues involved in the immune reactions. But it is very difficult, with the techniques now available, to suppress the immune reaction sufficiently to prevent rejection of the transplant without at the same time severely damaging the patient in other ways. In any such attempt, of course, the patient must be given constant massive doses of antibiotic drugs, because his own defenses against infection have been suppressed.

Basic to an understanding of immune reactions is greater knowledge of the way immunologic capabilities develop and of the way the body distinguishes between "self" and "not self." Various experiments have shown that the early embryo lacks immunologic capabilities, that these develop later, and at different times, some arising as late as several weeks after birth. If foreign substances are introduced into the embryo soon enough, before immunologic mechanisms against them develop, they are apparently considered "self"; thus a mouse injected as an early embryo with cells from another mouse will accept grafts from that mouse after it is born. The current view is that the mature organism identifies substances as "self" if they were present at the critical time during development and destroyed or suppressed the immunologic elements that started to develop against them; it identifies substances as "not self" if they were absent at the critical developmental moment and immunologic elements against them were consequently free to mature.

Occasionally the immunologic mechanisms get out of control and become sensitized to some part of the animal's own body, causing interference with or even destruction of that part. In such cases, the ability to distinguish between "self" and "not self" is impaired, and the body begins to destroy itself. Among the most intensively studied auto-immune diseases of man are rheumatoid arthritis, certain hemolytic anemias, rheumatic fever, ulcerative colitis, and glomerulonephritis. Auto-immune reactions have also been linked to some aspects of aging.

The role of the thymus.   In man, the thymus is a two-lobed, glandular-appearing structure located in the upper part of the chest just behind the sternum (breastbone) (Fig. 16.24); it often differs in size and shape in other vertebrates, but its cellular structure and location are much the same. In an embryo, or in a child under 10 or 12, the thymus is large and prominent, whereas in adults it is usually atrophied and difficult to locate.

Two principal functions have been attributed to the thymus. First, it is thought that very early in the development of the lymphoid system incipient lymphocytes move into the thymus from the bone marrow. After further development in the thymus, the lymphocytes migrate to such areas of the embryonic body as the lymph nodes and the spleen, settle down there, and give rise to the lymphoid elements that later make those areas immunologically competent. Apparently the thymus is essential to the development and maturation of immunologic capabilities, but once those capabilities have matured, they may last for a long time without further thymic intervention, declining only very

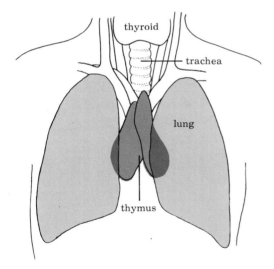

16.24. Location of the thymus in man. The thymus shown is that of a child; in an adult it would be proportionately much smaller.

slowly, unless they are destroyed by radiation or chemicals and must be re-established by the thymus. Thus removal of the thymus from newborn mice usually prevents them from developing immunologic competence, but removal when they are three weeks old has little detectable effect.

The second function attributed to the thymus is release of a hormone essential for functional maturation of the lymphocytes involved in such immune responses as rejection of tissue transplants. Evidence for the existence of such a hormone comes from experiments in which the thymus was removed from newborn mice, put into a container impermeable to whole cells, and then (while still in the container) reimplanted in the mice. Such mice developed essentially normal immunologic capabilities, an indication that some stimulus from the reimplanted thymus had enabled their lymphoid tissues to develop normally. Since the stimulus from the thymus could not involve transfer of whole cells, because of the container, it could be assumed that some diffusible chemical (a hormone) was involved.

REFERENCES

ABRAMOFF, P., and M. F. LA VIA, 1970. *Biology of the Immune Response.* McGraw-Hill, New York.

AREY, L. B., 1965. *Developmental Anatomy,* 7th ed. Saunders, Philadelphia.

BALINSKY, B. I., 1970. *An Introduction to Embryology,* 3rd ed. Saunders, Philadelphia.

BECKER, R. O., 1972. "Stimulation of Partial Limb Regeneration in Rats," *Nature,* pp. 109–111, January 14.

————, and D. G. MURRAY, 1970. "The Electrical Control System Regulating Fracture Healing in Amphibians," *Clinical Orthopaedics,* no. 73, pp. 169–198.

FLICKINGER, R. A., ed., 1966. *Developmental Biology.* Wm. C. Brown, Dubuque, Iowa. (A collection of 16 papers.)

GROSS, R. J., 1969. *Principles of Regeneration.* Academic Press, New York.

LEWIN, B. M., 1970. *The Molecular Basis of Gene Expression.* Wiley-Interscience, New York.

MARKERT, C. L., and H. URSPRUNG, 1971. *Developmental Genetics.* Prentice-Hall, Englewood Cliffs, N.J.

MAYER, A. M., and A. POLJAKOFF-MAYBER, 1963. *The Germination of Seeds.* Macmillan, New York.

NEEDHAM, A. E., 1964. *The Growth Process in Animals.* Van Nostrand, Princeton, N.J.

SISKIND, G. W., and B. BENACERRAF, 1969. "Cell Selection by Antigen in the Immune Response," in *Advances in Immunology,* vol. 10, ed. by F. J. Dixon and H. G. Kunkel. Academic Press, New York.

STENT, G. S., 1971. *Molecular Genetics.* Freeman, San Francisco. (See esp. Chapter 20.)

STEWARD, F. C., *et al.,* 1964. "Growth and Development of Cultured Plant Cells," *Science,* vol. 143, pp. 20–27.

THOMPSON, D'A. W., 1942. *On Growth and Form,* rev. ed. Macmillan, New York.

TRINKAUS, J. P., 1969. *Cells into Organs.* Prentice-Hall, Englewood Cliffs, N.J.

WARDLAW, C. W., 1968. *Morphogenesis in Plants,* 2nd ed. Methuen, London.

SUGGESTED READING

BEERMANN, W., and U. CLEVER, 1964. "Chromosome Puffs," *Scientific American,* April. (Offprint 180.)

EBERT, J. D., and I. SUSSEX, 1970. *Interacting Systems in Development,* 2nd ed. Holt, Rinehart & Winston, New York.

EDELMAN, G. M., 1970. "The Structure and Function of Antibodies," *Scientific American,* August. (Offprint 1185.)

FISCHBERG, M., and A. W. BLACKLER, 1961. "How Cells Specialize," *Scientific American,* September. (Offprint 94.)

FRIEDEN, E., 1963. "The Chemistry of Amphibian Metamorphosis," *Scientific American,* November. (Offprint 170.)

GURDON, J. B., 1968. "Transplanted Nuclei and Cell Differentiation," *Scientific American,* December. (Offprint 1128.)

HADORN, E., 1968. "Transdetermination in Cells," *Scientific American,* November. (Offprint 1127.)

HAYFLICK, L., 1968. "Human Cells and Aging," *Scientific American,* March. (Offprint 1103.)

LEVEY, R. H., 1964. "The Thymus Hormone," *Scientific American,* July. (Offprint 188.)

LOEWENSTEIN, W. R., 1970. "Intercellular Communication," *Scientific American,* May. (Offprint 1178.)

MILLER, J. F. A. P., 1964. "The Thymus and the Development of Immunologic Responsiveness," *Science,* vol. 144, pp. 1544–1551.

PTASHNE, M., and W. GILBERT, 1970. "Genetic Repressors," *Scientific American,* June. (Offprint 1179.)

REISFELD, R. A., and KAHAN, B. D., 1972. "Markers of Biological Individuality," *Scientific American,* June. (Offprint 1251.)

SINGER, M., 1958. "The Regeneration of Body Parts," *Scientific American,* October. (Offprint 105.)

STEWARD, F. C., 1963. "The Control of Growth in Plant Cells," *Scientific American,* October. (Offprint 167.)

SUSSMAN, M., 1964. *Growth and Development,* 2nd ed. Prentice-Hall, Englewood Cliffs, N.J.

WADDINGTON, C. H., 1953. "How Do Cells Differentiate?" *Scientific American,* September. (Offprint 45.)

————, 1966. *Principles of Development and Differentiation.* Macmillan, New York.

WESSELLS, N. K., and W. J. RUTTER, 1969. "Phases in Cell Differentiation," *Scientific American,* March. (Offprint 1136.)

WIGGLESWORTH, V. B., 1959. "Metamorphosis and Differentiation," *Scientific American,* February. (Offprint 63.)

WILLIAMS, C. M., 1950. "The Metamorphosis of Insects," *Scientific American,* April. (Offprint 49.)

PART IV

The biology of populations and communities

CHAPTER 17

Evolution

The topic of evolution was briefly introduced in Chapter 1 as a unifying principle for the study of many of the subjects that were to follow. That introduction was necessarily far from complete. But now these very subjects—the adaptations for the various life functions evolved by many kinds of organisms—have given us the background for a closer look at the mechanisms of evolution.

EVOLUTION AS CHANGE IN THE GENETIC MAKEUP OF POPULATIONS

Fundamental to the modern theory of evolution are two concepts —that the characteristics of living things change with time and that the change is directed by natural selection. The change we are here discussing is not change in an individual during its lifetime, although such change is a universal and important attribute of life; it is change in the characteristics of populations over the course of many generations. An individual cannot evolve, but a population can. The genetic makeup of an individual is set from the moment of conception; most of the changes during its lifetime are simply changes in the expression of the developmental potential inherent in its genes. But in populations, both the genetic makeup and the expression of the developmental potential can change. The former—change in the genetic makeup of a population in successive generations—is evolution.

Genetic variation as the raw material for evolution

A population is composed of many individuals. With rare exceptions, no two of these are exactly alike. In human beings, we are well aware of the uniqueness of the individual, for we are accustomed to recognizing different persons at sight, and we know from experience that each has distinctive anatomical and physiological characteristics, as well as distinctive abilities and behavior traits. We are also fairly well aware of individual variation in such common domesticated animals as dogs, cats, and horses. But we tend to overlook the similar individual variation in less familiar species such as robins, squirrels, earthworms, dandelions, and corn plants. Yet even though this variation may be less obvious to our unpracticed eye, it exists in all such species.

The members of a population, then, share some important features but differ from one another in numerous ways, some rather obvious, some very subtle. It follows that if there is selection against certain variants within a population and selection for other variants within it, the overall makeup of that population may change with time.

We discussed the main sources of genetic variation in Chapters 14 and 15. Ultimately, of course, all new genes arise by mutation. But once a variety of alleles is in existence, recombination becomes a mechanism that provides almost endless genotypic variation in the population. The variation resulting from meiosis and recombination at fertilization in biparental organisms has already been cited as of immense importance. Crossing-over and other chromosomal aberrations are further mechanisms of recombination. Together, mutation and recombination provide the genetic variability upon which natural selection can act to produce evolutionary change.

Exclusively phenotypic variations. Any phenotypic variation within a population may give rise to reproductive differentials between individuals, whether or not the variation reflects corresponding genetic differences. Thus variations produced by exposure to different environmental conditions during development, or produced by disease or accidents, are subject to natural selection. But even though the action of natural selection on all types of variations alters the immediate makeup of a population, it is only its action on variations reflecting genetic differences that has any long-term effect on the population. Variation that is exclusively phenotypic is not raw material for evolutionary change.

Lacking genetic data, many prominent biologists of the last century and of the early part of this century assumed that exclusively phenotypic variation could serve as evolutionary raw material; that it cannot is far from obvious to many nonbiologists even today. The theory of evolution by natural selection proposed by Darwin and Wallace had an influential rival during the nineteenth century in the concept of evolution by the inheritance of acquired characteristics—an old and widely held idea often identified with Jean Baptiste de Lamarck (1744–1829), who was one of its more prominent supporters in the early 1800's.

The Lamarckian hypothesis was that somatic characteristics acquired by an individual during its lifetime could be transmitted to its offspring. Thus the characteristics of each generation would be determined, in part at least, by all that had happened to the members of the preceding generations—by all the modifications that had occurred in them, including those caused by experience, use and disuse of body parts, and accidents. Evolutionary change would be the gradual accumulation of such acquired modifications over many generations. The classic example (though by now rather hackneyed) is the evolution of the long necks of giraffes. According to the Lamarckian view, ancestral giraffes with short necks tended to stretch their necks as much as they could to reach the tree foliage that served as a major part of their food. This frequent neck stretching caused their offspring to have slightly longer necks. Since these also stretched their necks, the next generation had still longer necks. And so, as a result of neck stretching to reach higher and higher foliage, each generation had slightly longer necks than the preceding generation. According to the modern theory of natural selection, on the other hand, ancestral giraffes probably had short necks, but the precise length of the neck must have varied from individual to individual because of their different genotypes. If the supply of food was somewhat limited, then individuals with longer necks had a better chance of surviving and leaving progeny than those with shorter necks. As a result, the proportion of individuals with genes for longer necks increased slightly with each succeeding generation.

Although the hypothesis of evolution by inheritance of acquired characteristics is rejected by most modern biologists, it was a logical and reasonable one when first proposed. It has simply not stood the test of further scientific research. In Lamarck's day (as in Darwin's), nothing was known about the mechanism of inheritance. Mendel had not yet performed his experiments on garden peas. It was not illogical, therefore, to assume that a change in any part of the body could be inherited. Now, however, it is known that somatic cells do not affect the genotype of the germ cells; immense alterations of the somatic cells can be brought about without in any way influencing the hereditary information in the gametes. The hypothesis of inheritance of acquired characteristics can therefore no longer be regarded as tenable.

The gene pool and factors that affect its equilibrium

The gene pool. To understand evolution as change in the genetic makeup of populations in successive generations, it is essential to know something about population genetics. Our study of the genetics of individuals in Chapter 14 was based on the concept of the genotype, which is the genetic constitution of an individual. Our study of the genetics of populations will be based in a similar manner on the concept of the gene pool, which is the genetic constitution of a population. The gene pool is the sum total of all the genes possessed by all the individuals in the population.

You will recall that the genotype of a diploid individual can contain a maximum of only two alleles of any given gene. But there is no such restriction on the gene pool of a population. It can contain any number of different allelic forms of a gene. The gene pool is characterized with regard to any given gene by the frequencies, or ratios, of the alleles of that gene in the population. Suppose, to use a simple example, that gene A occurs in only two allelic forms, A and a, in a particular sexually reproducing population. And suppose that allele A constitutes 90 percent of the total of both alleles while allele a constitutes 10 percent of the total. The frequencies of A and a in the gene pool of this population are therefore 0.9 and 0.1. If those frequencies were to change with time, the change would be evolution. When we say that evolution is change in the genetic makeup of populations, what we mean is that it is change in the gene frequencies (or genotype ratios) within gene pools. Therefore we can determine what factors cause evolution by determining what factors can produce a shift in gene frequencies.

Let us return to our hypothetical population. How can we calculate the frequencies of the genotypes that will be present in it? This calculation is not difficult if we assume that all possible genotypes have an equal chance of surviving. If allele A has a frequency of 0.9 in the population and allele a a frequency of 0.1, then 0.9 of the sperm cells and 0.9 of the egg cells will carry allele A, and 0.1 of the sperm cells and 0.1 of the egg cells will carry allele a. Using this information, we can set up a Punnett square much like the ones we used for crosses between two individuals:

		Sperms	
		0.9 A	0.1 a
Eggs	0.9 A	0.81 AA	0.09 Aa
	0.1 a	0.09 aA	0.01 aa

Notice that the only difference between this and a Punnett square for a cross between individuals is that here the sperms and eggs are not those produced by a single male and a single female but those produced by all the males and females in the population, with the frequency of each type of sperm and egg shown on the horizontal and vertical axes respectively. Filling in the boxes (by combining the indicated alleles and multiplying their frequencies) tells us that the frequency of the homozygous dominant genotype (AA) in the population will be 0.81, the frequency of the heterozygous genotype (Aa) will be 0.18, and the frequency of the homozygous recessive genotype (aa) will be 0.01.

We have thus fully characterized the gene pool of the present generation of our hypothetical population with regard to alleles A and a and the genotypes they form. Now we want to know whether the frequencies we have found will change in successive generations—in short, whether the population will evolve.

The Hardy-Weinberg Law and conditions for genetic equilibrium. It is easily assumed that the more frequent allele (in our hypothetical case, A) will automatically increase in frequency while the less frequent allele (a) will automatically decrease in frequency and eventually be lost from the population. But this assumption is incorrect. The rarity of a particular allele in a population does not automatically doom it to disappearance, as we can verify by using the known frequencies for one generation to compute the frequencies for the next generation. The gene frequencies in the gene pool of the present generation of our hypothetical population are 0.9 and 0.1, and the genotype ratios are 0.81, 0.18, and 0.01. If we use these figures to set up a Punnett square and compute the corresponding values for the next generation, we find that they are the same ones we started with. The gene frequencies of the next generation are also 0.9 and 0.1, and the genotype ratios are 0.81, 0.18, and 0.01. We could perform the same calculation for generation after generation, always with the same results; neither the gene frequencies nor the genotype ratios would change. We must conclude, therefore, that evolutionary change is not usually automatic, that it occurs only when something disturbs the genetic equilibrium. This was first recognized in 1908 by G. H. Hardy of Cambridge University and W. Weinberg, a German physician, working independently. According to the Hardy-Weinberg Law, *under certain conditions of stability both gene frequencies and genotype ratios remain constant from generation to generation in sexually reproducing populations.*

Let us examine the conditions of stability that the Hardy-Weinberg Law says must be met if the gene pool of a population is to be in genetic equilibrium. They are as follows:

1. The population must be large enough to make it highly unlikely that chance alone could significantly alter gene frequencies.
2. Mutations may not occur, or else there must be mutational equilibrium.
3. There may be no immigration or emigration.
4. Reproduction must be totally random.

With regard to the first condition, a population would have to be infinitely large for chance to be completely ruled out as a causal factor in the changing of gene frequencies. In reality, of course, no population is infinitely large, but many natural populations are large enough so that chance alone would not be likely to cause any appreciable alteration in the gene frequencies in their gene pools. Any breeding population with more than 10,000 members of breeding age is probably not significantly affected by random change. But gene frequencies in small isolated populations of, say, less than 100 breeding-age members are highly susceptible to random fluctuations, often called *genetic drift.*

The second condition for genetic equilibrium—no mutation or mutational equilibrium—is never met in any population. Mutations are always occurring. There is no known way of stopping them. As for mutational equilibrium, very rarely, if ever, are the mutations of alleles for the same character in exact equilibrium; i.e. reverse mutations do not completely cancel each other. There is thus a *mutation pressure* tending to cause a slow shift in the gene frequencies in the population. The more stable allele will tend to increase in frequency, the more mutable to decrease in frequency, unless some other factor offsets the mutation pressure. But even though mutation pressure is almost always present, it is probably seldom a major factor in producing changes in gene frequencies in a population. Acting alone, mutation would take an enormous amount of time to bring about any significant change (except in the origin of polyploidy, which we shall discuss later). Mutations increase variability and are thus the ultimate raw material of evolution, but they seldom determine the direction or nature of evolutionary change.

According to the third condition for genetic equilibrium, a gene pool cannot accept immigrants from other populations, for these would introduce new genes, and it cannot suffer loss of genes by emigration. Although most natural populations probably experience some gene migration, there are doubtless populations that experience none, and in many instances where migration does occur it is probably sufficiently slight to be essentially negligible as a factor causing shifts in gene frequencies.

The fourth condition for genetic equilibrium is that reproduction be totally random. Reproduction in this context does not mean simply the mating process, but the vast number of factors that contribute to the reproductive continuity of the population: selection of a mate, physical efficiency and frequency of the mating process, fertility, total number of zygotes produced at each mating, percentage of zygotes that lead to successful embryonic development and birth, survival of the young until they are of reproductive age, fertility of the young, etc. If reproduction is to be totally random, all these factors must be random; i.e. they must be independent of genotype. This condition is probably never met in any population, for the factors mentioned here are always correlated in part with genotype; indeed, there is probably no aspect of reproduction that is totally devoid of correlation with genotype. Nonrandom reproduction is the universal rule. And nonrandom reproduction, in the broad sense in which the term has been defined here, is synonymous with natural selection. Natural selection, then, is always operative in all populations.

In summary, of the four conditions necessary for genetic equilibrium (i.e. for no evolution) described by the Hardy-Weinberg Law, the first (large population size) is met reasonably often the second (no mutation) is never met, the third (no migration) is met sometimes, and the fourth (random reproduction) is never met. It follows that complete equilibrium in a gene pool is not expected. Evolutionary change is a fundamental characteristic of the life of all populations, including human populations.

The role of natural selection

Changes in individual gene frequencies caused by natural selection. Let us now return for a moment to our hypothetical population in which the initial frequencies of the alleles A and a are 0.9 and 0.1 and the genotype frequencies 0.81, 0.18, and 0.01. We have seen that although these frequencies will not change automatically with the passage of time—changing only when something disturbs the genetic equilibrium—mutation pressure (to a slight extent) and *selection pressure* (to a greater extent) are always disturbing this equilibrium. Suppose that natural selection acts against the dominant phenotype in our example, and that this negative selection pressure is strong enough to reduce the frequency of A in the present generation from 0.9 to 0.8 before reproduction occurs. (Of course there will be a corresponding increase in the frequency of a from 0.1 to 0.2, since the two frequencies must total 1.0). Now let us set up a Punnett square and calculate the genotype ratios that will be present in the zygotes of the second generation:

		Sperms	
		0.8 A	0.2 a
Eggs	0.8 A	0.64 AA	0.16 Aa
	0.2 a	0.16 aA	0.04 aa

We find that the genotype ratios of the zygotes in the second generation are different from those in the parental generation; instead of 0.81, 0.18, and 0.01, the ratios are 0.64, 0.32, and 0.04. If selection now acts against the dominant phenotype in this generation, and thereby again reduces the frequency of A, the genotype ratios in the third generation will be different from those of both preceding generations; the frequency of AA will be lower and that of aa will be higher. If this same selection pressure were to continue for many generations, the frequency of AA would fall very low and the frequency of aa would rise very high. Thus natural selection would have caused a change from a population in which 99 percent of the individuals showed the dominant phenotype and only one percent the recessive phenotype to a population in which very few showed the dominant phenotype and most showed the recessive phenotype. This evolutionary change of the phenotype most characteristic of the population would have occurred without the necessity of any new mutation, simply as a result of natural selection.

Rather than deal only with a hypothetical example, let us cite an actual situation in which selection has produced a radical shift in gene frequencies. Soon after the discovery of the antibiotic activity of penicillin, it was found that certain bacteria quickly developed resistance to the drug. Higher and higher doses of penicillin were necessary to kill the bacteria, and the resistant bacteria became a serious problem in hospitals. Clearly, under the influence of the strong selection exerted by the penicillin, the bacterial populations had evolved. Many studies have shown that the drug does not induce mutations for resistance; it simply selects against nonresistant bacteria. Apparently some genes determining metabolic pathways that confer resistance to penicillin were already present in low frequency in most populations, having arisen earlier as a result of random mutations. Individuals possessing these genes survived the antibiotic treatment, and, since it was they that reproduced and perpetuated the population (the nonresistant individuals having been killed), the next generation showed a marked resistance to penicillin. If such genes had not already been present in a population exposed to penicillin, no cells would have survived and the population would have been wiped out.

Evolution of drug resistance in bacteria is not entirely comparable to evolution in biparental organisms, because intense selection can change gene frequencies much more rapidly in haploid asexual organisms than in biparental ones. The recombination that occurs at every generation in a biparental species often re-establishes genotypes eliminated in the previous generation; this does not happen in asexual organisms. Nevertheless, even very small selection pressures can produce major shifts in gene frequencies in biparental populations when the time scale is one in which 50,000 years is a rather brief period.

The creative role of natural selection. So far, we have discussed situations in which we have posited only two clearly distinct phenotypes determined by two alleles of a single gene. But in reality the vast majority of characters on which natural selection acts are influenced by many different genes, most of which have multiple alleles in the population.

As environmental conditions shift over the course of many years, there is usually a corresponding evolutionary shift in the relative frequencies of complex phenotypes in the population. For example, if conditions become progressively wetter, we might find that in a certain plant species phenotypes specially suited for survival in wet habitats increase in frequency whereas phenotypes more suited to drier habitats decrease in frequency. If the shift in weather conditions is severe and if it lasts a long time, some wet-adapted phenotypes not present at all in the original population may become common. Where might they come from? One possibility would be that the new phenotypes result from new mutations. But mutation of any given gene is exceedingly rare and, if the phenotypes in question are influenced by many different genes, it is most unlikely that all would mutate at approximately the same time in a way that would make the plant better suited to the new environmental conditions. It is far more likely that the new phenotypes arise through the combination of old genes in new ways as a result of natural selection. Thus selection, even in the absence of new mutations, can produce new phenotypes. In a very real sense, selection can play a creative role in evolution.

In biparental populations, selection determines the direction of change largely by altering the frequencies of genes that arose through random mutation many generations before. It thus establishes new gene combinations and gene activities that produce new phenotypes. Mutation is not usually a major directing force in evolution; the principal evolutionary role of new mutations consists in replenishing the store of variability in the gene pool and thereby providing the potential upon which future selection can act.

The conservative role of natural selection. Natural selection also plays an extremely important conservative role. Each species, in the course of its evolution, comes to have a constellation of genes that interact in very precise ways in governing the developmental, physiological, and biochemical processes on which the continued existence of the species depends. Anything that disrupts the harmonious interaction of its genes is usually deleterious to the species. But in a sexually reproducing population, favorable groupings of genes tend to be dispersed and new groupings formed by the recombination that occurs when each generation reproduces. Most of these new groupings will be less adaptive than the original grouping (although a few may be more adaptive). And the vast majority of new mutations tend to disrupt rather than enhance the established harmonious relationships among the genes. If unchecked, recombination and random mutation would therefore tend to destroy the favorable gene groupings on which the fitness of the species rests. Selection, by constantly acting to eliminate all but the most favorable gene combinations, counteracts the disrupting, disintegrating tendency of recombination and mutation and is thus the chief factor maintaining stability where otherwise there would be chaos.

Effective selection pressure as the algebraic sum of numerous separate selection pressures. Until now, we have treated the selection pressure acting on a gene as though it were a simple, unitary factor.

But, in fact, most genes are subject to many separate selection pressures, some of them conflicting. Most and perhaps all genes have many different effects (pleiotropy), and it is most unlikely that all effects of any given gene will be advantageous. Whether a gene increases or decreases in frequency is determined by whether the sum of the various positive selection pressures produced by its advantageous effects is greater or smaller than the sum of the negative selection pressures produced by its harmful effects. If the algebraic sum (an addition taking into account plus and minus signs) of all the separate selection pressures is positive, the gene will increase in frequency, but if the algebraic sum is negative, the gene will decrease in frequency.

Many cases are known in which the effects of a given gene are more advantageous in the heterozygous than in the homozygous condition. In African Negroes the gene for sickle-cell anemia occurs much more frequently than we might expect in view of its highly deleterious effect when homozygous. This is because the gene, when heterozygous, confers on the possessor a partial resistance to malaria. The equilibrium frequency of the sickle-cell gene is thus determined by at least three separate selection pressures: the strongly negative selection pressure on the homozygotes, the weaker negative selection pressure on the heterozygotes as a result of their mild anemia, and the fairly strong positive selection pressure on the heterozygotes as a result of their resistance to malaria.

Effective selection pressure is determined by many separate selection pressures, whether the character is a simple one controlled by only one or a few genes or whether it is a complex one controlled by many different genes. As an example of the determination of a complex character, let us consider the selection pressures on plumage in dabbling ducks.

In most species of birds, it is the female that chooses the mate; the male displays until some female chooses him. Since it is more important, under these conditions, for the female to recognize the males of her species than for the male to recognize the females, natural selection in many species has favored an eye-catching exterior in the males that contrasts with the drab appearance of the females (see Pl. III-5A). This trend can be observed in a number of closely related species of dabbling ducks (Mallard, Pintail, Gadwall, etc.) that occur together in most of North America. Hybridization takes place among them because the females sometimes err in their selection of mates. Since the hybrids are apparently less viable than the parents, there has been strong selection for showy male plumage, distinctive for each species, that helps reduce the number of mating errors. No such selection pressure has operated on the females. These brownish nondescript birds closely resemble one another and are probably much less easily seen by predators than the males, whose bright plumage doubtless makes them more subject to predation—a liability that must cause strong selection against such plumage. But the positive selection resulting from reduced mating errors is apparently greater than the negative selection resulting from predation, and the showy plumage of the males has been maintained and even enriched. The situation is different, however, on some isolated islands where only one species of dabbling duck exists and where, therefore, no mating errors can occur. Here the negative selection predominates, and the males have lost their showy plumage and resemble the more protectively colored females.

ADAPTATION

Every organism is, in a sense, a complex bundle of immense numbers of adaptations. We have already examined a host of adaptations in earlier chapters, adaptations concerned with nutrient procurement, gas exchange, internal transport, regulation of body fluids, hormonal and nervous control, effector activity, reproduction, development, etc.

An adaptation can obviously take many forms. We can define it as any genetically controlled characteristic that aids an organism to survive and reproduce in the environment it inhabits. Adaptations may be structural, physiological, or behavioral. They may be genetically simple or complex. They may involve individual cells or subcellular components, or whole organs or organ systems. They may be highly specific, of benefit only under very limited circumstances, or they may be general, of benefit under many and varied circumstances.

Let us look now at a few particularly striking examples of adaptation, which will help clarify the processes by which adaptations come into being.

Growth habit in pasture plants. A good example of the rapidity with which a population may become adapted to changed environmental conditions is provided by a study published in 1937 by W. B. Kemp of the Maryland Agricultural Experiment Station. The owner of a pasture in southern Maryland had seeded the pasture with a mixture of grasses and legumes. Then he divided the pasture into two parts, allowing one to be heavily grazed by cattle, while protecting the other from the livestock and leaving it to produce hay. Three years after this division, Kemp obtained specimens of blue grass, orchard grass, and white clover from each part of the pasture and planted them in an experimental garden where all the plants were exposed to the same environmental conditions. He found that the specimens of all three species from the heavily grazed half of the pasture exhibited dwarf, rambling growth, while specimens of the same three species from the ungrazed half exhibited vigorous upright growth. In only three years' time, the two populations of each species, known to have been identical initially because one batch of seed was used for the entire pasture, had become markedly different in their genetically determined growth pattern. Apparently the grazing cattle in the one half of the pasture had devoured most of the upright plants, and only plants low enough to be missed had survived and set seed. There had been, in short, intense selection against upright growth in this half of the pasture and correspondingly intense selection for the adaptively superior dwarf, rambling growth. By contrast, in the other half of the pasture, where there was no grazing, upright growth was adaptively superior, and dwarf plants were unable to compete effectively.

Adaptations of flowers for pollination. The flowering plants depend on external agents to carry pollen from the male parts in the flowers of one plant to the female parts in the flowers of another plant.

The flowers of each species are adapted in shape, structure, color, and odor to the particular pollinating agents on which they depend, and they provide a particularly clear illustration of the adaptiveness of evolution.

Bee flowers have showy, brightly colored petals that are usually blue or yellow but seldom red (bees can see blue or yellow light well, but they cannot see red at all); they usually have a sweet, aromatic, or minty fragrance, to which bees are attracted; they are generally open only during the day (bees are active only then); and they often have a special protruding lip on which the bees can land. Flowers pollinated primarily by hummingbirds are usually red or yellow (hummingbirds can see red well but blue only poorly); are nearly odorless (the birds have a very poor sense of smell); and lack any protruding landing platform, which the birds don't need, since they ordinarily hover in front of the flowers while sucking the nectar (see Pl. IV-1A). The bases of the petals of flowers pollinated by bees, birds, or moths are often fused to form a tube whose length corresponds closely to the length of the tongue or bill of the particular species most important as the pollinator of that plant (Fig. 17.1).

An especially dramatic example of adaptation for pollination is seen in some species of orchids, where the flowers resemble in both shape and color the females of certain species of wasps or bees. The male insect is stimulated to attempt to copulate with the flower and becomes covered with pollen in the process. When he later attempts to copulate with another flower, some of the pollen from the first flower is deposited on the second. So complete is the deception that sperms have actually been found inside the orchid flowers after a visit by the male insect.

Cryptic coloration in animals. The fact that many animals blend into their surroundings so well as to be nearly undetectable has been recognized for many years, and recent careful studies have confirmed that, as had been assumed, cryptic coloration (Pls. IV-2, 3) is an adaptive characteristic that helps animals escape predation.

One of the most extensively studied cases of cryptic coloration is the so-called industrial melanism of moths. Certain populations of moths exhibiting *polymorphism*—two or more distinct forms of a genetically determined character—exist in two forms, light and dark. Since the mid-1800's the originally predominant light forms have given way, in industrial areas, to the dark (melanic) forms. For example, in the Manchester area, in England, the first black specimens of the species *Biston betularia* were caught in 1848; by 1895 melanics constituted about 98 percent of the total population in the area. It has been calculated that for such a remarkable shift in frequency to have occurred in so short a time the melanic form must have had at least a 30 percent advantage over the light form.

In 1937 E. B. Ford of Oxford University proposed an explanation for this striking evolutionary change, one confirmed some twenty years later by the field experiments of H. B. D. Kettlewell, also of Oxford. The species of moths exhibiting the rapid shift to melanism, though unrelated to one another, all habitually rest during the day in an exposed position on tree trunks or rocks, being protected from predation

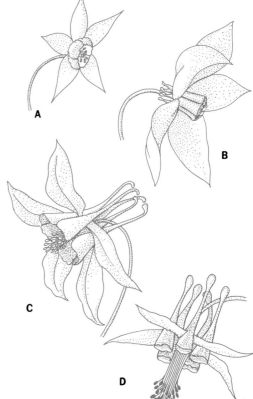

17.1. Characters of columbine flowers correlated with their pollinators. (A) *Aquilegia ecalcarata,* pollinated by bees. (B) *A. nivalis,* pollinated by long-tongued bees. (C) *A. vulgaris,* pollinated by long-tongued bumblebees. (D) *A. formosa,* pollinated by hummingbirds. The length and curvature of the nectar tubes of the flowers are correlated with the length and curvature of the bees' tongues and the hummingbirds' bills. [Adapted from V. Grant, "The Fertilization of Flowers," *Sci. Am.,* June, 1951. Copyright © 1951 by Scientific American, Inc. All rights reserved.]

17.2. Cryptic coloration of peppered moths. Top: Light and dark forms of *Biston betularia* at rest on a lichen-covered tree trunk in unpolluted countryside. The light moth is very difficult to see (it is slightly below and to the right of the dark moth). Bottom: Light and dark moths on a soot-covered tree trunk near Birmingham, England. Here the light form is the easier to see.   [From the experiments of Dr. H. B. D. Kettlewell, Oxford University.]

only by their close resemblance to their background. In former years, the tree trunks and rocks were rather light-colored and often covered with light-colored lichens. Against this background, the light forms of the moths were astonishingly difficult to see, whereas the melanic forms were quite conspicuous (Fig. 17.2). It would be expected, therefore, that predators such as birds would have captured melanics far more easily than the cryptically colored light moths. The light forms would thus have been strongly favored, and they would have occurred in much higher frequency than melanics. But with the advent of extensive industrialization, tree trunks and rocks were blackened by soot, and the lichens, which are particularly sensitive to such pollution, disappeared. In this altered environment, the melanic moths would have resembled the background more closely than the light moths. Thus selection would have been reversed and would now have favored the melanics, which would consequently have increased in frequency.

Warning coloration and mimicry in animals.   Whereas some animals have evolved cryptic coloration, others (particularly insects) have evolved colors and patterns that contrast boldly with their background and thus render them clearly visible to potential predators. Many of these animals are in some way disagreeable to predators; they may taste bad, or smell bad, or sting, or secrete poisonous substances. Such animals benefit by being gaudily colored and conspicuous because predators that have experienced their unpleasant features learn to recognize and avoid them more easily in the future (see Pl. IV-5). Their flashy appearance is protective because it warns potential predators that they should stay away. In fact, the warning is sometimes so effective that the protection extends to other species. After unpleasant experiences with one or two warningly colored insects, some vertebrate predators simply avoid all flashily colored insects, whether or not they resemble the ones they encountered earlier. One investigator offered over 200 different species of insects to an insectivorous monkey; the monkey accepted 83 percent of the cryptically colored insects but only 16 percent of the warningly colored ones, even though many of the insects belonged to species the monkey had probably not previously encountered.

Species not naturally protected by some unpleasant character of their own may closely resemble (mimic) in appearance and behavior some warningly colored unpalatable species (Fig. 17.3). Such a resemblance is adaptive; the mimic species suffers little predation because predators cannot distinguish it from its unpleasant models. This phenomenon is called **Batesian mimicry.** In some cases of Batesian mimicry, the unpalatable species has a whole group of mimics (see Pl. IV-4).

A second kind of mimicry, called **Müllerian mimicry,** involves the evolution of a similar appearance by two or more distasteful species. In this type of mimicry, each species is both model and mimic. Each species has some defensive mechanism, but if each had its own characteristic appearance, the predators would have to learn to avoid each species separately; the learning process would thus be a more demanding one, and some individuals of each prey species would have to be sacrificed to the learning process. If, however, several protected species evolve more and more toward one appearance type, they come to constitute a single prey group from the standpoint of the predators and avoidance is more easily learned.

SPECIES AND SPECIATION

We have so far discussed only one major aspect of evolution, the gradual change of a given population through time. Now we must turn to another of its major aspects, the processes whereby a single population may split, giving rise to two or more different descendent populations. But before we can discuss this topic meaningfully, we must examine more carefully the populations that we have so far casually taken for granted. With reference to sexually reproducing organisms, we can define a *population* as a group of individuals that share a common gene pool (i.e. that interbreed to a larger or smaller extent).

Units of population

Demes.   A deme is a small local population, such as all the deer mice or all the red oaks in a certain woodlot or all the perch in a given pond. Although no two individuals in a deme are exactly alike, the members of a deme do usually resemble one another more closely than they resemble the members of other demes, for two reasons: (1) They are more closely related genetically, because pairings occur more frequently between members of the same deme than between members of different demes; and (2) they are exposed to more similar environmental influences and hence to more nearly the same selection pressures.

It must be emphasized that demes are not clear-cut permanent units of population. Although the deer mice in one woodlot are more likely to mate among themselves than with deer mice in the next woodlot down the road, there will almost certainly be occasional matings between mice from different woodlots. Similarly, although the female parts of a particular red oak tree are more likely to receive pollen from another red oak tree in the same woodlot, there is an appreciable chance that they will sometimes receive pollen from a tree in another nearby woodlot. And the woodlots themselves are not permanent ecological features. They have only a transient existence as separate and distinct ecological units; neighboring woodlots may fuse after a few years, or a single large woodlot may become divided into two or more separate smaller ones. Such changes in ecological features will produce corresponding changes in the demes of deer mice and red oak trees. Demes, then, are usually temporary units of population that intergrade with other similar units.

Species.   Notice that intergradation is between "similar" demes. We expect some interbreeding between deer mice from adjacent demes, but we do not expect interbreeding between deer mice and house mice or between deer mice and black rats or between deer mice and gray squirrels. Nor do we expect to find crosses between red oaks and sugar maples or even between red oaks and pin oaks, even if they occur together in the same woodlot. In short, we recognize the existence of units of population larger than demes and both more distinct from each other and longer-lasting than demes. One such unit of population is that

17.3. An example of Batesian mimicry. Top: The Monarch butterfly, an ill-tasting species. Bottom: The Viceroy, a species that mimics the Monarch. Species in the group to which the Viceroy belongs ordinarily have a quite different appearance.

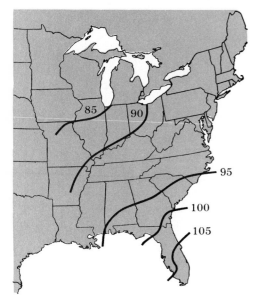

17.4. Clinal variation. Map showing by means of isophene lines (lines connecting equal values) the geographic variation in the mean number of subcaudal scales of the snake *Coluber constrictor* (the racer). [Modified from W. Auffenberg, *Tulane Stud. Zool.*, vol. 2, 1955.]

containing all the demes of deer mice. Another is that containing all the demes of red oaks. These larger units are known as species.

For centuries, it has been recognized that the variation between living organisms does not form a continuum—that there are, instead, many discontinuities in the variation, that plants and animals seem to be divided naturally into many separate and distinct "kinds," or species. This does not mean that all the individuals of any one species are precisely alike—far from it: any two individuals are probably distinguishable from each other in a variety of ways. But it does mean that all the members of a single species share certain biologically important attributes and that, as a group, they are genetically separated from other such groups.

Although the existence of discrete clusters of living things that can be called species has long been recognized, the concept of what a species is has changed many times in the course of history. One idea widely held by nonbiologists and once popular among biologists as well is that each species is a static, immutable entity typified by some ideal form, of which all the real individuals belonging to that species are rough approximations. This static, typological concept contradicts all that we have learned about evolution. The modern concept of species rejects the notion that there is some immutable ideal type for every species. A species, in the modern view, is a genetically distinctive group of natural populations (demes) that share a common gene pool and that are reproductively isolated from all other such groups. Or, to word it another way, a species is the largest unit of population within which effective *gene flow* (exchange of genetic material) occurs or can occur.

Notice that the modern definition of species says nothing about how different from each other two populations must be to qualify as separate species. Admittedly, most species can be separated on the basis of fairly obvious anatomical, physiological, or behavioral characters, and biologists frequently rely on these in determining species. But the final criterion is always reproduction—whether or not there is actual or potential gene flow. If there is complete intrinsic reproductive isolation between two outwardly almost identical populations (i.e. if there can be no gene flow between them), then those populations belong to different species despite their great similarity. On the other hand, if two populations show striking differences, but there is effective gene flow between them, then those populations belong to the same species. Anatomical, physiological, or behavioral characters simply serve as clues toward the identification of reproductively isolated populations; they are not in themselves regarded as determining whether a population constitutes a species.

Intraspecific variation. We have already discussed the sorts of variation that may occur between individuals of a single deme as a result of mutation and recombination, particularly the latter. This variation may involve almost imperceptible and intergrading differences or striking polymorphic discontinuities. We must now consider another sort of intraspecific variation, that between the demes of a single species, i.e. variation correlated with geographical distribution.

There is usually so much gene flow between adjacent demes of the same species that differences between them are slight. Thus the

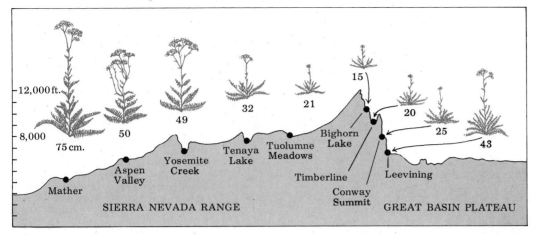

-12,000 ft.

49

32

21

15

20

25

-8,000

50

Bighorn Lake

43

75 cm.

Aspen Valley

Yosemite Creek

Tenaya Lake

Tuolumne Meadows

Timberline

Leevining

Mather

SIERRA NEVADA RANGE

Conway Summit

GREAT BASIN PLATEAU

17.5. Altitudinal cline in height of the milfoil *Achillea lanulosa.* The higher the altitude, the shorter the plants. This variation was shown to be genetic (i.e. not merely a phenotypic response) by collecting seeds from the locations indicated and planting them in a test garden at Stanford where all were exposed to the same environmental conditions. The differences in height were still evident in the plants grown from these seeds. [Modified from J. Clausen, D. D. Keck, and W. M. Hiesey, *Carnegie Inst. Wash. Publ.*, no. 581, 1948.]

ratio of alleles *A* and *a* may be 0.90 to 0.10 in one deme and 0.89 to 0.11 in the adjacent deme. But the farther apart two demes are, the smaller is the chance of direct gene flow between them, and hence the greater is the probability that the differences between them will be more marked. Some of this geographic variation may reflect chance events such as genetic drift or the occurrence in one deme of a mutation that would be favorable in all demes, but that has not yet spread to them. But much of the geographic variation probably reflects differences in the selection pressures operating on the populations as a result of different environmental conditions in their respective ranges. In other words, much geographic variation is adaptive. Each local population or deme tends to evolve adaptations to the specific environmental conditions in its own small portion of the species range. Such geographic variation is found in the vast majority of animal and plant species.

Environmental conditions often vary geographically in a more or less regular manner. There are changes in temperature with latitude, or with altitude on mountain slopes, or changes in rainfall with longitude, as in many parts of the western United States, or changes in topography with longitude, as from the Atlantic coast to the Central States. Such environmental gradients are usually accompanied by genetic-variation gradients, or ***clines,*** in the species of animals and plants that inhabit the areas involved. Most species show north–south clines in many characters (Fig. 17.4), and east–west clines are not uncommon. Altitudinal clines in various characters are also often found (Fig. 17.5). Many mammals and birds exhibit north–south clines in average body size, being larger in the colder climates farther north and smaller in the warmer climates farther south. Similarly, many mammalian species show north–south clines in the size of such extremities as the tails and ears, these parts being smaller in the demes farther north. The adaptive significance of these clines is obvious.

Sometimes geographically correlated genetic variation is not as gradual as in the clines discussed above. There may be a rather abrupt shift in some character in a particular part of the species range. Suppose, for example, that the average height of a certain species of plant decreases very gradually as one moves northward from Florida to Virginia, then decreases very rapidly in the counties of southen Virginia, and decreases only slightly as one continues to move northward (Fig.

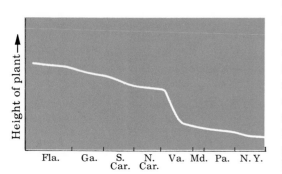

17.6. Graph showing stepped clinal variation in hypothetical plant. The horizontal spans correspond roughly to the north–south distances through the states. See text for discussion.

17.6). Such variation forms a stepped cline; i.e. the abrupt shift in height in southern Virginia breaks the otherwise gradual geographic variation and constitutes a step in the north–south cline. When such an abrupt shift in a genetically determined character occurs in a geographically variable species, the populations on the two sides of the step are sometimes regarded as subspecies or races. These terms are also sometimes applied to more isolated populations—such as those on different islands or in separate mountain ranges or, as in fish, in separate rivers—when the populations are recognizably different genetically but are believed potentially capable of interbreeding freely. *Subspecies* or *races* (the two terms, as used here, are equivalent) may be defined, then, as groups of natural populations within a species that differ genetically and that are partly isolated from each other reproductively because they have different ranges.

Note that two subspecies of the same species cannot, by definition, long occur together geographically, because it is only the limitation on interbreeding imposed by distance that keeps them genetically distinctive. If they occurred together, they would interbreed freely and any distinction between them would quickly disappear. Some biologists have argued against the formal recognition of subspecies, on these grounds: (1) The distinctions between them are often made arbitrarily on the basis of only one character; the fact that other characters may form entirely different patterns of variation is ignored (Fig. 17.7). (2) Most units so recognized probably have only a transitory existence as separate entities and do not, as was once thought, go on to become fully separate species.

Speciation

How do new species arise? Our exploration of the modern concept of species has provided some clues to the most usual method of speciation. Since a species is a group of populations sharing a common gene pool, and since populations tend to evolve adaptations to local conditions, some populations might, by one process or another, evolve characteristics sufficiently different for interbreeding between them and the other populations to become impossible, and they might thus come to constitute a new species. The question then becomes: By what processes do two sets of populations that initially shared a common gene pool acquire completely separate gene pools?

The role of geographic isolation. Most biologists agree that in the vast majority of cases (excluding speciation by polyploidy) the initiating factor in speciation is geographic separation. As long as all the populations of a species are in direct or indirect contact, gene flow will continue throughout the system and no splitting can occur, although various populations within the system may diverge in numerous characters and thus give rise to much intraspecific variation of the sorts discussed above. But if the initially continuous system of populations is divided by some geographic feature that constitutes a barrier to the dispersal of the species, then the separated population systems will no longer be able to exchange genes and their further evolution will therefore be independent. Given sufficient time, the two separate population systems will become more and more unlike each other as each evolves in its own way. At first, the only reproductive isolation between them will be geographic—isolation by physical separation—and they will potentially still be capable of interbreeding; according to the modern concept of species, they will still belong to the same species. Eventually, however, they may become genetically so different that there would be no effective gene flow between them even if they should again come into contact. When this point in their gradual divergence has been reached, the two population systems constitute two separate species.

There are at least three compelling reasons why geographically separated population systems will diverge in time.

First, chances are that the two systems will have somewhat different initial gene frequencies. Because most species exhibit geographic variation, it is most unlikely that a geographic barrier would divide a variable species into portions exactly alike genetically. It would be much more likely to separate populations already genetically different, such as the terminal portions of a cline. Separation can occur in other ways than through the splitting of a once-continuous distribution by a new geographic barrier. When, as often happens, a small number of individuals manage to cross an already existing barrier and found a new geographically isolated colony, these founders will, of course, carry with them in their own genotypes only a small percentage of the total genetic variation present in the gene pool of the parental population, and the new colony will thus have gene frequencies very different from those of the parental population. Obviously, if from the moment of their separation two populations have different genetic potentials, their future evolution is likely to follow different paths.

A second reason why separated population systems will diverge in the course of their evolution is that they will probably experience different mutations. Mutations are random (though some are more probable than others), and the chances are good that some mutations will occur in one of the populations and not in the other, and vice versa. Since there is no gene flow between the populations, a new mutant gene arising in one of them cannot spread to the other.

The third reason for evolutionary divergence of isolated populations is that they will almost certainly be exposed to different environmental selection pressures, since they occupy different ranges. The chances that two separate ranges will be identical in every significant environmental factor are essentially nil.

In addition to these three reasons for divergence, a fourth—genetic drift—would be important in small populations, as when a few founder individuals start a new colony.

The barriers that can cause the initial spatial separation leading to speciation are of many different types. A barrier is any physical or ecological feature that prevents the movement across it of the species in question. What is a barrier for one species may not be a barrier for another. Thus a prairie is a barrier for forest species but not, obviously, for prairie species. A mountain range is a barrier to species that can live only in lowlands, a desert is a barrier to species that require a moist environment, and a valley is a barrier to montane species. On a grander scale, oceans and glaciers have played a role in the speciation of many plants and animals. Let us look at a few actual examples of geographic isolation leading to speciation.

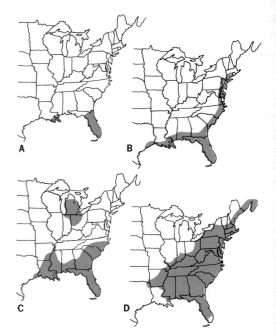

17.7 Discordant geographic variation in four characters of the snake *Coluber constrictor* in the eastern United States. No two of the characters vary together. (A) Areas where red eyes are found in juveniles. (B) Areas where red ventral spots are found on juveniles. (C) Areas where the loreal scale is in contact with the first supralabial scale in at least 10 percent of the specimens. (D) Areas where black adults are found. [Redrawn from W. Auffenberg, *Tulane Stud. Zool.*, vol. 2, 1955.]

One of the most frequently cited examples is that of two populations of squirrels separated by the Grand Canyon. The Kaibab squirrel on the north side and the Abert squirrel on the south side are clearly very closely related and doubtless evolved from the same ancestor, but they almost never interbreed at present because they do not cross the Grand Canyon. Biologists are not agreed whether these two squirrels have reached the level of full species or whether they should be considered well-marked geographic variants of a single species, but the fact remains that the Grand Canyon has acted as a barrier separating the two sets of populations, and that those populations have, as a result, evolved divergently until they have at least approached the level of fully distinct species.

On islands of the Pacific, in many instances, two closely related species of snails, clearly descended from the same ancestral population, live in valley woodlands separated by treeless ridges that the snails apparently cannot cross. Blind cave beetles (genus *Pseudanophthalmus*) living in different caves in the eastern United States have often diverged to the level of full species. Two river systems only a few miles apart but with no interconnections often have their own species of minnows.

Intrinsic reproductive isolation. According to the model of divergent speciation outlined above, the initial factor preventing gene flow between two closely related population systems is ordinarily an extrinsic one—geography. Then, the model says, as the two populations diverge, they accumulate differences that will lead, given enough time, to the development of intrinsic isolating mechanisms—biological characteristics that prevent the two populations from occurring together or from interbreeding effectively when (or if) they again occur together. In other words, speciation is initiated when through external barriers the two population systems become entirely *allopatric* (come to have different ranges), but is not completed until the populations have evolved intrinsic mechanisms that will keep them allopatric or that will

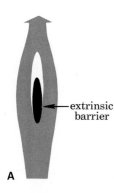

17.8. Model of geographic speciation. (A) An extrinsic barrier (geographic) divides a population, but the barrier breaks down before the two subpopulations have been isolated long enough to have evolved intrinsic reproductive isolating mechanisms; hence the populations fuse back together. (B) Two populations are isolated by an extrinsic barrier long enough to have evolved incomplete intrinsic reproductive isolating mechanisms. When the extrinsic barrier breaks down, some hybridization occurs. But the hybrids are not as well adapted as the parental forms; hence there is a strong selection pressure favoring forms of intrinsic isolation that prevent mating, and the two populations diverge more rapidly until mating between them is no longer possible. This rapid divergence is called character displacement. (C) Two populations are isolated by a geographic barrier so long that by the time the barrier breaks down they are too different to interbreed.

Table 17.1. Intrinsic isolating mechanisms

Mechanisms that prevent mating	1. Ecogeographic isolation 2. Habitat isolation 3. Seasonal isolation 4. Behavioral isolation 5. Mechanical isolation	Mechanisms operative in the parents, preventing fertilization
Mechanisms that prevent production of hybrid young after mating	6. Gametic isolation 7. Developmental isolation	
Mechanisms that prevent perpetuation of hybrids	8. Hybrid inviability 9. Hybrid sterility 10. Selective hybrid elimination	Mechanisms operative in the hybrids, preventing their success

keep their gene pools separate even when they are *sympatric* (have the same range) (Fig. 17.8). Let us now examine the various kinds of intrinsic isolating mechanisms that may arise. One way of classifying them is shown in Table 17.1.

1. *Ecogeographic isolation.* Two population systems, initially separated by some extrinsic barrier, may in time become so specialized for different environmental conditions that even if the original extrinsic barrier is removed they may never become sympatric, because neither can survive under the conditions where the other occurs. In other words, they may evolve genetic differences that will maintain their geographic separation.

2. *Habitat isolation.* When two sympatric populations occupy different habitats within their common range, the individuals of each population will be more likely to encounter and mate with members of their own population than with members of the other population. Their genetically determined preference for different habitats thus helps keep the two gene pools separate.

3. *Seasonal isolation.* If two closely related species are sympatric but breed during different seasons of the year, interbreeding between them will be effectively prevented.

4. *Behavioral isolation.* In Chapter 12 we discussed the immense importance of behavior in courtship and mating. And we emphasized the fundamental role of behavior patterns in species recognition among animals. We saw, for example, that ducks frequently have elaborate courtship displays, usually combined with striking color patterns in the males (see Pl. III-5D), and that these function in minimizing the chances that the females will select as a mate a male of the wrong species.

5. *Mechanical isolation.* If structural differences between two closely related species make it physically impossible for matings between males of one species and females of the other to occur, the two populations will obviously not exchange genes.

6. *Gametic isolation.* Even if individuals of two different animal species mate or the pollen from one plant species gets onto the stigma of another, actual fertilization may not take place. For example, if cross-insemination occurs between *Drosophila virilis* and *D. americana,* the sperms are rapidly immobilized by the unsuitable environment in the reproductive tract of the female and they never reach the egg cells.

7. *Developmental isolation.* Even when cross-fertilization occurs, the development of the embryo is often irregular and may cease before birth.

8. *Hybrid inviability.* Hybrids are often weak and malformed and frequently die before they reproduce; hence there is no actual gene flow through them from the gene pool of the one parental species to the gene pool of the other parental species. An example of hybrid inviability is seen in certain tobacco hybrids, which form tumors in their vegetative parts and die before they flower.

9. *Hybrid sterility.* Some interspecific crosses produce vigorous but sterile hybrids. The best-known example is, of course, the cross between the horse and the donkey, which produces the mule. Mules have many characteristics superior to those of both parental species, but they are sterile. No matter how many mules are produced, the gene pools of horses and donkeys remain distinct, because there is no gene flow between them.

10. *Selective hybrid elimination.* The members of two closely related populations may be able to cross and produce fertile offspring. If those offspring and their progeny are as vigorous and well adapted as the parental forms, then the two original populations will not remain distinct for long if they are sympatric, and it will no longer be possible to regard them as full species. But if the fertile offspring and their progeny are less well adapted than the parental forms, then they will soon be eliminated. There will be some gene flow between the two parental gene pools via the hybrids, but not much. The parental populations are consequently regarded as separate species. Usually, if they are sympatric, they will rapidly evolve more effective isolating mechanisms. The reason is clear. Since the hybrids are inferior and tend to die out within a generation or two, individuals of the parent species that tend to mate with members of their own species will, in the long run, have more descendants than individuals that tend to mate with members of the wrong species. In other words, there will be selection for correct mating and selection against wrong mating. Gene combinations that lead to correct mate selection will increase in frequency, and combinations that lead to incorrect selection will decrease, until eventually all hybridization ceases. The tendency of closely related sympatric species to diverge rapidly in characteristics that reduce the chances of hybridization and/or minimize competition between them is called *character displacement* (see Fig. 17.8B).

Situations in which only one of the ten isolating mechanisms discussed above is operative are extremely rare. Ordinarily two, three, four, or more all contribute to keeping two species apart. For example, closely related sympatric plant species often exhibit habitat and seasonal isolation in addition to some form of hybrid incapacity. In general, sympatric species, whether plant or animal, tend rapidly to evolve one or more of the forms of isolation that prevent mating (habitat, seasonal, behavioral, mechanical) rather than depend only on those forms that prevent the birth or perpetuation of hybrids. The reasons are similar to those given in the preceding paragraph: Individuals that tend to mate with members of the wrong species will leave fewer descendants than those that mate with members of their own species. Wrong matings produce gamete wastage, whether fertilization takes place or not, or whether the hybrids are viable or not.

Speciation by polyploidy. The model of speciation discussed above involves the gradual divergence of geographically separated populations. Among plants, there is another way in which new species may arise—an almost instantaneous process that makes it entirely possible for a parent to belong to one species and its offspring to belong to a different species. This process is speciation by polyploidy, which is the occurrence in cell nuclei of more than two complete sets of chromosomes (i.e. the cells are triploid, or tetraploid, or hexaploid, etc.).

Sometimes, as a result of nondisjunction of the chromosomes during meiosis, a diploid gamete is produced. If by chance it should unite in fertilization with another diploid gamete, also produced by aberrant meiosis, a tetraploid zygote results. Production of polyploid individuals in this manner has apparently been common in plants but very rare in animals. Polyploids are fertile and can breed with each other, but cannot cross with the diploid species from which they arose. Hence polyploid populations fulfill the requirements of the modern definition of species: They are genetically distinctive and they are reproductively isolated.

Adaptive radiation. One of the most striking aspects of life is its extreme diversity. A bewildering array of different species now occupies this globe. And the fossil record shows that of the species that have existed at one time or another those now living represent only a tiny fraction. Clearly, then, divergent evolution—the evolutionary splitting of species into many separate descendent species—has been an exceedingly frequent occurrence. Is it possible to account for such a degree of evolutionary radiation by the models outlined above? In particular, could opportunities for geographic isolation have been sufficient to lead to all the speciation not caused by polyploidy? After all, it is not unusual for a complex of four, five, or more closely related species to occur within a quite limited area. For example, 28 different species of the milliped genus *Brachoria,* many of them sympatric, are confined to a small portion of the deciduous forests of the eastern United States. How could so much speciation have occurred in so small an area if geographic isolation was a necessary factor? In an attempt to answer such questions, let us turn to a particularly instructive and historically important example—the finches on the Galápagos Islands, which played a major role in leading Charles Darwin to formulate his theory of evolution by natural selection.

The Galápagos Islands lie astride the equator in the Pacific Ocean roughly 600 miles west of the coast of Ecuador (Fig. 17.9). These islands have never been connected to South America or to any other land mass; nor have they been connected to each other. They apparently arose from the ocean floor as volcanoes somewhat more than a million years ago. At first, of course, they were completely devoid of life, and were thus an environment open to exploitation by whatever species from the South American mainland might chance to reach them. Relatively few species ever did so. The only land vertebrates present on

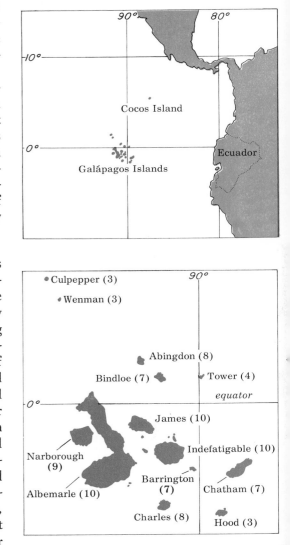

17.9. The Galápagos Islands. Top: The islands are located about 600 miles off the coast of Ecuador. Cocos Island is about 600 miles northeast of the Galápagos. Bottom: The islands shown in greater detail. The number in parentheses after each name indicates the number of species of Darwin's finches that occur on the island. The island names shown here are the English ones, which Darwin used. The Ecuadorian government has renamed the majority of the islands, but most of the biological literature continues to use the older English names. [Modified from D. Lack, *Darwin's Finches,* Cambridge University Press, 1947.]

the islands before man got there were six species of reptiles, two species of mammals, and a limited number of birds, including the famous Darwin's finches.

The 14 species of Darwin's finches constitute a separate subfamily found nowhere else in the world. They are believed to have evolved on the Galápagos Islands from some unknown finch ancestor that colonized the islands from the South American mainland. We can readily understand how the descendants of the geographically isolated colonizers would have undergone so much evolutionary change as to become, in time, very unlike their mainland ancestors. More perplexing at first glance is the manner in which the descendants of the original immigrants split into the separate populations that gave rise to today's 14 different species. The point to remember is that we are dealing not with a single island but with a cluster of more than 15 separate islands. The finches will not readily fly across wide stretches of water, and they show a strong tendency to remain near their home area. Hence a population on any one of the islands is effectively isolated from the populations on the other islands. We can suppose that the initial colony was established on some one of the islands where the colonizers chanced to land. Later, stragglers from this colony wandered or were blown to other islands and started new colonies. In time, the colonies on the different islands diverged for the reasons already outlined in our model of geographic speciation (different initial gene frequencies, different mutations, different selection pressures, and, in such small populations as some of these must have been, genetic drift). What we might expect, therefore, is a different species, or at least a different race, on each of the islands. But this is not what has actually been found; most of the islands have more than one species of finch, and the larger islands have ten (Fig. 17.9, bottom). How can we account for this?

Let us suppose that form A evolved originally on Indefatigable Island and that the closely related form B evolved on Charles Island. If, later, form A had spread to Charles Island before the two forms had been isolated long enough to evolve any but minor differences, the two forms might have interbred freely and merged with each other. But if A and B had been separated long enough to have evolved major differences before A invaded Charles Island, then A and B might have been intrinsically isolated from each other (i.e. been full species), and they might have been able to coexist on the same island without interbreeding (Fig. 17.10). If they formed occasional hybrids, those hybrids might well have been less viable than the parental forms. Accordingly, natural selection would have favored individuals that mated only with their own kind, and this selection pressure would have led rapidly to more effective intrinsic isolating mechanisms preventing the gamete wastage involved in cross-matings. It has been shown, in fact, that Darwin's finches readily recognize members of their own species and show little interest in members of a different species.

We have now arrived at a point in our hypothetical example where Indefatigable is occupied by species A and Charles is occupied by both A and B. It would be highly unlikely that A and B could coexist indefinitely if they utilized the same food supply or the same nesting sites; the ensuing competition would be very severe, and the less well adapted species would tend to be eliminated by the other unless it

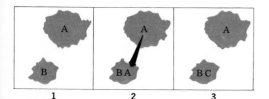

17.10. Model of speciation on the Galápagos Islands. (1) An ancestral form colonized two islands, and the two populations, being isolated from each other, eventually evolve into separate species A and B. (2) Some individuals of A colonize B's island. The two species coexist, but intense competition between them leads to rapid divergent evolution. (3) This rapid evolution of the population of A on B's island causes it to become more and more different from the original species A, until eventually it is sufficiently distinct to be considered a full species, C, in its own right.

evolved differences that minimized the competition. In other words, wherever two or more closely related species occurred together, natural selection would favor the evolution of different feeding and nesting habits, the phenomenon we have called character displacement. This is precisely what we find in Darwin's finches. The 14 species form four groups (genera) (Fig. 17.11). One group includes six species that live primarily on the ground; of these, some feed primarily on seeds and others feed mostly on cactus flowers. Of the species that feed on seeds, some feed on large seeds, some on medium-sized seeds, and some on small seeds. The second group contains six species that live primarily in trees. Of these, one is a vegetarian and the others eat insects, but the insect eaters differ from one another in the size of their prey and in the way they catch them. A third group contains only one species, which has become very unfinchlike and strongly resembles the warblers of the mainland. The fourth group contains only one species, restricted to Cocos Island. Correlated with the differences in diet among the species are major differences in the size and shape of their beaks. These characteristics of the beak are apparently the principal means by which the birds recognize other members of their own species.

Now, if on Charles Island selection favored character displacement between species A and B, the population of species A on Charles Island would become less and less like the population of species A on Indefatigable Island. Eventually these differences might become so great that the two populations would be intrinsically isolated from each other and would thus be separate species. We might now designate as species C the Charles population derived from species A. The geographic separation of the two islands would thus have led to the evolution of three different species (A, B, and C) from a single original species. The process of island hopping followed by divergence could continue indefinitely and produce many additional species. It was doubtless such a process, involving initial divergence on separate islands followed by intensification of differences when sympatry later developed, that led to the formation of the 14 species of Darwin's finches.

Now let us apply the principles learned from Darwin's finches to the case of the 28 species of *Brachoria* millipeds confined to a small area in the eastern United States. These animals live in the humus layer on the floor of deciduous forests. They are rather sluggish and seldom move very far. It would have been easy for populations to become isolated in local forested areas separated by less hospitable regions. Such allopatric populations could have become sufficiently different so that, when conditions changed and they became sympatric again, they would behave as full species. Thus the same sorts of processes as seen on islands can account for radiation in a continental area. And on a somewhat larger geographic scale, the same processes can account for the observed adaptive radiation in insects, fish, reptiles, birds, mammals, and many plant groups. In short, adaptive radiation on islands, like that of Darwin's finches, is dramatic and lends itself particularly well to analysis, but it does not differ in principle from adaptive radiation under other circumstances. Thus it helps show that our model of speciation can account for the great amount of divergence that was necessary to produce the immense diversity among living things. No other mechanism, except polyploidy in some plant groups, seems to have been required.

17.11. Darwin's finches. The finches numbered 1–6 are the ground finches (*Geospiza*). Those numbered 7–12 are the tree finches (*Camarhynchus*). Number 13 is the Warbler Finch (*Certhidea olivacea*). The Cocos Finch (*Pinaroloxias inornata*) is not shown. 1. Large Cactus Ground Finch (*G. conirostris*). 2. Large Ground Finch (*G. magnirostris*). 3. Medium Ground Finch (*G. fortis*). 4. Cactus Ground Finch (*G. scandens*). 5. Sharp-beaked Ground Finch (*G. difficilis*). 6. Small Ground Finch (*G. fuliginosa*). 7. Woodpecker Finch (*Camarhynchus pallidus*). 8. Vegetarian Tree Finch (*C. crassirostris*). 9. Large Insectivorous Tree Finch of Charles Island (*C. pauper*). 10. Large Insectivorous Tree Finch (*C. psittacula*). 11. Small Insectivorous Tree Finch (*C. parvulus*). 12. Mangrove Finch (*C. heliobates*). [*Biological Science: Molecules to Man,* Houghton Mifflin, 1963. Reprinted by permission of the Biological Sciences Curriculum Study.]

THE CONCEPT OF PHYLOGENY

Evolution implies that many unlike groups of organisms have a common ancestor and that all forms of life probably stem from the same remote beginnings. Thus one of the tasks of biology is to discover the relationships among the species alive today and to trace the ancestors from which they descended. But because so many different kinds of evidence must be weighed, reconstructing the evolutionary history—the phylogeny—of any group of organisms always entails a strong element of speculation (Fig. 17.12).

Determining phylogenetic relationships

Sources of data. When a biologist sets out to reconstruct the phylogeny of a group of species that he thinks are related, he usually has before him only the species living today. He cannot observe their phylogenetic history. To reconstruct it as closely as possible, he must make inferences based on any observational and experimental data that appear relevant. These data can usually be interpreted in several different ways, and only experience and good judgment can help the biologist choose among them.

The usual procedure in reconstructing phylogenies is to examine as many different characters of the species in question as possible and to determine in which characters they differ and in which they are alike. The assumption is that the differences and resemblances will reflect, at least in part, their true phylogenetic relationships. Ordinarily, as many different types of characters as possible are used in the hope that misleading data from any single character will be detected by a lack of agreement with the data from other characters.

The most easily studied and widely utilized characters pertain to morphology—including external morphology, internal anatomy and

17.12. A phylogenetic tree for Darwin's finches. Biologists often resort to phylogenetic trees in attempting to show the hypothetical connections between related organisms. No phylogenetic tree should be taken too seriously, however, because the evidence is usually sufficient only to indicate the broad outlines of the tree, and much guesswork (educated guesswork, preferably) goes into the adding of details. [Redrawn from D. Lack, *Darwin's Finches*, Cambridge University Press, 1947.]

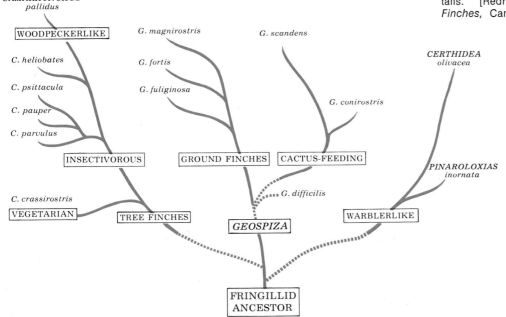

histology, and the morphology of the chromosomes in cell nuclei. It is particularly helpful, of course, when morphological characters of living species can be compared with those of fossil forms. The fossil record is the most direct source of evidence about the stages through which past forms of life passed, but unfortunately that record is usually very incomplete and is subject to the same sorts of errors of interpretation as the characters of living species. For many groups of organisms there is no suitable fossil record for working out the relationships between species; at best, the fossils may suggest the broad outlines of the evolution of major groups. Even in those groups where fossils are abundant, only the hard parts of the organisms' bodies have usually been preserved. Nevertheless, in some groups, of which the horses are an outstanding example, the fossil record has provided much phylogenetic information that could have been obtained from no other source.

Another frequently used source of information is embryology. Morphological characters are often easier to interpret if the manner in which they develop is known. For example, if it can be shown that a particular structure in organism A and a structure of quite different appearance in organism B both develop from the same embryonic primordium, then the resemblances and differences between those structures in A and B take on a phylogenetic significance that they would not have if they developed from entirely different embryonic primordia. Embryological evidence often allows biologists to trace the probable evolutionary changes that have occurred in important structures and helps them reconstruct the probable chain of evolutionary events that led to the modern forms of life. For example, the fact that pharyngeal gill pouches appear briefly during the early embryology of mammals, including man, is thought to indicate that the distant ancestors of land vertebrates were aquatic.

Life histories have also played an important role in phylogenetic studies. The stages through which plants pass during their life cycles are particularly important sources of information, as we shall see when we examine the algae and the fungi, for example.

The morphology of the adult and embryo, combined when possible with information from the fossil record and from life histories, has traditionally been the basic source of data on which phylogenetic hypotheses have been based. In addition, comparative physiology, comparative behavior, and comparative ecology have supplied valuable information and will doubtless increase in importance as the effort to interrelate form and function grows. More recently, techniques have been developed for comparing the proteins of different species (e.g. those in the blood) by means of chromatography or electrical separation (electrophoresis), and the use of such techniques will probably become more widespread in the future. A technique has even been developed for determining the degree of similarity between the DNA of different species.

Whenever new techniques for studying organisms are devised, the data they produce provide another source of information for the systematist (or taxonomist). In modern biology, it has become his task to gather together and correlate all that is known about the organisms he studies, and to try to reconstruct, in the light of modern evolutionary theory, some sort of intelligible picture of the organisms and their rela-

DIVERGENT
EVOLUTION

PARALLEL
EVOLUTION

CONVERGENT
EVOLUTION

17.13. Patterns of evolution. In divergent evolution, one stock splits into two, which become more and more unlike as time passes. In parallel evolution, two related species evolve in much the same way for a long period of time, probably in response to similar environmental selection pressures. Convergent evolution occurs when two groups that are not closely related come to resemble each other more and more as time passes, usually because they occupy similar habitats and have adopted similar environmental roles.

tionships with one another. The systematist is thus just what the term implies. He is the one who tries to fit together into an orderly system all of the information gathered about the organisms by anatomists, paleontologists, cytologists, physiologists, geneticists, embryologists, ethologists, ecologists, biochemists, and still other specialists.

The problem of convergence. Whether an investigator is using traditional morphological data or is making DNA comparisons, whether he is obtaining information from physiology or from behavior or from life histories, he is still faced with the problem of interpreting the similarities and differences that he finds. He must always ask himself whether close similarities in a particular character really indicate close phylogenetic relationship or whether they simply reflect similar adaptation to the same environmental situation. The latter phenomenon is common in nature and is a frequent source of confusion in phylogenetic studies. When organisms that are not closely related become more similar in one or more characters because of independent adaptation to similar environmental situations, they are said to have undergone convergent evolution, and the phenomenon is called convergence (Fig. 17.13). Whales, which are mammals descended from terrestrial ancestors, have evolved flippers from the legs of their ancestors; those flippers superficially resemble the fins of fish, but the resemblances are due to convergence and they do not indicate a close relationship between whales and fish. Both arthropods and terrestrial vertebrates have evolved jointed legs and hinged jaws, but these similarities do not indicate that arthropods and vertebrates have evolved from a common ancestor that also had jointed legs and hinged jaws; there is good reason to think that these two groups of animals evolved their legs and jaws independently and that their legless ancestors were not closely related. The "moles" of Australia are not true moles but marsupials (mammals whose young are born at an early stage of embryonic development and complete their development in a pouch on the mother's abdomen); they occupy the same habitat in Australia as do the true moles in other parts of the world and have, as a result, convergently evolved many startling similarities to the true moles (Fig. 17.14).

The preceding discussion makes it evident that when systematists find similarities between two species, they must try to determine whether the similarities are probably *homologous* (inherited from a common ancestor) or merely *analogous* (similar in function and often in superficial structure but of different evolutionary origins). Thus the wings of robins and those of bluebirds are considered homologous; i.e. the evidence indicates that both derive from the wings of a common avian ancestor. But the wings of robins and the wings of butterflies are only analogous, because, though functionally similar structures, they were not inherited from a common ancestor but evolved independently and from different ancestral structures.

It is always important to indicate in what sense two structures are considered homologous or analogous. Thus the wings of birds and the wings of bats are not homologous as wings, for they evolved independently, but they contain homologous bones—both types of wings having evolved from the forelimbs of ancient vertebrates that were ancestors to both birds and mammals. Similarly, the flippers of whales

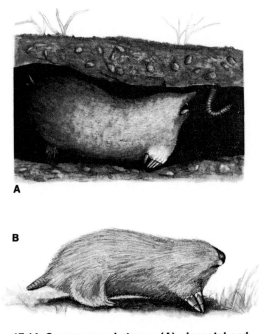

A

B

17.14. Convergence between (A) placental and (B) marsupial moles. These two animals are not related, but they have similar habits and have convergently evolved many startling similarities.

and seals evolved independently of each other, but both evolved from the front legs of land-mammal ancestors. Thus the flippers are homologous in the sense that both are forelimbs, with the same basic bone structure as other vertebrate forelimbs, but the modifications that make them flippers are analogous, not homologous.

Phylogeny and classification

To deal with the vast array of organic diversity—over a million known species of animals and over 350,000 known species of plants—biologists obviously need some sort of system by which species can be classified in a logical and meaningful manner. Many different kinds of classifications are possible. We could, for example, classify flowering plants according to their color: all white-flowered species in one group, all red-flowered species in a second group, all yellow-flowered species in a third group, etc. Or we could classify these same plants according to their average height. Or we could classify the plants according to the environment in which they grow. Each of these three classification systems, and many others that could be devised, would impose a measure of order, but how meaningful, how informative, would they be? Obviously the information they would convey—about flower color, about average height, about habitat—is of an incidental kind that would fail to set apart fundamentally different organisms.

The classification system used in biology today, by contrast, conveys morphological information to the morphologist, physiological information to the physiologist, ecological information to the ecologist, and so forth. It is a system based on phylogenetic relationships that automatically encodes information about every aspect of the organisms it classifies.

The classification hierarchy. Suppose you had to classify all the people on earth on the basis of where they live. You would probably begin by dividing the entire world population into groups based on country. This subdivision separates inhabitants of the United States from the inhabitants of France or those of Argentina, but it still leaves very large groups that must be further subdivided. Next, you would probably subdivide the population of the United States by states, then by counties, then by city or village or township, then by street, and finally by house number. You could do the same thing for Mexico, England, Australia, and all the other countries (using whatever political subdivisions in those countries correspond to states, counties, etc. in the United States). This procedure would enable you to place every individual in an orderly system of hierarchically arranged categories, as follows:

Country
State
County
City
Street
Number

Note that each level in this hierarchy is contained within and is partly determined by all levels above it. Thus, once the country has been determined as the United States, a Mexican state or a Canadian province is excluded. Similarly, once the state has been determined as Pennsylvania, a county in New York or one in California is excluded.

The same principles apply to the classification of living things on the basis of phylogenetic relationships. Again a hierarchy of categories is used, as follows:

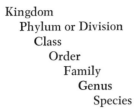

Kingdom
Phylum or Division
Class
Order
Family
Genus
Species

Each category in this hierarchy is a collective unit containing one or more groups from the next-lower level in the hierarchy. Thus a genus is a group of closely related species, a family a group of related genera, an order a group of related families, a class a group of related orders, etc. The species in any one genus are believed to be more closely related to each other than to species in any other genus; the genera in any one family are believed to be more closely related to each other than to genera in any other family; and the families in any one order are believed to be more closely related to each other than to families in any other order; etc.

Table 17.2 gives the classification of six different species. Notice that the table shows us immediately that the six species are not closely related, but that man and wolf are more closely related to each other, both being Mammalia, than either is to a bird such as the Herring Gull. And it shows us that the mammals and the bird are more closely related to each other than to the housefly, which is in a different phylum, or to the moss or the red oak, which are in a different kingdom. These relationships are correlated with similarities and differences in the morphology, physiology, and ecology of the six species. The particular species used in this table are all well known, and the relationships between them are probably intuitively clear to you. But many species are not so well known, and the relationships between them and other species not so clear. Much research may be necessary before they can be fitted into the classification system with any degree of certainty, and their assignment to genus or family (or even order) may have to be changed as more is learned about them.

Hierarchical classification systems similar to the one in current use have been employed by naturalists for many centuries. The current system dates from the work of the great Swedish naturalist Carolus Linnaeus (1707–1778), who wrote extensively on the classification of both plants and animals. Now, the rationale upon which Linnaeus based his system was necessarily very different from the phylogenetic one employed today. He worked a century before Darwin, and he had no idea of evolution, doubtless conceiving of each species as an immutable entity, the product of a divine creation. He was simply grouping organisms according to similarities, primarily morphological. That his results were so similar to those obtained today is a reflection of the fact that morphological characters, being products of evolution, tell us much about evolutionary relationships. The Linnaean system, however, produces results quite different from those of the modern phylogenetic system whenever it has to deal with cases of convergence or cases where gross morphological similarities are a poor indicator of phylogenetic relationships.

An outline of a modern classification of living things is given on pp. 546–549.

Nomenclature. The modern system of naming species also dates from Linnaeus. Before him, there had been little uniformity in the designation of species. Some species had a one-word name, others had two-word names, and still others had names consisting of long descriptive phrases. Linnaeus simplified things by giving each species a name consisting of two words: first, the name of the genus to which the species belongs and, second, a designation for that particular species. Thus the honeybee is called *Apis mellifera*. A certain species of carnation is called *Dianthus caryophyllus;* other species in the genus *Dianthus* have the same first word in their names, but each has its own specific designation (e.g. *Dianthus prolifer, Dianthus barbatus, Dianthus deltoides*). No two species can have the same name. Notice that the names are always Latin (or Latinized) and that the genus name is capitalized while the specific name is not. Both names are customarily printed in italics (underlined if handwritten or typed). The correct name for any species, according to the present rules, is usually the oldest validly proposed name.

The same Latin scientific names are used throughout the world. This uniformity of usage ensures that each scientist will know exactly which species another scientist is discussing. There would be no such assurance if common names were used; not only does a given species have a different common name in each language, but it often has two or three names in the same language.

Table 17.2. Classification of six species

Category	Haircap moss	Red oak	Housefly	Herring Gull	Wolf	Man
Kingdom	Plantae	Plantae	Animalia	Animalia	Animalia	Animalia
Phylum or Division	Bryophyta	Tracheophyta	Arthropoda	Chordata	Chordata	Chordata
Class	Musci	Angiospermae	Insecta	Aves	Mammalia	Mammalia
Order	Bryales	Fagales	Diptera	Charadriiformes	Carnivora	Primates
Family	Polytrichaceae	Fagaceae	Muscidae	Laridae	Canidae	Hominidae
Genus	*Polytrichum*	*Quercus*	*Musca*	*Larus*	*Canis*	*Homo*
Species	*commune*	*rubra*	*domestica*	*argentatus*	*lupus*	*sapiens*

REFERENCES

ALLEE, W. C., A. E. EMERSON, O. PARK, T. PARK, and K. P. SCHMIDT, 1949. *Principles of Animal Ecology.* Saunders, Philadelphia. (See esp. Section V, "Ecology and Evolution.")

DARWIN, C., 1859. *The Origin of Species by Means of Natural Selection.* John Murray, London. (Many modern editions are available, e.g. Modern Library, 1948, or, in paperback, New American Library, 1958.)

DOBZHANSKY, T., 1951. *Genetics and the Origin of Species,* 3rd ed. Columbia University Press, New York.

EATON, T. H., 1970. *Evolution.* Norton, New York.

EHRLICH, P. R., and R. W. HOLM, 1963. *The Process of Evolution.* McGraw-Hill, New York.

FORD, E. B., 1970. *Ecological Genetics,* 3rd ed. Barnes & Noble, New York.

GRANT, V., 1963. *The Origin of Adaptations.* Columbia University Press, New York.

MAYR, E., 1942. *Systematics and the Origin of Species.* Columbia University Press, New York.

———, 1969. *Principles of Systematic Zoology.* McGraw-Hill, New York.

———, 1970. *Populations, Species, and Evolution.* Belknap Press of Harvard University Press, Cambridge, Mass.

MOODY, P. A., 1970. *Introduction to Evolution,* 3rd ed. Harper & Row, New York.

SIMPSON, G. G., 1967. *The Meaning of Evolution,* 2nd ed. Yale University Press, New Haven, Conn.

STEBBINS, G. L., 1950. *Variation and Evolution in Plants.* Columbia University Press, New York.

SUGGESTED READING

BROWER, L. P., 1969. "Ecological Chemistry," *Scientific American,* February. (Offprint 1133.)

CAVALLI-SFORZA, L. L., 1969. " 'Genetic Drift' in an Italian Population," *Scientific American,* August. (Offprint 1154.)

EISELEY, L. C., 1956. "Charles Darwin," *Scientific American,* February. (Offprint 108.)

GRANT, V., 1951. "The Fertilization of Flowers," *Scientific American,* June. (Offprint 12.)

KETTLEWELL, H. B. D., 1959. "Darwin's Missing Evidence," *Scientific American,* March. (Offprint 842.)

LACK, D., 1947. *Darwin's Finches.* Cambridge University Press, New York. (Paperback edition, Harper Torchbooks, 1961.)

———, 1953. "Darwin's Finches," *Scientific American,* April. (Offprint 22.)

SAVAGE, J. M., 1969. *Evolution,* 2nd ed. Holt, Rinehart & Winston, New York.

STEBBINS, G. L., 1971. *Processes of Organic Evolution,* 2nd ed. Prentice-Hall, Englewood Cliffs, N.J.

WALLACE, B., and A. M. SRB, 1964. *Adaptation,* 2nd ed. Prentice-Hall, Englewood Cliffs, N.J.

WILLS, C., 1970. "Genetic Load," *Scientific American,* March. (Offprint 1172.)

CHAPTER 18

Ecology

Ecology is usually defined as the study of the interactions between organisms and their environment. "Environment" is given a very broad meaning here; it is taken to embrace all those things extrinsic to the organism that in any way impinge on it—not only light, temperature, rainfall, humidity, and topography, but also parasites, predators, mates, and competitors. Anything not an integral part of a particular organism is considered part of that organism's environment.

Much of the first half of this book dealt with life on the molecular, cellular, tissue, organ, organ-system, or individual level. Although ecology is often concerned with phenomena on these levels (e.g. the osmotic interactions between an organism and its environmental medium), the present chapter will deal primarily with three higher levels of organization: *populations,* which are groups of individuals belonging to the same species; *communities,* which are units composed of all the populations living in a given area; and *ecosystems,* which are communities and their physical environments considered together. Each of these designations may be applied to a small local entity or to a large widespread one. Thus the sycamore trees in a given woodlot may be regarded as a population, or all the sycamore trees in the eastern United States may be so regarded. Similarly, a small pond and its inhabitants or the forest in which the pond is located may be treated as an ecosystem.

The various ecosystems are linked to one another by biological, chemical, and physical processes. Energy in various forms, gases, inorganic chemicals, and organic compounds can cross ecosystem boundaries through meteorological factors such as wind and precipitation, geological ones such as running water, and biological ones such as the movement of animals. Thus the entire earth is itself a true ecosystem, in that no part is fully isolated from the rest. The global ecosystem is ordinarily called the *biosphere.* Except for energy, the biosphere is self-sufficient; all other requirements for life, such as water, oxygen, and nutrients, are supplied by utilization and recycling of materials already contained within the system.

THE ECONOMY OF ECOSYSTEMS

The flow of energy

Life depends, ultimately, on radiation from the sun. The familiarity of this statement should not blunt one to its significance. With the exception of the relatively unimportant chemosynthetic organisms, all forms of life obtain their high-energy organic nutrients, either directly or indirectly, from photosynthesis. Photosynthesis, however, utilizes less than one tenth of one percent of the solar energy reaching the surface of the earth, and of this fraction it is estimated that plants use from 15 to 50 percent, depending on the community, for their own metabolism. What is left over is known as net photosynthesis or *net primary productivity.* The total net primary productivity of the biosphere, which is estimated at about 6×10^{20} gram calories of energy per year,[1] constitutes the energy base for heterotrophic life on earth. Heterotrophic organisms—animals and some plants—obtain the energy they need by eating green plants or by eating other heterotrophic organisms that ate green plants.

The sequence of organisms through which energy may move in a community is customarily called a *food chain.* In most real communities, there are so many different possible food chains, so complexly intertwined, that together they form a community *food web* (Fig. 18.1). No matter how long a food chain or how complex a food web may be, however, certain basic characteristics are always present. Every food chain or web begins with the autotrophic organisms (green plants in the vast majority of cases) that are the *producers* for the community. And every food chain or web ends with *decomposers,* the organisms of decay—generally bacteria and fungi that release simple substances re-usable by the producers. The links between the producers and the decomposers are more variable. The producers may die and be acted upon directly by the decomposers, in which event there are no intermediate links. Or the producers may be eaten by *primary consumers,* the herbivores. These, in turn, may be either acted upon directly by decomposers or fed upon by *secondary consumers* such as carnivores or parasites or scavengers (Fig. 18.2).

Ecologists speak of the successive levels of nourishment in the food chains of a community as *trophic levels.* Thus all the producers together constitute the first trophic level; the primary consumers (herbivores) constitute the second trophic level; the herbivore-eating carnivores constitute the third trophic level; and so on. The species that comprise each trophic level differ from one community to another, but in general the pattern is the same. At each successive trophic level there is loss of energy from the system, a loss predictable from the Second

---

[1] A gram calorie (cal.) is the amount of heat energy required to raise the temperature of one gram of water from 14.5°C to 15.5°C. One kilocalorie (kcal. or Cal.) is equal to 1,000 gram calories. Food Calories for human diets are usually measured in kilocalories.

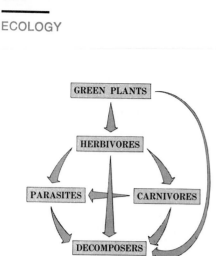

18.2. Diagram of the relationships between the principal trophic levels in an ecosystem. The green plants are the producers, which are eaten by the herbivores, the primary consumers. The primary consumers may in turn be eaten by parasites or carnivores, the secondary consumers. Producers or consumers may die and become food for decomposer organisms.

Law of Thermodynamics, which states that every energy transformation involves loss of some usable energy (see p. 72). Because of this unavoidable loss, the total amount of energy at each trophic level is lower than at the preceding level, usually much lower. There is less energy in the herbivores of a community than in the plants of that community, and there is less energy in the carnivores than in the herbivores, etc. Thus the distribution of energy within a community can be represented by a pyramid, with the first trophic level (producers) at the base and the last consumer trophic level at the apex (Fig. 18.3). Related to this *pyramid of energy* is the *pyramid of biomass.* In general, the decrease of energy at each successive trophic level means that less biomass can be supported at each level. Thus the total mass of carnivores in a given community is almost always less than the total mass of herbivores.

Cycles of materials

We have seen that energy is steadily drained from the ecosystem as it is passed along the links of a food chain. The system cannot continue

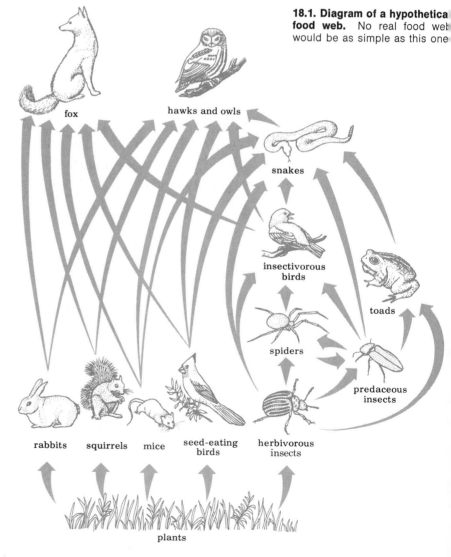

18.1. Diagram of a hypothetical food web. No real food web would be as simple as this one

functioning without a constant input of energy from the outside. In other words, there is no such thing as an energy cycle. But this is not the case with materials. The same materials can and must be used over and over again, and hence can be passed round and round through the ecosystem indefinitely. We can, therefore, speak of cycles of materials. Let us examine several such cycles.

The carbon cycle. The carbon dioxide contained in the atmosphere and dissolved in water constitutes the reservoir of inorganic carbon from which almost all organic carbon is derived. It is photosynthesis, largely by green plants, that extracts the carbon from this inorganic reservoir and incorporates it into the complex organic molecules characteristic of living substance (Fig. 18.4). Some of these organic molecules are soon broken down again by the plants, which release the carbon as CO_2 in the course of their own respiration. But much of the carbon remains in the plant bodies until they die or are eaten by animals. The carbon obtained from plants by animals may be released as CO_2 during respiration, or it may be eliminated in more complex compounds in the body wastes, or it may remain in the animals until they die. Usually the wastes from animals and the dead bodies of both plants and animals are broken down (respired) by the decomposers, and the carbon is released as CO_2. Notice that whether the carbon follows a short pathway involving only one or two trophic levels or a longer pathway involving three or more, most of it eventually returns as CO_2 to the air or water whence it started. This is, then, a true cycle (or rather a complex of interlocking cycles); carbon is constantly moving from the inorganic reservoir to the living system and back again.

The pathways just outlined are all pathways through which carbon moves rather rapidly. Complete passage through the system may take only minutes or hours or at most a few years. There are alternative pathways, however, that take much longer. The dead bodies of organisms occasionally fail to be decomposed promptly and are converted instead into coal, oil, gas, rock (particularly limestone), or diamond. Carbon in these forms may be removed from circulation for very long periods, perhaps permanently; but some of it may eventually return to the inorganic reservoir if the coal, oil, and gas (the fossil fuels) are burned or

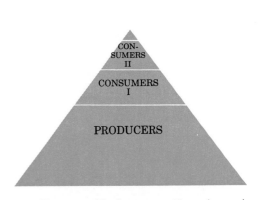

18.3. The pyramid of energy. There is much more energy at the producer level in an ecosystem than at the consumer levels, and there is more at the primary consumer level than at the secondary consumer level.

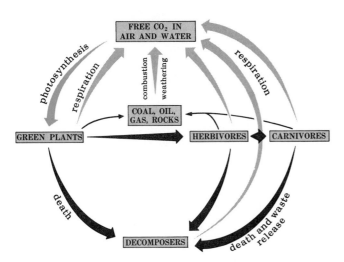

18.4. The carbon cycle.

if the rocks are sufficiently weathered. Man has of course greatly accelerated the return of such carbon to the active cycle. Many ecologists are concerned that the consequent increase in atmospheric CO_2 levels will cause drastic changes in the temperature at the earth's surface.

The nitrogen cycle. Another critical element in community metabolism is nitrogen, a constituent of the amino acids in proteins and of the nucleotides in nucleic acids. The reservoir of inorganic nitrogen is the gaseous N_2, which constitutes roughly 78 percent of the atmosphere. But N_2 has very little biological activity. It enters the bodies of all organisms, but in most cases comes out again without having played any significant role in their life processes. Some microorganisms, however—a few bacteria, the blue-green algae, and a few fungi—can use N_2 in the synthesis of substances usable by other organisms. This process is known as *nitrogen fixation.* Although some nitrogen fixation may also occur as a result of electrical discharges, such as lightning, the amount is minimal, and it is biological nitrogen fixation by microorganisms that provides most of the usable nitrogen for the earth's ecosystems (Fig. 18.5).[2]

Some of the nitrogen-fixing bacteria live in a close symbiotic relationship with the roots of higher plants, where they occur in prominent *nodules* (Fig. 18.6). The legumes (plants belonging to the pea family —beans, clover, alfalfa, lupine, etc.) are particularly well known for their numerous root nodules, but plants of some other families have them also. Other nitrogen-fixing microorganisms live free in soil or water. All of these nitrogen-fixing microorganisms can reduce N_2 to ammonia (NH_3).

The symbiotic bacteria in root nodules promptly release much of the fixed nitrogen they produce into the host plant's cytoplasm, primarily in the form of amino acids. Consequently legumes can grow well in soils very poor in nitrogen. The bacteria in their nodules not only supply

[2] Nitrogen is also provided by industrial processes. The amount is rising steadily as use of commercial fertilizers becomes more widespread.

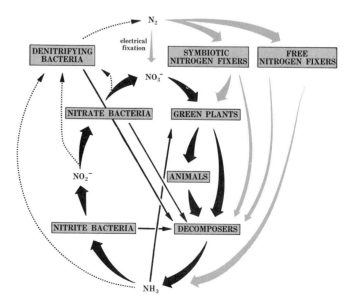

18.5. The nitrogen cycle. Gray arrows indicate paths of nitrogen fixation, by which nitrogen from the atmosphere is added to the main soil–organism part of the cycle. Dotted arrows indicate the paths of denitrification, by which nitrogen is removed from the main soil–organism cycle and returned to the atmosphere.

the plants with all the fixed nitrogen they need, but actually produce a surplus, some of which is excreted into the soil. Thus legumes tend to increase the fertility of the soil in which they grow.

The nitrogen-fixing microorganisms that live free in the soil or water release ammonia into the surrounding medium. When they die, the fixed nitrogen in their cells is broken down to ammonia by decomposer organisms. The decomposers act in the same way upon the organic nitrogen compounds in the bodies of green plants or animals or other microorganisms when they die, and upon the nitrogen compounds in the urine and feces of animals. Some of this free ammonia is picked up as ammonium ions (NH_4^+) by the roots of higher plants and incorporated into more complex compounds. But most flowering plants utilize nitrate in preference to ammonia.

Nitrate is produced from ammonia in the soil by nitrifying bacteria. The process of *nitrification* is usually accomplished by two different groups of bacteria, working in sequence. The first group converts ammonium ions into nitrite (NO_2^-), and the second group converts this nitrite into nitrate (NO_3^-) and releases it into the soil, where it can be picked up by the roots of plants.

Notice that nitrogen can cycle repeatedly from plants to decomposers to nitrifying bacteria to plants without having to return to the gaseous N_2 state in the atmosphere. In this respect, the nitrogen cycle differs from the carbon cycle, where every turn of the cycle includes a return of CO_2 to the atmosphere. Although nitrogen need not return to the atmosphere at every turn of the cycle, there is a steady drain of some of it away from the soil or water and back to the atmosphere. This is because some bacteria carry out a process of *denitrification,* converting ammonia or nitrite or nitrate into N_2 and releasing it. In short, the denitrifying bacteria remove nitrogen from the soil–organism part of the nitrogen cycle and return it to the atmosphere, while the nitrogen-fixing microorganisms do the reverse: They take nitrogen from the atmosphere and add it to the soil–organism part of the cycle.

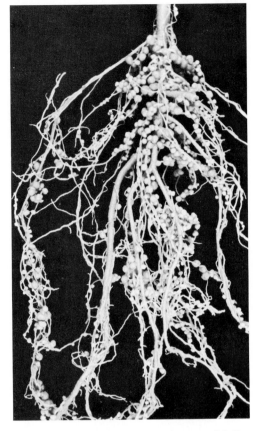

18.6. Photograph of roots of a legume (bird's-foot trefoil), showing nodules. [Courtesy Nitragin Co., Milwaukee, Wis.]

The phosphorus cycle. Another mineral essential to life is phosphorus. Like nitrogen, it is one of the chief ingredients in commercial fertilizers. Unlike carbon and nitrogen, for which the reservoir is the atmosphere, phosphorus has its reservoir in rocks (Fig. 18.7).

Under natural conditions, much less phosphorus than nitrogen is available to organisms; in natural waters, for example, the ratio of phosphorus to nitrogen is about 1 to 23. However, man, by mining roughly three million tons each year, has greatly accelerated the movement of this mineral from the rocks to the water–organism part of the cycle. One result has been a startling increase in algal populations, for which phosphorus had previously been the principal limiting resource in many ponds and streams. The proliferation of algae, in turn, has led to other major changes in the ecology of our fresh waters, as we shall see later in connection with ecological succession.

This brief summary of some of the biogeochemical cycles will suggest the complexity of the movement of materials through an ecosystem, the interdependence of the different species within it, and, in particular, the fundamental and essential role played by microorganisms.

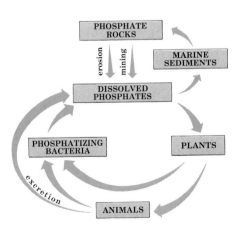

18.7. A simplified version of the phosphorus cycle. The phosphatizing bacteria produce the inorganic phosphate needed by plants—e.g. in both cyclic and noncyclic photophosphorylation.

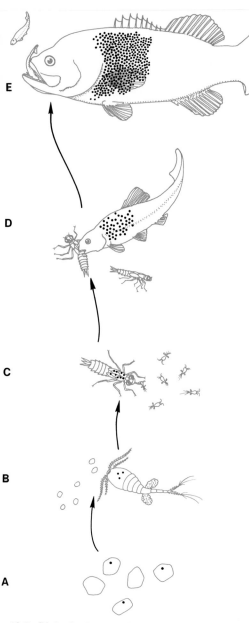

18.8. Biological magnification. (A) Some in-dividuals of a single-celled plant species at the bottom of a food chain have picked up a small amount of a stable nonexcretable chemical (black dots). (B) *Cyclops,* a small crustacean, incorporates into its own tissues the chemical from all the infected plants it eats. (C) A dragonfly nymph stores all the chemical ac-quired from the numerous *Cyclops* it eats. (D) Further magnification occurs when a minnow eats many of the infected dragonfly nymphs. (E) When a bass, the top predator in this food chain, eats many infected minnows, the result is a very high concentration of a chemical that was much less concentrated in the organisms lower in the chain.

Because these are not observed as easily as the larger plants and ani-mals, we have a tendency to forget about them, or to think only of the harmful microorganisms, especially those that cause diseases; and thus we fail to recognize that our continued existence depends on other microorganisms.

Man-made chemicals. Modern industry and agriculture have been releasing vast quantities of new or previously rare chemicals into the environment. The U.S. Food and Drug Administration has esti-mated the number of such compounds now in the environment at about half a million, with 400–500 new ones created each year. The path-ways through ecosystems are known for only a few. Some will probably be incorporated into the natural biogeochemical cycles and degraded to harmless simpler substances. But many are so different from any natu-rally occurring substances that we have no idea as yet what their eventual fate or effects on the biosphere may be. Many of the by-products of industrial processes have so far not even been fully characterized chemi-cally. Will some of these prove harmful to life? The answer is surely yes. But which will be harmful, and how harmful, and to what organ-isms? It is with these questions that much ecological research in the future must deal.

The matter is complicated by the fact that a substance may not be harmful in the form in which it is released, but may be changed by microorganisms, or by natural physical processes, into some other sub-stance with vastly different properties. A case in point is the mercury pollution of bodies of water near some plastics factories. The mercury was originally released in an insoluble and nontoxic form that was thought to be stable. When it settled in the bottom mud, however, microorganisms converted it into methyl mercury, a water-soluble com-pound that accumulates in organisms. The mercury poisoning that resulted was most severe in man and other so-called *top predators,* which are predators at or near the top of long food chains.

The harmful effect on top predators of the mercury from plastics factories (and of the mercury, principally from fungicides, that has recently caused the fish in many streams and lakes to be declared unsafe for human consumption) is an example of a common phenomenon called *biological magnification.*[3] If a persistent chemical,[3] when in-gested, is retained in the body rather than excreted, then that chemical will tend to become more and more concentrated as it is passed up the food chain. Figure 18.8 shows why this is so.

DDT (a persistent man-made compound that has become so pervasive in the biosphere that it can now be found in the fatty tissues of nearly every living organism) has had more severe effects on preda-tory birds such as the Bald Eagle and Peregrine Falcon than on seed-eating birds because of biological magnification. Some investigators have reported that the reproductive rate of eagles and falcons has been calamitously reduced because DDT interferes with deposition of cal-cium in the eggshells, with the result that the thin-shelled eggs are easily broken and few birds hatch.

[3] A persistent chemical is one comparatively stable under natural biological and environmental conditions.

The physical environment

Climate, weather, and microclimate.   We are all familiar with broad regional differences in temperature, humidity, rainfall, and other meteorological factors. Temperatures range between the extremes of hot tropical regions and permanently frozen polar areas; rain may fall several times a day on a rain forest or only a few times a decade on some deserts. The average pattern of such meteorological factors over an extended period is called the climate of a region. The climate influences the kind of vegetation and animal life that may inhabit the region, and the vegetation, in turn, influences the climate.

Although the climate of a desert is dry, at any given time rain may be falling. Thus we distinguish between climate, defined above, and weather, which is the immediate pattern of meteorological factors. Weather has obvious day-to-day effects on living organisms. Thus a late frost after a warm spell can destroy the potential productivity of a plant seedling; the reproductive success of many animals will be very different in unusually dry and unusually wet years. In other words, the climate of an area may play an important role in determining whether a given species will occur in the area, but the weather is what is important to the individuals making up that species at any particular moment.

Weather variations can be very local. Someone living only a few miles from the weather station serving his community may hear over the radio that the temperature is 20°F. Checking his own outdoor thermometer, he may find that it registers 12°F—8° less than the temperature at the station. This difference, which could easily be greater, may reflect a difference in microclimate. Within a single lot, variations of several degrees may occur between sheltered and exposed places, between points near the ground and points at various elevations above the ground. The same sorts of highly local variations may be found in humidity, wind velocity, amount of sunlight, soil type, etc. Accordingly, one should not expect to find exactly the same kinds of organisms living at all points within even a very small area. Plants and animals don't live under generalized regional conditions; they live under microclimatic conditions that may vary radically over a given region. It is the microclimatic conditions that ecologists must analyze if they want to understand fully the conditions important to the organisms they study.

The medium and the substratum.   Sea water is the most stable medium in which organisms can live. It undergoes remarkably little fluctuation in salt content, oxygen and carbon dioxide content, pH, or temperature. Only near the surface of the ocean or near the shore are fluctuations likely to be appreciable, and even there they are usually quite slow. Fresh water undergoes much greater fluctuations, particularly in very small ponds. In the course of a year, its temperature may vary by many degrees. And since temperature has a marked effect on the solubility of oxygen in water, the oxygen content may also vary greatly. The amount and type of soil and debris carried into streams and lakes by rainwater running off the neighboring land may profoundly alter the mineral content and the pH of the streams and lakes, and they may radically change the depth to which light can penetrate. Air, by contrast, seldom undergoes significant changes in chemical makeup, but it is subject to often extreme and rapid fluctuations in temperature and humidity. The amount of oxygen is much greater in air than in water.

A few organisms spend much of their lives suspended in air, and many spend their entire lives suspended in water. But most land organisms and many aquatic ones spend much of their time attached to or moving upon a solid surface, or substratum. The characteristics of that substratum are important components of the organisms' physical environment. Plants, for example, are often very sensitive to small differences in the depth, physical properties, and chemical content of soils —particularly of topsoil, a usually well-aerated layer that, containing much organic material, is favorable to root growth.

Most soils are a complex of mineral particles, organic material, water, soluble chemical compounds, and air. In this complex, the dominant components by far are the mineral particles, which consist largely of compounds of silicon and aluminum. They vary in size from tiny clay particles through silt to coarse sand grains. The proportions of clay, silt, and sand particles in any given soil determine many of its other characteristics. For example, very sandy soils, which contain less than 20 percent of silt and clay particles, have many air-filled spaces, but they are so porous and their particles have so little affinity for water that water rapidly drains through them and they are unsuitable for growth of many kinds of plants. As the percentage of clay particles increases, the water retention of the soil also increases until, in excessively clayey soils, the drainage is so poor and the water is held so tightly to the particles that the air spaces become filled with water; few plants can grow in such waterlogged soil. Although different species of plants are adapted to different soil types, most do best in soils of the type known as *loams,* which contain fairly high percentages of each size of particle. In such a soil, there is good but not excessive drainage and there is good aeration; the soil particles are surrounded by (or contain) a shell of water, but there are numerous air-filled spaces between them.

Loams usually also contain considerable amounts of organic material (roughly 3–10 percent), often in the form of **humus,** which consists mostly of breakdown products from cellulose and lignin. As the organic material decomposes, it releases inorganic substances required for good plant growth and thus contributes to soil fertility. Since humus usually has a rather porous spongy texture, it helps loosen clayey soils and increase the proportion of pore spaces, thus promoting drainage and aeration. It has the opposite effect on sandy soils, where it tends to reduce pore size by binding the sand grains together, thereby increasing the amount of water held in the soil.

Chemical analyses that give the total amount of the various nutrient ions present in soils can be misleading. A certain proportion of these ions may not be available to the plants. If, for example, water percolates downward very rapidly and in large quantities, as it tends to do in soils with a low proportion of clay particles, it will leach away many important ions; nitrate ions are especially susceptible to leaching, and sulfate, calcium, and potassium ions may all quickly be removed from the soil. More fundamentally, the availability of ions to plants depends on a complex equilibrium between ions free in the soil water, and hence available to the plants, and ions adsorbed on the surface of colloidal clay and organic particles, and hence not available to them. This equilibrium is shifted by a variety of factors, particularly acidity.

Acidity not only influences the availability of nutrient ions—iron, manganese, phosphate, and others; it also affects the activity of soil organisms, many of which are inhibited by high acidity. Though many plants grow best in slightly acid soils, most do not do well in strongly acid ones.

Among the many factors that may influence the acidity, and hence the ionic makeup, of the soil are atmospheric conditions. When rain contains sulfuric acid, as it may through industrial discharge of sulfur dioxide, the acid tends to displace nutrient ions, such as calcium, with the result that they are leached more rapidly from the soil into streams and lakes.

Soils cannot be fully understood without considering the effects of the plants and animals living in and on them. Plant roots break up the soil, remove substances from it, and add other substances. Plant shoots, by shielding the soil beneath them, alter the patterns of rainfall, humidity, light, and wind to which the soil is subject. And when the plants die, their substance adds organic material to the soil, changing both its physical and chemical makeup. Microorganisms in the soil alter its composition profoundly. Soil animals, such as earthworms, constantly work the soil, breaking down its organic components and moving materials between different soil layers. Thus we see that organisms are not only influenced by their physical environment but that they, in turn, modify that environment.

In a study at Hubbard Brook, New Hampshire, investigators dramatically demonstrated some of the effects vegetation may have on soil. They first obtained accurate measurements of the nutrient input and output of a particular watershed over a period of several years and then cleared the watershed of all its vegetation. When they again monitored input and output, they found not only a marked rise in the volume of runoff water (during one period it was actually 418 percent greater), but also an extraordinary loss of soil fertility. The runoff output of nitrate was as much as 45 times higher than in undisturbed watersheds; net losses of potassium were 21 times greater, of calcium 10 times greater. Apparently removal of the vegetation had so altered the chemistry of the soil that nutrients were bound less tightly to soil particles and hence were rapidly leached away. It can be seen, then, that man's wholesale destruction of vegetation results not only in increased erosion by wind and water, a long-familiar consequence, but also in severe loss of fertility in the soil that remains. The stability of the physical part of an ecosystem clearly depends on the production and decomposition of organic matter, and on an orderly flow of nutrients between the living and the nonliving components of the system.

Irrigation has been viewed as a way of greatly increasing the productivity of dry areas. In many cases, however, it leads simply to accelerated erosion; in the United States, for example, an estimated 2,000 irrigation dams are now useless impoundments of silt, sand, and gravel. In other cases, irrigation leads to rapid salinization of the soil, until eventually there is so much salt that plants cannot grow. The salinization may occur because adding water to land overlying a salty water table causes the ground water to rise, the salt thus being carried into the topsoil, or because salts, originally present in low concentrations in the irrigation water, accumulate in the soil as the water evaporates. The Indus valley of West Pakistan, the largest irrigated region in the world—over 23 million acres watered by canals—fell victim to salinization. The agricultural capabilities of the Nile valley may similarly be destroyed through the irrigation system made possible by the new Aswan high dam.

This brief discussion of soils and what can happen to them illustrates how complex, and at the same time how fragile, the earth's ecosystems are. Getting to know how ecosystems function, and using this knowledge, have virtually become ethical imperatives if human beings are to continue to inhabit the Spaceship Earth.

POPULATIONS AS UNITS OF STRUCTURE AND FUNCTION

Whatever aspect of ecology is being considered—whether it is the flow of energy and materials in the ecosystem, or symbiosis, or competition, or community structure and change—the reference is generally to species, or local subdivisions of species, rather than to individuals. Hence the species and the local populations of which it is composed are levels of biological organization to which we must direct special attention.

Population density and its regulation

The density of a population can be expressed in number of individuals per unit area or volume (e.g. 50 pine trees per acre) or in terms of biomass per unit area or volume (e.g. 4 tons of clover per acre). Variable as population densities are usually found to be, there is a theoretical upper limit, one determined by the total energy flow in the ecosystem (i.e. its productivity), the trophic level to which the species in question belongs, and the size and metabolic rate of the individuals. In other words, there is just so much energy available at a particular trophic level in any given ecosystem, and there can be no more biomass than that amount of energy can support. But actual densities often fluctuate at levels well below the theoretical maximum. Study of the factors that influence the densities of real populations in nature is thus a major aspect of modern ecology.

Spacing. Within any given area, the individuals of a population could be distributed uniformly, randomly, or in clumps. Uniform distributions are rare. They occur only where environmental conditions are fairly uniform throughout the area and where, in addition, there is intense competition or antagonism between individuals. Random distributions are also relatively rare. They occur only where the environmental conditions are uniform, where there is no intense competition or antagonism between individuals, and where, moreover, there is no tendency for the individuals to aggregate. Clumping is by far the most common distribution pattern for both plants and animals in nature, for several reasons: (1) The environmental conditions are seldom uniform throughout even a relatively small area. (2) Reproductive patterns frequently favor clumping. (3) Animals often exhibit behavior patterns that lead to active congregation in loose groups or in more organized colonies, schools, flocks, or herds.

A clumped distribution may increase competition for nutrients, food, space, or light, but this deleterious effect is often offset by some beneficial ones. For example, trees growing together in a hedgerow on a wide plain may compete more intensely for nutrients and light than if they were widely separated, but they may be better able to withstand strong winds; and the clump, which has less surface area in proportion to mass than an isolated tree, may be better able to conserve moisture. Aggregations of animals often reduce the rate of temperature change in their midst—an effect particularly important in cold weather. A group of animals may also have an advantage in locating food and in withstanding attacks by predators. Thus the optimum density for population growth and survival is often an intermediate one; undercrowding may be as deleterious as overcrowding.

Births, deaths, and survivorship. The immediate determiners of population size are the numbers of births and deaths per unit time and the average survival time. It is clear that these three population parameters are governed in part by environmental influences, for the actually observed values for these parameters are quite different from the ones theoretically possible.

The customary standard for evaluating actually observed birth rates is the maximum potential birth rate, which is the theoretical maximum under ideal conditions. It can be estimated by determining the birth rate per female of reproductive age (or sometimes simply per 1,000 individuals without regard to sex or age) under the best conditions that can actually be found, usually in the laboratory. The birth rates encountered in nature ordinarily fall well below the maximum, an indication that one way environmental factors can influence population size is to cause changes in the rate of production of new individuals.

The theoretical minimum mortality for a population is the number of deaths expected if all deaths were the result of "old age," i.e. if all individuals lived to the end of their potential life span. Of course, the average longevity in any real population will be far below the potential physiological longevity—an indication, once again, that environmental factors influence population dynamics. Since minimum mortality is not usually found in nature, it is often useful to determine the mortality rates for the various age groups in a population. Such data show us what stages in the life cycle are most susceptible to environmental control.

The curves in Fig. 18.9 illustrate several different survivorship patterns. Curve A approaches the pattern that would be expected if all the individuals in the population realized the average physiologically possible longevity. There would be full survival through all the early age intervals, and then all the individuals would die more or less at once and the curve would fall suddenly and precipitously. Curve D approaches the other extreme, where the mortality is exceedingly high among the very young but where any individual surviving the earliest life stages has a good chance of surviving for a long time thereafter. Between these two extremes is the condition represented by curve C, where the mortality

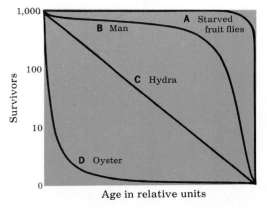

18.9. Four different types of survivorship curves. For an initial population of 1,000 individuals, the curves show the number of survivors (vertical axis) at different ages (horizontal axis), from 0 to the maximum possible age for that species. The vertical coordinate has a logarithmic scale. For explanation, see text. [Modified from E. P. Odum, *Fundamentals of Ecology*, Saunders, 1959, after Deevey.]

rate at all ages is constant. The survivorship curves for most wild-animal populations are probably intermediate between curves C and D, and the curves for most plant populations are probably near the extreme of D. In other words, high mortality among the young is the general rule in nature.

Changes in environmental conditions may radically alter the shape of the survivorship curve for any given population, and the altered mortality rates in turn may have profound effects on the dynamics of the population and on its future size. For example, the enormous increase in the population of human beings has been due mainly to a great reduction in mortality during the early life stages as a result of improvements in sanitation, nutrition, and medical care. These improvements have caused a shift in the human survivorship curve from one intermediate between curves C and D in primitive societies to one approaching A in the most advanced societies. There is no evidence that the shift is due to a rise in the human birth rate (in fact, the birth rate has almost certainly fallen) or, despite all the advances of modern medicine, to an increase in the potential life span.

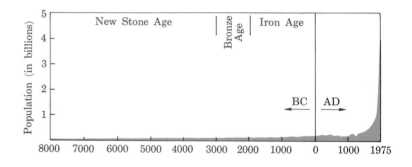

18.10. Growth of the human population of the world. The growth was slow for many thousands of years, but has become very rapid in the last century.

As Fig. 18.10 shows, the human population of the world increased very slowly for many thousands of years, even though the birth rate was probably high (large families ensured continuity despite very high mortality rates, particularly in infants). It has been estimated that there were about 5 million people on the earth ten thousand years ago, about 250 million by the year one, and about 500 million by 1650 A.D. Until about three hundred years ago, then, the human population seems to have doubled approximately every 1,600 years. Now, by contrast, it is doubling every 35 years; the present total is nearly 4 billion. How much longer the human species can continue with such a rate of increase, with such an imbalance between births and deaths, is one of the most pressing questions of our time. Some have argued, in fact, that it is *the* most important question; that all other aspects of the so-called ecological crisis—hunger, poverty, crowding, pollution, accumulation of wastes, destruction of the environment on which all life depends—are the inexorable consequences of a continuing rise in the number of human beings.

Biotic potential and environmental resistance. We have seen that the maximum possible birth rate for a population almost always far exceeds the actual ecological birth rate, and that the potential physiological longevity is usually far greater than the average ecological longevity. In other words, under natural conditions, the number of births is lower and the number of early deaths higher than would be expected under ideal conditions. One way to understand the dynamics of real populations is to find out what to expect of a population under ideal conditions and then try to determine how actual conditions modify this expected pattern. Let us assume, then, that we can study a population that is stationary, has a stable age distribution, faces no predation, parasitism, or competition, and exists in an environment with unlimited resources. The maximum population growth rate under such ideal conditions is called the *biotic potential.*

Every species has an enormous biotic potential. For example, one pair of houseflies starting breeding in April could have 191,010,000,-000,000,000,000 descendants by August if all their eggs hatched and if all the resulting young survived to reproduce. Such calculations played an important role in leading Charles Darwin to formulate his theory of natural selection. He himself calculated the reproductive potential of elephants, concluding that "after a period of from 740 to 750 years there would be nearly nineteen million elephants alive descended from the first pair." Someone has extended Darwin's calculations to show that in 100,000 years one pair of elephants would have so many living descendants that they would fill the visible universe.

That such astounding population explosions do not in fact occur is evidence enough that there are always limiting factors on any actual population, which prevent it from realizing its full biotic potential. The difference between the biotic potential and the actual rate of increase may be considered a measure of the *environmental resistance,* which is simply the sum total of the environmental limiting factors acting on the population in question.

Density-independent and density-dependent limiting factors. If a population were not subject to any limiting factors and could thus realize its full biotic potential, its growth would continue in an exponential fashion (Fig. 18.11A); the growth curve would rapidly approach the vertical. Real growth curves do start out exponentially, but exponential growth cannot long continue, because environmental limiting factors soon come into play. Sometimes the limiting factors are hardly operative until late in the increase, when they suddenly become very effective, usually causing a rapid decline in population density (Fig. 18.12). Such a population growth pattern is characteristic of some small insects with short life cycles; the population grows very rapidly during a period of favorable weather and then falls off abruptly when the weather changes. Notice that the limiting factor here is independent of the population density; the change in the weather is not caused by the increase in the population, and its limiting effect would be as severe, essentially, on a small population as on a large one.

Figure 18.11B illustrates a different sort of growth curve. Here the increase is slow at first, then becomes very rapid, and finally slows down again as the population size approaches an equilibrium position at or somewhat below the *carrying capacity* of the environment, which is the maximum population that the environment can support for an extended period of time.

An S-shaped curve such as 18.11B is produced when the limiting factors become increasingly effective as the density of the population rises, i.e. when the limiting factors are at least partly density-dependent.

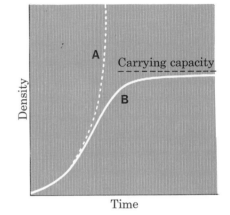

18.11. An unimpeded (exponential) growth curve (A) and the more normal S-shaped growth curve (B).

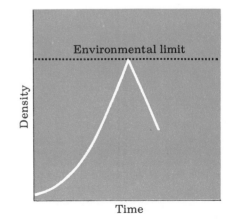

18.12. Growth curve produced when the environmental limiting factors do not become effective until late, and then bring about a sudden sharp decline.

The distinction between density-dependent and density-independent limiting factors is not always clear-cut. Strictly speaking, a density-independent factor is one that exerts a constant influence regardless of population density, but it is customary to consider factors that exert an influence of constant percentage as also density-independent. For example, if a hunter killed five rabbits every year regardless of the density of the rabbit population, he would be a density-independent limiting factor in the strict sense, but if he killed 5 percent of the local rabbit population each year, he would also frequently be considered a density-independent factor. A density-dependent limiting factor, by contrast, would be a hunter who killed 5 percent of the rabbits when the density was low, 20 percent when the density was intermediate, and 50 percent when the density was high. In other words, a density-dependent limiting factor is one whose intensity is at least partly determined by the density of the very population it helps limit.

The action of gross climatic factors and astronomical events like the diurnal light cycle and the annual changes in photoperiod is ordinarily density-independent, whereas the action of biotic factors (e.g. parasitism, predation, competition, endocrinological and behavioral changes) is usually density-dependent.

Parasitism, predation, and competition as density-dependent limiting factors. Parasitism and predation are usually density-dependent. As the population of the host or prey increases, a higher percentage is usually victimized. One reason is that the individuals—having perhaps been forced into less favorable situations or being weaker on account of the greater drain on available resources—become easier to find and attack. Another reason, apparently, is that as the relative densities of prey shift, predators that take a variety of prey species tend to alter their hunting patterns and concentrate on the most common species.

As the density of a prey species increases, the density of the predators feeding on it often increases also. This increase in predators, together with their increased concentration on the particular prey species, may cause the density of the prey to fall again. But as the density of the prey falls, there is usually a corresponding, but slightly later, fall in the density of the predator. The result is often a series of density fluctuations like those shown in Fig. 18.13, where the fluctuations of the predator (the lynx in this case) closely follow those of the prey (the hare), but with a characteristic time lag. Such linked fluctuations of predator and prey seem to indicate that the major limiting factor for the predator is

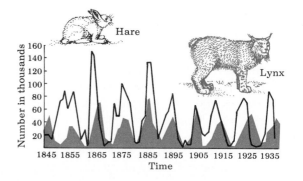

18.13. Fluctuations in abundance of lynx and showshoe hare over a period of ninety years. The oscillations of the lynx population (color) seem to follow those of the hare population. The index of abundance used here (vertical axis) is the number of pelts received by the Hudson's Bay Company. [Redrawn from D. A. MacLulich, *Univ. Toronto Studies,* Biol. Ser., no. 43, 1937.]

the availability of its food and that predation is probably one important limiting factor for the prey.

In stable predator-prey systems, predation is often decidedly beneficial to the prey population, even though it is destructive to individuals. This cardinal fact of ecology is frequently overlooked. When people set out to protect the prey from their "enemies" by killing the predators, the results are often very different from the ones they expected. A classic example is that of the deer on the Kaibab Plateau in Arizona. Prior to 1907 there was a stable population of about 4,000 deer on the plateau; the population was kept at this level, which was well below the estimated carrying capacity of the vegetation, by healthy populations of pumas and wolves. Then between 1907 and 1923 a concerted effort was made to free the deer of their "enemies" by exterminating the pumas and wolves. As a result—and perhaps also because of changing patterns of human land use— the deer population increased to 100,000 by 1925. But this number was far beyond the carrying capacity of the range, and it was not long before the area was stripped of most of the vegetation on which deer feed. In the two following years, over half the deer starved to death and the population continued to fall for many years. Not only had extermination of the predators thoroughly disrupted the previously stable deer population, but the huge population explosion of the deer disrupted the previously stable vegetation of the range and seems actually to have lowered the carrying capacity of the range to a level below what it had been in 1907 when the whole misguided effort began.

Similar difficulties have been brought on by the inadvertent destruction of predator-prey stability through pesticides. For example, application of certain insecticides to strawberries in an attempt to destroy cyclamen mites that were damaging the berries killed both the cyclamen mites and the carnivorous mites that preyed on them. The treatment permanently wiped out the predatory mites, but the cyclamen mites quickly reinvaded the strawberry fields, where, now free of their natural predators, they rapidly increased in density and did more damage than if the insecticides had never been applied.

One of the chief density-dependent limiting factors is competition. The continued healthy existence of most organisms depends on utilization of some environmental resources in limited supply, such as food, water, space, or light. As the population density increases, the competition for these limited resources becomes more intense and its deleterious results more and more effective in limiting the population; this is true of both intraspecific and interspecific competition. To take a familiar example, if flowers are planted too close together in a flower bed the plants will be weak and spindly and will produce few if any blooms; only if they are thinned, either artificially or by the natural death of the weakest individuals, will they grow well. The same sort of competition for space, light, water, and nutrients operates in a forest and keeps down the density of the trees.

Behavioral ecology

Individuals belonging to the same species utilize the same resources. If those resources are limited, the individuals must compete with one another. Thus intraspecific competition is an important aspect of the dynamics of an ecosystem. But, to some extent at least, members of the same species must also cooperate. The cooperation may be very extensive, as in colonial organisms or in animals that habitually live together in herds or flocks. The extreme of intraspecific cooperation is seen, of course, in animals that form complex societies (e.g. man and a few insects). In some species of animals, however, and in many plant species, the individuals pass much of their lives without any cooperative activity; their intraspecific interactions are restricted largely to antagonistic ones. Yet even in these species cooperation must sometimes occur if the species is to be perpetuated; the individuals must mate with each other (unless the species is completely asexual).

Within a species, then, two opposing sets of forces influence relationships between individuals—disruptive forces, particularly competition, which tend to drive the individuals apart, and cohesive ones, particularly reproduction but often, too, increased protection from predators and from destructive weather, which tend to hold the individuals together. Organisms have evolved a variety of behavioral patterns that help balance the conflicting disruptive and cohesive forces. We have already discussed the role of intraspecific communication, particularly pheromones and visual and vocal displays, in this regard. We shall here examine some aspects of behavior that directly influence the spacing, density, and survivorship of the species.

Physiological and behavioral mechanisms as density-dependent limiting factors. Some ecologists have suggested that physiological and behavioral phenomena may be important in limiting populations—that "within broad limits set by the environment, density-dependent mechanisms have evolved within the animals themselves to regulate population growth and curtail it short of the point of suicidal destruction of the environment."[4]

It has long been known that very dense populations often experience severe disease epidemics. There is now evidence that this proneness to epidemics is not due solely to the greater ease of spread of the pathogens; a density-induced change in host resistance is apparently also involved.

In numerous laboratory experiments with mice, it has been observed that increasing population density is accompanied by hypertrophy of the adrenal cortex and by degeneration of the thymus. Somatic growth is suppressed, sexual maturation delayed (or even totally inhibited at very high densities), and reproduction by mature mice diminished. The effects on reproduction include delayed spermatogenesis in males and, in females, prolonged estrous cycles, reduced rate of uterine implantation, and inadequate lactation. There is also evidence of increased intrauterine mortality of the embryos. Furthermore, it has been found that crowding of female mice before pregnancy can result in permanent behavioral disturbances in the young they later produce. There would seem, then, to be an endocrine feedback mechanism that can help regulate and limit population size by altering the reproductive rate.

It has not been established, however, that such endocrine changes, observed in the laboratory, act as major density-dependent limiting fac-

[4] J. J. Christian and D. D. Davis, "Endocrines, Behavior, and Population," *Science,* vol. 146, 1964, p. 1550.

tors in nature. The hypothesis has been severely criticized by some workers on the grounds that it is largely based on laboratory work with densities much higher than those that would actually occur in nature.

Some writers have suggested that endocrinological changes resulting from increased crowding might help stabilize human population densities. This seems improbable, because even if the stress of crowding induces hormonal changes in humans these would be unlikely to become severe or widespread enough to have significant effects on population dynamics before the densities reached extremes that would prove fatal for other reasons. We must remember that a highly social animal like man must have evolved a far greater tolerance to crowding than most animal species.

In some animals, crowding induces physiological and behavioral changes that result in increased emigration from the crowded region. Such changes can be observed in many species of aphids (small insects that suck the juices of plants). During seasons of the year when conditions are favorable, wingless females that reproduce parthenogenetically predominate. But when conditions deteriorate and competition becomes intense, sexually reproducing winged females develop and these may move out of the area in which they were born.

One of the best-known examples of physiological and behavioral changes induced by crowding is that of the solitary and migratory phases seen in several species of locusts, particularly in Eurasia. Individuals of the migratory phase have longer wings, a higher fat content, a lower water content, and a darker color than solitary-phase individuals, and they are much more gregarious and more readily stimulated to marching and flying by the presence of other individuals. The solitary phase is characteristic of low-density populations, and the migratory phase of high-density ones. As the density of a given population rises, the proportion of individuals developing into the migratory phase also rises; the sight and smell of other locusts seem to play an important role in triggering this line of development. When the proportion of migratory-phase individuals has risen sufficiently high, enormous swarms emigrate from the crowded area, consuming nearly all the vegetation in their path and often completely devastating agricultural crops.

How seriously are we to take the suggestion that emigration to other planets might be the solution to man's earthly population problem? Discounting for the moment the inhospitable conditions on planets close enough to be reasonably considered, simple calculations show that this proposal is not in fact a viable one. Bruce Wallace of Cornell University has pointed out that just to maintain the world's human population at its present level would require that rockets with a payload of 50 million tons depart each day. Building the number of rockets necessary to stabilize the population at this level for a full year might well require all the matter the earth is made of, which would leave those of us who stay at home with nothing to stand on!

Territoriality and home range. Little if any social organization can be observed in the simplest sorts of animal populations. But many aggregations, particularly in the vertebrates and in the more complex invertebrates, show considerable internal organization and integration. One frequent aspect of such organization is *territoriality,* a behavior in which an individual or a pair defends a definite area from intrusion by other individuals (particularly males of the same species). We discussed the role of displays in territorial defense in Chapter 12. Territoriality seems to function in spacing individuals in such a way as to minimize competition and individual antagonism and thus improve social stability.

Territoriality also plays a role in controlling the size of the breeding population. Although territories tend to be larger when population pressure is low and smaller when population pressure is high, there is usually a minimum requisite territory below which successful reproduction will not occur. The size of the minimum territory varies greatly from species to species, being rather large in animals that are only mildly social and sometimes quite small in highly social animals such as nesting gulls.

Even though most animals travel frequently beyond the borders of their defended territory, particularly in search of food, they tend to remain within a fairly limited area for long periods, sometimes for their entire lives. This area is called the *home range.* For whitetail deer, for example, the home range is about 0.5–2 square miles, for cottontail rabbits about 14 acres, and for meadow mice about a fifteenth of an acre.

Social hierarchies. Another aspect of organization common in animal aggregations, particularly those of vertebrates, is a social hierarchy. For example, a newly established flock of hens will form within a few days a series of dominance-subordination relationships that will effectively order the individuals within the flock. These relationships are based on the first few hostile encounters between each pair of birds. Bird A may be particularly large and aggressive, and may win her encounters with all other individuals of the flock. She thus becomes the dominant bird, and in future encounters the other hens will often move away from her without attempting to resist. Similarly, bird B may become established as dominant over all other birds in the flock except A. In other words, B can peck C or D without their pecking back, but B submits to pecking from A without pecking back. In this way, a flock of ten hens might come to have a *peck order* that could be diagramed as follows:

$$A \rightarrow B \rightarrow C \rightarrow D \rightarrow E \rightarrow F \rightarrow G \rightarrow H \rightarrow I \rightarrow J$$

where A is dominant over all the other birds, and has the "right" to peck them without being pecked back, and J is dominant over none.

A social hierarchy, once established, tends to give order and stability to the relationships within the group. There is less tension, and less fighting and disturbance. For example, when a flock of chickens is kept disrupted by frequent additions of new members and removal of old ones, the birds eat less, gain less weight, lay fewer eggs, fight more, and suffer more wounds than birds in a flock with a well-established peck order.

Societies. There may be some division of labor even in relatively loose social aggregations. For example, one individual may serve as a lookout for a prairie-dog town, or one crow may stay in a tall tree as a sentinel while the rest of the flock is feeding in a cornfield. Or, in a flock of migrating geese, one bird may act as leader, flying at the point of the V formation (Pl. III-1A). But such division of labor is primitive in comparison with that characteristic of some animal societies.

Societies are long-lasting animal aggregations with extensive division of labor and internal organization. They have evolved independently in several groups of animals, notably the insects and the mammals. Among insects, all termites and ants and some bees and wasps live in societies. It is tempting to draw analogies between insect and human societies, but the resemblances are actually superficial and the differences fundamental.

All insect societies are large, highly specialized families. Typically, a single inseminated female, the queen (Pl. III-1B), sometimes accompanied by workers, as in honeybees, founds a new social unit by laying eggs that hatch into larvae and subsequently develop into workers and, in ants and termites, into soldiers. Thus the division of labor in insect societies is based on major biological differences between individuals— in short, on the presence of different castes. In Hymenoptera (ants, bees, and wasps), fertilized eggs hatch either into a queen (a fertile female) or into workers (sterile females); in ants they may also hatch into soldiers. The size of the cell in which a given female larva develops and the amount and type of food it is given are evidently important elements in determining whether it will become a queen or a worker; pheromones may also be involved. In most cases, hymenopteran males, whose sole function is insemination of a new queen, hatch from unfertilized eggs and are haploid.

Based on the caste system, the organization and integration of insect societies depend on primarily instinctive behavior patterns subject to little modification. In the course of their development, however, individuals may perform a variety of tasks. For example, worker honeybees usually serve as nurse bees for roughly the first two weeks after metamorphosis, first incubating the brood and preparing brood cells, and later feeding the larvae. Then they may become house bees for a week or two, acting as storekeepers, housecleaners, wax secreters, or guards. Finally, they may become field bees for four or five weeks, foraging for nectar and pollen. Individuals have no choice of roles in this system; their role is determined by a combination of caste, stage of development, and the conditions of the hive.

In human societies, by contrast, family groups represent only one of many kinds of organizational units. A human society is typically an aggregation of many individuals and family groups, in which most adult members take part in reproduction; there is no setting aside of one or two individuals to serve as reproductive machines for the entire group while the other individuals remain sterile. Division of labor is not primarily based on biological differences, or castes, and behavior patterns are greatly modified by learning. In short, individuals have a considerable choice of roles, and they may change from one group to another.

INTERSPECIFIC INTERACTIONS

We have already seen that the various species of organisms that live in the same area and are thus part of the same community and ecosystem affect one another in a variety of ways. The herbivores depend on the green plants for their high-energy carbon compounds and organic nitrogen compounds. Carnivores depend on herbivores for these same materials. The green plants depend on microorganisms for nitrogen fixation and for nitrification of ammonia. All organisms, plant and animal, depend on the decomposers to rid the environment of the dead bodies and excreta that would otherwise soon prohibit life. Tall plants shade the ground below them and change the wind patterns and humidity to which the organisms living beneath them are exposed. Plants provide shelter and nesting sites for animals. And so on. We could continue this list of interactions almost indefinitely. The point is simply that a community is not just a collection of different species that happen to be able to live under the prevailing conditions in the area; it is an integrated system of species to a greater or lesser extent dependent on one another.

It is much beyond the scope of this book to examine all the various kinds of interspecific interactions that exist in most communities. Here we shall discuss only a few of the main interactions not covered elsewhere.

Symbiosis

Etymologically, symbiosis means simply "living together," without any implication of mutual benefit. This is the meaning the term was given when it was first introduced into biology, and this is the meaning it will have in this book. We shall, however, recognize three categories of symbiosis. The first is *commensalism,* a relationship in which one species benefits while the other receives neither benefit nor harm. The second is *mutualism,* where both species benefit. The third is *parasitism,* where one species benefits and individuals of the other species are harmed. We can summarize the distinctions as follows (a plus sign means benefit, a minus sign harm, and a zero no significant effect):

Relationship	Species A	Species B
Commensalism	+	0
Mutualism	+	+
Parasitism	+	−

A word of caution is in order: The division of symbiosis into three categories is in many ways arbitrary. Commensalism, mutualism, and parasitism grade into each other; it is often a question, for example, whether a given relationship should be considered commensalistic or parasitic. Each case of symbiosis is different from all others and must be studied and analyzed on its own merits.

Commensalism. The advantage derived by the commensal species from its association with the host frequently involves shelter, support, transport, or food, or several of these. For example, in tropical forests numerous small plants, called epiphytes, usually grow on the branches of the larger trees or in forks of their trunks. These commensals, among which species of orchids and bromeliads are prominent, are not parasites. They use the host trees only as a base of attachment and do not obtain nourishment from them. They apparently do no harm to the host except when so many of them are on one tree that they stunt its growth or cause limbs to break.

Sometimes it is difficult to tell what benefit is involved in a commensal relationship. For example, certain species of barnacles occur nowhere except attached to the backs of whales, and other species of barnacles occur nowhere except attached to the barnacles that are attached to whales. Just what advantages either of these groups of barnacles enjoys is not clear. In some cases of commensalism, however, the benefit is dramatically obvious. For example, certain species of fish regularly live in association with sea anemones, deriving protection and shelter from them and sometimes stealing some of their food. These fish swim freely among the tentacles of the anemones, even though those tentacles quickly paralyze other fishes that touch them. The anemones regularly feed on fish; yet the particular species that live as commensals with them sometimes actually enter the gastrovascular cavity of their host, emerging later with no apparent ill effects.

Mutualism.   Symbiotic relationships beneficial to both species are common. Two such mutualistic relationships are illustrated in Pl. III-8. Other examples are the relationship between a flowering plant and its insect or bird pollinators (Pl. IV-1); that between a legume and the nitrogen-fixing bacteria in its root nodules; and that between a human being and the bacteria synthesizing vitamin B_{12} in his intestine. The plants we call lichens are actually composites of an alga and a fungus united in such close mutualistic symbiosis that they give the appearance of being one plant. Apparently the fungus benefits from the photosynthetic activity of the alga, and the alga benefits from the water-retaining properties of the fungal walls.

Parasitism.   There is no sharp boundary between parasitism and predation. The usual distinction is that a predator eats its prey quickly and then goes on its way, while a parasite passes much of its life on or in the body of a living host, from which it derives food in a manner harmful to the host.

Parasites are customarily divided into two types: external and internal. External parasites live on the outer surface of their host, usually feeding on hair, feathers, scales, or skin, or else sucking blood. Internal parasites may live in the various tubes and ducts of the host's body, e.g. the digestive tract or respiratory passages or urinary ducts; or they may bore into and live embedded in tissues such as muscle or liver; or, in the case of viruses and some bacteria and protozoans, they may actually live inside the individual cells of their host.

Internal parasitism is usually marked by much more extreme specializations than external parasitism. The habitats available inside the body of another living organism are completely unlike those outside, and the unusual problems they pose have resulted in evolutionary adaptations quite different from those seen in free-living forms. For example, internal parasites have often lost organs or whole organ systems that would be essential in a free-living species. Tapeworms, for instance, have no digestive system. They live in their host's intestine, where they are bathed by the products of the host's digestion, which they can absorb directly across their body wall without having to carry out any digestion themselves.

Because of their frequent evolutionary loss of structures, internal parasites are often said to be degenerate. "Degenerate," of course, implies no value judgment, but simply refers to the lack, common in parasites, of many structures present in their free-living ancestors. From an evolutionary point of view, loss of structures useless in a new environment is an instance of positive adaptation. Such a loss is just as much an evolutionary advance, a specialization, as the development of increased complexity in some other environment. Specialization does not necessarily mean increased structural complexity; it only means the evolution of characteristics particularly suited to some special situation or way of life.

Perhaps the most striking of all adaptations of internal parasites pertain to their life histories and reproduction. Individual hosts don't live forever. If the parasitic species is to be perpetuated, therefore, a mechanism is needed for changing hosts. At some point in their life cycles, then, all internal parasites move from one host individual to another. But this is seldom simple. Rarely can a parasite move directly from one host to another of the same species. Consider the life cycle of a beef tapeworm. The eggs of such a tapeworm living in a man's intestine are shed in the host's feces. A cow eats plants contaminated with human feces, and the tapeworm eggs hatch in the cow's intestine. The young larvae bore through the wall of the cow's intestine, enter a blood vessel, and are carried by the blood to a muscle, where they encyst (become surrounded by a bladderlike case and lie inactive). If a man then eats the raw or insufficiently cooked beef, the tapeworm larvae become activated in his intestine, their heads attach to the intestinal wall, and mature worms develop. The beef tapeworm thus passes through two hosts during its life cycle: an intermediate host (cow), in which it undergoes some of its early development, and a final host (man), in which it matures. As life cycles of internal parasites go, the one just described is rather simple. It is not unusual for a life cycle to include two or three intermediate hosts and/or a free-living larval stage.

In the course of their evolution, parasites usually develop special features of behavior and physiology that improve both their adjustment to the particular characteristics of their host and their efficiency at competing with other parasites. This means that they often tend to become more and more specific. Where an ancestral organism may have parasitized all species in a particular family, each of its various descendants may parasitize only one species of host at each stage in its development. Parasites often tend to become more specific also with regard to the part of the host's body they can inhabit.

It must be remembered that as the parasites evolve the host species is also evolving, and that there is strong selection pressure for its evolution of more effective defenses against the ravages of its parasites. There is thus a constant interplay between host and parasite. As the one evolves better defenses, the other evolves ways of counteracting them—these counteractions leading to pressure on the host to evolve still better defenses, against which the parasite may then evolve new means of surviving, and so forth. Although this sort of mutual evolution will continue as long as the host-parasite relationship exists, a dynamic balance is usually reached eventually, the host surviving without being seriously damaged and the parasite prospering moderately well too. It is, in fact, decidedly disadvantageous to the parasite to kill its host. We have said that many parasites are host-specific. If they should cause the extinction of their host, then they themselves would also become extinct. Therefore the balance eventually reached is not due solely to the evolution of better defenses by the host; it is also due to the parasite's evolving in such a way that it becomes better adjusted to its host and causes less serious disturbance. Where serious disturbance does result, the reason may be that the host-parasite relationship is a relatively new one, or that a new and virulent form of the parasite has recently arisen, or that the host is not the main host of the parasite.

Predation

We shall here regard a predator as a free-living organism that feeds on other living organisms. This is an admittedly broad definition, which embraces both carnivores (predators that eat other animals) and herbivores (predators that eat plants). Many carnivorous predators and some herbivorous ones kill their prey, but a few carnivores (e.g. mosquitoes, biting flies) and many herbivores devour only a portion of their prey and the prey generally recovers. Predators are usually much less prey-specific than parasites.

The effect of a predatory species on its prey species varies. Sometimes the predator severely limits the numbers or distribution of its prey, often to a point far below what the environmental resources could support. In other cases, the predator plays only a minor role in the life of the prey, usually because it is much rarer than the prey or because it commonly utilizes some other food source. There is, of course, a continuous gradient of possibilities between these two extremes. Representing a fairly common middle ground are those situations in which the predator helps hold down the prey population, but not so severely as to endanger the continued existence of the species. This degree of predation is, in fact, often decidedly beneficial to the prey species (though not, of course, to the individuals eaten); it helps regulate the population size of the prey and prevent it from outrunning its resources, as we saw in the interaction between the Kaibab deer and their predators.

As in parasitism, long-established predator-prey relationships in a stable ecosystem tend to evolve toward a dynamic balance in which the predation is an important regulatory influence in the life of the prey species but not a real threat to its survival. This sort of balance is, of course, beneficial to the predator as well as to the prey, because extinction or severe depression of the prey population would diminish the future resources of the predator and in some cases might even lead to its own extinction as well. Factors involved in the predator-prey balance include the relative numbers and sizes of the two species, the vulnerability of the prey to the predator, the extent to which the predator can (or does) utilize other food sources, and the amount of energy obtained by the predator from consuming one prey individual.

Interspecific competition

Competition is an interaction in which both parties are harmed. If we were to fit the concept into the list on p. 412, it would be symbolized by minus-minus. Ordinarily interspecific competition occurs when two or more species utilize the same limited resource, such as food, water, sunlight, shelter, space, or nesting sites. The more similar the requirements of the species involved, the more intense the competition.

Or, as biologists often put it, the more the niches of the species overlap, the more intense the interspecific competition.

The concept of niche.   Niche, an important concept in ecology, defined in a variety of ways, is understood here as the functional role and position of an organism in the ecosystem. It should not be confused with *habitat,* which is the physical place where the organism lives. The characteristics of the habitat help define the niche, but they alone do not suffice. Also involved is what the organism eats; how and where it finds and captures its food; what extremes of heat and cold, dry and wet, sun and shade, and other climatic factors it can withstand, and what values of these factors are optimal for it; at what time of year and what time of day it is most active; what its parasites and predators are; where, how, and when it reproduces; etc. In short, every aspect of an organism's existence helps define that organism's niche. It must be emphasized that niche is an abstraction and as such can never be fully measured. The most we can do is measure certain of the more important parameters of an organism's niche.

The following graphic method of representing the niche of a species has been proposed. Consider one environmental variable, X_1, such as temperature; determine the high and low extremes the species in question can tolerate; plot these on a coordinate, and connect them by a line. Do the same for a second environmental variable, X_2, and draw its line at right angles to the first one. These two lines together determine a rectangular surface.

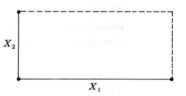

Now determine the values for a third environmental variable, X_3, and connect them by a line oriented at right angles to the other two. The three lines together now determine a volume within which every point corresponds to some combination of values for the three variables that would permit the species to survive.[5]

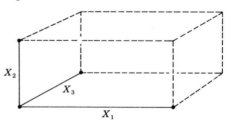

If you were to continue this procedure, adding more and more variables each expressed by a different dimension, you would obtain a multidimensional hypervolume. If you included a dimension for every variable relevant to the species in question, the resulting multidimensional hypervolume would represent the niche of the species.

[5] There is an important complication that partly invalidates this procedure. The environmental variables are seldom completely independent. Thus, for example, the temperature extremes a species can tolerate may be different at different humidities.

A

IV-1   ADAPTED TO ATTRACT

Flowers adapted for pollination by hummingbirds, such as the fuchsia-flowered goose-
berry (A), are often red, a color the birds see
well, have elongate corollas into which the
birds can insert their beaks to obtain nectar,
and generally lack a landing platform, which
the birds do not need because they hover in
midair while probing flowers. The goose-
berry blossoms have long sexual organs that
brush pollen onto the hummingbird's head
while it is getting nectar. Contrast these
flowers with those of *Eupatorium* (B), a herb
of the family Compositae, which are adapted
for pollination by butterflies, in this case a
Viceroy.

A, David G. Allen. B, courtesy Carolina Biological
Supply Co.

B

IV-2C

IV-2A
IV-2B

IV-2, IV-3   ADAPTED TO DECEIVE

Many animals have adaptations of color, shape, or behavior that make it difficult for predators to see or catch them. The geometrid moth larvae in 2A mimic juniper stems; note the markings on the larva at the upper left, which resemble those of the stem to its left. The closely related larvae in 2B mimic juniper leaves; the one inching up the stem is fairly easy to see, but the other one can hardly be distinguished from the surrounding leaves. The larva in 2C camouflages itself with pieces of its food plant, picking them up with its mouth and impaling them on hooks on its back. The four-eye butterfly fish (3A) and the leaf-tailed gecko lizard (3B) have adaptations that make it hard to tell which is the head end, and hence in which direction the animals are likely to move. The fish has a fake eyespot near the base of its tail, while its real eye is obscured by a black band. The gecko has an expanded tail resembling a head. The frog *Hyla versicolor* (3C) is well known for its ability to change color to match its background. The brownish frog has been on the tree trunk for some time. The green one has just been moved there from a pond, where it was among green duckweed; it has not had time to change color.

2A–C, J. G. Franclemont. 3A, Marineland of Florida. 3B, Australian News and Information Bureau. 3C, Verne N. Rockcastle.

IV-3A

IV-3B

IV-3C

IV-4 A MIMICRY COMPLEX

Many animals derive protection, not from concealment or the possession of efficient weapons, but from imitation of other animals that do have efficient weapons, which make predators avoid them. The Pipevine Swallowtail (top), a poisonous species avoided by birds, is mimicked by the females of the Black Swallowtail (2), the Tiger Swallowtail (3), and the Diana Fritillary (6), but not by the males of these species (1, 4, 5). Both sexes of the Red-spotted Purple (7), which occurs in the Pipevine's range, are mimics, whereas the Banded Purple (8), another race of the same species, which has a range different from the Pipevine's, looks quite unlike it. It is not known why, in this group, the females should all mimic and the males (with one exception) not.

Photo, J. G. Franclemont.

IV-5   ADAPTED TO INTIMIDATE

No bluff, the fierce mien of this dragonfish. Glands in its dorsal, anal, and pelvic spines are loaded with a powerful poison. Although its feathery fins are adapted for concealment among the coral growths where it lives, it also benefits from its highly idiosyncratic appearance, which would-be predators readily identify with dangerous experience. Not surprisingly, most of them give the dragonfish a wide berth.

Photo, Marineland of Florida.

IV-6A

IV-6B

IV-6, IV-7 DIVERSITY OF BIOMES

Among the world's major plant formations, or biomes, are (6A) the taiga, dominated by coniferous trees and typically inhabited by moose—a female is here shown feeding; (6B) temperate deciduous forests, whose leaves may take on spectacular coloring just before they drop; (7A) the Arctic tundra, dotted by ponds, and with a plant life dominated by lichens, grasses, and low-growing herbs such as alpine azalea (inset); (7B) the desert, where the dominant plants are annuals that grow only after heavy rains, thornbushes whose small thick leaves may be shed during prolonged dry spells, and succulent perennials like cactuses, which can store much water; the brilliant blooms belong to a hedgehog cactus. Abandoned woodpecker holes like those in the tall cactus in the center are often appropriated by other desert animals.

6A, Leonard Lee Rue III from National Audubon Society. 6B, Alice Kadmon. 7A, with inset, Arthur A. Allen. 7B, Harry and Ruth Crockett from National Audubon Society; inset, Bruce Hunter, courtesy of the American Museum of Natural History.

IV-7A
IV-7B

A

B

IV-8   COMMUNITY SUCCESSION

What is dry land today may have been a swamp
a few decades ago. What is now bare rock will
perhaps have disappeared under a cover of
vegetation a few years from now. In time, the
biological communities that plants and animals
form in any given area tend to displace one
another, until a relatively stable community becomes established. In the Okefinokee Swamp
in Georgia, a pond is slowly being invaded by
vegetation (A); the roots of the plants help hold
silt and thus abet the buildup of soil and organic
matter that will eventually produce dry land
there. First to gain a foothold on a rock surface
are often lichens (shown bearing fruiting bodies
in B); chemicals that they produce corrode the
rock and thus help prepare the way for later
successional stages.

A, David G. Allen. B, courtesy Carolina Biological Supply
Co.

Consequences of intense competition. Most (though not all) ecologists subscribe to the *competitive exclusion principle,* which says that two species cannot for long simultaneously occupy the same niche in the same place. But it is implicit in our definition of niche that no two species could ever occupy the same niche. To do so, they would have to be identical in every respect and hence they would be one species, not two. What, then, is the value of this principle? Its value is much like that of the Hardy-Weinberg Law, which also describes a situation that would never really occur. Both the Hardy-Weinberg Law and the competitive exclusion principle establish null conditions—points of departure, so to speak—and thus facilitate our understanding of what will happen when the null conditions do not hold. Thus, for example, the competitive exclusion principle helps us see why the species living together in a stable community ordinarily occupy quite distinct niches, and what sorts of interactions will result when two species with very similar (though not identical) niches occur together.

The more similar the two niches are, the more likely it is that both species will utilize in common and in the same way at least one limited resource (food, shelter, nesting sites, etc.). They will therefore be competing for that limited resource. Such competition usually leads to one (or two) of three possible outcomes: (1) The competitive superiority of one of the rival species may be such that the other is driven to extinction. (2) One species may be competitively superior in some regions, and the other may be superior in other regions with different environmental conditions, with the result that one is eliminated in some places and the other is eliminated in other places; i.e. sympatry disappears but both species survive in allopatric ranges. (3) The two species may rapidly evolve in divergent directions under the strong selection pressure resulting from their intense competition; in other words, they may undergo character displacement, rapidly evolving greater differences in their niches. Whether extinction, range restriction, character displacement, or a combination of the last two, will be the outcome in any given case of intense interspecific competition is determined by a host of factors too complex to discuss here.

Let us look at a few well-studied examples of competition and its consequences. G. F. Gause of the University of Moscow worked with two closely related species of protozoans, *Paramecium caudatum* and *Paramecium aurelia,* in laboratory cultures. When cultured separately, the population curve for each species had a typical S shape (Fig. 18.14). But when the two species were cultured together, the population growth rate of *P. aurelia* was slower than normal and *P. caudatum* failed to survive. Similar results were obtained by Thomas Park and his colleagues at the University of Chicago. They worked with flour beetles of the genus *Tribolium.* When *T. confusum* and *T. castaneum* were kept together in the same container of flour, one or the other species always became extinct. The conditions of temperature and humidity under which the competition took place greatly influenced which species would prevail. Thus *T. castaneum* usually prevailed under hot-wet conditions, while *T. confusum* usually prevailed under cool-dry conditions.

It is not always easy to detect the differences between the niches of two or more closely related sympatric species. At first glance, the species may appear to be occupying the same niche in a stable way and thus to discredit the exclusion principle. But closer study usually reveals

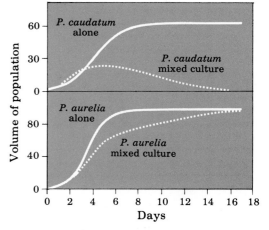

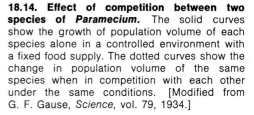

18.14. Effect of competition between two species of *Paramecium*. The solid curves show the growth of population volume of each species alone in a controlled environment with a fixed food supply. The dotted curves show the change in population volume of the same species when in competition with each other under the same conditions. [Modified from G. F. Gause, *Science*, vol. 79, 1934.]

differences of fundamental importance. When Robert MacArthur, then at the University of Pennsylvania, studied a community where several closely related species of warblers (small insect-eating birds) occurred together, he found that their feeding habits were significantly different. Myrtle Warblers fed predominantly among the lower branches of spruce trees; Black-throated Green Warblers fed in the middle portions of the trees; and Blackburnian Warblers fed toward the tops of the same trees and on the outer tips of the branches. We noted similar differences in feeding habits between sympatric Galápagos finches.

THE BIOTIC COMMUNITY

Darwin noted the influence of the number of cats on the seed production of clover in a certain area. It was mice, predators on bees' nests, that provided the link.

It is not always so easy to discover the links among species in a biotic community, the web of interactions usually being highly complex. Generally a great diversity of plant and animal species occupy the same general area, sharing some resources and competing for others. The biotic community they form can be considered a fundamental unit of life, with its own characteristic structure and functional interrelationships.

Dominance, species diversity, and community stability

The various species of plants and animals in a community are not all present in equal numbers. On the basis of productivity or biomass, it is usual to find that a few species are dominant. In a community with over one hundred kinds of producing plants, from 30 to well over 90 percent of the productivity may be concentrated in a single species. In complex communities, a common pattern is a few markedly dominant species, a great many species of intermediate dominance, and a few rare species.

Simple ecosystems, with only a few species, often show strong dominance by a single species. In more complex communities, with intricate food webs, a diversity of predators and consumers act to prevent the emergence of one highly dominant species. The suppression of a single strongly dominant species allows a more efficient energy flow to the other organisms and maintenance of a greater number of species.

Efforts to eliminate undesirable species from a community often reveal hidden linkages to other organisms, and may dramatically demonstrate the complex interactions upon which community stability rests. Some years ago the World Health Organization (WHO) began a campaign to eradicate malaria-carrying mosquitoes in the Borneo states of Malaysia, where as many as 90 percent of the population of some areas suffered from the disease. Mosquito control was achieved by spraying the interiors of the village huts with DDT and Dieldrin, two powerful contact insecticides, and malaria was indeed eradicated. But soon the villagers became aware that the thatch roofs of their huts were rotting and beginning to collapse. Investigation showed that the deterioration, which occurred only in sprayed huts, was due to the larvae of a moth that normally lives in small numbers in the thatch roofs. Whereas the thatch-eating moth larvae avoided food sprayed with the insecticides, the moth's natural enemy, a parasitic wasp, was very sensitive to them. The net result was a substantial increase in the population of the larvae eating the thatch.

That would have been an interesting story in itself, but there was yet another side effect potentially more serious. Cockroaches and a small house lizard, the gecko, are two normal inhabitants of the village huts. Insecticide-contaminated cockroaches were eaten by the geckos, which were in turn eaten by house cats (as were some cockroaches). The cats, poisoned by the accumulation of insecticide, died. The result was a population explosion of rats, which are potential carriers of such diseases as leptospirosis, typhus, and plague. In an attempt to restore the cat population, WHO and the Royal Air Force undertook a remarkable venture, "Operation Cat Drop," in which they parachuted cats into the villages. With the cat population restored, the rats and the consequent threat of serious disease subsided.

Fortunately, not every pest-control effort entails such complications. Insecticides rapidly broken down in the environment and more selective in their toxicity have been developed, and these are less disruptive of biological communities. Much to be preferred, however, is biological control, e.g. the use of the natural predators, parasites, or pathogens of the pest.

A consistent effect of pollution, whatever the kind, is the reduction of local species diversity. Top predators are frequently the first species to disappear. With the predators gone, one or more prey species will multiply unchecked and further degrade the habitat, often making it unfit for other species and sometimes even for the original population of the exploding species itself, as in the case of the Kaibab deer population we looked at earlier. Disturbances of many sorts—chemical pollution, wildfire, earthquakes, massive radiation, or simply the interference of man—all seem to have approximately the same effect on an ecosystem; they reduce species diversity, shorten food chains, and increase instability.

It is generally accepted among ecologists that complex communities are more stable than simple ones, but most of their evidence is indirect, coming as it does from laboratory experiments with predator-prey systems or with systems of several competing species. Such laboratory-culture systems virtually always result in the extinction of at least one of the species (see Fig. 18.14). By making the environment more diverse, however, as by adding tunnels, blind alleys, and other obstacles to travel, the wild fluctuations and oscillations typical of laboratory predator-prey systems can be reduced in amplitude and extinction postponed. Thus both species diversity and structural diversity of the environment seem to favor stability. Many other factors, notably climate, also play a role under natural conditions. Among the most difficult tasks of modern ecology is assessing the stability of natural communities—determining whether a community is stable enough to resist a proposed disturbance by man without catastrophic consequences.

Ecological succession

The characteristics of succession. Succession is a more or less orderly process of community change. It involves replacement, in the course of time, of the dominant species within a given area by other species. A farmer's field is allowed to lie fallow. A crop of annual weeds grows in it during the first year. Many perennial herbs appear in the second year and become even more common in the third year. Soon, however, these are superseded as the dominant vegetation by woody shrubs, which, in turn, may eventually be replaced by trees. Or a lake dries up, and its sandy bottom becomes covered with grass, which later gives way to trees.

But why the change? Why does succession occur? The cause cannot be climate, because succession will occur even if the climate remains the same. Climate may be a major factor in determining what sorts of species will follow one another, but the succession itself must result from other changes. The most important of these are the modifications of the physical environment produced by the community itself. Most successional communities tend to alter the area in which they occur in such a way as to make it less favorable for themselves and more favorable for other communities. In effect, each community in the succession sows the seeds of its own destruction. Consider the alterations initiated by pioneer communities on land. Usually these communities will produce a layer of litter on the surface of the soil. The accumulation of litter affects the runoff of rainwater, the soil temperature, and the formation of humus. The humus, in turn, contributes to soil development and thus alters the availability of nutrients, the water relations, the pH and aeration of the soil, and the sorts of soil organisms that will occur. But the organisms characteristic of the pioneer communities that produced these changes may not prosper under the new conditions, and they may be replaced by invading competitors that do better in an area with the new type of soil.

Though successions in different places and at different times are not identical—the species involved are often completely different— some ecologists have nevertheless formulated generalizations that they have reason to think hold true in most cases where both autotrophs and heterotrophs are involved:

1. The species composition changes continuously during the succession, but the change is usually more rapid in the earlier stages than in the later ones.
2. The total number of species represented increases initially and then becomes more or less stabilized in the older stages. This is particularly true of the heterotrophs, whose variety is usually much greater in the later stages of the succession.
3. Both the total biomass in the ecosystem and the amount of nonliving organic matter increase during the succession until a more stable stage is reached.
4. The food webs become more complex, and the relations between species in them become better defined or more specialized.
5. Although the amount of new organic matter synthesized by the producers remains approximately the same, except at the very beginning, the percentage utilized at the various trophic levels rises.

In summary, if these ecologists are right, the trend of most successions is toward a more complex ecosystem in which less energy is wasted and hence a greater biomass can be supported without an increase in the supply of energy.

Some examples of succession.   A much-studied type of succession is that in ponds (Pl. IV-8A). Sediments washed from the surrounding land begin to fill the pond, and the dead bodies of planktonic organisms add organic material. Soon pioneer submerged vascular plants appear in the shallower water near the margins of the pond. Their roots hold the silt, and the lake bottom is built up faster where they are growing. As these plants die, their bodies accumulate faster than decomposers can break them down. Soon the water is shallow enough for broad-leaved floating pondweeds, such as water lilies, to displace the submerged species, which now become established in a zone farther out in the pond, where conditions are more favorable for them. But as the bottom continues to build up, the floating pondweeds are in their turn displaced by emergent species (plants that have their roots in the mud of the bottom but their shoots extending into the air above the water), such as cattails, bulrushes, and reeds. These plants grow very close together and hold the sediment tightly, and their great bulk results in rapid accumulation of organic material. Soon conditions are dry enough for a few terrestrial plants to gain a foothold. Now an area that was formerly part of the pond is newly formed dry land. This entire sequence can sometimes be seen as a nearly continuous series of zones girdling a pond or lake. With the passage of the years, the pond becomes smaller and smaller as the zones move nearer and nearer its center. Eventually nothing of the pond remains.

Orderly succession in lakes and ponds is easily disturbed by the activities of man. For example, both the clearing of vegetation and the turning of soil in agricultural cultivation greatly increase the amount of soil in the runoff water in streams, which in turn leads to faster siltation of lakes. Perhaps more important is the tremendous increase of nutrients in lakes that results from runoff of inorganic fertilizers and from release of sewage and detergents. The normal limiting resource for algae in many fresh-water lakes is phosphorus, but this mineral is now being poured into the aquatic environment in enormous quantities. One consequence is extensive algal blooms that cover the surface of the water with scum and foul the shores with stinking masses of rotting organic matter. Now, the increased photosynthetic productivity associated with the algal blooms might be expected to make more food available for higher links in food webs and thus to be of benefit to the biotic community. But excessive growth of algae actually causes destruction of many of the higher links in the food webs. At the end of the growing season, many of the algae die and sink to the bottom, where they stimulate massive growth of bacteria the following year. There is so much bacteria-produced decomposition that the oxygen of the deeper colder layers of the lake becomes depleted, with the result that cold-water fish such as trout, whitefish, pike, and sturgeon die and are replaced by less valuable species such as carp and catfish. Deoxygenation of the water also causes chemical changes in the bottom mud that produce increased quantities of odorous, sometimes toxic, gases. These changes also further accelerate the *eutrophication,* or aging process, of the lake.[6]

[6] The term "eutrophication" was originally applied to the accumulation of nutrients and increase in organic matter that are a natural part of the succession of lakes. Recently it has been applied not so much to the natural successional process as to the greatly accelerated one resulting from human interference.

Extensive use of tertiary treatment of sewage, which few municipalities now carry out, combined with a reduction of the phosphate content of detergents, would slow the death of our lakes, but would not prevent it, because at least 30 percent of the polluting phosphorus comes from agricultural sources, where this mineral is essential to the production of the large crops needed to feed our burgeoning population. Furthermore, the change from phosphate to nitrogen detergents, advocated in the name of "better ecology," may simply shift the pollution problem from one place to another. Although the tertiary sewage treatments devised so far are quite effective in removing phosphates, they are much less so in removing nitrates, which may be the natural limiting nutrient in some fresh waters and probably are the major natural limiting nutrient in estuaries. Unless new detergents free of nutrients that are difficult to remove can be developed, or more effective sewage treatment devised, a return to soaps may become imperative.

Successions need not begin with land reclaimed from lakes or ponds. Consider a bare rock surface (Pl. IV-8B). The first pioneer plants may be lichens, which grow during the brief periods when the rock surface is wet and lie dormant during periods when the surface is dry. The lichens release acids and other substances that corrode the rock. Dust particles and bits of dead lichen may collect in the tiny crevices thus formed, and pioneer mosses may gain anchorage there. The mosses grow in tufts or clumps that trap more dust and debris and gradually form a thickening mat. A few fern spores or seeds of grasses and annual herbs may land in the mat of soil and moss and germinate. These may be followed by perennial herbs. As more and more such plants survive and grow, they catch and hold still more mineral and organic material, and the new soil layer thus becomes thicker. Later, shrubs and even trees may start to grow in the soil that now covers what once was a bare rock surface.

Succession on abandoned croplands, plowed grasslands, or cutover forests often proceeds relatively quickly in its initial stages, because the effects of the previous communities have not been wholly erased and the physical conditions are not as bleak as on a beach or a bare rock surface. Suppose, for example, that a cornfield in Georgia is abandoned. The very first year it will be covered with annual weeds, such as ragweed, horseweed, and crabgrass. In the second year, ragweed, goldenrod, and asters will probably be common, and there will be much tall grass. The grass will usually be dominant for several years, and then more and more shrubs and tree seedlings will appear. The first tree seedlings to grow well in the unshaded field will be pines, and eventually a pine forest will replace the grass and shrubs. But pine seedlings do not grow well in the shade of older pines. Seedlings of oaks, hickories, and other deciduous trees are more shade-tolerant, and these trees will gradually develop in the lower strata of the forest beneath the old pines, eventually replacing them. The deciduous forest thus formed is more stable and will ordinarily maintain itself for a very long time.

Climax and biome

The concept of climax.   If man or some other disruptive factor doesn't interfere, most successions eventually reach a stage that is much more stable than those that preceded it. The community of this stage is called the *climax community.* It has much less tendency than earlier successional communities to alter its environment in a manner injurious to itself. In fact, its more complex organization, larger organic structure, and more balanced metabolism enable it to buffer its own physical environment to such an extent that it can be self-perpetuating. Consequently it may persist for centuries, not being replaced by another stage so long as climate, physiography, and other major environmental factors remain essentially the same. It must be emphasized, however, that a climax community is not static; it does slowly change, and will change rapidly if there are major shifts in the environment, either physical or biotic. For example, fifty years ago chestnut trees were among the dominant plants in the climax forests of much of eastern North America, but they have been almost completely eliminated by a fungal blight, and the present-day climax forests of the region are dominated by other species. Thus there can be no absolute distinction between climax and the other stages of succession; the difference between them is relative.

Since the beginning of the century, some American ecologists have held that all successions in a given large climatic region will converge to the same climax type—that there is only one type of climax community for the region and that any sites dominated by other communities have not yet reached climax, no matter how stable and long-lasting they may seem. Thus a beech-maple forest has been considered the climax for most of the northeastern United States, and a white spruce—balsam fir forest has been considered the climax for much of Canada.

Many modern ecologists, however, argue that since each species is distributed according to its own particular biological potentialities, climax has meaning only in relation to the individual site and its environmental conditions. In their view, the climax for a given spot should be determined, not by referring to some theoretical regional climax, but by actually observing what populations replace others and then maintain themselves in a stable condition.

Biomes. Even though the present trend is away from definition of regional climaxes in terms of aggregations of particular dominant and subdominant species, most biologists find it convenient to recognize a limited number of major climax formations called biomes.

Let us briefly survey some of the world's major biomes. In the far northern parts of North America, Europe, and Asia is the *tundra.* The tundra is the most continuous of the earth's biomes, forming a circumpolar band interrupted only narrowly by the North Atlantic and the Bering Sea. It corresponds roughly to the region where the subsoil is permanently frozen. The land has the appearance of a gently rolling plain, with many lakes, ponds, and bogs in the depressions (Pl. IV-7A). "Tundra" is a Siberian word meaning north of the timberline. There are, in fact, a few trees on the tundra, but they are small, widely scattered, and clearly not the dominant vegetation except locally. Much of the ground is covered by mosses (particularly sphagnum), lichens (particularly so-called reindeer moss), and a few species of grasses. There are numerous small perennial herbs, which are able to withstand frequent freezing and which grow rapidly during the brief cool summers, often carpeting the tundra with brightly colored flowers. Reindeer, caribou, arctic wolves, arctic foxes, arctic hares, and lemmings are among the principal mammals; polar bears are common on parts of the tundra near the coast. Vast numbers of birds, particularly shorebirds (sandpipers, plovers, etc.) and waterfowl (ducks, geese, etc.), nest on the tundra in summer, but they are not permanent residents and migrate south for the winter. Insects, particularly flies (including mosquitoes), are incredibly abundant. In short, far from being a barren lifeless land as many people think, the tundra teems with life. It is true, however, that though the number of individual organisms on the tundra is often very large, the number of different species is quite limited.

South of the tundra, in both North America and Eurasia, is a wide zone dominated by coniferous forests. This is the *taiga* (Pl. IV-6A). Like the tundra, it is dotted by countless lakes, ponds, and bogs. And like the tundra, it has very cold winters. But it has longer and somewhat warmer summers, during which the subsoil thaws and vegetation grows abundantly. The number of different species living in the taiga is larger than on the tundra, but it is considerably smaller than in biomes farther south. Though conifers (including spruce, fir, and tamarack) are the most characteristic of the larger plants in the taiga, some deciduous trees (e.g. paper birch) are also common. Moose, black bear, wolves, lynx, wolverines, martens, squirrels, and many smaller rodents are important mammals in the taiga communities. Birds are abundant in summer.

The biomes south of the taiga do not form such definite circum-global belts as the tundra and the taiga. There is more variation in the amount of rainfall at this latitude, and consequently more longitudinal variation in the types of climax communities that predominate. In those parts of the temperate zone where rainfall is abundant and the summers are relatively long and warm, as in most of the eastern United States, most of central Europe, and part of eastern Asia, the climax communities are frequently dominated by broad-leaved trees. Such areas, in which the foliage changes color in autumn and drops, constitute the *deciduous-forest biomes* (Pl. IV-6B). They characteristically include many more species than the taiga to the north.

Tropical areas with abundant rainfall are (or, more correctly, were) usually covered by *tropical rain forests,* which include some of the most complex communities there are on earth. The diversity of species is enormous; a temperate forest is composed of two or three, or at most ten, dominant tree species, but a tropical rain forest may be composed of a hundred or more. One may actually have difficulty finding any two trees of the same species within an area of many acres. The dominant trees are usually very tall, and their interlacing tops form dense canopies that intercept much of the sunlight, leaving the forest floor only dimly lit even at midday. The canopy likewise breaks direct fall of rain, but water drips from it to the forest floor much of the time, even when no rain is actually falling. It also shields the lower levels from wind and hence greatly reduces the rate of evaporation. The lower levels of the forest are consequently very humid. Temperatures near the forest floor are nearly constant. The pronounced differences in the microenvironmental conditions at different levels within such a forest result in a striking degree of vertical stratification; many species of animals and epiphytic plants (plants growing on the large trees) occur only in the canopy, others occur only in the middle strata, and still others occur only on the forest floor. Some vertical stratification is found in any community, particularly any forest community, but nowhere is it so extensively developed as in a tropical rain forest.

Huge areas in both the temperate and tropical regions of the world are covered by *grassland* biomes. These are typically areas where either relatively low total annual rainfall (10–20 inches) or uneven seasonal occurrence of rainfall makes conditions inhospitable for forests but suitable for often luxurious growth of grasses. Temperate and tropical grasslands are remarkably similar in appearance, although the particular species they contain may be very different. In both cases, there are usually vast numbers of large and conspicuous herbivores, often including ungulates (e.g. bison and pronghorn antelope in the United States). Burrowing rodents or rodentlike animals are often common (e.g. prairie dogs in the western United States).

In places where rainfall is very low, often less than 10 inches per year, not even grasses can survive as the dominant vegetation, and *desert* biomes occur (Pl. IV-7B). Deserts are subject to the most extreme temperature fluctuations of any biome type; during the day they are exposed to intense sunlight, and the temperature of both air and soil may rise very high (to 105°F or higher for air temperature and to 160°F or higher for surface temperature), but in the absence of the moderating influence of abundant vegetation, heat is rapidly lost at night, and a short while after sunset searing heat has usually given way to bitter cold. Some deserts, such as parts of the Sahara, are nearly barren of vegetation, but more commonly there are scattered drought-resistant shrubs (e.g. sagebrush, greasewood, creosote bush, and mesquite) and succulent plants that can store much water in their tissues (e.g. cactuses in New World deserts and euphorbias in Old World deserts). In addition, there are often many small rapid-growing annual herbs with seeds that will germinate only when there is a hard rain; once they germinate, the young plants shoot up, flower, set seed, and die, all within a few days. Most desert animals are active primarily at night or during the brief periods in early morning and late afternoon when the heat is not so intense. Most show numerous remarkable physiological and behavioral adaptations for life in their hostile environment.

We have seen that, moving north or south on the earth's surface, one may pass through a series of different biomes. The same thing is true if one moves vertically on the slopes of tall mountains. Climatic conditions change with altitude, and the biotic communities change correspondingly (Fig. 18.15). Thus arms, or isolated pockets, of the taiga extend far south in the United States on the slopes of the Appalachian Mountains in the east and of the Rockies and Coast Ranges in the west. There are even tundralike spots on the highest peaks.

BIOGEOGRAPHY

The geological or ecological zones that intervene between any two regions are apt to render dispersal considerably more difficult for some species than for others. A wide expanse of ocean may prohibit movement of horses or elephants, but coconut palms may cross it in fair numbers, because their large water-resistant seeds can float in sea water for many weeks without harm. A grassland separating two forested areas may be an almost insuperable barrier for some forest animals and prevent them from moving from one forest to the other. Other forest animals may have difficulty crossing the grassland but may do so occasionally, and still others may cross the grassland freely. In short, what is a barrier to dispersal for one species may be a possible but difficult route for another or an easily negotiated path for a third. The sorts of routes and barriers that are effective for different species help explain the distribution patterns of organisms on the earth's surface.

For a complete understanding of these distribution patterns, it would be necessary to know the ecology of all living species—including their physiological potential for survival in other habitats, the ecological opportunities they would find in those habitats, the ways in which they can be physically dispersed, and the routes and barriers they would be likely to encounter—as well as the detailed geography of the earth. But there is also a historical element to be considered. The earth itself and the organisms on it are constantly changing, and present distributions are in very large part the result of past conditions—conditions often

18.15. The correspondence between latitudinal and altitudinal life zones in eastern North America.

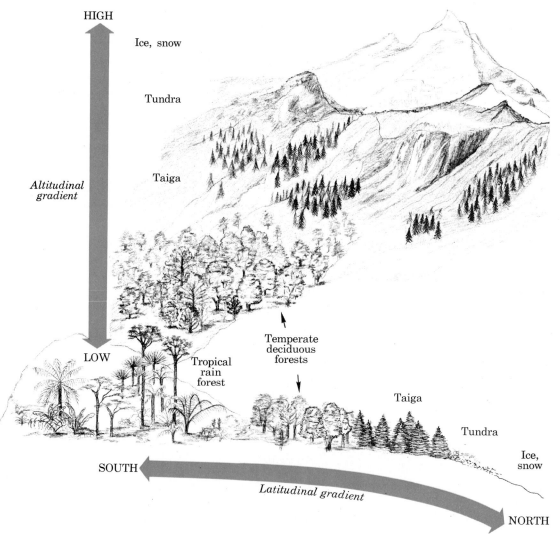

HIGH

Ice, snow

Tundra

Taiga

Altitudinal gradient

LOW

Tropical rain forest

Temperate deciduous forests

Taiga

Tundra

Ice, snow

SOUTH

Latitudinal gradient

NORTH

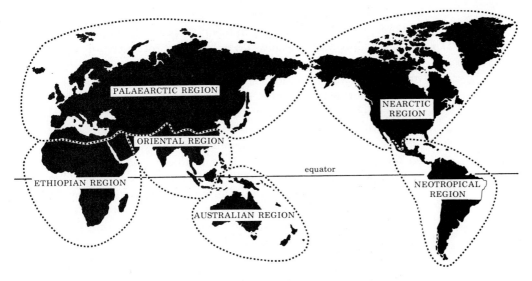

18.16. Biogeographic regions of the world.

very different from those now prevailing. For example, a knowledge of present conditions alone would be insufficient to explain why certain mammalian groups occur in South America, Africa, and southern Asia but nowhere else. Only by combining knowledge of present conditions with evidence from the fossil record and with geological evidence of the past shapes, connections, and climates of the earth's land masses and oceans can we hope to gain insight into the present geography of life.

With these ideas in mind, let us briefly consider some of the major regional patterns of distribution, as these have been described and classified by biologists over the past century (Fig. 18.16).

The island continents. The biota (flora and fauna) of the *Australian region* (Australia, New Zealand, and adjacent islands) is by all odds the most unusual found on any of the earth's major land masses (see Pls. III-2C, 5C, IV-3B). Many species common in Australia occur nowhere else. Conversely, many species widespread in the rest of the world are absent from Australia. This biological evidence, together with convincing geological evidence, indicates that Australia has not been connected to the Eurasian land mass for a very long time, if ever. The ancestors of most, if not all, organisms now living in the Australian region must have crossed a water barrier to get there. But since that water barrier was dotted by islands (the East Indies, presently the nation of Indonesia), it is not necessary to assume that a great expanse of ocean was crossed at any one time; organisms could have spread from one island to the next over a period of millions of years. Furthermore, most of the western islands of the East Indies were probably interconnected as an extension of the Asian land mass several times in the past; hence the distance from Asia to Australia has not always been as great as it is today.

Perhaps the best-known aspect of Australia's curious biota is its mammalian fauna, which is completely unlike that of any other continent. Other than wild dogs (dingoes) and pigs, both of which were probably brought to the region by prehistoric man, the only placental mammals (those in which the entire embryonic development takes place in the mother's uterus) present in the Australian region before European explorers landed there were a number of species of rodents belonging to a single family and a variety of bats. Bats can, of course, fly across water barriers and would be expected to reach most oceanic islands. The rodents of Australia are apparently relatively recent arrivals, which came from Asia by island-hopping through the East Indies. Most of the ecological niches that on other continents would be filled by placental mammals are in Australia filled by marsupials (mammals whose young develop only a short while in the uterus, are then born, move to a pouch on the mother's abdomen, attach to a nipple, and complete their embryonic development in the pouch). Apparently the marsupials reached Australia very early and, encountering no competition from placental mammals, underwent extensive evolutionary radiation. Since they were filling niches similar to those filled in the rest of the world by placentals and were thus subject to similar selection pressures, they evolved striking convergent similarities to the placentals. They variously resemble placental shrews, jumping mice, weasels, wolverines, wolves, anteaters, moles (see Fig. 17.14, p. 387), rats, flying squirrels, groundhogs, bears, etc. The uninitiated visitor to an Australian zoo finds it hard to believe when first looking at an assemblage of these animals that he is not seeing close relatives of the mammals familiar to him from other parts of the world. Not all marsupials look like the ecologically equivalent placentals, however; though some of the kangaroos play an ecological role very similar to that of horses and other large placental grazers, they present a markedly different appearance.

The South American continent, known to biologists as the **Neotropical region** (meaning "new tropics"), has had a history similar to that of Australia. It too has been an island continent unconnected to the other major land masses of the world through much of its history. And it too had an early mammalian fauna that included a great variety of marsupials, as the fossil record shows. But it also had a variety of placentals that probably reached it during a short period of connection to North America via a Central American land bridge during the early part of the Age of Mammals (about 60 million years ago). After this land bridge disappeared, both the marsupials and the placentals evolved in isolation many characteristics convergent to those evolved by other placentals on the main World Continent. There were times during this period of isolation lasting some 60 million years when the water barrier between South America and what is now northern Central America was not so wide. A few additional placental mammals chanced to get across into South America at such times; among these were the ancestors of the modern New World monkeys and a number of rodents. Finally, not so very long ago (a few million years), connection to North America was re-established and many additional immigrants arrived in South America. Some species also moved in the opposite direction, from South America to North America; the opossum, the porcupine, and the armadillo are examples. Central America has never been more than a narrow bridge, however, a filter route, along which some species but not others could pass. Its climate and rugged terrain have not been hospitable to many northern species, which, accordingly, have never been able to move between the North and South American continents. Thus South America, like Australia, has been an island continent through much of its history, but its nearness to North America (Mexico) and its occasional direct connections to North America via the Central American land bridge have given it a more diverse biota with more similarities to that of the World Continent.

The World Continent.   Europe, Asia, Africa, and North America have formed a relatively continuous land mass, the so-called World Continent, throughout most of geological time. Their biotas are consequently more alike in many aspects than they are like those of the two island continents. Nevertheless, biologists customarily divide the World Continent into four biogeographical regions: the *Nearctic* ("new northern"), which is North America; the *Palaearctic* ("old northern"), which is Europe and northern Asia; the *Oriental,* which is southern Asia; and the *Ethiopian,* which is Africa south of the Sahara.

The geological feature that looks like the most obvious barrier between the Palaearctic and Ethiopian regions is the Mediterranean Sea, but this is not really the important barrier. Many species can move between Europe and North Africa by circling around the eastern end of the Mediterranean. The real barrier to species dispersal is the Sahara Desert. Africa north of the Sahara is part of the Palaearctic region.

The Oriental region—tropical Asia—is separated from Palaearctic northern Asia in most places by east–west mountain ranges, of which the Himalayas between India and China are part. It is interesting to note that these east–west mountains constitute important breaks between climatic regions, and that they therefore act as both topographic and climatic barriers between cold-adapted and warm-adapted species. By contrast, north–south mountains such as those in North America tend to facilitate mixing of cold-adapted and warm-adapted species.

That the Palaearctic, Oriental, and Ethiopian regions constitute an essentially continuous land mass is obvious, but you may wonder why the Nearctic is considered part of the same land mass. The answer is that North America and Asia have been connected by a Siberian land bridge through much of their geological history. Part of this bridge, between what are now Alaska and Siberia, is beneath water at the present time. Because the climate of Alaska and Siberia is so forbidding, you might not expect many organisms to use a bridge in that region. But again present conditions are misleading. The climate of that area has not always been so severe. Fossils of many temperate and even subtropical species of plants and animals are abundant in Alaska. Indeed all the evidence indicates that on many occasions during the past 60 million years or so there has been much movement between Asia and North America via the Siberian land bridge. In fact, the Nearctic and Palaearctic regions are biologically so similar that many biologists regard them as a single region, which they call the Holarctic.

The history of the major land masses and the two principal land bridges helps account for the disjunct distribution referred to earlier. When a mammalian group occurs in South America, Africa, and southern Asia, but not elsewhere, we may assume that it was once widespread and moved between the New World and the Old World via the Siberian land bridge, and between North and South America via Central America, but that it then became extinct in the north, either because of climatic changes or because of intense competition. The fossil record shows that this pattern of dispersal between southern regions via the north has occurred again and again. For example, members of the camel family are found today in South America (llamas, alpacas, vicuñas, etc.), northern Africa, and central Asia, but fossils indicate that the family originated in North America, spread to South America via Central America and to the Old World via Siberia, and later became extinct in North America; hence the disjunct distribution we see today.

REFERENCES

ALLEE, W. C., A. E. EMERSON, O. PARK, T. PARK, and K. P. SCHMIDT, 1949. *Principles of Animal Ecology.* Saunders, Philadelphia.

ANDREWARTHA, H. G., and L. C. BIRCH, 1954. *The Distribution and Abundance of Animals.* University of Chicago Press, Chicago.

CHRISTIAN, J. J., and D. D. DAVIS, 1964. "Endocrines, Behavior, and Population," *Science,* vol. 146, pp. 1550–1560.

DARLINGTON, P. J., 1957. *Zoogeography.* Wiley, New York.

EHRLICH, P. R., and A. H. EHRLICH, 1972. *Population, Resources, Environment,* 2nd ed. Freeman, San Francisco.

ELTON, C. S., 1958. *The Ecology of Invasions by Animals and Plants.* Methuen, London.

FARVAR, M. T., ed., 1971. *Careless Technology: Ecology and International Development.* Doubleday Natural History Press, New York.

HAZEN, W. E., ed., 1970. *Readings in Population and Community Ecology,* 2nd ed. Saunders, Philadelphia. (A collection of 23 important articles by prominent ecologists.)

KENDEIGH, S. C., 1961. *Animal Ecology.* Prentice-Hall, Englewood Cliffs, N.J.

McINTOSH, R. P., 1958. "Plant Communities," *Science,* vol. 128, pp. 115–120.

NICHOLSON, A. J., 1957. "The Self-Adjustment of Populations to Change," *Cold Spring Harbor Symposium on Quantitative Biology,* vol. 22, pp. 153–172.

ODUM, E. P., 1971. *Fundamentals of Ecology,* 3rd ed. Saunders, Philadelphia.

OOSTING, H. J., 1956. *The Study of Plant Communities,* 2nd ed. Freeman, San Francisco.

ROGERS, W. P., 1962. *The Nature of Parasitism.* Academic Press, New York.

SLOBODKIN, L. B., 1961. *Growth and Regulation of Animal Populations.* Holt, Rinehart & Winston, New York.

WHITTAKER, R. H., 1953. "A Consideration of Climax Theory: The Climax as a Population and Pattern," *Ecological Monographs,* vol. 23, pp. 41–78.

————, 1970. *Communities and Ecosystems.* Macmillan, New York.

SUGGESTED READING

ANDREWARTHA, H. G., 1971. *Introduction to the Study of Animal Populations,* 2nd ed. University of Chicago Press, Chicago. (Paperback edition, Phoenix, Chicago, 1971.)

BILLINGS, W. D., 1969. *Plants, Man, and the Ecosystem,* 2nd ed. Wadsworth, Belmont, Calif.

"The Biosphere," special issue of *Scientific American,* September, 1970.

BORMANN, F. H., and G. E. LIKENS, 1970. "The Nutrient Cycles of an Ecosystem," *Scientific American,* October. (Offprint 1202.)

BOUGHEY, A. S., 1971. *Man and the Environment.* Macmillan, New York.

BRESLER, J. B., ed., 1968. *Environments of Man.* Addison-Wesley, Reading, Mass.

CARSON, R., 1961. *The Sea Around Us,* rev. ed. Oxford University Press, New York. (Paperback edition, Signet, New York, 1954.)

————, 1955. *The Edge of the Sea.* Houghton Mifflin, Boston. (Paperback edition, Signet, New York, 1971.)

————, 1970. *Silent Spring,* rev. ed. Fawcett World, New York. (Paperback.)

CLARK, J. R., 1969. "Thermal Pollution and Aquatic Life," *Scientific American,* March. (Offprint 1135.)

COLE, L. C., 1958. "The Ecosphere," *Scientific American,* April. (Offprint 144.)

COOPER, C. F., 1961. "The Ecology of Fire," *Scientific American,* April.

DEEVEY, E. S., 1958. "Bogs," *Scientific American,* October. (Offprint 840.)

DUBOS, R., 1965. *Man Adapting.* Yale University Press, New Haven.

HOLDREN, J. P., and P. R. EHRLICH, eds., 1971. *Global Ecology,* Harcourt Brace Jovanovich, New York.

KORMONDY, E. J., ed., 1965. *Readings in Ecology.* Prentice-Hall, Englewood Cliffs, N.J. (A collection of passages from 60 important papers.)

LEOPOLD, A. S., 1961. *The Desert.* Time Inc., New York.

ODUM, E. P., 1963. *Ecology.* Holt, Rinehart & Winston, New York.

ODUM, H. T., 1971. *Environment, Power, and Society.* Wiley, New York.

PLASS, G. N., 1959. "Carbon Dioxide and Climate," *Scientific American,* July. (Offprint 823.)

RAY, J. D., and G. E. NELSON, eds., 1971. *What a Piece of Work Is Man.* Little, Brown, Boston.

SMITH, R. L., 1966. *Ecology and Field Biology.* Harper & Row, New York.

WAGNER, R. H., 1971. *Environment and Man.* Norton, New York.

WECKER, S. C., 1964. "Habitat Selection," *Scientific American,* October. (Offprint 195.)

WENT, F. W., 1955. "The Ecology of Desert Plants," *Scientific American,* April. (Offprint 114.)

WOODWELL, G. M., 1967. "Toxic Substances and Ecological Cycles," *Scientific American,* March. (Offprint 1066.)

PART V

The genesis and diversity of organisms

The origin and early evolution of life

Few questions have exercised the human imagination like that of the origin of life. Religion, mythology, philosophy have proposed a great variety of answers to it, but different though these answers have been, most of them share the assumption that the phenomenon must be attributed to an agency outside nature, a creator. In the same way, the diversity of species was conceived as resulting from separate, deliberate acts of creation. It was not until the latter part of the nineteenth century that the theory of evolution was able to account for the origin of species without invoking a supernatural agency. Can twentieth-century science do the same for the origin of life itself?

THE ORIGIN OF LIFE

We discussed the principle of biogenesis—that life can arise only from life—in Chapter 3. Pasteur's work effectively put to rest, as far as most biologists were concerned, the long-held idea of spontaneous generation. No longer could men of science seriously entertain the notion that the maggots in decaying meat arise *de novo* from the meat, or that earthworms arise from the soil during heavy rains, or that microorganisms appear spontaneously in spoiling broth. It may seem strange, then, that in the mid-twentieth century spontaneous generation should be a topic of major interest in biology. The modern theorists do not suggest that life can arise spontaneously under present conditions on earth; indeed, most of them are convinced that it cannot. Their position is that life could and did arise spontaneously from nonliving matter under the conditions prevailing on the early earth, and that it is from such beginnings that all present earthly life has descended.

One purpose of this chapter is to outline for you a now widely held theory of the origin of life, which was first enunciated clearly and forcefully by the Russian biochemist A. I. Oparin in 1936. Although the broad outlines of the theory have wide support, many of the details are disputed. After all, we have no direct evidence concerning the origin of life; we can only gather indirect evidence to show how life could have arisen and how it probably arose.

Formation of the earth and its atmosphere. We are not certain how the solar system originated. The hypothesis most widely held today is that the sun and its planets formed between four and a half and five billion years ago from a cloud of cosmic dust and gas. Most of this material began to condense rapidly into a more compact mass. The condensation produced enormous heat and pressure, which initiated thermonuclear reactions and converted the main condensed mass into the sun. Within the remainder of the dust and gas cloud, which now formed a disc held in the gravitational field of the newborn sun, lesser centers of condensation began to form. These became the planets, of which the earth is one.

As the earth condensed, a stratification of its components took place, the heavier materials, such as iron and nickel, sinking into the core and the lighter substances becoming more concentrated nearer the surface. Among these lighter materials must have been hydrogen, nitrogen, oxygen, and carbon, all of which were to play crucial roles in the later origin of life.

The present atmosphere of the earth is an oxidizing one, containing about 78 percent molecular nitrogen (N_2), 21 percent molecular oxygen (O_2), and 0.033 percent carbon dioxide (CO_2), plus traces of rarer gases such as helium and neon. The primordial atmosphere, by contrast, appears to have been a reducing one, i.e. one containing a significant proportion of hydrogen, which reduced much of the available nitrogen, oxygen, and carbon. Thus the nitrogen in the atmosphere was probably in the form of ammonia (NH_3), the oxygen in the form of water vapor (H_2O), and the carbon primarily in the form of methane (CH_4). Note that methane is a simple hydrocarbon—an organic compound; thus the atmosphere of the early earth contained organic molecules long before there were any organisms. This sort of atmosphere (containing large percentages of methane, ammonia, and hydrogen) is still found on Jupiter and Saturn.

Initially, most of the earth's water was probably present as vapor in the atmosphere, forming dense clouds many miles thick. Whenever any of this vapor condensed and fell to the surface as rain, the moderately high temperatures of the earth's crust doubtless caused prompt evaporation of the water. But eventually the crust cooled sufficiently for liquid water to remain on it. Torrential rains must have fallen, filling the low places on the crust with water and giving rise to the first oceans. As rivers rushed down the slopes, they must have dissolved away and carried with them salts and minerals of various sorts, which slowly accumulated in the seas. Some of the methane and much of the ammonia (as ammonium ions, NH_4^+) from the atmosphere probably also dissolved in the waters of the newly formed oceans.

Formation of small organic molecules. If the early earth had a reducing atmosphere, as Oparin suggested and as many astronomers and chemists agree, and the primitive warm seas contained a mixture of salts, ammonium, and methane, how were the more complex organic molecules formed? Methane may be organic, but it is far from being a sufficient base on which to build living things. A mixture of ammonia, methane, water, and hydrogen is thermodynamically stable. There is no tendency for these materials to react with each other to form other compounds. Yet for life to have arisen it would seem at the very least that the critical building-block materials—particularly amino acids and purines and pyrimidines, the nitrogenous bases in nucleic acids—would have been necessary. How might these compounds have been formed on the abiotic primitive earth?

If more complex organic compounds were produced by reactions in the stable mixture of ammonia, methane, water, and hydrogen, clearly some external source of energy must have been acting on the mixture. One possible source might have been solar radiation, including visible light, X rays, and, probably most important, ultraviolet light. A second possibility is energy from electrical discharges, such as lightning. A third is heat. Can it be demonstrated that one of these sources of energy or a combination of them can cause reactions that produce complex organic compounds from a mixture of ammonia, methane, water, and hydrogen? An answer to this question was provided in 1953 by Stanley L. Miller, then a graduate student at the University of Chicago.

Miller set up an airtight apparatus in which the four gases could be circulated past electrical discharges from tungsten electrodes. He kept the gases circulating continuously in this way for one week, and then analyzed the contents of his apparatus. He found that an amazing variety of organic compounds had been synthesized. Among these were some of the biologically most important amino acids and also such substances as urea, hydrogen cyanide, acetic acid, and lactic acid. Lest it be objected that microorganisms had contaminated his gas mixture and synthesized the compounds, Miller in another experiment circulated the gases in the same way but without any electrical discharges; no significant yield of complex organic compounds resulted. In still another experiment, he prepared the apparatus with the gas mixture inside and then sterilized it at 130°C for 18 hours before starting the sparking. The yields of complex compounds were the same as in his first experiments; a great variety of organic compounds were formed. Clearly, the synthesis was not brought about by microorganisms but was abiotic—a synthesis in the absence of any living organisms, a synthesis under conditions presumably similar to those on the primordial earth. This experiment by Miller, which gave the first conclusive evidence that some of the steps hypothesized by Oparin could really occur, marked a turning point in the scientific approach to the problem of how life began.

In the years since 1953, many investigators have synthesized a great variety of organic compounds (including purines, pyrimidines, and simple sugars) from reducing mixtures of gases in which the only initial carbon source was methane. Some, using ultraviolet light or heat or both as energy sources, have also obtained large yields; these results are important, because on the early earth ultraviolet light was probably more available as a source of energy than lightning discharges. Significantly, the amino acids most easily synthesized in all these experiments performed under abiotic conditions were the very ones that are most abundant in proteins today.

The wide variety of conditions under which abiotic synthesis of the organic compounds essential to life has now been demonstrated makes it safe to conclude that, even if conditions on the primitive earth were only roughly similar to the ones postulated, synthesis of such compounds on the early earth was not only possible but highly probable.

Even if organic compounds were synthesized abiotically on the primordial earth, would they not have been destroyed too fast to accumulate in quantities sufficient for the later origin of living things? After all, most of these organic compounds are known to be highly perishable. But why are they perishable? First, because they tend to react slowly with molecular oxygen and become oxidized. Second, because they are broken down by organisms of decay, primarily microorganisms. But the prebiotic atmosphere contained virtually no free oxygen, and there were no organisms of any kind. Therefore neither oxidation nor decay would have destroyed the organic molecules, and they could have accumulated in the seas over hundreds of millions of years. No such accumulation would be possible today.

Formation of polymers.   Let us suppose, then, that a variety of hydrocarbons, fatty acids, amino acids, purine and pyrimidine bases, simple sugars, and other relatively small organic compounds slowly accumulated in the ancient seas. This is still not a sufficient basis for the beginning of life. Macromolecules are needed, particularly polypeptides and nucleic acids. How might these polymers have formed from the building-block substances present in the "hot soup" of the ancient oceans? This question is not easy to answer, and several hypotheses are currently supported by different investigators.

Some think that the concentration of organic material in the seas was high enough for chance bondings between simpler molecules to give rise, over a period of hundreds of millions of years, to considerable quantities of macromolecules. They point out that even though each such polymerization reaction is rather unlikely in the absence of protein enzymes, nevertheless on the time scale here involved enough rare and unlikely events may occur to produce, collectively, a major change. As George Wald of Harvard University has said, "Given so much time, the 'impossible' becomes possible, the possible probable, and the probable virtually certain."

The occurrence of polymerization might have been favored by adsorption of the building-block compounds on surfaces such as those of clay minerals. Or if small amounts of dilute solution of building-block compounds had collected in puddles on the beaches of lagoons and ponds, the heat of the sun would have evaporated much of the water, thus concentrating the organic chemicals, and provided energy for polymerization reactions. The resulting polymers might then have been washed back into the pond. A process such as this could slowly have built up a supply of macromolecules in the pond. The hypothesis seems a reasonable one, for Sidney W. Fox of the University of Miami has shown that if a nearly dry mixture of amino acids is heated, poly-peptide molecules are indeed synthesized (particularly if phosphates are present). Alternatively, after condensation by evaporation, the energy for polymerization reactions in the puddles might have come from ultraviolet radiation rather than heat.

Although various concentrating mechanisms may well have played a role in prebiotic polymerization reactions, too much may have been made of the difficulty of incorporating building-block compounds into polymers, at least in the case of amino acids. In experiments of the Miller type, various researchers have observed that spherules yielding amino acids upon hydrolysis are formed after only 48 hours, whereas free amino acids do not appear in appreciable quantities until much later. Similarly, in some of their experiments on thermal synthesis, Fox and his associates obtained polymers first, and amino acids only later, *after hydrolysis* of the polymers. In short, abiotic synthesis may differ fundamentally from the more familiar biochemical syntheses in that polymers may be formed first, rather than the expected monomers.

Formation of molecular aggregates and primitive cells. We have now reached a point in our model for the origin of life where the "hot soup" of the ancient seas, or at least the "soup" in some estuaries and lagoons, contains a mixture of salts and organic molecules, includ-ing polymers such as polypeptides and perhaps nucleic acids. How could the orderliness that characterizes living things emerge from this mix-ture?

Oparin pointed out that, under appropriate conditions of tempera-ture, ionic composition, and pH, colloids of macromolecules tend to give rise to complex units called *coacervate* droplets. Each such droplet is a cluster of macromolecules surrounded by a shell of water in which the individual water molecules are rigidly oriented relative to the col-loidal particles. There is thus a definite demarcation or interface be-tween the coacervate droplet and the liquid in which it floats. In a sense, the shell of oriented water molecules forms a membrane around the droplet. Now, coacervate droplets have a marked tendency to adsorb and incorporate various substances from the surrounding solution; sometimes this tendency, which is a selective one, is so pronounced that the droplets may almost completely remove some materials from the medium. In this way, the droplets may grow at the expense of the surrounding liquid. And coacervate droplets have a strong tendency toward formation of definite internal structure; i.e. the molecules within the droplet tend to become arranged in an orderly manner in-stead of being randomly scattered. As more and more different materials are incorporated into the droplet, a membrane consisting of surface-active substances may form just inside the shell of oriented water mole-cules, the permeability of the boundary of the droplet thus becoming even more selective than before. Thus, although coacervate droplets are not alive in the usual sense of the word, they do exhibit many properties ordinarily associated with living organisms.

Fox, like Oparin, has envisioned the prebiological systems (pre-bionts) that led to development of the first cells as microscopic multi-molecular droplets, but he suggests proteinoid microspheres rather than coacervate droplets. Fox's microspheres are droplets that form spontaneously when hot aqueous solutions of polypeptides are cooled. The microspheres may exhibit many properties characteristic of cells, including swelling in a hypoosmotic medium and shrinking in a hyperosmotic one, formation of a double-layered outer boundary, an internal movement reminiscent of cytoplasmic streaming, nonrandom movement of the whole unit if the microsphere is asymmetric and contains ATP, growth in size and increase in complexity, budding in a manner superficially similar to that seen in yeasts, and a tendency to aggregate in clusters of various types resembling those seen in many bacteria.

Since either type of droplet, complex coacervate or proteinoid microsphere, is structurally organized and sharply separated from the external medium, the chemical reactions that take place within the droplet depend not only on the conditions of the medium but also on the physicochemical organization of the droplet itself. Because various substances may be more concentrated in the droplet, the probability of their taking part in chemical reactions is increased; and because of the organization within the droplet, each reaction that takes place will influence other reactions in ways that are most unlikely when the substances are free in the external medium. Furthermore, catalytic activity of both inorganic substances such as metallic compounds and organic ones such as proteins is enhanced by the regular spatial arrangement of molecules within the droplet. In short, the special conditions within the droplet will exert selective and regulative influence over the chemical reactions taking place within the droplet.

Vast numbers of different prebiological systems of this kind may have arisen in the seas of the early earth. Most would probably have been too unstable to last long. But a few may have contained particularly favorable combinations of materials, especially complexes with catalytic activity, and may thus have developed unusually harmonious interactions between the reactions occurring within them. They might thus have survived longer and undergone relatively coordinated growth. As such droplets increased in size, they would have been more susceptible to physical fragmentation, which would have produced new smaller droplets with composition and properties essentially similar to those of the original droplet. These, in turn, would have grown and fragmented again. This primitive reproduction might not initially have been under the control of nucleic acids, even though these compounds could have been synthesized under abiotic conditions and may well have been incorporated into some of the prebionts. The nucleic acids would, of course, have been able to reproduce themselves exactly if a sufficient pool of nucleotides was present and if there were appropriate catalysts (even weak ones).

Clearly, if the new droplets formed on fragmentation of the most stable prebionts could have acquired more exact replicates of the favorable features of the parent droplet, their chances of survival would have been greater than without a genetic control system. In other words, chances of survival would have been increased if the sequence of nucleotides in nucleic acids could have come to code not only for more nucleic acid but for other components of the droplets, particularly proteins. As far as is known, nucleic acids have no catalytic properties and proteins have no duplicative properties; hence the establishment of a functional relation between the two was crucial to the origin of the type of life we know today. Just how a correlation between nucleotide sequences in nucleic acids and amino acid sequences in proteins would have arisen remains an unsolved problem, but the presumption is that it slowly came about and thus made possible much more accurate duplication during the reproductive process, and also more precise control over the chemical reactions taking place within the droplets. It is proposed, at any rate, that a small percentage of prebionts with particularly favorable characteristics slowly developed into the first primitive cells. Notice that there was almost certainly no abrupt transition from "nonliving" prebionts to "living" cells. The attributes that we normally associate with life were acquired gradually. At this stage, the boundary between living and nonliving is an arbitrary one.

Not all biologists accept the sequence outlined above. Some think it more likely that the first "living" things were self-replicating macromolecules such as nucleic acids—"naked genes," if you will. The first cells would then have arisen as these macromolecules slowly accumulated a shell of other substances (a primitive cytoplasm) around themselves. In other words, while the first model for the genesis of cells suggests that prebionts capable of a very primitive and imprecise form of reproduction arose first and then slowly developed a genetic control system, the suggestion here is that the control system arose first and that cytoplasm and a membrane then developed around it.

Viruses are the closest thing to naked genes found in nature today. They are simply nucleic acids with a thin shell of protein around them, and might thus be modern representatives of one of the earliest stages in the evolution of cells from naked genes. This possibility has been disputed on the ground that all modern viruses are obligatorily parasitic —that they can reproduce only inside living cells—and consequently could not have existed before there were any cells. Today, it is true, the raw materials necessary for virus reproduction can be found only inside living cells (which is why viruses are obligate parasites), but these substances would have been available in the "hot soup" hypothesized for the prebiological earth; hence viruses or something like them could have been free-living as long as the "hot soup" existed.

Whether either of these models for the origin of the first cells approximates the facts can perhaps never be settled with finality. But further attempts to duplicate in the laboratory the developmental sequences these models suggest may cast more light on this intriguing subject in the future.

Evolution of more complex biochemical pathways by primitive organisms. According to the view now most widely accepted, the earliest organisms were heterotrophs. They used as nutrients the carbohydrates, amino acids, and other organic compounds free in the environment in which they lived. In other words, they depended on previous abiotic synthesis of organic compounds. But as the organisms became more abundant and more efficient at removing nutrients from the medium, they must have begun to deplete the supply of nutrients. The rate of abiotic formation of organic matter was most likely never very high, and it must have taken many millions of years for a moderate supply of nutrients to accumulate. Now that supply was probably being used up at an accelerating rate. With consequently increased competition, forms inefficient at obtaining nutrients doubtless perished; those more efficient survived in greater frequency. Natural selection would have favored any new mutation that enhanced the ability of its possessor to obtain or process food.

At first, the primitive organisms probably carried out relatively few complex biochemical transformations. They could obtain most of the materials they needed ready-made. But it would have been these very materials—materials that could be utilized directly with little alteration—that would have dwindled most rapidly. Hence there would have been strong selection for any organisms that could utilize alternative nutrients. Suppose, for example, that compound A, which was necessary for the life of cells, was initially available in the medium but that its supply was being rapidly exhausted. If some cells possessed a mutant gene that coded for an enzyme, a, that catalyzed synthesis of A from another compound, B, in greater supply in the medium, then those cells would have had an adaptive advantage over cells that lacked the mutant gene. They could survive even when A was no longer available in the medium by carrying out the reaction

$$B \xrightarrow{\quad a \quad} A$$

But then there would have been increasing demand for free B, and the rate of its utilization would soon have exceeded the rate of its abiotic synthesis. Thus the supply of B would have dwindled, and there would have been strong selection for any cells possessing a second mutant gene that coded for an enzyme, b, catalyzing synthesis of B from C. These cells would not have been dependent on a free supply of either A or B because they could make both A and B for themselves as long as they could obtain sufficient C:

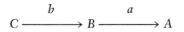

$$C \xrightarrow{\quad b \quad} B \xrightarrow{\quad a \quad} A$$

This process of evolution of synthetic ability might have continued until eventually most cells made all the A they required by carrying out a long chain of chemical reactions.

Evolution of autotrophy. Even though the primitive heterotrophs probably evolved biochemical pathways that enabled them to utilize a greater variety of the organic compounds free in their environment, and even though some of them probably evolved other methods of feeding, such as saprophytism, parasitism, and predation, nevertheless life would eventually have ceased if all nutrition had remained heterotrophic. Not only must nutrients have been used up much faster than they were being synthesized, but also the organisms themselves must have been altering the environment in ways that decreased the rate of abiotic synthesis of organic compounds. For example, their respiration, which would have been fermentative in the absence of molecular oxygen, would have released carbon dioxide into the atmosphere. Abiotic synthesis of complex organic compounds from CO_2 is much less likely than from methane.

That life did not become extinct as the supply of free organic compounds dwindled is attributable to the evolution of photosynthetic pathways by some of the primitive organisms. The first such pathway may well have been that of cyclic photophosphorylation, in which light of visible wavelengths is used as an energy source in synthesis of ATP by cells. Later, the much more complex pathways of noncyclic photo-phosphorylation and CO_2 fixation, in which energy from sunlight is used in synthesis of carbohydrate from carbon dioxide and water (or some other hydrogen source) would have arisen. From this time on-ward, the continuation of life on earth depended on the activity of the photosynthetic autotrophs.

The evolution of photosynthesis probably administered the coup de grâce to significant abiotic synthesis of complex organic compounds. An important by-product of plant photosynthesis is molecular oxygen. The oxygen released by photosynthesis must have helped convert the atmosphere from a reducing into an oxidizing one, particularly since much of the free hydrogen present initially must by this time have escaped from the planet into space. Once the layer of ozone (O_3) now present high in the atmosphere had been formed by some of the oxygen, it effectively screened out most of the ultraviolet radiation from the sun and allowed very little high-energy radiation to reach the earth's surface. In other words, living organisms, once they arose, changed their environment in a way that destroyed the conditions that had made possible the origin of life.

Once molecular oxygen became a major component of the atmo-sphere, both heterotrophic and autotrophic organisms could evolve the biochemical pathways of aerobic respiration, by which far more energy can be extracted from nutrient molecules than by fermentation alone.

The possibility of life on other planets.   If life could arise spon-taneously from nonliving matter on the primordial earth, might it also have arisen elsewhere in the universe? So far, no one knows the answer. Life may be unique to the planet Earth. But most scientists concerned with the problems of the origin of organisms are convinced that life has probably arisen many times in many places. They point out that no unduplicable event was necessary to the origin of life on earth. On the contrary, all the events now hypothesized and all the known character-istics of life seem to fall well within the general laws of the universe; i.e. they are natural phenomena susceptible of duplication. Indeed, according to some scientists, biochemical evolution (i.e. life) should be regarded as an inevitable part of the overall evolution of matter in the universe. Given the immense size of the universe, they argue, it would actually be unreasonable to think that life is restricted to one small planet in one minor solar system.

PRECAMBRIAN EVOLUTION

The oldest known fossils (as of this writing) are estimated to date back 3.1 billion years. They appear to be bacteria and blue-green algae —procaryotic cells, as would be expected of the oldest fossils, since these cells are assumed to be more primitive than eucaryotic ones. Most authorities think it probable that the first cellular organisms, living at a time when all nutrition was heterotrophic, were of the bacterial type, and that it was the evolution of blue-green algae from bacteria that

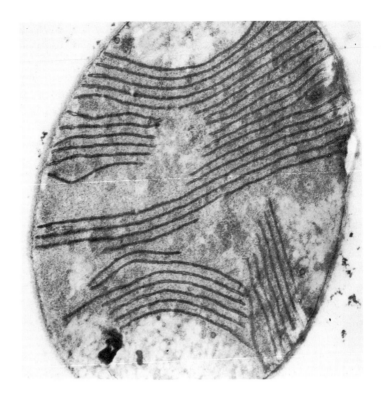

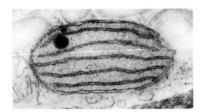

19.1. A blue-green alga (right), a procaryote, compared with the chloroplast of a red alga (above), a eucaryote. The photosynthetic lamellae in the two algae are arranged in a very similar way, but whereas they are loose in the cytoplasm of the procaryote, in the eucaryote they are contained in chloroplasts. Both photos × 30,000. [Left: Courtesy R. E. Lee, University of the Witwatersrand. Right: Courtesy M. Jost, Michigan State University.]

brought about the so-called oxygen revolution. These algae, like true plants but unlike photosynthetic bacteria, utilize water as the hydrogen source in noncyclic photophosphorylation and hence release molecular oxygen as a by-product.

More and more fossils of procaryotic organisms from the first two and a half billion years or so of life are being found, but fossils of higher forms of life do not appear in any quantity until the *Cambrian period,* which began about 600 million years ago (see Table 21.2, p. 478). Many of the Cambrian fossils are of relatively complex organisms—how complex is suggested by the fact that most of the animal phyla extant today are represented. The few Precambrian fossils of eucaryotic organisms are fairly complex too; hence we have very little evidence concerning a most fascinating evolutionary development—the early radiation of life and the origin of the major phyla.

In the absence of any direct evidence concerning the evolution of the first eucaryotic cells from procaryotic progenitors, study of the special properties of such characteristic eucaryotic organelles as chloroplasts, mitochondria, centrioles, and the basal bodies of cilia and flagella has shed some light on the possible origin of the complicated kind of cell that is the structural unit of higher plants and animals. These organelles contain some genetic material and are self-replicating; and they can carry out protein synthesis on their own ribosomes (see chloroplast ribosomes in Fig. 4.10, p. 84). In short, they have many features in common with free-living procaryotic organisms. Several investigators, most prominently Lynn Margulis of Boston University, have made the intriguing suggestion that these organelles might be modern descendants of ancient procaryotic cells that became obligate endosymbionts of other cells and evolved in concert with their hosts ever since.

It was noted as long ago as the nineteenth century that chloroplasts resemble certain free-living blue-green algae (Fig. 19.1). Might they have originated as such? A bacterial origin for mitochondria was first proposed in the 1920's. More recent research findings have tended to support these surmises regarding the origin of the two kinds of organelle: (1) The enzymes for synthesis of DNA, RNA, and protein in chloroplasts and mitochondria are qualitatively like those of procaryotic cells and correspondingly unlike those in the rest of the plant cell. (2) The nuclear DNA of eucaryotic cells is coated with protein, whereas the DNA of chloroplasts and mitochondria is naked like that of procaryotic cells. (3) The ribosomes of chloroplasts and mitochondria, like those of procaryotes, tend to be 15 percent smaller than the cytoplasmic ribosomes of eucaryotes. (4) Protein synthesis by the cytoplasmic ribosomes of eucaryotic cells is inhibited by certain drugs (e.g. cycloheximide) and not by others (e.g. chloramphenicol, erythromycin); protein synthesis in both procaryotic cells and chloroplasts and mitochondria shows exactly opposite specificity. In summary, a growing list of characteristics attests to a possible derivation of these two organelles from procaryotic ancestors.

A word of caution: Though the organelles discussed here contain DNA, and divide, grow, and differentiate partly on their own, they are not fully autonomous entities. Many of their proteins are specified by nuclear genes. Presumably the symbionts surrendered much of their genetic control to the host during the hundreds of thousands of years since they last lived as free cells.

Although we have been concerned with the origin of eucaryotic cells—literally, true nucleate cells—we have not so far mentioned the membrane-delimited nucleus. The fact is that evidence on the origin of the nuclear membrane and the associated endoplasmic reticulum is far weaker than that on the origin of chloroplasts and mitochondria. And it is true that these organelles are as fundamental to eucaryotic cells as the presence of a nuclear membrane.

THE KINGDOMS OF LIFE

With a fossil record nearly blank for the long span of time when the basic pattern of organismic diversity was coming into being, our ideas about the evolutionary relationships between the major phyla and divisions of organisms are rather vague. We have little evidence concerning the relationships among the major groups of algae. We don't know whether fungi evolved from photosynthetic green algae, or directly from heterotrophic organisms such as bacteria, or from some other stock. We are uncertain about the relationships of many Protozoa to multicellular plants or to multicellular animals.

Traditionally, men have assigned all living things to one or another of a few large categories called kingdoms. One of the oldest and most widely used such classifications recognizes only two kingdoms—one for plants and the other for animals. This dichotomy works well as long as the organisms to be classified are the generally familiar ones. Dandelions, grasses, daffodils, roses, and oak trees can easily be recognized as plants. Similarly, cats, horses, chickens, earthworms, and houseflies can easily be recognized as animals. Things become more difficult when the organisms in question are bread molds, sea anemones, or sponges. These don't fit quite so neatly within the common intuitive concept of plant and animal. Nevertheless, bread molds, when their characteristics are carefully examined, seem definitely more plantlike than animal-like, despite their lack of chlorophyll; and biologists can easily convince themselves that sea anemones and sponges are animals, despite their sedentary way of life.

But what about unicellular organisms? The ones zoologists have traditionally called Protozoa have been particularly troublesome to those who insist on a neat separation between plants and animals. This is true especially of the group of protozoans known as flagellates. These creatures have long flagella that enable them to swim actively in a manner intuitively felt to be animal-like. Yet some of them possess chlorophyll and carry out photosynthesis, a characteristic ordinarily considered decidedly plantlike. How should organisms such as these be classified?

Whatever the criteria chosen, it is impossible to make a clean separation between plants and animals that will not cause as many problems as it solves. The reason is obvious. Unicellular organisms (and some multicellular ones) are at an evolutionary level where it is essentially meaningless to talk about plants and animals. The kingdoms are artificial constructions, human efforts to cope with the tremendous diversity of the living world. They are not rules of nature; real living things don't come in two convenient categories labeled "plant" and "animal." At the lowest evolutionary levels, about the only distinction between plants and animals that stands up is: Plants are living things studied by people who say they are studying plants (botanists), and animals are living things studied by people who say they are studying animals (zoologists). Facetious as this distinction sounds, it is basically accurate. Since the unicellular green flagellates are studied by both botanists and zoologists, the first calling them algae and the second calling them Protozoa—without doing violence to any important biological principle—it is practical to classify green flagellates as both plants and animals. They will thus appear in two places in the classification used in this book—in the algal divisions Chlorophyta (green algae), Euglenophyta (euglenoid algae), or Chrysophyta (yellow-green and golden-brown algae) and in the Protozoa (see pp. 546 and 547).

Like green flagellates, bacteria and blue-green algae are difficult to classify as either plants or animals, but for a different reason: They are not closely related to either. Here is one instance where a clear and sharp break exists between major groups. Bacteria and blue-green algae are procaryotic, whereas all other organisms are eucaryotic. Biologists have traditionally put these two groups in the plant kingdom, but it is plain that this procedure is no longer acceptable. We shall follow the practice, which is rapidly gaining favor, of setting them apart as a separate kingdom, the **Monera.**

The viruses have not been mentioned in this discussion of kingdom classification because they differ in such fundamental ways from cellular organisms that biologists have been hesitant to regard them as living things. Yet they undeniably possess many properties in common with living organisms, and, living or not, they have come to be a major focus of interest in modern biology. Perhaps we should expand our definition of life to include noncellular entities that possess nucleic acid genes, in which case we could establish a separate kingdom for the viruses.

REFERENCES

KENYON, D. H., and G. STEINMAN, 1969. *Biochemical Predestination.* McGraw-Hill, New York.

MARGULIS, L., 1970. *Origin of Eukaryotic Cells.* Yale University Press, New Haven.

MILLER, S. L., 1953. "Production of Amino Acids Under Possible Primitive Earth Conditions," *Science,* vol. 117, pp. 528–529.

OPARIN, A. I., 1968. *Genesis and Evolutionary Development of Life.* Academic Press, New York.

RAVEN, P. H., 1970. "Multiple Origin for Plastids and Mitochondria," *Science,* vol. 169, pp. 641–646.

SHKLOVSKII, I. S., and C. SAGAN, 1966. *Intelligent Life in the Universe.* Holden-Day, San Francisco.

WHITTAKER, R. H., 1969. "New Concepts of Kingdoms of Organisms," *Science,* vol. 163, pp. 150–160.

SUGGESTED READING

EHRENSVARD, G., 1962. *Life: Origin and Development.* University of Chicago Press, Chicago.

KEOSIAN, J., 1968. *The Origin of Life,* 2nd ed. Reinhold, New York.

MARGULIS, L., 1971. "Symbiosis and Evolution," *Scientific American,* August. (Offprint 1230.)

OPARIN, A. I., 1938. *The Origin of Life on Earth.* Macmillan, New York. (Also available in paperback as *The Origin of Life,* Dover, New York, 1953.)

SAGAN, C., 1971. "Origin of Life," in *Topics in the Study of Life: The Bio Source Book,* ed. by V. G. Dethier *et al.* Harper & Row, New York.

SHAPLEY, H., 1963. *The View from a Distant Star.* Basic Books, New York.

Viruses and Monera

Although viruses are not ordinarily given a place in formal classifications of living organisms, both biologists and nonbiologists regularly use such phrases as "live virus vaccine" and "killed virus vaccine." The inconsistency is excusable. On the one hand, viruses lack all metabolic machinery and cannot reproduce in the absence of a host. On the other, they possess nucleic acid genes that encode sufficient information for the production of new viruses with the same characteristics as those that supply the templates; and reproduction with gene-controlled heredity is one of the most basic attributes of life.

Whether viruses are classed as living, as nonliving, or as something in between, it must be granted that modern study of procaryotic organisms—the bacteria and blue-green algae—is intimately bound to the study of viruses. Not only do viruses and procaryotes have similar genetic material, but their life cycles are intertwined. We shall therefore consider both these groups in the present chapter.

VIRUSES

The discovery of viruses. By the latter part of the nineteenth century, the idea had become firmly established that many diseases are caused by microorganisms. But for some diseases—among them smallpox, which it was known could be induced in a healthy person by something in the pus from a smallpox victim—no microbial agent could be found.

A suggestive experiment was performed in 1892 by a Russian biologist, Dmitri Iwanowsky, who was studying tobacco mosaic disease. The leaves of affected tobacco plants become mottled and wrinkled. If juice is extracted from an infected plant and rubbed on the leaves of a healthy one, that plant soon develops tobacco mosaic. If, however, the juice is heated nearly to boiling before it is rubbed on the healthy leaves, no disease develops. Concluding that the disease must be caused by bacteria in the plant juice, Iwanowsky passed juice from an infected tobacco plant through a very fine porcelain filter in order to remove the bacteria; he then rubbed the filtered juice on the leaves of healthy plants. Contrary to his expectation, the plants developed mosaic disease. During the next several decades, many other diseases of both plants and animals were found to be caused by infectious agents so small that they could pass through porcelain filters and could not be seen with even the best light microscopes. These microbial agents of disease came to be called filterable viruses, or simply viruses. They were at first assumed to be very small bacteria.

There were, however, a few hints that viruses might be something quite different from bacteria. First, all attempts to culture them on media customarily used for bacteria failed. Second, the virus material, unlike bacteria, could be precipitated from an alcoholic suspension without loss of its infectious power. But not until 1935 was it conclusively demonstrated that viruses and bacteria are two very different things. In that year, W. M. Stanley of the Rockefeller Institute isolated and crystallized tobacco mosaic virus. If the crystals were injected into tobacco plants, they again became active, multiplied, and caused disease symptoms in the plants. The fact that viruses could be crystallized showed that they were not cells but must be much simpler chemical entities.

Reproduction of viruses. Because viruses have no metabolic machinery of their own—they lack enzyme systems and they cannot generate ATP—and because they lack raw materials for synthesis, they cannot reproduce themselves in the same sense as true living organisms. It is the host cell, not the virus, that manufactures new virus particles when the genetic material of the old virus provides the instructions. Some types of virus, known as bacteriophage, appropriate the metabolic machinery of bacterial cells in order to reproduce; other types of virus make use of plant or animal cells. The viral genetic material is a single nucleic acid molecule, a few thousand to 250,000 nucleotides long. In bacteriophage the nucleic acid is DNA; in other types of virus it may be either DNA or RNA.

We have already discussed the reproduction of bacteriophage viruses (see p. 321). You will recall that a free phage particle becomes attached by the tip of its tail to the wall of a bacterial cell. The phage nucleic acid is injected into the host while the protein coat remains outside. Once inside the bacterial cell, the phage DNA provides genetic information for synthesis of new viral DNA and protein and for assembling these into complete new phage particles, which are then released when the cell lyses (ruptures). This series of events is called the *lytic cycle.*

Sometimes viral DNA does not immediately take control of the host cell's metabolic machinery and put it to work making new virus particles. Instead, it is integrated into the bacterial chromosome and is reproduced with the chromosome for an indefinite number of generations; it is then known as *provirus.* This so-called *lysogenic cycle* occurs when a viral gene induces production by the host of a repressor substance that blocks expression of the viral genes governing reproduction. As long as repression continues, the viral nucleic acid may remain integrated in the host's chromosome. The repressor substance also confers on the host cell immunity to virulent infection by other viruses of the same or similar types. Indeed, a standard test to determine whether cells are carrying provirus of a given type is to attempt to infect them with the virus in question.

The free-particle stage (sometimes called virion) of plant and animal viruses is usually simpler in structure than that of bacteriophage. In some cases the particle is rod-shaped (actually helical), in others polyhedral (Fig. 20.1). The absence of a tail like that of bacteriophage means that plant and animal viruses cannot inject their nucleic acid into the host cell. The whole virus particle penetrates into the cell, though only the nucleic acid component plays a role in viral multiplication, the protein coat being rapidly destroyed by host enzymes. Most plant viruses depend on insect vectors to inject them through the thick cell walls into the host-cell cytoplasm. Since the host cells of animal viruses have no walls, these viruses can come into direct contact with the cell membrane at special adsorption sites for which they have a high affinity, and then penetrate through the membrane into the cell interior. Much of the host specificity of such viruses results from their differing affinities for the various adsorption sites.

After sufficient quantities of new viral nucleic acid and new protein coats have been produced within the host cell, the two components are assembled into complete new viral particles ready for release from the host. As in bacteriophage, the release may be by lysis of the cell —in the case of plant cells, through a virus-induced enzyme that attacks the cell wall. In animal cells, however, the release is sometimes by extrusion, a process whereby the virus is enveloped in a small piece of cell membrane; the host cells may not be killed immediately.

The origin of viruses.   Viruses are not cells and thus, according to the cell theory, fail to meet one requisite that must be met by all living things. Can they be regarded as living things nevertheless? The answer may depend on which view of their origin one accepts. Each of the three hypotheses of viral origin outlined below has received serious consideration:

1. Viruses may be organisms that have reached the extreme of evolutionary specialization for parasitism. Loss of structures is commonly seen in internal parasites, and it is conceivable that intracellular parasites might lose everything but their nucleus. Since a virus particle resembles a cellular nucleus, viruses might have arisen from cellular ancestors by a process of gradual loss of all other cellular components. If this hypothesis is correct, then viruses should perhaps be regarded as degenerate cells and classified as living organisms.

2. The ancestors of modern viruses may have been free-living noncellular predecessors of cellular organisms. When the organic nutrients of the primordial "hot soup" seas disappeared, those of their descendants that remained noncellular survived by becoming parasites on the cellular organisms that had arisen by that time. According to this

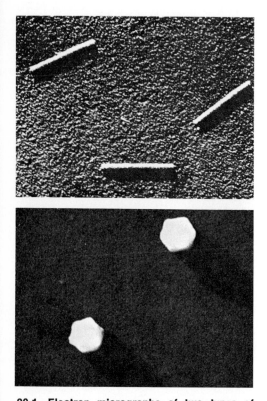

20.1. Electron micrographs of two types of virus. Top: Tobacco mosaic virus, a rod-shaped virus. Bottom: *Tipula* iridescent virus, a polyhedral virus that attacks the larvae of a species of crane fly. [Courtesy Virus Laboratory, University of California, Berkeley.]

view, modern viruses are representatives of an early "nearly living" stage in the origin of life.

3. Viruses may be fragments of genetic material detached from the chromosomes of cellular organisms. They might originally have existed as bare nucleic acid but might later have evolved the capacity of causing their host cells to synthesize a protein shell within which replicates of the nucleic acid could be enclosed. If this hypothesis is correct, viruses may have arisen as fragments of bacterial or plant or animal chromosomes. The host specificity of viruses might be a reflection of their origin; a given type of virus might be able to parasitize only species fairly closely related to the one from which the virus was originally derived.

Viral disease. Among the many human diseases caused by viruses are chicken pox, mumps, measles, smallpox, yellow fever, rabies, influenza, viral pneumonia, the common cold, poliomyelitis (infantile paralysis), fever blisters, several types of encephalitis, and infectious hepatitis. Unfortunately most viral infections do not respond to treatment with sulfur drugs or antibiotics, which have been so effective in bacterial diseases. But some of the most pernicious virus diseases, notably smallpox and polio, are preventable by means of vaccines; Edward Jenner demonstrated as early as 1796 that a person vaccinated with material from cowpox lesions develops an immunity to smallpox (see also p. 456).

Immune responses, though important in preventing reinfection, ordinarily appear too late to account for recovery from viral diseases. Recent evidence suggests that at least one important factor in recovery is a protein called *interferon*, which is produced by host cells in response to invading viruses. Apparently the interferon cannot save the cell in which it is produced, but when it is released from that cell and enters uninfected cells, it confers on them a resistance to viral infection, probably by inducing production of another protein that blocks replication of viral nucleic acid. In other words, the interferon acts as a messenger from infected cells to uninfected cells telling them to mobilize their defenses against viral infection.

Though interferon induced by one type of virus temporarily inhibits infection by virtually all other types in the host that produces it, it is not effective in other host species; in other words, interferon is host-specific, not virus-specific. This means that interferons produced by other animals are of no value in treating human beings. Since the quantities required for combating viral diseases are much greater than can be obtained from human donors, direct administration of interferon cannot be used in treatment. However, a group of researchers at the Merck Institute have succeeded in synthesizing an analogue of viral RNA, called poly I:C, which has proved a very effective inducer of interferon production when injected into patients, and has not so far revealed any serious side effects. Poly I:C and other similar experimental substances may come into general medical use in the next few years.

Evidence is accumulating that some, and perhaps all, cancers involve virus pathogens. Though free virus particles are seldom associated with the tumors, there is reason to think that the cells contain proviruses that have modified the cells' metabolic activity and caused them to become malignant. Viral genes probably activate the DNA-synthesizing machinery of the cells and thus initiate the proliferation characteristic of cancer. Other viral genes may alter the surface properties of the cells, and this may be one reason why they escape normal regulation. In some cases, cells may contain latent cancer viruses, acquired early in life, which do not become malignant until induced by an irritant, such as certain carcinogenic chemicals, tobacco smoke, frequent abrasion, or radioactivity.

MONERA

The kingdom Monera includes two divisions: the Schizomycetes (bacteria) and the Cyanophyta (blue-green algae). The cells of both groups are procaryotic; i.e. they lack a nuclear membrane, mitochondria, an endoplasmic reticulum, Golgi apparatus, and lysosomes.

Schizomycetes

Early studies of bacteria were based largely on species belonging to a group now called the eubacteria ("true" bacteria), all rather similar in their basic characteristics. But later other groups of organisms with quite different properties also came to be regarded as bacteria. Thus the modern Schizomycetes constitute an assemblage of diverse forms probably not very closely related to one another. The bacteria could easily be separated into three or more divisions as distinct from one another as each is from the blue-green algae.

Bacterial cells. Most bacteria are very tiny, far smaller than the individual cells in the body of a multicellular plant or animal. In fact, some bacteria are as small as some of the largest viruses. However, even the smallest bacteria are fundamentally different from viruses in that they are cellular. They always contain both RNA and DNA, whereas viruses contain one or the other but never both; they possess ribosomes, whereas viruses do not; they always contain many different proteins with enzymatic functions, whereas viruses carry no enzymatic proteins (except occasionally some concerned with attachment to and penetration into a host cell); they can generate ATP and use it in the synthesis of many other organic compounds, whereas viruses cannot; they provide both the raw material and the metabolic machinery for their own reproduction, whereas viruses do not. No entity truly transitional between viruses and cellular organisms is known to exist.

The cells of most bacteria have one of three fundamental shapes: spherical or ovoid (see Fig. 20.4), cylindrical or rod-shaped (Fig. 20.2), or helically coiled (Fig. 20.3). Spherical bacteria are called *cocci* (singular, coccus); rod-shaped ones are called *bacilli;* and helically coiled ones are called *spirilla.*

When cell division takes place, the daughter cells of some species remain attached and form characteristic aggregates. Thus cells of the bacterium that causes pneumonia are often found in pairs (diplococci). The cells of some spherical species form long chainlike aggregates (streptococci) (Fig. 20.4), while others form grapelike clusters (staphylococci). Each of the cells in a diplococcal, streptococcal, or staphylo-

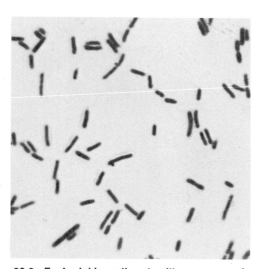

20.2. *Escherichia coli,* a bacillus common in the human digestive tract. [Courtesy Carolina Biological Supply Co.]

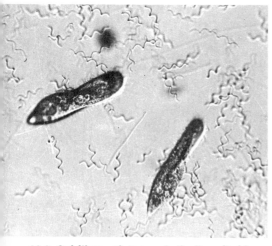

20.3. *Spirillum volutans,* a helically coiled bacterium. The two paramecians (which are Protozoa) are included for size comparison. [Courtesy General Biological Supply House, Inc., Chicago.]

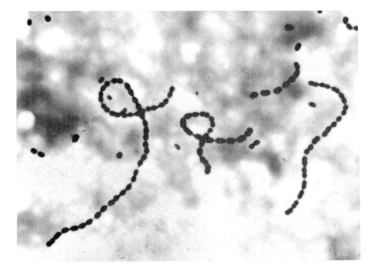

20.4. *Streptococcus lactis*, a bacterium common in milk. All streptococci are spherical (coccal) bacteria normally grouped in chainlike clusters. [Courtesy General Biological Supply House, Inc., Chicago.]

coccal aggregate is an independent organism, but some bacteria form multicellular filaments, in which adjacent cells have a common cell wall. Many Actinomycetes, among other filamentous forms, strikingly resemble molds (which are fungi), but this is a case of evolutionary convergence and not evidence of close relationship, because the cells of Actinomycetes, like those of other bacteria, are procaryotic whereas those of fungi are eucaryotic.

Like plant cells, most bacterial cells are enclosed in a cell wall, which protects the cell both from physical damage and from osmotic disruption. But the walls of eucaryotic cells derive their tensile strength largely from cellulose and related compounds (or from chitin in fungi), those of procaryotic cells from murein, a polymer composed of amino acids and amino sugars (and their derivatives, particularly muramic acid). This important difference between the chemical composition of procaryotic and eucaryotic cells is the basis for the selective activity of some drugs, such as penicillin. Nontoxic to plants and animals (or to resting bacterial cells), penicillin is toxic to growing bacteria because it inhibits formation of murein, and thus interferes with bacterial multiplication.

Differences in the relative amounts of certain components in the walls of different types of bacteria make the cells show characteristic reactions to a variety of stains. Since there are few visible morphological characters that can be used in identifying bacteria, diagnostic staining is an important laboratory tool.

Many bacterial cells secrete polysaccharide mucoid materials that accumulate on the outer surface of the cell wall and form a *capsule.* The capsule apparently makes the cell more resistant to the defenses of host organisms; hence encapsulated strains of a given bacterial species are more likely to cause disease than unencapsulated strains.

Some of the eubacteria (mostly rod-shaped ones) can form special resting cells called *endospores,* which enable them to withstand conditions that would quickly kill the normal active cell. Each small endospore develops inside a vegetative cell and contains DNA plus a limited amount of other essential materials from that cell (Fig. 20.5). It is enclosed in an almost indestructible spore coat. Once the endospore has

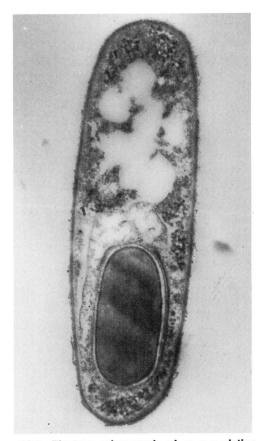

20.5. Electron micrograph of a sporulating bacillus. The spore is the dark oval in the lower end of the cell. The developing spore coat is clearly visible. The white areas in the upper end of the cell are not vacuoles but areas filled with fatty material. × 34,400. [Courtesy G. B. Chapman, *J. Bacteriol.,* vol. 71, 1956.]

fully developed, the remainder of the vegetative cell in which it formed may disintegrate. Because of their very low water content and refractile coats, spores of many species can survive an hour or more of boiling or an hour in a hot oven. They can be frozen for decades or perhaps for centuries without harm. They can survive long periods of drying. And they can even withstand treatment with strong disinfectant solutions. When conditions again become favorable, the spores may germinate, giving rise to normal vegetative cells that resume growing and dividing. Fortunately, few disease-causing bacteria can form endospores.

Many bacteria can move about actively. In some cases, the motion is produced by the beating action of flagella. Bacterial flagella are structurally very different from the flagella of eucaryotic cells. They are not enclosed within the cytoplasmic membrane, and they do not contain the nine peripheral and two central fibrils found in all flagella of eucaryotic cells. Some bacteria that lack flagella exhibit a peculiar gliding movement that does not involve any visible locomotor organelles; the mechanism of this movement has not yet been discovered.

Bacterial reproduction. Although bacteria were long thought to lack nuclei, newer techniques have revealed the presence of a nuclear area—one not surrounded by a membrane, however. Analysis has shown that there is DNA in the nuclear area, and that the bacterial genes are arranged in sequence along a single circular chromosome composed only of DNA.

Most bacteria reproduce by **binary fission,** a type of cell division in which two equal daughter cells with characteristics essentially like those of the parent cell are produced without mitosis. It has been established that when a bacterial cell undergoes fission, each daughter cell receives a full set of genes, i.e. a complete chromosome. Therefore nuclear division must have taken place. The details of nuclear division in procaryotic cells have been discovered only recently (Fig. 20.6). The DNA, which is attached to the cell membrane, is replicated, and the two chromosomes move into separate nuclear areas long before division of the cytoplasm occurs. When the plasma membrane grows inward, it simply partitions the binucleate parent cell into two daughter cells each of which already has its own nuclear area.

Bacteria commonly have an enormous reproductive potential. Many species may divide as often as once every 20 minutes under favorable conditions. If all the descendants of a cell of this type survived and divided every 20 minutes, the single initial cell would have about 500,000 descendants at the end of 6 hours, and by the end of 24 hours the total weight of its descendants would be about 4,000,000 pounds. Although increases of this magnitude do not actually occur, the real increases are frequently huge, which helps explain the rapidity with which food sometimes spoils or a disease develops.

Although the reproductive process itself is asexual, genetic recombination does occur occasionally, at least in some bacteria. Three mechanisms of recombination are known: **conjugation,** in which part of a chromosome is transferred from a donor cell to a recipient; **transformation,** in which a living cell picks up fragments of DNA released into the medium from dead cells (see p. 321); and **transduction,** in which fragments of DNA are carried from one cell to another by viruses.

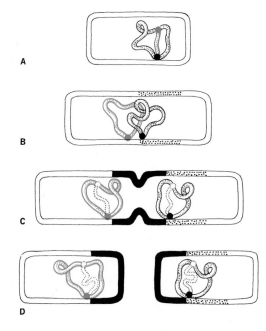

A

B

C

D

20.6. Binary fission of a procaryotic cell. (A) The circular chromosome of a young cell is attached to the plasma membrane near one end; it has already begun replication (the partially formed new chromosome is shown in color). (B) Chromosomal replication is complete, and the new chromosome now has an independent point of attachment to the membrane. During replication, additional membrane and wall (stippled) was formed, pushing the chromosomes nearer the midpoint of the cell. (C) More new membrane and wall (black) has formed between the points of attachment of the two chromosomes. Part of this growth forms invaginations that will give rise to a septum cutting the cell in two. Each chromosome has already begun replicating. (D) Fission is complete, and each of the daughter cells is at the same developmental stage as the cell in (A).

Bacterial nutrition.   Most bacteria are heterotrophic, being either saprophytes or parasites. Like animals, the majority of these bacteria are aerobic; i.e. they cannot live without molecular oxygen, which they use in the respiratory breakdown of carbohydrates and other food materials to carbon dioxide and water. Aerobic respiration is carried out with the help of a cytochrome electron-transport system, built (since bacteria lack mitochondria) into the inner surface of the cell membrane or its invaginations. However, to some bacteria that obtain all their energy by fermentation, oxygen is lethal; such bacteria are called *obligate anaerobes,* by contrast with *facultative anaerobes,* which can live either in the presence or in the absence of molecular oxygen. Some of the facultative anaerobes are simply indifferent to oxygen, obtaining all their energy from fermentation whether oxygen is present or not; others obtain their energy by fermentation when oxygen is not available, but carry out aerobic respiration (via the Krebs cycle) when oxygen is present, and may grow faster under these conditions. Besides lactic and alcoholic fermentations—the most common fermentative processes in living organisms—at least ten other types of fermentations occur in different groups of bacteria. Fermentation is, of course, a much less efficient energy-yielding process than aerobic respiration.

Bacteria differ considerably in the sorts of molecules they can use as energy sources and in the specific amino acids and vitamins they require. These differences provide valuable diagnostic characters for workers attempting to identify unknown bacteria. Samples of the organisms to be identified are placed on a variety of nutrient media and cultured at standard temperatures. By determining on which of the media the organisms will grow and on which not and then, when they grow, comparing the color, texture, etc. of the colony they produce with the same characters in known species, it is often possible to assign the unknown organisms to the proper group or even to the proper species.

Most bacteria, as we said, are heterotrophs, but some are either chemosynthetic or photosynthetic autotrophs. The chemosynthetic bacteria oxidize inorganic compounds, such as ammonia, nitrite, sulfur, hydrogen gas, or ferrous iron, and trap the released energy.

Bacteria as agents of disease.   Perhaps the bacteria best known to most people are the ones that cause diseases in man, his domesticated animals, or his cultivated plants. Most of the so-called "germs" are either bacteria or viruses (though a few are fungi, protozoans, or parasitic worms). The idea that bacteria can cause disease—often called the Germ Theory of Disease—was first developed by Louis Pasteur in the late nineteenth century. Though scorned at first, it soon gained the support of prominent scientists and physicians of the day.

The long list of human diseases caused by bacteria includes bubonic plague, cholera, diphtheria, syphilis, gonorrhea, leprosy, scarlet fever, tetanus, tuberculosis, typhoid fever, whooping cough, bacterial pneumonia, bacterial dysentery, meningitis, strep throat, boils, and abscesses. Equally long lists could be compiled of bacterial diseases of other animals or of bacterial diseases of plants.

Microorganisms cause disease symptoms in a variety of ways. Their immense numbers may place such a tremendous material burden on the host's tissues that they interfere with normal function. Or they may actually destroy cells and tissues, or produce poisons, called **toxins.** There are even cases, as we mentioned in Chapter 16, where the disease symptoms are not caused directly by the microorganisms but result from an excessive immune response to the microorganisms by the host's body.

There are several ways in which the human body resists attacks by pathogenic microorganisms. The first line of defense, once the pathogens have gotten into the body past the protective epidermal tissues, is phagocytic action by certain kinds of white blood cells; the pathogens are engulfed and destroyed. The second line of defense is production of antibodies that react with the antigens of the pathogens. Production of antibodies the first time a host individual is exposed to a particular antigen is a rather slow process, which may take days or even weeks. However, in most cases the immunity thus built up is relatively long-lasting. For example, a person who has once had whooping cough or chicken pox usually remains immune for life to further infection by the pathogens of those diseases. Immunity to some diseases, among them most respiratory infections, does not last for life, but may have a duration of several months.

Modern medicine often takes advantage of the body's antigen-antibody reaction to induce prophylactic immunity—immunity that prevents a first case of disease. The patient is inoculated with either a vaccine or an antiserum. A *vaccine* is material containing antigen from the pathogen. Sometimes the antigen consists of dead microorganisms, as in the Salk vaccine for poliomyelitis.[1] Sometimes it consists of microorganisms that are alive but have been treated in a manner that weakens them sufficiently to prevent their causing disease; the Sabin oral vaccine for poliomyelitis is an example. Vaccines also sometimes consist either of a small amount of active bacterial toxin (enough to induce formation of antibodies but not enough to produce disease) or of inactivated toxin, called toxoid; tetanus toxoid is an example. Vaccines, whatever the kind, induce active immunity in the patient; i.e. they stimulate the patient to produce his own antibodies. They are therefore rather slow-acting, but their effects are long-lasting. By contrast, inoculation with **antiserum** produces almost immediate immunity, but the immunity lasts only a short time, because an antiserum contains presynthesized antibodies instead of antigens, and hence produces passive rather than active immunity. An antiserum is made by injecting antigen into some other animal, usually a horse, waiting until the animal has produced antibodies specific for that antigen, and then removing blood serum containing the antibodies from the animal.

Beneficial bacteria. Contrary to the popular impression, beneficial bacteria outnumber harmful ones. In Chapter 18 we mentioned the importance of bacteria as organisms of decay—a process that not only prevents the accumulation of dead bodies and metabolic wastes but also converts materials such as the nitrogen of proteins into a form usable by other living things. We also mentioned the essential role of nitrogen-fixing bacteria. The human intestine contains bacteria that synthesize vitamins absorbed by the body and other bacteria that aid in the digestion of certain materials. Anyone who has been given such

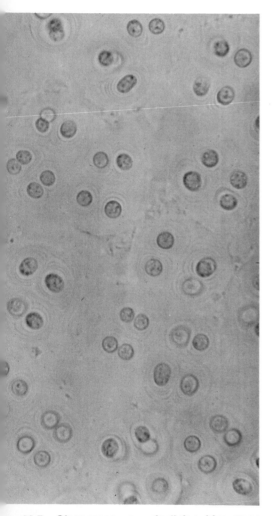

20.7. *Gloeocapsa,* a unicellular blue-green alga. Groups of spherical cells are enclosed in layers of gelatinous material (the gelatinous sheaths are faintly visible as concentric rings around the cells). The species occur on wet rocks and other damp objects, where they may form large masses of jelly. [Courtesy J. M. Kingsbury, Cornell University.]

[1] Poliomyelitis is a viral disease. The discussion of bodily response to infection and of vaccines and antiserums applies to both bacterial and viral diseases.

massive doses of antibiotics as to have his intestinal flora exterminated can testify to the ensuing disturbance in normal intestinal activity.

Bacteria are also of great importance in many industrial processes. Manufacturers often find it easier and cheaper to use cultured microorganisms in certain difficult syntheses than to try to perform the syntheses themselves. Among the many substances manufactured commercially by means of bacteria are acetic acid (vinegar), acetone, butanol, lactic acid, and several vitamins. Bacteria are also used in the retting of flax and hemp, a process that decomposes the pectin material holding the cellulose fibers together; the fibers, once freed, may be used in making linen, other textiles, and rope. Commercial preparation of skins for making leather goods often involves use of bacteria.

Many branches of the food industry depend on bacteria. You are probably aware of the central role of bacteria in the making of dairy products, such as butter and the various kinds of cheeses; the characteristic flavor of Swiss cheese, for example, is due in large part to propionic acid produced by bacteria.

Many farmers depend on bacterial action in the making of silage for use as cattle feed. Also of considerable interest to farmers is the possibility that bacteria pathogenic for destructive insects may eventually be usable in lieu of chemical insecticides.

Particularly interesting is the modern use of bacteria in the production of antibiotics that can help control other bacteria. In fact, most of the antibiotic drugs in use today (but not penicillin) are produced by bacteria or, if synthesized artificially, were discovered in these organisms. Among these drugs are streptomycin, Aureomycin, Terramycin, and neomycin.

Cyanophyta

The Cyanophyta, or blue-green algae, are procaryotic unicellular or filamentous organisms. The unicellular forms are either rods or spheres, which may occur singly or as colonies embedded in a gelatinous matrix (Fig. 20.7). The filamentous forms are multicellular in the sense that adjacent cells have common end walls (Fig. 20.8). In some species, plasmodesmata penetrate these walls, interconnecting the cytoplasm of the cells. Yet, for the most part, the individual cells remain the important units of function.

The cell walls of blue-green algae differ from those of most bacteria in that they usually contain cellulose, but like those of bacteria they also usually seem to contain some muramic acid. Outside the wall proper, there is often a layer of more or less firm gelatinous material called a sheath, composed of pectic materials.

The cytoplasm of blue-green algae, which seems to be composed of very dense colloidal material, is unusually viscid—so viscid, in fact, that no Brownian movement can be detected in it, and that high-speed centrifugation does not cause rearrangement of its constituents. The large cell vacuole characteristic of higher-plant cells is absent, as are mitochondria, endoplasmic reticulum, Golgi apparatus, and a nuclear membrane. Many ribosomes are present, however, and there are numerous proteinaceous granules and granules of a stored carbohydrate material called cyanophycean starch, which is very similar (perhaps identical) to the glycogen used as a storage product in animal cells. No blue-green

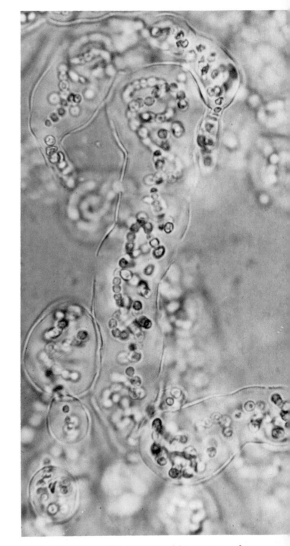

20.8. *Nostoc*, a filamentous blue-green alga. In *Nostoc*, a very common genus, several bead-like filaments may be grouped in a shared gelatinous matrix. [Courtesy J. M. Kingsbury, Cornell University.]

algae ever possess flagella. The peculiar gliding motion exhibited by many species is like that seen in some bacteria, and like it remains unexplained.

All blue-green algae possess photosynthetic pigments located in lamellar structures that appear to be flattened vesicles (see Fig. 19.1, p. 444); these structures are not contained within chloroplasts. The chlorophyll of the blue-green algae is chlorophyll *a*, the pigment also found in higher plants, rather than bacterial chlorophyll. In addition to chlorophyll and various carotenoids, these organisms contain phycocyanin (blue pigment) or sometimes phycoerythrin (red pigment). It is the presence of phycocyanin with the chlorophyll that gives these algae their characteristic blue-green color. However, not all "blue-green" algae are blue-green; black, brown, yellow, red, grass green, and other colors also occur. The periodic redness of the Red Sea is due to a species that contains a particularly large amount of phycoerythrin.

Cell division in blue-green algae is by binary fission, as in bacteria. Reproduction also frequently occurs by fragmentation of filaments. No form of genetic recombination has ever been unequivocally demonstrated in these organisms; they are the only major group in which all reproduction, so far as is known, is totally asexual. It must be admitted, however, that they have not been studied nearly as thoroughly as the bacteria, and it seems entirely possible that some form of recombination will eventually be found in them. Almost nothing is known about their genetics.

Blue-green algae occur in numerous and varied habitats—in fresh or salt water, in or on soil, on the bark of trees, on wet cliffs and ledges. Ponds or lakes containing a rich supply of organic matter, particularly nitrogenous compounds, often develop huge populations ("blooms," as they are called) of blue-green algae, which may make the water so green that objects only a few inches below the surface are invisible. Such blooms may give the water an objectionable odor, clog filters of water supplies, and even be toxic to livestock. Some species of blue-green algae live mutualistically with fungi in the compound plants called lichens. A few species live in habitats that rank among the most inhospitable known—the hot springs that occur in various parts of the world; these species can grow well at temperatures as high as 85°C (185°F).

One reason blue-green algae occur in environments so bleak that no other organisms can inhabit them is probably that some have the simplest nutritional requirements of any known organisms. Nitrogen-fixing species, for example, need only water, light, and minerals, in addition to N_2 and CO_2, which are readily available from the atmosphere.

REFERENCES

BROCK, T. D., ed., 1961. *Milestones in Microbiology.* Prentice-Hall, Englewood Cliffs, N.J. (55 papers by leading workers.)

————, 1970. *Biology of Microorganisms.* Prentice-Hall, Englewood Cliffs, N.J.

HAHON, N., ed., 1964. *Selected Papers on Virology.* Prentice-Hall, Englewood Cliffs, N.J. (40 papers by leading workers.)

Luria, S. E., and J. E. Darnell, 1967. *General Virology*. Wiley, New York.

Pankratz, H. S., and C. C. Bowen, 1963. "Cytology of Blue-Green Algae: I. The Cells of *Symploca muscorum*," *American Journal of Botany,* vol. 50, pp. 387–399.

Ris, H., and R. N. Singh, 1961. "Electron Microscope Studies on Blue-Green Algae," *Journal of Biophysical and Biochemical Cytology,* vol. 9, pp. 63–80.

Salton, M. R. J., 1964. *The Bacterial Cell Wall*. Elsevier, Amsterdam.

Sokatch, J. R., 1969. *Bacterial Physiology and Metabolism*. Academic Press, New York.

Stanier, R. Y., and C. B. van Niel, 1962. "The Concept of a Bacterium," *Archiv für Mikrobiologie,* vol. 42, pp. 17–35.

Stanier, R. Y., M. Doudoroff, and E. A. Adelberg, 1970. *The Microbial World,* 3rd ed. Prentice-Hall, Englewood Cliffs, N.J.

Stent, G. S., 1971. *Molecular Genetics*. Freeman, San Francisco.

Wolstenholme, G. E. W., and M. O'Connor, eds., 1968. *Interferon*. Churchill, London.

SUGGESTED READING

Braude, A. I., 1964. "Bacterial Endotoxins," *Scientific American,* March. (Offprint 177.)

DeKruif, P., 1926. *Microbe Hunters*. Harcourt, Brace, New York. (Paperback edition, Pocket Books, New York, 1959.)

Dulbecco, R., 1967. "The Induction of Cancer by Viruses," *Scientific American,* April. (Offprint 1069.)

Edgar, R. S., and R. H. Epstein, 1965. "The Genetics of a Bacterial Virus," *Scientific American,* February. (Offprint 1004.)

Hilleman, M. R., and A. A. Tytell, 1971. "The Induction of Interferon," *Scientific American,* July. (Offprint 1226.)

Horne, R. W., 1963. "The Structure of Viruses," *Scientific American,* January. (Offprint 147.)

Isaacs, A., 1961. "Interferon," *Scientific American,* May. (Offprint 87.)

Losick, R., and P. W. Robbins, 1969. "The Receptor Site for a Bacterial Virus," *Scientific American,* November. (Offprint 1161.)

Luria, S. E., 1970. "The Recognition of DNA in Bacteria," *Scientific American,* January. (Offprint 1167.)

Morowitz, H. J., and M. E. Tourtellotte, 1962. "The Smallest Living Cells," *Scientific American,* March. (Offprint 1005.)

Sharon, N., 1969. "The Bacterial Cell Wall," *Scientific American,* May. (Offprint 1142.)

Sistrom, W. R., 1969. *Microbial Life,* 2nd ed. Holt, Rinehart & Winston, New York.

Smith, I. M., 1968. "Death from Staphylococci," *Scientific American,* February.

Zinsser, H., 1935. *Rats, Lice and History*. Little, Brown, Boston.

The plant kingdom

The various divisions[1] of the plant kingdom have traditionally been separated into two groups: the **Thallophyta** and the **Embryophyta.** Once recognized as subkingdoms, these categories are without status in most current classifications, for some of the divisions they embrace are but uncertainly related. Although accorded little phylogenetic significance, the categories are useful in that they include groups of plants at similar levels of structural complexity. The Thallophyta are the more primitive plants, the Embryophyta the more advanced. Disregarding some special cases, we can summarize the chief distinctions between them as follows:

1. Thallophytes are either unicellular or multicellular. Embryophytes are always multicellular.
2. Multicellular thallophytes usually show little if any tissue differentiation; there is thus no anatomical basis for distinguishing roots, stems, or leaves, the entire plant being known as a **thallus.** There is far more differentiation in embryophytes; the higher embryophytes have distinct roots, stems, and leaves.
3. The reproductive structures of thallophytes are often unicellular. Those of embryophytes are multicellular.
4. The reproductive structures of thallophytes lack the protective jacket of sterile (i.e. nondividing) cells characteristic of the reproductive structures of embryophytes.
5. The zygotes of thallophytes do not develop into embryos until after they have been released from the female reproductive organs where they were produced. In embryophytes—and here

[1] "Division" in the plant kingdom is the equivalent of "phylum" in the animal kingdom.

is the source of their name—the early stages of embryonic development occur while the embryo is still contained within the female reproductive organ.

The thallophytes are customarily divided into two categories on the basis of the presence or absence of chlorophyll. Thus the photosynthetic thallophytes are called *algae,* and the nonphotosynthetic thallophytes are called *fungi.* Again, neither of these terms is accorded formal taxonomic recognition in most modern classifications, because it is now clear that no close relationship exists either among the various algal divisions or among the various fungal divisions. The names will therefore be used here simply as terms of convenience without phylogenetic significance.

A formal outline of the classification of the plant kingdom used in this book is given in the Appendix. The following list relates the traditional groupings to the formal classification.

"Thallophyta"
Division Euglenophyta
Division Chlorophyta
Division Chrysophyta
Division Pyrrophyta   "Algae"
Division Phaeophyta
Division Rhodophyta
Division Myxomycophyta
Division Eumycophyta   "Fungi"

"Embryophyta"
Division Bryophyta
Division Tracheophyta

Omitted from the discussion in this chapter are the algal divisions Chrysophyta and Pyrrophyta and the fungal division Myxomycophyta. The bacteria and blue-green algae, which have been placed in a separate kingdom, the Monera, are discussed in Chapter 20.[2]

EUGLENOPHYTA (The euglenoids)

The euglenoids are unicellular organisms that show a combination of plantlike and animal-like characteristics. They are plantlike in that many species have chlorophyll and are photosynthetic. They are animal-like in lacking a cell wall and being highly motile; and, like animals, the species that lack chlorophyll are heterotrophic. Zoologists have traditionally regarded the euglenoids as animals and placed them among the flagellated Protozoa. Botanists, on the other hand, have regarded them as plants and placed them among the algae. They seem to have no close relatives among the other algae, however, and for this reason botanical classifications usually put them in a division by themselves. Most live in fresh water; but a few are found in soil, on damp surfaces, or even in the digestive tracts of certain animals.

A representative genus is *Euglena.* A typical cell of *Euglena* is elongate ovoid, with an anterior invagination from which a long flagel-

[2] Classifications that recognize only two kingdoms—Plantae and Animalia—place the bacteria among the fungi as the division Schizomycetes (or Schizomycophyta), and the blue-green algae among the algae as the division Cyanophyta.

lum emerges (see Fig. 8.4, p. 157). Since the cell lacks a wall, it is fairly flexible, and its shape may change somewhat as it swims about; however, the pellicle by which it is bounded prevents excessive alterations of its shape.

CHLOROPHYTA (The green algae)

The green algae are of particular interest because they are generally regarded as the group from which the higher plants arose. They are thus probably the only algal division that has not been a phylogenetic dead end. The majority of green algae live in fresh water, but some live in moist places on land, and there are many marine species.

Many divergent evolutionary tendencies, all probably beginning with walled and flagellated unicellular organisms, can be perceived in the Chlorophyta: (1) the evolution of motile colonies; (2) a change to nonmotile unicells and colonies; (3) the evolution of extensive tubelike bodies with numerous nuclei but without cellular partitions (coenocytic organisms); (4) the evolution of multicellular filaments and even three-dimensional leaflike thalluses.

Chlamydomonas **as a representative unicellular green alga.** *Chlamydomonas* is a genus of modern unicellular green algae that probably resemble the ancestral organisms from which the rest of the plant kingdom arose. Its species are common in ditches, pools, and other bodies of fresh water and in soils. The oval haploid cell has a cellulose wall and two anterior flagella of equal length (Fig. 21.1). It contains a single large cup-shaped chloroplast that fills from one half to two thirds of the basal portion of the cell. The chlorophylls are of the same types (*a* and *b*) as in vascular plants. The only other pigments are the carotenoids, which are also found in higher plants. (This type of pigmentation is characteristic of all Chlorophyta; unlike many other algae, they have no special accessory pigments or unusual chlorophylls.) A conspicuous *pyrenoid* in the basal portion of the chloroplast of *Chlamydomonas* functions as the site of starch synthesis. A stigma, or eyespot, is also located inside the chloroplast. The cell has no large central vacuole such as is seen in mature cells of higher plants.[3]

Asexual reproduction is common in *Chlamydomonas* (Fig. 21.2). A vegetative cell resorbs its flagella; then mitotic division of the nucleus and longitudinal cytokinesis take place simultaneously. This process gives rise to two daughter cells, both of which lie within the wall of the original cell. In some species, the two daughter cells are promptly released by breakdown of the wall; in other species, the daughter cells themselves divide while still inside the wall of the parent cell. The daughter cells develop a wall and flagella just before they are released as motile *zoospores,* which are asexual reproductive cells, i.e. reproductive cells not specialized as gametes. In *Chlamydomonas* the zoospores are smaller than mature vegetative cells but otherwise indistinguishable from them; in many species of algae, however, there are noticeable morphological differences between the zoospores and the mature cells.

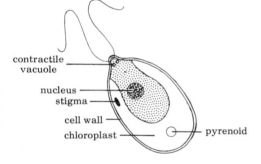

contractile vacuole
nucleus
stigma
cell wall
chloroplast
pyrenoid

21.1. Mature cell of *Chlamydomonas*.

---

[3] The descriptions of *Chlamydomonas* and other algae given in this chapter are based largely on material in J. M. Kingsbury, *Biology of the Algae,* published privately, Ithaca, N.Y., 1963.

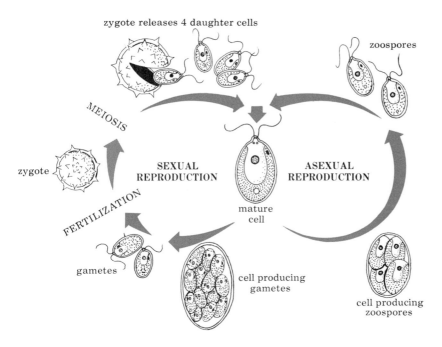

zygote releases 4 daughter cells

zoospores

MEIOSIS

zygote

SEXUAL
REPRODUCTION

ASEXUAL
REPRODUCTION

FERTILIZATION

mature
cell

gametes

cell producing
gametes

cell producing
zoospores

21.2. Life history of *Chlamydomonas.* Left: Diagram showing stages of both the sexual and asexual cycles. Below: Schematic representation of life cycle for comparison with the life cycles of other organisms. Note that the zygote is the only diploid stage. This type of life cycle was probably characteristic of the first sexually reproducing unicellular organisms, and it may thus be the type from which all other types arose.

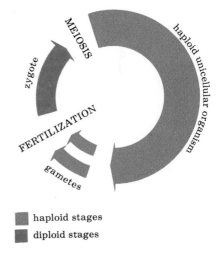

MEIOSIS

zygote

FERTILIZATION

haploid unicellular organism

gametes

■ haploid stages
■ diploid stages

The free zoospores soon grow to full size, completing the asexual reproductive cycle.

Under certain conditions, *Chlamydomonas* may reproduce sexually. A mature haploid vegetative cell divides mitotically to produce several gamete cells, which develop walls and flagella and are released from the parent cell. The gametes are attracted to each other and form large clumps. Eventually the clumped cells, which have shed their walls, move apart in pairs. The members of a pair lie side by side, and their cytoplasms slowly fuse. Finally, their nuclei unite in the process of fertilization, which produces a single diploid cell, the zygote. The zygote sheds its flagella, sinks to the bottom, and develops a thick protective wall. It can withstand unfavorable environmental conditions, such as the drying up of the pond or the cold of winter. When conditions are again favorable, it germinates, dividing by meiosis to produce four new haploid cells, which are released into the surrounding water. The new cells quickly mature, thus completing the sexual reproductive cycle.

Sexual reproduction in most species of *Chlamydomonas* is at a very simple level, and thus gives us insight into the way sexuality probably arose. There are no separate male and female individuals. Furthermore, the gametes are all alike; they cannot be separated into male gametes (sperms) and female gametes (eggs). Such a condition, where all gametes are alike, is called *isogamy;* it is probably the primitive (ancestral) condition in plants. The isogametes of *Chlamydomonas* are indistinguishable from vegetative cells; they may be viewed simply as small vegetative cells that tend to fuse and act as gametes under certain conditions. This, too, is probably the primitive condition; the specialization of gametes as morphologically distinctive cells—a characteristic of most higher plants and animals—is surely a later evolutionary development.

Notice that the haploid stages of the life cycle of *Chlamydomonas* are the dominant ones; the only diploid stage is the zygote. Dominance of the haploid stages is characteristic of most very primitive plants, and it seems clear that this was the ancestral condition.

The volvocine series. As an example of one of the evolutionary tendencies that can be traced in the Chlorophyta, let us examine the so-called volvocine or motile-colony series. This is a series of genera showing a gradual progression from the unicellular condition of *Chlamydomonas* to an elaborate colonial organization.

Gonium may be taken as an example of the simplest colonial stage. Each colony of *Gonium* is made up of 4, 16, or 32 cells (depending on the species), each of which is morphologically similar to *Chlamydomonas*. The cells are embedded in a mucilaginous matrix. In some species, delicate cytoplasmic strands run between the cells; these may provide a route for direct interaction and coordination between the cells of the colony (see Pl. I-3C). That some sort of coordination does indeed exist is shown by the organized fashion in which the flagella of all the cells beat together and thus enable the colony to swim as a unit.

Pandorina is a genus of colonial forms slightly more complex than *Gonium.* Each colony is a hollow sphere in which 8, 16, or 32 cells are arranged in a single layer about the periphery, their flagella oriented to the outside of the sphere (Fig. 21.3). Three main advances over *Gonium* are noticeable: (1) The colony shows some regional differentiation; it has definite anterior and posterior halves (detectable both by the orientation of the colony when it is swimming and by the larger size of the stigma in the anterior cells than in the posterior ones). (2) The vegetative cells of the colony are so dependent upon one another that they cannot live apart from the colony, and the colony itself cannot survive if disrupted or broken. (3) Sexual reproduction is *heterogamous;* i.e. it involves two different kinds of gametes—small male gametes and larger female gametes, both of which, however, have flagella and are free-swimming.

A still more advanced genus is *Pleodorina* (Fig. 21.4), whose spherical colonies are composed of 32 to 128 cells. These large colonies exhibit considerable division of labor. The anterior cells are purely vegetative, never participating in reproduction. The posterior cells, which function in both asexual and sexual reproduction, are much larger. Sexual reproduction is heterogamous; in fact, in some cases the large female gametes lose their flagella and thus become true nonmotile egg cells. This type of advanced heterogamy, where only the male gamete is motile and the female gamete is a nonflagellated nonmotile egg cell, is called *oögamy.*

The culmination of the evolutionary series here being traced is represented by the genus *Volvox* (Fig. 21.5; Pls. I-2, 3). Its spherical colonies are very large, consisting of about 500–50,000 cells. Most of these cells are exclusively vegetative. A few cells (between 2 and 50) scattered in the posterior half of the colony are much larger than the others and are specialized for reproduction. Each of the female reproductive cells can give rise to an entire new daughter colony. Sexual reproduction is always oögamous.

We can summarize the major lines of evolutionary changes manifest in this series as follows: (1) a change from unicellular to colonial

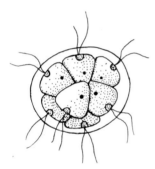

21.3. *Pandorina* colony. The cells are embedded in a gelatinous matrix.

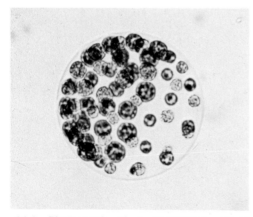

21.4. Photograph of a *Pleodorina* colony.
[Courtesy J. M. Kingsbury, Cornell University.]

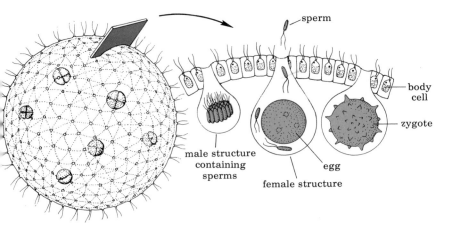

body
cell

zygote

male structure
containing
sperms

egg

female structure

21.5. *Volvox*. Left: The colony is very large, containing 500 to 50,000 vegetative cells. Six daughter colonies at various stages of development can be seen still embedded in the matrix of the parent colony. Right: Section through the surface of a colony showing male and female reproductive structures. Sperms released by the male structure enter the female structure and fertilize the egg. After a period of inactivity, the zygote divides meiotically, and the haploid cells thus formed then divide mitotically, producing a new daughter colony, which is eventually released.

life, and a tendency for the number of cells in the colonies to increase; (2) increasing coordination of activity among the cells; (3) increasing interdependence among the vegetative cells, so that they cannot live apart from the colonies and the colonies cannot survive if disrupted; (4) increasing division of labor, particularly between vegetative and reproductive cells; (5) a gradual change from isogamy to simple heterogamy to the extreme form of heterogamy called oögamy.

In tracing this *Chlamydomonas–Gonium–Pandorina–Pleodorina–Volvox* series, we do not mean to imply that each genus evolved from the preceding one; the available evidence will not allow us to decide whether it did or not. But it does seem likely that each of these genera evolved from an ancestor that resembled in many important ways the modern genus placed just before it in this series, and therefore that the actual evolutionary progression from some unicellular ancestor to *Volvox* involved a series of stages similar to those represented by the modern genera discussed here. Study of this series consequently suggests how complex colonial forms may have evolved, and indicates one possible way in which the type of multicellularity characteristic of most animals may have arisen. After all, it is largely an arbitrary decision whether one calls *Volvox* colonial or multicellular. Though multicellular animals certainly did not evolve from *Volvox* or any of the other genera discussed here, a similar evolutionary series, beginning with a nonwalled unicellular organism, might well have been the beginning of multicellularity in the animal kingdom.

Some multicellular green algae. Many green algae have a multicellular stage in their life cycle. In most cases, this stage is a filamentous thallus, which may be either nonbranching or branching, depending upon the species (Pl. I-4A). Let us take *Ulothrix* as a first example.

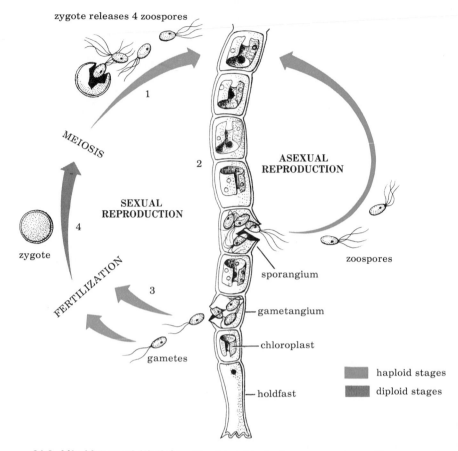

zygote releases 4 zoospores

MEIOSIS

SEXUAL
REPRODUCTION

ASEXUAL
REPRODUCTION

zygote

FERTILIZATION

gametes

zoospores

sporangium

gametangium

chloroplast

holdfast

haploid stages
diploid stages

21.6. Life history of *Ulothrix*. The haploid plant may reproduce either asexually or sexually (though a single filament would never reproduce in both ways at once as shown here). Asexual reproduction is the more common; certain cells of the filament develop into sporangia and produce zoospores, which settle down and develop into new filaments. When a filament begins reproducing sexually, a cell becomes specialized as a gametangium, which produces isogametes. Two of these may then fuse in fertilization, producing a zygote, which divides meiotically and releases zoospores.

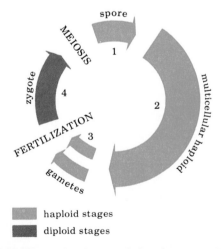

spore

MEIOSIS

zygote

FERTILIZATION

gametes

multicellular haploid

haploid stages
diploid stages

21.7. Life cycle characteristic of most multi-cellular green algae. Note that multicellularity is present only in the haploid phase.

The species of *Ulothrix* are unbranched filamentous forms, most of which live in fresh water although a few are marine. The filament of each plant is a very small threadlike structure attached to the substratum by a specialized cell called a ***holdfast*** (Fig. 21.6). Except for the holdfast cell, all the cells of the filament are identical and are arranged end to end in a single series. The filament increases in length as its cells divide horizontally and as the new cells thus added to the chain grow to mature size. Adjacent cells have common end walls—a basic step in the evolution of multicellularity in algae. Each cell contains a single nucleus and a single large chloroplast.

Ulothrix may reproduce by fragmentation (each fragment growing into a complete plant), by asexually produced zoospores, or by sexual processes. In asexual reproduction, any cell of the filament except the holdfast may act as a ***sporangium*** (spore-producing structure), producing zoospores, each of which has four flagella. After the zoospores are released, they swim about for a short while, and then settle down and give rise to a new filament.

Under certain environmental conditions, a filament may cease reproducing asexually and begin reproducing sexually. A cell becomes specialized as a **gametangium** (gamete-producing structure) and eventually releases biflagellate isogametes (stage 3 in Fig. 21.6). The zygote (stage 4) formed by the union of two of these gametes develops a thick wall and functions as a resting stage capable of withstanding unfavorable environmental conditions. At germination, the zygote divides by meiosis, producing haploid zoospores (stage 1), each of which will grow into a new filament (stage 2). The main difference, then, between this life cycle and that of *Chlamydomonas* is the addition of the haploid multicellular stage (stage 2). As in *Chlamydomonas,* the only diploid stage is the zygote. This type of life cycle is diagramed in a more generalized form in Fig. 21.7.

Ulva, or sea lettuce, is a green alga with an expanded leaflike thallus two cells thick (Fig. 21.8). Its life cycle (and that of its close relatives) is more complex than those of the other green algae discussed here in that it includes both a multicellular haploid and a multicellular diploid stage (stages 2 and 5 in Fig. 21.9). The entire cycle can be summarized as follows: Haploid zoospores (stage 1) divide mitotically to produce the haploid multicellular thalluses of stage 2. These may reproduce either asexually by means of zoospores or sexually by means of gametes (stage 3). Fusion of pairs of gametes (fertilization) produces diploid zygotes (stage 4). Upon germination, the zygotes divide mitotically (not meiotically as in the green algae previously discussed), producing diploid multicellular thalluses (stage 5). Eventually certain reproductive cells (sporangia) of these diploid plants divide by meiosis, producing haploid zoospores, which begin a new cycle. A life cycle of this type is said to exhibit **alternation of generations** in that a haploid multicellular plant alternates with a diploid multicellular plant. The haploid multicellular stage is customarily called a **gametophyte** (meaning that it is a plant that can produce gametes), and the diploid multicellular stage is called a **sporophyte** (meaning that it is a plant that can reproduce only by spores).

We have seen that multicellularity in plants arose first in the gametophyte; many green algae have no sporophyte stage. *Ulva* shows a more advanced life cycle in that both gametophyte and sporophyte stages are present. They are equally prominent, moreover, being nearly equal in duration and almost identical in appearance; in this plant, then, the haploid portion of the life cycle is no longer dominant over the diploid.

PHAEOPHYTA (The brown algae)

The brown algae are almost exclusively marine, the few freshwater species being quite rare. Many of the plants called seaweeds are members of this division. They are most common along rocky coasts of the cooler parts of the oceans, where they normally grow attached to the bottom in the littoral (intertidal) and upper sublittoral zones. They may be seen forming great stringy mats over the rocks exposed at low tide along the New England coast. A few species occur in warmer seas, and some of these differ from the majority of brown algae in being able to live and grow when detached from the substratum; e.g. some species of

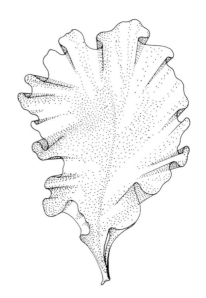

21.8. *Ulva,* a marine green alga with a three-dimensional leaflike thallus.

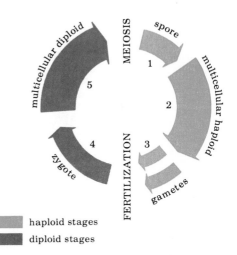

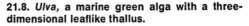

haploid stages
diploid stages

21.9. Life cycle of *Ulva* and *Ectocarpus.* The gametophyte (multicellular haploid) and sporophyte (multicellular diploid) stages are equally prominent. *Ulva* and its close relatives are unusual among the green algae in having a life cycle of this sort; most of the Chlorophyta have no alternation of generations, the sporophyte stage being absent. *Ectocarpus,* a brown alga, is also unusual in having a life cycle of this type, but for a different reason: It is one of the few members of the Phaeophyta in which the sporophyte is not more prominent than the gametophyte.

21.10. Photograph of part of a *Sargassum* thallus. Note the characteristic bladders; they are filled with gas and act as floats. [Courtesy J. M. Kingsbury, Cornell University.]

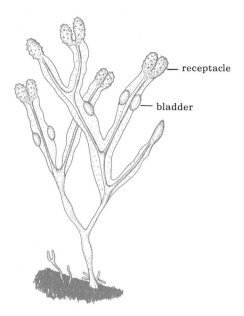

— receptacle

— bladder

Sargassum (Fig. 21.10) form dense floating mats that cover much of the surface of the so-called Sargasso Sea, which occupies some two and a half million square miles of ocean between the West Indies and North Africa.

All brown algae are multicellular, and most are macroscopic, some growing as long as 150 feet or more. The thallus (plant body) may be a filament, or it may be a large and rather complex three-dimensional structure. The individual cells are much like those of higher land plants, having cellulose cell walls with pits through which plasmodesmata pass, large central vacuoles, usually several plastids, and no pyrenoids. However, unlike the cells of most higher land plants, they usually have centrioles.

Like all photosynthetic plants, the Phaeophyta possess chlorophyll *a*. However, they have chlorophyll *c* instead of the chlorophyll *b* found in euglenoids, green algae, and higher land plants. Large amounts of a carotenoid called **fucoxanthin** are also present, and it is this that gives the characteristic brownish color to these algae. A variety of unusual carbohydrates are synthesized and used as structural or storage materials.

Reproduction may be either asexual or sexual. The latter often involves specialized multicellular sex organs called **antheridia** if they produce male gametes and **oögonia** if they produce female gametes (or gametangia if all the gametes are alike, i.e. if the plant is isogamous). The sex organs are ordinarily not enclosed by a protective layer of sterile jacket cells. The life cycle is usually characterized by alternation of gametophyte (haploid) and sporophyte (diploid) multicellular generations. In many forms, such as *Ectocarpus,* the gametophyte and sporophyte (stages 2 and 5 in Fig. 21.9) are essentially similar in structure and neither can be said to be dominant. In other forms, such as *Laminaria,* the haploid gametophyte is reduced and the diploid sporophyte is much larger and more prominent. In a few, such as *Fucus* (Fig. 21.11; Pl. I-5A), reduction of the haploid stages has progressed so far that there is no longer any multicellular haploid gametophyte and the only haploid cells in the life cycle are the gametes (Fig. 21.12); such a life cycle, very rare in plants, is essentially the sort seen in animals.

Laminaria is an example of the group of brown algae commonly called kelps. The sporophyte thallus is large (6 feet long or more in *Laminaria;* 150 feet long or more in some kelps) and consists of a root-like **holdfast,** a stemlike **stipe,** and an expanded leaflike **blade** (Fig. 21.13). Although thallophyte plants usually lack tissue differentiation, the stipe of some kelps has an outer surface tissue (epidermis), a middle tissue (cortex) containing many plastids, and a central core tissue (medulla); it may even have a meristematic layer similar to the cambium of higher vascular plants and, in a few genera, a phloemlike conductive tissue in the medulla. In short, these brown algae are complex plants that have convergently evolved many similarities to the vascular plants.

21.11. *Fucus,* often called rockweed, a brown alga common along northern coasts. Each thallus is flattened and characterized by repeated dichotomous branching. Each younger axis consists of a midrib and thin paired wings. In some species (including the one shown here), there are bladders (floats) at intervals along the wings. The tips of fertile thalluses develop swollen reproductive structures called receptacles, whose surface is pocked by numerous tiny openings that lead into cavities (conceptacles) where the sex organs are located.

However, none of them have a protective layer of sterile jacket cells around their reproductive organs; none develop multicellular embryos while these are still inside the oögonia; none have a cuticle; and none have xylem.

RHODOPHYTA (The red algae)

The red algae are mostly marine seaweeds (Fig. 21.14; Pl. I-4C), but a few live in fresh water or on land. They often occur at greater depths than the brown algae. Most are multicellular and are attached to the substratum, but a few species are unicellular.

In addition to chlorophyll *a*, which is found in all photosynthetic organisms except the photosynthetic bacteria, the Rhodophyta often possess chlorophyll *d*, which is not found in any other group of plants. They also contain phycocyanins and phycoerythrins. It is the phycoerythrins that give many of these algae their characteristic reddish color. It should be emphasized, however, that "red algae" are not always red; many are black.

The accessory pigments of the Rhodophyta play an important role in absorbing light for photosynthesis. The wavelengths preferentially absorbed by chlorophyll *a* for use in photosynthesis, which are among those at the ends of the visible spectrum, do not penetrate to the depths at which the red algae grow, partly because of the imperfect transparency of the water, partly because they are selectively absorbed by pigmented phytoplankton. The wavelengths that do penetrate deep enough are mostly those of the central portion of the spectrum, which are not readily absorbed by chlorophyll *a*. But these wavelengths can be absorbed by the accessory pigments of the Rhodophyta, which then pass the energy to chlorophyll *a*. Thus the accessory pigments make it possible for red algae to live at depths where other algae, lacking these pigments, cannot survive.

EUMYCOPHYTA (The true fungi)

Fungi are by definition thallophyte plants that lack chlorophyll. Thus if bacteria are regarded as plants, then they must be designated as

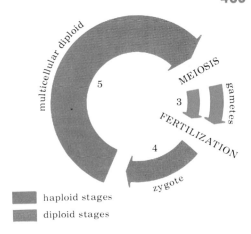

■ haploid stages
▨ diploid stages

21.12. Life cycle of *Fucus*. *Fucus* and its close relatives have a very unusual life cycle. They are the only multicellular plants in which the multicellular haploid stage (the gametophyte) is completely absent. In this respect, their life cycle is like that of animals.

21.13. *Laminaria digitata*. This specimen was detached from the substratum on which it was growing and spread out for photographing. Note the rootlike holdfast, the stemlike stipe, and the leaflike blade. [Courtesy J. M. Kingsbury, Cornell University.]

21.14. Photographs of representative red algae. Left: *Polysiphonia*. Right: *Agardhiella*. [Courtesy J. M. Kingsbury, Cornell University.]

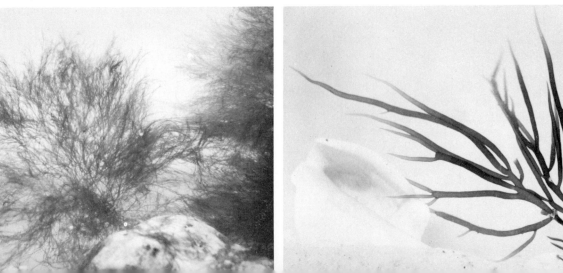

fungi. The term "fungus," however, is often used in a more restricted sense to refer only to the members of the Eumycophyta, the so-called true fungi. This is a large group containing thousands of parasitic and saprophytic species, many of economic importance. Some are parasitic on or in animals, including man; many skin diseases, including "ring-worm" and athlete's foot, are caused by fungi, and there are several serious fungal diseases of the lungs. Other fungi are parasitic on plants, and some of these may cause losses of millions of dollars when they attack agricultural crops. Still others cause spoilage of bread, fruit, vegetables, and other foodstuffs, and deterioration of leather goods, fabrics, paper, lumber, and other valuable products. However, the numerous pathogenic or destructive fungi should not cause us to forget the many others that are beneficial. Yeasts are used extensively in the manufacture of alcoholic products and to make bread dough "rise." The antibiotic penicillin is obtained from a fungus (see Pl. I-5B). Fungi are important in the manufacture of many cheeses, and certain mushrooms are regularly used as food (Pl. I-5C). And fungi, together with bacteria, decompose vast quantities of dead organic material that would otherwise rapidly accumulate and make the earth uninhabitable.

The body of a fungus is either unicellular or filamentous. The individual filaments are called *hyphae,* and a mass of hyphae is called a *mycelium.* The cell walls sometimes contain cellulose, but more often chitin is their most important component. Most saprophytic fungi secrete digestive enzymes onto their food material and absorb the products of the extracellular digestion (see p. 104). Parasitic fungi may carry out extracellular digestion, or they may directly absorb materials produced by the body of their host. Reproduction may be either asexual or sexual, but in both cases the haploid stages are usually dominant.

The division Eumycophyta may be divided into four classes, each of which is given full divisional status in some classifications. We shall discuss three of them: Phycomycetes, Ascomycetes, and Basidiomycetes. Most of the characteristics that distinguish these groups are related to their sexual reproduction.

Phycomycetes (The algal fungi)

The hyphae of the Phycomycetes characteristically lack cross walls although they contain many haploid nuclei (i.e. they are coenocytic). Cross walls appear only during the formation of reproductive structures. Except for their lack of chlorophyll, these fungi resemble the coenocytic green algae, and for this reason they are often called the algal fungi.

As an example of a member of this class, let us examine the common black bread mold, *Rhizopus.* The hyphae of this mold form a whitish or grayish mycelium on the bread. If the mycelium is examined carefully, it can be seen to include three types of hyphae: hyphae (called stolons) that form a network on the surface of the bread; root-like hyphae (called rhizoids) that penetrate into the bread and function both in anchoring the plant and in absorbing nutrients; and hyphae (called sporangiophores) that grow upright from the surface and bear globular sporangia on their ends (Fig. 21.15A). Thousands of asexual spores are produced in each sporangium. The spores, which have no flagella, are very tiny and light, and when liberated at maturity (by

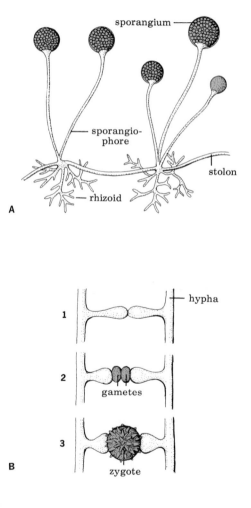

21.15. Rhizopus. (A) Hyphae with sporangia. (B) Sexual reproduction. (1) Short branches from two different hyphae meet. (2) The tips of the branch hyphae are cut off as gametes. (3) The gametes fuse in fertilization to form a zygote with a thick spiny wall.

disintegration of the wall of the sporangium) they may be carried long distances by air currents. If a spore lands in a suitable location, where conditions are warm and moist, it germinates and soon gives rise to a new mass of hyphae, thus completing the asexual cycle.

In the sexual reproduction of *Rhizopus,* short branches from two different hyphae (which must be of different mating types or sexes) contact each other at their tips (Fig. 21.15B). Cross walls soon form just back of the tips of these hyphal branches, thus delimiting gamete cells, which fuse to form a zygote. The zygote develops a thick protective wall and enters a period of dormancy usually lasting from one to three months. At germination, the nucleus of the zygote undergoes meiosis, and a short hypha grows from the zygote. This haploid hypha promptly produces a sporangium, which releases asexual spores that grow into new mycelia. Note that the only diploid stage in the entire sexual cycle is the zygote (Fig. 21.16).

Ascomycetes (The sac fungi)

The members of this large class are very diverse, varying all the way from unicellular yeasts through powdery mildews and cottony molds to complex cup fungi. These last form a cup-shaped structure composed of many hyphae tightly packed together (Fig. 21.17). The vegetative hyphae of Ascomycetes, unlike those of Phycomycetes, are septate (i.e. they possess cross walls) and are thus long multicellular filaments rather than coenocytic tubes (though cells with two nuclei each are formed during sexual reproduction).[4]

Though their vegetative structures differ, all Ascomycetes resemble each other in forming a reproductive structure called an ***ascus*** during their sexual cycle. An ascus is a sac in which haploid spores (usually eight, but sometimes four) are produced; all the spores in an ascus are derived from a single parent cell. The events leading to the formation of a mature ascus are explained in Fig. 21.17.

[4] Actually the septa (cross walls) are often incomplete, having large holes in their centers. The cytoplasm of adjacent cells is thus frequently continuous.

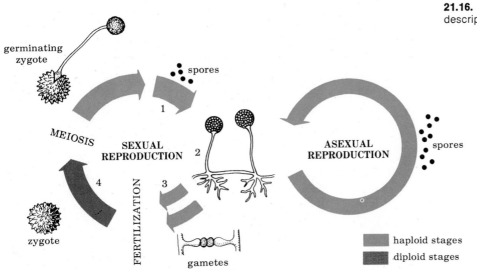

21.16. Life cycle of *Rhizopus*. See text for description.

germinating zygote

spores

MEIOSIS

SEXUAL REPRODUCTION

ASEXUAL REPRODUCTION

spores

FERTILIZATION

zygote

gametes

haploid stages
diploid stages

Most Ascomycetes also reproduce asexually by means of special spores called *conidia* (Fig. 21.18). Conidia are produced in chains at the end of conidiophore hyphae (but not inside sporangia). Each conidium can grow into a new fungal plant.

It may seem strange that yeasts are considered members of the Ascomycetes. They are unicellular, and their asexual reproduction is by budding, not by conidia formation. However, under certain conditions a single yeast cell may function as an ascus, producing four spores. The spores are more resistant to unfavorable environmental conditions than vegetative cells, and they may enable yeasts to survive temperature extremes or periods of prolonged drying.

In previous chapters, we mentioned the *lichens,* plants composed of a fungus and an alga growing together in a complex symbiotic relationship (Fig. 21.19; Pl. I-5E). The fungal component of most lichens is an ascomycete, though in a few tropical forms the fungus is a member of the Basidiomycetes, discussed below. The algal components are usually Chlorophyta, but may be Cyanophyta.

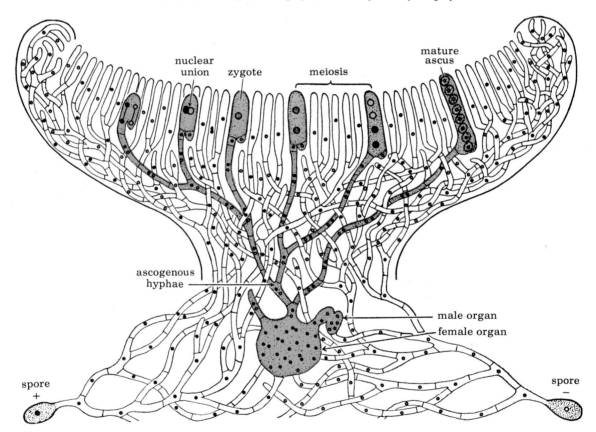

21.17. Diagram of a magnified section through a cup fungus. Hyphae of two haploid mycelia, one derived from a plus (female) spore and the other from a minus (male) spore, participate in forming the cup structure and in producing the spores. The plus mycelium bears a female organ (ascogonium), and the minus mycelium bears a male organ (antheridium). A tube grows from the female organ to the male organ, and then minus nuclei (small circles), acting as male nuclei, move into the female organ and become associated with plus nuclei (black dots). Next, hyphae grow from the female organ; each cell in these hyphae contains two associated nuclei, one minus and one plus. The terminal cells of these hyphae eventually become elongate, and their nuclei unite, forming a zygote nucleus. The zygote nucleus promptly divides meiotically, and each of the four haploid nuclei thus formed then divides mitotically, producing a total of eight small spore cells, which are still contained within the wall of the old zygote cell, now called an ascus. When the mature ascus ruptures, the spores are released and may grow into new mycelia. [From L. W. Sharp, *Fundamentals of Cytology*, McGraw-Hill Book Co., 1943. Used by permission.]

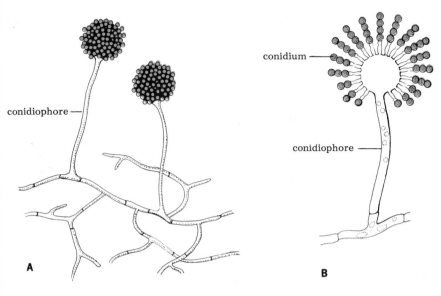

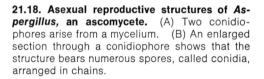

21.18. Asexual reproductive structures of *Aspergillus*, an ascomycete. (A) Two conidiophores arise from a mycelium. (B) An enlarged section through a conidiophore shows that the structure bears numerous spores, called conidia, arranged in chains.

Basidiomycetes (The club fungi)

Many of the largest and most conspicuous fungi—puffballs, mushrooms, toadstools, and bracket fungi—are Basidiomycetes (see Pl. I-5C, D, E). Though the above-ground portion of these plants looks like a solid mass of tissue, and in some is differentiated into a stalk and a prominent cap (Pl. I-5D), it is nevertheless composed of hyphae, as are all fungi (Fig. 21.20). It should be emphasized that the above-ground portion, or fruiting body, of many mushrooms is only a small part of the total plant; there is an extensive mass of hyphae in the soil. The hyphae are septate.

The members of this class are distinguished by their possession of club-shaped reproductive structures called *basidia*, which are functionally equivalent to the asci of the Ascomycetes. The events leading to the formation of a mature basidium are explained in Fig. 21.20.

THE MOVEMENT ONTO LAND

Let us now turn to the Embryophyta—plants that have evolved numerous adaptations for life on land. We have seen that life probably arose in water, and that many plants, notably the algae, are still largely restricted to the aquatic environment; the few algae that live on land are not truly terrestrial, occurring, as they do, only in very moist places and actually living in a film of moisture. The evolutionary move from an aquatic to a terrestrial existence was not a simple one, for the terrestrial environment is in many ways hostile to life. Among the many problems faced by a land plant are the following:

1. Obtaining enough water when fluid no longer bathes the entire surface of the plant body.
2. Transporting water and dissolved substances from restricted areas of intake to other parts of the plant body, and transporting the products of photosynthesis to those parts of the plant that no longer carry out this process for themselves.
3. Preventing excessive loss of water by evaporation.

21.19. Photographs of two kinds of lichens. Top: Pixie-cup lichens growing in a clump of moss. Bottom: Common rock lichen growing in grayish scaly patches on a walk. [Courtesy Verne N. Rockcastle, Cornell University.]

4. Maintaining a sufficiently extensive moist surface for gas exchange when the surrounding medium is air instead of liquid.

5. Supporting a large plant body against the pull of gravity when the buoyancy of an aqueous medium is no longer available.

6. Carrying out reproduction when there is little water through which flagellated sperms may swim and when the zygote and early embryo are in severe danger of desiccation.

7. Withstanding the extreme fluctuations in temperature, humidity, wind, light, and other environmental parameters to which terrestrial organisms are often subjected.

Much of the evolution of the embryophyte plants can best be understood in terms of adaptations that help solve these problems.

As we indicated at the beginning of this chapter, embryophyte plants characteristically have multicellular reproductive organs with an outer layer of sterile (i.e. nondividing) jacket cells that help protect the enclosed gametes from desiccation; male reproductive organs of this type are known as antheridia, female reproductive organs as *archegonia*.[5]

[5] Like "oögonium," "archegonium" denotes an organ producing female gametes, but most botanists restrict "archegonium" to the jacketed female reproductive organs of embryophytes, "oögonium" to the unjacketed ones of thallophytes.

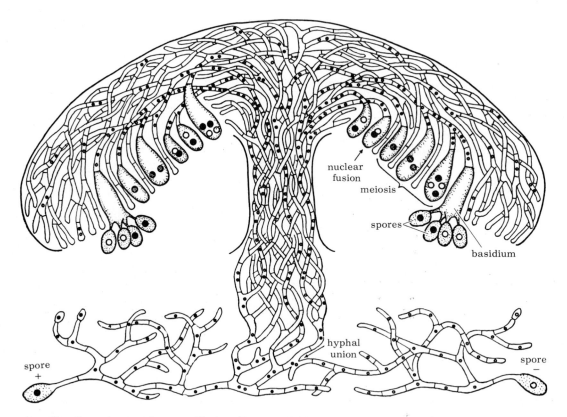

21.20. Diagram of section through a mushroom. Hyphae from two uninucleate haploid mycelia—one of the plus (female), the other of the minus (male) mating type—unite and give rise to binucleate hyphae, which develop into the above-ground portion of the mushroom. Certain terminal cells of the hyphae become zygotes when their nuclei fuse in fertilization. The zygote then forms a club-shaped cell known as a basidium. Its diploid nucleus divides by meiosis, producing four new haploid nuclei. Four small protuberances develop on the end of the basidium, and the haploid nuclei migrate into these. The tip of each protuberance then becomes walled off as a spore, which falls or is ejected from the basidium. Each spore may give rise to a new mycelium. [From L. W. Sharp, *Fundamentals of Cytology*, McGraw-Hill Book Co., 1943. Used by permission.]

The sporangia in embryophytes are also multicellular, and they too have a layer of jacket cells. All embryophytes are oögamous, and the egg cells are fertilized while they are still contained within the archegonia. Each zygote develops into a multicellular diploid embryo while still inside the archegonium. The embryo obtains some of its water and nutrients from the parent plant and is thus a parasite on it. This type of embryonic development, clearly an adaptation permitting the stages of development most susceptible to desiccation to take place in a favorably moist microenvironment, is strongly reminiscent of the internal gestation of mammals.

The surfaces of the aerial parts of the plant bodies of embryophytes are usually covered by a waxy cuticle, which waterproofs the epidermis and helps prevent excessive water loss.

The principal pigments in embryophytes are chlorophylls *a* and *b*, and the reserve material is starch (Table 21.1). In other words, these plants are biochemically similar to the Chlorophyta, from which they almost certainly arose.

BRYOPHYTA (The liverworts, hornworts, and mosses)

The bryophytes are relatively small plants that grow in moist places on land—on damp rocks and logs, on the forest floor, in swamps or marshes, or beside streams and pools. Some species can survive periods of drought, but only by becoming dormant and ceasing to grow. In short, the bryophytes live on land, but they have never become fully

Table 21.1. A comparison of the major plant divisions

Characteristics	Chloro-phyta	Phaeo-phyta	Rhodo-phyta	Eumyco-phyta	Bryo-phyta	Tracheo-phyta
Usually, flagellated sperms	+	+	−	+ or −	+	+ or −
Chlorophyll *a*	+	+	+	−	+	+
Chlorophyll *b*	+	−(*c* instead)	−(*d* instead)	−	+	+
Principal reserve material usually starch	+	−	−	−	+	+
Sporophyte equal or dominant to gametophyte in most species	*	+	+	−	−	+
Usually, multicellular reproductive organs with jacket cells	−	−	−	−	+	+
Embryo development within archegonium	−	−	−	−	+	+
Cuticle usually present	−	−	−	−	+	+
Both xylem and phloem present	−	−	−	−	−	+

* Many Chlorophyta species whose full life cycles are known have a dominant gametophyte, but some have a dominant sporophyte. Since many species have not yet been studied, it is not possible to say which condition is the more usual.

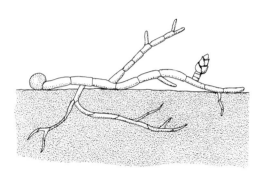

21.21. A young moss plant. The spore (round object on surface of soil, at left) gives rise to a filamentous plant (called a protonema) that strikingly resembles a green alga. The protonema develops into the mature moss plant. [Modified from H. J. Fuller and O. Tippo, *College Botany*, Holt, Rinehart and Winston, Inc., 1954.]

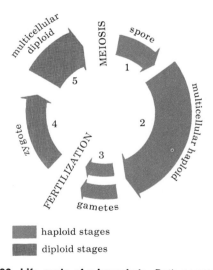

haploid stages
diploid stages

21.22. Life cycle of a bryophyte. Both gametophyte (stage 2) and sporophyte (stage 5) are present; the former is dominant.

emancipated from their ancestral aquatic environment, and they have therefore never become a dominant group of plants. Their great dependence upon a moist environment is linked to two characteristics: They retain flagellated sperm cells, which must swim to the egg cells in the archegonia, and they lack vascular tissues, and hence lack efficient long-distance internal transport of fluids. The absence of xylem, which functions as a major supportive tissue in vascular plants, has probably also limited the size they can attain.

The bryophytes are thought to have arisen from filamentous green algae. Indeed, a very young moss plant, called a protonema (Fig. 21.21), often closely resembles a green algal filament. As the plant grows, it forms some branches (rhizoids) that enter the ground and function like roots, anchoring the plant and absorbing water and nutrients. Other branches form upright shoots with stemlike and leaflike parts. Note that we say "like" roots, stems, and leaves. Since bryophytes lack vascular tissue, the anatomical criteria for distinguishing true roots, stems, and leaves cannot be applied to them.

We saw earlier that among the larger and more complex algae, most of which exhibit alternation of generations, there seems to have been an evolutionary tendency toward reduction of the gametophyte (multicellular haploid) stage and increasing emphasis of the sporophyte (multicellular diploid) stage. Thus in both brown algae and red algae the sporophyte is at least as prominent as the phylogenetically older gametophyte, and it is often much more prominent. As we shall see later, the same evolutionary tendency is found in the vascular plants. But this tendency is not apparent in the bryophytes, where the haploid gametophyte (stage 2 in Fig. 21.22) is clearly the dominant stage in the life cycle. The "leafy" green moss plant or liverwort is the gametophyte. These plants bear antheridia and archegonia in which gametes (stage 3) are produced. The flagellated male gametes (sperms) are released from the antheridia and swim through a film of moisture, such as rain or heavy dew, to archegonia, where they fertilize the egg cells, producing zygotes (stage 4). Each zygote then divides mitotically, producing a diploid sporophyte (stage 5).

In a moss, this sporophyte is a relatively simple structure consisting of three parts: a foot embedded in the "leafy" green gametophyte, a stalk, and a distal capsule, or sporangium (Fig. 21.23, Pl. I-4D). The sporophyte has a few chloroplasts and carries out some photosynthesis, but it also obtains nutrients parasitically from the gametophyte to which it is attached. Meiosis occurs within the mature capsule of the sporophyte, producing haploid spores (stage 1), which are released. These may germinate, developing into protonemata and eventually into mature gametophyte plants (stage 2), thus completing the life cycle.

The gametophyte plants of some liverworts resemble mosses except that the "leaves" are scaly in appearance, and the "stem" is prostrate. Other liverworts grow as flat green structures lying on the substratum (Fig. 21.24). The life cycle of liverworts is much like that of mosses except that the sporophyte is even simpler. Asexual reproduction sometimes occurs by production of special cells called gemmae, usually borne in cuplike structures located on the surface of the flat gametophyte. When detached from the parent plant, the gemmae can grow into new gametophytes.

TRACHEOPHYTA (The vascular plants)

We have seen that, though most bryophytes live on land, in a sense they are not fully terrestrial. The tracheophytes, by contrast, have evolved a host of adaptations to the terrestrial environment that have enabled them to invade all but the most inhospitable land habitats. In the process, they have diverged sufficiently from one another for botanists to classify them in five subdivisions as follows:

Division Tracheophyta
    Subdivision Psilopsida (psilopsids)
    Subdivision Lycopsida (club mosses)
    Subdivision Sphenopsida (horsetails)
    Subdivision Pteropsida (ferns)
    Subdivision Spermopsida (seed plants)

All members of this division (with a few minor exceptions) possess four important attributes lacking in even the most advanced and complex algae: a protective layer of sterile jacket cells around the reproductive organs; multicellular embryos, which are retained within the archegonia; cuticles on the aerial parts; and xylem (see Table 21.1). All four are obviously fundamental adaptations for a terrestrial existence. Many other such adaptations, absent in the earliest tracheophytes, appear in more advanced members of the division; a history of the evolution of these adaptations is a history of the increasingly extensive exploitation of the terrestrial environment by vascular plants. Let us briefly trace this history of adaptation to life on land.

21.23. Gametophyte and sporophyte stages of a moss. The lower "leafy" plant is the gametophyte. The sporophyte plant (brown stalk and capsule) is attached to the gametophyte.

21.24. Liverworts (*Marchantia*). The gametophyte is a flat green organism with a scaly appearance. The two cup-shaped structures on the plants in the center are gemmae cups. [Courtesy Verne N. Rockcastle, Cornell University.]

Table 21.2.   The geologic time scale

Era	Period	Epoch	Millions of years (approx.) from start of period to present	Plant life	Animal life
CENOZOIC	Quaternary	Recent	0.01	Increase in number of herbs	Rise of civilizations
		Pleistocene	2		First *Homo*
	Tertiary	Pliocene	12	Dominance of land by angiosperms	First men
		Miocene	25		
		Oligocene	36		Dominance of land by mammals, birds, and insects
		Eocene	58		
		Paleocene	63		

— Building of ancestral Rocky Mountains —

Era	Period	Millions of years	Plant life	Animal life
MESOZOIC	Cretaceous	135	Angiosperms expand as gymnosperms decline	Last of the dinosaurs; second great radiation of insects
	Jurassic	181	Gymnosperms (esp. cycads and conifers) still dominant; last of the seed ferns; a few angiospermlike plants appear	Dinosaurs abundant; first mammals and birds
	Triassic	230	Dominance of land by gymnosperms; decline of lycopsids and sphenopsids	First dinosaurs

— Building of ancestral Appalachian Mountains —

Era	Period	Millions of years	Plant life	Animal life
PALEOZOIC	Permian	280	Forests of lycopsids, sphenopsids, seed ferns, and conifers	Great expansion of reptiles; decline of amphibians; last of the trilobites
	Carboniferous*	345	Great coal forests, dominated at first by lycopsids and sphenopsids, and later also by ferns and seed ferns; first conifers	Age of Amphibians; first reptiles; first great radiation of insects
	Devonian	405	Expansion of primitive tracheophytes; first liverworts	Age of Fishes; first amphibians and insects
	Silurian	425	Invasion of land by primitive tracheophytes	Invasion of land by a few arthropods
	Ordovician	500	Marine algae abundant	First vertebrates (Agnatha)
	Cambrian	600	Primitive marine algae (esp. Cyanophyta and probably Chlorophyta)	Marine invertebrates abundant (including representatives of most phyla)

— Interval of Great Erosion —

PRECAMBRIAN		Primitive marine life

* In North America, the Lower Carboniferous is often called the Mississippian period, and the Upper Carboniferous is called the Pennsylvanian period.

Psilopsida

The oldest undisputed fossil representatives of the vascular plants are from the Silurian period, which means that they lived more than 405 million years ago (Table 21.2). They are classified among the Psilopsida and the Lycopsida (see below). Most of the Psilopsida lived during the Devonian period and then became extinct. Two living genera, *Psilotum* and *Tmesipteris,* have been regarded as members of this ancient group, but there is recent evidence to suggest that they may actually be very primitive ferns.

The sporophyte was the dominant stage in the Psilopsida. It was a simple dichotomously branching plant that possessed vascular tissue but lacked leaves and had no true roots, though it did have underground stems with unicellular rhizoids similar to root hairs. The tips of some of the aerial branches bore sporangia.

Botanists, noting the resemblance between the psilopsids and certain branching filamentous green algae, have suggested that it was from such algae that these primitive vascular plants arose. Since it is the psilopsid sporophyte that resembles the green algae, it must be concluded that, if the vascular plants did indeed evolve from green algae, either those algae had a prominent sporophyte stage or the first vascular plants rapidly evolved such a stage.

Lycopsida (The club mosses)

The first representatives of the subdivision Lycopsida also appeared in the Silurian period and may be as old as or older than the Psilopsida. During the Devonian and Carboniferous periods, these were among the dominant plants on land. Some of them were very large trees that formed the earth's first forests. Toward the end of the Paleozoic era, however, the group was displaced by more advanced types of vascular plants, and only five genera are alive today. One of these, *Lycopodium* (often called running pine or ground pine), is common in many parts of the United States, and is frequently used in Christmas decorations (Fig. 21.25).

Unlike the psilopsids, lycopsids have true roots. It is generally supposed that these arose from branches of the ancestral algae that penetrated the soil and ramified underground. Lycopsids also have true leaves, which are thought to have arisen as simple scalelike outgrowths (emergences) from the outer tissues of the stem. Certain of the leaves that have become specialized for reproduction bear sporangia on their surfaces. Such reproductive (fertile) leaves are called **sporophylls.** In many lycopsids, the sporophylls are congregated on a short length of stem and form a conelike structure, or strobilus (Fig. 21.26). The strobili are rather club-shaped; hence the name "club mosses" for the lycopsids (note, however, that lycopsids are not related to the true mosses, which are bryophytes).

The spores produced by *Lycopodium* are all alike, and each can give rise to a gametophyte that will bear both archegonia and antheridia. However, some lycopsids have two types of sporangia: one that produces very large spores called **megaspores,** which develop into female gametophytes bearing archegonia; and one that produces small spores called **microspores,** which develop into male gametophytes bearing antheridia.

21.25. A *Lycopodium* plant with strobili. [Courtesy Verne N. Rockcastle, Cornell University.]

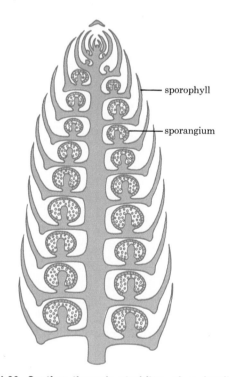

21.26. Section through strobilus of a fossil lycopsid. [Modified from H. N. Andrews, *Ancient Plants and the World They Lived In,* copyright 1947, Comstock Publishing Co. Used by permission of Cornell University Press.]

sporophyll

sporangium

21.27. Carboniferous swamp forest. Note the tree sphenopsids with their jointed stems and whorls of leaves. The trunk at right is a lycopsid. [Portion of group in Carnegie Museum, Pittsburgh. Used by permission.]

Plants that produce only one kind of spore, and hence have only one kind of gametophyte that bears both male and female organs, are said to be **homosporous.** Plants that produce both megaspores (female) and microspores (male) are said to be **heterosporous.**

Sphenopsida (The horsetails)

The sphenopsids first appear in the fossil record in the Devonian period. They became a major component of the land flora during the late Paleozoic era, and then declined. Specimens of the one living genus, *Equisetum,* are commonly called horsetails or scouring rushes (Pl. I-4B). Though most of these are small (under 3 feet), some of the ancient sphenopsids were large trees (Fig. 21.27). Much of the coal we use today was formed from the dead bodies of these plants.

Like the lycopsids, sphenopsids have true roots, stems, and leaves. The stems are hollow and are jointed. Whorls of leaves occur at each joint. Many of the extinct sphenopsids had cambium, and hence secondary growth, but the modern species do not. Spores are borne in terminal cones (Fig. 21.28). In *Equisetum,* all spores are alike (i.e. the plants are homosporous) and give rise to small gametophytes that bear both archegonia and antheridia (i.e. the sexes are not separate).

Pteropsida (The ferns)

In the opinion of many botanists, the ferns evolved from the Psilopsida. They first appeared in the Devonian period and greatly increased in importance during the Carboniferous. Their decline after the Paleozoic era was much less severe than that of the psilopsids, lycopsids, and sphenopsids, and, as you know, there are many modern species.

The ferns are fairly advanced plants with a very well developed vascular system and with true roots, stems, and leaves. The leaves are

21.28. Reproductive stems of *Equisetum.* Left: Portions of two stems bearing immature terminal cones. The hollow stems can easily be pulled apart at the joints. The leaves, arranged in whorls at the joints, are very small and inconspicuous. Right: Portion of a stem bearing a mature cone. The whorls of leaves along the stem are more evident. [Courtesy Verne N. Rockcastle, Cornell University.]

thought to have arisen in a different manner from those of the lycopsids. Instead of being emergences, they are probably flattened and webbed branch stems; i.e. a group of small branches probably became planated (arranged in one plane), and the interstices filled with tissue.[6] The leaves are sometimes simple, but more often they are compound, being divided into numerous leaflets that may give the plant a lacy appearance.[7] In a few ferns (e.g. the large tree ferns of the tropics), the stem is upright, forming a trunk. But in most modern ferns, especially those of temperate regions, the stems are prostrate on or in the soil, and the large leaves are the only parts normally seen.

The large leafy fern plant is the diploid sporophyte phase. Spores are produced in sporangia located in clusters on the underside of some leaves (sporophylls) (Fig. 21.29). In some species, the sporophylls are relatively little modified and look like the nonreproductive leaves. In other species, the sporophylls look quite different from vegetative leaves; sometimes they are so highly modified that they do not look like leaves at all, forming spikelike structures instead (Fig. 21.30).

Most modern ferns are homosporous; i.e. all their spores are alike. After germination, the spores develop into gametophytes that bear both archegonia and antheridia (Fig. 21.31). These gametophytes are tiny (seldom more than one fourth of an inch wide), thin, and often more or less heart-shaped. Although most people are familiar with the sporophytes of ferns, few have ever seen a gametophyte, and even fewer would guess that it had anything to do with a fern. Small and obscure as it is,

---

[6] Leaves arising as emergences are called microphylls. Those arising as planated and webbed branch systems are called megaphylls.

[7] When a fern leaf, or frond, is divided into leaflets, the leaflets are called pinnae. The pinnae may themselves be subdivided into pinnules.

21.29. Undersurface of fertile leaf (sporophyll) of polypody fern. Each of the round dots is a sorus, which is a cluster of many tiny sporangia. [Courtesy Verne N. Rockcastle, Cornell University.]

21.30. Leaves of sensitive fern. The leaf on the left is a mature sporophyll, which bears little resemblance to the sterile leaf on the right. The leaf in the middle is an intermediate type. [Courtesy Verne N. Rockcastle, Cornell University.]

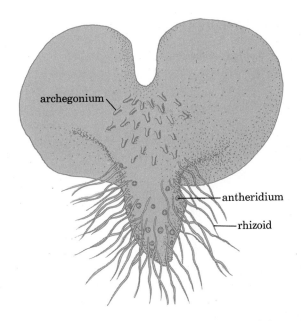

archegonium

antheridium

rhizoid

21.31. Fern gametophyte. This is a much-magnified view of the undersurface of the tiny heart-shaped organism. [Modified from H. J. Fuller and O. Tippo, *College Botany,* Holt, Rinehart and Winston, Inc., 1954.]

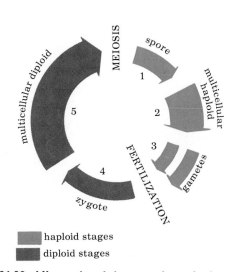

■ haploid stages
■ diploid stages

21.32. Life cycle of ferns and seed plants.
Both gametophyte (stage 2) and sporophyte
(stage 5) are present; the latter is much the more
prominent. Compare this life cycle with that of
bryophytes shown in Fig. 21.22.

however, the fern gametophyte is an independent photosynthetic organism. Here, then, is a life cycle in which all five principal stages are present, but in which the multicellular haploid stage has been much reduced and the multicellular diploid stage emphasized (Fig. 21.32).

In some respects, the ferns (and also the three primitive groups of vascular plants discussed above) are no better adapted for life on land than the bryophytes. Their vascularized sporophytes can live in drier places and grow bigger, but for a number of reasons—because their nonvascularized free-living gametophytes can survive only in moist places, because their sperms are flagellated and must have a film of moisture through which to swim to the egg cells in the archegonia, and because the young sporophyte develops directly from the zygote without passing through any protected seedlike stage—these plants are most successful in habitats where there is at least a moderate amount of moisture.

Spermopsida (The seed plants)

The seed plants have been by far the most successful in fully exploiting the terrestrial environment. They first appeared in the Carboniferous and soon replaced the lycopsids and sphenopsids as the dominant land plants, a position they still hold today. In these plants, the gametophytes are even more reduced than in the ferns—they are not photosynthetic or free-living—and the sperms of most modern species are not independent free-swimming flagellated cells. In addition, the young embryo, together with a supply of nutrients, is enclosed within a desiccation-resistant seed coat and can remain dormant for extended periods if environmental conditions are unfavorable. In short, the aspects of the reproductive process that are most vulnerable in more primitive vascular plants have been eliminated in the seed plants (Table 21.3).

The seed plants have traditionally been divided into two classes, the Gymnospermae and the Angiospermae. It is becoming increasingly

Table 21.3.   A comparison of the subdivisions of Tracheophyta

Characteristics	Psilop-sida	Lycop-sida	Sphenop-sida	Pterop-sida	Spermopsida Gymno-sperms	Angio-sperms
Vascular tissue	+	+	+	+	+	+
True roots and leaves	−	+	+	+	+	+
Megaphyllous leaves	−	−	−	+	+	+
Gametophyte retained in sporophyte archegonium	−	−	−	−	+	+
Sperm cells without flagella	−	−	−	−	+ (− in primitive groups)	+
Production of seeds	−	−	−	−	+	+
Flowers and fruit	−	−	−	−	−	+

clear, however, that the relationships between the five groups put together as the gymnosperms are not particularly close and that these groups differ from one another at least as much as they differ from the angiosperms. Consequently many modern classifications recognize each of the gymnospermous groups as a separate class. We have adopted this procedure in the technical classification given in the Appendix and outlined below, but shall discuss the gymnospermous groups together.

Subdivision Spermopsida
    Class Pteridospermae ⎫
    Class Cycadae   ⎪
    Class Ginkgoae   ⎬ "Gymnosperms"
    Class Coniferae   ⎪
    Class Gneteae   ⎭
    Class Angiospermae

The gymnosperms. The first gymnosperms appear in the fossil record in the early Carboniferous, but the group probably arose in the Devonian. Many of those first seed plants had bodies that closely resembled those of the ferns, and indeed for many years their fossils were thought to be fossils of ferns. Slowly, however, evidence accumulated that some of the "ferns" that were such important components of the coal-age forests produced seeds, not spores. Today these fossil plants— usually called the seed ferns—are grouped together as the class Pteridospermae of the subdivision Spermopsida. No members of this class survive today.

Another ancient group, the cycads and their relatives (class Cycadae), may have arisen from the seed ferns. These plants first appeared in the Permian period and became very abundant during the Mesozoic era. They had large palmlike leaves; the palmlike plants so often shown in pictures of the dinosaur age are usually cycads, not true palms. The cycads declined after the rise of the angiosperms in the Cretaceous period, but nine genera containing over a hundred species are in existence today. They are generally called sago palms and are fairly common in some tropical regions.

The Ginkgoae are still another group that was once widespread but is now nearly extinct. There is only one living species, the ginkgo or maidenhair tree, which is often planted as a lawn tree but is almost unknown in the wild.

By far the best-known group of gymnosperms is that of the conifers (class Coniferae), which includes such common species as pines, spruces, firs, cedars, hemlocks, yews, and larches. The leaves of most of these are small evergreen needles or scales (Pl. II-7A). This group first arose in the Permian period and was very common during the Mesozoic era. It remains an important part of the earth's flora.

Let us examine in some detail the life cycle of a pine tree as an example of the seed method of reproduction. The large pine tree is the diploid sporophyte stage (stage 5 in Fig. 21.32). This tree produces reproductive structures called *cones*. A cone is a spiral cluster of modified leaves (sporophylls) on a short section of stem. Each sporophyll (often called a cone scale) bears on its surface two sporangia in which haploid spores (stage 1) are produced by meiosis. There are two kinds of cones: large female cones whose sporangia produce megaspores and

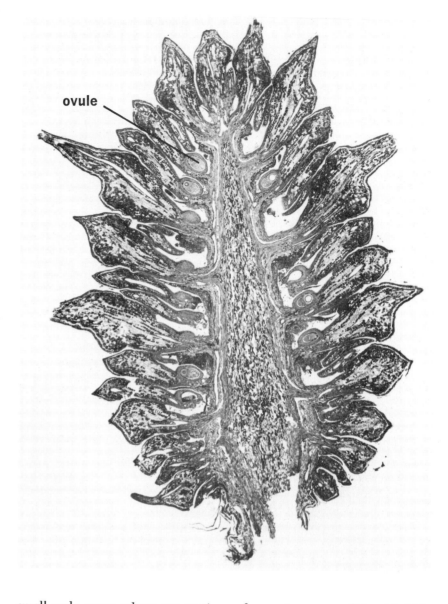

ovule

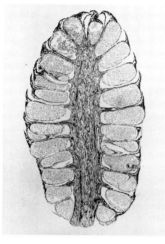

21.33. Sections of male and female pine cones.
Left: Male cone. Each sporophyll (cone scale) bears a large sporangium that becomes a pollen sac. Right: Female cone. Ovules can be seen on the surface of the sporophylls, near their base. [Left: Courtesy Thomas Eisner, Cornell University. Right: Courtesy Carolina Biological Supply Co.]

small male cones whose sporangia produce microspores (Fig. 21.33). (Production of two kinds of spores—heterospory—is characteristic of all seed plants, both gymnosperms and angiosperms.)

Each scale of a female cone bears two sporangia on its upper (adaxial) surface (Fig. 21.34B). Each sporangium is encased in an integument with a small opening, the *micropyle*, at one end (Fig. 21.34C). Meiosis takes place inside the sporangium, producing four haploid megaspores, three of which soon disintegrate. Next, the single remaining megaspore gives rise, by repeated mitotic divisions, to a multicellular mass, which is the female gametophyte (megagametophyte). When mature, the female gametophyte produces two to five tiny archegonia at its micropylar end. Egg cells develop in the archegonia. Note that the megaspore is never released from the sporangium, and that the female gametophyte derived from it remains embedded in the sporangium, which is still attached to the cone scale. The composite structure consisting of integument, sporangium, and female gametophyte is called an *ovule.*

Each microspore produced by the sporangium of a male cone becomes a *pollen grain.* It develops a thick coat, which is highly resistant to loss of water, and winglike structures on each side, which doubtless aid its dispersal by wind. Within the pollen grain, the haploid nucleus divides mitotically several times, and walls develop around each nucleus. In this manner, the pollen grain becomes four-celled (Fig. 21.34A); two of the cells soon degenerate. The mature pollen grain is released from the cone when the sporangium bursts. A single male cone may release millions of tiny pollen grains, which may be carried many miles (sometimes as many as a hundred) by the wind. Note that the pollen grains are multicellular haploid structures (if four cells may be said to be "multi") and that they constitute the male gametophyte (micro-gametophyte) (stage 2 in Fig. 21.32).

Most of the millions of pollen grains released by a pine tree fail to reach a female cone. But of the few that sift down between the scales of a female cone, some land in a sticky secretion near the open micro-pylar end of an ovule. As this secretion dries, it is drawn through the micropyle, carrying the pollen grains with it. The arms of the integument around the micropyle then swell and close the opening. When a pollen grain comes into contact with the end of the sporangium just inside the micropyle, it develops a tubular outgrowth, the *pollen tube.* The two functional nuclei of the pollen grain enter the tube, and one of them divides. One of the daughter nuclei thus produced then divides again, producing two sperm nuclei.[8] Thus a germinated pollen grain contains four active nuclei plus the two nuclei of the degenerate cells; this six-nucleate condition is as far as the male gametophyte of pine ever develops.

The pollen tube grows down through the tissue of the sporangium and penetrates into one of the archegonia of the female gametophyte.[9] There it discharges its sperm nuclei, one of which fertilizes the egg cell. The resulting zygote (stage 4) then divides mitotically to produce a tiny embryo sporophyte. The embryo is still contained in the female gametophyte, which is itself contained in the sporangium. Finally, the entire ovule is shed from the cone as a *seed,* which consists of three main components: a seed coat derived from the old integument, stored food material derived from the tissue of the female gametophyte, and an embryo.

The main advances of the pine life cycle over the fern life cycle can be summarized as follows:

1. There are two types of sporangia, which produce two types of spores—microspores (male) and megaspores (female).
2. Two different kinds of gametophytes are derived from the two types of spores (i.e. the sexes are separate in the gametophyte stage).

[8] The two nuclei that enter the tube are the generative and tube nuclei. The generative cell divides, producing a stalk cell and a body cell, and the body cell then divides to form two sperms.

[9] Since an ovule contains several archegonia, several embryos may begin development, but usually only one completes it.

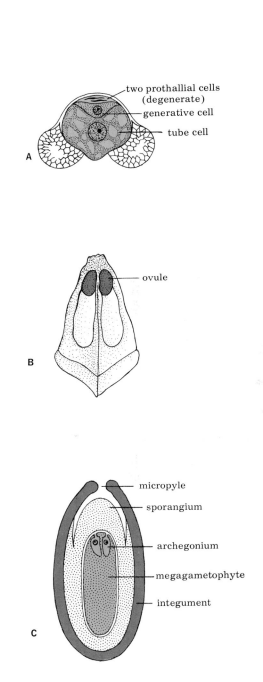

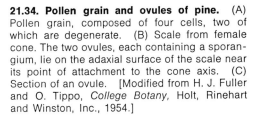

21.34. Pollen grain and ovules of pine. (A) Pollen grain, composed of four cells, two of which are degenerate. (B) Scale from female cone. The two ovules, each containing a sporangium, lie on the adaxial surface of the scale near its point of attachment to the cone axis. (C) Section of an ovule. [Modified from H. J. Fuller and O. Tippo, *College Botany,* Holt, Rinehart and Winston, Inc., 1954.]

3. The gametophytes are much further reduced than those of ferns; they do not possess chlorophyll and are not free-living. A male gametophyte consists only of the six-nucleate pollen grain and tube. A female gametophyte is only a mass of haploid tissue that remains in the sporangium and is parasitic on it.
4. There are usually no flagellated sperm cells.[10]
5. The young embryo is contained within a seed.

The angiosperms. There are a few angiospermlike fossils from the Triassic and Jurassic periods, but the first undisputed representatives of this group are from the Cretaceous. Great expansion of the angiosperms occurred in the Cretaceous, and these plants became the dominant land flora of the Cenozoic era, as they are today.

We have seen that the reproductive structures of gymnosperms are cones (or structures derived from cones) and that the ovules, which later become the seeds, are borne naked on the surface of the sporophylls. In the angiosperms, by contrast, the reproductive structures are flowers, and the ovules are enclosed within modified leaves called carpels.

A flower, like a cone, is a short length of stem with modified leaves attached to it. The modified leaves of a typical flower (Fig. 21.35) occur in four sets attached to the enlarged end (receptacle) of the flower stalk: (1) The *sepals* enclose and protect all the other floral parts during the bud stage. They are usually small, green, and leaflike, but in some species they are large and brightly colored. All the sepals together form the *calyx.* (2) Internal to the sepals are the *petals,* which together form the *corolla.* In flowers pollinated by insects, birds, or other animals, the petals are usually quite showy, but in those pollinated by wind they are often reduced or even absent. (3) Just inside the circle of the corolla are the *stamens,* which are the male reproductive organs; i.e. they are the sporophylls that produce the microspores. Each stamen consists of a stalk, called a *filament,* and a terminal ovoid pollen-producing structure called an *anther.* (4) In the center of the flower is the female reproductive organ, the *pistil* (some species have more than one pistil per flower). Each pistil consists of an *ovary* at its base, a slender stalk (more than one in some species) called a *style,* which rises from the ovary, and an enlarged apex called a *stigma.* The pistil is derived from one or more sporophylls, which in flowers are called *carpels.*[11] The four kinds of floral organs—sepals, petals, stamens, and pistils—are all present in so-called complete flowers, but some flowers, which are said to be incomplete, lack one or more of them.

Within the ovary are one or more (at least one for each carpel) sporangia, called ovules, which are attached by short stalks to the wall of the ovary. Meiosis occurs once in each ovule, producing four haploid megaspores, three of which usually soon disintegrate. The remaining megaspore then divides mitotically several times, producing, in most species, a structure composed of seven cells, one of which is much larger than the others and contains two nuclei, called polar nuclei (Fig.

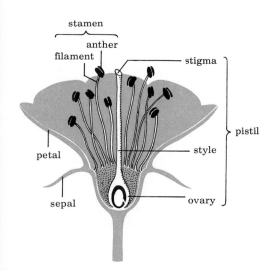

21.35. The parts of a flower.

stamen
anther
filament
stigma
petal
style
sepal
ovary
pistil

[10] The pollen tubes of primitive gymnosperms, such as ginkgoes and cycads, produce flagellated sperm cells, but these have only a very short distance to swim to reach the egg cells, the pollen grains having already been carried to the ovules by the wind.
[11] A simple pistil is composed of only one sporophyll, or carpel. A compound pistil is composed of several fused carpels.

21.36A). This haploid seven-celled eight-nucleate structure is the much-reduced female gametophyte (often called an embryo sac). One of the cells located near the micropylar end will act as the egg cell.

Each anther has four sporangia in which meiosis occurs repeatedly, producing numerous haploid microspores. The wall of each microspore thickens, and the nucleus divides mitotically, producing a generative nucleus and a tube nucleus. The resulting thick-walled two-nucleate structure is a pollen grain—a male gametophyte—which is released from the anther when the mature sporangium splits open.

A pollen grain germinates when it falls (or is deposited) on the stigma of a pistil, which is usually rough and sticky. A pollen tube begins to grow, and the two nuclei of the pollen grain move into it. The generative nucleus then divides, giving rise to two sperm nuclei (Fig. 21.36B). The pollen tube grows down through the tissues of the stigma and style and enters the ovary (Fig. 21.37). When the tip of the pollen tube reaches an ovule, it enters the micropyle and then discharges the two sperm nuclei into the female gametophyte (embryo sac). One of the sperm nuclei fertilizes the egg cell, and the zygote thus formed develops into an embryo sporophyte. By the time fertilization occurs, the two polar nuclei of the female gametophyte have combined to form a diploid *fusion nucleus,* with which the second sperm nucleus unites to form a triploid nucleus. This nucleus undergoes a series of divisions, and a triploid tissue, called *endosperm,* is formed. The endosperm functions in the seed as a source of stored food for the embryo.

After fertilization, the ovule matures into a seed, which, as in pine, consists of seed coat, stored food (endosperm), and embryo. However, the angiosperm seed differs from that of pine in being enveloped by the ovary. It is the ovary that develops into the *fruit,* usually enlarging greatly in the process. Sometimes other structures associated with the ovary, such as the receptacle, are also incorporated into the fruit. The ripe fruit may burst, expelling the seeds, as in peas (where the pod is the fruit). Or the ripe fruit with the seeds still inside may fall from the plant, as in tomatoes, squash, cucumbers, apples, peaches, and acorns. The fruit not only helps protect the seeds from desiccation during their early development, before they have fully ripened, but often also helps disperse them—as when an animal, attracted by the fruit, carries it to other locations or eats both fruit and seeds and later releases the unharmed seeds in its feces.

The main features in which the angiosperm life cycle differs from that of gymnosperms can be summarized as follows:

1. The reproductive structures are flowers instead of cones. The sporophylls (stamens and pistils) of flowers are more extensively modified and less leaflike than the sporophylls (scales) of cones.
2. The ovules are embedded in the tissues of the female sporophylls instead of lying bare on their surface.
3. The gametophytes are even more reduced than those of gymnosperms. The male gametophyte (pollen grain and tube) has only three nuclei. The female gametophyte usually has only eight nuclei.
4. In pollination, the pollen grains are not deposited close to the

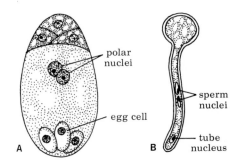

21.36. Gametophytes of an angiosperm. (A) Female gametophyte (embryo sac), which is composed of seven cells. One cell is much larger than the others and contains the two polar nuclei. (B) Male gametophyte (pollen grain and tube). [Modified from H. J. Fuller and O. Tippo, *College Botany,* Holt, Rinehart and Winston, Inc., 1954.]

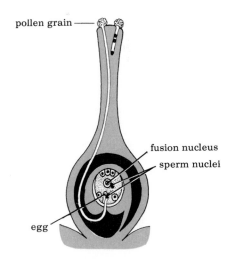

21.37. Fertilization of an angiosperm. Pollen grains land on the stigma and give rise to pollen tubes that grow downward through the style. One of the pollen tubes shown here has reached the ovule in the ovary and discharged its sperm nuclei into it. One sperm nucleus will fertilize the egg cell, and the other will unite with the diploid fusion nucleus (derived from the two polar nuclei) to form a triploid nucleus, which will give rise to endosperm.

21.38. Easter lily flower. This is a monocot, and all the flower parts are in threes or multiples of three. There are three sepals (the outer three white structures, which resemble petals), three petals, six stamens (one is nearly hidden behind the pistil), and a three-lobed stigma (which shows that the single pistil is compound, formed by fusion of three primordial pistils). [Courtesy Verne N. Rockcastle, Cornell University.]

opening of the ovule, as in gymnosperms, but are deposited on the stigma instead. The pollen tube thus has much farther to grow in angiosperms.

5. Angiosperms have "double fertilization," one sperm fertilizing the egg cell and the other uniting with the fusion nucleus to give rise to triploid endosperm. Gymnosperms, by contrast, have single fertilization; one sperm fertilizes the egg, and the other soon deteriorates. The stored food in the seed of gymnosperms is the haploid tissue of the female gametophyte and is thus developmentally quite different from the triploid endosperm of the angiosperms.

6. The seeds of angiosperms are enclosed in fruits that develop from the ovaries and associated structures; gymnosperms have no fruit.

The class Angiospermae is customarily divided into two subclasses, the Dicotyledoneae and the Monocotyledoneae. The dicots include oaks, maples, elms, willows, roses, beans, clovers, tomatoes, asters, and dandelions. The monocots include grasses, corn, wheat, rye, daffodils, irises, lilies, and palms. There are certain basic differences between the two groups:

1. As the names imply, the embryos of dicots have two cotyledons, whereas those of monocots have only one.

2. Dicots often have cambium and secondary growth; monocots usually do not.

3. The vascular bundles in the stems of young dicots are arranged in a circle or fused to form a tubular vascular cylinder; monocots have more scattered vascular bundles.

4. The leaves of dicots usually have net venation; those of monocots usually have parallel venation.

5. Dicot leaves generally have petioles; monocots generally do not. Most monocots can be recognized by the way the leaf base clasps the stem (as in corn).

6. The flower parts of dicots occur in fours or fives or multiples of these (e.g. four sepals, four petals, four stamens); those of monocots occur in threes or multiples of three (Fig. 21.38).

REFERENCES

ALEXOPOULOS, C. J., 1962. *Introductory Mycology,* 2nd ed. Wiley, New York.

ANDREWS, H. N., 1961. *Studies in Paleobotany.* Wiley, New York.

BIERHORST, D. W., 1971. *Morphology of Vascular Plants.* Macmillan, New York.

BOLD, H. C., 1967. *Morphology of Plants,* 2nd ed. Harper & Row, New York.

CHAPMAN, V. J., 1962. *The Algae.* St. Martin's Press, New York.

DAWSON, E. Y., 1966. *Marine Botany.* Holt, Rinehart & Winston, New York.

DELEVORYAS, T., 1962. *Morphology and Evolution of Fossil Plants.* Holt, Rinehart & Winston, New York.

ESAU, K., 1965. *Plant Anatomy,* 2nd ed. Wiley, New York.

FOSTER, A. F., and E. M. GIFFORD, 1959. *Comparative Morphology of Vascular Plants.* Freeman, San Francisco.

KINGSBURY, J. M., 1963. *Biology of the Algae.* Published privately, Ithaca, N.Y.

LEWIN, R. A., 1962. *Physiology and Biochemistry of Algae.* Academic Press, New York.

PRESCOTT, G. W., 1968. *The Algae: A Review.* Houghton Mifflin, Boston.

SCAGEL, R. F., *et al.,* 1965. *An Evolutionary Survey of the Plant Kingdom.* Wadsworth, Belmont, Calif.

SMITH, G. M., 1955. *Cryptogamic Botany,* 2nd ed. Vol. 1: *Algae and Fungi;* vol. 2: *Bryophytes and Pteridophytes.* McGraw-Hill, New York.

SUGGESTED READING

AHMADJIAN, V., 1963. "The Fungi of Lichens," *Scientific American.* February.

ANDREWS, H. N., 1947. *Ancient Plants and the World They Lived In.* Comstock, Ithaca, N.Y.

————, 1963. "Early Seed Plants," *Science,* vol. 142, pp. 925–937.

BOLD, H. C., 1970. *The Plant Kingdom,* 3rd ed. Prentice-Hall, Englewood Cliffs, N.J.

BONNER, J. T., 1956. "The Growth of Mushrooms," *Scientific American,* May.

CHRISTENSEN, C. M., 1965. *The Molds and Man,* 3rd ed. University of Minnesota Press, Minneapolis. (Paperback edition, McGraw-Hill, New York, 1965.)

DELEVORYAS, T., 1966. *Plant Diversification.* Holt, Rinehart & Winston, New York.

DOYLE, W. T., 1970. *Non-Seed Plants: Form and Function,* 2nd ed. Wadsworth, Belmont, Calif.

EMERSON, R., 1952. "Molds and Men," *Scientific American,* January. (Offprint 115.)

FULLER, H. J., and O. TIPPO, 1954. *College Botany,* rev. ed. Holt, New York. (See esp. Part III.)

GRANT, V., 1951. "The Fertilization of Flowers," *Scientific American,* June. (Offprint 12.)

LAMB, I. M., 1959. "Lichens," *Scientific American,* October. (Offprint 111.)

WILSON, C. L., and W. E. LOOMIS, 1967. *Botany,* 4th ed. Holt, Rinehart & Winston, New York.

The animal kingdom

We now turn to the phyla of the animal kingdom, with many of which you already have some familiarity from our previous discussion of such of their vital processes as nutrient procurement, gas exchange, internal transport, excretion, coordination, and development. In our earlier discussions, we paid particular attention to representatives of the Protozoa, Coelenterata, Platyhelminthes (flatworms), Mollusca, Annelida (segmented worms), Arthropoda, Echinodermata, and Chordata (especially vertebrates). We shall again concentrate on these large and important groups, but shall also briefly mention some smaller phyla we previously ignored. Our aim here is to bring together the various phases of animal life discussed in other chapters and to suggest both the immense diversity within the kingdom and possible evolutionary patterns.

A formal outline of the classification of the animal kingdom used in this book is given in the Appendix. We must stress that this classification, though a widely used one, is not accepted by all biologists. Much is uncertain regarding the evolution of animal groups, and different biologists interpret the data in different ways.

PROTOZOA

Protozoans are usually said to be unicellular. However, as the great student of invertebrate animals Libbie Henrietta Hyman, of the American Museum of Natural History, has pointed out, "Each protozoan is to be regarded not as equivalent to a cell of a more complex animal but as a complete organism with the same properties and characteristics as cellular animals." Although they "necessarily lack tissues and organs, since these are defined as aggregations of differentiated cells," many do exhibit "a remarkable degree of functional differentiation." Instead of organs, they have functionally equivalent subcellular structures called *organelles.* In recognition of the complexity of Protozoa, which often far exceeds that of other individual cells, Hyman and many other biologists prefer to call them *acellular* organisms, i.e. organisms whose bodies do not exhibit the usual construction of cells.

Protozoans occur in a great variety of habitats, including the sea, fresh water, soil, and the bodies of other organisms—in fact, wherever there is moisture. Most are solitary, but some are colonial. Many are free-living, but others are commensalistic, mutualistic, or parasitic. The great majority are heterotrophic, but some of the flagellates possess chlorophyll and are photosynthetic autotrophs; these plantlike forms, which include *Euglena, Chlamydomonas, Gonium, Pandorina,* and *Volvox,* are classified as algae by botanists (see pp. 461–465). The heterotrophic forms usually digest food particles in food vacuoles (see p. 105 and Figs. 5.6 and 5.7). There are no special organelles for gas exchange, the general cell membrane serving as the exchange surface. Many species, particularly those living in hypoosmotic media such as fresh water, possess contractile vacuoles, which function primarily in osmoregulation (see p. 157 and Fig. 8.4). Small amounts of nitrogenous waste may also be expelled by the contractile vacuoles, but most of it is released as ammonia by diffusion across the general cell surface. Locomotion is by formation of pseudopodia or by means of beating cilia or flagella. Where a single individual has many cilia, their action is coordinated by a system of fibrils connecting their bases; these fibrils seem to have conductile properties rather like those of nerves in multicellular animals. Reproduction is sometimes asexual and sometimes sexual. Most fresh-water and parasitic protozoans can encyst when conditions are unfavorable; they secrete a thick resistant case around themselves and become dormant.

The Protozoa have traditionally been divided into four classes: Flagellata, Sarcodina, Sporozoa, and Ciliata. Each of these is a heterogeneous assemblage of structurally similar organisms that are probably not closely related. Furthermore, the relationships of the classes to one another are unclear; this is particularly true of the Ciliata. For these reasons, some biologists choose to divide the Protozoa into two or more separate phyla. Although a more conservative system is used here, the two subphyla treated below (Plasmodroma and Ciliophora) might well be raised to the rank of full phyla.

Subphylum Plasmodroma

Class Flagellata. As their name implies, the Flagellata are protozoans that possess flagella as the principal locomotor organelles. They appear to be the most primitive of all the Protozoa, and it seems likely that some (and possibly all) of the other protozoan groups arose from them. As indicated in Chapter 21, it also seems likely that the various groups of algae (and through them the higher plants) evolved from flagellates; indeed, many of the modern algae (including all Euglenophyta and many members of the Chlorophyta) are classified as flagellate Protozoa by zoologists. Most zoologists (though not all) believe that the flagellates were also the ancestors of the multicellular animals. There is good reason to think, then, that this group of organisms played a key role in the evolution of life on earth, probably giving rise to other Protozoa, to most (and perhaps all) of the plant kingdom, and to the animal kingdom. In saying this, we are not suggesting that any flagellate species still living today was the ancestor of these other organisms, but simply that the ancestral flagellates from which modern flagellates are descended probably also gave rise to other Protozoa and to multicellular plants and animals. This relationship might be diagramed as follows:

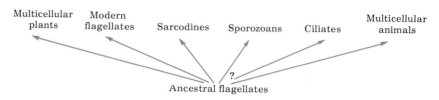

We discussed representatives of the plantlike photosynthetic flagellates in Chapter 21; we have as yet said nothing about the animal-like heterotrophic flagellates. A few of these are free-living in salt or fresh water, but most live as symbionts in the bodies of higher plants or animals. Several species, for instance, are found in the gut of termites, where they participate in the digestion of the cellulose consumed by the termite (Fig. 22.1); this is an example of mutualism.

Class Sarcodina. The Sarcodina are the amoeboid Protozoa. They are thought to be more closely related to the flagellates than to the other protozoan classes, because some flagellates undergo amoeboid phases, and, conversely, some Sarcodina have flagellated stages. The most familiar sarcodines are the naked fresh-water species of the genera *Amoeba* (Fig. 22.2) and *Pelomyxa,* which have asymmetrical bodies that constantly change shape as new pseudopodia are formed and old ones retracted. These pseudopodia, which are large and have rounded or blunt ends, function both in locomotion (see p. 235) and in feeding by phagocytosis (see p. 45 and Fig. 3.9; also p. 105). The food consists of small algae, other protozoans, and even some small multicellular animals such as rotifers and nematode worms.

Also included in the Sarcodina are several groups of protozoans that secrete shells around themselves, which, often quite elaborate and complex, can be used in species identification. The pseudopodia of the shelled Sarcodina are usually thin and pointed. Two groups, the Foraminifera and the Radiolaria, have played major roles in the geologic history of the earth. Both are extremely abundant in the oceans, and when the individuals die their shells become important components of the bottom mud. The shells of foraminiferans, which are calcareous, are especially prevalent in the mud at depths of 8,000 to 15,000 feet; the bottom ooze in deeper parts of the ocean is composed chiefly of the siliceous shells of radiolarians (Fig. 22.3). Much of the limestone and chalk now present on the earth was formed from deposits of foraminiferal shells, and radiolarian shells have contributed to the formation of siliceous rocks such as chert.

Class Sporozoa. The Sporozoa lack special locomotor organelles (except in the male gametes). They are parasites and usually exhibit complex life cycles. Among human diseases caused by sporozoans are coccidiosis and malaria. As you probably know, malaria, which is caused

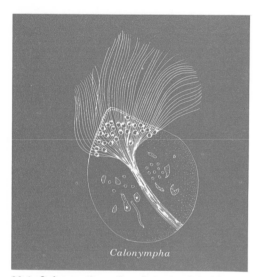

22.1. *Calonympha,* a flagellate that inhabits the gut of termites. [Redrawn from C. Janicki, *Z. Wiss. Zool.,* vol. 112, 1915.]

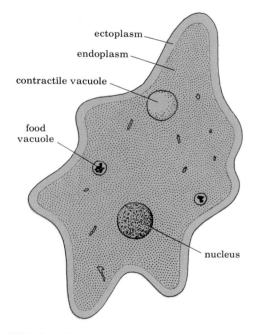

22.2. *Amoeba,* a common representative of the Sarcodina.

by species of the genus *Plasmodium,* is transmitted from host to host by female *Anopheles* mosquitoes.

Subphylum Ciliophora

Class Ciliata. The class Ciliata is the largest of the four protozoan classes and also the most homogeneous. Its relationship to the other classes is unclear, and it differs so markedly from them that it is put in a subphylum of its own.

As their name implies, ciliates possess numerous cilia as locomotor organelles (Pl. III-8A). In most species, the cilia are present throughout life, but in a few (Suctoria) they are absent in the adult stages. The ciliates exhibit the greatest elaboration of subcellular organelles of any Protozoa (see Pl. I-1), and it is to them that the term "acellular" applies best. As we saw in our discussion of *Paramecium* (p. 105), they may have a special oral groove and cytopharynx into which food particles are drawn in currents produced by beating cilia, and they often have an anal "pore" through which indigestible wastes are expelled from food vacuoles. Conductile fibrils connect the bases of the cilia, and there may

22.3. Some sarcodines with thin pointed pseudopodia. *Actinophrys* is a heliozoan. *Acanthometra* is a radiolarian. *Textularia* and *Globigerina* are foraminiferans. [Modified from L. H. Hyman, *The Invertebrates,* McGraw-Hill Book Co., 1940. Used by permission.]

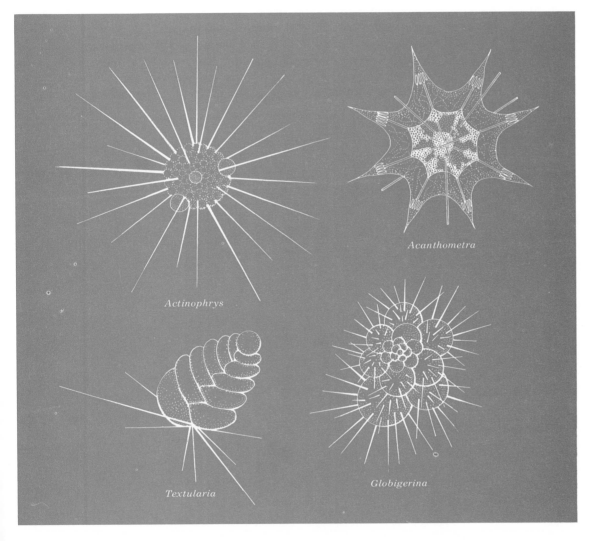

Acanthometra

Actinophrys

Textularia

Globigerina

be a system of contractile fibers (sometimes striated) analogous to the muscular system of multicellular animals. Stiffened plates occasionally found in the pellicle of the cell together constitute a "skeleton." In some species, there is a long stalk by which the individual may attach itself to the substratum (Fig. 22.4). A few species have tentacles for capture of prey. Some can discharge toxic threadlike darts called trichocysts that resemble the nematocysts of coelenterates; these may function in defense against predators, in capturing prey, or in anchoring the organism to the substratum during feeding.

Ciliates differ from all other members of the animal kingdom in having two quite different types of nuclei: a large **macronucleus** and one or more small **micronuclei** (see Fig. 5.6, p. 105). The macronucleus controls the normal metabolism of the cell. The micronuclei are concerned only with reproduction and with giving rise to the macronucleus.

PORIFERA (The sponges)

Sponges are aquatic, mostly marine animals. Although the larvae are ciliated and free-swimming, the adults, which tend to be colonial, are always sessile and are usually attached to rocks or shells or other submerged objects. They are multicellular, but show few of the features ordinarily associated with multicellular animals. For example, they have no digestive system, no nervous system,[1] and no circulatory system. In fact, they have no organs of any kind, and even their tissues are not well defined. They thus represent a very low grade of organization.

The body of a sponge is rather like a perforated sac. Its wall is

[1] In some sponges, a few cells of the body wall possess elongate processes that may have special conductile properties.

22.4. Some representative ciliates. Note the variation in shape and in the arrangement of organelles. *Vorticella* has a long stalk by which it attaches to the substratum. [Modified from L. H. Hyman, *The Invertebrates,* McGraw-Hill Book Co., 1940. Used by permission.]

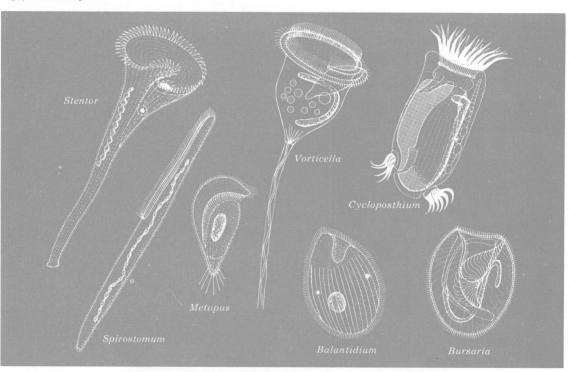

composed of three layers: an outer layer of flattened epidermal cells, a gelatinous middle layer with wandering amoeboid cells, and an inner layer of flagellated cells (Fig. 22.5). These last are unusual in that the base of the flagellum is encircled by a delicate collar; such cells are called *collar cells* (or choanocytes).

The wall of a sponge is perforated by numerous pores, each surrounded by a single pore cell. The beating of the flagella of the collar cells produces water currents that flow through the pores into the central cavity (spongocoel) and out through a larger opening (osculum) at the end of the body. Microscopic food particles brought in by these currents adhere to the collar cells and are engulfed; the food may be digested in food vacuoles of the collar cells themselves or passed to the amoeboid cells for digestion. The water currents also bring oxygen to the cells and carry away carbon dioxide and nitrogenous wastes (largely ammonia).

Sponges characteristically possess an internal skeleton secreted by the amoeboid cells. This skeleton is composed of crystalline *spicules* or proteinaceous fibers or both. The spicules are made of calcium carbonate or siliceous material; their chemical composition and shape are the basis for sponge classification. The fibrous skeletons of the bath sponges (*Spongia*) are cleaned and sold for many uses. A living bath sponge, which looks rather like a piece of raw liver, bears little resemblance to the familiar commercial object.

Among the free-living flagellated Protozoa are some collared organisms that closely resemble the collar cells of sponges. Such cells are found in no other organisms. For this reason, many biologists think that the Porifera evolved from collared flagellates. Other biologists disagree, pointing out that larval sponges have no collar cells, and suggest that sponges arose instead from a hollow free-swimming colonial flagellate. In either case, it seems very likely that sponges arose from the Protozoa independently of the other multicellular animals. The phylum Porifera would then stand as an evolutionary development entirely separate from the rest of the animal kingdom, and it must be concluded that multicellular animals evolved at least twice from the flagellates.

Because the Porifera differ so greatly from other multicellular animals, and probably arose independently, they are often regarded as constituting a separate subkingdom, the Parazoa.

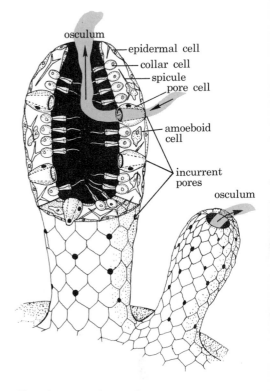

22.5. **Anatomy of part of a sponge colony.**

THE RADIATE PHYLA

The phyla Coelenterata and Ctenophora (of which the first will be discussed in more detail below) comprise radially symmetrical animals with bodies at a relatively simple grade of construction. These animals have definite tissue layers, but no distinct internal organs, no head, and no central nervous system, though they possess nerve nets (see Fig. 10.2, p. 203). There is a digestive cavity, but it has only one opening, which must serve as both mouth and anus; i.e. it is a gastrovascular cavity. Tentacles are usually present. There is no coelom or other internal space between the wall of the digestive cavity and the outer body wall.

It was once thought that the bodies of these animals consist of only two layers of cells—an outer epidermis (ectoderm) and an inner gastrodermis (endoderm)—but it is now known that a third layer called mesoglea (mesoderm) usually occurs between these two (see Fig. 5.8,

p. 107), just as it does in higher multicellular animals. However, this mesodermal layer is not as well developed in the radiate phyla as in higher phyla. Usually gelatinous, it has a few scattered cells, which may be amoeboid or fibrous.

Coelenterata

The phylum Coelenterata contains a variety of aquatic organisms, among which are the hydras mentioned so frequently in earlier chapters, jellyfishes, sea anemones, and corals. The hydras live in fresh water, but most other coelenterates are marine.

The coelenterate body shows some cell specialization and division of labor. Thus the outer epidermis contains sensory–nerve cells, gland cells, special cells that produce nematocysts, small interstitial cells, and epitheliomuscular cells. These last are the main structural elements of the epidermis and consist of a columnar cell body with several contractile basal extensions. In other words, the contractile elements of coelenterates are parts of cells that also have other important functions; there are no cells specialized exclusively for contraction—no separate muscle cells. Note also that these contractile elements are ectodermal, not mesodermal as in most higher animals. The gastrodermis also contains contractile elements; these are basal extensions of cells whose cell bodies constitute the bulk of the lining of the gastrovascular cavity and function in digestion. Here again a single cell performs two functions that in higher animals are performed by separate elements. In short, there is some division of labor among cells in coelenterates, but it is never as complete as in most bilateral multicellular animals; and most functions performed by tissues derived from mesoderm in other animals are performed by ectodermal or endodermal cells in coelenterates.

The phylum Coelenterata is divided into three classes: Hydrozoa, Scyphozoa, and Anthozoa.

Class Hydrozoa. The best-known members of this class are the fresh-water hydras. We have discussed their feeding (p. 106), gas exchange (p. 120), nervous control (p. 202), and locomotion (p. 236). Little more need be said about them here. In many ways, however, hydras are not typical members of their class. Many hydrozoans are colonial and have a complex life cycle, in which a sedentary hydralike *polyp* stage alternates with a free-swimming jellyfishlike *medusa* stage (Fig. 22.6, Pl. I-7C). By contrast, hydras are solitary and have only a polyp stage (which is not completely sedentary).

Let us examine *Obelia* as an example of a colonial hydrozoan (Fig. 22.7). Much of the life of *Obelia* is passed as a sedentary branching colony of polyps. The colony arises from an individual hydralike polyp by asexual budding; the buds fail to separate, and the new polyps remain attached by hollow stemlike connections. The gastrovascular cavities of all the polyps are interconnected via the cavity in the stems. The cells lining the stem cavity have long flagella that circulate the fluid in the cavity. Partly digested food can be passed from one polyp to another in this moving fluid. Both the stems and the polyps (except the mouth and tentacles) are enclosed in a hard chitinous case secreted by the ectoderm. Rings or joints in the case at intervals along the stems permit some flexibility for the colony.

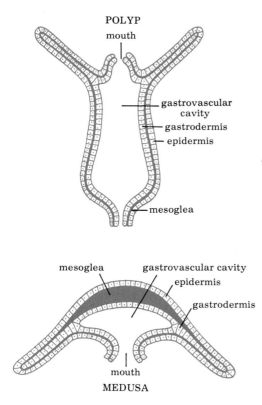

22.6. Diagram contrasting polyp and medusa. The basic structure of these two forms is the same. A medusa is like a flattened polyp turned upside down.

A mature *Obelia* colony consists of two kinds of polyps: feeding polyps with tentacles and nematocysts, and reproductive polyps without tentacles.[2] The reproductive polyps regularly bud off tiny transparent free-swimming medusas. Each medusa is umbrella- or bell-shaped, with numerous tentacles hanging from the margin of the bell. A tube with a mouth at its end hangs from the middle of the undersurface of the bell (where the clapper of a real bell would be). The medusas are the dispersal and sexual stage in the life cycle. They swim feebly by alternately contracting and relaxing the contractile cells in the bell, but much of their movement results from drifting with the water currents.

Certain cells in mature medusas undergo meiosis and develop into either sperms or eggs, which are released into the surrounding water. Fertilization takes place, and the resulting zygote develops into a hollow blastula. Gastrulation does not take place by the invagination process described in Chapter 16. Instead, endodermal cells proliferated by the wall (ectoderm) of the blastula wander into the blastocoel until they completely fill it. This solid gastrula then develops into an elongate ciliated larva called a *planula.* The planula eventually settles to the bottom, attaches by one end to some object, and develops a mouth and tentacles at the other end, becoming a polyp that gives rise to a new colony. The life cycle is thus completed. Note that the alternation of polyp and medusa stages in a coelenterate like *Obelia* differs from the alternation of generations in plants in that both polyp and medusa are diploid; as in all multicellular animals, the only haploid stage in the life cycle is the gametes.

[2] In view of this division of labor—the reproductive polyps are nourished with food captured by the feeding polyps and passed to them through the common gastrovascular cavity in the connecting stems—and in view of the structural continuity between the polyps, it might be argued that the so-called colony is not really a colony, but a complex individual.

22.7. Life cycle of *Obelia*, a colonial hydrozoan. Since the medusas are of separate sexes, the eggs and sperms are produced by different individuals. See text for description.

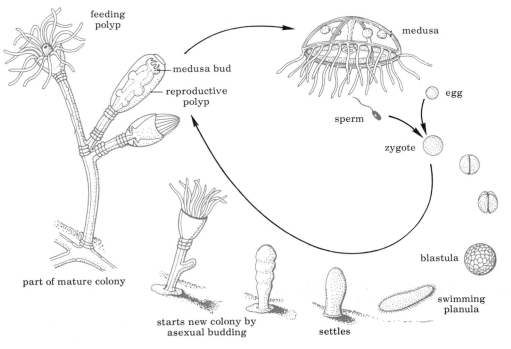

feeding polyp

medusa bud

reproductive polyp

part of mature colony

starts new colony by asexual budding

settles

medusa

sperm

egg

zygote

blastula

swimming planula

Class Scyphozoa. The scyphozoans are the true jellyfishes. In these animals, the medusa is the dominant and conspicuous stage in the life cycle, and the polyp stage is restricted to a small larva. This larva, which develops from the planula, promptly produces medusas by budding (Fig. 22.8). Scyphozoan medusas resemble the hydrozoan medusas already described except that they are usually much larger and have long oral arms (endodermal tentacles) arising from the margin of the mouth; the marginal tentacles may be much reduced as in *Aurelia* (shown in Fig. 22.8), or they may be quite long.

Class Anthozoa. The class Anthozoa includes sedentary polypoid forms such as sea anemones (Pls. II-3C, III-8B), sea fans, and corals (Pl. I-8A). All are marine. There is no trace of a medusa stage in their life cycle. They are the most advanced members of the Coelenterata, and their body structure is much more complex than that of simple polyps like the hydras. They possess a tubular pharynx leading into the gastrovascular cavity, which is divided into numerous radiating compartments by longitudinal septa; their mesoderm (mesoglea) is much thicker than that of other coelenterates and is often elaborated into a fibrous connective tissue; and their muscles are much better developed.

The corals, anthozoans that secrete a hard limy skeleton, have played a very important role in the geologic history of the earth, particularly in tropical oceans. As their skeletons have accumulated over the ages, they have formed many reefs, atolls, and islands, especially in the South Pacific.

Origin of the Metazoa

Speculation concerning the origin of the Metazoa (all multicellular animals except the Porifera, or sponges, and another group of extremely

22.8. Life cycle of *Aurelia*, a jellyfish. The polyps are shown much enlarged. Since the medusas are of separate sexes, the eggs and sperms are produced by different individuals. See text for description.

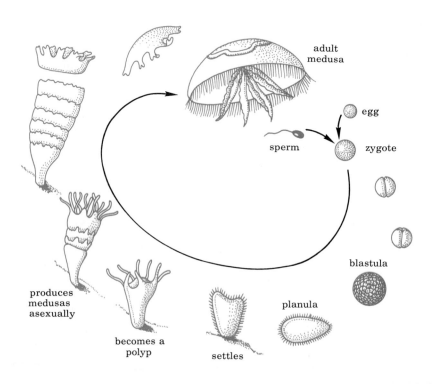

simple multicellular organisms, the Mesozoa) has long centered on the radiate phyla—not only because the radiates, particularly the hydrozoan coelenterates, seem to be among the simplest metazoans, but also because their saclike, essentially two-layered bodies strikingly resemble embryonic gastrulas. Now, a spherical colonial flagellate like *Volvox* certainly resembles an embryonic blastula. As long ago as 1874 Ernst Haeckel suggested that the ancestor of multicellular animals was a hollow-sphere colonial flagellate,[3] and that the coelenterates arose from this hypothetical ancestor, called a **blastaea,** by a process of invaginating gastrulation. The higher animals would then have arisen from the early two-layered *gastraea* ancestor of the modern coelenterates by assuming a creeping mode of life and slowly becoming bilateral. According to this hypothesis, the blastula and gastrula stages in the embryonic development of higher animals are recapitulations of early steps in the evolution of these animals from the Protozoa.

It was soon pointed out that gastrulation in coelenterates does not take place by invagination. Instead, as we have already seen, the endoderm arises by inwandering of cells produced at the inner surface of the ectoderm, and a solid gastrula, which develops into a planula larva, is formed. Both the hollowing out of the interior to form the gastrovascular cavity and the breaking through of a mouth occur later, when the larva develops into a polyp. Thus it seems likely that the invagination type of gastrulation was a later evolutionary development and not the original method of formation of endoderm. Consequently, although the idea of a hollow blastaea ancestor as the starting point was retained in later versions of Haeckel's hypothesis, the idea of a gastraea stage was abandoned and a *planuloid* stage hypothesized instead.

The idea of metazoan origin from flagellates via blastaea and planuloid steps is widely accepted among biologists today, but other hypotheses have been proposed, of which the most prominent is that multicellular animals arose from multinucleate ciliates, not from flagellates. According to this hypothesis, the bodies of the earliest metazoans were coenocytic; i.e. they contained many nuclei, each controlling the cytoplasm around it, but had no cellular partitions. Formation of cellular membranes around each nucleus and its associated cytoplasm would have produced a typical multicellular animal, probably resembling some of the simplest flatworms. Note that this hypothesis begins with a bilaterally symmetrical[4] ciliate and hence assumes that the first metazoans were bilateral; it considers the radial symmetry of the Coelenterata a secondary characteristic.

THE ACOELOMATE BILATERIA

There are two phyla—the Platyhelminthes and the Nemertina—that contain what biologists generally regard as the most primitive bilaterally symmetrical animals. In both phyla, the body is composed of three well-developed tissue layers—ectoderm, mesoderm, and endoderm

[3] The colonial flagellate ancestral to multicellular animals would probably have differed from *Volvox* in having cells without walls and without chlorophyll.

[4] Bilateral symmetry is the property of having two similar sides. A bilaterally symmetrical animal has definite dorsal (upper) and ventral (lower) surfaces and definite anterior (head) and posterior (tail) ends.

—and the mesoderm is a solid mass that fills what once was the embryonic blastocoel. In other words, there is no coelom—no cavity between the digestive tract and the body wall (see Fig. 22.16A). For this reason, these two phyla are known as the acoelomate bilateria.

Platyhelminthes (The flatworms)

The flatworms, as their name implies, are dorsoventrally flattened, elongate animals.[5] Their digestive cavity (not always present) resembles that of coelenterates; i.e. it is a gastrovascular cavity—a cavity with a single opening that must serve as both mouth and anus. However, there is a muscular pharynx leading into the cavity, and the cavity itself is often profusely branched, especially in the free-living species (see p. 107 and Fig. 5.9). As in coelenterates, the amount of extracellular digestion is limited, most of the food particles being phagocytized and digested intracellularly by the cells of the wall of the gastrovascular cavity. Respiratory and circulatory systems are absent. However, there is a flame-cell excretory system (see p. 157 and Fig. 8.5), and there are well-developed reproductive organs (usually both male and female in each individual). That both the excretory system, with its flame bulbs and tubules, and the reproductive organs should be present signifies that the flatworms have advanced beyond the tissue level of construction seen in the radiate phyla to an organ level of construction. The more extensive development of mesoderm, leading to greater division of labor, was probably a major factor in making this advance possible. Mesodermal muscles are well developed. Several longitudinal nerve cords running the length of the body and a tiny "brain" ganglion located in the head constitute a central nervous system (see p. 203 and Fig. 10.3).

The phylum is divided into three classes: Turbellaria, Trematoda, and Cestoda. The last two are entirely parasitic.

Class Turbellaria. The members of this class, of which the fresh-water planarians often mentioned in earlier chapters are examples, are free-living flatworms ranging from microscopic size to a length of several inches. The body is clothed by an epidermal layer, which is usually ciliated (at least in part). Although a few turbellarians live on land, most are aquatic (the majority marine).

Turbellarians usually have a gastrovascular cavity, but most members of one small order, the Acoela, do not (Fig. 22.9). For a variety of reasons (not just the absence of a digestive cavity), some biologists consider the Acoela the most primitive bilateral animals, and suggest that a primitive *acoeloid* organism might well have arisen from a planuloid ancestor. According to these biologists, both the more complex flatworms and the other metazoan phyla probably evolved from such an acoeloid organism. This account of the early evolution of the Metazoa can be diagramed as shown at left.

Proponents of the ciliate hypothesis of metazoan origin also assume an acoeloid stage at the root of the Metazoa, but they derive the acoeloid directly from a multinucleate ciliate, not from a planuloid. They point out that acoels are about the same size as some ciliates, are ciliated, and

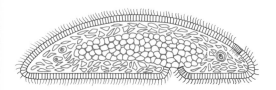

22.9. An acoel flatworm. There is a ventral mouth but no digestive cavity. The entire interior of the animal is filled with an almost solid mass of tissue (color).

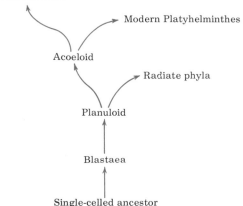

[5] The term "worm" is applied to a great variety of unrelated animals. It is a descriptive, not a taxonomic, term that denotes possession of a slender elongate body, usually without legs or with very short ones.

sometimes have poorly developed cellular partitions. They think the primitive acoeloid organism was the ancestor of the radiate phyla as well as of the bilateral phyla. Their ideas can be diagrammed as shown at right.

Class Trematoda (The flukes). The flukes are parasitic flatworms. They lack cilia, and in place of the cellular epidermis of their turbellarian ancestors they have a thick cuticle secreted by the cells below. This cuticle is highly resistant to enzyme action and is thus an important adaptation for a parasitic way of life. Flukes characteristically possess suckers, usually two or more, by which they attach themselves to their host (Fig. 22.10). They have a two-branched gastrovascular cavity, which does not ramify throughout the body like that of turbellarians. Much of their volume is occupied by reproductive organs, including two or more large testes, an ovary, a long much-coiled uterus in which eggs are stored prior to laying, and yolk glands.

The members of one order of flukes are ectoparasites (external parasites) on the gills or skin of fresh-water and marine fishes. A few of these flukes sometimes wander into the body openings of their hosts, and it is probably from such a beginning that the endoparasitism (internal parasitism) typical of the members of the other two orders arose.

The endoparasitic flukes frequently have very complicated life cycles involving two to four different hosts. The blood fluke, *Schistosoma japonicum,* common in China, Japan, Taiwan, and the Philippines, is a species with two hosts. The adult worm inhabits blood vessels near the intestine of a human being. When ready to lay its eggs, it pushes its way into one of the very small blood vessels in the wall of the intestine. There it deposits so many eggs that the vessel ruptures, discharging the eggs into the intestinal cavity, whence they are carried to the exterior in the feces. If there is a modern sewage system, that is the end of the story. But in many Asiatic countries, human feces are regularly used as fertilizer. Thus the eggs get into water in rice fields, irrigation canals, or rivers, where they hatch into tiny ciliated larvae. A larva swims about until it finds a snail of a certain species; it dies if it cannot soon find the correct species. When it finds such a snail, it bores into the body of the snail and feeds on its tissues. It then reproduces asexually, and the new individuals thus produced leave the snail and swim about until they come in contact with the skin of a human being, such as a farmer wading in a rice paddy or a boy swimming in a pond. They attach themselves to the skin and digest their way through it and into a blood vessel. Carried by the blood to the heart and lungs, they eventually reach the vessels of the intestine, where they settle down, mature, and lay eggs, thus initiating a new cycle. Schistosomas in the human body cause a serious disease called schistosomiasis, which is characterized initially by a cough, rash, and body pains, followed by severe dysentery and anemia. The disease so saps the strength of its victims that they become weak and emaciated and often die of other diseases to which their weakened condition makes them susceptible.

Class Cestoda (The tapeworms). Adult tapeworms (Fig. 22.11) live as internal parasites of vertebrates, almost always in the intestine. However, the life cycle usually involves one or two intermedi-

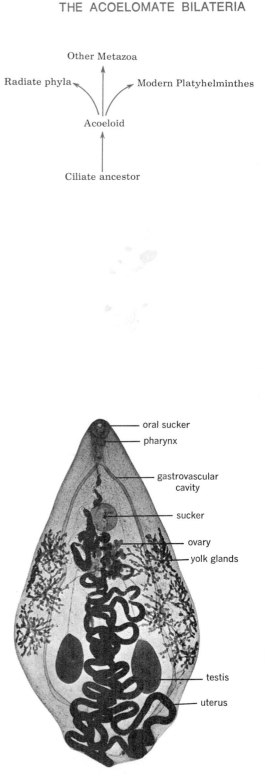

22.10. *Prosthogonimus macrorchis,* a fluke that parasitizes the oviducts of the domestic hen. [Courtesy General Biological Supply House, Inc., Chicago.]

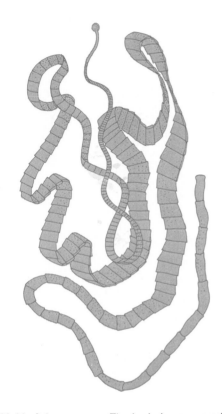

22.11. A tapeworm. The body is composed of a small head and neck, followed by a large number of segments called proglottids. As the proglottids ripen, they break off and pass with the host's feces to the outside. New proglottids are produced just back of the neck.

ate hosts, which may be invertebrate or vertebrate, depending on the species. The life cycle of the beef tapeworm, in which the intermediate host is a cow and the final host human, was outlined on p. 414.

Tapeworms exhibit many special adaptations for their parasitic way of life. Like the flukes, they have a resistant cuticle instead of the epidermis of their free-living ancestors. And they have neither mouth nor digestive tract. Bathed by the food in their host's intestine, they absorb predigested nutrients across their general body surface. Diffusion, probably augmented by active transport, suffices to provision all the cells, because none are far from the surface.

The head of a tapeworm is a small knoblike structure called a *scolex,* which usually bears suckers, and often also hooks, by which the worm attaches itself to the wall of the host's intestine (Fig. 22.12, right). Immediately behind the scolex is a neck region, which is followed by a very long ribbonlike body (beef tapeworms occasionally grow to 75 feet, fish tapeworms to 60 feet, and pork tapeworms to 25 feet). This long body is usually divided by transverse constrictions into a series of segments called *proglottids* (Fig. 22.12, left).

Each proglottid is essentially a reproductive sac, containing both male and female organs. Sperm cells, usually from a more anterior proglottid of the same animal, enter the genital pore and fertilize the egg cells, which are then combined with yolk from a yolk gland and enclosed in a shell. The fertilized eggs, already undergoing development, are stored in a uterus, which may become so engorged with eggs that it occupies most of the volume of a mature proglottid. Eventually all the sexual organs except the uterus degenerate, and the proglottid, which is now "ripe," detaches from the worm and passes out of the host's body with the feces. As ripe proglottids are released from the end of the worm, new ones are produced just back of the neck. A single ripe proglottid may contain more than 100,000 eggs, and the annual output of one worm may be more than 600 million.

If an appropriate intermediate host eats food contaminated with feces containing tapeworm eggs, its enzymes digest the shells of the eggs. The embryos thus released bore through the wall of the host's intestine, enter a blood vessel, and are carried by the blood to the muscles, where they encyst. If a human being eats the raw or "rare" meat of this intermediate host (e.g. beef, pork, or fish), the walls of the cysts are digested away; the young tapeworms attach themselves to the intestinal wall and, nourished by an abundant supply of food, begin to grow and produce eggs, thus starting a new cycle.

Nemertina (The proboscis worms)

The members of the phylum Nemertina are long slender worms characterized by a very long eversible muscular proboscis enclosed in a tubular cavity at the anterior end of the body. This proboscis, used in capturing prey and also in defense, is often two or more times the length of the worm's body and is somewhat coiled when enclosed in its sheath. The worms are common along both the Atlantic and Pacific shores of the United States. They are usually found sheltered under stones, shells, or seaweeds, or burrowing in the sand or mud in shallow water.

Although nemertines resemble turbellarian flatworms (their probable ancestors) in many ways, they differ from them in two important characteristics not encountered in the animals considered thus far. First, they have a *complete digestive system*—one that has two openings, a mouth and an anus. Such a system makes possible specialization of sequentially arranged chambers for different functions and thus permits an assembly-line processing of food, as we saw in Chapter 5. Second, they have a simple blood circulatory system, which presumably facilitates transport of materials from one part of the body to another.

DIVERGENCE OF THE PROTOSTOMIA AND DEUTEROSTOMIA

We saw in Chapter 16 that the embryonic cavity called the archenteron, formed during gastrulation, becomes the digestive tract of an adult animal. But the archenteron has only one opening to the outside, the blastopore. In animals like coelenterates and flatworms, where the digestive tract is a gastrovascular cavity, the blastopore becomes the combined mouth and anus. But in nemertine worms and the other higher animals that have complete digestive systems, does the blastopore become the mouth, or does it become the anus? Embryologists have shown that in nemertines the site of the embryonic blastopore becomes the mouth and that the anus is an entirely new opening. This is also the case in many other animals, including nematode worms, molluscs, and annelids. But in a few phyla, among them two large and important ones—the Echinodermata and the Chordata—the situation is reversed: The embryonic blastopore becomes the anus, and the mouth is the new opening.

This fundamental difference in embryonic development suggests that a major split occurred in the animal kingdom soon after the origin of a bilateral ancestor. One evolutionary line led to all the phyla in which the blastopore becomes the mouth; these phyla are often called the

22.12. Proglottids and scolex of *Taenia pisiformis*, the dog tapeworm. Left: Mature proglottids. Right: Scolex. Note the hooks and the suckers. [Courtesy General Biological Supply House, Inc., Chicago.]

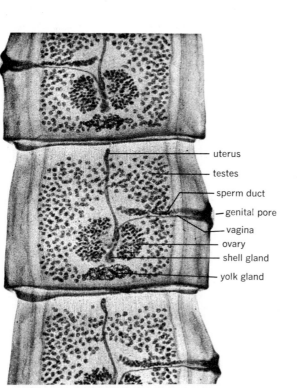

— uterus
— testes
— sperm duct
— genital pore
— vagina
— ovary
— shell gland
— yolk gland

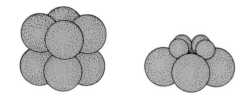

22.13. Radial and spiral cleavage patterns.
Left: Radial cleavage, characteristic of deuterostomes. The cells of the two layers are arranged directly above each other. Right: Spiral cleavage, characteristic of protostomes. The cells in the upper layer are located in the angles between the cells of the lower layer.

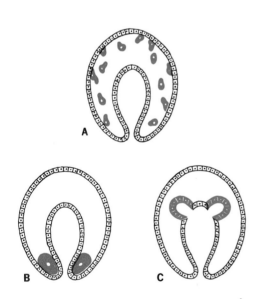

22.14. Different modes of origin of mesoderm.
(A) In the radiate phyla, mesoderm (colored cells) arises from inwandering cells derived from the ectoderm. A small amount of the mesoderm of the protostome phyla arises in this way also. (B) In most protostome phyla, the bulk of the mesoderm arises from initial cells located near the blastopore, at the junction between the ectoderm and the endoderm. (C) In the deuterostome phyla, the mesoderm arises as pairs of pouches from the endodermal wall of the archenteron.

Protostomia (from the Greek *protos,* first, and *stoma,* mouth). The other evolutionary line led to the phyla in which the blastopore becomes the anus and a new mouth is formed; these phyla are called the **Deuterostomia** (from the Greek *deuteros,* second, later, and *stoma*).

As might be expected if the Protostomia and Deuterostomia diverged at a very early stage of their evolution, they differ in a number of other fundamental characters besides the mode of formation of mouth and anus. A further essential difference between them is that the early cleavage stages are determinate in protostomes and indeterminate in deuterostomes; i.e. the developmental fates of the first few cells of a protostome embryo are already at least partly determined, and if these cells are separated no one of them can form a complete individual, whereas the first few cells of a deuterostome embryo are not determined, and each cell, if separated, can develop into a normal individual. Furthermore, the two groups exhibit strikingly different patterns of cleavage; the early cleavages in protostomes are usually oblique to the polar axis[6] of the embryo and thus give rise to a spiral arrangement of cells, whereas the early cleavages in deuterostomes are either parallel or at right angles to the polar axis and thus give rise to a so-called radial arrangement of cells (Fig. 22.13). The basic larval types are also different in the two groups, as we shall see later.

Another fundamental difference between the protostome and deuterostome phyla is seen in the method of origin of the mesoderm in the embryo. The mesoderm of the radiate phyla arises from inwandering cells derived from the ectoderm. A small amount of mesoderm forms this way in the protostomes also, though not in the deuterostomes. Most of the mesoderm in protostomes and all of it in deuterostomes is derived from endoderm instead of ectoderm. However, in protostomes this mesoderm arises as a solid ingrowth of cells from a single initial cell located near the blastopore (Fig. 22.14), whereas in deuterostomes (vertebrates excepted) it arises by a saclike outfolding of the gut wall, as we saw in Chapter 16.[7]

Still another difference, correlated with the preceding one, has to do with the method of formation of the coelom, if one is present. A true coelom is defined as a cavity enclosed entirely by mesoderm and located between the digestive tract and the body wall. In the coelomate protostomes, this cavity usually arises as a split in the initially solid mass of mesoderm. In the deuterostomes, by contrast, the coelom arises as the cavity in the mesodermal sacs as they evaginate from the wall of the archenteron.[8]

To summarize, the Protostomia and the Deuterostomia differ most conspicuously in the following aspects of their development: the fate of the embryonic blastopore; the determinateness and pattern of the initial cleavages; the mode of origin of the mesoderm and of the coelom (if one is present); and type of larva.

---

[6] The polar axis runs between the animal and vegetal poles.

[7] There are actually a variety of other ways in which mesoderm may arise from endoderm, but embryologists usually interpret them as variants of one or the other of the two processes described here.

[8] A coelom that arises as a split in an initially solid mass of mesoderm is called a schizocoelom. One that forms as the cavity in a pouch of mesoderm is called an enterocoelom. The coelomate protostomes are sometimes called the schizocoelous phyla, and the deuterostomes the enterocoelous phyla.

One way of diagraming the relationships among the animal phyla, taking into consideration the split between the protostomes and the deuterostomes, is shown in Fig. 22.15. This figure, set up in the traditional fashion of a phylogenetic tree, shows only one of many possible interpretations of evolutionary relationships.

THE PSEUDOCOELOMATE PROTOSTOMIA

In several protostome phyla, the body cavity is functionally analogous to a coelom but differs from a true coelom, which is entirely enclosed by mesoderm, in being partly bounded by ectoderm and endo-

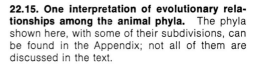

22.15. One interpretation of evolutionary relationships among the animal phyla. The phyla shown here, with some of their subdivisions, can be found in the Appendix; not all of them are discussed in the text.

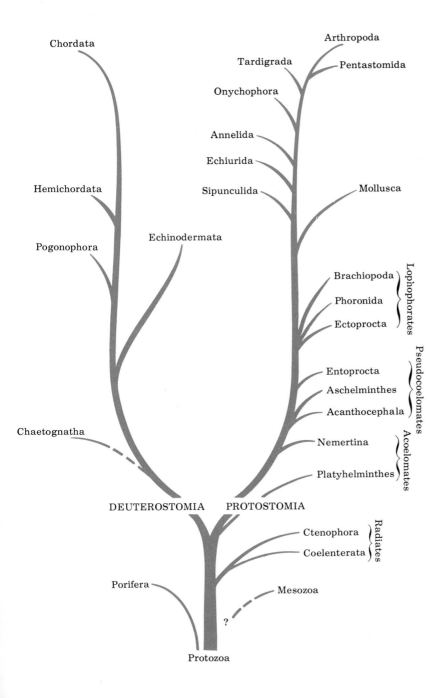

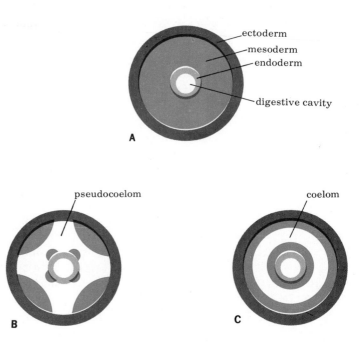

22.16. Diagrams of acoelomate, pseudocoelomate, and coelomate body types.
(A) Acoelomate body. There is no body cavity, the entire space between the ectoderm and endoderm being filled by a solid mass of mesoderm.   (B) Pseudocoelomate body. There is a functional body cavity, but it is not entirely bounded by mesoderm.   (C) Coelomate body. The body cavity is completely bounded by mesoderm.

derm. Such a cavity is called a *pseudocoelom* (Fig. 22.16). It is
actually only the remnant of the embryonic blastocoel.

The largest and most important of the three phyla of pseudocoelomate protostomes is the phylum Aschelminthes.

Aschelminthes

The Aschelminthes are generally small, wormlike animals that lack
a definitely delimited head and have a straight or slightly curved complete digestive tract and a cuticle. There is no respiratory or circulatory
system. A flame-cell excretory system occurs in most classes but not in
nematodes, which have a special type of excretory system unique with
them.

The phylum is divided into five classes, each recognized as a separate phylum by some biologists. We shall discuss only two: Rotifera and
Nematoda.

Class Rotifera.   The rotifers—or wheel animalcules, as they are
commonly called—are microscopic, usually free-living aquatic animals
with a crown of cilia at the anterior end (Fig. 22.17). The cilia are
generally arranged in a circle, and when beating they often give the
appearance of a rotating wheel; hence the name of the class. When
feeding, rotifers attach themselves by a tapering posterior "foot," and
the beating cilia draw a current of water into the mouth. In this manner, very small protozoans and algae are swept into a complicated
muscular pharynx, where they are ground up by seven hard jawlike
structures.

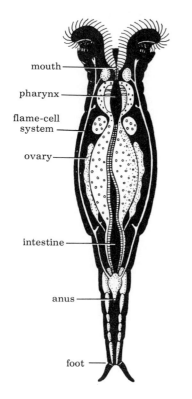

22.17. Section of a rotifer.

The fresh-water rotifers are extremely abundant. Anyone examining a drop of water for Protozoa under a microscope is likely to see one or more of these interesting animals. In fact, most of them are no larger than protozoans, and it is often difficult, when encountering them for the first time, to realize that they are multicellular.

Class Nematoda (The roundworms). Nematode worms have round elongate bodies that usually taper nearly to a point at both ends. Unlike flatworms, they have no cilia. The body is enclosed in a tough cuticle. Just under the epidermal layer of the body wall are bundles of longitudinal muscles; there are no circular muscles. The lack of circular muscles and the stiff cuticle severely limit the types of movements possible for the worms, and they usually thrash about in what appears to be a random and inefficient manner.

Nematodes are extremely abundant, and occur in almost every type of habitat. Of the many free-living in soil or water, most are very tiny, often microscopic. A single spadeful of garden soil may contain a million or more, and a bucket of water from a pond usually contains comparable numbers. Many other nematodes are internal parasites of both plants and animals; these also are often small, but some may attain a length of 3 feet or more.

Nematodes parasitic on cultivated plants cause an annual loss of millions of dollars. Others parasitic on human beings cause some serious diseases. *Trichinella spiralis,* for example, causes the disease called trichinosis, often contracted by eating insufficiently cooked pork. Adult *Trichinella* worms inhabit the small intestine of numerous species of mammals, among them hogs. Impregnated females bore through the wall of the intestine and deposit young larvae (which hatched from the eggs while still inside the uterus of the female) in the lymphatic vessels of their host. The larvae are carried by the lymph and blood to all parts of the body. They then bore out of the vessels, eventually entering every organ and tissue. However, only those that bore into skeletal muscles (especially the muscles of the diaphragm, ribs, tongue, and eyes) survive. In the muscles, they grow in size (to about one millimeter) and then curl up and encyst (Fig. 22.18); the thick wall of the cyst is formed by the host's tissues. If insufficiently cooked pork containing such cysts is eaten by a human being, the walls of the cysts are digested away and the worms complete their development in the man's intestine. The adult worms then deposit larvae in the lymph vessels in the wall of the intestine, and the larvae move through the human body as they do in hogs, eventually encysting in muscles.

Most of the damage of trichinosis occurs during the migration of the larvae, when half a billion or more may simultaneously bore through the body after one infection. Symptoms include excruciating muscular pains, fever, anemia, weakness, and sometimes localized swellings. Some victims die, and those that do not may sustain permanent muscular damage. Prevention of the disease is simple: Pork must be thoroughly cooked to kill the encysted larvae. It is well to remember that one ounce of infected pork in the center of a large chunk of meat where heat does not reach it may easily contain as many as 100,000 encysted worms, each of which, when mature, may produce 1,500 young larvae in the body of a new host.

22.18. Photograph of *Trichinella spiralis* encysted in muscle. [Courtesy Bausch and Lomb, *Focus.*]

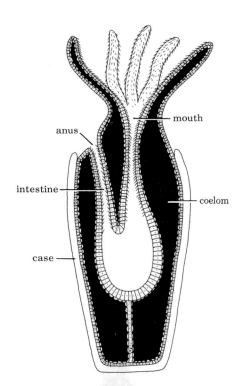

mouth

anus

intestine

coelom

case

**22.19. Section through the body of an ecto-
proct, showing the U-shaped digestive tract.**
The anus is located outside the ring of tentacles.

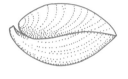

22.20. Brachiopods. The shells are reminiscent of old Roman lamps. The animals usually
attach themselves to the substratum by means
of a stalk (right).

THE COELOMATE PROTOSTOMIA

All the protostome phyla except the ones discussed above possess true coeloms that in most groups arise as a split in an initially solid mass of mesoderm. All have a complete digestive tract, and most have well-developed circulatory, excretory, and nervous systems.

The lophophorate phyla

There are three small phyla, Phoronida, Ectoprocta (Fig. 22.19), and Brachiopoda (Fig. 22.20), that resemble one another in having a *lophophore*—a fold, usually horseshoe-shaped, that encircles the mouth and bears numerous ciliated tentacles. The lophophore is a feeding device; its tentacular cilia create water currents that sweep plankton and tiny particles of detritus into a groove leading to the mouth.

All members of the lophophorate phyla are aquatic, and most are marine. Adults are usually sessile and secrete a protective case, tube, or shell around themselves, but the larvae are ciliated and free-swimming. The digestive tract is U-shaped in Phoronida and Ectoprocta and in some Brachiopoda; the anus lies outside the crown of tentacles.

Mollusca

The phylum Mollusca is the second-largest in the animal kingdom. Among its best-known members are snails and slugs, clams and oysters, squids and octopuses.

The various groups of molluscs may differ considerably in outward appearance (Fig. 22.21), but most have fundamentally similar body plans. The soft body consists of three principal parts: (1) a large ventral muscular *foot,* which can be extruded from the shell (if one is present) and functions in locomotion; (2) a *visceral mass* above the foot, which contains the digestive system, the excretory organs (nephridia), the heart, and other internal organs; and (3) a heavy fold of tissue called the *mantle,* which covers the visceral mass and which in most species contains glands that secrete a shell. The mantle often overhangs the sides of the visceral mass, thus enclosing a *mantle cavity,* which often accommodates gills (see Fig. 22.23).

Molluscs have an open circulatory system; i.e. during part of each circuit the blood is in large open sinuses where it bathes the tissues directly. Blood drains from the sinuses into vessels that run into the gills, where the blood is oxygenated. From the gills, the blood goes to the heart, which pumps it into vessels that lead it back to the sinuses; a typical circuit, then, is heart–sinuses–gills–heart.

Most marine molluscs pass through one or more ciliated free-swimming larval stages, but fresh-water and land snails complete the corresponding developmental stages while still in the egg and hatch as miniature editions of the adult.

The Mollusca are customarily divided into six classes: Amphineura, Monoplacophora, Gastropoda, Scaphopoda, Pelecypoda, and Cephalopoda.

Class Amphineura. Exclusively marine animals, the amphineurans are generally regarded as the most primitive living members of the Mollusca, and are thus a suitable group with which to begin study of the basic molluscan body plan.

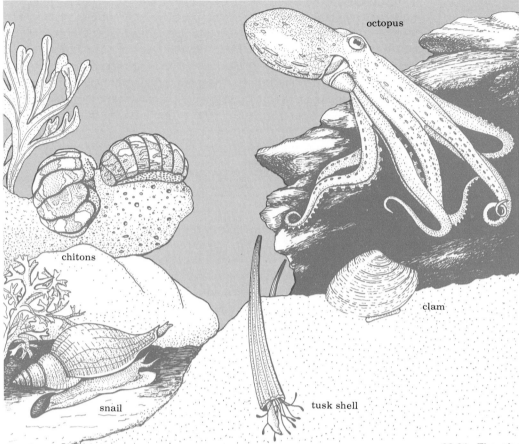

The best-known Amphineura are the chitons, which have an ovoid bilaterally symmetrical body with an anterior mouth and a posterior anus (Fig. 22.22). The coelom is reduced to a small cavity surrounding the heart.[9] The shell consists of eight serially arranged dorsal plates.

Chitons lead a sluggish, nearly sessile life. They creep about on the surface of rocks in shallow water, rasping off fragments of algae with a horny toothed organ called a *radula.* Their broad flat foot can develop tremendous suction, and when disturbed they clamp down so tenaciously to the rock that they can hardly be pried loose.

[9] The lumina of the gonads and nephridia are also thought to be remnants of the coelom.

22.21. The classes of Mollusca. Chitons are members of the Amphineura. Snails are in the class Gastropoda. Clams are members of the Pelecypoda. Tusk shells belong to the Scaphopoda. The octopus is a representative of the Cephalopoda. (Cutaway at lower right shows clam and tusk shell partly buried in a ridge of sand.)

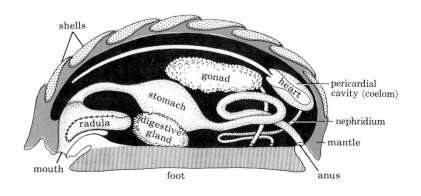

22.22. Lateral view of a section of a chiton.

Class Monoplacophora.   Members of this class, long known as fossils, had been assumed to be extinct for about 350 million years when, in 1952, ten living specimens (genus *Neopilina*) were dredged from a deep trench in the Pacific Ocean off the coast of Costa Rica. These specimens sparked a lively debate on the ancestry of the Mollusca, because they showed some internal segmentation, a characteristic seen in no other members of the phylum. Since it was already known that the early cleavage pattern and larval type of molluscs show striking similarities to the corresponding developmental stages in the segmented worms (Annelida), the segmentation of *Neopilina* led many biologists to conclude that the ancestral molluscs were segmented animals, perhaps primitive annelids. Other biologists, however, are convinced that the segmentation of *Neopilina* is secondary, not primitive, and that the original molluscan body was unsegmented. Whichever view is correct, it seems clear that the Mollusca and the Annelida are fairly closely related.

Class Gastropoda (The snails and their relatives).   Most gastropods have a coiled shell. In some cases, however, the coiling is minimal. Some species—e.g. the nudibranchs—have lost the shell (see Pl. I-6B).

The early larva in gastropods is bilateral, but as the animal develops, the digestive tract bends downward and forward until the anus comes to lie close to the mouth. Then the entire visceral mass rotates through an angle of 180 degrees, so that it comes to lie dorsal to the head in the anterior part of the body. Most of the visceral organs on one side (usually the left) atrophy, and growth proceeds asymmetrically, producing the characteristic spiral.

Except for the peculiar twisting and coiling of their body, gastropods are thought to be rather like the ancestral molluscs. They have a distinct head with well-developed sense organs. Most have a well-developed radula and feed on bits of plant or animal tissue that they grate or brush loose with this organ.

Gastropods occur in a great variety of habitats. The majority are marine, and their often large and decorative shells are among the most prized finds on a beach, but there are also many fresh-water species and some that live on land. The land snails are one of the few groups of fully terrestrial invertebrates. In most of them, the gills have disappeared, but the lining of the mantle cavity has become very highly vascularized and functions as a lung.

Class Scaphopoda (The tusk shells).   Scaphopods have a long tubular shell, open at both ends (Fig. 22.21). One end is usually smaller than the other, and the shell thus has a tusklike or toothlike appearance. All scaphopods are marine, living buried in mud or sand.

Class Pelecypoda (The bivalve molluscs).   As the term "bivalve" indicates, these animals have a two-part shell. The two parts, or valves, are usually similar in shape and size and are hinged on one side (the animal's dorsum) (Fig. 22.23). The animals open and shut them by means of large muscles. Among the more common bivalves are clams, oysters, scallops, cockles, file shells (Pl. I-6A), and mussels. Most lead rather sedentary lives as adults, though scallops sometimes swim about

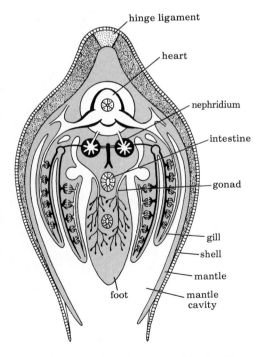

22.23. Cross section of a clam.   [Modified from W. Stempell, *Zoologie im Grundriss*, Borntraeger, 1926.]

hinge ligament

heart

nephridium

intestine

gonad

gill

shell

mantle

mantle cavity

foot

by rapidly opening and shutting the valves of their shell. Lacking a radula, pelecypods are filter feeders, straining tiny food particles from the water flowing across their gills.

Class Cephalopoda (Squids, octopuses, and their relatives). Many of the cephalopods bear little outward resemblance to other molluscs. Unlike their sedentary relatives, they are often specialized for rapid locomotion and a predatory way of life—for killing and eating large prey, such as fishes or crabs. Though fossil cephalopods often had large shells, these are much reduced or absent in most modern forms.[10] The body is elongate, with a large and well-developed head encircled by long tentacles.

Some species attain large size, often being several feet long. The giant squids of the North Atlantic are the largest living invertebrates; the biggest recorded individual was 55 feet long (including the tentacles) and weighed approximately 2 tons. Octopuses (Pl. II-3B) never grow anywhere near this size (except in Hollywood).

Among cephalopods, squids, particularly, have convergently evolved many similarities to vertebrates: internal cartilaginous supports analogous to the vertebrate skeleton; a cartilaginous braincase rather like a skull; an exceedingly well-developed nervous system with a large and complex brain; and, perhaps most striking of all, large camera-type image-forming eyes, which work exactly the way ours do.

Annelida

The Annelida, or segmented worms, have received attention in various connections throughout this book. We have considered their digestive system (p. 108), gas exchange (p. 120), closed circulatory system (p. 137), nephridia (p. 158), nervous system (p. 204), and their hydrostatic skeleton, muscle arrangement, and methods of locomotion (p. 236). The discussion here will therefore be brief.

The phylum is usually divided into three classes: Polychaeta, Oligochaeta, and Hirudinea.

Class Polychaeta. Polychaetes are marine annelids with a well-defined head bearing eyes and antennae. Each of the numerous serially arranged body segments usually bears a pair of lateral appendages called *parapodia* that function in both locomotion and gas exchange (see Fig. 6.5, p. 120). There are numerous stiff bristles called setae or chaetae on the parapodia (the name "Polychaeta" means many chaetae).

Some polychaetes swim or crawl about actively; others are more sedentary, usually living in tubes they construct in the mud or sand of the ocean bottom (Fig. 22.24). The beating parapodia keep water currents flowing through the tubes; these currents bring oxygen, and in some cases food particles, to the worm. Many of the tube dwellers are beautiful animals, particularly the fanworms and peacock worms, which have a crown of colorful fanlike or featherlike processes that they wave in the water at the entrance to their tube (Pl. I-8B).

All the segments of the body are usually much alike. The coelom of each is partly separated from the coeloms of adjacent segments by

[10] *Nautilus,* a modern form with a well-developed shell, is an exception.

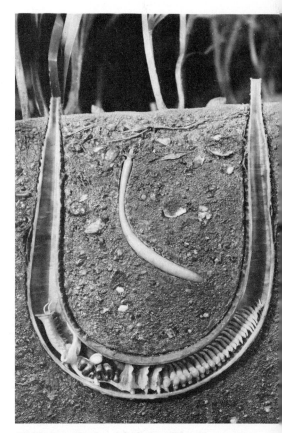

22.24. Polychaete and sipunculid. The parchment worm (*Chaetopterus pergamentacus*) is a polychaete that lives in a U-shaped tube (here shown in section). Between the arms of the tube is a sipunculid (*Phascolosoma gouldii*); the Sipunculida are a small phylum of worms related to the Annelida. [Courtesy American Museum of Natural History.]

membranous intersegmental partitions; the partitions of many poly-chaetes are not complete, however, and in some species they have been entirely lost. Each segment generally has its own ventral ganglion and its own pair of nephridia.

The sexes are separate in the majority of species. In primitive polychaetes most segments produce gametes, but in more advanced species gamete production is restricted to a few specialized segments. The gametes are usually shed into the coelom and leave the body through the nephridia. Fertilization is external. In many species, development includes a ciliated free-swimming larval stage called a *trochophore* (see Fig. 22.38).

Class Oligochaeta. The class Oligochaeta contains the earth-worms and many fresh-water species. They differ from polychaetes in that they lack a well-developed head and parapodia, have fewer setae ("Oligochaeta" means few chaetae or setae), usually combine male and female organs in the same individual, and usually have more complete intersegmental partitions. We have described most of the important characteristics of earthworms in earlier chapters (see Fig. 5.10, p. 108; Fig. 7.12, p. 137; Fig. 8.6, p. 158; and Fig. 10.4, p. 204).

Class Hirudinea (The leeches). The leeches, which probably evolved from oligochaetes, are the most specialized annelids. Their body is dorsoventrally flattened and often tapered at both ends. The first and last segments are modified to form suckers, of which the posterior one is much the larger. Leeches show almost no internal segmentation; the intersegmental partitions have been completely lost except in a few very primitive species.

Most leeches are bloodsuckers. When such a leech attacks a host, it selects a thin area of the host's integument, attaches itself by its posterior sucker, applies the anterior sucker very tightly to the skin, and either painlessly slits the skin with small bladelike jaws or dissolves an opening by means of enzymes. It then secretes a substance into the wound that prevents coagulation of the blood, and begins to suck the blood, usually consuming an enormous quantity at one feeding and then not feeding again for a fairly long time.

Arthropoda

The phylum Arthropoda is by far the largest of the phyla. More than 800,000 species have been described, and there are doubtless hun-dreds of thousands (perhaps millions) more yet to be discovered. Prob-ably more than 80 percent of all the animal species on the earth belong to this phylum.

Arthropods are characterized by jointed chitinous exoskeletons and jointed legs. The exoskeleton, which is secreted by the epidermis, func-tions both as a point of attachment for muscles and as a protective armor, but it imposes limitations on growth and must be periodically molted if the animal is to undergo much increase in size.

Together with their elaborate exoskeleton, arthropods have evolved a complex musculature quite unlike that of most other invertebrates. It comprises not only longitudinal and circular bands, as in so many inver-tebrates, but also separate muscles that, running in myriad directions, make possible an extensive repertoire of movements. Most of the muscles are striated.

The nervous system is very well developed. Of a similar organization as the annelid nervous system, it consists of a dorsal brain and a ventral double nerve cord. Primitively there were ganglia in each segment, but in many groups the ganglia have tended to move forward and fuse into larger ganglionic masses. Sensory organs are many and varied. Like the nervous system, discussed more extensively in Chapter 10 (p. 204), hormonal control, too, is well developed in arthropods.

As we have seen (p. 138, Fig. 7.13), arthropods have an open circulatory system. There is usually an elongate dorsal vessel called the heart, which pumps the blood forward into arteries (the extent of these arteries varies greatly among the various groups of arthropods). From the arteries, the blood goes into open sinuses, where it bathes the tissues directly. Eventually the blood returns to the posterior portion of the heart.

The body spaces through which the blood moves constitute the *hemocoel*—not a true coelom but a cavity derived from the embryonic blastocoel. Although arthropods almost certainly descended from an annelidlike ancestor with well-developed coelomic cavities, and although such cavities develop in arthropod embryos, they are not retained as the functional body cavity in the adult.

In most aquatic arthropods (excluding secondarily aquatic ones), excretion of nitrogenous wastes (primarily ammonia) is principally by way of the gills. Aquatic species usually also have special saclike glands, located near or in the head and equipped with ducts leading to the outside, that play a minor role in excretion. The excretory organs in most groups of terrestrial arthropods are Malpighian tubules (see Fig. 8.9, p. 161).

The sexes are usually separate. Fertilization is internal in all terrestrial and in most aquatic forms.

It is generally held that arthropods evolved either from a polychaete annelid or from the ancestor of the polychaetes. The arthropod body plan may be viewed as an elaboration and specialization of the segmented body of that annelid ancestor. The evidence indicates that the first arthropods had long wormlike bodies composed of many nearly identical segments, each bearing a pair of legs. All the legs were alike. Among the host of modifications of this ancestral body plan that have arisen in the various groups of arthropods during the millions of years of their evolution, four tendencies can be recognized: (1) reduction in the total number of segments; (2) grouping of segments into distinct body regions, such as a head and trunk, or a head, thorax, and abdomen; (3) increasing cephalization, i.e. incorporation of more segments into the head and concentration of nervous control and sensory perception in or just behind the head; (4) specialization of the legs of some segments for a variety of functions other than locomotion, and complete loss of legs from many other segments.

Subphylum Trilobita. Arthropods were very abundant in the Paleozoic seas, and fossils from that era are plentiful. Particularly common in rocks of the first half of the Paleozoic are the fossils of an extinct group—the Trilobita (Fig. 22.25)—characterized by a usually oval

22.25. A fossil of a trilobite. [Courtesy N. F. Snyder, Cornell University.]

and flattened shape and three body regions: a head, apparently composed of four fused segments, that bore a pair of slender antennae and, often, compound eyes; a thorax consisting of a variable number of separate segments; and an abdomen composed of several fused segments. It is not to this tripartite division, however, that the name "Trilobita" refers, but to a division of the body into a median lobe and two lateral lobes by two prominent longitudinal furrows running along the dorsum.

Trilobites, though certainly different from the first arthropods, and exhibiting specializations of their own (e.g. the longitudinal furrows and the fusion of the abdominal segments), nevertheless approach the hypothetical arthropod ancestor more closely than any other known group. One primitive character stands out—the lack of specialization and structural differentiation of the appendages. The fossils show that every segment bore a pair of legs, and that all these legs, including those of the four head segments, were nearly identical. There were thus no appendages specialized as mouthparts.

Trilobites are so markedly different from all other arthropods that they are often regarded as a separate subphylum. Two other subphyla—Chelicerata and Mandibulata—are usually recognized. In both these groups, the tendency toward specialization of some appendages and loss of others is quite evident; thus, in both, the appendages of the most anterior segments have been modified as mouthparts and no longer function in locomotion.

Subphylum Chelicerata. The chelicerate body is usually divided into two regions: a cephalothorax and an abdomen. There are no antennae. The pincerlike or fanglike mouthparts, called *chelicerae,* are believed to be derived from the first pair of postoral legs of the ancestral arthropods. The cephalothorax usually bears five other pairs of appendages besides the chelicerae; in some groups these are all walking legs, while in others only the last four pairs are legs, the first pair being modified as feeding devices called *pedipalps,* which are often much longer than the chelicerae (see Fig. 22.27). The legs of the abdominal segments have been either lost or modified into respiratory or sexual structures.

The subphylum Chelicerata includes four classes. One (Eurypterida) consists entirely of animals extinct since the Paleozoic era, and the members of another (Pycnogonida, the sea spiders) are very rare marine animals.

Members of a third class (Xiphosura) are familiar to anyone who has spent some time on the Atlantic beaches of North America (or the coast of Asia from Japan and Korea to Malaysia and Indonesia). These are the horseshoe crabs (Fig. 22.26), which are not really crabs at all but living relics of an ancient chelicerate class most members of which have been extinct for millions of years.

Members of the fourth class of chelicerates—Arachnida—are familiar to everyone. These are the spiders, ticks, mites, daddy longlegs, scorpions, whipscorpions (Fig. 22.27), and their relatives. Though the various groups of arachnids differ structurally in many ways, most have two body regions, a cephalothorax and an abdomen (these are not distinguishable in ticks, mites, or daddy longlegs). There are often simple

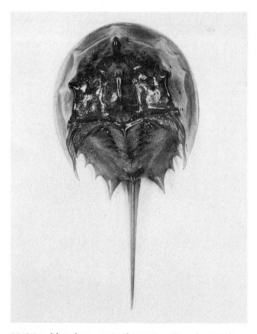

22.26. *Limulus polyphemus,* the horseshoe crab. Dorsal view. [Courtesy American Museum of Natural History.]

eyes on the cephalothorax, but never any compound eyes or antennae. The cephalothorax bears six pairs of appendages: a pair of chelicerae, a pair of pedipalps, and four pairs of walking legs. In most groups, prey is seized and torn apart by the pedipalps. The chelicerae, too, may function in manipulating prey, or they may be modified as poison fangs, as in spiders.

Subphylum Mandibulata. The members of this subphylum differ from chelicerates in having antennae and in having *mandibles* instead of chelicerae as their first pair of mouthparts. Mandibles are modified from the basal segment of the ancestral legs and function in biting and chewing (though in some species they are secondarily modified for piercing and sucking). They are never clawlike or pincerlike, as chelicerae frequently are. In most mandibulates, there are two additional pairs of mouthparts called *maxillae.*

The subphylum comprises six classes, including the Chilopoda (centipeds; Pl. I-7A); the Diplopoda (millipeds); and the Crustacea and Insecta, described below.

Class Crustacea. Some representatives of this class, such as crayfish, lobsters, shrimps, and crabs (Fig. 22.28, Pl. III-8B), are well known to most people. But there are many other species of Crustacea, which bear little superficial resemblance to these familiar animals; among them are fairy shrimps, water fleas, brine shrimps, sand hoppers, barnacles, and sow bugs; many of these are very small odd-looking creatures (Fig. 22.29).

Crustacea characteristically have two pairs of antennae, a pair of mandibles, and two pairs of maxillae. But the rest of the appendages vary greatly from group to group, and whatever could be said about those of one group, such as crayfish and lobsters, would have little

22.27. A whipscorpion, *Mastigoproctus giganteus.* Whipscorpions and other arachnids have six pairs of appendages: a pair of fanglike chelicerae (not visible in the photograph); a pair of stout toothed pincerlike pedipalps; and four pairs of legs (the first pair, which are long and slender, have a sensory–tactile function and are not used in walking). The posterior knob with its "whip" has slits through which the animal can spray a poisonous secretion. [Courtesy Thomas Eisner, Cornell University.]

22.28. The land crab *Geocarcinus.* Three types of jointed appendages can be seen here: mouthparts (folded against underside of head), large chelipeds (pincers), and walking legs. Notice the eyes on mobile stalks. [Walter Dawn from National Audubon Society.]

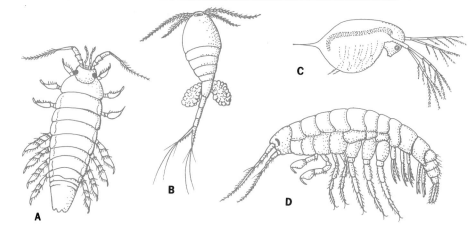

22.29. Some representative small Crustacea.
(A) Marine sow bug (*Idotea*). (B) Fresh-water copepod (*Cyclops*). (C) Water flea (*Daphnia*). (D) Sand hopper (*Gammarus*). [Based on drawings by Louise G. Kingsbury.]

relevance to those of other groups. In fact, the Crustacea are an enormously diverse assemblage of animals that can hardly be characterized in any simple way. Some have a cephalothorax and an abdomen; others have a head and a trunk, or a head, thorax, and abdomen, or even a unified body. Most are free-living, but some are parasitic. Most are active swimmers, but some, like barnacles, secrete a shell and are sessile. The majority are marine, but there are many fresh-water species, and a few, such as sow bugs, are terrestrial and have a simple tracheal system. We could go on listing divergences, but the point of the amazing diversity of this group has been made. This is a class in which the basic arrangement of a segmented body with numerous jointed appendages has been modified and exploited in countless ways as the members of the class have diverged into different habitats and adopted different modes of life.

Class Insecta. This is an enormous group of diverse animals that occupy almost every conceivable habitat on land and in fresh water. If numbers are the criterion by which to judge biological success, then the insects are the most successful group of animals that has ever lived; there are more species of insects than of all other animal groups combined. But there is one qualification to their dominant role—they do not occur in the sea; the role played by insects on land is played in the sea by Crustacea.

There are a few insect fossils from the Devonian, but it was in the Carboniferous and Permian periods that insects took their place as one

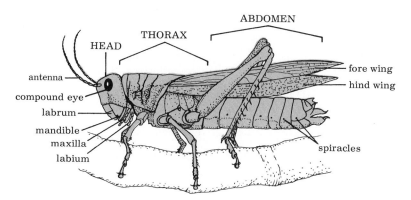

22.30. A grasshopper.

of the dominant groups of animals (see Table 21.2, p. 478). By the end of the Paleozoic era, many of the modern orders had appeared, and the number of species was enormous. A second great period of evolutionary radiation began in the Cretaceous and continues to the present time; this second radiation is correlated with the rise of flowering plants.

The insect body is divided into three regions: a head, a thorax, and an abdomen (Fig. 22.30). The head segments are completely fused, and in adults their boundaries cannot be distinguished. The head bears numerous sensory receptors, usually including compound eyes; one pair of antennae; and three pairs of mouthparts derived from ancestral legs. The mouthparts include a pair of mandibles, a pair of maxillae, and a lower lip, or *labium,* formed by fusion of the two second maxillae (Fig. 22.31). The upper lip, or *labrum,* which has not traditionally been classified as a mouthpart, may also be derived from ancestral legs.

The thorax is composed of three segments, each of which bears a pair of walking legs. In many insects (but not all), the second and third thoracic segments each bear a pair of wings.

The abdomen is composed of a variable number of segments (12 or fewer). Abdominal segments are devoid of legs, but highly modified remnants of the ancestral appendages may be present at the posterior end, where they function in mating and egg laying.[11]

We have already discussed the insects at considerable length in other chapters and need not repeat ourselves here. See the discussion of the insect tracheal system (p. 124, Pl. II-5D), the circulatory system (p. 138, Fig. 7.13), the Malpighian excretory organs (p. 161, Fig. 8.9), nervous control and sensory perception (pp. 204, 219, Pl. II-5A), the exoskeleton and muscles (p. 236, Fig. 11.2A), behavior (Chapter 12, Pls. III-1B, IV-2), and development (p. 346, Fig. 16.16, Pl. II-8).

The insects are classified in approximately 25 orders. Representatives of nine of them are shown in Fig. 22.32.

THE DEUTEROSTOMIA

There are only five phyla in the Deuterostomia, and only two of these—Echinodermata and Chordata—are major groups. We shall discuss these and a small phylum—Hemichordata—that is important from an evolutionary standpoint. The other two phyla—Chaetognatha (arrow worms) and Pogonophora (beard worms)—contain only a few, exclusively marine species and are not important for our purposes here.

Echinodermata

The echinoderms, exclusively marine, mostly bottom-dwelling animals, include the sea stars, brittle stars, sea urchins, sand dollars, sea cucumbers, and sea lilies. Almost all members of the phylum possess an internal skeleton composed of numerous calcareous plates embedded in the body wall. These plates may be separate, or they may be fused to form a rigid boxlike structure. The skeleton frequently bears many bumps or spines that project from the surface of the animal (these are particularly noticeable in sea urchins; see Pl. I-7B). It is this character-

[11] A few primitive insects retain vestiges of appendages on many abdominal segments. These may have a sensory function.

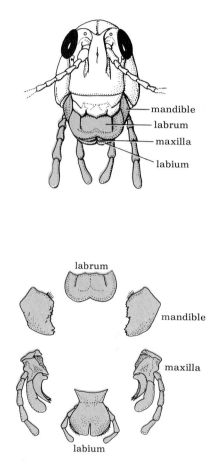

22.31. The mouthparts of a grasshopper. Top: Front view of head, with mouthparts in place. Bottom: The mouthparts, including the labrum, removed from the head but kept in their proper relative positions. The mandibles and probably the labrum (upper lip) are derived from the basal segments of ancestral legs; all the other segments of those legs have been lost. The maxillae and labium (lower lip) retain more of the segments of the legs from which they are derived; the basal segments are enlarged, but the distal segments form slender leglike structures called palps, which bear many sensory receptors. [Modified from T. I. Storer and R. L. Usinger, *General Zoology,* McGraw-Hill Book Co., 1957. Used by permission.]

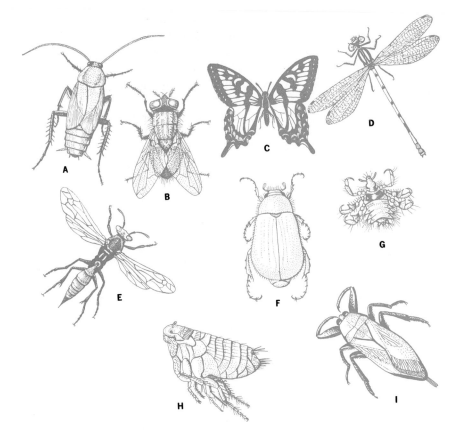

22.32. Some representatives of the major insect orders. (A) Cockroach (order Orthoptera). (B) Fly (Diptera). (C) Butterfly (Lepidoptera). (D) Damselfly (Odonata). (E) Wasp (Hymenoptera). (F) Beetle (Coleoptera). (G) Louse (Anoplura). (H) Flea (Siphonaptera). (I) Bug (Hemiptera). The insects are not drawn to a uniform scale.

istic that gives the animals the name "Echinodermata" (from the Greek *echino-,* spiny, and *derma,* skin). Although the adults are radially symmetrical, the larvae are bilateral, and it is generally held that echinoderms evolved from bilateral ancestors.

Echinoderms have a well-developed coelom in which the various internal organs are suspended. The complete digestive system is the most prominent of the organ systems. There is no special excretory system, and the blood circulatory system, though present, is poorly developed. The nervous system is radially organized, consisting of nerve networks that connect to ringlike ganglionated nerve cords running around the body of the animal (there are often three of these cords); there is no brain.

A characteristic unique in echinoderms is their **water-vascular system.** This is a system of tubes (usually called canals) filled with watery fluid. Water can enter the system through a sievelike plate, called a madreporite, on the surface of the animal. A tube from this plate leads to a ring canal that encircles the esophagus (Fig. 22.33). Five radial canals branch off the ring canal and run along symmetrically spaced grooves or bands on the surface of the animal. Many short side branches from the radial canals lead to hollow **tube feet** that project to the exterior. Each tube foot is a thin-walled hollow cylinder, with a sucker on its end. At the base of each tube foot is a muscular ampulla. When the ampulla contracts, the fluid in it is prevented by a valve from flowing into the radial canal; consequently it is forced into the tube foot, which is thereby extended. The foot attaches to the substratum by its sucker, and then longitudinal muscles in its wall contract, shortening it and pulling the animal forward (while forcing the water back into the ampulla). This cycle of events, repeated rapidly by the many tube feet of an animal like a sea star, enables it to get about, albeit slowly. The tube feet also enable it to hold tightly to a rock or other object by applying suction, or to pull open the valves of the shell of a clam or oyster, on which it will feed.

The sexes are usually separate. Eggs and sperms are shed into the surrounding water, where fertilization occurs. Cleavage is radial and indeterminate. The free-swimming ciliated larva has a complete digestive tract, with the anus derived from the embryonic blastopore, the mouth being a new opening.

Class Asteroidea (The sea stars). The body of a sea star (starfish) consists of a central *disc* and usually of five rays, or *arms*,[12] each with a groove bearing rows of tube feet running along the middle of its lower surface. The outer surface of the animal is studded with many short spines and numerous tiny skin gills, which are thin fingerlike evaginations of the body wall that protrude to the outside between the plates of the endoskeleton (see Fig. 6.4, p. 120). The cavity of each skin gill is continuous with the general coelom. The madreporite is on the upper surface (but not in the center; in this respect, the radial symmetry of the animal is not perfect).

The mouth is located in the center of the lower surface of the disc and the anus in the center of the upper surface (thus the lower surface is the morphological anterior end of the animal, and the upper surface the morphological posterior end). The digestive tract of a sea star is straight and very short, consisting of a short esophagus, a broad stomach that fills most of the interior of the disc, and a very short intestine. The stomach is divided by a constriction into two parts: a large eversible part

[12] Though most species of sea stars have five arms, some have more.

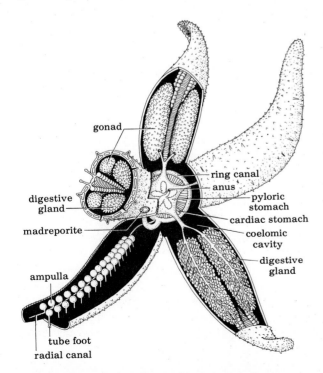

22.33. Dissection of a sea star (dorsal view).

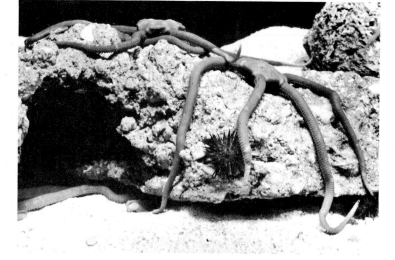

22.34. Brittle stars. The body disc is smaller than that of sea stars. Loco-motion is by rapid lashing of the long slender arms. Note the sea urchin be-tween the arms of one of the brittle stars. [New York Zoological Society photo.]

22.35. A sea cucumber. Unlike other echinoderms, a sea cucumber has a much reduced endoskeleton and a leathery body, and it lies on its side rather than on the oral surface. It has five rows of tube feet (two of them visible here) corresponding to the five arms of a sea star, and the mouth is sur-rounded by tentacles. [Courtesy Carolina Biological Supply Co.]

22.36. A fossil sea lily. By contrast with other echinoderms, most sea lilies are sessile, having a long stalk by which they attach to the sub-stratum, and their mouth is on the upper side. Long feathery arms surround the mouth. [Cour-tesy American Museum of Natural History.]

(cardiac stomach) at the esophageal (lower) end and a smaller non-eversible part (pyloric stomach) at the intestinal (upper) end. At-tached to the noneversible part are five pairs of large digestive glands; each pair of glands lies in the coelomic cavity of one of the arms (Fig. 22.33).

When the sea star feeds, it pushes the lower part of the stomach out through the mouth, turning it inside out and placing it over food material such as the soft body of a clam of oyster. The stomach secretes digestive enzymes onto the food, and digestion begins. The partly digested food is then taken into the upper part of the stomach and into the digestive glands, where digestion is completed and the products are absorbed.

Other echinoderm classes. There are four other classes of echino-derms: (1) Ophiuroidea, the brittle stars (Fig. 22.34), serpent stars, and basket stars; (2) Echinoidea, the sea urchins (Pl. I-7B), sand dol-lars, and heart urchins; (3) Holothuroidea, the sea cucumbers (Fig.

22.35); and (4) Crinoidea, the sea lilies (Fig. 22.36), which are the oldest and most primitive group. Although members of these four classes often show little superficial resemblance to sea stars, their structure is fundamentally similar. For example, sea urchins and sea cucumbers lack the five arms of sea stars, but they do have five bands of tube feet and thus show the same basic pentaradiate symmetry.

Hemichordata

The hemichordates, many of which are called acorn worms (or tongue worms), are marine animals often found living in U-shaped burrows in sand or mud along the coast. They are fairly large worms, ranging from 3½ to 17 inches in length. Their body consists of an anterior conical proboscis (thought by some to resemble an acorn— hence their name), a collar, and a long trunk (Fig. 22.37). The mouth is situated ventrally, at the junction between the proboscis and the collar. A particularly important feature is a series of *gill slits* in the wall of the pharynx. Water drawn into the mouth is forced back into the pharynx and out through these slits. Oxygen is removed from the in-drawn water and carbon dioxide released into it by blood in beds of capillaries in the septa between the slits. Another important characteristic of hemichordates is the occurrence during development of a ciliated larval stage that strikingly resembles the larvae of some echinoderms.

The relationships between echinoderms, hemichordates, and chordates

It may seem strange that the Echinodermata form the major phylum generally considered most closely related to our own phylum, the Chordata. But as we saw earlier when we discussed the differences between the Protostomia and Deuterostomia, certain characteristics seem to link echinoderms, hemichordates, and chordates and set them apart from all the protostome phyla. These characteristics include formation of the anus from the embryonic blastopore, radial and indeterminate cleavage, origin of the mesoderm as pouches, and formation of the coelom as the cavities in the mesodermal pouches.

The Hemichordata have long held special interest for zoologists

22.37. An adult acorn worm.

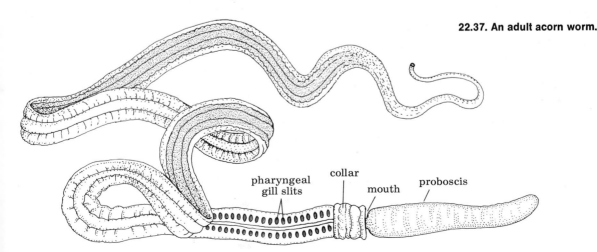

pharyngeal gill slits   collar   mouth   proboscis

Table 22.1. A comparison of some of the major animal phyla

Phylum	Symmetry	Cleavage	Body cavity	Digestive tract	Circulatory system	Ciliated larva	Segmentation
Coelenterata	Radial		None	Gastrovascular cavity	Absent	Planula	Absent
Platyhelminthes	Bilateral	Determinate	None	Gastrovascular cavity	Absent	Trochophorelike in some	Absent or correlated with reproduction
Aschelminthes	Bilateral	Determinate	Pseudocoelom	Complete, with mouth from blastopore	Absent	None or a unique type	Absent
Mollusca	Bilateral	Determinate	Coelom much reduced	Complete, with mouth from blastopore	Open	Trochophore	Absent (except in *Neopilina*)
Annelida	Bilateral	Determinate	Coelom	Complete, with mouth from blastopore	Closed	Trochophore	Present
Arthropoda	Bilateral	Determinate	Hemocoel (coelom, degenerate)	Complete, with mouth from blastopore	Open	None	Present
Echinodermata	Secondarily radial	Indeterminate	Coelom	Complete, with anus from blastopore	A special type; often poorly developed	Dipleurula	Absent
Hemichordata	Bilateral	Indeterminate	Coelom	Complete, with anus from blastopore	Open	Dipleurula	Absent
Chordata	Bilateral	Indeterminate	Coelom	Complete, with anus from blastopore	Closed (except in tunicates)	None	Present

because their apparent affinities to both the Echinodermata and the Chordata seem to provide additional evidence of a relationship between those two large and important phyla. The ciliated larvae of hemichordates are so much like those of some echinoderms that they were mistaken for echinoderms when first discovered. This larval type, sometimes called a *dipleurula* (Fig. 22.38), is found only in the echinoderms and hemichordates. It has a band of cilia that forms a ring encircling the mouth. It thus differs from the trochophore larva found in many protostomes (including some turbellarian Platyhelminthes, the lophophorate phyla, Mollusca, and Annelida), which has a band of cilia encircling the body anterior to the mouth. The similar larvae of hemichordates and echinoderms, as well as the similarities in early embryology mentioned above, indicate that these two groups must stem from a common ancestor. In view of the complicated metamorphosis that in echinoderms produces a radial adult from a bilateral larva, it seems likely that echinoderms have deviated greatly from the ancestral type and that hemichordates are probably nearer that ancestral type.

The most obvious resemblance of hemichordates to chordates is their possession of pharyngeal gill slits, which are found in all chordates but nowhere else in the animal kingdom. The hemichordates also have a dorsal nerve cord that is sometimes hollow and resembles the dorsal hollow nerve cord characteristic of chordates. Because of these resemblances, the hemichordates were regarded for many years as primitive members of the phylum Chordata. Though they are now generally regarded as a separate phylum, which may actually be closer to the echinoderms than to the chordates, recognition of their ties with both Chordata and Echinodermata has helped clarify the phylogenetic relationship between these two major groups. Note that there is no suggestion here that chordates evolved from echinoderms, but simply that the two groups diverged from a common ancestor at some remote time.

Some of the important characteristics of the major animal phyla are compared in Table 22.1.

Invertebrate Chordata

Throughout this book, we have used the terms "vertebrate" and "invertebrate," and have assigned all the animals discussed to one or the other of the categories they designate. But this division of the animal kingdom is in many respects an odd one, because neither category coincides with any phylum or group of phyla. Indeed, the term "vertebrate" designates only a part of one phylum; the rest of that phylum and all the other phyla then fall under the heading "invertebrate." The phylum that contains both invertebrate and vertebrate members is Chordata.

The phylum Chordata is customarily divided into three subphyla: Urochordata, Cephalochordata, and Vertebrata. These share three important characteristics: (1) All have, at least during embryonic development, a structure called a *notochord* (whence the name "Chordata"). This is a flexible supportive rod running longitudinally through the dorsum of the animal just ventral to the nerve cord. (2) All have pharyngeal gill slits (or pouches) at some stage in their development. (3) All have a dorsal hollow nerve cord.

The Urochordata and the Cephalochordata are both invertebrate; i.e. they have no backbone.

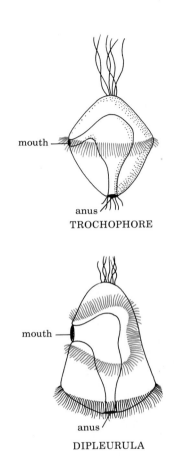

mouth —

anus
TROCHOPHORE

mouth —

anus
DIPLEURULA

22.38. Trochophore and dipleurula larval types. The band of cilia of the trochophore is located anterior to the mouth, whereas the corresponding band of the dipleurula encircles the mouth.

Subphylum Urochordata (The tunicates). In the best-known class of tunicates (sometimes called sea squirts), the adults are sessile marine animals that little resemble other chordates except in having pharyngeal gill slits (Fig. 22.39).[13] These structures function in both gas exchange and feeding, acting as a strainer for removing small food particles from the water flowing through them. According to one hypothesis, the pharyngeal gill slits, which are so distinctive a trait of chordates, first evolved as an adaptation for this sort of filter feeding and only later came to function in gas exchange as well.

Larval tunicates, which are motile, show much more resemblance to the other chordates. With their elongate bilaterally symmetrical bodies and long tails, they look rather like tadpoles. They possess a well-developed dorsal hollow nerve cord and a notochord beneath it in the tail region (Fig. 22.40). When the larvae settle down and undergo metamorphosis to the adult form, the notochord and most of the nerve cord are lost.

Some biologists hold that the tunicates and vertebrates descended from a common ancestor that was free-swimming and resembled a modern tunicate larva. If this is so, then the sessile structure of modern adult tunicates is a later specialization. An alternative hypothesis is that the common ancestor was sessile, more like adult tunicates, and that vertebrates evolved from its motile larva; in other words, in the line

[13] Members of two smaller classes of tunicates are free-swimming planktonic organisms.

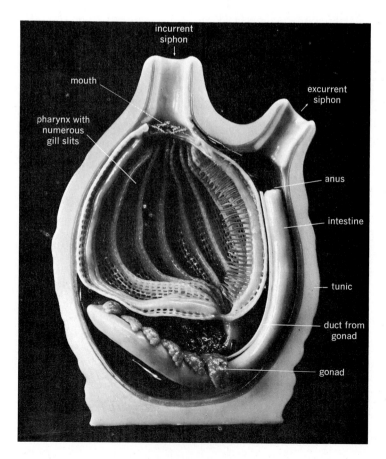

22.39. Cutaway model of an adult tunicate.
[Courtesy American Museum of Natural History.]

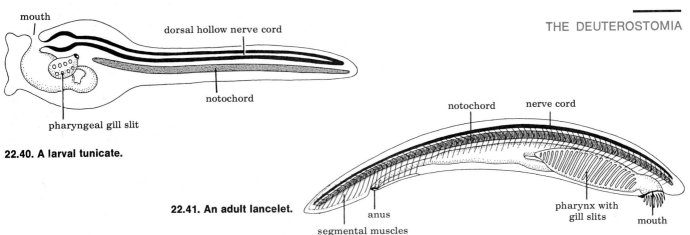

22.40. A larval tunicate.

22.41. An adult lancelet.

leading to the vertebrates, the larval stage increased in importance and duration, until finally it could reproduce without undergoing metamorphosis and the ancestral sessile stage dropped out of the life cycle entirely.

Subphylum Cephalochordata (The lancelets). There are about 30 species of these small marine animals. Though capable of swimming, they spend most of their time buried tail down in sand in shallow water, with only their anterior end exposed. They are filter feeders, taking in water through the mouth and straining it through the pharyngeal gill slits.

The body of a typical lancelet (usually called amphioxus) is about 2 inches long, translucent, and shaped rather like a fish (Fig. 22.41). Both the dorsal hollow nerve cord and the notochord are well developed and are retained throughout life. A feature not seen in tunicates but characteristic of both cephalochordates and vertebrates is segmentation. In lancelets, this segmentation is most noticeable in the muscles, which are in V-shaped, segmentally arranged bundles.

Vertebrate Chordata

As the name "Vertebrata" implies, the animals in this group are characterized by an endoskeleton that includes a backbone composed of a series of vertebrae. The vertebrae develop around the notochord, which in most vertebrates is present in the embryo only. The serial arrangement of the vertebrae and the organization of the muscles are the principal tokens of segmentation.

We discussed the anatomy, physiology, behavior, and development of vertebrates at length in other parts of this book. Here we shall be concerned primarily with the evolutionary history of the group.

Class Agnatha. The vertebrates are one of the few major animal groups not represented among the Cambrian fossils. The oldest vertebrate fossils are from the Ordovician period, which began some 500 million years ago (see Table 21.2, p. 478). Those first vertebrate fossils are of bizarre fishlike animals covered by thick plates of bony material. Though they had a skeleton, they lacked an important character found

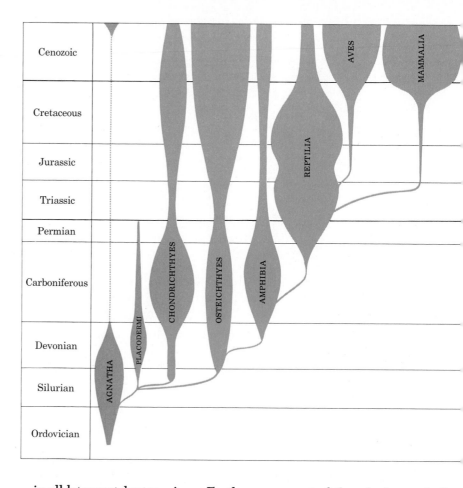

22.42. Evolution of the vertebrate classes.

in all later vertebrates—jaws. Furthermore, most of them had no paired fins. These ancient fishes constitute the class Agnatha (which means jawless). Most were probably filter feeders, straining food material from mud and water flowing through their gill system.

The Agnatha continued as an important group through the Silurian period, sharing the seas with the already abundant sponges, coelenterates, brachiopods (far more numerous then than now), molluscs (particularly gastropods and cephalopods), trilobites and eurypterids, and echinoderms. But by the end of the Silurian the Agnatha had begun to decline, and they disappear from the fossil record by the end of the Devonian (Fig. 22.42).

A few peculiar species living today, the lampreys (Fig. 22.43)

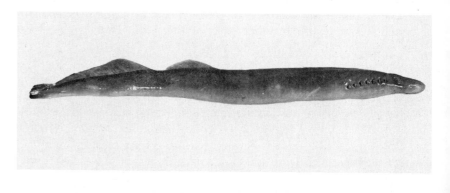

22.43. The sea lamprey (*Petromyzon*). Right: Lateral view. Note the pharyngeal gill slits. Above: Ventral view, showing mouth sucker. [Right: Courtesy Carolina Biological Supply Co. Above: New York Zoological Society photo.]

and the hagfishes, are generally classified as Agnatha, although they are quite unlike the Paleozoic armored species. They have a soft body without either armor or scales; they have a cartilaginous skeleton, having lost all trace of bone; and their jawless mouth is modified as a round sucker that is lined with many horny teeth and accommodates a rasping tongue. They feed by attaching themselves by their sucker to other fishes, rasping a hole in the skin of the prey and sucking blood and other body fluids. The lampreys have a larval filter-feeding stage that strikingly resembles amphioxus.

Class Placodermi. The decline of the ancient Agnatha coincided with the rise of a second class of armored fishes—the Placodermi—which first appeared in the Silurian, having probably arisen from primitive agnaths. The Placodermi were an important group during the Devonian, but most became extinct by the end of that period; a few survived until the Permian, when they too disappeared.

The Placodermi mark a notable advance in vertebrate evolution in their possession of hinged jaws. The acquisition of hinged jaws was one of the most important events in the history of vertebrates, because it made possible a revolution in the method of feeding and hence in the entire mode of life of early fishes. They became more active and wide-ranging animals, usually with paired fins. Many became ferocious predators. Even those that remained mud feeders were evidently adaptively superior to the ecologically similar agnaths, which they gradually replaced.

Anatomical and embryological studies have convinced biologists that the hinged jaws of the placoderms developed from a set of gill-support bars (Fig. 22.44). Notice that hinged jaws arose independently in two important animal groups, the arthropods and the vertebrates, but that, although they are functionally analogous structures, they arose in entirely different ways—in the one case from ancestral legs and in the other from skeletal elements in the wall of the pharyngeal region.

Though the Placodermi themselves have been extinct at least 230 million years, two other classes that arose from them in the Devonian are still important elements of our fauna. These are the Chondrichthyes (cartilaginous fishes) and the Osteichthyes (bony fishes).

Class Chondrichthyes. The modern Chondrichthyes (sharks, skates, rays, and their relatives) are distinguished by their cartilaginous skeletons; bone is unknown in the group. Though a cartilaginous skeleton might at first be taken as a primitive trait, it is not thought to be one in Chondrichthyes. Their ancestors among the Placodermi probably had bony skeletons, and loss of the bone must be regarded as an evolutionary specialization.

Class Osteichthyes. The other class that arose from the Placodermi—the Osteichthyes—is a large one that includes most of the fishes familiar to you. Its members are the dominant vertebrates in both fresh water and the oceans, as they have been since the Devonian—the so-called Age of Fishes.

The earliest Osteichthyes probably lived in fresh water. In addition to gills, they had lungs, which they probably used as supplementary

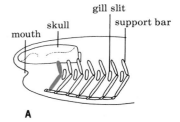

A

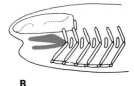

B

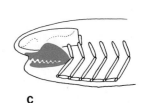

C

22.44. Evolution of the hinged jaws of vertebrates. (A) The earliest vertebrates had no jaws: The structures (color) that in their descendants would become jaws were gill-support bars. (B) A pair of gill-support bars has been modified into weak jaws. (C) The jaws have become larger and stronger. [Modified from A. S. Romer, *The Vertebrate Body*, Saunders, 1949.]

22.45. *Latimeria,* **a modern lobe-finned fish.**

22.46. The movement of vertebrates onto land.
A Devonian lobe-finned fish (*Eusthenopteron*),
which probably pulled itself out of the water
onto mud flats and sandbars. [Courtesy American Museum of Natural History.]

gas-exchange devices when the water was stagnant and deficient in oxygen. There are still a few living relict species with lungs.

Soon after the Osteichthyes arose from the Placodermi, the class split into two divergent groups. One underwent great evolutionary radiation, giving rise to nearly all the bony fishes alive today. The other radiated considerably in the late Paleozoic, but today is represented by only six relict species—five species of lungfishes (one in Australia, three in Africa, and one in South America) and one species of "lobe-fin" known only from deep waters off the southeast coast of Africa. Despite its rarity in the present fauna, this second group of Osteichthyes is of special evolutionary interest, because it is thought to have been ancestral to the land vertebrates.

Let us look more closely at the ancient lobe-fin fishes. This group has long been known from fossils, but was thought entirely extinct for some 75 million years, until, in 1939, a specimen was caught off the east coast of South Africa; since then, additional specimens of these living fossils (genus name *Latimeria*) have been caught and studied (Fig. 22.45). While these lobe-fins are not the particular ones thought to be the ancestors of land vertebrates, they are believed to resemble those ancestral forms in many ways.

In addition to lungs, the lobe-fins had another important preadaptation for life on land—the large fleshy bases of their paired pectoral and pelvic fins. At times, especially during droughts, lobe-fins living in fresh water probably used these leglike fins to pull themselves onto sandbars and mud flats (Fig. 22.46). They may even have managed, albeit with great difficulty, to crawl to a new pond or stream when the one they were in dried up.

Now, by the Devonian period the land had already been colonized by plants (see Table 21.2, p. 478), but was still nearly devoid of animal life. Hence any animal that could survive on land would have had a whole new range of habitats open to it without competition. Any lobe-fin fishes that had appendages slightly better suited for land locomotion than those of their fellows would have been able to exploit these habitats more fully; through selection pressure exerted over millions of years, the fins of these first vertebrates to walk (or, rather, crawl) on land would slowly have evolved into legs. Thus, by the end of the Devonian, with a host of other adaptations for life on land evolving at the same time, one group of ancient lobe-fin fishes must have given rise to the first amphibians.

Classes Amphibia and Reptilia. Numerous fossils indicate that, as would be expected, the first amphibians were still quite fishlike. In fact, they probably spent most of their time in the water. But as they progressively exploited the ecological opportunities open to them on land, they slowly became a large and diverse group. So numerous were they during the Carboniferous that that period is often called the Age of Amphibians, just as the period before it, the Devonian, is called the Age of Fishes. The amphibians were still abundant in the Permian, but during that period they slowly declined as the members of a new class, the Reptilia, replaced them.

The end of the Permian, which also marked the end of the Paleozoic era, was a time of great change, both geological and biological. The ancestral Appalachian Mountains were built up; the last trilobites and the last placoderms disappeared; the once common brachiopods declined; and older types of corals, molluscs, echinoderms, crustaceans, and fishes were replaced by more modern representatives of those groups. This so-called Permo-Triassic crisis also witnessed the extinction of most groups of amphibians. By the end of the Triassic, the only members of this class that survived were the immediate ancestors of the few small groups of modern Amphibia—the salamanders (order Urodela), the apodes (order Apoda), and the frogs and toads (order Anura) (Fig. 22.47).

The first reptiles had evolved from primitive amphibians by the late Carboniferous. The class expanded during the Permian, replacing its amphibian predecessors, and became a huge and dominant group during the Mesozoic era, which is often called the Age of Reptiles.

One might well wonder why the reptiles were so effectively able to displace the once dominant amphibians. There were doubtless many reasons. But surely one of the most compelling was that the reptiles were terrestrial in the fullest sense of the word while the amphibians were not. Amphibians continued to use external fertilization and to lay fishlike eggs—eggs that had no amnion or shell and hence had to be deposited either in water or in very moist places on land lest they dry up. Larval development remained aquatic. Amphibians were thus bound to the ancestral fresh-water environment by the necessities of their mode of reproduction. Furthermore, even adult amphibians probably had thin moist skin and were in danger of desiccation if conditions became very dry. But in both these respects, reptiles had completed the transition to land life; they used internal fertilization, laid amniotic shelled eggs, had no larval stage, and had dry, scaly, relatively impermeable skin. Evolution of the amniotic egg—often called the "land egg"—which provides a fluid-filled chamber in which the embryo can develop even when the egg itself is in a dry place, was an advance as important in the conquest of land as the evolution of legs by the Amphibia.

The Reptilia had many other characteristics that made them better suited for terrestrial life than the Amphibia. The legs of the ancient amphibians were small, weak, attached far up on the sides of the body, and oriented laterally; hence they were unable to support much weight, and the belly of the animal often dragged on the ground; walking was doubtless slow and labored, as it is in salamanders today. The legs of reptiles were usually larger and stronger and could thus support more weight and effect more rapid locomotion; in many (though not all) species, they were also attached lower on the sides and oriented more vertically, so that the animal's body cleared the ground. Whereas the lungs of amphibians were poorly developed and inefficiently ventilated, those of reptiles were fairly well developed, and greater rib musculature made their ventilation more efficient. Whereas the amphibian heart was three-chambered (two atria and one ventricle), that of reptiles was four-chambered (though the partition between the ventricles was seldom complete); hence there was little mixing of oxygenated and deoxygenated blood.

The class Reptilia is represented in our modern fauna by members of four groups: turtles (order Chelonia), crocodiles and alligators (order Crocodylia), lizards and snakes (order Squamata), and the tuatara

22.47. Modern Amphibia. Above: Spotted salamander (*Ambystoma maculatum*). Below: American toad (*Bufo americanus*) calling. Note the huge expanded vocal sac. [Courtesy Verne N. Rockcastle, Cornell University.]

(order Rhynchocephalia) (Fig. 22.48). The last, the tuatara (*Sphenodon punctatum*), is the sole surviving member of its ancient order; it is found only on a few islands off the coast of New Zealand. Members of the other three orders are fairly abundant, totaling about 6,000 living species.

All the living reptiles except the crocodilians are directly descended from an important Permian group called the stem reptiles (cotylosaurs). This group also gave rise to several other lineages, including two (ichthyosaurs and plesiosaurs) that returned to the aquatic environment, one (therapsids) that ultimately led to the mammals, and one (thecodonts) that in its turn gave rise to crocodilians, birds, the flying reptiles called pterosaurs, and the great assemblage of reptiles called dinosaurs (Fig. 22.49). The dinosaurs were extremely abundant and varied during the Jurassic and Cretaceous periods.

22.48. Some modern Reptilia. Left top: Black racer (*Coluber constrictor*), a common snake in the eastern United States. Left middle: Snapping turtle (*Chelydra serpentina*). Bottom: Salt-marsh crocodile (*Crocodylus palustris*). Right: Fence lizard (*Sceloporus undulatus*). [Left top and middle: New York Zoological Society photos. Bottom: Courtesy American Museum of Natural History. Right: Courtesy Verne N. Rockcastle, Cornell University.]

22.49. Reconstitutions of some extinct Reptilia.
Top: Well-adapted for marine life, the plesiosaurs (left) and ichthyosaurs (right) fed on fish. Middle: Many of the therapsids, the ancestors of the mammals, must have been very active animals. They often had well-differentiated teeth and vertically placed legs that raised their bodies high above the ground. Bottom: The dinosaurs were a highly diversified group. *Triceratops* (left) was a herbivore, *Tyrannosaurus* a giant carnivore about 47 feet long and 19 feet high. [Top and bottom: From murals by Charles R. Knight. Courtesy Field Museum of Natural History, Chicago. Middle: Courtesy American Museum of Natural History.]

By the end of the Cretaceous (which was also the end of the Mesozoic era), all the plesiosaurs, pterosaurs, and dinosaurs had disappeared. Only members of the four groups of modern reptiles remained. It is true that the decline of the dinosaurs was not as sudden as is often supposed; it took tens of millions of years. But it was a dramatic event in the history of life on earth nonetheless. Why previously successful animals should have died out on such a scale has never been satisfactorily explained.

Class Aves. By the late Triassic or early Jurassic, two different lineages of reptiles, descended from the thecodonts, had developed the power of flight. One of these lineages, the pterosaurs, included animals with wings consisting of a large membrane of skin stretched between the body and the enormously elongated arm and fourth finger; some species had wingspreads as great as 26 feet. The pterosaurs were common for a time, but eventually became extinct. The other lineage developed wings of an entirely different sort, in which many long feathers, derived from scales, were attached to the modified forelimbs. This line eventually became so different from the other reptiles that we designate it as a separate class—Aves—the birds.

The oldest known fossil bird (*Archaeopteryx*), from the middle Jurassic, still had many reptilian characters, e.g. teeth and a long jointed tail. Neither of these traits is present in modern birds, which have a beak instead of teeth and only a tiny remnant of the ancestral tail bones (the tail of a modern bird consists only of feathers).

Along with wings, birds evolved a host of other adaptations for their very active way of life. One of the most important was warm-bloodedness (homeothermy)—the ability to maintain a high and constant metabolic rate, and hence great activity, despite fluctuations in environmental temperature. An anatomical feature that helped make possible the metabolic efficiency necessary for homeothermy was the complete separation of the two ventricles of the heart; birds have completely four-chambered hearts. The insulation against heat loss provided by the body feathers plays an important role in temperature regulation; in modern birds, all the scales except those of the feet are modified as feathers (see Pl. II-1A). Among other adaptations for flight are light hollow bones and an extensive system of air sacs attached to the lungs (Fig. 6.10, p. 123). Birds also have very keen senses of vision, hearing, and equilibrium.

The newly hatched young of birds are usually not yet capable of complete temperature regulation, and they cannot fly. In many species, in fact, they are featherless, blind, and almost entirely helpless. Accordingly, most birds exhibit elaborate nest-building and parental-care behavior (Pls. III-6, 7).

Class Mammalia. Both birds and mammals evolved from reptiles, and both became homeothermic and highly successful organisms, but the two groups did not arise from the same ancestral reptilian stock. Precisely at what point therapsid reptiles (Fig. 22.49, middle) ceased and mammals began, it is impossible to say; there was no sudden transformation of reptile into mammal, no dramatic event to mark the appearance of the first member of our class. The characters that distinguish modern mammals from stem reptiles arose not only very gradually but over different spans of time.

Let us briefly review the outstanding characteristics of mammals. They have a four-chambered heart and are homeothermic. They have a diaphragm, which increases breathing efficiency. There is increased separation (by the palate) of the respiratory and alimentary passages. The body is covered with an insulating layer of hair. The limbs are oriented ventrally and lift the body high off the ground. The lower jaw is composed of only one bone (compared with six or more in most reptiles), and the teeth are complexly differentiated for a variety of functions. There are three bones in the middle ear (compared with one in reptiles and birds). The brain, particularly the neocortex, is much larger than in reptiles, and behavior is more easily modifiable by experience. No eggs are laid (except in monotremes); embryonic development occurs in the uterus of the mother, and the young are born alive. After birth, the young are nourished on milk secreted by the mammary glands of the mother.

As indicated above, one small group—the monotremes—is fundamentally different from all other members of the class. These mammals lay eggs; yet they secrete milk. In many other ways, they are a curious blend of reptilian traits, mammalian traits, and traits peculiar to themselves. It seems clear that they were a very early offshoot of the mammalian lineage and were not ancestral to the other mammals. Some biologists think they should be considered mammal-like reptiles rather than reptilelike mammals. The only living monotremes are the spiny anteater and the duck-billed platypus, both found in Australia (Fig. 22.50).

22.50. Duck-billed platypus, an egg-laying mammal. [New York Zoological Society photo.]

The main stem of mammalian evolution split into two parts very early, one leading to the marsupials and the other to the placentals. The characteristic difference between them is that marsupial embryos remain in the uterus for a relatively short time and then complete their development while attached to a nipple in an abdominal pouch of the mother (Fig. 22.51), whereas placental embryos complete their development in the uterus.

The living placental mammals are classified in approximately 16 orders, several of which contain species familiar to almost everyone. A few of the most important orders are listed below:

INSECTIVORA.   Moles, shrews
CHIROPTERA.   Bats
PRIMATES.   Lemurs, monkeys, apes, men
EDENTATA.   Sloths, anteaters, armadillos
LAGOMORPHA.   Rabbits, hares, pikas
RODENTIA.   Rats, mice, squirrels, gophers, beavers, porcupines
CETACEA.   Whales, dolphins, porpoises
CARNIVORA.   Cats, dogs, bears, raccoons, weasels, skunks, minks, badgers, otters, hyenas, seals, walruses
PROBOSCIDEA.   Elephants
PERISSODACTYLA.   Odd-toed ungulates (hoofed animals): horses, zebras, tapirs, rhinoceroses
ARTIODACTYLA.   Even-toed ungulates: pigs, hippopotamuses, camels, deer, giraffes, antelopes, cattle, sheep, goats, bison

22.51. Euro kangaroo with young in pouch. [New York Zoological Society photo.]

The oldest fossils identified as placental mammalian ones are from the Jurassic. They are of small, probably secretive creatures that are thought to have fed primarily on insects. They remained a relatively unimportant part of the fauna until the end of the Mesozoic. Of the modern orders, Insectivora is closest to this ancient group. The great radiation from the insectivore ancestors dates from the beginning of the Cenozoic era, as the mammals rapidly filled the many niches left open by the demise of the dinosaurs. The Cenozoic, which includes the present, is aptly termed the Age of Mammals.

Evolution of the Primates

As members of the mammalian order Primates, we naturally have a special interest in its evolutionary history, and, in particular, in that part of its history that concerns the origin of man.

Fossil evidence indicates that the primates arose from an arboreal stock of small shrewlike insectivores very early in the Cenozoic. The groups soon split into several evolutionary lines that have had independent histories ever since. Though the modern representatives of these evolutionary lines are a rather heterogeneous lot, most of them share the following characteristics: (1) retention of the clavicle (collarbone), which is greatly reduced or lost in many other mammals; (2) development of a shoulder joint permitting relatively free movements in all directions, and an elbow joint permitting some rotational movement; (3) retention of five functional digits on each foot; (4) enhanced individual mobility of the digits, especially the first digits (thumb and big toe), which are usually apposable; (5) modification of the claws into flattened nails; (6) development of sensitive tactile pads on the digits; (7) abbreviation of the snout or muzzle; (8) elaboration of the visual apparatus and development of binocular vision; (9) expansion of the brain, particularly the cerebral cortex; (10) usually only two mammae; (11) usually only one young per pregnancy. Most of these traits are correlated with an arboreal way of life.

In quadrupedal terrestrial mammals, the limbs function as props and as instruments of propulsion for running and galloping; they have tended to evolve toward greater stability at the expense of freedom of movement. Think of the forelimbs of a dog or a horse: The clavicle is greatly reduced or lost; the two limbs are positioned close together under the animal, and their movement is restricted largely to one plane (i.e. they can move easily back and forth but cannot be spread far to the side like human arms). By contrast, in an animal leaping about in the branches of a tree, the limbs function in grasping and swinging; mobility at the shoulder, elbow, and digit joints facilitates such activities, as does attachment of the limbs (braced by the clavicle) far apart at the sides of the body instead of underneath.

The eyes of many quadrupedal terrestrial mammals (e.g. horses, cows, dogs) are located on the sides of the head. As a result, they can survey a very wide total visual field, but the fields of the two eyes overlap only slightly; i.e. the animals do not have binocular stereoscopic (three-dimensional) vision. But stereoscopic vision aids in localizing near objects, and an animal jumping from limb to limb obviously must be able to detect very accurately the position of the next limb. Hence the arboreal way of life of the early primates doubtless led to selection for

22.52. Tarsier. Note the huge eyes of this nocturnal animal and the highly developed digits. [New York Zoological Society photo.]

stereoscopic vision and, consequently, for eyes directed forward rather than laterally. This change, in turn, would have led to the distinctive flattened forward-directed face of most higher primates.

Now, hands capable of grasping the next limb and keen eyes with broadly overlapping fields of vision would not by themselves have met the requirements of an arboreal way of life. Essential, too, would have been neural and muscular mechanisms capable of very precise eye-hand coordination. This need was doubtless one of the factors that led to the early expansion of the primate brain.

We could continue in this manner, relating other characteristics of primates to the demands of arboreal life, but the point has been made: Many of the traits most important to us as human beings first evolved because our distant ancestors lived in trees.

The prosimians and the monkeys.   The living primates are usually classified in two suborders: Prosimii and Anthropoidea. The first, the prosimians ("pre-monkeys"), are a miscellaneous group of more or less primitive primates, including the lemurs, aye-ayes, lorises, pottos, galagos, and tarsiers (Fig. 22.52).

The first members of the suborder Anthropoidea had diverged from a prosimian stock by the Oligocene epoch. Actually, two lines of anthropoids probably arose at about the same time from closely related prosimians. One of these led to the New World monkeys, and the other, which soon split, led to the Old World monkeys and to the apes. We may diagram these hypothetical relationships as shown at right.

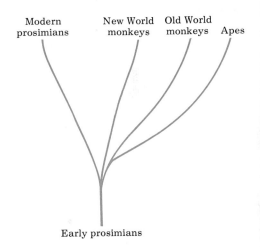

The following is the formal classification within the anthropoid suborder:

Suborder Anthropoidea
    Superfamily Ceboidea
        Family Cebidae, New World monkeys
        Family Callithricidae, marmosets
    Superfamily Cercopithecoidea
        Family Cercopithecidae, Old World monkeys and baboons
    Superfamily Hominoidea
        Family Pongidae, apes
        Family Hominidae, men

The apes.   The living great apes (Pongidae) fall into four groups: gibbons, orangutans, gorillas, and chimpanzees. All are fairly large animals that have no tail, a relatively large skull and brain, and very long arms. All have a tendency, when on the ground, to walk semi-erect.

The gibbons, of which several species are found in southeast Asia, represent a lineage that probably split from the others soon after the pongid line itself arose. They are the smallest of the apes (about 3 feet tall when standing). Their arms are exceedingly long, reaching the ground even when the animal is standing erect. The gibbons are amazing arboreal acrobats and spend almost all their time in trees.

The one living species of orangutan is native to Sumatra and Borneo. Though the orangs are fairly large, and their movements slow and deliberate, they nevertheless spend most of their time in trees and only rarely descend to the ground.

Gorillas, of which there are two forms in Africa, are the largest of

22.53. Chimpanzee. [New York Zoological Society photo.]

the apes; wild adult males may weigh as much as 450 pounds (up to 600 pounds in zoos) and stand 6 feet tall. Their arms, while proportionately much longer than those of man, are not as long as those of gibbons and orangs. Unlike gibbons and orangs, gorillas spend most of their time on the ground. Despite their fierce appearance, they are not usually aggressive.

Chimpanzees, which are native to tropical Africa, have been used extensively in psychological experiments. In general appearance, they are the most manlike of the living apes (Fig. 22.53). They are about the same size as orangs, but their arms are shorter. Although they spend most of their time in trees, they descend to the ground more frequently than orangs, and sometimes even adopt a bipedal position (their usual locomotion, however, is quadrupedal, with the knuckles of the hand used for support). They are quite intelligent and can learn to perform a variety of tasks.

The evolution of man

The earliest members of the family Hominidae (men) probably arose from the same pongid stock that produced the gorillas and chimpanzees. Both paleontological evidence and biochemical and serological data indicate that gorillas, chimpanzees, and man are more closely related, in terms of recency of common ancestry, than any one of them is to orangutans or gibbons. We can diagram the relationships as follows:

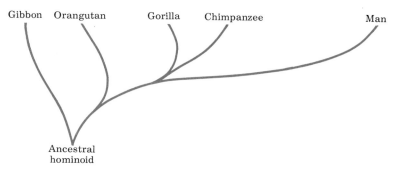

Gibbon   Orangutan   Gorilla   Chimpanzee   Man

Ancestral
hominoid

The current conception of the common ancestor of modern apes (at least of gorillas and chimpanzees) and man is based largely on fossils of several species assigned to the genus *Dryopithecus,* which ranged widely over Europe, Africa, and Asia during the Miocene epoch.[14] These animals had a skull with a low rounded cranium, moderate supraorbital ridges, and moderate forward projection of face and jaws. The arms were only modestly specialized for brachiation—swinging from branch to branch—and the feet indicate some tendency toward bipedal posture.

Only a few of the many anatomical changes that occurred in the course of evolution from ape ancestor to modern man can be mentioned here: (1) The jaw became shorter (making the muzzle shorter), and the teeth became smaller. (2) The point of attachment of the skull to the vertebral column shifted from the rear of the braincase to a position under the braincase, so that the skull was balanced more on top of the vertebral column. (3) The braincase became much larger, and, as it did, a prominent vertical forehead developed. (4) The eyebrow ridges

[14] Included in *Dryopithecus,* as here used, are the African forms sometimes called *Proconsul.*

and other keels on the skull were reduced as the muscles that once attached to them became smaller. (5) The nose became more prominent, with a distinct bridge and tip. (6) The arms became shorter. (7) The feet became flattened, and then an arch developed. (8) The big toe moved back into line with the other toes and ceased being apposable. The various fossil men are intermediate in these characteristics.

Several traits that man inherited from his ape ancestors were important in leading to the development of one of his prime abilities— the making and use of tools. The long widely spaced arms of apes with their special shoulder and elbow joints, while ideal for moving about in trees, were not well suited for terrestrial locomotion. Hence, when on the ground, the animals tended to adopt an upright or semi-upright posture, thus freeing their forelimbs for the performance of functions other than walking. And the hands, with their apposable thumbs originally evolved for grasping tree branches, were preadapted for manipulative actions. Again we see that our "human" traits owe much to our ape ancestors' arboreal existence.

Fossil men.   The first human fossil bones were found in 1856 in Germany. In the years since, many bones of ancient men have been discovered. At first, the tendency of anthropologists was to erect both a new genus and a new species for each new find, without regard to biological criteria for erecting such categories. The result was a very long list of names that gave no indication of the probable relationships of the organisms they designated. More recently, however, the modern biological ideas concerning speciation and intraspecific variation have been increasingly applied to the study of fossil man, and this, together with the discovery of many new fossils, particularly in Africa, has begun to improve our understanding of human evolution.

The oldest fossils now assigned to the family Hominidae are of *Ramapithecus punjabicus*[15] from the late Miocene (about 14 million years ago). This was an apelike creature, but one that exhibited early stages in reduction of the size of the incisors and canines—a change that may have been correlated with increased use of the hands rather than the teeth in obtaining food. Unfortunately the only fossils of *Ramapithecus* found to date are teeth and jaw fragments; we know nothing, therefore, of its cranial capacity or stance.

The first truly manlike hominids are usually assigned to the genus *Australopithecus* (from *australis,* southern, and *pithecus,* ape—"southern," because the first specimens were found in South Africa). Until recently, the emergence of man was dated less than two million years ago, in the Pleistocene epoch, to which the oldest known fossils of *Australopithecus* were assigned. But with the discovery in 1969 and 1971 of far more ancient remains of *Australopithecus* in Ethiopia and Kenya, that estimate had to be revised; it now appears that the first hominids lived at least four million years ago, in the Pliocene epoch.

The *Australopithecus* species, often called South African ape men, were apparently fully bipedal, and they probably used simple bone tools. They had large jaws but almost no forehead or chin, and their cranial capacity was only about 450–700 cubic centimeters (compared with

---

[15] Included in *Ramapithecus punjabicus* are the fossils formerly called *Kenyapithecus.*

350–450 for normal chimpanzees and 1,200–1,600 for normal modern men) (Fig. 22.54). The fossils probably represent at least two species, which apparently lived contemporaneously for some three million years. One, which had considerably larger teeth than the other, was probably primarily herbivorous. The other was probably a predator, and is thought to be the more likely ancestor of later human lineages.[16]

A later stage in human evolution is represented by fossils that may be classified as *Homo erectus*[17] (often called Java man) (Fig. 22.55). Specimens have been found in Asia and Africa, and perhaps Europe. This species first appeared about 750,000 years ago. The cranial capacity was considerably larger than that of *Australopithecus*, normally ranging from 775 to 1,100 cubic centimeters. However, the facial features remained primitive, with a projecting massive jaw, large teeth, almost no chin, a receding forehead, heavy bony eyebrow ridges, and a broad low-bridged nose. Not only did *Homo erectus* make and use tools, but he also used fire. Casts of the interior of the skulls indicate the presence of the speech areas of the brain; of course we have no way of knowing whether or how language was used.

Modern man is given the Latin name *Homo sapiens* (wise man). Early representatives of this species, which probably first appeared about 40,000 years ago (the date is disputed), are often called Cro-Magnon man (Fig. 22.55). Unfortunately we are uncertain where the early evolutionary stages of *Homo sapiens* occurred. The species appeared rather suddenly in Europe, probably having migrated from Africa or southern Asia.

When *Homo sapiens* first settled in Europe, he coexisted for a time with a related form already established there. This form, called Neanderthal man (*Homo neanderthalensis*), is the one to which the bones discovered in 1856 belonged (Fig. 22.55). Neanderthal man was about 5–5½ feet tall, and had a receding forehead, prominent eyebrow ridges, and a receding chin, but his brain was as big as that of modern man (perhaps a little bigger). He made many kinds of tools and even buried his dead, which has been interpreted as a capacity on his part for abstract and religious thought. Neanderthal man disappeared soon after *Homo sapiens* arrived in his range, presumably eliminated by either combat or competition.[18]

The races of man. As we saw in Chapter 17, widespread species often tend to become subdivided into geographic races. Man is no exception. *Homo sapiens* is an extremely variable species, and regional populations are often recognizably different (or were before the great

---

[16] The smaller-toothed species, *Australopithecus africanus*, includes specimens from Olduvai Bed I (in Tanzania) called *Homo habilis* by Leakey. The larger-toothed species, usually called *Australopithecus robustus* (though *paleojavanicus*, the older name, may be technically the valid one), includes the forms originally described in *Paranthropus*, *Meganthropus*, and *Zinjanthropus*.

[17] *Homo erectus* includes the specimens from Olduvai Bed II called *Homo habilis* by Leakey, and also the forms originally described in *Pithecanthropus*, *Sinanthropus*, *Telanthropus*, and *Atlanthropus*.

[18] Neanderthal man is here regarded as a separate species. Some workers, who consider this form a subspecies of *Homo sapiens* and designate it *Homo sapiens neanderthalensis*, have suggested that Neanderthal man's disappearance may not have been so much an elimination as an absorption into *Homo sapiens sapiens* as a result of interbreeding.

22.54. Reconstruction of *Australopithecus africanus*. [Courtesy American Museum of Natural History.]

22.55. Prehistoric men. Top: Restorations of (left to right) Java man, Neanderthal man, and Cro-Magnon man. Bottom: Lateral view of restored skulls of the same three types. Note differences in size of braincase, height of forehead, size of eyebrow ridge, length of jaw, size of teeth, etc. [Courtesy American Museum of Natural History.]

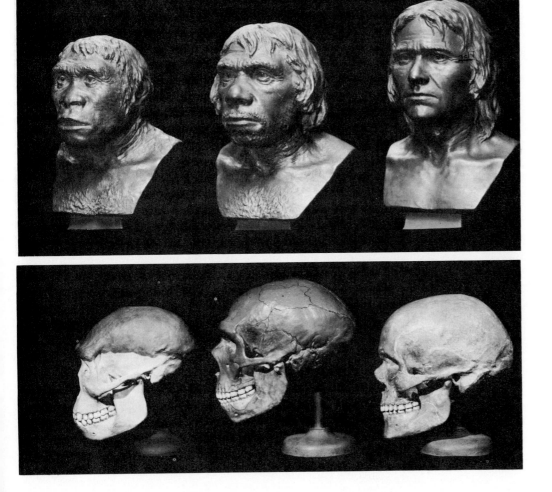

mobility of the last few centuries began). Thus Scandinavians tend to have blue eyes and a fair complexion, while south Europeans tend to have brown eyes and a darker complexion. Eskimos look different from Mohawk Indians, and they in turn look different from Apaches. Pygmies of the Congo are obviously different from their taller neighbors. Many of the differences probably reflect adaptations to different environmental conditions. Thus, for example, the prevalence of darker skin in tropical and subtropical regions may be a protective adaptation against damaging ultraviolet solar radiation.

Now, races, by definition, are regional populations that differ genetically but have no effective intrinsic isolating mechanisms. There are seldom sharp boundaries between them, and they intergrade over wide areas. Designation of races in most species is thus an arbitrary matter, and there is no such thing as a "pure" race. An almost unlimited number of different races of the snake *Coluber constrictor* could be erected, depending on which of the characters whose distributions are illustrated in Fig. 17.7 (p. 377) are chosen for emphasis. The same considerations apply to man. Some authorities have chosen to recognize as many as 30 different races, while others recognize only three: the traditional Caucasoid, Mongoloid, and Negroid. Another widely used system recognizes five: the traditional three, plus American Indians and Australian aborigines.

No one of these systems has any more biological validity than the others, since races, as categories, are human inventions. What is biologically real is the geographic variation within the species *Homo sapiens,* a variation that will surely tend to break down as men move about more and more. The main barriers to interbreeding in many parts of the world are now cultural or social rather than geographic, and it seems very unlikely that such barriers will even approach the potency of the original geographic barriers. Hence, whatever races are recognized now, it seems probable that they will become less and less distinct as time goes on. This, too, is a phenomenon that has occurred countless times in other species.

The interaction of cultural and biological evolution. One of the most interesting discoveries of recent years is that early hominids used tools long before their brains were much larger than those of apes. Thus the old idea that a large brain and high intelligence were necessary prerequisites for the use of tools has been discredited. Early man's use of tools may, in fact, have been an important factor in leading to evolution of higher intelligence. Once the use and making of tools began, the more adept would surely have had an advantage over their less talented fellows, with a resulting strong selection for neural mechanisms favoring improved fashioning and use of tools. Thus perhaps, instead of considering culture the crown of man's fully evolved intelligence, we should regard early cultural development and increasing intelligence as two faces of the same coin; in a sense, the highly developed brain of modern man may be as much a consequence as a cause of culture.

Cultural evolution can proceed at a far more rapid pace than biological evolution. Words are much more effective as units of inheritance than genes in spreading new developments and giving dominance to new approaches originated by a few talented individuals. But the two types of evolution continue to be interwoven just as they were in the use of tools; they may well be even more so in the future.

Man, by his unrivaled ability to alter his environment, is influencing in profound ways the evolution of all species with which he comes in contact. Thus, as we have seen, new strains of bacteria have evolved in response to man's use of antibiotics; many species of insects have evolved new physiological and behavioral traits as a consequence of the intense selection resulting from use of insecticides; the clearing of forests for agricultural purposes has led to drastic decline in the population densities of some species and to increase in others; long-established balances between prey species and their predators and parasites have been destroyed, often with far-reaching consequences for the entire ecosystem. Since disruption of the ecosystem will unavoidably grow with the ever wider demand for better living standards, the great challenge to man is to use the resources of his knowledge and technology to direct change in ways that will benefit both his own species and the other organisms around him.

Not only can man influence the evolution of other species, but he can now deliberately alter some aspects of the future evolution of his own species. Thus modern medicine, by saving people with gross genetic defects that would once have been fatal, permits perpetuation of genes that natural selection would formerly have eliminated. Does this mean that man should practice eugenics—deliberately restrict, by law or by social pressure, the perpetuation of some genetic traits and encourage the perpetuation of others? Some persons, apprehensive over what they see as the inevitable "decline" of the species, have urged just that. Others have pointed out that, although there may indeed be an increase in traits that would once have been maladaptive, modern man lives in an environment of his own design, in which those traits are no longer so deleterious.

Man's ability to control his own evolution may eventually no longer depend primarily on regulating reproduction. Now that the genetic code has been deciphered, the day will surely come when the DNA of genes can deliberately be altered in order to design, at least in part, new human beings. When that day comes, how do we decide what to design? What shall we look for in human beings? We might all agree to rid our species of the genes for muscular dystrophy or sickle-cell anemia (at least in areas where there is no malaria), but once the techniques for achieving these apparently worthy ends are mastered, suggestions will surely be put forward for other alterations to which we cannot all agree.

More immediately pressing, perhaps, is the problem of regulating the size of human populations now that we have interfered with the action of the many former natural regulating factors. The current population problem is the result of major advances in the technology of death control—prevention and cure of once fatal diseases, improved nutrition—which societies have generally been as eager to accept as they have been reluctant to accept compensatory birth control. No species can continue indefinitely with its birth and death rates unbalanced. There are but two possible solutions, which many find equally repellent: more birth control or less death control.

These complex and unnerving problems—at once biological, economic, political, and moral—must be faced now. Men have already gone too far toward modifying biological evolution to pull back. How the next few generations meet these problems may well have as profound an influence on the future of life as anything that has happened since the first cells appeared in the primordial seas.

REFERENCES

BARNES, R. D., 1968. *Invertebrate Zoology,* 2nd ed. Saunders, Philadelphia.

BORROR, D. J., and D. M. DELONG, 1971. *An Introduction to the Study of Insects,* 3rd ed. Holt, Rinehart & Winston, New York.

CLARK, W. E. LEGROS, 1959. *The Antecedents of Man.* Edinburgh University Press, Edinburgh. (Paperback edition, Harper Torchbooks, New York, 1963.)

COMSTOCK, J. H., 1940. *An Introduction to Entomology,* 9th ed. Comstock, Ithaca, N.Y.

HYMAN, L. H., 1940–1967. *The Invertebrates* (6 vols.). Vol. 1: *Protozoa Through Ctenophora,* 1940; vol. 2: *Platyhelminthes and Rhynchocoela,* 1951; vol. 3: *Acanthocephala, Aschelminthes, and Entoprocta,* 1951; vol. 4: *Echinodermata,* 1955; vol. 5: *Smaller Coelomate Groups,* 1959; vol. 6: *Mollusca I,* 1967. McGraw-Hill, New York.

IMMS, A. D., 1957. *A General Textbook of Entomology,* 9th ed. Methuen, London.

KUDO, R. R., 1971. *Protozoology,* 5th ed. Charles C Thomas, Springfield, Ill.

LAGLER, K. F., J. E. BARDACH, and R. R. MILLER, 1962. *Ichthyology.* Wiley, New York.

ROMER, A. S., 1966. *Vertebrate Paleontology,* 3rd ed. University of Chicago Press, Chicago.

———, 1970. *The Vertebrate Body,* 4th ed. Saunders, Philadelphia.

STORER, T. I., and R. L. USINGER, 1972. *General Zoology,* 5th ed. McGraw-Hill, New York.

VILLEE, C. A., W. F. WALKER, and F. E. SMITH, 1968. *General Zoology,* 3rd ed. Saunders, Philadelphia.

WELTY, J. C., 1962. *The Life of Birds.* Saunders, Philadelphia.

YOUNG, J. Z., 1957. *The Life of Mammals.* Oxford University Press, New York.

———, 1962. *The Life of Vertebrates,* 2nd ed. Oxford University Press, New York.

SUGGESTED READING

AUSTIN, O. L., 1961. *Birds of the World.* Golden Press, New York.

BUCHSBAUM, R., 1972. *Animals Without Backbones,* rev. 2nd ed. University of Chicago Press, Chicago.

———, and L. J. MILNE, 1960. *The Lower Animals: Living Invertebrates of the World.* Doubleday, Garden City, N.Y.

CLARK, J. D., 1958. "Early Man in Africa," *Scientific American,* July. (Offprint 820.)

COCHRAN, D. M., 1961. *Living Amphibians of the World.* Doubleday, Garden City, N.Y.

FINGERMAN, M., 1969. *Animal Diversity.* Holt, Rinehart & Winston, New York.

GILLIARD, E. T., 1958. *Living Birds of the World.* Doubleday, Garden City, N.Y.

GLAESSNER, M. F., 1961. "Pre-Cambrian Animals," *Scientific American,* March. (Offprint 837.)

HANSON, E. D., 1972. *Animal Diversity,* 3rd ed. Prentice-Hall, Englewood Cliffs, N.J.

HASELDEN, K., and P. HEFNER, eds., 1968. *Changing Man: The Threat and the Promise.* Doubleday, Garden City, N.Y.

HERALD, E. S., 1961. *Living Fishes of the World.* Doubleday, Garden City, N.Y.

KLOTS, A. B., and E. B. KLOTS, 1959. *Living Insects of the World.* Doubleday, Garden City, N.Y.

MILLOT, J., 1955. "The Coelacanth," *Scientific American,* December. (Offprint 831.)

NEWELL, N. D., 1963. "Crises in the History of Life," *Scientific American,* February. (Offprint 867.)

PILBEAM, D. R., and E. L. SIMONS, 1965. "Some Problems of Hominid Classification," *American Scientist,* vol. 53, pp. 237–259.

ROMER, A. S., 1971. *The Vertebrate Story,* rev. ed. (titled *Man and the Vertebrates* in older editions). University of Chicago Press, Chicago.

SANDERSON, I. T., 1955. *Living Mammals of the World.* Doubleday, Garden City, N.Y.

SCHMIDT, K. P., and R. F. INGER, 1957. *Living Reptiles of the World.* Doubleday, Garden City, N.Y.

SIMONS, E. L., 1964. "The Early Relatives of Man," *Scientific American,* July. (Offprint 622.)

WASHBURN, S. L., 1960. "Tools and Human Evolution," *Scientific American,* September. (Offprint 601.)

WECKLER, J. E., 1957. "Neanderthal Man," *Scientific American,* December. (Offprint 844.)

APPENDIX

A classification of living things

The classification given here is one of many in current use. Some other systems recognize more or fewer divisions and phyla, and combine or divide classes in a variety of other ways; but compared with the large areas of agreement, the differences between the various classifications are minor. Chapters 19–22 discuss certain of the points at issue between advocates of different systems.

Most classes are listed here, and for some classes (e.g. Insecta and Mammalia) orders are also given. Except for a few extinct groups of particular evolutionary importance (e.g. Placodermi), only groups with living representatives are included. A few of the better-known genera are mentioned as examples in each of the taxons.

KINGDOM MONERA

DIVISION SCHIZOMYCETES. Bacteria[*]

CLASS MYXOBACTERIA. *Myxococcus, Chondromyces*

CLASS SPIROCHETES. *Leptospira, Cristispira, Spirocheta, Treponema*

CLASS EUBACTERIA. *Staphylococcus, Escherichia, Salmonella, Pasteurella, Streptococcus, Bacillus, Spirillum, Caryophanon*

CLASS RICKETTSIAE. *Rickettsia, Coxiella*

CLASS BEDSONIA. *Chlamydia*

DIVISION CYANOPHYTA. Blue-green algae. *Gloeocapsa, Microcystis, Oscillatoria, Nostoc, Scytonema*

[*] There is no generally accepted classification for bacteria at the class level. The system used here is based on R. Y. Stanier, M. Doudoroff, and E. A. Adelberg, *The Microbial World* (3rd ed.; Prentice-Hall, 1970), though these authors do not formally designate their categories as classes. Other systems usually recognize more classes, some of them very small.

KINGDOM PLANTAE

DIVISION EUGLENOPHYTA. Euglenoids. *Euglena, Eutreptia, Phacus, Colacium*

DIVISION CHLOROPHYTA. Green algae

CLASS CHLOROPHYCEAE. True green algae. *Chlamydomonas, Volvox, Ulothrix, Spirogyra, Oedogonium, Ulva*

CLASS CHAROPHYCEAE. Stoneworts. *Chara, Nitella, Tolypella*

DIVISION CHRYSOPHYTA

CLASS XANTHOPHYCEAE. Yellow-green algae. *Botrydiopsis, Halosphaera, Tribonema, Botrydium.*

CLASS CHRYSOPHYCEAE. Golden-brown algae. *Chrysamoeba, Chromulina, Synura, Mallomonas*

CLASS BACILLARIOPHYCEAE. Diatoms. *Pinnularia, Arachnoidiscus, Triceratium, Pleurosigma*

DIVISION PYRROPHYTA. Dinoflagellates. *Gonyaulax, Gymnodinium, Ceratium, Gloeodinium*

DIVISION PHAEOPHYTA. Brown algae. *Sargassum, Ectocarpus, Fucus, Laminaria*

DIVISION RHODOPHYTA. Red algae. *Nemalion, Polysiphonia, Dasya, Chondrus, Batrachospermum*

DIVISION MYXOMYCOPHYTA. Slime molds

CLASS MYXOMYCETES. True slime molds. *Physarum, Hemitrichia, Stemonitis*

CLASS ACRASIAE. Cellular slime molds. *Dictyostelium*

CLASS PLASMODIOPHOREAE. Endoparasitic slime molds. *Plasmodiophora*

CLASS LABYRINTHULEAE. Net slime molds. *Labyrinthula*

DIVISION EUMYCOPHYTA. True fungi

CLASS PHYCOMYCETES. Algal fungi. *Rhizopus, Mucor, Phycomyces*

CLASS OÖMYCETES. Water molds, white rusts, downy mildews. *Saprolegnia, Phytophthora, Albugo*

CLASS ASCOMYCETES. Sac fungi. *Neurospora, Aspergillus, Penicillium, Saccharomyces, Morchella, Ceratostomella*

CLASS BASIDIOMYCETES. Club fungi. *Ustilago, Puccinia, Coprinus, Lycoperdon, Psalliota, Amanita*

DIVISION BRYOPHYTA

CLASS HEPATICAE. Liverworts. *Marchantia, Conocephalum, Riccia, Porella*

CLASS ANTHOCEROTAE. Hornworts. *Anthoceros*

CLASS MUSCI. Mosses. *Polytrichum, Sphagnum, Mnium*

DIVISION TRACHEOPHYTA. Vascular plants

Subdivision Psilopsida. *Psilotum, Tmesipteris*

Subdivision Lycopsida. Club mosses. *Lycopodium, Phylloglossum, Selaginella, Isoetes, Stylites*

Subdivision Sphenopsida. Horsetails. *Equisetum*

Subdivision Pteropsida. Ferns. *Polypodium, Osmunda, Dryopteris, Botrychium, Pteridium*

Subdivision Spermopsida. Seed plants

CLASS PTERIDOSPERMAE. Seed ferns. No living representatives

CLASS CYCADAE. Cycads. *Zamia*

CLASS GINKGOAE. *Ginkgo*

CLASS CONIFERAE. Conifers. *Pinus, Tsuga, Taxus, Sequoia*

CLASS GNETEAE. *Gnetum, Ephedra, Welwitschia*

CLASS ANGIOSPERMAE. Flowering plants

SUBCLASS DICOTYLEDONEAE. Dicots. *Magnolia, Quercus, Acer, Pisum, Taraxacum, Rosa, Chrysanthemum, Aster, Primula, Ligustrum, Ranunculus*

SUBCLASS MONOCOTYLEDONEAE. Monocots. *Lilium, Tulipa, Poa, Elymus, Triticum, Zea, Ophyrys, Yucca, Sabal*

KINGDOM ANIMALIA

SUBKINGDOM PROTOZOA

PHYLUM PROTOZOA. Acellular animals

Subphylum Plasmodroma

CLASS FLAGELLATA (or Mastigophora). Flagellates. *Trypanosoma, Calonympha, Chilomonas* (also *Euglena, Chlamydomonas,* and other green flagellates included in Plantae as well)

CLASS SARCODINA. Protozoans with pseudopodia. *Amoeba, Pelomyxa, Globigerina, Textularia, Acanthometra*

CLASS SPOROZOA. *Plasmodium, Monocystis*

Subphylum Ciliophora

CLASS CILIATA. Ciliates. *Paramecium, Opalina, Stentor, Vorticella, Spirostomum*

SUBKINGDOM PARAZOA

PHYLUM PORIFERA. Sponges

CLASS CALCAREA. Calcareous (chalky) sponges. *Scypha, Leucosolenia, Sycon, Grantia*

CLASS HEXACTINELLIDA. Glass sponges. *Euplectella, Hyalonema, Monoraphis*

CLASS DEMOSPONGIAE. *Spongilla, Euspongia, Axinella*

SUBKINGDOM MESOZOA

PHYLUM MESOZOA. *Dicyema, Pseudicyema, Rhopalura*

SUBKINGDOM METAZOA

SECTION RADIATA

PHYLUM COELENTERATA (or Cnidaria)

CLASS HYDROZOA. Hydrozoans. *Hydra, Obelia, Gonionemus, Physalia*

CLASS SCYPHOZOA. Jellyfishes. *Aurelia, Pelagia, Cyanea*

CLASS ANTHOZOA. Sea anemones and corals. *Metridium, Pennatula, Gorgonia, Astrangia*

PHYLUM CTENOPHORA. Comb jellies

CLASS TENTACULATA. *Pleurobrachia, Mnemiopsis, Cestum, Velamen*

CLASS NUDA. *Beroe*

SECTION PROTOSTOMIA

PHYLUM PLATYHELMINTHES. Flatworms

CLASS TURBELLARIA. Free-living flatworms. *Planaria, Dugesia, Leptoplana*

CLASS TREMATODA. Flukes. *Fasciola, Schistosoma, Prosthogonimus*

CLASS CESTODA. Tapeworms. *Taenia, Dipylidium, Mesocestoides*

PHYLUM NEMERTINA (or Rhynchocoela). Proboscis worms. *Cerebratulus, Lineus, Malacobdella*

PHYLUM ACANTHOCEPHALA. Spiny-headed worms. *Echinorhynchus, Gigantorhynchus*

PHYLUM ASCHELMINTHES

CLASS ROTIFERA. Rotifers. *Asplanchna, Hydatina, Rotaria*

CLASS GASTROTRICHA. *Chaetonotus, Macrodasys*

CLASS KINORHYNCHA (or Echinodera). *Echinoderes, Semnoderes*

CLASS NEMATODA. Round worms. *Ascaris, Trichinella, Necator, Enterobius, Ancylostoma, Heterodera*

CLASS NEMATOMORPHA. Horsehair worms. *Gordius, Paragordius, Nectonema*

PHYLUM ENTOPROCTA. *Urnatella, Loxosoma, Pedicellina*

PHYLUM PRIAPULIDA. *Priapulus, Halicryptus*

PHYLUM ECTOPROCTA (or Bryozoa). Bryozoans, moss animals

CLASS GYMNOLAEMATA. *Paludicella, Bugula*

CLASS PHYLACTOLAEMATA. *Plumatella, Pectinatella*

PHYLUM PHORONIDA. *Phoronis, Phoronopsis*

PHYLUM BRACHIOPODA. Lamp shells

CLASS INARTICULATA. *Lingula, Glottidia, Discina*

CLASS ARTICULATA. *Magellania, Neothyris, Terebratula*

PHYLUM MOLLUSCA. Molluscs

CLASS AMPHINEURA

SUBCLASS APLACOPHORA. Solenogasters. *Chaetoderma, Neomenia, Proneomenia*

SUBCLASS POLYPLACOPHORA. Chitons. *Chaetopleura, Ischnochiton, Lepidochiton, Amicula*

CLASS MONOPLACOPHORA. *Neopilina*

CLASS GASTROPODA. Snails and their allies (univalve molluscs). *Helix, Busycon, Crepidula, Haliotis, Littorina, Doris, Limax*

CLASS SCAPHOPODA. Tusk shells. *Dentalium, Cadulus*

CLASS PELECYPODA. Bivalve molluscs. *Mytilus, Ostrea, Pecten, Mercenaria, Teredo, Tagelus, Unio, Anodonta*

CLASS CEPHALOPODA. Squids, octopuses, etc. *Loligo, Octopus, Nautilus*

PHYLUM SIPUNCULIDA. *Sipunculus, Phascolosoma, Dendrostomum*

PHYLUM ECHIURIDA. *Echiurus, Urechis, Thalassema*

PHYLUM ANNELIDA. Segmented worms

CLASS POLYCHAETA (including Archiannelida). Sandworms, tubeworms, etc. *Nereis, Chaetopterus, Aphrodite, Diopatra, Arenicola, Hydroides, Sabella*

CLASS OLIGOCHAETA. Earthworms and many freshwater annelids. *Tubifex, Enchytraeus, Lumbricus, Dendrobaena*

CLASS HIRUDINEA. Leeches. *Trachelobdella, Hirudo, Macrobdella, Haemadipsa*

PHYLUM ONYCHOPHORA. *Peripatus, Peripatopsis*

PHYLUM TARDIGRADA. Water bears. *Echiniscus, Macrobiotus*

PHYLUM PENTASTOMIDA. *Cephalobaena, Linguatula*

PHYLUM ARTHROPODA

Subphylum Trilobita. No living representatives

Subphylum Chelicerata

CLASS EURYPTERIDA. No living representatives

CLASS XIPHOSURA. Horseshoe crabs. *Limulus*

CLASS ARACHNIDA. Spiders, ticks, mites, scorpions, whipscorpions, daddy longlegs, etc. *Archaearanea, Latrodectus, Argiope, Centruroides, Chelifer, Mastigoproctus, Phalangium, Ixodes*

CLASS PYCNOGONIDA. Sea spiders. *Nymphon, Ascorhynchus*

Subphylum Mandibulata

CLASS CRUSTACEA. *Homarus, Cancer, Daphnia, Artemia, Cyclops, Balanus, Porcellio*

CLASS CHILOPODA. Centipeds. *Scolopendra, Lithobius, Scutigera*

CLASS DIPLOPODA. Millipeds. *Narceus, Apheloria, Polydesmus, Julus, Glomeris*

CLASS PAUROPODA. *Pauropus*

CLASS SYMPHYLA. *Scutigerella*

CLASS INSECTA. Insects

ORDER COLLEMBOLA. Springtails. *Isotoma, Achorutes, Neosminthurus, Sminthurus*

ORDER PROTURA. *Acerentulus, Eosentomon*

ORDER DIPLURA. *Campodea, Japyx*

ORDER THYSANURA. Bristletails, silverfish, firebrats. *Machilis, Lepisma, Thermobia*

ORDER EPHEMERIDA. Mayflies. *Hexagenia, Callibaetis, Ephemerella*

ORDER ODONATA. Dragonflies, damselflies. *Archilestes, Lestes, Aeshna, Gomphus*

ORDER ORTHOPTERA (including Isoptera). Grasshoppers, crickets, walking sticks, mantids, cockroaches, termites, etc. *Schistocerca, Romalea, Nemobius, Megaphasma, Mantis, Blatta, Periplaneta, Reticulitermes*

ORDER DERMAPTERA. Earwigs. *Labia, Forficula, Prolabia*

ORDER EMBIARIA (or Embiidina or Embioptera). *Oligotoma, Anisembia, Gynembia*

ORDER PLECOPTERA. Stoneflies. *Isoperla, Taeniopteryx, Capnia, Perla*

ORDER ZORAPTERA. *Zorotypus*

ORDER CORRODENTIA. Book lice. *Ectopsocus, Liposcelis, Trogium*

ORDER MALLOPHAGA. Chewing lice. *Cuclotogaster, Menacanthus, Menopon, Trichodectes*

ORDER ANOPLURA. Sucking lice. *Pediculus, Phthirius, Haematopinus*

ORDER THYSANOPTERA. Thrips. *Heliothrips, Frankliniella, Hercothrips*

ORDER HEMIPTERA (including Homoptera). Bugs, cicadas, aphids, leafhoppers, etc. *Belostoma, Lygaeus, Notonecta, Cimex, Lygus, Oncopeltus, Magicicada, Circulifer, Psylla, Aphis*

ORDER NEUROPTERA. Dobsonflies, alderflies, lacewings, mantispids, snakeflies, etc. *Corydalus, Hemerobius, Chrysopa, Mantispa, Agulla*

ORDER COLEOPTERA. Beetles, weevils. *Copris, Phyllophaga, Harpalus, Scolytus, Melanotus, Cicindela, Dermestes, Photinus, Coccinella, Tenebrio, Anthonomus, Conotrachelus*

ORDER HYMENOPTERA. Wasps, bees, ants, sawflies. *Cimbex, Vespa, Glypta, Scolia, Bembix, Formica, Bombus, Apis*

ORDER MECOPTERA. Scorpionflies. *Panorpa, Boreus, Bittacus*

ORDER SIPHONAPTERA. Fleas. *Pulex, Nosopsyllus, Xenopsylla, Ctenocephalides*

ORDER DIPTERA. True flies, mosquitoes. *Aedes, Asilus, Sarcophaga, Anthomyia, Musca, Chironomus, Tabanus, Tipula, Drosophila*

ORDER TRICHOPTERA. Caddisflies. *Limnephilus, Rhyacophila, Hydropsyche*

ORDER LEPIDOPTERA. Moths, butterflies. *Tinea, Pyrausta, Malacosoma, Sphinx, Samia, Bombyx, Heliothis, Papilio, Lycaena*

SECTION DEUTEROSTOMIA

PHYLUM CHAETOGNATHA. Arrow worms. *Sagitta, Spadella*

PHYLUM ECHINODERMATA

CLASS CRINOIDEA. Crinoids, sea lilies. *Antedon, Ptilocrinus, Comactinia*

CLASS ASTEROIDEA. Sea stars. *Asterias, Ctenodiscus, Luidia, Oreaster*

CLASS OPHIUROIDEA. Brittle stars, serpent stars, basket stars, etc. *Asteronyx, Amphioplus, Ophiothrix, Ophioderma, Ophiura*

CLASS ECHINOIDEA. Sea urchins, sand dollars, heart urchins. *Cidaris, Arbacia, Strongylocentrotus, Echinanthus, Echinarachnius, Moira*

CLASS HOLOTHUROIDEA. Sea cucumbers. *Cucumaria, Thyone, Caudina, Synapta*

PHYLUM POGONOPHORA. Beard worms. *Siboglinum, Lamellisabella, Oligobrachia, Polybrachia*

PHYLUM HEMICHORDATA

CLASS ENTEROPNEUSTA. Acorn worms. *Saccoglossus, Balanoglossus, Glossobalanus*

CLASS PTEROBRANCHIA. *Rhabdopleura, Cephalodiscus*

PHYLUM CHORDATA. Chordates

Subphylum Urochordata (or Tunicata). Tunicates

CLASS ASCIDIACEA. Ascidians or sea squirts. *Ciona, Clavelina, Molgula, Perophora*

CLASS THALIACEA. *Pyrosoma, Salpa, Doliolum*

CLASS LARVACEA. *Appendicularia, Oikopleura, Fritillaria*

Subphylum Cephalochordata. Lancelets, amphioxus. *Branchiostoma, Asymmetron*

Subphylum Vertebrata. Vertebrates

CLASS AGNATHA. Jawless fishes. *Cephalaspis,** *Pteraspis,** *Petromyzon, Entosphenus, Myxine, Eptatretus*

CLASS PLACODERMI. No living representatives

CLASS CHONDRICHTHYES. Cartilaginous fishes. *Squalus, Hyporion, Raja, Chimaera*

CLASS OSTEICHTHYES. Bony fishes
  SUBCLASS SARCOPTERYGII
    ORDER CROSSOPTERYGII (or Coelacanthiformes). Lobe-fins. *Latimeria*
    ORDER DIPNOI (or Dipteriformes). Lungfishes. *Neoceratodus, Protopterus, Lepidosiren*
  SUBCLASS BRACHIOPTERYGII. Bichirs. *Polypterus*
  SUBCLASS ACTINOPTERYGII. Higher bony fishes. *Amia, Cyprinus, Gadus, Perca, Salmo*

CLASS AMPHIBIA
    ORDER ANURA. Frogs and toads. *Rana, Hyla, Bufo*
    ORDER URODELA. Salamanders. *Necturus, Triturus, Plethodon, Ambystoma*
    ORDER APODA. *Ichthyophis, Typhlonectes*

CLASS REPTILIA
    ORDER CHELONIA. Turtles. *Chelydra, Kinosternon, Clemmys, Terrapene*
    ORDER RHYNCHOCEPHALIA. Tuatara. *Sphenodon*
    ORDER CROCODYLIA. Crocodiles and alligators. *Crocodylus, Alligator*
    ORDER SQUAMATA. Snakes and lizards. *Iguana, Anolis, Sceloporus, Phrynosoma, Natrix, Elaphe, Coluber, Thamnophis, Crotalus*

CLASS AVES. Birds. *Anas, Larus, Columba, Gallus, Turdus, Dendroica, Sturnus, Passer, Melospiza*

* Extinct.

CLASS MAMMALIA. Mammals
  SUBCLASS PROTOTHERIA
    ORDER MONOTREMATA. Egg-laying mammals. *Ornithorhynchus, Tachyglossus*
  SUBCLASS THERIA. Marsupial and placental mammals
    ORDER MARSUPIALIA. Marsupials. *Didelphis, Sarcophilus, Notoryctes, Macropus*
    ORDER INSECTIVORA. Insectivores (moles, shrews, etc.). *Scalopus, Sorex, Erinaceus*
    ORDER DERMOPTERA. Flying lemurs. *Galeopithecus*
    ORDER CHIROPTERA. Bats. *Myotis, Eptesicus, Desmodus*
    ORDER PRIMATES. Lemurs, monkeys, apes, man. *Lemur, Tarsius, Cebus, Macacus, Cynocephalus, Pongo, Pan, Homo*
    ORDER EDENTATA. Sloths, anteaters, armadillos. *Bradypus, Myrmecophagus, Dasypus*
    ORDER PHOLIDOTA. Pangolin. *Manis*
    ORDER LAGOMORPHA. Rabbits, hares, pikas. *Ochotona, Lepus, Sylvilagus, Oryctolagus*
    ORDER RODENTIA. Rodents. *Sciurus, Marmota, Dipodomys, Microtus, Peromyscus, Rattus, Mus, Erethizon, Castor*
    ORDER CETACEA. Whales, dolphins, porpoises. *Delphinus, Phocaena, Monodon, Balaena*
    ORDER CARNIVORA. Carnivores. *Canis, Procyon, Ursus, Mustela, Mephitis, Felis, Hyaena, Eumetopias*
    ORDER TUBULIDENTATA. Aardvark. *Orycteropus*
    ORDER PROBOSCIDEA. Elephants. *Elephas, Loxodonta*
    ORDER HYRACOIDEA. Coneys. *Procavia*
    ORDER SIRENIA. Manatees. *Trichechus, Halicore*
    ORDER PERISSODACTYLA. Odd-toed ungulates. *Equus, Tapirella, Tapirus, Rhinoceros*
    ORDER ARTIODACTYLA. Even-toed ungulates. *Pecari, Sus, Hippopotamus, Camelus, Cervus, Odocoileus, Giraffa, Bison, Ovis, Bos*

Glossary

Interalphabetized with the vocabulary are the main prefixes and combining forms used in biology. Generally of Greek or Latin origin, many of them have acquired a new meaning in biology (see *blasto-, -cyte, caryo-, -plasm*). Familiarity with these forms will make it easier for you to learn and remember the numerous terms in which they are incorporated.

Some units of measurement

meter (m.)	39.37 inches
centimeter (cm.)	0.39 inch
millimeter (mm.)	0.039 inch
micron (μ)	one thousandth mm.
angstrom (Å)	one ten millionth mm.
gram (g.)	0.035 ounce
liter (l.)	1.057 quarts
milliliter (ml.)	0.034 fluid ounce

a-. Without, lacking.

ab-. Away from, off.

abdomen [L belly]. In mammals, the portion of the trunk posterior to the thorax, containing most of the viscera except heart and lungs. In other animals, the posterior portion of the body.

acellular. Not constructed on a cellular basis.

acid [L *acidus* sour]. A substance that increases the concentration of hydrogen ions when dissolved in water, that has a pH lower than 7.

ACTH. *See* adrenocorticotrophic hormone.

action potential. *See* potential.

active transport. Movement of a substance across a membrane by a process requiring expenditure of energy by the cell.

ad-. Next to, at.

adaptation. Any genetically controlled characteristic that aids an organism to survive and reproduce in the environment it inhabits.

adenosine diphosphate (ADP). A doubly phosphorylated organic compound that can be further phosphorylated to form "energy currency."

adenosine triphosphate (ATP). A triply phosphorylated organic compound that functions as "energy currency" for organisms.

adipose [L *adeps* fat]. Fatty.

ADP. *See* adenosine diphosphate.

adrenal [L *renes* kidneys]. An endocrine gland of vertebrates located near the kidneys.

adrenalin. A hormone produced by the adrenal medulla that stimulates "fight-or-flight" reactions.

adrenocorticotrophic hormone (ACTH). A hormone produced by the pituitary that stimulates the adrenal cortex.

adsorb [L *sorbēre* to suck up]. Hold on a surface.

advanced. New, unlike the ancestral condition.

aerobic [L *aer* air]. With oxygen.

agonistic [Gk *agōnistēs* combatant]. Having to do with attack, escape, or fear.

alcohol. Any of a class of organic compounds in which one or more OH groups are attached to a carbon backbone.

all-, allo- [Gk *allos* other]. Other, different.

allele. Any of several alternative gene forms at a given chromosomal locus.

allopatric [L *patria* homeland]. Having different ranges.

all-or-none. The property of responding maximally or not at all.

alveolus [L little hollow]. A small cavity, especially one of the microscopic cavities that are the functional units of lungs.

amino acid. An organic acid carrying an amino group ($—NH_2$); the building-block compound of proteins.

amnion [Gk caul]. An extraembryonic membrane that forms a fluid-filled sac containing the embryo in reptiles, birds, and mammals.

amoeboid [Gk *amoibē* change]. Amoebalike in the tendency to change shape by protoplasmic flow.

amylase [L *amylum* starch]. A starch-digesting enzyme.

an-. Without.

anaerobic [L *aer* air]. Without oxygen.

analogous. Of characters in different organisms: similar in function and often in superficial structure but of different evolutionary origins.

angio-, -angium [Gk *angeion* vessel]. Container, receptacle.

anion. A negatively charged ion.

anterior. Toward the front end.

antheridium [Gk *anthos* flower]. Male reproductive organ of a plant; produces sperm cells.

antibody. A protein that destroys or inactivates an antigen.

antigen. A foreign substance, usually a protein, that stimulates the body to produce antibodies against it.

anus [L ring]. Opening at the posterior end of the digestive tract, through which indigestible wastes are expelled.

aorta. The main artery of the systemic circulation.

apical. At, toward, or near the apex or tip of a structure such as a plant shoot.

arch- [Gk *archein* to begin]. Primitive, original.

archegonium [Gk *archegonos* the first of a race]. Female reproductive organ of a higher plant; produces egg cells.

archenteron [Gk *enteron* intestine]. The cavity in an early embryo that becomes the digestive cavity.

arteriole. A small artery.

artery. A blood vessel that carries blood away from the heart.

artifact. A by-product of scientific manipulation rather than an inherent part of the thing observed.

ascus [Gk *askos* bag]. The elongate spore sac of an ascomycete fungus.

asexual. Without sex.

atom [Gk *atomos* indivisible]. The smallest unit of an element, not divisible by ordinary chemical means.

ATP. *See* adenosine triphosphate.

auto-. Self, same.

autonomic nervous system. A portion of the vertebrate nervous system, comprising motor neurons not normally under direct voluntary control that innervate internal organs.

autosome [Gk *sōma* body]. Any chromosome other than a sex chromosome.

autotrophic [Gk *trophē* food]. Capable of manufacturing organic nutrients from inorganic raw materials.

auxin [Gk *auxein* to grow]. Any of a class of plant hormones that promote cell elongation and can diffuse into a decapitated plant from an agar block, causing it to bend in the dark (Went test).

axon [Gk *axōn* axis]. A fiber of a nerve cell that conducts impulses away from the cell body and can release, but cannot itself be stimulated by, transmitter substance.

bacteriophage [Gk *phagein* to eat]. A virus that attacks bacteria, *abbrev.* phage.

basal. At, near, or toward the base, i.e. the point of attachment, of a structure such as a limb.

base. A substance that increases the concentration of hydroxyl ions when dissolved in water, that has a pH higher than 7.

basidium. The spore-bearing structure of basidiomycetes (club fungi).

bi-. Two.

bilateral symmetry. The property of having two similar sides, with definite upper and lower surfaces and definite anterior and posterior ends.

binary fission. Reproduction by the division of a cell into two essentially equal parts by a nonmitotic process.

bio- [Gk *bios* life]. Life, living.

biogenesis [Gk *genesis* source]. Origin of living organisms from other living organisms.

biological magnification. Increasing concentration of relatively stable chemicals as they are passed up a food chain from initial consumers to top predators.

biomass. The total weight or volume of all the organisms, both plant and animal, in a given area.

biome. A major climax plant formation.

biotic. Pertaining to life.

biotic potential. The maximum population growth rate under ideal conditions.

blasto- [Gk *blastos* bud]. Embryo.

blastocoel [Gk *koilos* hollow]. The cavity of a blastula.

blastopore [Gk *poros* passage]. The opening from the cavity of the archenteron to the exterior in a gastrula.

blastula. An early embryonic stage in animals, preceding the delimitation of the three principal tissue layers; frequently spherical and hollow.

Brownian movement. Random movement of microscopic particles suspended in a gas or liquid.

buffered. Resisting change in pH.

caecum [L *caecus* blind]. A blind diverticulum of the digestive tract.

cambium [L *cambiare* to exchange]. The principal lateral meristem of vascular plants; gives rise to most secondary tissue.

cAMP. *See* cyclic adenosine monophosphate.

capillary [L *capillus* hair]. A tiny blood vessel across whose walls exchange of materials between the blood and the tissues takes place; receives blood from arteries and carries it to veins.

carbohydrate. Any of a class of organic compounds composed of carbon, hydrogen, and oxygen. Examples: sugar, starch, cellulose.

carboxyl group. A carbon, hydrogen, oxygen group, —COOH, characteristic of organic acids.

cardiac [Gk *kardia* heart]. Pertaining to the heart.

carnivore [L *carnis* of flesh; *vorare* to devour]. An organism that feeds on animals.

carotenoid [L *carota* carrot]. Any of a group of red, orange, and yellow accessory pigments of plants, found in plastids.

carrying capacity. The maximum population that a given environment can support indefinitely.

cartilage. A specialized type of dense fibrous connective tissue with a rubbery intercellular matrix.

caryo- [Gk *karyon* kernel]. Nucleus.

Casparian strip. A lignified and waterproofed thickening in the radial and end walls of endodermal cells of plants.

catalysis [Gk *katalyein* to dissolve]. Acceleration of a chemical reaction by a substance that is not itself permanently changed by the reaction.

catalyst. A substance that produces catalysis.

cation. A positively charged ion.

caudal [L *cauda* tail]. Pertaining to the tail.

cell sap. *See* sap.

cellulose [L *cellula* cell]. A complex polysaccharide that is a major constituent of most plant cell walls.

central nervous system. A portion of the nervous system containing association neurons and exerting some control over the rest of the nervous system. In vertebrates, the brain and spinal cord.

centriole [L *centrum* center]. A cytoplasmic organelle, located just outside the nucleus, that organizes the spindle during mitosis and meiosis.

centromere [L *centrum;* Gk *meros* part]. A special region on a chromosome that attaches to a spindle fiber during mitosis or meiosis.

cephalization [Gk *kephalē* head]. Localization of neural coordinating centers and sense organs at the anterior end of the body.

cerebellum [L small brain]. A part of the hindbrain of vertebrates that controls muscular coordination.

cerebrum [L brain]. Part of the forebrain of vertebrates, the chief coordination center of the nervous system.

character displacement. The rapid divergent evolution in sympatric species of characters that minimize competition and/or hybridization between them.

chemosynthesis. Autotrophic synthesis of organic materials, energy for which is derived from inorganic molecules.

chitin [Gk *chitōn* tunic]. Polysaccharide that forms part of the hard outer covering of insects, crustaceans, and other invertebrates; also occurs in the cell walls of fungi.

chlorophyll [Gk *chlōros* greenish yellow; *phyllon* leaf]. The green pigment of plants necessary for photosynthesis.

chloroplast. A plastid containing chlorophyll.

chrom-, -chrome [Gk *chrōma* color]. Colored; pigment.

chromatid. A single chromosomal strand.

chromatography. Process of separating substances by adsorption on media for which they have different affinities.

chromosome [Gk *sōma* body]. A filamentous structure in the cell nucleus, along which the genes are located.

cilium [L eyelid]. A short hairlike locomotory organelle on the surface of a cell.

cistron. The unit of genetic function, probably many nucleotides long.

cleavage. Division of a zygote.

climax (ecological). A relatively stable stage reached in some ecological successions.

cline [Gk *klinein* to lean]. Gradual variation, correlated with geography, in a character of a species.

cloaca [L sewer]. Common chamber that receives materials from the digestive, excretory, and reproductive systems.

clone [Gk *klōn* twig]. A group of cells or organisms derived asexually from a single ancestor and hence genetically identical.

co-. With, together.

codon. The unit of genetic coding, probably three nucleotides long.

coel-, -coel [Gk *koilos* hollow]. Hollow, cavity; chamber.

coelom. A body cavity bounded entirely by mesoderm.

coenocytic [Gk *koinos* common]. Having more than one nucleus in a single mass of cytoplasm.

coenzyme. A substance necessary for activation of an enzyme.

coleoptile [Gk *koleon* sheath; *ptilon* feather]. A sheath around the young shoot of grasses.

collenchyma [Gk *kolla* glue]. A supportive tissue in plants in which the cells usually have thickenings at the angles of the walls.

colloid [Gk *kolla*]. A stable suspension of particles that, though larger than in a true solution, do not settle out.

colon. The large intestine.

com-. Together.

commensalism [L *mensa* table]. A symbiosis in which one party benefits and the other is neither benefited nor harmed.

community. In ecology, a unit composed of all the populations living in a given area.

competition. In ecology, utilization by two or more individuals, or by two or more populations, of the same limited resource; an interaction where both parties are harmed.

condensation reaction. A reaction joining two compounds with resultant formation of water.

conditioning. The associating, as a result of reinforcement, of a response with a stimulus with which it was not previously associated.

conjugation [L *jugare* to join, marry]. Process of genetic recombination between two organisms (e.g. bacteria, algae) through a cytoplasmic bridge between them.

connective tissue. A type of animal tissue whose cells are embedded in an extensive intercellular matrix; connects, supports, or surrounds other tissues and organs.

contractile vacuole. An excretory and/or osmoregulatory vacuole in some cells, which, by contracting, ejects fluids from the cell.

cork [L *cortex* bark]. A waterproofed tissue, derived from the cork cambium, that forms at the outer surfaces of the older stems and roots of woody plants; the outer bark or peridem.

corpus luteum [L yellow body]. A yellowish structure in the ovary, formed from the follicle after ovulation, that secretes estrogen and progesterone.

cortex [L bark]. In plants, tissue between the epidermis and the vascular cylinder of stems and roots. In animals, the outer barklike tissue of some organs, as. *cerebral cortex*, *adrenal cortex*, etc.

cotyledon [Gk *kotylē* cup]. A "seed leaf," a food-digesting and -storing part of a plant embryo.

covalent bond. A chemical bond resulting from the sharing of a pair of electrons.

crossing-over. Exchange of parts of two homologous chromosomes.

cross section. *See* section.

cryptic [Gk *kryptos* hidden]. Concealing.

cuticle [L *cutis* skin]. A waxy layer on the outer surface of leaves, insects, etc.

cyclic adenosine monophosphate (**cyclic AMP** or **cAMP**). Compound, synthesized in living cells from ATP, that functions as an intracellular mediator of hormonal action; also plays a part in neural transmission and some other kinds of cellular control systems.

cyst [Gk *kystis* bladder, bag]. (1) A saclike abnormal growth. (2) Capsule that certain organisms secrete around themselves and that protects them during resting stages.

-cyte, cyto- [Gk *kytos* container]. Cell.

cytochrome. Any of a group of iron-containing pigments important in the electron transport of oxidative phosphorylation and photophosphorylation.

cytokinesis (Gk *kinēsis* motion]. Division of the cytoplasm of a cell.

cytoplasm. All of a cell except the nucleus.

deamination. Removal of an amino group.

deciduous [L *decidere* to fall off]. Shedding leaves each year.

dehydration reaction. A condensation reaction.

deme [Gk *dēmos* population]. A local unit of population.

dendrite [Gk *dendron* tree]. A fiber of a nerve cell that leads impulses toward the cell body; can be stimulated by transmitter substances but does not secrete them.

deoxyribonucleic acid (**DNA**). A nucleic acid found especially in the cell nucleus, thought to be the genetic material.

-derm [Gk *derma* skin]. Skin, covering; tissue layer.

di-. Two.

dicot. A member of a subclass of the angiosperms, or flowering plants, distinguished mainly by the presence of two cotyledons in the embryo; *cf.* monocot.

differentiation. The process of developmental change from an immature to a mature form, especially in a cell.

diffusion. The movement of dissolved or suspended particles from one place to another as a result of their heat energy (thermal agitation).

digestion. Hydrolysis of complex nutrient compounds into their building-block units.

diploid [Gk *diploos* double]. Having two of each type of chromosome.

disaccharide. A double sugar, i.e. one composed of two simple sugars.

distal [L *distare* to stand apart]. Situated away from some reference point (usually the main part of the body).

diverticulum [L *devertere* to turn aside]. A blind sac branching off a cavity or canal.

DNA. *See* deoxyribonucleic acid.

dominant. Of an allele: exerting its full phenotypic effect despite the presence of another allele of the same gene, whose phenotypic expression it blocks. *Dominant phenotype, dominant character:* one caused by a dominant allele.

dorsal [L *dorsum* back]. Pertaining to the back.

DPN. *See* nicotinamide adenine dinucleotide.

duodenum [L *duodeni* twelve each (i.e. finger's breadths)]. The first portion of the small intestine of vertebrates, into which ducts from the pancreas and gallbladder empty.

ecosystem [Gk *oikos* habitation]. The sum total of physical features and organisms occurring in a given area.

ecto-. Outside, external.

ectoderm. The outermost tissue layer of an animal embryo. Also, tissue derived from the embryonic ectoderm.

effector. The part of an organism that produces a response, e.g. muscle, cilium, flagellum.

egg. An egg cell or female gamete. Also a structure in which embryonic development takes place, especially in birds and reptiles; consists of an egg cell, various membranes, and often a shell.

electron. A negatively charged primary subatomic particle.

embryo. A plant or animal in an early stage of development; generally still contained within the seed, egg, or uterus.

emulsion [L *emulsus* milked out]. Suspension, usually as fine droplets, of one liquid in another.

-enchyma [Gk *parenchein* to pour in beside]. Tissue.

end-, endo-. Within, inside; requiring.

endergonic [Gk *ergon* work]. Energy-absorbing.

endocrine [Gk *krinein* to separate]. Pertaining to ductless glands that produce hormones.

endoderm. The innermost tissue layer of an animal embryo.

endodermis. A plant tissue, especially prominent in roots, that surrounds the vascular cylinder; all endodermal cells have Casparian strips.

endoplasmic reticulum [L *reticulum* network]. A system of membrane-bounded channels in the cytoplasm.

endoskeleton. An internal skeleton.

endosperm [Gk *sperma* seed]. A nutritive material in the seeds of flowering plants.

entropy. Measure of the unusable or unavailable energy in a system, hence also a measure of the disorder of a system.

enzyme [Gk *zymē* leaven].   A protein that acts as a catalyst.

epi-.   Upon, outer.

epicotyl.   The portion of the axis of a plant embryo above the point of attachment of the cotyledons; forms the shoot.

epidermis [Gk *derma* skin].   The outermost portion of the skin or body wall of an animal.

episome [Gk *sōma* body].   Genetic element at times free in the cytoplasm, at other times integrated into a chromosome.

epithelium.   An animal tissue that forms the covering or lining of all free body surfaces, both external and internal.

erythrocyte [Gk *erythros* red].   A red blood cell, i.e. a blood cell containing hemoglobin.

esophagus [Gk *phagein* to eat].   An anterior part of the digestive tract; in mammals it leads from the pharynx to the stomach.

estrogen [L *oestrus* frenzy].   Any of a group of vertebrate female sex hormones.

estrous cycles [L *oestrus*].   In mammals, the higher primates excepted, a recurrent series of physiological and behavioral changes connected with reproduction.

eu- [Gk *eus* good].   Most typical, true.

eucaryotic cell.   A cell containing a distinct membrane-bounded nucleus, characteristic of all organisms except bacteria and blue-green algae.

evaginated [L *vagina* sheath].   Folded or protruded outward.

eversible [L *evertere* to turn out].   Capable of being turned inside out.

evolution [L *evolutio* unrolling].   Change in the genetic makeup of a population with time.

ex-, exo-.   Out of, outside; producing.

excretion.   Release of metabolic wastes and excess water.

exergonic [Gk *ergon* work].   Energy-releasing.

exoskeleton.   An external skeleton.

extrinsic.   External to, not a basic part of; as in *extrinsic isolating mechanism*.

fauna.   The animal life of a given area or period.

feces [L *faeces* dregs].   Indigestible wastes discharged from the digestive tract.

feedback.   The process by which a control mechanism is regulated through the very effects it brings about.

fermentation.   Anaerobic production of alcohol, lactic acid, or some similar compound from carbohydrate via the glycolytic pathway.

fertilization.   Fusion of nuclei of egg and sperm.

fetus [L *fetus* pregnant].   An embryo in its later development, still in the egg or uterus.

fixation.   (1) Conversion of a substance into a biologically more usable form, as the conversion of CO_2 into carbohydrate by photosynthetic plants or the incorporation of N_2 into more complex molecules by nitrogen-fixing bacteria. (2) Process of treating living tissue for microscopic examination.

flagellum [L whip].   A long hairlike locomotory organelle on the surface of a cell.

flora.   The plant life of a given area or period.

follicle [L *follis* bag].   A jacket of cells around an egg cell in an ovary.

follicle-stimulating hormone (FSH).   A gonadotrophic hormone of the anterior pituitary that stimulates growth of follicles in the ovaries of females and function of the seminiferous tubules in males.

food chain.   Sequence of organisms, including producers, consumers, and decomposers, through which energy and materials may move in a community.

free energy.   Usable energy in a chemical system; energy available for producing change.

fruit.   A mature ovary (or cluster of ovaries).

FSH.   *See* follicle-stimulating hormone.

gamete [Gk *gametē(s)* wife, husband].   A sexual reproductive cell that must usually fuse with another such cell before development begins; an egg or sperm.

gametophyte [Gk *phyton* plant].   A haploid plant that can produce gametes.

ganglion [Gk tumor].   A structure containing a group of cell bodies of neurons.

gastr-, gastro- [Gk *gastēr* belly].   Stomach; ventral; resembling the stomach.

gastrovascular cavity.   An often branched digestive cavity, with only one opening to the outside, that conveys nutrients throughout the body; found only in animals without circulatory system.

gastrula.   A two-layered, later three-layered, animal embryonic stage.

gastrulation.   The process whereby a blastula develops into a gastrula, usually by an involution of cells.

gel.   Colloid in which the suspended particles form a relatively orderly arrangement; *cf.* sol.

-gen; -geny [Gk *genos* birth, race].   Producing; production, generation.

gene [Gk *genos*].   A unit of inheritance.

gene flow.   The movement of genes from one part of a population to another, or from one population to another, via gametes.

gene pool.   The sum total of all the genes of all the individuals in a population.

generator potential.   *See* potential.

genetic drift.   Change in the gene pool purely as a result of chance, and not as a result of selection, mutation, or migration.

genotype.   The particular combination of genes present in the cells of an individual.

germ cell.   A sexual reproductive cell; an egg or sperm.

gibberellin.   A plant hormone one of whose effects is stem elongation in some plants.

gill.   An evaginated area of the body wall of an animal, specialized for gas exchange.

gizzard.   A chamber of an animal's digestive tract specialized for grinding food.

glucose [Gk *glykys* sweet].   A six-carbon sugar; plays a central role in cellular metabolism.

glycogen [Gk *glykys*].   A polysaccharide that serves as the principal storage form of carbohydrate in animals.

glycolysis [Gk *glykys*].   Anaerobic respiration of carbohydrates to pyruvic acid.

Golgi apparatus.   Membranous subcellular structure that plays a role in storage and modification particularly of secretory products.

gonadotrophic.   Stimulatory to the gonads.

gonads [Gk *gonos* seed].   The testes or ovaries.

granum [L grain].   A stacklike grouping of photosynthetic membranes in a chloroplast.

guard cell.   A specialized epidermal cell that regulates the size of a stoma of a leaf.

habit [L *habitus* disposition].   In biology, the characteristic form or mode of growth of an organism.

habitat [L it lives].   The kind of place where a given organism normally lives.

haploid [Gk *haploos* single].   Having only one of each type of chromosome.

hemoglobin [Gk *haima* blood].   A red iron-containing pigment in the blood that functions in oxygen transport.

hepatic [Gk *hēpar* liver].   Pertaining to the liver.

herbaceous [L *herbaceus* grassy].   Having a stem that remains soft and succulent; not woody.

herbivore [L *herba* grass; *vorare* to devour].   An animal that eats plants.

hetero- [Gk *heteros* other].   Other, different.

heterogamy [Gk *gamos* marriage].   The condition of producing gametes of two or more different types.

heterotrophic [Gk *trophē* food].   Incapable of manufacturing organic compounds from inorganic raw materials, therefore requiring organic nutrients from the environment.

heterozygous [Gk *zygōtos* yoked].   Having two different alleles of a given gene.

hilum.   Region where blood vessels, nerves, ducts, enter an organ.

hist- [Gk *histos* web].   Tissue.

histology.   The structure and arrangement of the tissues of organisms; the study of these.

homeo-, homo- [Gk *homoios* like].   Like, similar.

homeostasis.   The tendency in an organism toward maintenance of physiological stability.

homeothermic [Gk *thermē* heat].   Capable of self-regulation of body temperature; warm-blooded.

home range.   An area within which an animal tends to confine all or nearly all its activities for a long period of time.

homologous.   Of chromosomes: bearing genes for the same characters. Of characters in different organisms: inherited from a common ancestor.

homozygous [Gk *zygōtos* yoked].   Having two doses of the same allele of a given gene.

hormone [Gk *horman* to set in motion].   A control chemical secreted in one part of the body that affects other parts of the body.

hybrid.   In evolutionary biology, a cross between two species. In genetics, a cross between two genetic types.

hydro- [Gk *hydōr* water].   Water; fluid; hydrogen.

hydrocarbon.   Any compound containing only carbon and hydrogen.

hydrolysis [Gk *lysis* loosing].   Breaking apart of a molecule by addition of water.

hydrostatic [Gk *statikos* causing to stand].   Pertaining to the pressure and equilibrium of fluids.

hydroxyl ion.   The OH^- ion.

hyper-.   Over, overmuch.

hyperosmotic.   Of a substance separated from another by a semipermeable membrane: having a higher osmotic pressure than it.

hypertrophy [Gk *trophē* food].   Abnormal enlargement, excessive growth.

hypha [Gk *hyphē* web].   A fungal filament.

hypo-.   Under, lower.

hypocotyl.   The portion of the axis of a plant embryo below the point of attachment of the cotyledons; forms the root.

hypoosmotic.   Of a substance separated from another by a semipermeable membrane: having a lower osmotic pressure than it.

hypothalamus.   Part of the posterior portion of the vertebrate forebrain, containing important centers of the autonomic nervous system and centers of emotion.

inducer.   In embryology, a substance that stimulates differentiation of cells or development of a particular structure. In genetics, a substance that activates particular genes.

inorganic compound.   A chemical compound not based on carbon.

in situ [L in place].   In its natural or original position.

insulin [L *insula* island].   A hormone produced by the islets of Langerhans in the pancreas that helps regulate carbohydrate metabolism, especially conversion of glucose into glycogen.

inter-.   Between; e.g. *interspecific,* between two or more different species.

intra-.   Within; e.g. *intraspecific,* within a single species.

intrinsic.   Inherent in, a basic part of; as in *intrinsic isolating mechanism.*

invaginated [L *vagina* sheath].   Folded or protruded inward.

invertebrate [L *vertebra* joint].   Lacking a backbone, hence an animal without bones.

in vitro [L in glass].   Not in the living organism, in the laboratory.

in vivo [L in the living].   In the living organism.

ion.   An electrically charged atom.

iso-.   Equal, uniform.

isogamy [Gk *gamos* marriage].   The condition of producing gametes of only one type, no distinction existing between male and female.

isolating mechanism.   An obstacle to interbreeding, either extrinsic, such as a geographical barrier, or intrinsic, such as structural or behavioral incompatibility.

isosmotic. Of a substance separated from another by a semipermeable membrane: having the same osmotic pressure as it.

isotope [Gk *topos* place]. An atom differing from another atom of the same element in the number of neutrons in its nucleus.

kin-, kino- [Gk *kinēma* motion]. Motion, action.

lactic acid. A three-carbon organic acid produced in animals and some microorganisms by fermentation.

lamella. A thin platelike structure; a fairly straight intracellular membrane.

larva [L ghost, mask]. Immature form of some animals that undergo radical transformation to attain the adult form.

lateral. Pertaining to the side.

lenticel. A porous region in the periderm of a woody stem through which gases can move.

leukocyte [Gk *leukos* white]. A white blood cell.

LH. *See* luteinizing hormone.

ligament [L *ligare* to bind]. A type of connective tissue linking two bones at a joint.

lignin [L *lignum* wood]. An organic compound in wood that makes cellulose harder and more brittle.

linkage. The location of two or more genes on the same chromosome, which, in the absence of crossing-over, causes the characters they control to be inherited together.

lip- [Gk *lipos* fat]. Fat or fatlike.

lipase. A fat-digesting enzyme.

lipid. Any of a variety of compounds insoluble in water but soluble in ethers and alcohols; includes fats, oils, waxes, phospholipids, and steroids.

locus [L place]. A particular location on a chromosome, hence often used synonymously with gene.

lumen [L light, opening]. The space or cavity within a tube or sac.

lung. An internal chamber specialized for gas exchange in an animal.

luteinizing hormone (LH). A gonadotrophic hormone of the pituitary that stimulates conversion of a follicle into a corpus luteum and secretion of progesterone by the corpus luteum; also stimulates secretion of sex hormone by the testes.

lymph [L *lympha* water]. A fluid derived from tissue fluid and transported in special lymph vessels to the blood.

lymphocyte. White blood cell that, upon stimulation by an antigen, gives rise to plasma cells, which produce antibody.

-lysis, lyso- [Gk *lysis* loosing]. Loosening, decomposition.

lysogenic. Of bacteria: carrying bacteriophage capable of lysing, i.e. destroying, other bacterial cells.

lysosome. A subcellular organelle in which digestive enzymes are stored.

macro-. Large.

Malpighian tubule. An excretory diverticulum of the digestive tract in insects and some other arthropods.

medulla [L marrow, innermost part]. (1) The inner portion of an organ, e.g. *adrenal medulla*. (2) The *medulla oblongata*, a portion of the vertebrate hindbrain that connects with the spinal cord.

medusa [*after* Medusa, mythological monster with snaky locks]. The free-swimming stage in the life cycle of a coelenterate.

mega-. Large; female.

megaspore. A spore that will germinate into a female plant.

meiosis [Gk *meiōsis* diminution]. A process of nuclear division in which the number of chromosomes is reduced by half.

memory trace. A change in one or more cells in a neuronal circuit hypothesized to account for the persistence of a memory.

meristematic tissue [Gk *meristos* divisible]. A plant tissue that functions primarily in production of new cells by mitosis.

meso-. Middle.

mesoderm. The middle tissue layer of an animal embryo.

mesophyll [Gk *phyllon* leaf]. The parenchymatous middle tissue layers of a leaf.

meta-. Posterior, later; change in.

metabolism [Gk *metabolē* change]. The sum of the chemical reactions within a cell (or a whole organism), including the energy-releasing breakdown of molecules and the synthesis of complex molecules and new protoplasm.

metamorphosis [Gk *morphē* form]. Transformation of an immature animal into an adult. More generally, change in the form of an organ or structure.

micro-. Small; male.

microorganism. A microscopic plant, animal, or virus, especially bacteria, protozoans, and viruses.

microspore. A spore that will germinate into a male plant.

middle lamella. A layer of substance deposited between the walls of adjacent plant cells.

mineral. In biology, any inorganic substance.

mitochondrion [Gk *mitos* thread; *chondrion* small grain]. A subcellular organelle in which aerobic respiration takes place.

mitosis [Gk *mitos*]. A process of nuclear division, characterized by complex movements of chromosomes along a spindle, that results in new nuclei with the same number of chromosomes as the original nucleus.

mold. Any of many fungi that produce a cottony or furry growth.

molecule. A chemical unit consisting of two or more atoms bonded together.

monocot. A member of a subclass of the angiosperms, or flowering plants, distinguished mainly by the presence of a single cotyledon in the embryo; *cf.* dicot.

-morph, morpho- [Gk *morphē* form]. Form, structure.

morphogenesis. The establishment of shape and pattern in an organism.

morphology. The form and structure of organisms or parts of organisms; the study of these.

motivation. The internal state of an animal that is the immediate cause of its behavior.

motor neuron. A neuron leading from the central nervous system toward an effector.

mouthparts. Structures or appendages near the mouth used in manipulating food.

mucosa. Any membrane secreting mucus (a slimy protective substance). Specifically the membrane lining the stomach and intestine.

muscle [L *musculus* small mouse, muscle]. A contractile tissue of animals.

mutation [L *mutatio* change]. Any relatively stable heritable change in the genetic material.

mutualism. A symbiosis in which both parties benefit.

mycelium [Gk *mykēs* fungus]. A mass of hyphae forming the body of a fungus.

myo- [Gk *mys* mouse, muscle]. Muscle.

NAD. *See* nicotinamide adenine dinucleotide.

NADP. *See* nicotinamide adenine dinucleotide phosphate.

natural selection. Differential reproduction in nature, leading to an increase in the frequency of some genes or gene combinations and to a decrease in the frequency of others.

navigation. The initiation and/or maintenance of movement toward a goal by means other than recognition of landmarks.

nematocyst [Gk *nēma* thread; *kystis* bag]. A specialized stinging cell in coelenterates; contains a hairlike structure that can be ejected.

neo-. New.

neocortex. Portion of the cerebral cortex in mammals, of relatively recent evolutionary origin; often greatly expanded in the higher primates and dominant over other parts of the brain.

nephr- [Gk *nephros* kidney]. Kidney.

nephridium. An excretory organ consisting of an open bulb and a tubule leading to the exterior; found in many invertebrates, e.g. segmented worms.

nephron. The functional unit of a vertebrate kidney, consisting of Bowman's capsule, convoluted tubule, and loop of Henle.

nerve. A bundle of neuron fibers.

nerve net. A nervous system without any central control, as in coelenterates.

neuron. A nerve cell.

neutron. An electrically neutral subatomic particle with approximately the same mass as a proton.

niche. The functional role and position of an organism in the ecosystem.

nicotinamide adenine dinucleotide (NAD). An organic compound that functions as an electron acceptor in respiration. Also known as diphosphopyridine nucleotide (DPN).

nicotinamide adenine dinucleotide phosphate (NADP). An organic compound that functions as an electron acceptor in photosynthesis and respiration. Also known as triphosphopyridine nucleotide (TPN).

nitrogen fixation. Incorporation of nitrogen from the atmosphere into substances more generally usable by organisms.

node (of plant) [L *nodus* knot]. Point on a stem where a leaf or bud is (or was) attached.

notochord [Gk *nōtos* back; *chordē* string]. In the lower chordates and in the embryos of the higher vertebrates, a flexible supportive rod running longitudinally through the back just ventral to the nerve cord.

nucleic acid. Any of several organic acids that are polymers of nucleotides and function in transmission of hereditary traits, in protein synthesis, and in control of cellular activities.

nucleolus. A dense body within the nucleus, composed largely of RNA.

nucleotide. A chemical entity consisting of a five-carbon sugar with a phosphate group and a purine or pyrimidine attached; building-block unit of nucleic acids.

nucleus (of cell) [L kernel]. A large membrane-bounded organelle containing the chromosomes.

nutrient [L *nutrire* to nourish]. A substance usable in metabolism.

nymph [Gk *nymphē* bride, nymph]. Immature stage of insect that undergoes gradual metamorphosis.

olfaction [L *olfacere* to smell]. The sense of smell.

omnivorous [L *omnis* all; *vorare* to devour]. Eating a variety of foods, including both plants and animals.

ontogeny [Gk *ōn* being]. The course of development of an individual organism.

o-, oö- [Gk *ōion* egg]. Egg.

oögamy. A type of heterogamy in which the female gametes are large nonmotile egg cells.

oögonium. Unjacketed female reproductive organ of a thallophyte plant.

oral [L *oris* of the mouth]. Relating to the mouth.

organ [Gk *organon* tool]. A body part usually composed of several tissues grouped together into a structural and functional unit.

organelle. A well-defined subcellular structure.

organic compound. A chemical compound containing carbon.

organism. An individual living thing.

orientation. The act of turning or moving in relation to some external feature, e.g. a source of light.

osmoregulation. Regulation of the osmotic concentration of body fluids in such a manner as to keep them relatively constant despite changes in the external medium.

osmosis [Gk *ōsmos* thrust]. Movement of a solvent (usually water in biology) through a semipermeable membrane.

osmotic pressure. The pressure that must be exerted on a solution or colloid to keep it in equilibrium with pure water when it is separated from the water by a semipermeable membrane; hence a measure of the tendency of the solution or colloid to take in water.

ov-, ovi- [L *ovum* egg]. Egg.

ovary. Female reproductive organ in which egg cells are produced.

ovulation. Release of an egg from the ovary.

ovule. A plant structure, composed of an integument, sporangium, and megagametophyte, that develops into a seed after fertilization.

ovum. A mature egg cell.

oxidation. Energy-releasing process involving removal of electrons from a substance; in biological systems, generally the removal of hydrogen (or sometimes the addition of oxygen).

pancreas. In vertebrates, a large glandular organ located near the stomach that secretes digestive enzymes into the duodenum and also produces hormones.

papilla [L nipple]. A small nipplelike protuberance.

para-. Alongside of.

parapodium [Gk *podion* little foot]. One of the paired segmentally arranged lateral flaplike protuberances of polychaete worms.

parasitism [Gk *parasitos* eating with another]. A symbiosis in which one party benefits at the expense of the other.

parasympathetic nervous system. One of the two parts of the autonomic nervous system.

parathyroids. Small endocrine glands of vertebrates located near the thyroid.

parenchyma. A plant tissue composed of thin-walled, loosely packed, relatively unspecialized cells.

parthenogenesis [Gk *parthenos* virgin]. Production of offspring without fertilization.

pathogen [Gk *pathos* suffering]. A disease-causing organism.

pectin. A complex polysaccharide that partially fills the spaces between the fibrils in a plant cell wall and is a major constituent of the middle lamella.

pellicle [L *pellis* skin]. A thin skin or membrane.

pepsin [Gk *pepsis* digestion]. A protein-digesting enzyme of the stomach.

peptide bond. A bond between two amino acids resulting from a condensation reaction between the amino group of one acid and the acidic group of the other.

peri-. Surrounding.

pericycle. A layer of cells inside the endodermis but outside the phloem of roots and stems.

periderm. The corky outer bark of older stems and roots.

peristalsis [Gk *stalsis* contraction]. Alternating waves of contraction and relaxation passing along a tubular structure such as the digestive tract.

permeable. Of a membrane: permitting other substances to pass through.

petiole [L *pediculus* small foot]. The stalk of a leaf.

PGAL. *See* phosphoglyceraldehyde.

pH. Symbol for the logarithm of the reciprocal of the hydrogen ion concentration; hence a measure of acidity. A pH of 7 is neutral; lower values are acidic, higher values alkaline (basic).

phage. *See* bacteriophage.

phagocytosis [Gk *phagein* to eat]. The active engulfing of particles by a cell.

pharynx. Part of the digestive tract between the oral cavity and the esophagus; in vertebrates, also part of the respiratory passage.

phenotype [Gk *phainein* to show]. The physical manifestation of a genetic trait.

pheromone [Gk *pherein* to carry + hormone]. A substance that, secreted by one organism, influences the behavior or physiology of other organisms of the same species.

phloem [Gk *phloios* bark]. A plant vascular tissue that transports organic materials.

-phore [Gk *pherein* to carry]. Carrier.

phosphoglyceraldehyde (**PGAL**). A three-carbon phosphorylated carbohydrate, important in both photosynthesis and glycolysis.

phospholipid. A compound composed of glycerol, fatty acids, a phosphate group, and a nitrogenous group.

phosphorylation. Addition of a phosphate group.

photo- [Gk *phōs* light]. Light.

photon. One of the discrete units or packets into which radiant energy is subdivided.

photoperiodism. A response by an organism to the duration and timing of the light and dark conditions.

photosynthesis. Autotrophic synthesis of organic materials in which the source of energy is light.

phototropism [Gk *tropos* turn]. A turning response to light.

-phyll [Gk *phyllon* leaf]. Leaf.

phylogeny [Gk *phylē* tribe]. Evolutionary history of an organism.

physiology [Gk *physis* nature]. The life processes and functions of organisms; the study of these.

-phyte, phyto- [Gk *phyton* plant]. Plant.

phytochrome. A protein pigment of plants sensitive to red and far-red light.

pinocytosis [Gk *pinein* to drink]. The active engulfing by cells of liquid or of very small particles.

pistil. The female reproductive organ of a flower, composed of one or more megasporophylls.

pith. A tissue (usually parenchyma) located in the center of a stem (rarely a root), internal to the xylem.

pituitary. An endocrine gland located near the brain of vertebrates; known as the master gland because it secretes hormones that regulate the action of other endocrine glands.

placenta [Gk *plax* flat surface]. An organ in mammals, made up of fetal and maternal components, that aids in exchange of materials between the fetus and the mother.

plasm-, plasmo-, -plasm [Gk *plasma* something formed or molded]. Formed material; plasma; cytoplasm.

plasma. Blood minus the cells and platelets.

plasma membrane. The outer membrane of a cell.

plasmodesma [Gk *desma* bond]. A delicate cytoplasmic connection between adjacent plant cells.

plasmolysis. Shrinkage of a plant cell away from its wall when in a hyperosmotic medium.

plastid [Gk *plastos* formed]. Relatively large organelle in plant cells that functions in photosynthesis and/or nutrient storage.

pleiotropic [Gk *pleiōn* more]. Of a gene: having more than one phenotypic effect.

poikilothermic [Gk *poikilos* various; *thermē* heat]. Incapable of precise self-regulation of body temperature, dependent on environmental temperature; cold-blooded.

pollen grain [L *pollen* flour dust]. A microgametophyte of a seed plant.

poly-. Many.

polymer [Gk *meros* part]. A large molecule consisting of a chain of small molecules bonded together by condensation reactions or similar reactions.

polymorphism [Gk *morphē* form]. The simultaneous occurrence of several discontinuous phenotypes in a population.

polyp [Gk *polypous* many-footed]. The sedentary stage in the life cycle of a coelenterate.

polypeptide chain. A chain of amino acids linked together by peptide bonds.

polyploid. Having more than two complete sets of chromosomes.

polysaccharide. Any carbohydrate that is a polymer of simple sugars.

population. In ecology, a group of individuals belonging to the same species.

portal system [L *porta* gate]. A blood circuit in which two beds of capillaries are connected by a vein (e.g. *hepatic portal system*).

posterior. Toward the hind end.

potential. Short for *potential difference*, the electrification of one point or structure relative to the electrification of some other point or structure. *Resting p.:* a relatively steady potential difference across a cell membrane, particularly of a nonfiring nerve cell or a relaxed muscle cell. *Action p.:* a sharp change in the potential difference across the membrane of a nerve or muscle cell that is propagated along the cell; in nerves, identified with the nerve impulse. *Generator p.:* a change in the potential difference across the membrane of a sensory cell that, if it reaches a threshold level, may trigger an action potential along the associated neural pathway.

predation [L *praedatio* plundering]. The feeding of free-living organisms on other organisms.

primitive [L *primus* first]. Old, like the ancestral condition.

primordium [L *primus; ordiri* to begin]. Rudiment, earliest stage of development.

pro-. Before.

proboscis [Gk *boskein* to feed]. A long snout; an elephant's trunk. In invertebrates, an elongate, sometimes eversible process originating in or near the mouth that often serves in feeding.

procaryotic cell. A type of cell that lacks a membrane-bounded nucleus; found only in bacteria and blue-green algae.

progesterone [L *gestare* to carry]. One of the principal female sex hormones of vertebrates.

prot-, proto-. First, primary.

protein. A long polypeptide chain.

proteolytic. Protein-digesting.

proton. A positively charged primary subatomic particle.

protoplasm. Living substance, the material of cells.

provirus. Viral nucleic acid integrated into the genetic material of a host cell.

proximal. Near some reference point (often the main part of the body).

pseudo-. False; temporary.

pseudocoelom. A functional body cavity not entirely enclosed by mesoderm.

pseudopod [L *podium* foot]. A transitory cytoplasmic protrusion of an amoeba or an amoeboid cell.

pulmonary [L *pulmones* lungs]. Relating to the lungs.

purine. Any of several double-ringed nitrogenous bases important in nucleic acids.

pyloric [Gk *pylōros* gatekeeper]. Referring to the junction between the stomach and the intestine.

pyrimidine. Any of several single-ringed nitrogenous bases important in nucleic acids.

pyruvic acid. A three-carbon compound produced by anaerobic respiration.

race. A subspecies.

radial symmetry. A type of symmetry in which the body parts are arranged regularly around a central line (in animals, running through the oral-anal axis) rather than on the two sides of a plane.

radiation. As an evolutionary phenomenon, divergence of members of a single lineage into different niches or adaptive zones.

recessive. Of an allele: not expressing its phenotype in the presence of another allele of the same gene, therefore expressing it only in homozygous individuals. *Recessive character, recessive phenotype:* one caused by a recessive allele.

rectum [L *rectus* straight]. The terminal portion of the intestine.

reduction. Energy-storing process involving addition of electrons to a substance; in biological systems, generally the addition of hydrogen (or sometimes the removal of oxygen).

reflex [L *reflexus* bent back]. A functional unit of the nervous system, involving the entire pathway from receptor cell to effector.

reinforcement (psychological). Reward for a particular behavior.

releaser. A structure, action, sound, etc., that gives rise to releasing stimuli, i.e. to stimuli particularly effective in triggering behavioral responses.

renal [L *renes* kidneys]. Pertaining to the kidney.

respiration [L *respiratio* breathing out]. (1) The release of energy by oxidation of fuel molecules. (2) The taking in of O_2 and release of CO_2; breathing.

resting potential. *See* potential.

retina. The tissue in the rear of the eye that contains the sensory cells of vision.

rhizoid [Gk *rhiza* root]. Rootlike structure.

ribonucleic acid (RNA). Any of several nucleic acids in which the sugar component is ribose and one of the nitrogenous bases is uracil.

ribosome. A small cytoplasmic organelle that functions in protein synthesis.

RNA. *See* ribonucleic acid.

salt. Any of a class of generally ionic compounds that may be formed by reaction of an acid and a base, e.g. table salt, NaCl.

sap. Water and dissolved materials moving in the xylem; less commonly, solutions moving in the phloem. *Cell sap:* the fluid content of a plant-cell vacuole.

saprophyte [Gk *sapros* rotten]. A heterotrophic plant or bacterium that lives on dead organic material.

sclerenchyma [Gk *scléros* hard]. A plant supportive tissue composed of cells with thick secondary walls.

section. *Cross* or *transverse s.:* section at right angles to the longest axis. *Longitudinal s.:* section parallel to the longest axis. *Radial s.:* longitudinal section along a radius. *Sagittal s.:* vertical longitudinal section along the midline of a bilaterally symmetrical animal.

seed. A plant reproductive entity consisting of an embryo and stored food enclosed in a protective coat.

segmentation. The subdivision of an organism into more or less equivalent serially arranged units.

selection pressure. In a population, the force for genetic change resulting from natural selection.

semipermeable. Permeable to some substances but not to others.

sensory neuron. A neuron leading from a receptor cell to the central nervous system.

septum [L barrier]. A partition or wall.

sessile [L *sessilis* of sitting, low]. Of animals, sedentary. Of plants, without a stalk.

sex-linked. Of genes: located on the X chromosome.

shoot. A stem with its leaves, flowers, etc.

sieve element. A conductile cell of the phloem.

sol. Colloid in which the suspended particles are dispersed at random; *cf.* gel.

solute. Substance dissolved in another (the solvent).

solution [L *solutio* loosening]. A homogeneous molecular mixture of two or more substances.

solvent. Medium in which one or more substances (the solute) are dissolved.

-soma, somat-, -some [Gk *sōma* body]. Body, entity.

somatic. Pertaining to the body; to all cells except the germ cells; to the body wall; to that part of the nervous system that is at least potentially under control of the will and whose reflex arcs include one sensory and one motor neuron (*cf.* autonomic nervous system).

specialized. Adapted to a special, usually rather narrow, function or way of life.

speciation. The process of formation of new species.

species [L kind]. The largest unit of population within which effective gene flow occurs or could occur.

sperm [Gk *sperma* seed]. A male gamete.

sphincter [Gk *sphinktēr* band]. A ring-shaped muscle that can close a tubular structure by contracting.

spindle. A fibrillar structure with which the chromosomes are associated in mitosis and meiosis.

sporangium. A plant structure that produces spores.

spore [Gk *spora* seed]. An asexual reproductive cell, often a resting stage adapted to resist unfavorable environmental conditions.

sporophyll [Gk *phyllon* leaf]. A modified leaf that bears spores.

sporophyte [Gk *phyton* plant]. A diploid plant that produces spores.

stamen [L thread]. A male sexual part of a flower; a microsporophyll of a flowering plant.

starch. A glucose polymer, the principal polysaccharide storage product of vascular plants.

stele [Gk *stēlē* upright slab]. The vascular cylinder in the center of a root or stem, bounded externally by the endodermis.

stereo- [Gk *stereos* solid]. Solid; three-dimensional.

steroid. Any of a number of complex, often biologically important compounds (e.g. some hormones and vitamins), composed of four interlocking rings of carbon atoms.

stimulus. Any environmental change that is detected by a receptor.

stoma [Gk mouth]. An opening, regulated by guard cells, in the epidermis of a leaf or other plant part.

subspecies. A genetically distinctive geographic subunit of a species.

substrate. (1) The base on which an organism lives, e.g. soil. (2) In chemical reactions, a substance acted upon, as by an enzyme.

succession. In ecology, progressive change in the plant and animal life of an area.

sucrose. A double sugar composed of a unit of glucose and a unit of fructose; table sugar.

suspension. A heterogeneous mixture in which the particles of one substance are kept dispersed by agitation.

sym-, syn.- Together.

symbiosis [Gk *bios* life]. The living together of two organisms in an intimate relationship.

sympathetic nervous system. One of the two parts of the autonomic nervous system.

sympatric [L *patria* homeland]. Having the same range.

synapse [Gk *haptein* to fasten]. A juncture between two neurons.

synapsis. The pairing of homologous chromosomes during meiosis.

synergistic [Gk *ergon* work]. Acting together with another substance or organ to achieve or enhance a given effect.

systemic circulation. The part of the circulatory system supplying body parts other than the gas-exchange surfaces.

taxonomy [Gk *taxis* arrangement]. The classification of organisms on the basis of their evolutionary relationships.

tendon [L *tendere* to stretch]. A type of connective tissue attaching muscle to bone.

territory. A particular area defended by an individual against intrusion by other individuals, particularly of the same species.

testis. Primary male sex organ in which sperms are produced.

thalamus [Gk *thalamos* inner chamber]. Part of the rear portion of the vertebrate forebrain, a center for integration of sensory impulses.

thallus [Gk *thallos* young shoot]. A plant body exhibiting relatively little tissue differentiation and lacking true roots, stems, and leaves.

thorax [Gk *thōrax* breastplate]. In mammals, the part of the trunk anterior to the diaphragm, which partitions it from the abdomen. In insects, the body region between the head and the abdomen, bearing the walking legs and wings.

thymus [Gk *thymos* warty excrescence]. Glandular organ essential to the development of most immunologic capabilities in vertebrates.

thyroid [Gk *thyreoeidēs* shield-shaped]. An endocrine gland of vertebrates located in the neck region.

thyroxin. A hormone, produced by the thyroid, that stimulates a speedup of metabolism.

tissue [L *texere* to weave]. An aggregate of cells, usually similar in both structure and function, that are bound together by intercellular material.

TPN. *See* nicotinamide adenine dinucleotide phosphate.

trachea. In vertebrates, the part of the respiratory system running from the pharynx into the thorax; the "windpipe." In land anthropods, an air duct running from an opening in the body wall to the tissues.

tracheid. An elongate thick-walled tapering conductile cell of the xylem.

trans-. Across; beyond.

transduction [L *ducere* to lead]. The transfer of genetic material from one host cell to another by a virus.

transformation. The incorporation by bacteria of fragments of DNA released into the medium from dead cells.

translocation. In botany, the movement of organic materials from one place to another within the plant body, primarily through the phloem.

transpiration. Release of water vapor from the aerial parts of a plant, primarily through the stomata.

-trophic [Gk *trophē* food]. Nourishing; stimulatory.

tropism [Gk *tropos* turn]. A turning response to a stimulus, primarily by differential growth patterns in plants.

turgid [L *turgidus* swollen]. Swollen with fluid.

turgor pressure [L *turgēre* to be swollen]. The pressure exerted by the contents of a cell against the cell membrane or cell wall.

tympanic membrane [Gk *tympanon* drum]. A membrane of the ear that picks up vibrations from the air and transmits them to other parts of the ear.

unit membrane. A membrane composed of two layers of protein with two layers of lipid between them.

urea. The nitrogenous waste product of mammals and some other vetebrates, formed in the liver by combination of ammonia and carbon dioxide.

ureter. The duct carrying urine from the kidney to the bladder in higher vertebrates.

urethra. The duct leading from the bladder to the exterior in higher vertebrates.

uric acid. An insoluble nitrogenous waste product of most land arthropods, reptiles, and birds.

uterus. In mammals, the chamber of the female reproductive tract in which the embryo undergoes much of its development; the womb.

vaccine [L *vacca* cow]. Drug containing an antigen, administered to induce active immunity in the patient.

vacuole [L *vacuus* empty]. A membrane-bounded vesicle or chamber in a cell.

vascular tissue [L *vasculum* small vessel]. Tissue concerned with internal transport, such as xylem and phloem in plants and blood and lymph in animals.

vaso- [L *vas* vessel]. Blood vessel.

vector [L *vectus* carried]. Transmitter of pathogens.

vegetative. Of plant cells and organs: not specialized for reproduction. Of reproduction: asexual. Of bodily functions: involuntary.

vein [L *vena* blood vessel]. A blood vessel that transports blood toward the heart.

vena cava [L hollow vein]. One of the two large veins that return blood to the heart from the systemic circulation of vertebrates.

ventral [L *venter* belly]. Pertaining to the belly or underparts.

vessel cell. A highly specialized cell of the xylem, with thick secondary walls and extensively perforated end walls.

villus [L shaggy hair]. A highly vascularized fingerlike process from the intestinal lining.

virus [L slime, poison]. A submicroscopic noncellular, obligatorily parasitic particle, composed of a protein shell and a nucleic acid core, that exhibits some properties normally associated with living organisms, including the ability to mutate and to evolve.

viscera [L]. The internal organs, especially those of the great central body cavity.

vitamin [L *vita* life]. An organic compound, necessary in small quantities, that a given organism cannot synthesize for itself and must obtain prefabricated in the diet.

X chromosome. The female sex chromosome.

xylem [Gk *xylon* wood]. A vascular tissue that transports water and dissolved minerals upward through the plant body.

Y chromosome. The male sex chromosome.

yolk. Stored food material in an egg.

zoo- [Gk *zōion* animal]. Animal; motile.

zoospore. A ciliated or flagellated plant spore.

zygote [Gk *zygōtos* yoked]. A fertilized egg cell.

Index

Boldface identifies pages with illustrations, which may also include relevant text discussion. *Italics* identifies definitions or the main discussion of a topic.

maxilla, *515, 516, 517*
maze running, 254, 255
measles, 451
Mecoptera, 548
Medicago, **127**
medulla oblongata, **227, 228**
medusa, **496, 497, 498**, Pl. I-7C
megagametophyte, 484, **485**
Meganthropus, 538*n*
megaphyll, *481n*
megaspore, *479–480*, 483, 484, 485, 486
meiosis, 284–289, **286**
    first division of, compared with mitosis, 288
    in life cycle, 289–290
melanin, 301, 302
melanism, industrial, 371–372
melanophore, 185
melatonin, 185–186
membrane. *See* cell membrane; endoplasmic reticulum; nuclear membrane; postsynaptic membrane; semipermeable membrane; unit membrane
memory, 230–231
Mendel, G., 292–296, 298, 312, 333
meningitis, 455
menopause, 194
menstrual cycle, 191–195, **192, 193**
menstruation, 191
mercury pollution, 398
Mering, J. von, 177
meristem, apical, 58, 129, 165, *338, 339*
meristematic tissue, *58*
mesoderm, *342, 342, 343*, 344
    origin of, 342, 504, **506**
mesoglea, 106, **107**, 495, **496**
mesophyll, **83**, 118
Mesozoa, 499, **505**, 547
Mesozoic, 478, 529, 532, 534
mesquite, 425
metabolic rate, 92
metabolism, *91n*
metacarpal, **238**
metamorphosis, **346**, Pl. II-8
metaphase
    in meiosis, **286**, 287, 288
    in mitosis, **281**, 282, **283**, 288
metatarsal, **238**
Metazoa, 498–499, *500–501*, 547–549
methane, **22**, 23, 436, 437
Metopus, **494**
microclimate, *399*
microgametophyte, 485
micronucleus, **105**, 494
micronutrient, 97
microorganism, 396–398. *See also* bacteria; germ; Protozoa; virus
microphyll, *481n*
micropyle, *484, 485*
microscope, **36**
microspore, *479–480*, 484, 485, 487
microvillus, 112, **113**
midbrain, **227, 228**
middle lamella, *47*, 284
midget, 183
midgut, **161**

migration, in bird, 270
mildew, 471
milfoil, clinal variation in, **375**
Miller, S. L., 437
milliped, 69, 515
    speciation in, 381, 383
mimicry, 372, **373**, Pls. IV-2, 4
mineral nutrition
    in green plant, 96–97, 136, 400–401
    in heterotroph, 103–104
mineralocorticoids, 182–183
Minkowski, O., 177
minnow, geographic isolation of, 378
Miocene, 478, 536, 537
mite, 407, 514
mitochondrion, **49, 50**, 52, *90–91*, 243
    origin of, 444–445
    in procaryotic vs. eucaryotic cell, 55
    self-replication of, 444
mitosis, *280–284, 281, 283*
    compared with first division of meiosis, 288
    duration of, 280
mold, 67
    bread, 104, 446, **470–471**
molecule, *19*
moles, convergent evolution of, **387**
mollusc (Mollusca), 69, 508–511, 522, 547
    ancestry of, **505**, 510
    body plan of, 508
    circulatory system of, 138, 508
    classes of, **509**, 547
    determinate cleavage in, 347
    eye of, 219
    as filter feeder, 109
    respiratory pigment of, 147
molt, of arthropod, 346, 352, 512
molybdenum, for heterotrophs, 103
Monarch butterfly, 373
Monera, 446, 452–458, 546
monkey, 428, 535
    rhesus, 305
    and warning coloration, 373
monocot (Monocotyledoneae), 68, 488, 546
    compared with dicot, 488
    root, 100
    seed, **336**, 337
    stem, **128**, 129
Monod, J., 353
monohybrid inheritance, 292–299
Monoplacophora, 508, 510, 547
monosaccharide, *24*
Monotremata, 533, 549
moose, 424, Pl. IV-6A
Morgan, T. H., 310, 313, 314, 333
morphogenesis, *341*
morphogenetic movement, **341**, 344, 345, 346
morphology, and phylogeny, 386
morula, *341, 342*
mosquito, 418, 424, 493
    behavior of male, **258**, 260
moss, 67, 422, 423, **476, 477**, Pl. I-4D
    haircap, classification of, 390
moth
    antenna of, Pl. II-5C

moth (*Cont.*)
    hearing of, 223, 263
    industrial melanism of, 371–372
    larva, mimicry by, Pl. IV-2
    metamorphosis of, 346, Pl. II-8
    as pollinator, 372
    sense of smell of, 218, 264
    sex attractant pheromone of, 264
    sex determination in, 309*n*
motivation, 256–257, 258–259
motor end plate, **214**
Mougeotia, Pl. I-4A
mouse, 533
    effect of thymus removal on, 357
    estrous periods of, 191
    meadow, home range of, 410
    population, regulation of, 265, 408
    primer pheromones of, 265
mouth, *108*
    embryonic development of, 503–504
    human, **109**
mouthparts, **517**
mRNA. *See* RNA, messenger
mucoprotein, in cell membrane, **38**
mucosa, 111
mucus membrane, 111
mudpuppy, cones from, **220**
mule, sterility of, 380
Müllerian mimicry, 372
mumps, 451
muramic acid, 55, 453, 457
murein, 453
Musci, 546
muscle, 62, 66, 238–239
    antagonist, 238
    of arthropod, 512–513
    in carbohydrate storage, 179
    cardiac, 62, 66, **239**
    contraction, 240–246
        molecular basis of, 241–244
        stimulus for, 244–246
    embryonic development of, 344
    fatigue, 89, 241
    fiber, *238, 239*
    of insect, **236**, 239
    of invertebrate, 236, 239
    response to stimulus, **240–241**
    skeletal, 62, 66, 238–240, **239**, 242, 243
    smooth, 62, 66, 238–240, **239**
    striated. *See* muscle, skeletal
    synergist, 238
    tone, 241
    visceral. *See* muscle, smooth
mushroom, 67, 470, 473, **474**, Pl. I-5C, I-5D
mussel, 510
mutation, *307–308*
    in deciphering genetic code, 331
    gene, as unit of, 333–334
    and genetic equilibrium, 365, 366
    and natural selection, 307, 308, 368
    pressure, 365
muton, **334**
mutualism, *412*, 413, 492
mycelium, *470, 472, 473, 474*
myelin sheath, **205–206**, 214
myofibril, *242, 243*, 245